Advanced Organic Chemistry

Part B: Reactions and Synthesis

ADVANCED ORGANIC CHEMISTRY

PART A: Structure and Mechanisms

PART B: Reactions and Synthesis

Advanced Organic Chemistry

Part B: Reactions and Synthesis

Francis A. Carey
and Richard J. Sundberg
University of Virginia, Charlottesville, Virginia

A PLENUM/ROSETTA EDITION

Library of Congress Cataloging in Publication Data

Carey, Francis A   1937-
   Advanced organic chemistry.

   "A Plenum/Rosetta edition."
   Includes bibliographical references and index.
   CONTENTS: pt. A. Structure and mechanisms. pt. B. Reactions and synthesis.
   1. Chemistry, Organic. I. Sundberg, Richard J., 1938-   II. Title.
[QD251.2.C36 1977]   547   76-54956
ISBN 0-306-25004-7 (pt. B)

First Paperback Printing — April 1977
Second Paperback Printing — November 1977

A Plenum/Rosetta Edition
Published by Plenum Publishing Corporation
227 West 17th Street, New York, N.Y. 10011

© 1977 Plenum Press, New York
A Division of Plenum Publishing Corporation
227 West 17th Street, New York, N.Y. 10011

Printed in the United States of America

Preface to Part B

In Part A, the structural and mechanistic groundwork of organic chemistry was considered. Part B assumes that the student possesses a mastery of these areas and emphasizes the synthetic application of organic reactions. Mechanisms are discussed in sufficient detail to allow the student to understand the basis for the selectivity of the reaction and its stereochemistry, but fine points of mechanistic detail are not emphasized. Many of the most general synthetic reactions are illustrated by referenced examples included in the schemes.

The organization is along the lines of reaction type rather than functional groups. The first nine chapters discuss most of the important reactions presently in use in organic synthesis. Although the emphasis here is on synthesis, the reactions that are discussed in each chapter are usually members of related mechanistic families. Chapter 10 discusses synthetic tactics and strategy in general. Chapter 11 considers some of the special features of macromolecular synthesis.

As in Part A, the majority of the references are to English language journals that are widely accessible. References have been chosen primarily because they are illustrative of a given point or are useful leading references. No attempt has been made to consider authors' priority in the selection of references.

A number of problems are given with each chapter. An attempt has been made in those dealing with synthesis to make the problems cumulative in the sense that reactions discussed in preceding chapters may be involved, while reactions that have yet to be discussed are avoided. Of course, synthetic problems have numerous "correct answers"; therefore, although literature references to the synthetic problems are given, there may in many instances be other, perhaps preferable, alternatives. Many of the problems will be quite challenging, and the student should not feel discouraged at not being able to match the solutions to synthetic challenges reported in the literature. Indeed, it may be most useful to treat the more difficult problems as takeoff points for in-class discussion and analysis.

Contents of Part B

List of Figures

List of Tables

List of Schemes

Contents of Part A

Advanced Organic Chemistry

Chemistry

Part B: Reactions and Synthesis

Alkylation of Carbon via Enolates and Enamines

The alkylation of carbon nucleophiles by S_N2-type processes is an important transformation in the synthesis of organic compounds. The generation and alkylation of such nucleophiles are described in this chapter. Alkylation and acylation of nucleophilic carbon species by other mechanisms are discussed in Chapter 2.

1.1. Generation of Carbon Nucleophiles by Proton Abstraction

The most general means of generating carbon nucleophiles involves removal of a proton from a carbon atom. The anions thus generated are referred to as *carbanions*. The removal of a proton from a carbon atom is greatly facilitated by substituent groups that can stabilize the resulting negative charge. Carbonyl groups are especially important in this function, and the carbanions that are formed are often called *enolates*. Several typical examples of proton abstraction are listed in Scheme 1.1. Scheme 1.2 illustrates the delocalization of negative charge possible in the resulting carbanions.

The position of the equilibrium in these acid–base reactions will depend on the relative acidity of the carbon acid and of the species BH (or, conversely, the basic strengths of B^- and the carbanion). Some approximate pK values for typical carbon acids and the conjugate acids of some species commonly employed as bases are shown in Table 1.1. The numerical values recorded are approximate, since there is no method of accurately establishing absolute acidity in a single solvent medium for

Scheme 1.1. Generation of Carbon Nucleophiles by Proton Abstraction

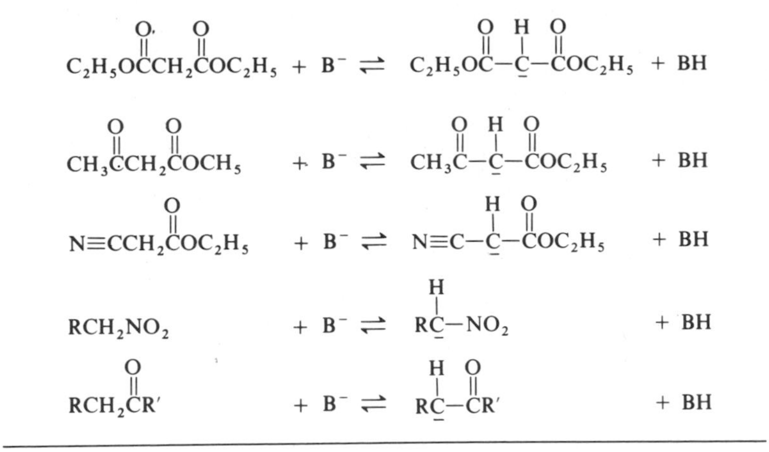

$$C_2H_5OCCH_2COC_2H_5 + B^- \rightleftharpoons C_2H_5OC-\overset{H}{\underset{\cdot}{C}}-COC_2H_5 + BH$$

$$CH_3CCH_2COCH_5 + B^- \rightleftharpoons CH_3C-\overset{H}{\underset{\cdot}{C}}-COC_2H_5 + BH$$

$$N\equiv CCH_2COC_2H_5 + B^- \rightleftharpoons N\equiv C-\overset{H}{\underset{\cdot}{C}}-COC_2H_5 + BH$$

$$RCH_2NO_2 + B^- \rightleftharpoons R\overset{H}{\underset{\cdot}{C}}-NO_2 + BH$$

$$RCH_2CR' + B^- \rightleftharpoons R\overset{H}{\underset{\cdot}{C}}-CR' + BH$$

substances of widely varying acidic strength. Furthermore, there can be significant shifts in relative acidity from solvent to solvent.

From the pK values reported in Table 1.1, an approximate ordering of some substituents with respect to their ability to stabilize carbanions can be established. The order suggested is $NO_2 > COR > CN \approx CO_2R > SOR > Ph > R$.

By comparing the approximate pK values of the conjugate acids of the basic catalysts with those of the carbon acid of interest, it is possible to estimate the position of the acid–base equilibrium for a given reactant–base combination. If we consider the case of a simple alkyl ketone, for example, it can be seen that hydroxide ion, and primary alkoxide ions, will convert only a fraction of such a ketone to its anion:

$$RCCH_3 + RCH_2O^- \longleftrightarrow R\overset{O^-}{C}=CH_2 + RCH_2OH$$

The slightly more basic tertiary alkoxides are comparable to the enolates in basicity,

Scheme 1.2. Resonance in Some Carbanions

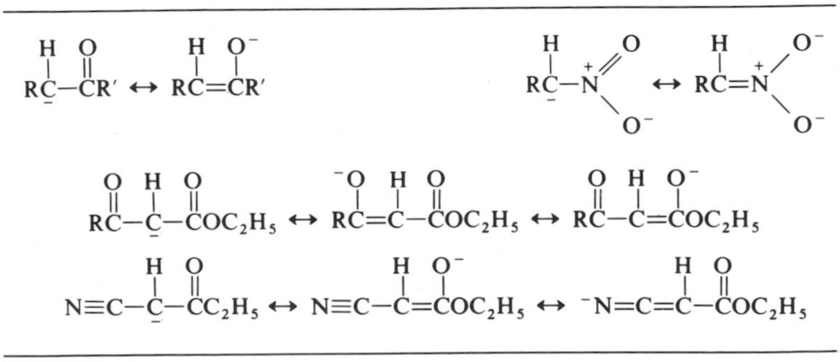

$$R\overset{H}{\underset{\cdot}{C}}-\overset{O}{C}R' \leftrightarrow R\overset{H}{C}=\overset{O^-}{C}R' \qquad R\overset{H}{\underset{\cdot}{C}}-\overset{+}{N}\overset{O}{\underset{O^-}{\diagup}} \leftrightarrow R\overset{H}{C}=\overset{+}{N}\overset{O^-}{\underset{O^-}{\diagup}}$$

$$R\overset{O}{C}-\overset{H}{\underset{\cdot}{C}}-\overset{O}{C}OC_2H_5 \leftrightarrow R\overset{^-O}{C}=\overset{H}{C}-\overset{O}{C}OC_2H_5 \leftrightarrow R\overset{O}{C}-\overset{H}{C}=\overset{O^-}{C}OC_2H_5$$

$$N\equiv C-\overset{H}{\underset{\cdot}{C}}-\overset{O}{C}C_2H_5 \leftrightarrow N\equiv C-\overset{H}{C}=\overset{O^-}{C}OC_2H_5 \leftrightarrow {}^-N=C=\overset{H}{C}-\overset{O}{C}OC_2H_5$$

Table 1.1. Approximate pK Values for Some Carbon Acids and Basic Catalysts

Carbon acid	pK^a	Common bases[a,b]	pK^c
$O_2NCH_2NO_2$	3.6	$CH_3CO_2^-$	4.2
$CH_3COCH_2NO_2$	5.1		
$C_2H_5NO_2$	8.6		
$CH_3COCH_2COCH_3$	9		
$PhCOCH_2COCH_3$	9:6	PhO^-	9.9
CH_3NO_2	10.2		
$CH_3COCH_2CO_2C_2H_5$	10.7	$(C_2H_5)_3N$	10.7
$\underset{\textstyle \mid}{CH_3}$			
$CH_3COCHCOCH_3$	11	$(C_2H_5)_2NH$	11
$NCCH_2CN$	11.2		
$C_2H_5O_2CCH_2CO_2C_2H_5$	12.7		
Cyclopentadiene	14.5		
$\underset{\textstyle \mid}{C_2H_5}$			
$C_2H_5O_2CCHCO_2C_2H_5$	15	HO^-	15.7
		CH_3O^-	16
		$C_2H_5O^-$	18
CH_3COCH_3	20	$(CH_3)_3CO^-$	19
Fluorene	20.5		
CH_3CN	25		
Ph_3CH	33	NH_2^-	35
		$CH_3SOCH_2^-$	35
		$(C_2H_5)_2N^-$	36
		$CH_3(CH_2)_3Li$	>45

a. D. J. Cram, *Fundamentals of Carbanion Chemistry*, Academic Press, New York, NY, 1965, pp. 8–20, 41.
b. H. O. House, *Modern Synthetic Reactions*, Second Edition, W. A. Benjamin, Menlo Park, CA, 1972, p. 494.
c. pK of the conjugate acid.

and more complete proton transfer will occur in the presence of such bases:

$$\underset{\displaystyle \parallel}{RCCH_3} \ + \ R_3CO^- \ \rightleftharpoons \ \underset{\displaystyle \mid}{RC}=CH_2 \ + \ R_3COH$$

Stronger bases, such as the amide ion, the hydride ion, the dimethyl sulfoxide anion, and the triphenylmethyl anion, are capable of essentially complete conversion of a ketone to its enolate:

$$\underset{\displaystyle \parallel}{RCCH_3} \ + \ Ph_3CLi \ \rightleftharpoons \ \underset{\displaystyle \mid}{RC}=CH_2 \ + \ Ph_3CH$$

For any of the other carbon acids, it is also possible to estimate the position of the acid–base equilibrium with a given base. It is important to bear in mind the position of such equilibria as other aspects of reactions of carbanions are considered.

1.2. Kinetic Versus Thermodynamic Control in Formation of Enolates

When a methylene or methinyl group carries two electron-attracting substituents, its hydrogens are more rapidly removed by bases than those activated by a single substituent. The resulting carbanion is also more stable than isomeric anions in which only one of the substituents can delocalize the negative charge. Thus, in ethyl acetoacetate, for example, a hydrogen in the 2-position is removed in preference to one of the less acidic protons at C-4:

$$CH_3\overset{\overset{\displaystyle O}{\|}}{C}CH_2CO_2C_2H_5 \overset{B^-}{\longrightarrow} CH_3\overset{\overset{\displaystyle O}{\|}}{C}\underset{}{CH}CO_2C_2H_5 \longleftrightarrow \ ^-CH_2\overset{\overset{\displaystyle O}{\|}}{C}CH_2CO_2C_2H_5$$
(major)   (very minor)

$$PhCH_2\overset{\overset{\displaystyle O}{\|}}{C}CH_2CH_3 \overset{B^-}{\longrightarrow} Ph\underset{}{CH}\overset{\overset{\displaystyle O}{\|}}{C}CH_2CH_3 \longleftrightarrow PhCH_2\overset{\overset{\displaystyle O}{\|}}{C}\underset{}{CH}CH_3$$
(major)   (minor)

If a single activating group is present, the competition for formation of the isomeric enolates is usually much more closely balanced. An important case to consider is the situation in ketones with nonidentical alkyl branches. Studies in this area have led to some insight into the factors that determine which of the possible enolates will be formed.

The composition of the enolate mixture may be governed by kinetic or thermodynamic factors. In the former case, the product composition is governed by the relative rates of two competing proton-abstraction reactions:

$$R_2CH\overset{\overset{\displaystyle O}{\|}}{C}CH_2R'$$

$$R_2C=\overset{\overset{\displaystyle O^-}{|}}{C}CH_2R' \qquad\qquad R_2CH\overset{\overset{\displaystyle O^-}{|}}{C}=CHR'$$
A   B

$$\frac{[\mathbf{A}]}{[\mathbf{B}]} = \frac{k_a}{k_b}$$

The ratio of the products is governed by *kinetic control.* On the other hand, if enolates **A** and **B** can be interconverted rapidly, equilibrium will be established and the product composition will reflect the relative thermodynamic stability of the enolates:

$$R_2CH\overset{\overset{\displaystyle O}{\|}}{C}CH_2R'$$

$$R_2C=\overset{\overset{\displaystyle O^-}{|}}{C}CH_2R \overset{K}{\rightleftharpoons} R_2CH\overset{\overset{\displaystyle O^-}{|}}{C}=CHR'$$
A   B

$$\frac{[\mathbf{A}]}{[\mathbf{B}]} = K$$

In this case, the product composition is governed by *thermodynamic control.*

By adjusting the conditions under which an enolate mixture is formed from a ketone, it is possible to establish either kinetic or thermodynamic control. Kinetic control will be observed when the enolates, once formed, are interconverted only slowly. This situation is observed when a very strong base—for example triphenylmethyllithium—is used in an aprotic solvent and no excess ketone is present. Use of lithium as the metal ion also favors kinetic control. Protic solvents and excess ketone must be excluded, since their presence would permit equilibration by proton-transfer reactions subsequent to enolate formation. The small lithium

$$\underset{\text{A}}{R_2C\!\!=\!\!\overset{\overset{\displaystyle O^-}{|}}{C}CH_2R'} + R_2CH\overset{\overset{\displaystyle O}{\|}}{C}CH_2R' \rightleftharpoons R_2CH\overset{\overset{\displaystyle O}{\|}}{C}CH_2R' + \underset{\text{B}}{R_2CH\overset{\overset{\displaystyle O^-}{|}}{C}\!\!=\!\!CHR'}$$

cation is tightly coordinated with the oxygen atom of the enolate ion and tends to decrease the rate of proton-exchange reactions. House and his associates have studied the composition of enolates formed under conditions of kinetic control by adding the ketone to a solution of triphenylmethyllithium in the aprotic solvent dimethoxyethane. Equilibration (thermodynamic control) occurs when excess ketone is present.[1] The enolate composition was determined by allowing the enolates to react with acetic anhydride. Rapid formation of enol acetates occurs, and subsequent determination of the ratio of enol acetates reveals the ratio of enolates present in the solution.

$$R_2C\!\!=\!\!\overset{\overset{\displaystyle O^-}{|}}{C}CH_2R' + R_2CH\overset{\overset{\displaystyle O^-}{|}}{C}\!\!=\!\!CHR' \xrightarrow{(CH_3CO)_2O} R_2C\!\!=\!\!\overset{\overset{\displaystyle O\overset{\displaystyle\|}{C}CH_3}{|}}{C}CH_2R' + R_2CH\overset{\overset{\displaystyle O\overset{\displaystyle\|}{C}CH_3}{|}}{C}\!\!=\!\!CHR'$$

The enol acetate mixture can be analyzed by gas chromatography or by nmr analysis. Table 1.2 shows the data obtained for several ketones. A consistent relationship is found in these and related data. Conditions of kinetic control usually favor the less substituted enolate, as is true in each of the cases shown in Table 1.2. The principal reason for this result is probably that removal of the less hindered hydrogen is more rapid, for steric reasons, than removal of more hindered protons, and this more rapid reaction leads to the less substituted enolate. Similar results were obtained when an amine anion, lithium diisopropylamide, was used instead of triphenylmethyllithium.[2] On the other hand, at equilibrium it is the more substituted enolate that is usually the dominant species. The stability of carbon–carbon double bonds increases with increasing substitution, and it is this substituent effect that leads to the greater stability of the more substituted enolate.

Proton abstraction from α,β-unsaturated ketones occurs preferentially from the γ-carbon atom to afford the more stable enolate:

$$R_2CH\overset{\overset{\displaystyle O}{\|}}{C}CH\!\!=\!\!CHCH_2R' \xrightarrow{B^-} \underset{\text{(more stable)}}{R_2CH\overset{\overset{\displaystyle O^-}{|}}{C}\!\!=\!\!CH\!-\!CH\!\!=\!\!CHR'} \gg \underset{\text{(less stable)}}{R_2C\!\!=\!\!\overset{\overset{\displaystyle O}{|}}{C}CH\!\!=\!\!CHCH_2R'}$$

1. H. O. House and B. M. Trost, *J. Org. Chem.* **30**, 1341 (1965).
2. H. O. House, M. Gall, and H. D. Olmstead, *J. Org. Chem.* **36**, 2361 (1971).

These isomeric enolates differ in stability because the one system is fully conjugated, whereas cross-conjugation is present in the second case. The cross-conjugated isomer restricts the delocalization of the negative charge to the oxygen and α-carbon, whereas in the conjugated system the oxygen, α'-carbon, and β'-carbon all bear part of the negative charge.

The terms "kinetic control" and "thermodynamic control" are applicable to other reactions besides enolate formation; this concept was covered in general terms in Part A, Section 4.8. In discussions of other reactions in this chapter, it may be stated that a given reagent or set of conditions favors the "thermodynamic product." This statement means that the mechanism operating is such that the various possible products are equilibrated after initial formation. When this is true, the dominant product can be predicted by considering the relative stabilities of the various possible products. On the other hand, if a given reaction is under "kinetic control," prediction or interpretation of the relative amounts of products must be made by analyzing the competing rates of formation of the products.

Table 1.2. Compositions of Enolate Mixtures[a]

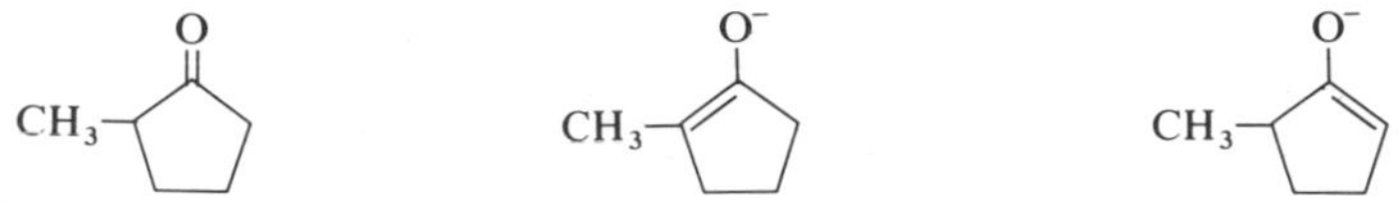

	Kinetic: 28	72
	Thermodynamic: 94	6

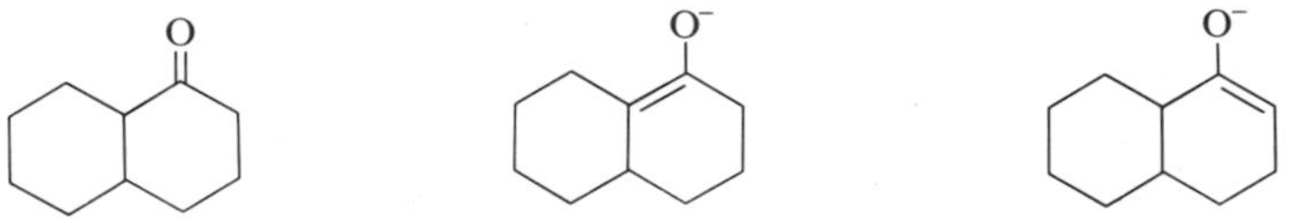

	Kinetic: 10	90
	Thermodynamic: 66	34

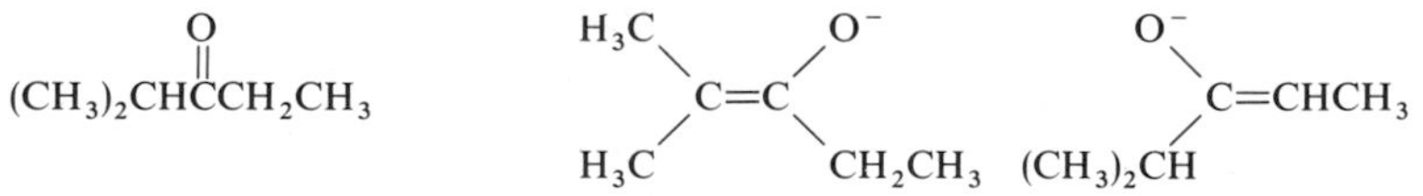

	Kinetic: <1	>99

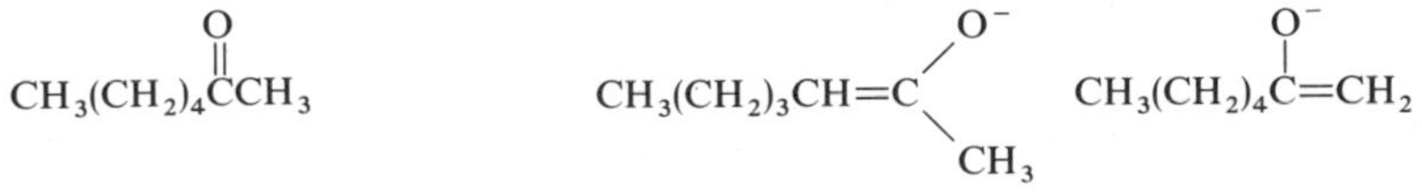

	Kinetic: ~25	~75
	Thermodynamic: ~87	~13

a. From H. O. House and B. M. Trost, *J. Org. Chem.* **30**, 1341 (1965).

Scheme 1.3. Generation of Specific Enolates 7

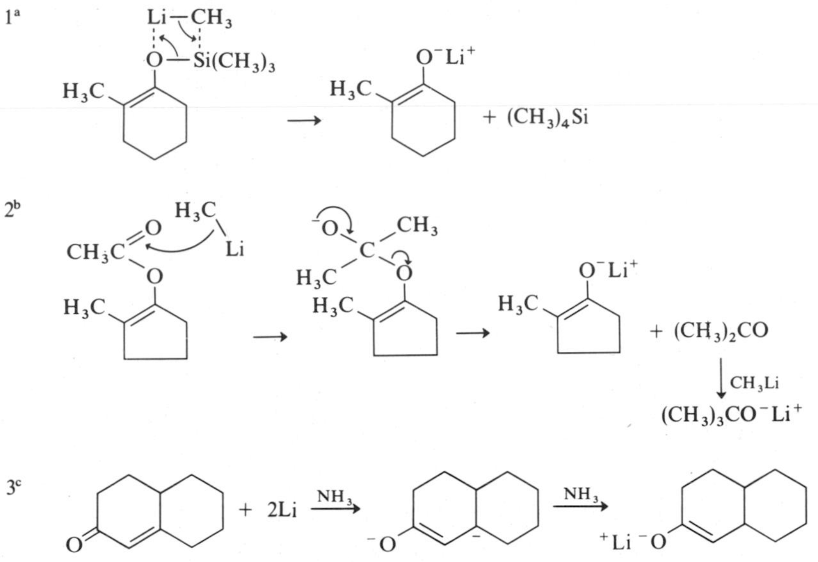

a. G. Stork and P. E. Hudrlik, *J. Am. Chem. Soc.* **90**, 4464 (1968); H. O. House, L. J. Czuba, M. Gall, and H. D. Olmstead, *J. Org. Chem.* **34**, 2324 (1969).
b. H. O. House and B. M. Trost, *J. Org. Chem.* **30**, 2502 (1965).
c. G. Stork, P. Rosen, N. Goldman, R. V. Coombs, and J. Tsuji, *J. Am. Chem. Soc.* **87**, 275 (1965).

1.3. Other Means of Generating Enolates

The development of conditions under which lithium enolates do not equilibrate with other possible isomeric enolates has permitted the use of reactions that are more specific than proton abstraction to generate specific enolates. Three such methods and the reaction mechanisms involved are shown in Scheme 1.3. The synthetic use of solutions containing specific enolate species is described in the following section.

Cleavage of enol trimethylsilyl ethers or enol acetates by methyllithium (entries 1 and 2, Scheme 1.3) as a route to specific enolate formation is limited by the availability of these materials. Preparation of the enol trimethylsilyl ethers and enol acetates from the corresponding ketones usually affords a mixture of the two possible derivatives, which must be then separated. It is sometimes possible to find conditions that favor the formation of one isomer; for example, reaction of 2-methyl-cyclohexanone with lithium diisopropylamide and trimethylchlorosilane affords the less highly substituted enol ether preferentially by 99 : 1 over the more highly substituted one (kinetically controlled conditions).[3]

3. H. O. House, L. J. Czuba, M. Gall, and H. D. Olmstead, *J. Org. Chem.* **34**, 2324 (1969).

Lithium–ammonia reduction of α,β-unsaturated ketones (entry 3, Scheme 1.3) provides a more generally useful method for generating specific enolates, since the desired starting material is often readily available by the use of various condensation reactions (Chapter 2).

1.4. Alkylations of Enolates

The alkylation of substances such as β-diketones, β-ketoesters, and esters of malonic acid can be carried out in alcoholic solvents using metal alkoxides as bases. The presence of two electron-withdrawing substituents favors formation of a single enolate by abstraction of a hydrogen from the carbon situated between them. Alkylation then occurs by an S_N2 process.

Scheme 1.4. Alkylations of Relatively Acidic Carbon Acids

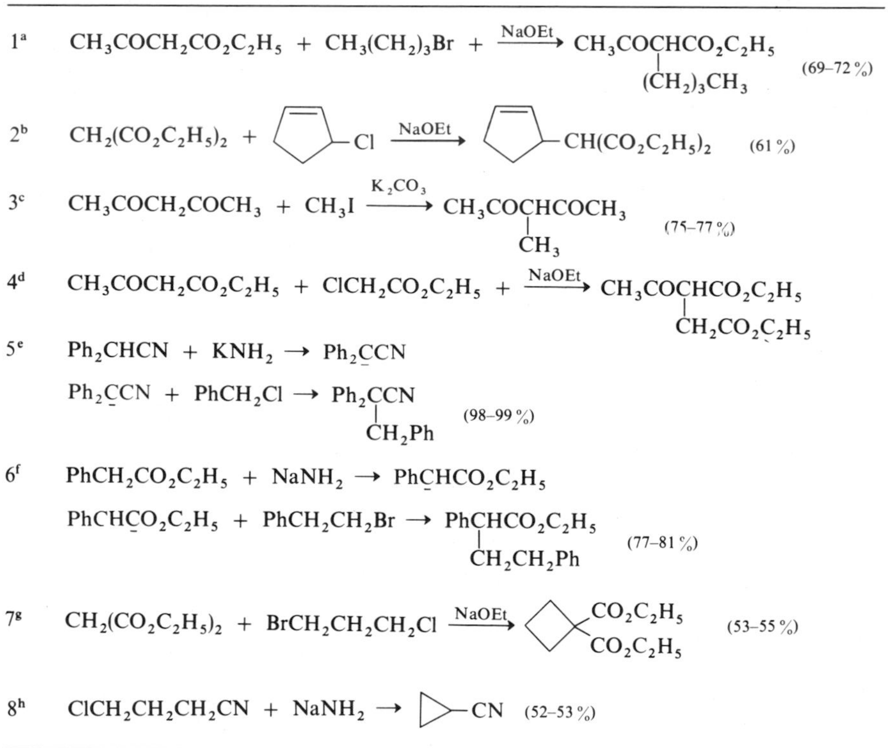

1[a] $CH_3COCH_2CO_2C_2H_5 + CH_3(CH_2)_3Br + \xrightarrow{NaOEt} CH_3COCHCO_2C_2H_5$
$\quad\quad (CH_2)_3CH_3$ (69–72%)

2[b] $CH_2(CO_2C_2H_5)_2 + $ $-Cl \xrightarrow{NaOEt}$ $-CH(CO_2C_2H_5)_2$ (61%)

3[c] $CH_3COCH_2COCH_3 + CH_3I \xrightarrow{K_2CO_3} CH_3COCHCOCH_3$
$\quad\quad CH_3$ (75–77%)

4[d] $CH_3COCH_2CO_2C_2H_5 + ClCH_2CO_2C_2H_5 + \xrightarrow{NaOEt} CH_3COCHCO_2C_2H_5$
$\quad\quad CH_2CO_2C_2H_5$

5[e] $Ph_2CHCN + KNH_2 \rightarrow Ph_2\underset{\cdot}{C}CN$

$Ph_2\underset{\cdot}{C}CN + PhCH_2Cl \rightarrow Ph_2CCN$
$\quad\quad CH_2Ph$ (98–99%)

6[f] $PhCH_2CO_2C_2H_5 + NaNH_2 \rightarrow Ph\underset{\cdot}{C}HCO_2C_2H_5$

$Ph\underset{\cdot}{C}HCO_2C_2H_5 + PhCH_2CH_2Br \rightarrow PhCHCO_2C_2H_5$
$\quad\quad CH_2CH_2Ph$ (77–81%)

7[g] $CH_2(CO_2C_2H_5)_2 + BrCH_2CH_2CH_2Cl \xrightarrow{NaOEt}$ $\begin{array}{l}CO_2C_2H_5\\CO_2C_2H_5\end{array}$ (53–55%)

8[h] $ClCH_2CH_2CH_2CN + NaNH_2 \rightarrow$ $-CN$ (52–53%)

a. C. S. Marvel and F. D. Hager, *Org. Synth.* **I**, 248 (1941).
b. R. B. Moffett, *Org. Synth.* **IV**, 291 (1963).
c. A. W. Johnson, E. Markham, and R. Price, *Org. Synth.* **42**, 75 (1962).
d. H. Adkins, N. Isbell, and B. Wojcik, *Org. Synth.* **II**, 262 (1943).
e. C. R. Hauser and W. R. Dunnavant, *Org. Synth.* **IV**, 962 (1963).
f. E. M. Kaiser, W. G. Kenyon, and C. R. Hauser, *Org. Synth.* **47**, 72 (1967).
g. R. P. Mariella and R. Raube, *Org. Synth.* **IV**, 288 (1963).
h. M. J. Schlatter, *Org. Synth.* **III**, 223 (1955).

Scheme 1.5. Synthesis of Ketones and Carboxylic Acid Derivatives via Alkylation Techniques

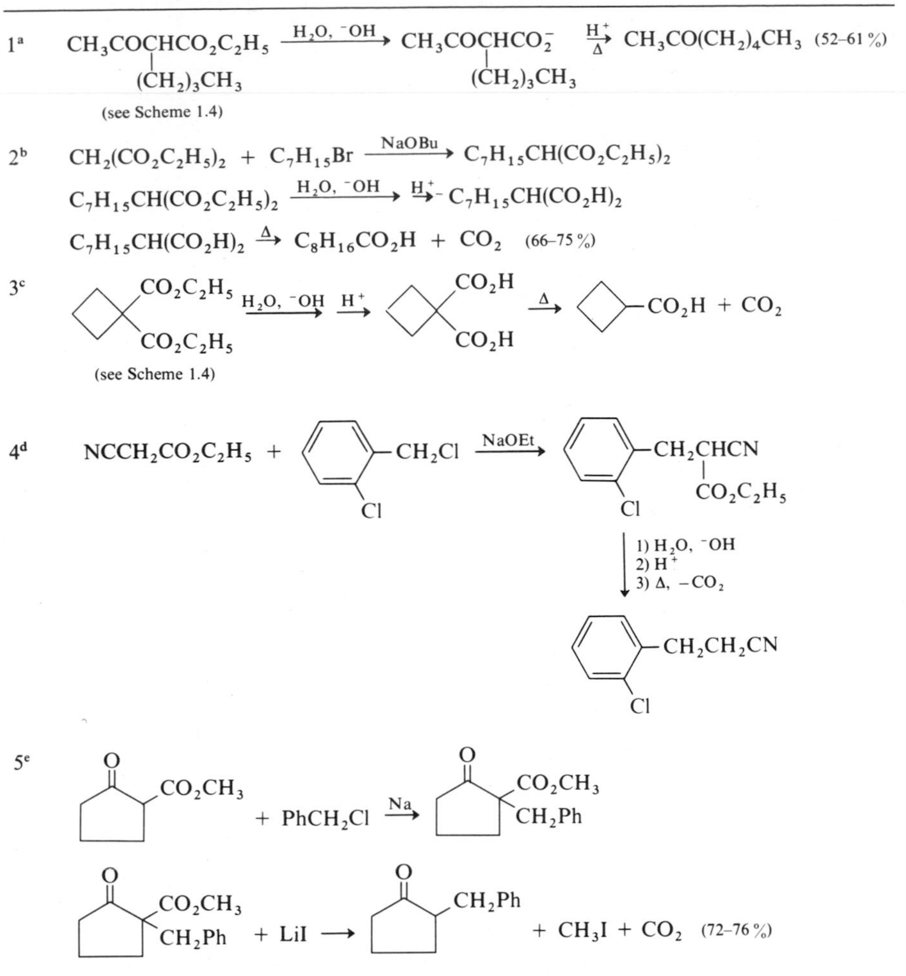
1[a] CH3COCHCO2C2H5 (CH2)3CH3 H2O, -OH → CH3COCHCO2- (CH2)3CH3 H+, Δ → CH3CO(CH2)4CH3 (52–61%)
(see Scheme 1.4)
2[b] CH2(CO2C2H5)2 + C7H15Br NaOBu → C7H15CH(CO2C2H5)2
C7H15CH(CO2C2H5)2 H2O, -OH → H+ → C7H15CH(CO2H)2
C7H15CH(CO2H)2 Δ → C8H16CO2H + CO2 (66–75%)
3[c] CO2C2H5 CO2C2H5 H2O, -OH H+ → CO2H CO2H Δ → CO2H + CO2
(see Scheme 1.4)
4[d] NCCH2CO2C2H5 + CH2Cl Cl NaOEt → CH2CHCN CO2C2H5 Cl
1) H2O, -OH 2) H+ 3) Δ, -CO2
CH2CH2CN Cl
5[e] O CO2CH3 + PhCH2Cl Na → O CO2CH3 CH2Ph
O CO2CH3 CH2Ph + LiI → O CH2Ph + CH3I + CO2 (72–76%)

a. J. R. Johnson and F. D. Hager, *Org. Synth.* **I**, 351 (1941).
b. E. E. Reid and J. R. Ruhoff, *Org. Synth.* **II**, 474 (1943).
c. G. B. Heisig and F. H. Stodola, *Org. Synth.* **III**, 213 (1955).
d. J. A. Skorcz and F. E. Kaminski, *Org. Synth.* **48**, 53 (1968).
e. F. Elsinger, *Org. Synth.* **45**, 7 (1965).

Some examples of the more important alkylation reactions with relatively acidic carbon acids are included in the reactions shown in Scheme 1.4. These reactions are important means of synthesizing a variety of ketones, carboxylic acids, and related compounds, as illustrated in Scheme 1.5. These reactions share a common mechanism involving base-catalyzed formation of a carbanion followed by nucleophilic attack via an S_N2 mechanism. The alkylating agent must be a suitable substrate for an S_N2 reaction. Primary halides and sulfonates are the best substrates. Secondary systems usually give poorer yields because of competition from elimination

Scheme 1.6. Alkylation of Some Specific Enolates

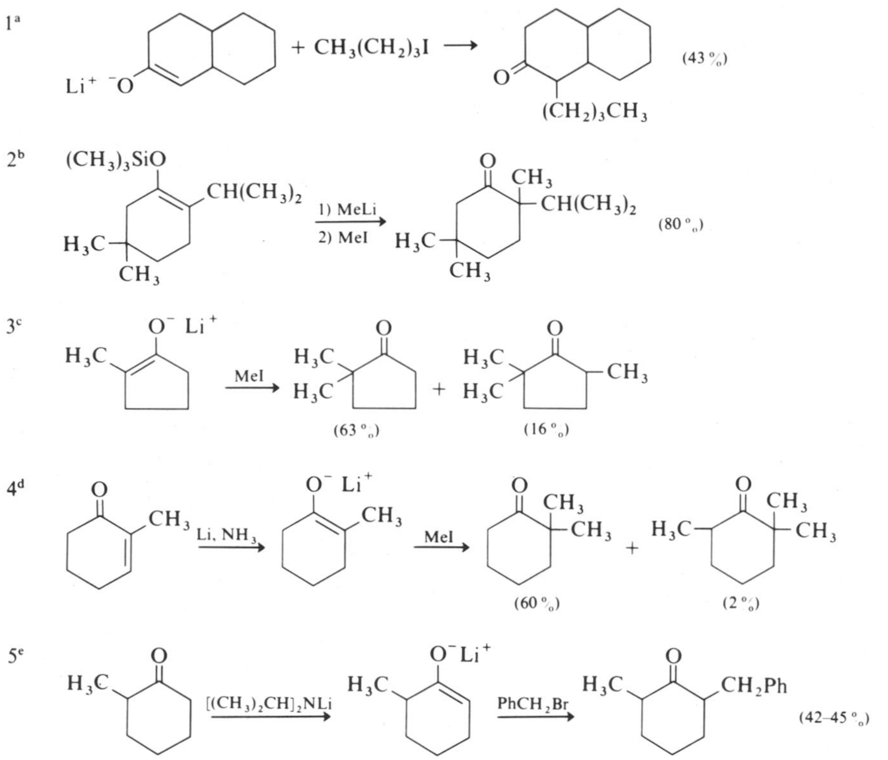

1[a]

+ CH$_3$(CH$_2$)$_3$I ⟶ (43 %)

2[b]

1) MeLi
2) MeI

(80 %)

3[c]

MeI ⟶ (63 %) + (16 %)

4[d]

Li, NH$_3$ ⟶ MeI ⟶ (60 %) + (2 %)

5[e]

[(CH$_3$)$_2$CH]$_2$NLi ⟶ PhCH$_2$Br ⟶ (42–45 %)

a. G. Stork, P. Rosen, N. Goldman, R. V. Coombs, and J. Tsugji, *J. Am. Chem. Soc.* **87**, 275 (1965).
b. G. Stork and P. E. Hudrlik, *J. Am. Chem. Soc.* **90**, 4464 (1968).
c. H. O. House and B. M. Trost, *J. Org. Chem.* **30**, 2502 (1965).
d. H. A. Smith, B. J. L. Huff, W. J. Powers, III, and D. Caine, *J. Org. Chem.* **32**, 2851 (1967).
e. M. Gall and H. O. House, *Org. Synth.* **52**, 39 (1972).

reactions. Tertiary halides or sulfonates are unsatisfactory because elimination rather than substitution occurs.

Methylene groups can be dialkylated if sufficient amounts of base and alkylating agent are used. Dialkylation can be an undesirable side reaction if the monoalkyl derivative is the desired product. Use of dihalides as the alkylating reagents leads to ring formation, as illustrated by the diethyl cyclobutanedicarboxylate synthesis (entry 7) shown in Scheme 1.4. Intramolecular reactions are also possible, as illustrated by the synthesis of cyclopropanecarbonitrile (entry 8, Scheme 1.4).

If only a single electron-withdrawing substituent is present, as with simple ketones, esters, and nitriles, the formation of alkyl derivatives in high yield requires careful control of reaction conditions. Use of bases that are strong enough to effect only partial conversion of the substrate to its anion can result in aldol-condensation reactions with ketones and Claisen condensations with esters (see Chapter 2 for discussion of these reactions). This problem can be partially avoided by use of very strong bases such as the amide, hydride, or triphenylmethyl anions.

It was pointed out that with unsymmetrical ketones, proton-abstraction reactions normally lead to mixtures of the various possible enolates. The composition of the mixture is dependent on the extent of kinetic or thermodynamic control. Perhaps the simplest cases to consider are enolates generated by the methods discussed in Section 1.3. Here a single enolate is the dominant species, and subsequent alkylation leads mainly to its alkylation product. Scheme 1.6 shows some examples of alkylation of specific enolates.

Entries 3 and 4 in Scheme 1.6 illustrate the formation of by-products by dialkylation. The dialkylation occurs because the monoalkylated product, once formed, can react with remaining unreacted enolate to form the enolate of the alkylated ketone:

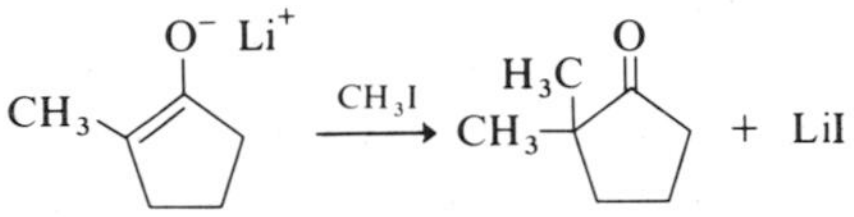

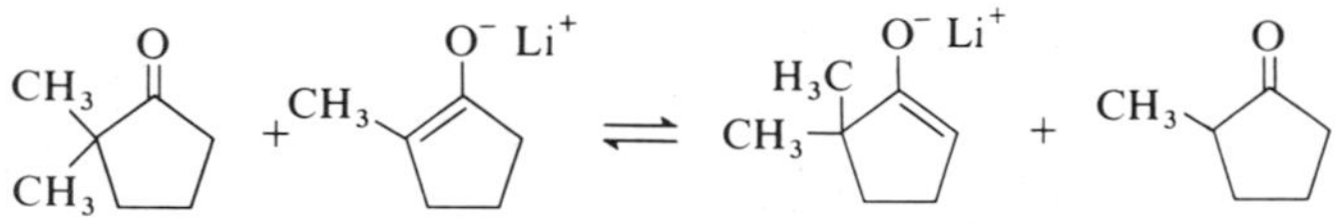

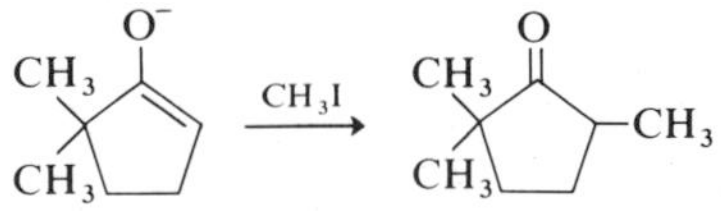

These processes can occur very rapidly in protic solvents or when a weakly coordinating cation is used. It is for this reason that base-catalyzed alklyations of ketones in protic solvents seldom give good yields of monoalkylation products. The use of specific lithium enolates in aprotic solvents minimizes these difficulties.[4]

The stereochemistry of enolate alkylations has been studied by determining the stereochemistry of products from alkylation of cyclic ketones. The stereochemistry of alkylation of the enolates 1 and 2 has been determined. While 1 shows no

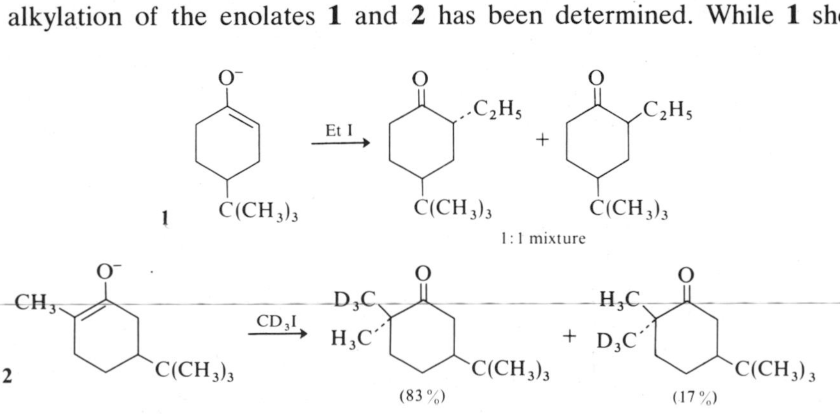

4. H. O. House, *Rec. Chem. Prog.* **28**, 98 (1967).
5. H. O. House, B. A. Terfertiller, and H. D. Olmstead, *J. Org. Chem.* **33**, 935 (1968).
6. H. O. House and M. J. Umen, *J. Org. Chem.* **38**, 1000 (1973).

preference for the two possible approaches by alkylating agent, **2** gives more of the product corresponding to introduction of the new alkyl group in the axial orientation. In general, high stereoselectivity with respect to the two faces of an enolate system is not to be expected unless one side of the enolate is sterically hindered. Then, the alkylation can be quite stereoselective.

1.5. Generation and Alkylation of Dianions

In the presence of a very strong base, such as amide ion or an organolithium reagent, it is possible to convert dicarbonyl compounds to their dianions. Subsequent alkylation of such dianions leads to alkylation at the more strongly basic enolate site, rather than at the carbon atom between the two carbonyl carbons. The more acidic methylene group activated by two carbonyl substituents is the preferred site in the monoanion, as discussed earlier. The ability to determine the site of monoalkylation by choice of the amount and nature of the basic catalyst has significantly expanded the synthetic utility of enolate alkylations.[7] Scheme 1.7 gives some examples of formation and alkylation of dianions.

1.6. Solvent Effects in Enolate Alkylations

Certain aprotic polar solvents, including dimethylformamide, dimethyl sulfoxide, and hexamethylphosphoramide, have been found to markedly accelerate enolate alkylation reactions.[8] The relative rates of alkylation of the sodium enolate of diethyl *n*-butylmalonate by butyl bromide are shown in Table 1.3. The greatly enhanced rates in dimethylformamide and dimethyl sulfoxide illustrate the rate enhancement by polar aprotic solvents.

The nature of the solvent determines the environment immediately around the carbanion. In nonpolar solvents such as hydrocarbons and simple ethers, carbanions

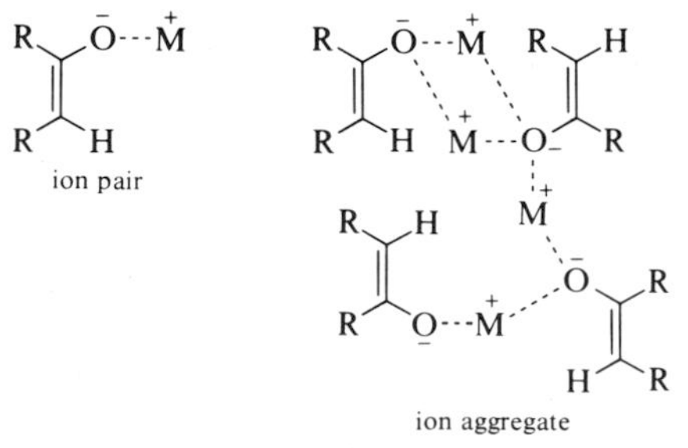

ion pair

ion aggregate

7. T. M. Harris and C. M. Harris, *Org. React.* **17**, 155 (1969).
8. A. J. Parker, *Chem. Rev.* **69**, 1 (1969).

Scheme 1.7. Generation and Alkylation of Dianions

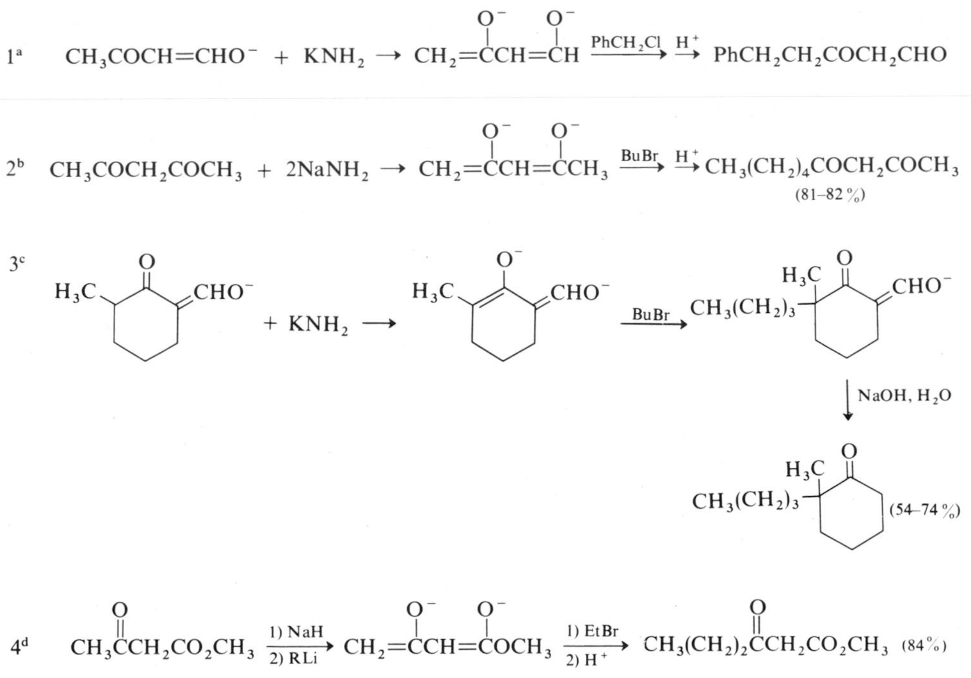

13

SECTION 1.6.
SOLVENT
EFFECTS IN
ENOLATE
ALKYLATIONS

1^a $CH_3COCH=CHO^- + KNH_2 \rightarrow CH_2=\overset{\overset{O^-}{|}}{C}CH=\overset{\overset{O^-}{|}}{C}H \xrightarrow{PhCH_2Cl} \xrightarrow{H^+} PhCH_2CH_2COCH_2CHO$

2^b $CH_3COCH_2COCH_3 + 2NaNH_2 \rightarrow CH_2=\overset{\overset{O^-}{|}}{C}CH=\overset{\overset{O^-}{|}}{C}CH_3 \xrightarrow{BuBr} \xrightarrow{H^+} CH_3(CH_2)_4COCH_2COCH_3$
(81–82 %)

3^c

NaOH, H$_2$O

(54–74 %)

4^d $CH_3\overset{\overset{O}{||}}{C}CH_2CO_2CH_3 \xrightarrow[2) RLi]{1) NaH} CH_2=\overset{\overset{O^-}{|}}{C}CH=\overset{\overset{O^-}{|}}{C}OCH_3 \xrightarrow[2) H^+]{1) EtBr} CH_3(CH_2)_2\overset{\overset{O}{||}}{C}CH_2CO_2CH_3$ (84%)

a. T. M. Harris, S. Boatman, and C. R. Hauser, *J. Am. Chem. Soc.* **85**, 3273 (1963), S. Boatman, T. M. Harris, and C. R. Hauser, *J. Am. Chem. Soc.* **87**, 82 (1965); K. G. Hampton, T. M. Harris, and C. R. Hauser, *J. Org. Chem.* **28**, 1946 (1963).
b. K. G. Hampton, T. M. Harris, and C. R. Hauser, *Org. Synth.* **47**, 92 (1967).
c. S. Boatman, T. M. Harris, and C. R. Hauser, *Org. Synth.* **48**, 40 (1968).
d. S. N. Huckin and L. Weiler, *J. Am. Chem. Soc.* **96**, 1082 (1974).

are closely associated with metal ions in ion pairs or ion aggregates. Solubility in these solvents is also limited. Nonpolar solvents are unable to separate and solvate the enolate and metal ions. In hydroxylic solvents, the polar hydroxyl groups promote dissociation of ion aggregates and ion pairs by solvation of both the anion

Table 1.3. Relative Alkylation Rates of Sodium Diethyl *n*-Butylmalonate in Various Solvents[a]

Solvent	Relative rate
Benzene	1
Tetrahydrofuran	14
Dimethoxyethane	80
Dimethylformamide	970
Dimethyl sulfoxide	1420

a. From H. E. Zaugg, *J. Am. Chem. Soc.* **83**, 837 (1961).

and the cation. Hydrogen bonds are formed between the solvent and enolate anion:

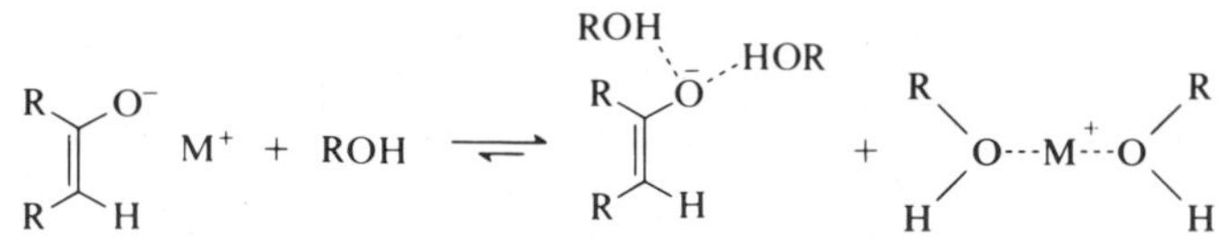

Solvents such as dimethylformamide have relatively large dipole moments. The oxygen atoms are sites of high electron density and are capable of coordinating

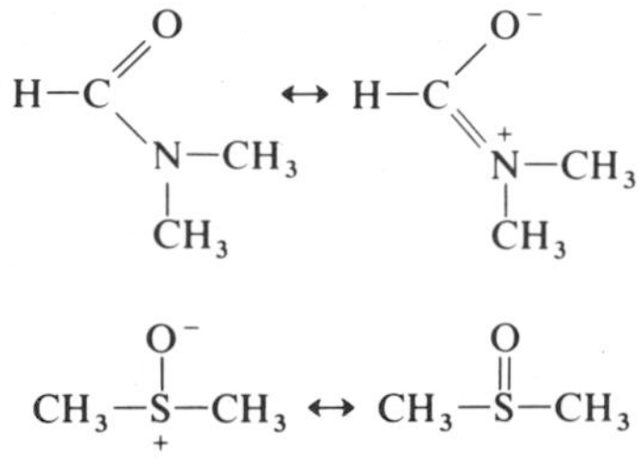

positively charged metal ions. The positive sites in these solvents are less effective at solvating anions, probably, at least in part, because the positive charge is shielded by the methyl groups in each molecule. The alkali metal enolates exist as ion pairs even

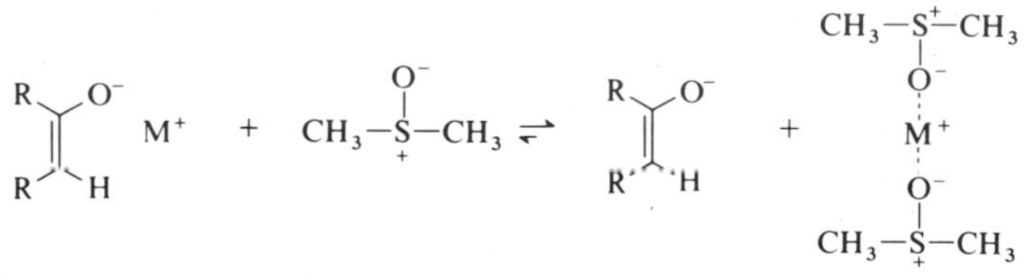

in these polar solvents, but the extensive solvation of the cation decreases the strength of the attraction between the ions.[9]

It is clear that the differing environments around the enolate anion in these various media will strongly affect the reactivity of the enolate anion. Solvation around the enolate anion must be disrupted, at least partially, to permit alkylation to take place. The high reactivity in polar aprotic solvents is due in large part to the fact

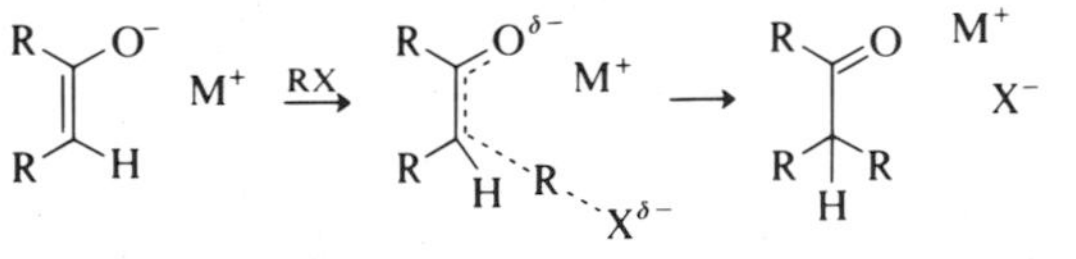

that the anion is largely unsolvated, and little energy needs to be expended in breaking down ion-pairing or solvation. Stated in another way, the unsolvated enolate reactant is relatively higher in energy than a solvated ion, and the activation energy for the reaction is therefore lower. This difference in reactivity is illustrated in Figure 1.1.

9. H. E. Zaugg, J. F. Ratajczyk, J. E. Leonard, and A. D. Schaefer, *J. Org. Chem.* **37**, 2249 (1972).

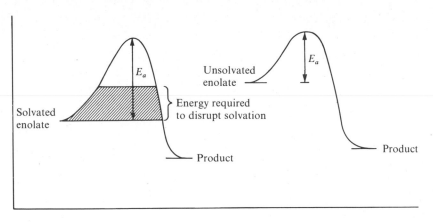

Figure 1.1. Enhanced reactivity of unsolvated enolates.

1.7. Oxygen Versus Carbon as the Site of Alkylation

Enolate anions present two possible sites for alkylation. Nucleophiles with more than one potential site for electrophilic attack are referred to as *ambident nucleophiles*. When the alkylating agent is an alkyl halide, carbon alkylation is normally

$$\text{C-alkylation} \qquad R\overset{O^-}{\underset{|}{C}}=CH_2 \ + \ R'-X \ \rightarrow \ R\overset{O}{\underset{\|}{C}}CH_2R'$$

$$\text{O-alkylation} \qquad R\overset{O^-}{\underset{|}{C}}=CH_2 \ + \ R'-X \ \rightarrow \ R\overset{O-R'}{\underset{|}{C}}=CH_2$$

the dominant process. Let us consider the factors involved in the preference for C-alkylation; three interrelated factors must be taken into account. As discussed in the preceding section, the extent of solvation of the enolate anion has a strong effect on anion reactivity. The detailed structure of the solvated anion can affect the ratio of oxygen to carbon alkylation. If there is strong solvation in the vicinity of the oxygen atom and a looser solvation at carbon, relatively greater reactivity at carbon would be expected. Since the charge distribution in an enolate places much of the negative charge on oxygen, it is not unreasonable to expect the strongest solvation by hydrogen bonding at the oxygen.

Second, and as discussed in Part A, Chapter 5, nucleophilicity in S_N2 reactions is associated with polarizability. The more easily a nucleophile's electronic cloud can be distorted to permit bond formation, the stronger an S_N2 nucleophile it will be. Comparison of the oxygen and carbon ends of an ambident enolate ion with regard to nucleophilicity leads to the conclusion that the less electronegative carbon atom is more polarizable and to the prediction that the carbon end of the anion will be more nucleophilic.

Finally, the structures of the transition states for carbon and oxygen alkylation must be compared:

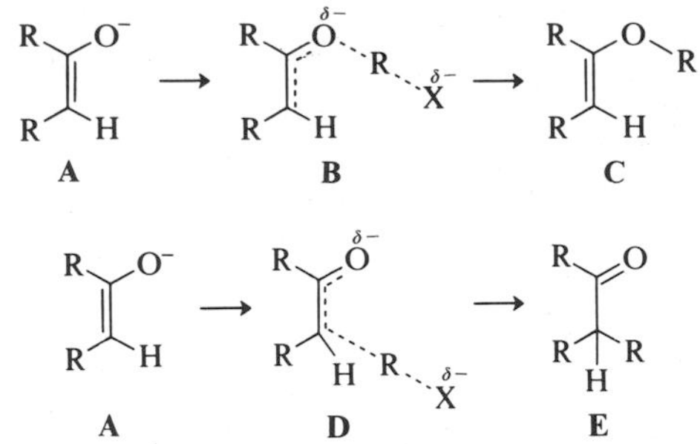

Comparison of the bonds present in the O-alkylation product with those in the C-alkylated ketone show that the ketone is considerably more stable because of the strength of the carbonyl bond:

C—O	79		C=O	173
C—O	79		C—C	80
C=C	145		C—C	80
	303 kcal/mol			333 kcal/mol

Under the conditions of most alkylation reactions, the products of O- and C-alkylation are usually not interconvertible, so it is not valid to predict C-alkylation directly on the basis of thermodynamic considerations. But, the transition state **B** and **D** resemble the products **C** and **E**, respectively, to a significant extent. It is therefore expected that the transition state for C-alkylation will be lower in energy, anticipating to some extent the greater stability of the product. This is depicted in Figure 1.2. The competition between O- and C-alkylation will then depend upon the interplay among (1) solvation effects, (2) nucleophilicity of the C and O ends of the enolate ion, and (3) transition-state structure.

In the case of enolate anions, C-alkylation usually dominates, except with very reactive alkylating agents. When the alkylating species is highly reactive, the transition state resembles the reactant enolate ion rather than product, and the high charge density at oxygen leads to O-alkylation. Oxygen alkylation is unusual in base-catalyzed alkylations. Such reagents as diazomethane and the triethyloxonium ion, however, tend to react at oxygen.[5] Alkyl sulfates and sulfonates tend to give

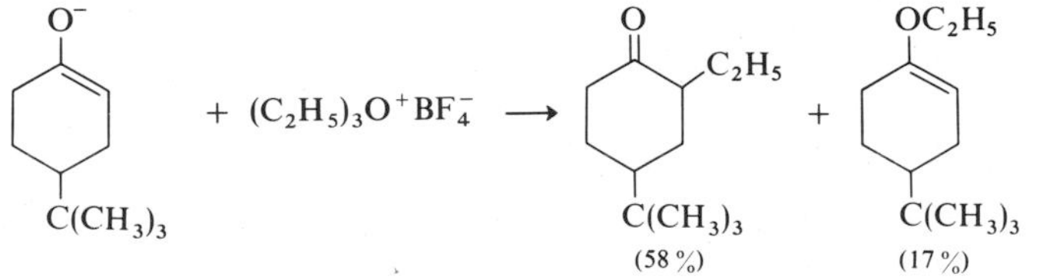

5. See p. 11.

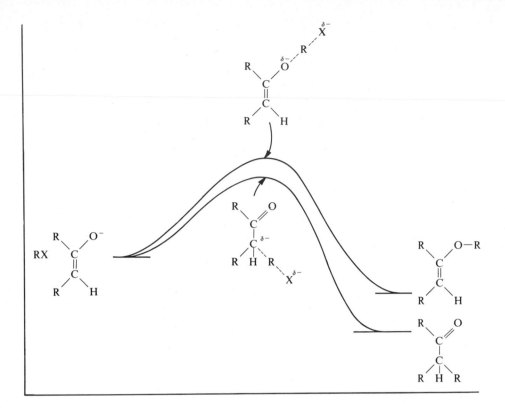

Figure 1.2. O- versus C-alkylation.

more O-alkylation product than the corresponding halides. Chloromethyl methyl ether, an especially reactive halide, gives an exceptionally high percentage of O-alkylation.[10]

Enolates in which the carbon atom is relatively hindered may also give extensive O-alkylation, especially in polar aprotic solvents. For example, diphenyl-acetophenone has been found to give much O-alkylation product in the presence of dimethyl sulfoxide.[11] Hexamethylphosphoramide, another dipolar aprotic solvent, gives higher O : C alkylation ratios for alkylation of the ethyl acetoacetate anion than are observed in dimethyl sulfoxide and is perhaps the solvent of choice if O-alkylation is desired.[12]

10. R. M. Coates and J. E. Shaw, *J. Org. Chem.* **35**, 2597, 2601 (1970).
11. H. D. Zook, T. J. Russo, E. F. Ferrand, and D. S. Stotz, *J. Org. Chem.* **33**, 2222 (1968).
12. W. J. LeNoble and H. F. Morris, *J. Org. Chem.* **34**, 1969 (1969).

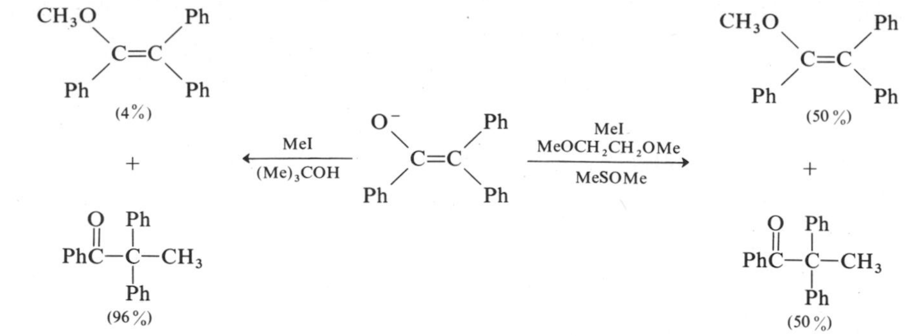

The competition between C- and O-alkylation is rather closely balanced in the alkylation of phenolate anions:

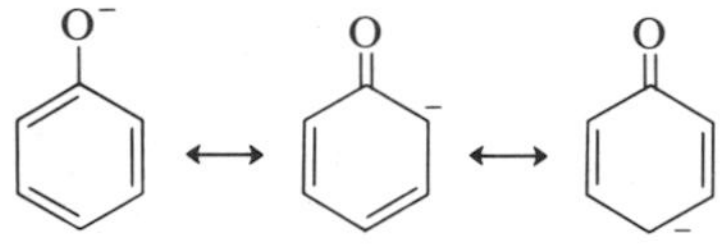

In this case, C-alkylation is burdened energetically by the fact that aromaticity is destroyed as C-alkylation proceeds:

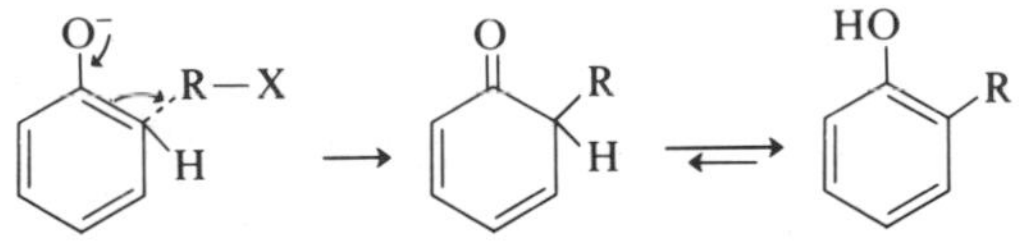

The effect of solvent on the site of alkylation has been clearly demonstrated for phenolate ions. In solvents such as dimethyl sulfoxide, dimethylformamide, ethers, and alcohols, O-alkylation is dominant. In water, phenol, and trifluoroethanol, however, extensive amounts of C-alkylation occur.[13] These latter solvents would be expected to form particularly strong hydrogen bonds with the oxygen atom of the phenolate anion. This strong solvation decreases the reactivity at oxygen and favors carbon alkylation.

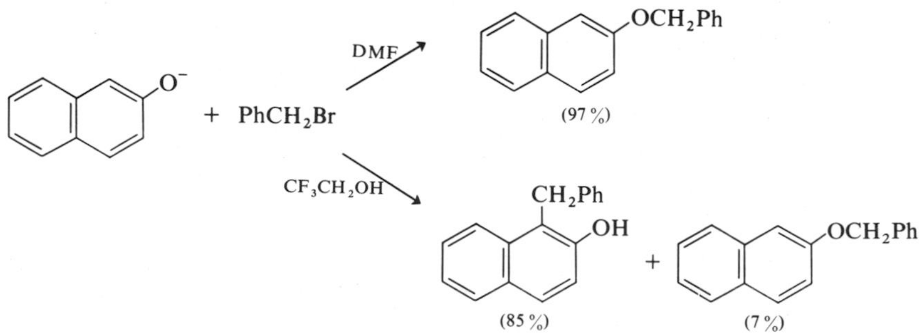

In enolates formed from α,β-unsaturated ketones by proton abstraction from

13. N. Kornblum, P. J. Berrigan, and W. J. LeNoble, *J. Am. Chem. Soc.* **85**, 1141 (1963); N. Kornblum, R. Seltzer, and P. Haberfield, *J. Am. Chem. Soc.* **85**, 1148 (1963).

the γ-carbon, there are three potential sites for attack by electrophiles: the oxygen, the α-carbon, and the γ-carbon. The kinetically preferred site for both protonation and alkylation is the α-carbon. Protonation of the enolate provides a method for

19

SECTION 1.8.
ALKYLATIONS OF
ALDEHYDES,
ESTERS, NITRILES,
AND NITRO
COMPOUNDS

converting α,β-unsaturated ketones and esters to the less stable β,γ-unsaturated isomers:

(major)   (minor)   Ref. 14

$$CH_3CH=CHCO_2C_2H_5 \xrightarrow{\text{LiNR}_2} \xrightarrow{\text{H}_2\text{O}} CH_2=CHCH_2CO_2C_2H_5 + CH_3CH=CHCO_2C_2H_5 \quad \text{Ref. 15}$$

(87%)   (13%)

Alkylation also takes place selectively at the α-carbon, but these enolates have a very strong tendency to undergo dialkylation, and introduction of a single alkyl group is difficult.[16]

1.8. Alkylations of Aldehydes, Esters, Nitriles, and Nitro Compounds

Among the groups of compounds having functionalities that can stabilize negative charge on carbon, the ketones have been the most widely studied. Base-catalyzed α-alkylations of aldehydes are rare. Aldehydes are very reactive toward base-catalyzed aldol condensations (Chapter 2), and this reaction is dominant with bases that convert aldehydes only partially to the conjugate base. Complete formation of the corresponding enolate with a very strong base followed by introduction of an alkylating agent might be a feasible means of aldehyde alkylation, but there are relatively few examples of such reactions. Alkylation via enamines or imine magnesium salts provides an indirect procedure for alkylation of aldehydes. These reactions will be discussed in Section 1.9.

Base-catalyzed alkylations of simple esters require strongly basic catalysts. Relatively weak bases such as alkoxides promote condensation reactions (Chapter 2). The techniques for successful formation of ester enolates have been developed

14. H. J. Ringold and S. K. Malhotra, *Tetrahedron Lett.*, 669 (1962); S. K. Malhotra and H. J. Ringold, *J. Am. Chem. Soc.* **85**, 1538 (1963).
15. M. W. Rathke and D. Sullivan, *Tetrahedron Lett.*, 4249 (1972).
16. G. Stork and J. Benaim, *J. Am. Chem. Soc.* **93**, 5938 (1971).

only recently. Highly hindered bases, particularly diisopropylamide anion, can successfully abstract an α-proton from esters and lactones at low temperatures without competing addition at the carbonyl group.[17] The resulting enolates can be successfully alkylated with alkyl bromides or iodides:

$$CH_3(CH_2)_4CO_2C_2H_5 \xrightarrow[\text{2) } CH_3I]{\text{1) } LiNR_2} CH_3(CH_2)_3\underset{\underset{CH_3}{|}}{CH}CO_2C_2H_5 \qquad \text{(83\%)}$$

Ref. 17a

Less hindered bases can be used to form the enolate, provided tertiary butyl esters are employed to retard reactions at the carbonyl group. For example, lithium amide in liquid ammonia has been successfully used for the alkylation of t-butyl acetate.[18]

Phenylacetonitriles are good substrates for carbon alkylation. The phenyl group enhances the acidity of the C–H bond and stabilizes the resulting anion:

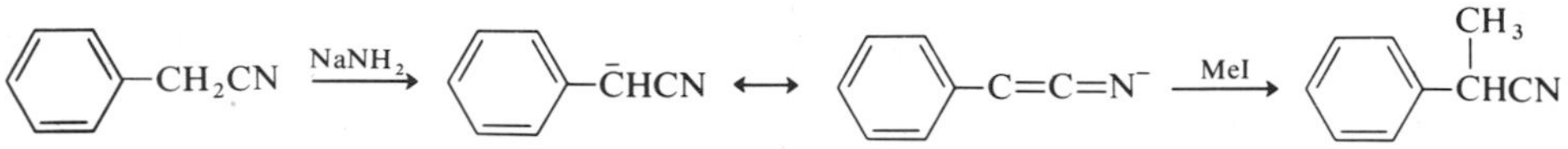

Aliphatic nitriles are less acidic, and stronger bases are required for complete anion formation. There have been too few studies of alkylations of simple nitriles to define the conditions for successful C-alkylation of nitriles.

Although aliphatic nitro compounds are more acidic than aldehydes, esters, ketones, or nitriles, the alkylation of the resulting anions almost always occurs predominantly at oxygen rather than carbon.[19] The O-alkylation products, which are

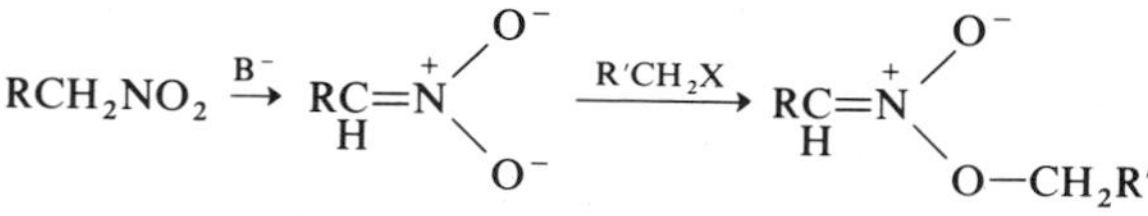

called *nitronic esters*, are very unstable and decompose in basic solution to give an aldehyde and an oxime:

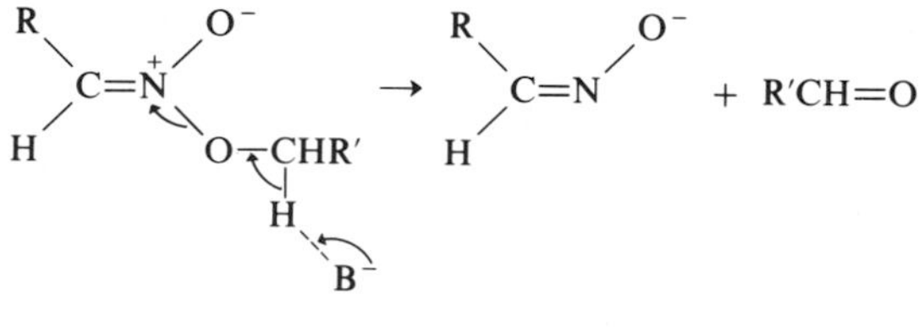

17a. M. W. Rathke and A. Lindert, *J. Am. Chem. Soc.* **93**, 2318 (1971).

b. R. J. Cregge, J. L. Herrmann, C. S. Lee, J. E. Richman, and R. H. Schlessinger, *Tetrahedron Lett.*, 2425 (1973).

c. J. L. Herrmann and R. H. Schlessinger, *Chem. Commun.*, 711 (1973).

18. W. R. Dunnavant and C. R. Hauser, *J. Org. Chem.* **25**, 1693 (1960); H. Sisido, K. Sei, and M. Nozaki, *J. Org. Chem.* **27**, 2681 (1962).

19. H. B. Hass and M. L. Bender, *J. Am. Chem. Soc.* **71**, 1767, 3482 (1949); N. Kornblum and R. A. Brown, *J. Am. Chem. Soc.* **86**, 2681 (1964); H. B. Hass and M. L. Bender, *Org. Synth.* **IV**, 932 (1963).

In some cases, this reaction has been useful for the synthesis of aldehydes from primary halides, but it seldom gives significant amounts of C-alkylation product.[20]

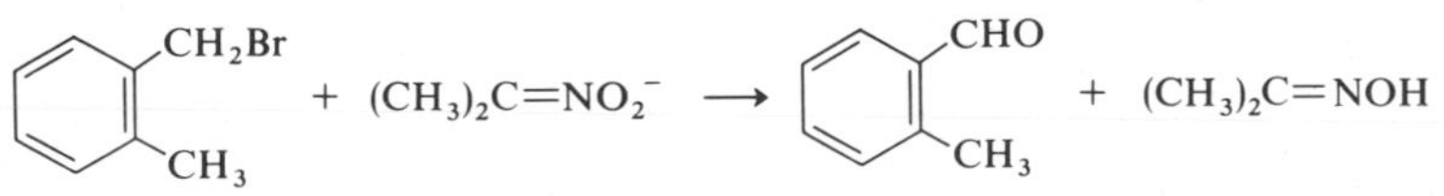

1.9. The Nitrogen Analogs of Enols and Enolates—Enamine Alkylations

The nitrogen analogs of ketones and aldehydes are known as *imines* or *azomethines*. These compounds can be prepared by condensation of amines with ketones and aldehydes[21]:

$$R-\overset{O}{\overset{\|}{C}}-R + RNH_2 \longrightarrow R-\overset{N-R}{\overset{\|}{C}}-R + H_2O$$

When secondary amines are heated with ketones and aldehydes in the presence of an acidic catalyst, a related condensation reaction occurs and can be driven to completion by removal of water. This is often accomplished by azeotropic distillation. The condensation product is a substituted vinylamine or *enamine*.

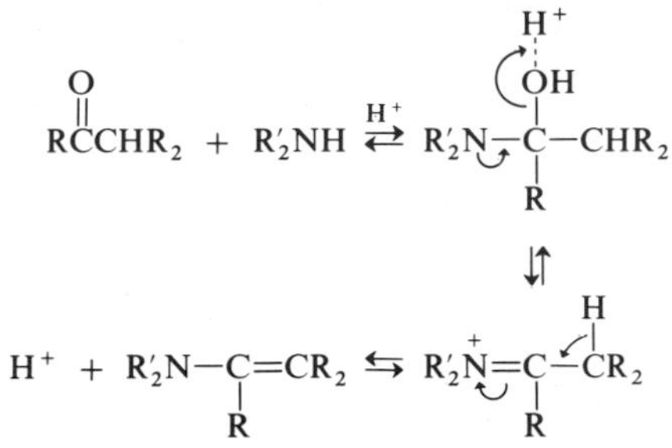

There are other methods for preparing enamines from ketones that utilize strong dehydrating reagents to drive the reaction to completion. For example, mixing carbonyl compounds and secondary amines followed by addition of titanium tetrachloride rapidly gives enamines. This method is applicable to hindered as well as to ordinary amines.[22] Another procedure involves converting the secondary amine to

$$2\,RCH_2\overset{O}{\overset{\|}{C}}R' + 6\,R_2NH + TiCl_4 \longrightarrow 2\,R\underset{NR_2}{\underset{|}{C}}H{=}CR' + 4\,R_2\overset{+}{N}H_2\ \overset{-}{Cl} + TiO_2$$

20. For an exception, see R. C. Kerber, G. W. Urry, and N. Kornblum, *J. Am. Chem. Soc.* **87**, 4520 (1965).
21. P. Y. Sollenberger and R. B. Martin, in S. Patai (ed.), *Chemistry of the Amino Group*, Chapter 7, Interscience, New York, NY, 1968.
22. W. A. White and H. Weingarten, *J. Org. Chem.* **32**, 213 (1967).

its trimethylsilyl derivative. Because of the higher affinity of silicon for oxygen than nitrogen, enamine formation is favored and takes place under mild conditions.[23]

$$R_2NSi(CH_3)_3 \ + \ RCH_2\overset{\overset{\displaystyle O}{\|}}{C}R' \ \rightarrow \ RCH=\overset{\overset{\displaystyle NR_2}{|}}{C}R' \ + \ (CH_3)_3SiOH$$

The β-carbon atom of an enamine is a nucleophilic site because of conjugation with the nitrogen atom. Indeed, acidification of enamines results in protonation at

$$R_2N-\underset{\underset{\displaystyle R}{|}}{C}=CR_2 \ \leftrightarrow \ R_2\overset{+}{N}=\underset{\underset{\displaystyle R}{|}}{C}-\overset{-}{C}R_2$$

$$R_2N-\underset{\underset{\displaystyle R}{|}}{C}=CR_2 \ \overset{H^+}{\rightarrow} \ R_2\overset{+}{N}=\underset{\underset{\displaystyle R}{|}}{C}-CHR_2$$

the carbon atom, giving an iminium ion. The nucleophilicity of the β-carbon atom of enamines can be utilized in certain synthetically useful alkylation reactions:

$$R_2N-\underset{\underset{\displaystyle R}{|}}{C}=\underset{\underset{\displaystyle R}{|}}{C}\overset{R}{\diagup} \quad R'-X \ \rightarrow \ R_2\overset{+}{N}=\underset{\underset{\displaystyle R}{|}}{C}-\underset{\underset{\displaystyle R}{|}}{\overset{\overset{\displaystyle R}{|}}{C}}-R' \ \overset{H_2O}{\rightarrow} \ R\overset{\overset{\displaystyle O}{\|}}{C}-\underset{\underset{\displaystyle R}{|}}{\overset{\overset{\displaystyle R}{|}}{C}}-R'$$

The enamines derived from cyclohexanones have been of particular interest. The enamine mixture formed from pyrrolidine and 2-methylcyclohexanone is predominantly **3**.[24] The tendency of pyrrolidine to provide the less substituted cyclohexanone

3 (90%)   **4** (10%)

enamine is quite general. A steric effect is responsible for the preference for the less substituted enamine. Maximum interaction of the π-orbital of the double bond and the nitrogen lone pair requires coplanarity of the nitrogen and carbon atoms. A severe destabilizing nonbonded repulsion arises in the more substituted isomer.

Because of the predominance of the less substituted enamine, alkylations occur

23. R. Comi, R. W. Franck, M. Reitano, and S. M. Weinreb, *Tetrahedron Lett.*, 3107 (1973).
24. W. D. Gurowitz and M. A. Joseph, *J. Org. Chem.* **32**, 3289 (1967).

Scheme 1.8. Enamine Alkylations

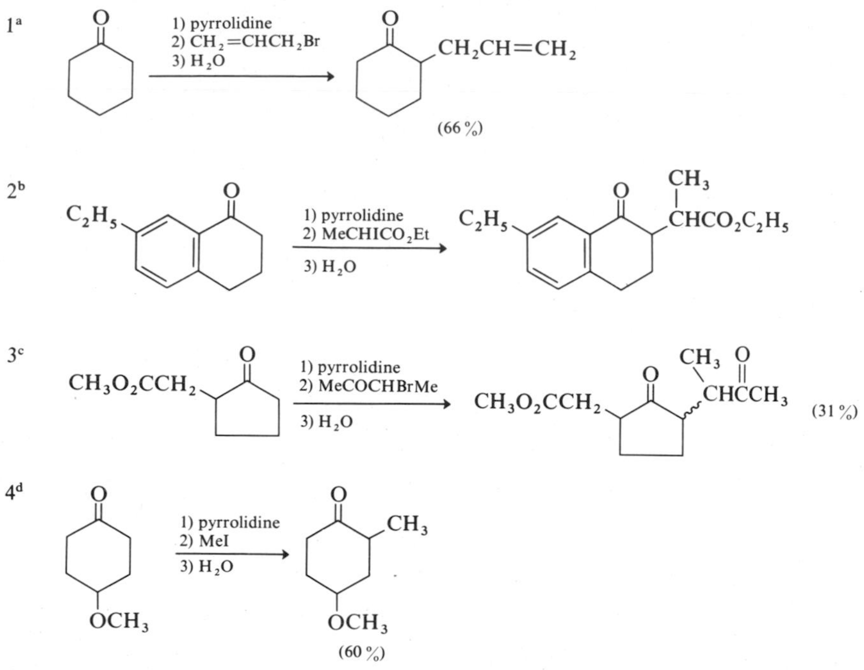

23

SECTION 1.9.
THE NITROGEN
ANALOGS OF
ENOLS AND
ENOLATES

a. G. Stork, A. Brizzolara, H. Landesman, J. Szmuszkovicz, and R. Terrell, *J. Am. Chem. Soc.* **85**, 207 (1963).
b. D. M. Locke and S. W. Pelletier, *J. Am. Chem. Soc.* **80**, 2588 (1958).
c. K. Sisido, S. Kurozumi, and K. Utimoto, *J. Org. Chem.* **34**, 2661 (1969).
d. G. Stork and S. D. Darling, *J. Am. Chem. Soc.* **86**, 1761 (1964).

primarily at the less hindered α-carbon. Synthetic advantage can be taken of this selectivity; typical reactions are shown in Scheme 1.8. The alkylated ketone is obtained by hydrolysis of the reaction mixture. Alkylation of the nitrogen in the enamine system is an important competing reaction. Hydrolysis of these N-

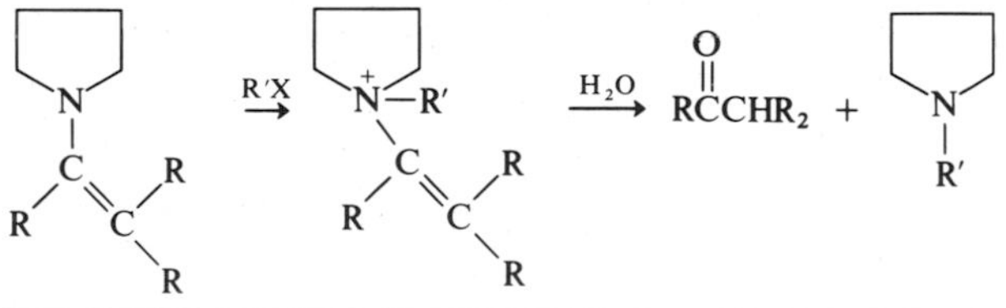

alkylation products leads to recovery of unreacted ketone. This competing reaction has thus far limited successful enamine alkylations to particularly reactive alkylating agents, such as methyl iodide, benzyl halides, α-haloketones, α-haloesters, and α-haloethers.

The nitrogen analogs of enolate ions can be prepared from imines and strong bases. The resulting anions are alkylated by alkyl halides:

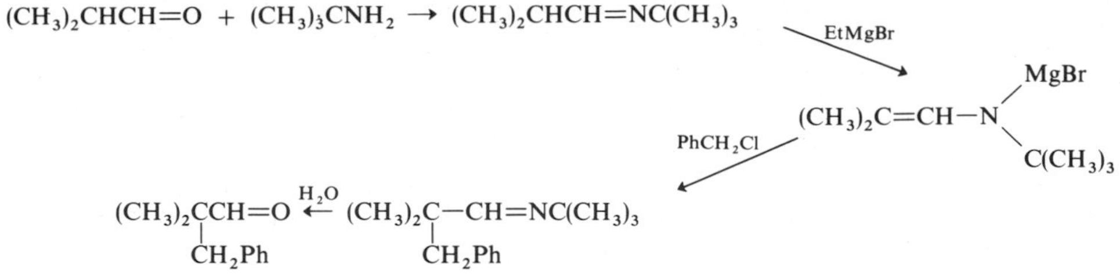

$(CH_3)_2CHCH=O + (CH_3)_3CNH_2 \rightarrow (CH_3)_2CHCH=NC(CH_3)_3$

$\xrightarrow{EtMgBr}$

$(CH_3)_2C=CH-N\overset{MgBr}{\underset{C(CH_3)_3}{\diagup\diagdown}}$

$\xrightarrow{PhCH_2Cl}$

$(CH_3)_2\underset{CH_2Ph}{\overset{|}{C}}-CH=NC(CH_3)_3 \xrightarrow{H_2O} (CH_3)_2\underset{CH_2Ph}{\overset{|}{C}}CH=O$

(80 % overall yield)

$\xrightarrow[\substack{2)\ BuI \\ 3)\ H_2O}]{1)\ EtMgBr}$

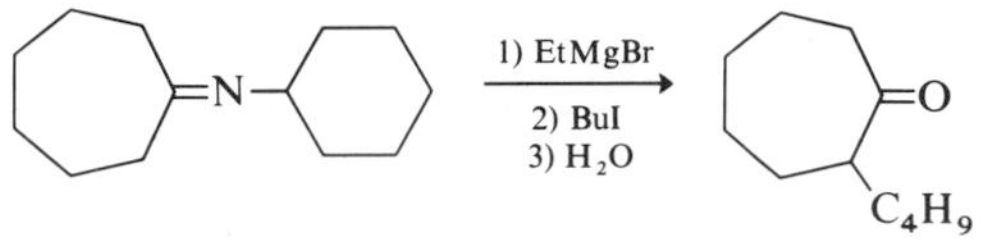

Such alkylation procedures are found to give alkylation primarily on the less hindered carbon in unsymmetrical ketones.[25]

$CH_3\overset{O}{\overset{||}{C}}CH(CH_3)_2 \xrightarrow[\substack{3)\ BuI \\ 4)\ H_2O}]{\substack{1)\ RNH_2 \\ 2)\ RMgX}} CH_3(CH_2)_4\overset{O}{\overset{||}{C}}CH(CH_3)_2$ (70 %)

Enamine chemistry is also the basis for the synthetic utility of dihydrooxazine derivatives. The quaternary salt **5** is converted to the cyclic enamine **6** on reaction with sodium hydride. Aldehydes are obtained after reduction and hydrolysis of the alkylation product.[26]

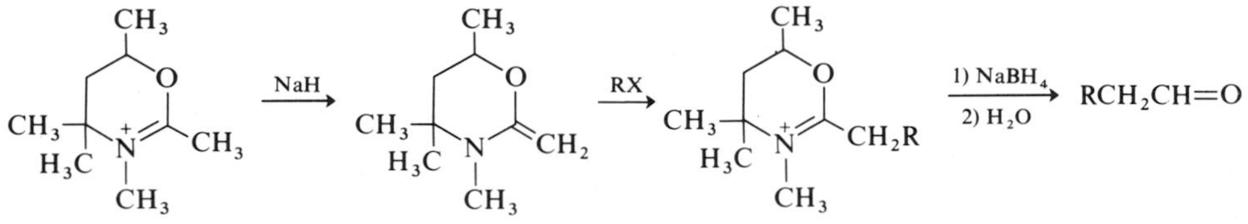

$\xrightarrow{NaH}$ ... $\xrightarrow{RX}$ ... $\xrightarrow[2)\ H_2O]{1)\ NaBH_4} RCH_2CH=O$

5   **6**

1.10. Alkylation of Carbon by Conjugate Addition

The previous sections have dealt primarily with reactions in which the new carbon–carbon bond is formed in an S_N2 reaction between the nucleophilic carbon species and the alkylating reagent. There is another general and important method for alkylation of carbon that should be discussed at this point. This reaction involves the addition of a nucleophilic carbon species to an electrophilic multiple bond. The reaction is applicable to a wide variety of enolates and enamines. The electrophilic

25. G. Stork and S. R. Dowd, *J. Am. Chem. Soc.* **85**, 2178 (1963).
26. A. I. Meyers and N. Nazarenko, *J. Am. Chem. Soc.* **94**, 3243 (1972); A. I. Meyers, A. Nabeya, H. W. Adickes, I. R. Politzer, G. R. Malone, A. C. Kovelesky, R. L. Nolen, and R. C. Portnoy, *J. Org. Chem.* **38**, 36 (1973).

Scheme 1.9. Alkylation of Carbon by Conjugate Addition

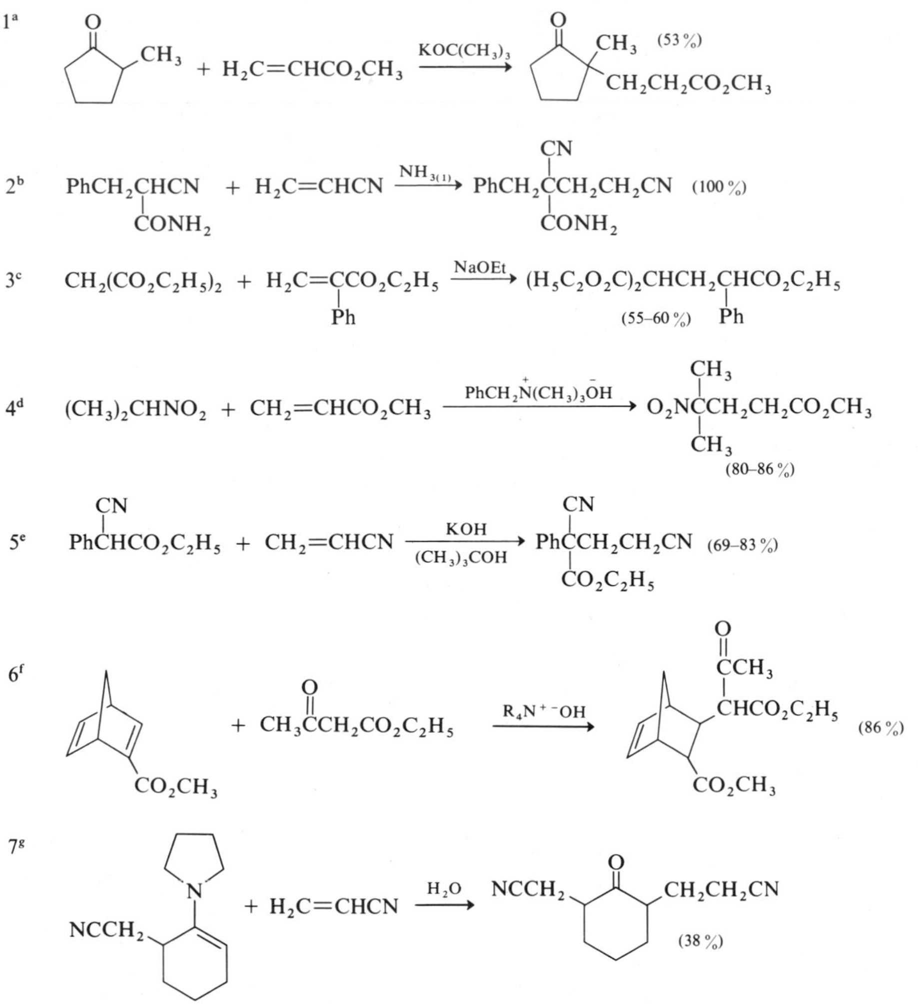

25

1[a]   2-methylcyclopentanone $+ H_2C=CHCO_2CH_3$ $\xrightarrow{KOC(CH_3)_3}$ product (53%) with CH_3 and $CH_2CH_2CO_2CH_3$

2[b] $PhCH_2\overset{|}{C}HCN$ (with $CONH_2$) $+ H_2C=CHCN$ $\xrightarrow{NH_{3(l)}}$ $PhCH_2\overset{CN}{\underset{CONH_2}{C}}CH_2CH_2CN$ (100%)

3[c] $CH_2(CO_2C_2H_5)_2 + H_2C=\overset{|}{\underset{Ph}{C}}CO_2C_2H_5$ $\xrightarrow{NaOEt}$ $(H_5C_2O_2C)_2CHCH_2\overset{|}{\underset{Ph}{C}}HCO_2C_2H_5$ (55–60%)

4[d] $(CH_3)_2CHNO_2 + CH_2=CHCO_2CH_3$ $\xrightarrow{PhCH_2\overset{+}{N}(CH_3)_3\overset{-}{O}H}$ $O_2N\overset{CH_3}{\underset{CH_3}{C}}CH_2CH_2CO_2CH_3$ (80–86%)

5[e] $Ph\overset{CN}{\underset{}{C}}HCO_2C_2H_5 + CH_2=CHCN$ $\xrightarrow[(CH_3)_3COH]{KOH}$ $Ph\overset{CN}{\underset{CO_2C_2H_5}{C}}CH_2CH_2CN$ (69–83%)

6[f]   bicyclic diene with CO_2CH_3 $+ CH_3\overset{O}{\underset{}{C}}CH_2CO_2C_2H_5$ $\xrightarrow{R_4N^{+-}OH}$ product with $\overset{O}{\overset{||}{C}}CH_3$, $CHCO_2C_2H_5$, CO_2CH_3 (86%)

7[g]   pyrrolidine enamine with NCCH$_2$ $+ H_2C=CHCN$ $\xrightarrow{H_2O}$ $NCCH_2$ cyclohexanone CH_2CH_2CN (38%)

a. H. O. House, W. L. Roelofs, and B. M. Trost, *J. Org. Chem.* **31**, 646 (1966).
b. S. Wakamatsu, *J. Org. Chem.* **27**, 1285 (1962).
c. E. M. Kaiser, C. L. Mao, C. F. Hauser, and C. R. Hauser, *J. Org. Chem.* **35**, 410 (1970).
d. R. B. Moffett, *Org. Synth.* **IV**, 652 (1963).
e. E. C. Horning and A. F. Finelli, *Org. Synth.* **IV**, 776 (1963).
f. K. Alder, H. Wirtz, and H. Koppelberg, *Justus Liebigs Ann. Chem.* **601**, 138 (1956).
g. L. Mandell, J. U. Riper, and K. P. Singh, *J. Org. Chem.* **28**, 3440 (1963).

reaction partners are typically α,β-unsaturated ketones, esters, or nitriles, but other electron-withdrawing substituents also activate the carbon–carbon double bond to nucleophilic attack. The reaction is called either the *Michael reaction* or *conjugate addition*.[27] The process can also occur with other nucleophiles, such as alkoxide ions or amines, but these reactions are outside the scope of the present discussion.

27. E. D. Bergmann, D. Ginsburg, and R. Pappo, *Org. React.* **10**, 179 (1959).

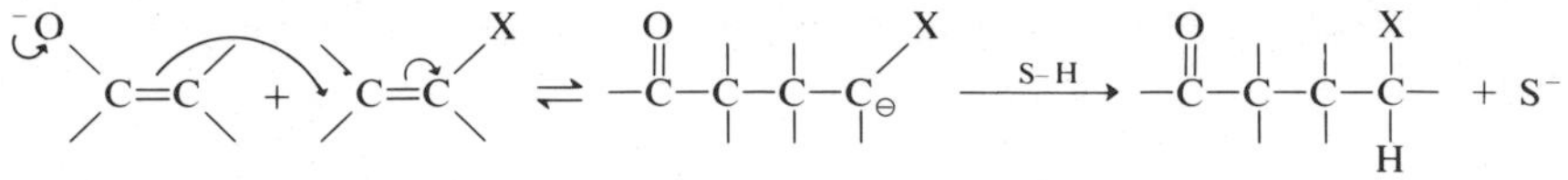

The reaction is usually catalyzed by base. All the steps are reversible, and protic solvents—particularly alcohols—that are capable of protonating the intermediate anionic adduct to give product are usually employed. The reaction can be thermodynamically or kinetically controlled, depending on the severity of the reaction conditions. The reaction is most familiar with such nucleophiles as the enolates of malonate esters or β-ketoesters, which are stabilized by two electron-attracting substituents. Good yields of alkylation products of simple ketones and nitroalkanes, however, have been obtained. Enamines are also reactive as nucleophiles in the Michael reaction. A wide variety of olefins having one or more electron-attracting groups in conjugation with the double bond have been employed as the acceptor molecule. Acetylenes can also function as the electrophilic species.

Some typical examples of Michael reactions are recorded in Scheme 1.9. The reaction is of very broad scope, and a large number of examples have been tabulated.[27]

Enamines of cyclic ketones react with acrolein to give bicyclic ketones[28]:

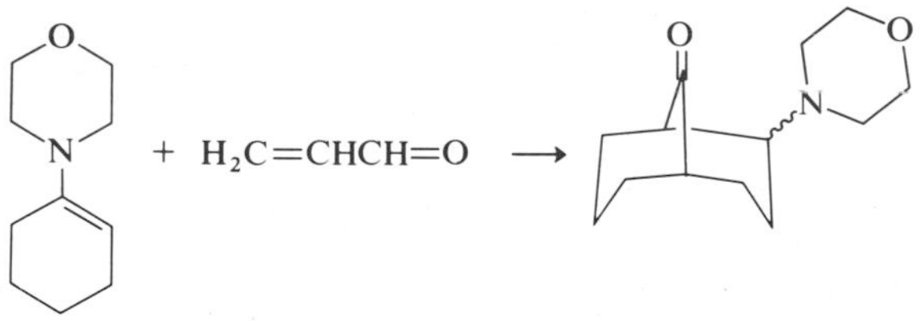

Ref. 28

Several steps are involved in this reaction, including the transfer of the enamine function to the aldehyde carbonyl. The reaction begins with conjugate addition:

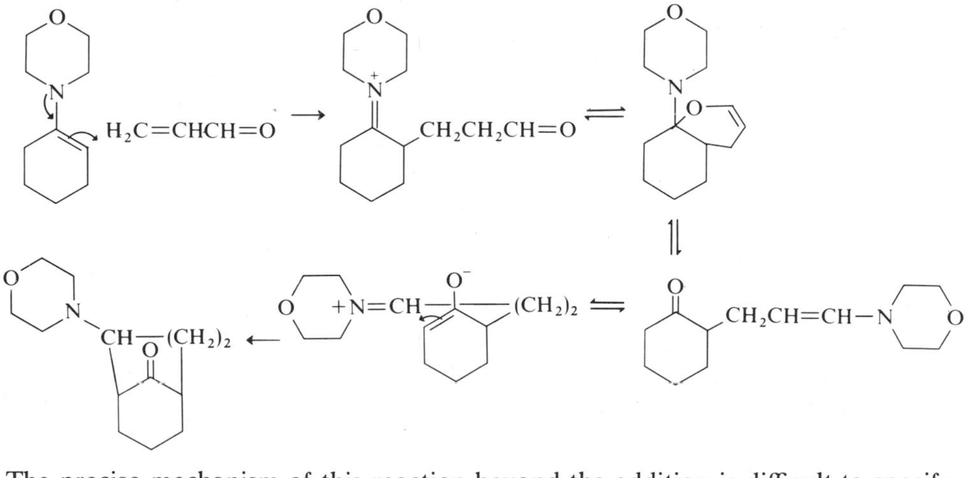

The precise mechanism of this reaction beyond the addition is difficult to specify.

27. See p. 25.
28. G. Stork and H. K. Landesman, *J. Am. Chem. Soc.* **78**, 5129 (1956).

Scheme 1.10. Addition of Cyanide Ion to Electrophilic Alkenes

27

SECTION 1.10.
ALKYLATION OF
CARBON BY
CONJUGATE
ADDITION

1[a]   PhCH=CPh $\xrightarrow[\text{CH}_3\text{OH, H}_2\text{O}]{\text{KCN}}$ PhCH—CHPh   (95–98 %)
         | | |
         CN   NC   CN

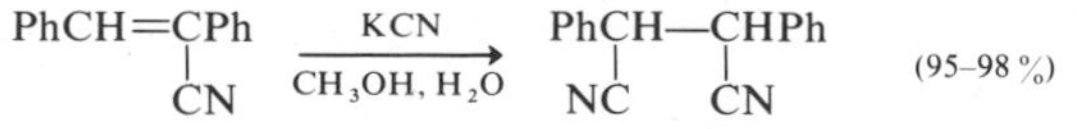

2[b]

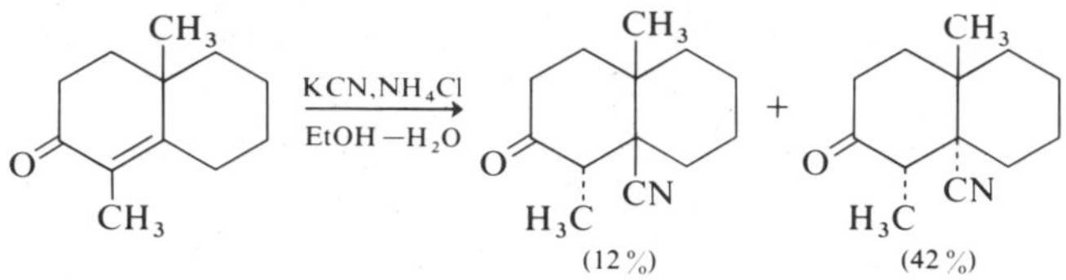

(12 %)   (42 %)

3[c]

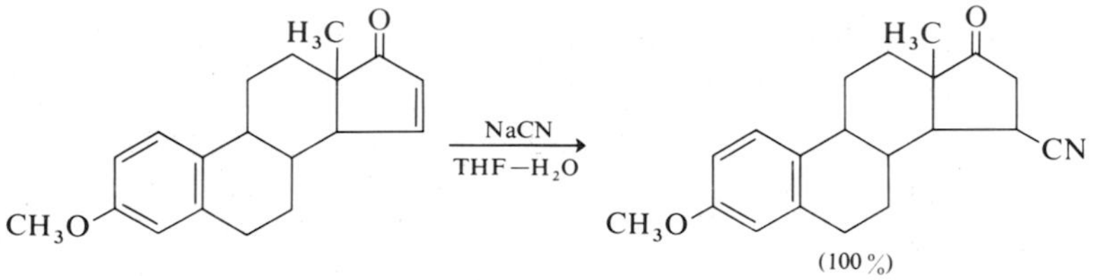

$\xrightarrow[\text{THF}-\text{H}_2\text{O}]{\text{NaCN}}$

(100 %)

4[d]

$\xrightarrow{\text{Et}_3\text{Al}-\text{HCN}}$

(92–93 %)

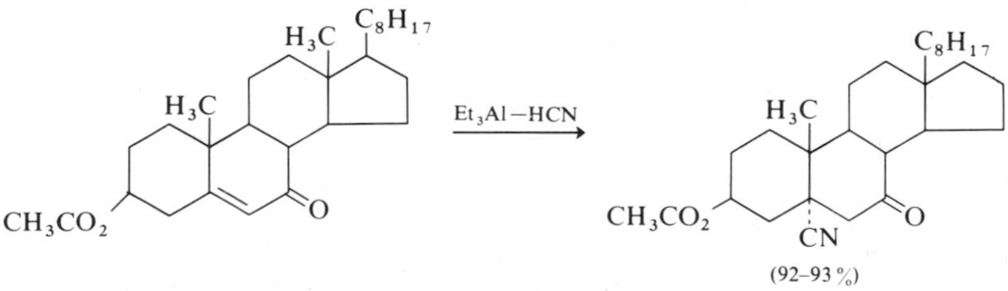

a. J. A. McRae and R. A. B. Bannard, *Org. Synth.* **IV**, 393 (1963).
b. O. R. Rodig and N. J. Johnston, *J. Org. Chem.* **34**, 1942 (1969).
c. E. W. Cantrall, R. Littell, and S. Bernstein, *J. Org. Chem.* **29**, 64 (1964).
d. W. Nagata and M. Yoshioka, *Org. Synth.* **52**, 100 (1972).

Presumably, the transfer of the amino function occurs by partial hydrolysis and re-formation of the enamine at the aldehyde group.[29] The method is a convenient way of constructing certain bicyclic ring systems.

Bicyclic rings can also be constructed from enamines using bifunctional reagents. The bromoester **7**, for example, reacts by both an S_N2 alkylation and a Michael addition, giving the bicyclic ring system **8**[30]:

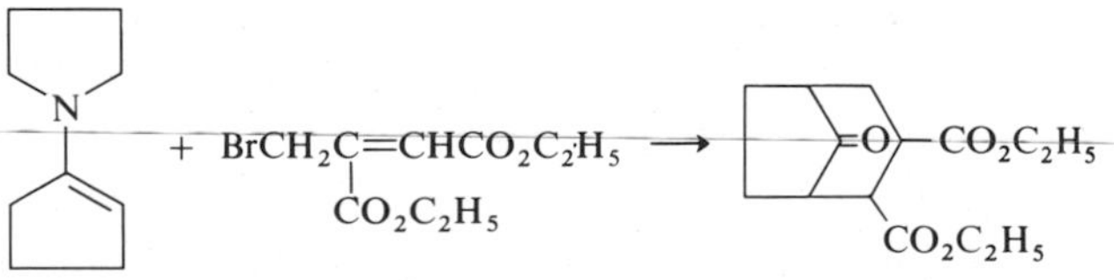

7   **8**

29. R. N. Schut and T. M. H. Liu, *J. Org. Chem.* **30**, 2845 (1965); R. D. Allen, B. G. Cordiner, and R. J. Wells, *Tetrahedron Lett.*, 6055 (1968).
30. R. P. Nelson and R. G. Lawton, *J. Am. Chem. Soc.* **88**, 3884 (1966).

The cyanide ion is a reactive carbon nucleophile toward electrophilic alkenes, leading to overall addition of hydrogen cyanide:

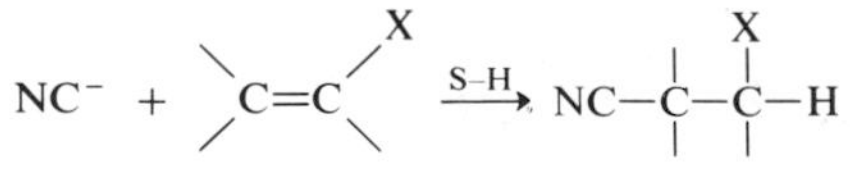

$$NC^- + \quad \overset{X}{\underset{}{C=C}} \quad \xrightarrow{S-H} \quad NC-\overset{X}{\underset{}{C}}-\overset{X}{\underset{}{C}}-H$$

The standard conditions for such reactions involve alcoholic solutions of potassium or sodium cyanide. Triethylaluminum-hydrogen cyanide and diethylaluminum-cyanide have been introduced for the same purpose.[31] These reagents have been successful in instances where the more standard procedures involving cyanide ion fail. These reagents also provide a degree of control over the stereochemistry of the reaction. The former reagent gives kinetically controlled product, while the latter leads to the thermodynamically more stable nitrile. Some examples of both methods are given in Scheme 1.10.

Another very important method for adding a carbon chain at the β-carbon atom of an α,β-unsaturated carbonyl system involves organometallic reagents. This reaction will be discussed in Chapter 5.

General References

Enolates

D. J. Cram, *Fundamentals of Carbanion Chemistry*, Academic Press, New York, NY, 1965.
H. O. House, *Modern Synthetic Reactions*, Second Edition, W. A. Benjamin, Menlo Park, CA, 1972, Chapter 9.

Enamines

A. G. Cook, *Enamines: Synthesis, Structure and Reactions*, Marcel Dekker, New York, NY, 1969.

Problems

(*References for these problems will be found on page 497.*)

1. Write the structure of all possible enolates for each ketone or aldehyde. Indicate which you would expect to be favored in a kinetically controlled deprotonation. Which would you expect to be the most stable enolate in each case?

(a)

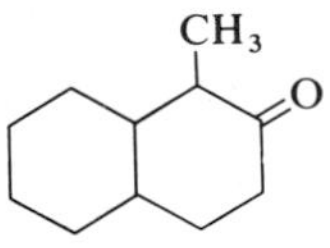

(b) $PhCH_2CH_2\overset{O}{\overset{\|}{C}}CH(CH_3)_2$

31. W. Nagata, M. Yoshioka, and S. Hirai, *J. Am. Chem. Soc.* **94**, 4635 (1972); W. Nagata, M. Yoshioka, and M. Murakami, *J. Am. Chem. Soc.* **94**, 4654 (1972).

(c)

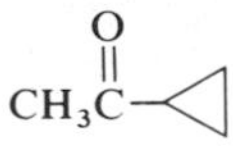

(d)

(e)

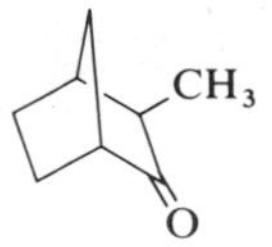

2. Show how each of the following compounds could be prepared using the Michael reaction:
   (a) 4,4-dimethyl-5-nitropentan-2-one
   (b) diethyl 2,3-diphenylglutarate
   (c) ethyl 2-benzoyl-3-(2-pyridyl)butyrate
   (d) 2-phenyl-3-oxocyclohexaneacetic acid

3. Intramolecular alkylation of enolates has been used to advantage in synthesis of bi- and tricyclic compounds. Indicate how such a procedure could be used to synthesize each of the following molecules, by drawing the structure of a suitable precursor:

(a)

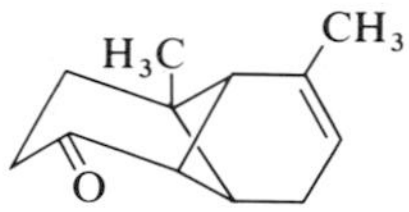

(c)

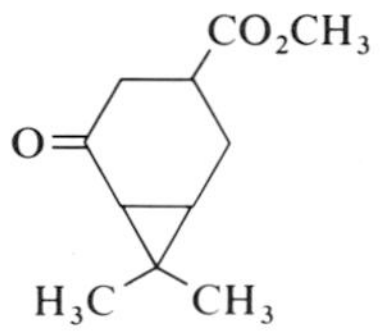

(b)

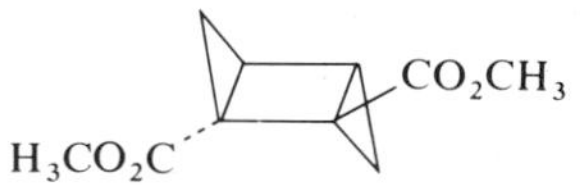

(d)

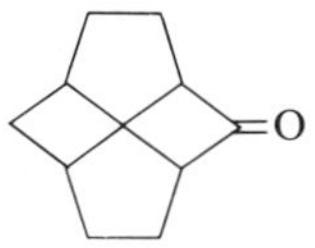

4. Predict the form of the rate expression that would be observed in base-catalyzed addition of 3,5-pentanedione to methyl vinyl ketone.

5. Treatment of 2,3,3-triphenylpropionitrile with one equivalent of potassium amide in liquid ammonia followed by addition of benzyl chloride affords 2-benzyl-2,3,3-triphenylpropionitrile in 97% yield. Use of two equivalents of potassium amide gives an 80% yield of 2,3,3,4-tetraphenylbutyronitrile under the same reaction conditions. Explain.

6. Substituted acetophenones react with ethyl phenylpropiolate under the conditions of the Michael reaction to give pyrones. Formulate a mechanism.

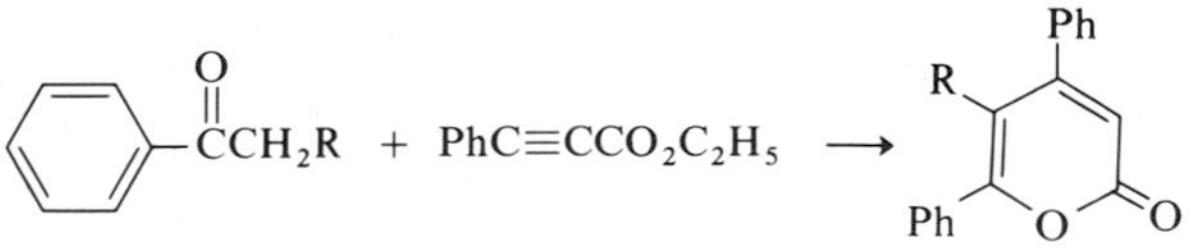

7. The reaction of simple ketones such as 2-butanone or phenylacetone with α,β-unsaturated ketones gives cyclohexenones when the reaction is effected by heating in methanol with potassium methoxide. Explain how the cyclohexenones are formed. What structures are possible for the cyclohexenones? Can you suggest means for distinguishing between possible isomeric cyclohexenones?

8. Suggest starting materials and reaction conditions for obtaining each of the following compounds by a procedure involving alkylation of nucleophilic carbon:

(a)

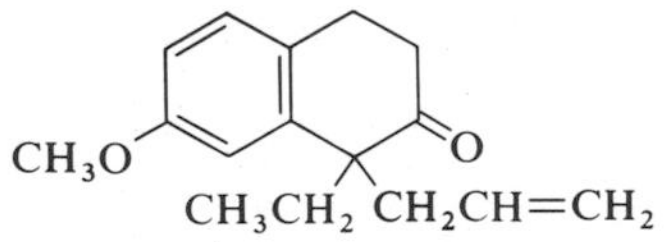

CH_3O

CH_3CH_2 $CH_2CH=CH_2$

(d)

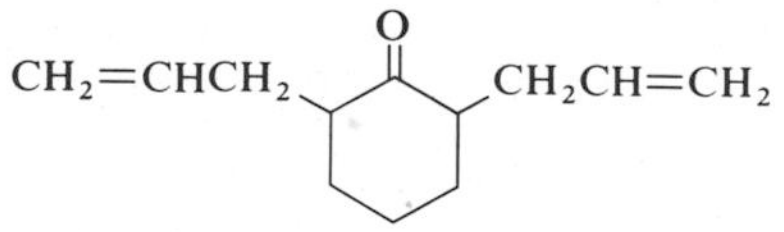

$CH_2=CHCH_2$ $CH_2CH=CH_2$

(b)

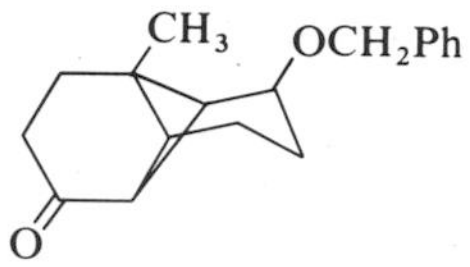

CH_3 OCH_2Ph

(e)

CN

$H_2C=CHCH_2CPh$

CNH_2

O

(c)

$PhCHCO_2H$

$PhCH_2$

(f)

O CH_3CO O

$CH_3CCH_2CH_2$

O

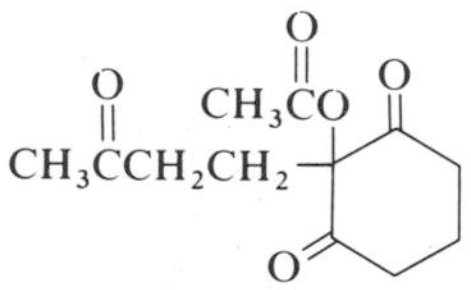

9. Suggest starting materials for synthesis of each of the following compounds, using Michael reactions:

(a)

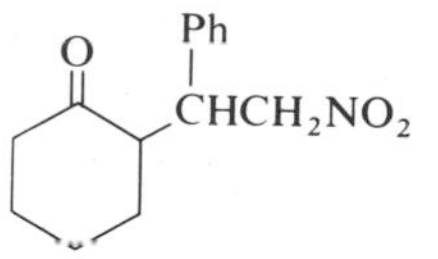

O Ph

$CHCH_2NO_2$

(c)

O OCH_3

O

Ph O $CHNO_2$

NO_2 CH_3

(b)

Ph O

$PhCHCHCH_2CCH_3$

CN

(d)

HO $CH=O$ O

$CH_2CH_2CCH_3$

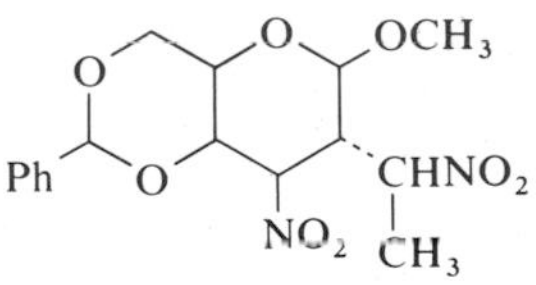

H_3C

O

O

O

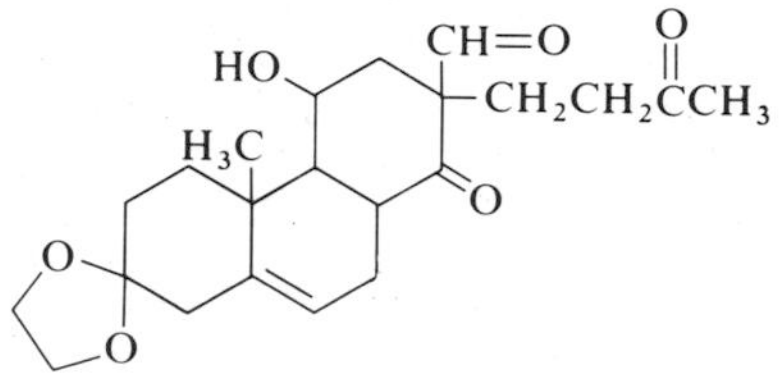

10. The Michael reaction of 2-methyl-5-*t*-butylcyclohexanone with ethyl acrylate has been studied. Two products, both of which are 2,2-disubstituted, rather than

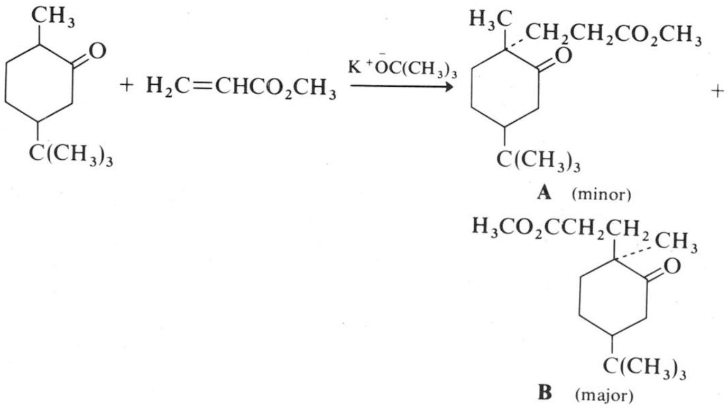

CH_3

O

$+ H_2C=CHCO_2CH_3$ $\xrightarrow{K^+OC(CH_3)_3}$

$C(CH_3)_3$

H_3C $CH_2CH_2CO_2CH_3$

O

$C(CH_3)_3$

A (minor)

$+$

$H_3CO_2CCH_2CH_2$ CH_3

O

$C(CH_3)_3$

B (major)

2,6-disubstituted, are found. The addition reaction was catalyzed by potassium *t*-butoxide in 1-butanol. In control experiments, it was shown that the two products did not interconvert under these conditions. Are the products of this reaction kinetically controlled or thermodynamically controlled? What factors do you feel lead to formation of **B** as the major product?

11. In a synthesis of diterpenes via compound **C**, a key intermediate **B** was obtained from carboxylic acid **A**. Suggest a series of reactions for obtaining **B** from **A**:

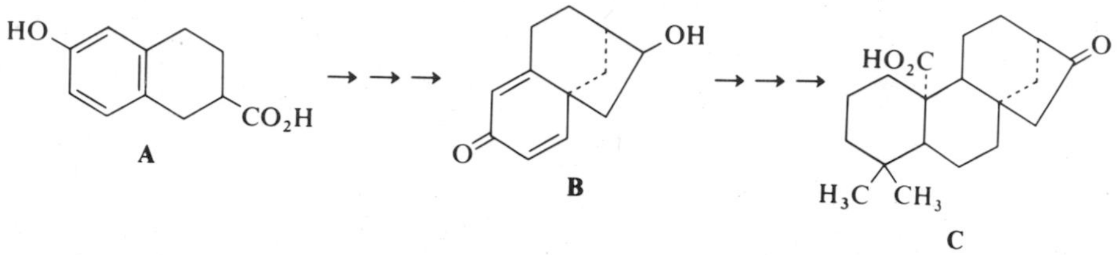

12. In a synthesis of the terpene longifolene, the tricyclic intermediate **A** was obtained from a bicyclic intermediate by an intramolecular Michael addition. Deduce the possible structure(s) of the bicyclic precursor.

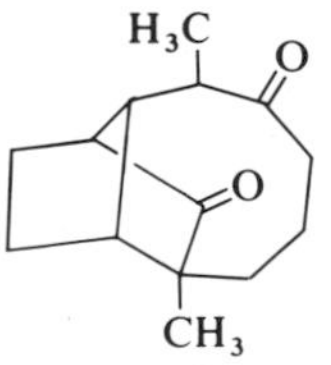

13. Suggest a sequence of reactions for accomplishing each of the following transformations:

(a)

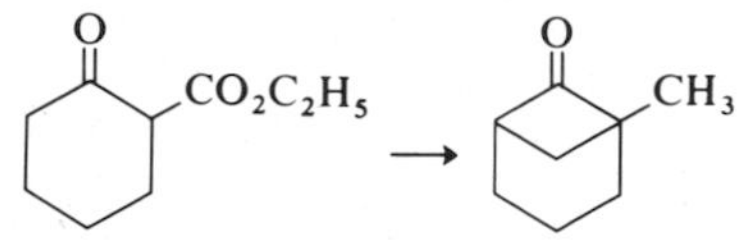

(b)

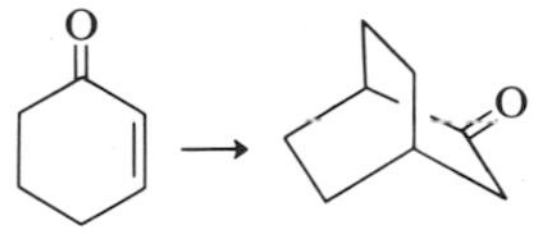

(c)

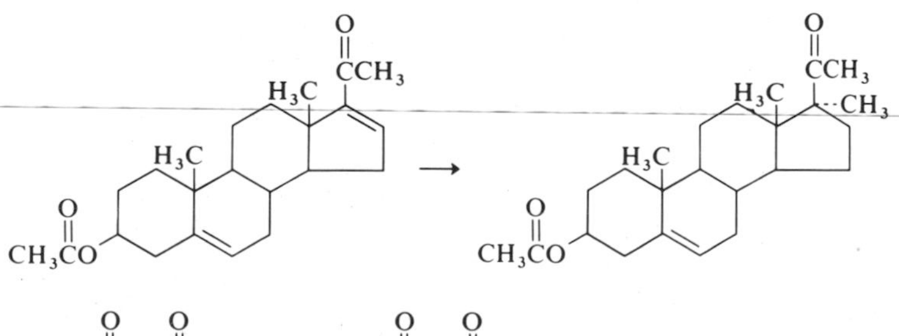

(d) $(CH_3O)_2PCH_2CCH_3 \rightarrow (CH_3O)_2PCH_2C(CH_2)_4CH_3$

(e) $PhCH_2CO_2C_2H_5 \rightarrow PhCH_2CH_2\underset{\underset{Ph}{|}}{C}HCO_2C_2H_5$

14. A method for specific alkylation of unsymmetrical succinic acids involves treatment of the monoester with 2 equivalents of amide ion in liquid ammonia, followed by the alkylating agent:

$$\underset{\underset{CH_2CO_2H}{|}}{R}CHCO_2C_2H_5 \xrightarrow[\text{2) R'X}]{\text{1) 2 NH}_2{}^-} \underset{\underset{CH_2CO_2H}{|}}{\overset{R'}{R}}CHCO_2C_2H_5$$

Explain why this technique is specific, whereas alkylation of the diester with 1 equivalent of base gives a mixture of products.

15. The alkylation of 3-methyl-2-cyclohexenone with several dibromides led to the products shown below. Discuss the course of each reaction and suggest an explanation for the dependence of the product structure on the identity of the dihalide.

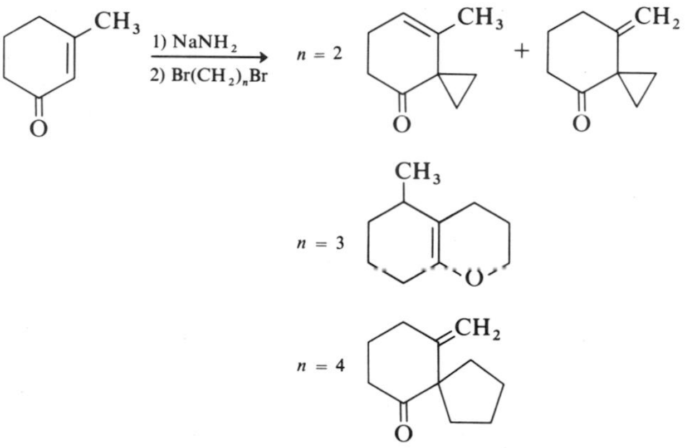

16. Arrange in order of increasing pK:

$$\underset{O}{\overset{O}{\|}}\text{PhCCH}_2\text{CH}_2\text{Ph}, \quad (CH_3)_3C\overset{O}{\overset{\|}{C}}CH_3, \quad (CH_3)_3C\overset{O}{\overset{\|}{C}}CH(CH_3)_2, \quad Ph\overset{O}{\overset{\|}{C}}CH_2CH_2CH_3$$

17. Predict the structure and stereochemistry of the product of alkylation of **A** with methyl iodide.

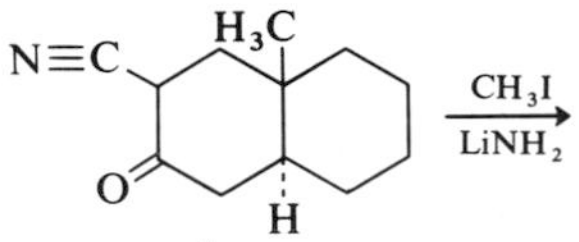

Reactions of Nucleophilic Carbon Species with Carbonyl Groups

The reactions discussed in this chapter are related mechanistically in that all begin by addition of a carbon nucleophile to a carbonyl or related unsaturated carbon site.

$$\overset{\diagdown}{\diagup}C^- \;+\; \overset{O}{\underset{\diagdown}{\overset{\|}{\underset{\diagup}{C}}}} \;\rightarrow\; -\overset{|}{\underset{|}{C}}-\overset{O^-}{\underset{|}{\overset{|}{C}}}-$$

The initial addition phase can then be followed by any of several reaction sequences. The discussion emphasizes those reactions that have attained synthetic importance.

2.1. Aldol Condensation

The aldol condensation reaction is the acid- or base-catalyzed self-condensation of a ketone or aldehyde.[1-3] Under certain conditions, the reaction product may undergo further transformations, especially dehydration. The reaction may also occur between two different carbonyl compounds, in which case the term *mixed aldol condensation* is applied. The mechanistic pattern of the reaction, involving attack by

1. A. T. Nielsen and W. J. Houlihan, *Org. React.* **16**, 1 (1968).
2. R. L. Reeves, in *Chemistry of the Carbonyl Group*, S. Patai (ed.), Interscience, New York, N.Y., 1966, pp. 580–593.
3. H. O. House, *Modern Synthetic Reactions*, Second Edition, W. A. Benjamin, Menlo Park, CA, 1972, pp. 629–682.

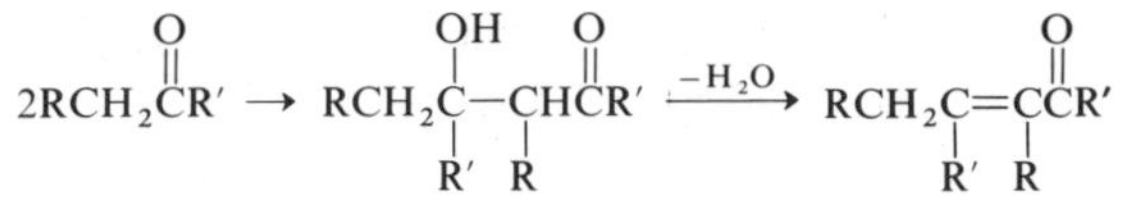

$$2RCH_2CR' \rightarrow RCH_2\overset{OH}{\underset{R'}{C}}-\overset{O}{\underset{R}{CHCR'}} \xrightarrow{-H_2O} RCH_2\overset{O}{\underset{R'}{C}}=\overset{}{\underset{R}{CCR'}}$$

R' = H, or alkyl or aryl

nucleophilic carbon at a carbonyl group, can be recognized in a number of other types of reactions, some of which will be discussed in later sections of this chapter.

Base-Catalyzed Mechanism

1. Addition phase
$$RCH_2COR' + B^- \rightleftharpoons RCH=\underset{O^-}{CHR'} + BH$$

$$\underset{\underset{\overset{|}{RCH_2CR'}}{\overset{||}{O}}}{RCH=CR'} \rightleftharpoons RCH_2\overset{||}{\underset{O^-}{CR'}} \underset{B^-}{\overset{BH}{\rightleftharpoons}} RCH_2\overset{O}{\underset{OH}{CR'}}$$

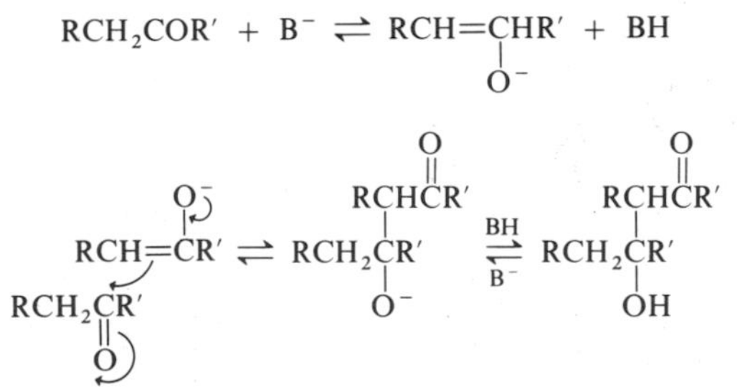

2. Dehydration phase
$$\underset{\underset{OH}{RCH_2CR'}}{RCHCR'} + B^- \rightleftharpoons RCH_2\overset{O}{\underset{OH}{C}}-R' \rightarrow$$

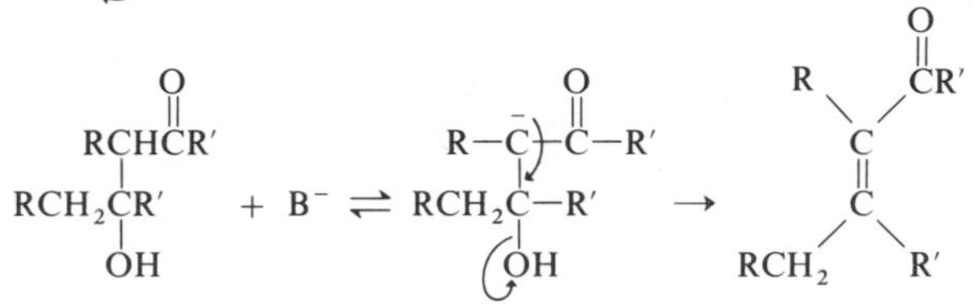

Acid-Catalyzed Mechanism

1. Addition phase
$$RCH_2\underset{O}{CR'} + HA \rightleftharpoons RCH_2\underset{+OH}{CR'} + A^-$$

$$RCH_2\underset{+OH}{CR'} \xrightarrow{-H^+} RCH=\underset{OH}{CR'}$$

$$\underset{\underset{O}{RCH_2CR'}}{RCH=CR'} \xrightarrow{OH} RCH_2\underset{OH}{CR'} + HA$$

HA

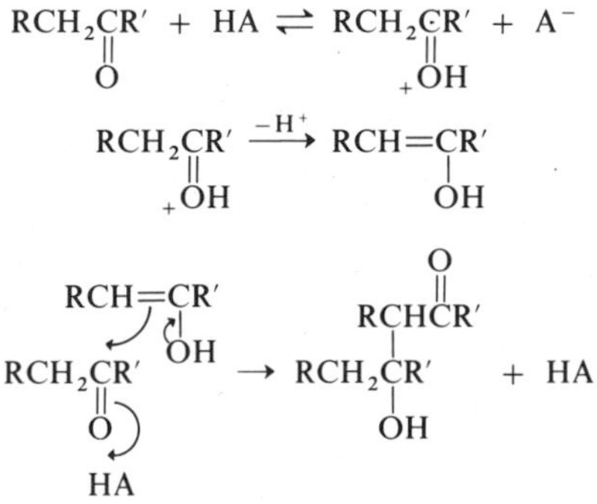

2. Dehydration phase
$$RCH_2\underset{OH}{CHCR'} \overset{HA}{\rightleftharpoons} RCH_2\underset{+OH}{CR'} \xrightarrow{-H_2O, -H^+}$$

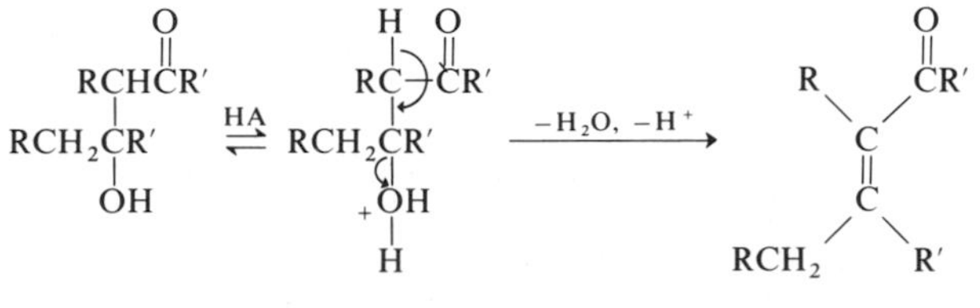

Because of this resemblance in reaction pattern, the term *generalized aldol condensation* has been applied to a broader group of reactions in which enolates or enols act as nucleophilic species in reactions with ketones and aldehydes. In general, the reactions in the addition phase of both base-catalyzed and acid-catalyzed aldol condensations are easily reversible.

The equilibrium constant for the formation of the addition product is not always favorable; for acyclic ketones, for example, it is usually unfavorable. A method for self-condensation of acetone has been developed to overcome this problem and it shows how special procedures can sometimes be used to drive a reaction with an unfavorable equilibrium constant to completion. The reaction is carried out in a Soxhlet extractor with barium hydroxide, an insoluble basic catalyst, in the thimble.[4] The acetone vapor is condensed and flows over the basic barium oxide catalyst. The product is formed (to a small extent) while acetone is in contact with the catalyst and then passes into the flask. Since the product has a much higher boiling point than acetone, the vapor is always nearly pure acetone. Because the reaction product is not in contact with the basic catalyst, its concentration builds up in the flask far above its normal equilibrium concentration, and the reverse reaction is very slow in the absence of catalyst.

The equilibrium constant for the dehydration phase of aldol condensations is usually favorable, largely because a conjugated α,β-unsaturated carbonyl system is formed. When the reaction conditions are sufficiently vigorous to cause dehydration, the overall reaction can go to completion even if the equilibrium constant for the addition phase is not favorable.

There are substantial difficulties involved in utilizing the aldol condensation of aldehydes from a synthetic point of view because both the starting materials and products are often sensitive to side reactions including polymerizations. Careful choice of conditions has provided numerous successful cases,[1] however, as illustrated by the first three entries in Scheme 2.1.

The self-condensation of ketones without dehydration is rare because of the unfavorable equilibria in the addition phase. Conditions which involve dehydration, however, can usually effect successful condensation. Many are recorded in the review of Nielsen and Houlihan.[1] Entries 4–6 in Scheme 2.1 are examples of such condensations.

Intramolecular aldol condensations are often more successful than the intermolecular type. A particularly important example of the synthetic use of intramolecular condensation is in the Robinson annelation, a procedure that constructs a new six-membered ring from a ketone that has an enolizable hydrogen.[5] The stages in which this alkylation–cyclization occurs are outlined below. The cyclization

4. J. B. Conant and N. Tuttle, *Org. Synth.* **I**, 199 (1941).
1. See p.33.
5. E. D. Bergmann, D. Ginsburg, and R. Pappo, *Org. React.* **10**, 179 (1950); J. W. Cornforth and R. Robinson, *J. Chem. Soc.*, 1855 (1949).

36

Scheme 2.1. Examples of the

CHAPTER 2
REACTIONS OF
NUCLEOPHILIC
CARBON
SPECIES

A. Aldehyde and Ketone Self-Condensation

1^a $CH_3CH_2CH_2CH{=}O$ $\xrightarrow{\text{KOH}}$ $CH_3CH_2CH_2\overset{\text{OH}}{\underset{\underset{C_2H_5}{|}}{C}}HCHCH{=}O$ (75 %)

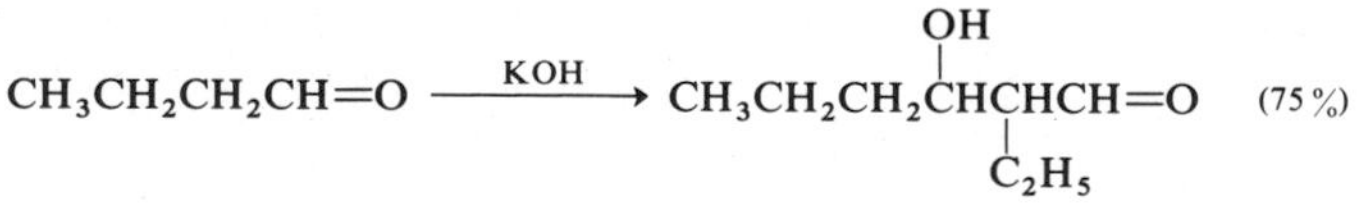

2^b $C_7H_{15}CH{=}O$ $\xrightarrow{\text{NaOEt}}$ $C_7H_{15}CH{=}\underset{\underset{C_6H_{13}}{|}}{C}HCH{=}O$ (79 %)

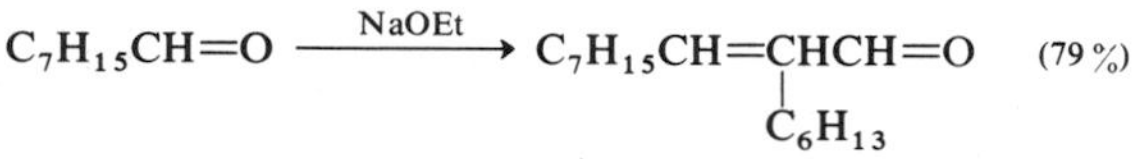

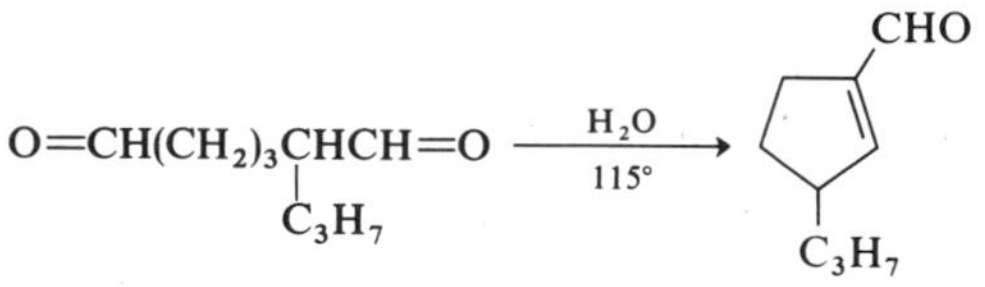

3^c $O{=}CH(CH_2)_3\underset{\underset{C_3H_7}{|}}{C}HCH{=}O$ $\xrightarrow[115°]{\text{H}_2\text{O}}$

4^d $(CH_3)_2CO$ $\xrightarrow[\text{resin}]{\text{Dowex-50}}$ $(CH_3)_2C{=}CH\overset{O}{\overset{||}{C}}CH_3$ (79 %)

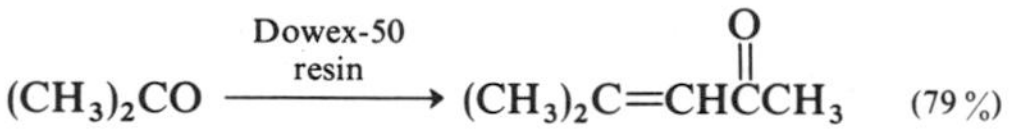

5^e

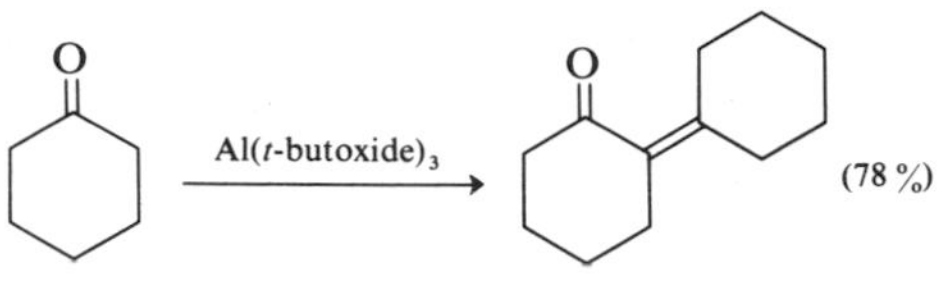

B. Mixed Condensations and Cyclizations

6^f

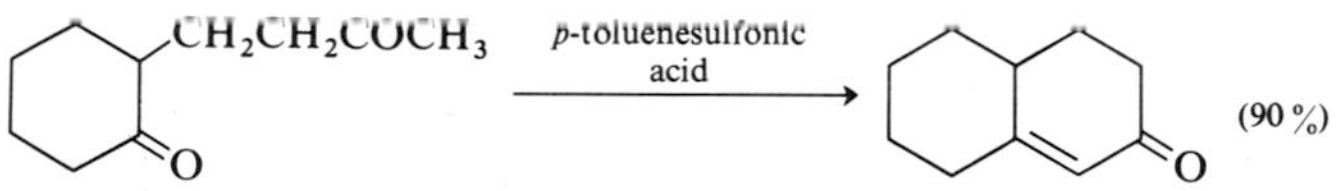

7^g $(CH_3)_3C\overset{O}{\overset{||}{C}}CH_3 \; + \; PhCHO \; \rightarrow \; (CH_3)_3C\overset{O}{\overset{||}{C}}CH{=}CHPh$ (90 %)

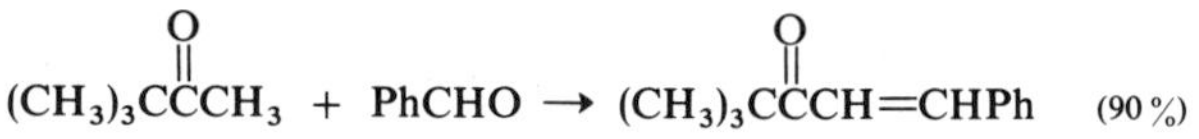

8^h

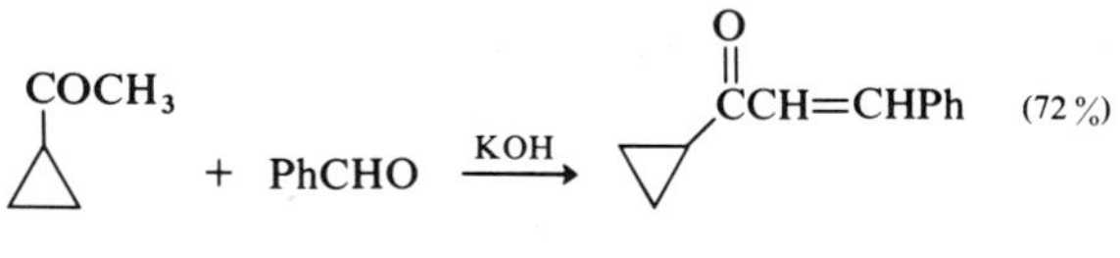

9^i

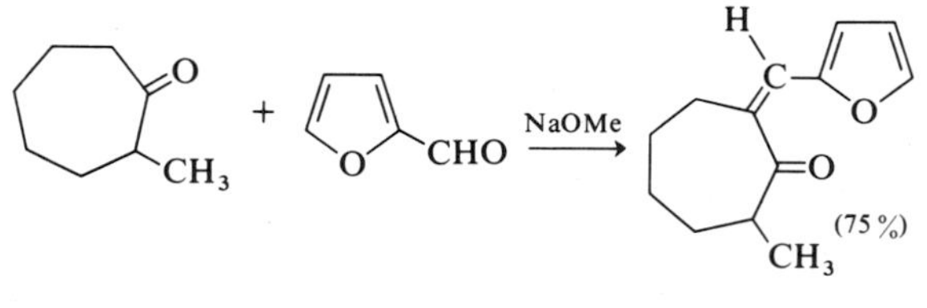

10^j

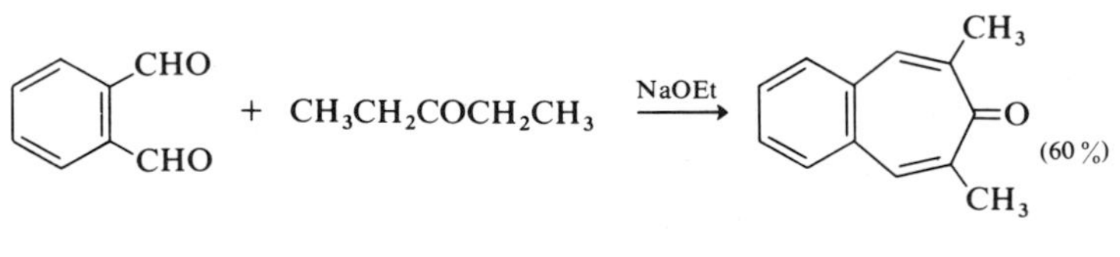

11^k

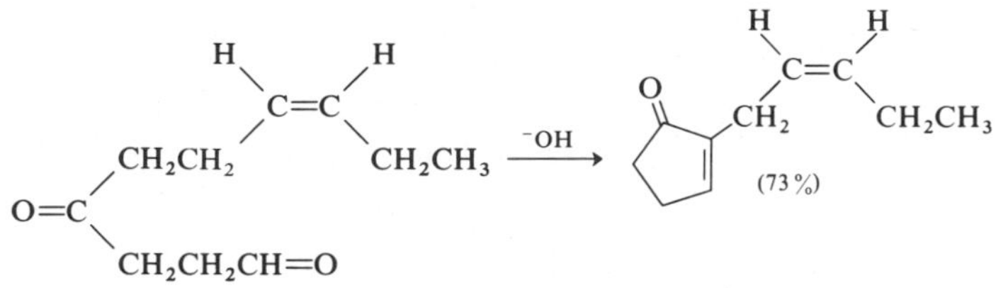

(73%)

C. Condensations Involving Ester Enolates

12^l $CH_3CO_2C_2H_5 + Ph_2CO \xrightarrow[\text{2) } NH_4Cl]{\text{1) } LiNH_2}$ $Ph_2\overset{\overset{\displaystyle OH}{|}}{C}CH_2CO_2C_2H_5$

(75–84%)

13^m $CH_3CO_2C_2H_5 \xrightarrow{(Me_3Si)_2NLi} LiCH_2CO_2C_2H_5 \longrightarrow$

$\xrightarrow{PhCH=CHCH=O}$ $PhCH=CH\overset{\overset{\displaystyle OH}{|}}{C}HCH_2CO_2C_2H_5$ (94%)

14^n $(CH_3)_2CHCO_2H \xrightarrow[R_2NLi]{\text{2 mol}}$ $(CH_3)_2\overset{\overset{\displaystyle}{|}}{\underset{\underset{\displaystyle Li}{|}}{C}}CO_2Li \longrightarrow$

$\xrightarrow{(CH_3CH_2)_2C=O}$ $HO-\overset{\overset{\displaystyle CH_2CH_3}{|}}{\underset{\underset{\displaystyle CH_2CH_3}{|}}{C}}-\overset{\overset{\displaystyle CH_3}{|}}{\underset{\underset{\displaystyle CH_3}{|}}{C}}-CO_2H$ (77%)

15° $CH_3CO_2C_2H_5 + LiN[Si(CH_3)_3]_2 \longrightarrow LiCH_2CO_2C_2H_5$

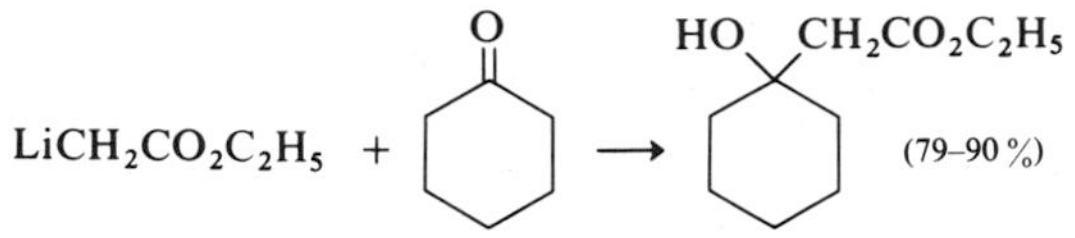

$LiCH_2CO_2C_2H_5 +$ $\longrightarrow$ (79–90%)

a. V. Grignard and A. Vesterman, *Bull. Chim. Soc. Fr.* **37**, 425 (1925).
b. F. J. Villani and F. F. Nord, *J. Am. Chem. Soc.* **69**, 2605 (1947).
c. J. English and G. W. Barber, *J. Am. Chem. Soc.* **71**, 3310 (1949).
d. N. B. Lorette, *J. Org. Chem.* **22**, 346 (1957).
e. W. Wayne and H. Adkins, *J. Am. Chem. Soc.* **62**, 3401 (1940).
f. D. J. Baisted and J. S. Whitehurst, *J. Chem. Soc.,* 4089 (1961).
g. G. A. Hill and G. Bramann, *Org. Synth.* **I**, 81 (1941).
h. S. C. Bunce, H. J. Dorsman, and F. D. Popp, *J. Chem. Soc.,* 303 (1963).
i. A. M. Islam and M. T. Zemaity, *J. Am. Chem. Soc.* **79**, 6023 (1957).
j. D. Meuche, H. Strauss, and E. Heilbronner, *Helv. Chim. Acta* **41**, 2220 (1958).
k. A. I. Meyers and N. Nazarenko, *J. Org. Chem.* **38**, 175 (1973).
l. W. R. Dunnavant and C. R. Hauser, *Org. Synth.* **44**, 56 (1964).
m. M. W. Rathke, *J. Am. Chem. Soc.* **92**, 3222 (1970).
n. G. W. Moersch and A. R. Burkett, *J. Org. Chem.* **36**, 1149 (1971).
o. M .W. Rathke, *Org. Synth.* **53**, 66 (1973).

phase will be recognized as an intramolecular aldol condensation. This annelation procedure is an important method for the construction of six-membered rings. Scheme 2.2 provides some examples of the Robinson annelation.

The initial version of the reaction (as illustrated by entry 1) used an *in situ* source of methyl vinyl ketone. The quaternary salt of 4-dimethylamino-2-butanone decomposes easily to the unsaturated ketone. In more recent procedures—for example, entries 2–4—the α,β-unsaturated ketone is added directly. Entry 3 illustrates the use of the enamine of a ketone as the reactive nucleophile in a Robinson annelation.

A recently introduced version of the Robinson annelation procedure involves the use of methyl 1-trimethylsilylvinyl ketone. The reaction follows the normal sequence of conjugate addition, aldol addition, and dehydration:

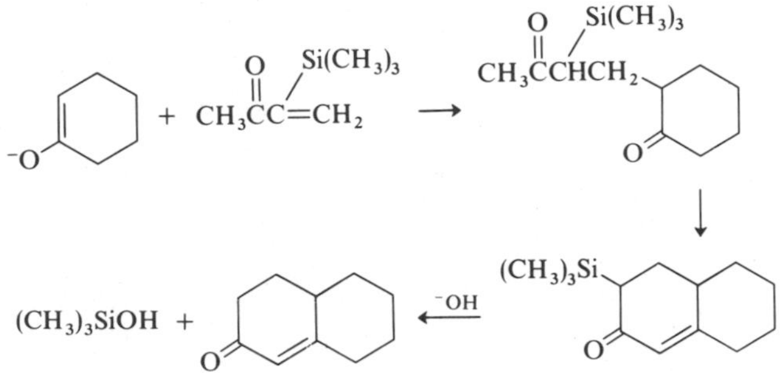

Ref. 6

The role of the trimethylsilyl group is to stabilize the intermediate carbanion formed by conjugate addition. The silyl group is removed under conditions similar to those required for the dehydration; the removal occurs by nucleophilic attack on silicon, resulting in displacement of the ketone. The advantage of the substituted methyl vinyl ketone is that it permits the annelation reaction to be carried out in aprotic solvents under conditions where enolate equilibration does not take place. The annelation of unsymmetrical ketones can therefore be controlled by using specific enolates generated by the methods described in Chapter 1.

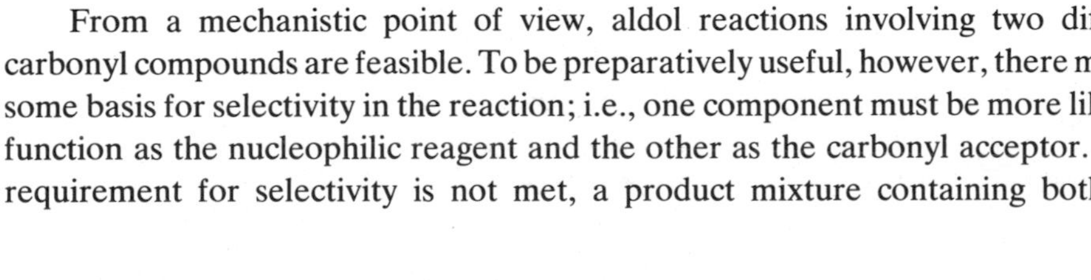

(69%)

Ref. 7

From a mechanistic point of view, aldol reactions involving two different carbonyl compounds are feasible. To be preparatively useful, however, there must be some basis for selectivity in the reaction; i.e., one component must be more likely to function as the nucleophilic reagent and the other as the carbonyl acceptor. If this requirement for selectivity is not met, a product mixture containing both self-

6. G. Stork and B. Ganem, *J. Am. Chem. Soc.* **95**, 6152 (1973); G. Stork and J. Singh, *J. Am. Chem. Soc.* **96**, 6181 (1974).
7. R. K. Boeckman, Jr., *J. Am. Chem. Soc.* **96**, 6179 (1974).

Scheme 2.2. The Robinson Annelation Reaction

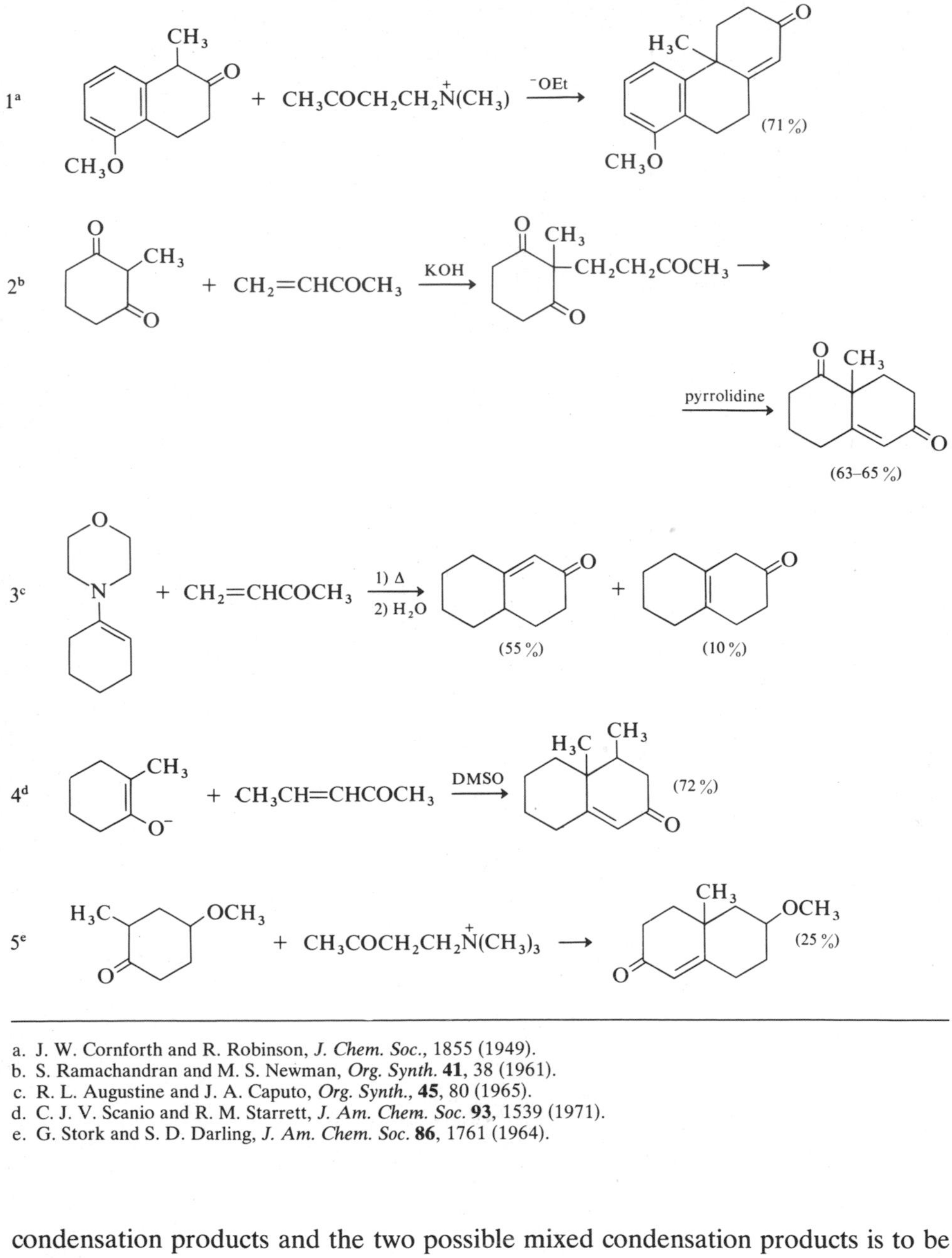

1[a]   (with CH$_3$, CH$_3$O structures)   + CH$_3$COCH$_2$CH$_2$$\overset{+}{N}$(CH$_3$) $\xrightarrow{\text{ }^-\text{OEt}}$ (product) (71 %)

2[b]   (structure with CH$_3$)   + CH$_2$=CHCOCH$_3$ $\xrightarrow{\text{KOH}}$ (structure with CH$_3$, -CH$_2$CH$_2$COCH$_3$) →

$\xrightarrow{\text{pyrrolidine}}$ (product with CH$_3$) (63–65 %)

3[c]   (morpholine enamine structure)   + CH$_2$=CHCOCH$_3$ $\xrightarrow[\text{2) H}_2\text{O}]{\text{1) }\Delta}$ (product) (55 %)   +   (product) (10 %)

4[d]   (structure with CH$_3$, O$^-$)   + CH$_3$CH=CHCOCH$_3$ $\xrightarrow{\text{DMSO}}$ (product with H$_3$C, CH$_3$) (72 %)

5[e]   (structure with H$_3$C, OCH$_3$)   + CH$_3$COCH$_2$CH$_2$$\overset{+}{N}$(CH$_3$)$_3$ →   (product with CH$_3$, OCH$_3$) (25 %)

a. J. W. Cornforth and R. Robinson, *J. Chem. Soc.*, 1855 (1949).
b. S. Ramachandran and M. S. Newman, *Org. Synth.* **41**, 38 (1961).
c. R. L. Augustine and J. A. Caputo, *Org. Synth.,* **45**, 80 (1965).
d. C. J. V. Scanio and R. M. Starrett, *J. Am. Chem. Soc.* **93**, 1539 (1971).
e. G. Stork and S. D. Darling, *J. Am. Chem. Soc.* **86**, 1761 (1964).

condensation products and the two possible mixed condensation products is to be expected.

One of the most important cases of mixed condensations involves the reaction of aromatic aldehydes with aliphatic ketones or aldehydes. An aromatic aldehyde cannot function as the nucleophilic species because, lacking an α-hydrogen atom, it is incapable of forming an enol or enolate. Dehydration is favored because it leads to a double bond conjugated with both the carbonyl group and the aromatic ring:

$$\text{ArCH}{=}\text{O} \; + \; \underset{\text{O}}{\overset{\text{O}}{\text{RCH}_2\overset{\|}{\text{C}}\text{R}'}} \; \rightleftharpoons \; \underset{\text{R}}{\text{Ar}\overset{\text{OH}}{\underset{|}{\text{CH}}}\overset{\text{O}}{\overset{\|}{\text{CH}}\overset{\|}{\text{C}}\text{R}'}} \; \xrightarrow{-\text{H}_2\text{O}} \; \text{ArCH}{=}\underset{\text{R}}{\overset{\overset{\text{O}}{\overset{\|}{\text{CR}'}}}{\text{C}}}$$

There are many examples of both acid- and base-catalyzed condensation reactions involving aromatic aldehydes. The name *Claisen–Schmidt condensation* is associated with this type of mixed aldol reaction. Entries 7–10 in Scheme 2.1 are a few of the hundreds of examples of this reaction that have been recorded.

There is a strong preference for formation of compounds having *trans* stereochemistry in the base-catalyzed condensation of aromatic aldehydes with methyl ketones. This stereoselectivity can be traced to the dehydration step. The transition states leading to *cis* and *trans* product are sketched in Figure 2.1. Coplanarity of the substituents on C-2 and C-3 is attained as the elimination proceeds. The steric compression that develops between the phenyl group and the ketone R group raises the energy of the transition state that leads to *cis* product, and therefore the *trans* product is preferentially formed.

Additional insight into the factors affecting product structure was obtained by studies on the Claisen–Schmidt condensation of 2-butanone with benzaldehyde.[8]

$$\underset{\text{CH}_3}{\text{PhCH}{=}\overset{\text{O}}{\overset{\|}{\text{C}}}\overset{|}{\underset{|}{\text{CCH}_3}}} \; \xleftarrow{\text{HCl}} \; \text{PhCHO} \; + \; \text{CH}_3\text{COCH}_2\text{CH}_3 \; \xrightarrow{\text{NaOH}} \; \text{PhCH}{=}\text{CH}\overset{\text{O}}{\overset{\|}{\text{C}}}\text{CH}_2\text{CH}_3$$

The results indicate how the interplay between the relative rates of the various reaction steps determines the identity of the reaction product. When catalyzed by base, 2-butanone reacts with benzaldehyde at the methyl group; under conditions of acid catalysis, the site of reaction is the methylene group. The reaction conditions do not permit the isolation of the intermediate ketols because the addition phase is rate-limiting, but these compounds have been prepared by an alternative route. They behave as shown in the following equations:

$$\text{Ph}\overset{\text{OH}}{\underset{|}{\text{CH}}}\text{CH}_2\overset{\text{O}}{\overset{\|}{\text{C}}}\text{CH}_2\text{CH}_3 \; \xrightarrow{\text{NaOH}} \; \text{PhCH}{=}\text{CH}\overset{\text{O}}{\overset{\|}{\text{C}}}\text{CH}_2\text{CH}_3 \; + \; \text{PhCH}{=}\text{O} \; + \; \text{CH}_3\text{CH}_2\overset{\text{O}}{\overset{\|}{\text{C}}}\text{CH}_3$$

$$\underset{\text{CH}_3}{\text{Ph}\overset{\text{OH}}{\underset{|}{\text{CH}}}\overset{|}{\text{CH}}\overset{\text{O}}{\overset{\|}{\text{C}}}\text{CH}_3} \; \xrightarrow{\text{NaOH}} \; \text{PhCH}{=}\text{O} \; + \; \text{CH}_3\text{CH}_2\overset{\text{O}}{\overset{\|}{\text{C}}}\text{CH}_3$$

These results establish that base-catalyzed dehydration is slow relative to the reverse of the addition phase for the branched-chain isomer. The reason for selective formation of the straight-chain product under conditions of base catalysis is then

8. M. Stiles, D. Wolf, and G. V. Hudson, *J. Am. Chem. Soc.* **81**, 628 (1959).

Figure 2.1. Transition states for base-catalyzed dehydration in Claisen–Schmidt condensations.

obvious. The straight-chain ketol is the only productive addition intermediate. Acid treatment of each of the intermediates gives the dehydration product having the corresponding carbon skeleton, along with some of the cleavage products. Under conditions of acid catalysis, then, either intermediate can be dehydrated. Under

$$\text{PhCHCH}_2\text{CCH}_2\text{CH}_3 \xrightarrow{\text{HCl}} \text{PhCH}=\text{CHCCH}_2\text{CH}_3 + \text{PhCHO} + \text{CH}_3\text{COCH}_2\text{CH}_3$$

$$\text{PhCHCHCCH}_3 \xrightarrow{\text{HCl}} \text{PhCH}=\text{CCCH}_3 + \text{PhCHO} + \text{CH}_3\text{COCH}_2\text{CH}_3$$

acid-catalyzed conditions, the addition phase is rate-determining, and the relative amounts of the two dehydrated products are determined by the amounts of intermediates formed. The more substituted enol is favored, and the branched chain isomer is therefore the major product in the acid-catalyzed reaction:

$$\text{CH}_3\text{COCH}_2\text{CH}_3 \xrightarrow{\text{H}^+} \text{CH}_3\text{C}=\text{CHCH}_3 + \text{CH}_2=\text{CCH}_2\text{CH}_3$$
(major) (minor)

$$\text{CH}_3\text{C}=\text{CHCH}_3 + \text{PhCHO} \xrightarrow{\text{slow}} \text{PhCHCHCCH}_3$$

$$\text{PhCHCHCCH}_3 \xrightarrow{\text{fast}} \text{PhCH}=\text{CCCH}_3$$

In any given system, the structure of the final product will depend upon the magnitude of the individual rate constants. In general, condensations of methyl ketones with aromatic aldehydes follow the pattern observed for 2-butanone; i.e., base catalysis favors the linear condensation product, while acid catalysis favors the branched product.

It may be possible to carry out mixed aldol condensations that would otherwise lead to product mixtures by preforming the enolate of one of the two carbonyl compounds. If the addition step is then faster than proton transfer between the enolate and carbonyl acceptor, a single product will be formed. Since the aldol reaction is reversible, some technique of trapping the adduct and preventing reversal must also be employed. This has been done by adding ZnCl_2 to the reaction mixture, which results in the formation of a stable chelate. To illustrate, the selective condensation of phenylacetone with butyraldehyde was achieved in 54% yield:

$$PhCH_2\overset{\overset{\displaystyle O}{\|}}{C}CH_3 \xrightarrow[\text{(dimethoxyethane)}]{NaH} PhCH=\overset{\overset{\displaystyle O^-}{|}}{C}CH_3$$

1) ZnCl$_2$
2) C$_3$H$_7$CH=O, 5–10°

$$\underset{\underset{\displaystyle C_3H_7CHOH}{|}}{Ph\overset{\overset{\displaystyle O}{\|}}{C}HCCH_3} \longleftarrow \underset{\underset{\displaystyle C_3H_7CH}{|}}{PhCHC}\overset{CH_3}{\underset{O}{\diagdown_{O\cdots}}Zn}$$

The stereochemistry of the dominant aldol addition product can be predicted by examining the structure of the zinc chelate. The major product is that corresponding to the maximum number of equatorial substituents in the chelate ring.[9]

Mixed condensations in which the nucleophilic enolate is derived from an ester have also been developed. Very strong bases have usually been used for enolate formation. For example, the lithium enolate of ethyl acetate is generated using lithium bis(trimethylsilyl)amide as the base. Condensation with carbonyl compounds proceeds readily (entry 13, Scheme 2.1) without apparent complications from proton-transfer reactions between the ester enolate and carbonyl compound. The dilithium salts of carboxylic acids can also add to carbonyl compounds (entry 14, Scheme 2.1).

2.2. Related Condensation Reactions

A number of preparatively useful reactions are variants of the mechanistic pattern established by the aldol condensation. One important group is a family of condensations that effect transformations quite similar to the aldol, but that are particularly effectively catalyzed by amines or buffer systems containing amines and the corresponding conjugate acid. These amine-catalyzed reactions are often referred to as *Knoevenagel condensations*.[10]

In several cases, it has been established that the amines do not function as simple bases, but, instead, are involved in prior reaction with the carbonyl compounds as well. Kinetic evidence in support of such a mechanism in the condensation of aromatic aldehydes with nitromethane has been reported.[11] The fact that such condensations are often most effectively catalyzed when a weak acid is present in addition to the amine suggests that amines do not function as simple base catalysts.

9. H. O. House, D. S. Crumrine, A. Y. Teranishi, and H. D. Olmstead, *J. Am. Chem. Soc.* **95**, 3310 (1973).

10a. G. Jones, *Org. React.* **15**, 204 (1967).
   b. R. L. Reeves, in *The Chemistry of the Carbonyl Group*, S. Patai (ed.), Interscience, New York, NY, 1966, pp. 593–599.

11. T. I. Crowell and D. W. Peck, *J. Am. Chem. Soc.* **75**, 1075 (1953).

$$ArCH{=}O \ + \ C_4H_9NH_2 \ \rightleftarrows \ ArCH{=}NC_4H_9$$

$$\overset{H^+}{ArCH{=}NC_4H_9} \rightarrow \ ArCHNHC_4H_9 \rightarrow \ \overset{H^+}{ArCH{-}NHC_4H_9} \rightarrow \ ArCH{=}CHNO_2$$

There have been few definitive mechanistic studies, however, so there is no reason to conclude that imines and iminium intermediates are involved in all amine-catalyzed condensations.

In preparative applications of the reaction, its principal utility has been condensation of ketones and aldehydes with easily enolizable compounds containing two activating groups. Malonic esters and cyanoacetic esters are the most common examples.[12] Usually, the product is the "dehydrated" compound, with the saturated intermediates being isolated only under especially mild conditions. Nitroalkanes are also effective nucleophilic substrates. The single, strongly electron-withdrawing nitro group sufficiently activates the α-hydrogens to permit formation of the nucleophilic nitronate anion under mildly basic conditions. A relatively highly acidic proton in the potential nucleophile is important for two reasons. First, weak bases, such as amines, can then provide a sufficient concentration of enolate for reaction, without causing deprotonation of the ketone or aldehyde. Self-condensation of the carbonyl component is thus minimized. Second, the highly acidic proton facilitates the elimination step that drives the condensation to completion.

$$R_2C{-}\underset{X}{\overset{H}{C}}\overset{CO_2R}{\underset{CN}{}} \longrightarrow R_2C{=}C\overset{CO_2R}{\underset{CN}{}}$$

X = OH or NR₂

A closely related variation of the reaction uses cyanoacetic acid or malonic acid, as opposed to the corresponding esters, as the potential nucleophile. The mechanism of the addition phase of the reaction under these circumstances is similar to the previously discussed cases. The addition intermediates, however, are susceptible to decarboxylation. In many instances, the decarboxylation and elimination phases may occur as a single concerted process.[13] Many of the decarboxylative condensations have been carried out in pyridine, and it has been shown that pyridinium ion can

$$\overset{O}{R\overset{\|}{C}R} \ + \ CH_2(CO_2H)_2 \rightarrow R_2C{-}CHCO_2H \rightarrow R_2C{=}CHCO_2H$$

12. A. C. Cope, C. M. Hofmann, C. Wyckoff, and E. Hardenbergh, *J. Am. Chem. Soc.* **63**, 3452 (1941).
13. E. J. Corey, *J. Am. Chem. Soc.* **74**, 5897 (1952).

catalyze the decarboxylation of arylidenemalonic acids.[14] This then provides an

$$ArCH=C(CO_2H)_2 \; + \; \underset{\overset{|}{N}}{\overset{\overset{H^+}{N}}{\bigcirc}} \;\longrightarrow\; \underset{\underset{\overset{|}{N}}{\bigcirc}}{\overset{\displaystyle ArCH-CHCO_2H}{\underset{\overset{|}{C=O}}{\overset{|}{\underset{O-H}{}}}}} \;\longrightarrow\; ArCH=CHCO_2H$$

alternative mechanism by which decarboxylation might occur.

Scheme 2.3 provides a few examples of condensation reactions of the Knoevenagel type.

2.3. The Mannich Reaction

The *Mannich reaction* is very closely related to the Knoevenagel condensation reaction in that it involves iminium intermediates. The reaction, which is carried out in mildly acidic solution, effects α-alkylation of ketones and aldehydes with dialkylaminomethyl groups. The electrophilic species is the iminium ion derived from

$$RCH_2\overset{\overset{\displaystyle O}{\|}}{C}R' \; + \; CH_2=O \; + \; HN(CH_3)_2 \;\rightarrow\; RCH\overset{\overset{\displaystyle O}{\|}}{C}R' \atop \underset{CH_2N(CH_3)_2}{|}$$

the amine and formaldehyde. The reaction is quite general for aldehydes and ketones

$$CH_2=O \; + \; HN(CH_3)_2 \;\rightleftharpoons\; HOCH_2N(CH_3)_2 \;\overset{H^+}{\rightleftharpoons}\; H_2O \; + \; CH_2=\overset{+}{N}(CH_3)_2$$

$$RCH_2\overset{\overset{\displaystyle O}{\|}}{C}R' \;\overset{H^+}{\rightleftharpoons}\; \underset{\underset{+}{CH_2=\overset{+}{N}(CH_3)_2}}{RCH=\overset{\overset{\displaystyle OH}{|}}{C}R'} \;\longrightarrow\; RCH\overset{\overset{\displaystyle O}{\|}}{C}R' \atop \underset{CH_2N(CH_3)_2}{|}$$

having at least one enolizable hydrogen. For practical preparative purposes, the reaction is limited to secondary amines, since dialkylation becomes a significant problem with primary amines. The dialkylation reaction, however, can be used advantageously in some ring closures.

$$\underset{CH_3O_2C\overset{|}{C}H}{\overset{CH_3CH_2}{|}}\!-\!\overset{\overset{\displaystyle O}{\|}}{C}\!-\!\underset{CHCO_2CH_3}{\overset{CH_2CH_3}{|}} \; + \; CH_2O \; + \; CH_3NH_2 \;\longrightarrow\; \overset{\overset{\displaystyle O}{\|}}{\underset{\underset{CH_3}{\overset{|}{N}}}{\underset{C_2H_5}{\diagdown}\;\;\diagup{C_2H_5}}{CH_3O_2C\diagdown \qquad \diagup CO_2CH_3}}$$

Ref. 15

The importance of the Mannich reaction stems from the synthetic utility of the resulting aminoketones. Thermal decomposition of the amines or the derived

14. E. J. Corey and G. Fraenkel, *J. Am. Chem. Soc.* **75**, 1168 (1953).
15. C. Mannich and P. Schumann, *Chem. Ber.* **69**, 2299 (1936).

Scheme 2.3. Amine-Catalyzed Condensations of the Knoevenagel Type

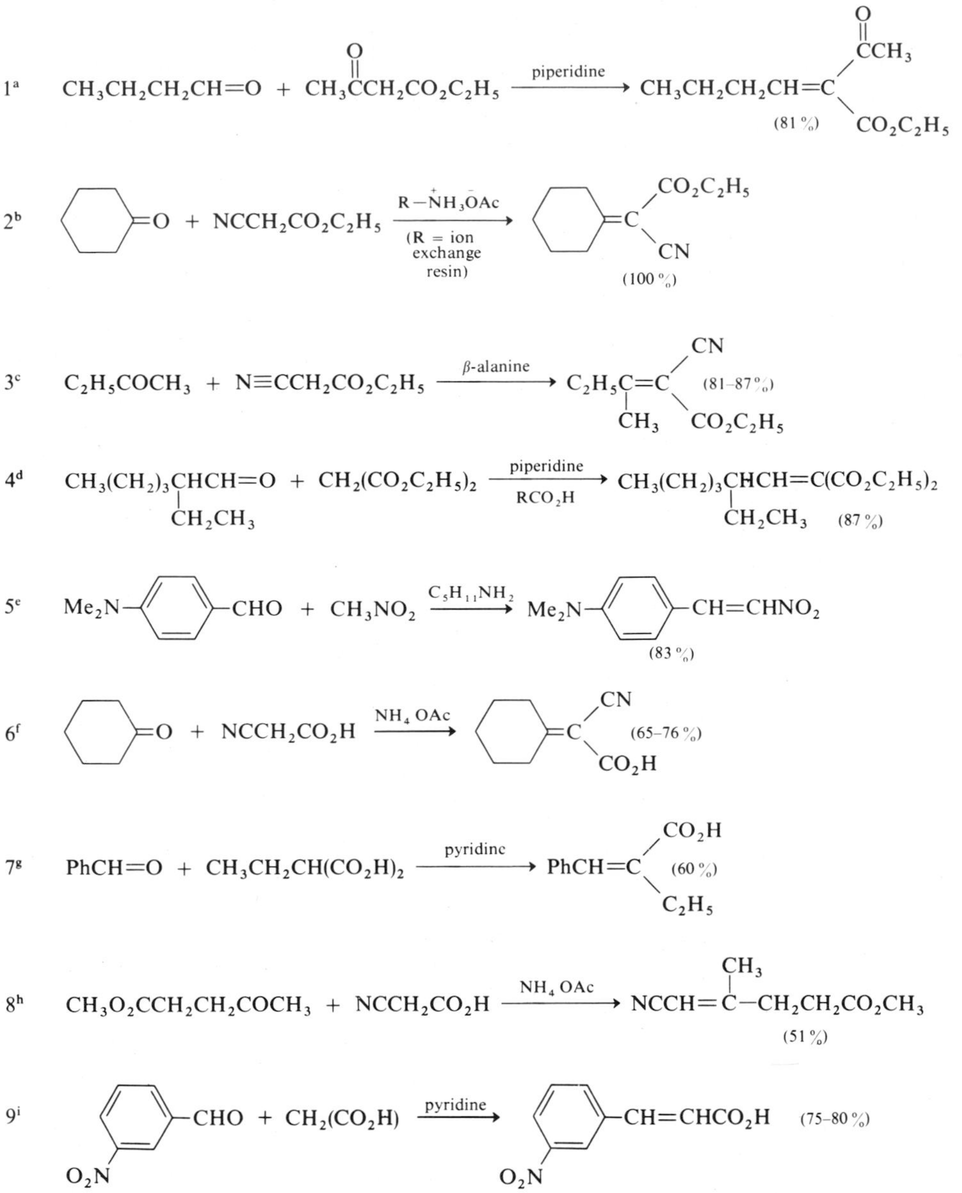

1^a $CH_3CH_2CH_2CH{=}O$ + $CH_3\overset{O}{\overset{\|}{C}}CH_2CO_2C_2H_5$ $\xrightarrow{\text{piperidine}}$ $CH_3CH_2CH_2CH{=}C\overset{\overset{O}{\overset{\|}{CCH_3}}}{\underset{CO_2C_2H_5}{}}$ (81%)

2^b ⬡$=O$ + $NCCH_2CO_2C_2H_5$ $\xrightarrow[\substack{(R = ion \\ exchange \\ resin)}]{R-\overset{+}{N}H_3OAc}$ ⬡$=C\overset{CO_2C_2H_5}{\underset{CN}{}}$ (100%)

3^c $C_2H_5COCH_3$ + $N{\equiv}CCH_2CO_2C_2H_5$ $\xrightarrow{\beta\text{-alanine}}$ $C_2H_5\underset{CH_3}{C}{=}C\overset{CN}{\underset{CO_2C_2H_5}{}}$ (81–87%)

4^d $CH_3(CH_2)_3\underset{CH_2CH_3}{CH}CH{=}O$ + $CH_2(CO_2C_2H_5)_2$ $\xrightarrow[RCO_2H]{\text{piperidine}}$ $CH_3(CH_2)_3\underset{CH_2CH_3}{CH}CH{=}C(CO_2C_2H_5)_2$ (87%)

5^e $Me_2N{-}$⬡$-CHO$ + CH_3NO_2 $\xrightarrow{C_5H_{11}NH_2}$ $Me_2N{-}$⬡$-CH{=}CHNO_2$ (83%)

6^f ⬡$=O$ + $NCCH_2CO_2H$ $\xrightarrow{NH_4OAc}$ ⬡$=C\overset{CN}{\underset{CO_2H}{}}$ (65–76%)

7^g $PhCH{=}O$ + $CH_3CH_2CH(CO_2H)_2$ $\xrightarrow{\text{pyridine}}$ $PhCH{=}C\overset{CO_2H}{\underset{C_2H_5}{}}$ (60%)

8^h $CH_3O_2CCH_2CH_2COCH_3$ + $NCCH_2CO_2H$ $\xrightarrow{NH_4OAc}$ $NCCH{=}\overset{CH_3}{\underset{}{C}}{-}CH_2CH_2CO_2CH_3$ (51%)

9^i $\underset{O_2N}{}$⬡$-CHO$ + $CH_2(CO_2H)$ $\xrightarrow{\text{pyridine}}$ $\underset{O_2N}{}$⬡$-CH{=}CHCO_2H$ (75–80%)

a. A. C. Cope and C. M. Hofmann, *J. Am. Chem. Soc.* **63**, 3456 (1941).
b. R. W. Hein, M. J. Astle, and J. R. Shelton, *J. Org. Chem.* **26**, 4874 (1961).
c. F. S. Prout, R. J. Hartman, E. P.-Y. Huang, C. J. Korpics, and G. R. Tichelaar, *Org. Synth.* **IV**, 93 (1063).
d. E. F. Pratt and E. Werble, *J. Am. Chem. Soc.* **72**, 4638 (1950).
e. D. E. Worrall and L. Cohen, *J. Am. Chem. Soc.* **66**, 842 (1944).
f. A. C. Cope, A. A. D'Addieco, D. E. Whyte, and S. A. Glickman, *Org. Synth.* **IV**, 234 (1963).
g. W. J. Gensler and E. Berman, *J. Am. Chem. Soc.* **80**, 4949 (1958).
h. R. Stevens, *J. Chem. Soc.* 1118 (1960).
i. R. H. Wiley and N. R. Smith, *Org. Synth.* **IV**, 731 (1963).

Scheme 2.4. Synthesis and Utilization of Mannich Bases

1[a] $PhCOCH_3 + CH_2O + (CH_3)_2\overset{+}{N}H_2Cl^- \longrightarrow PhCOCH_2CH_2\overset{H_+}{N}(CH_3)_2Cl^-$ (70%)

2[b] $CH_3COCH_3 + CH_2O + (CH_3CH_2)_2\overset{+}{N}H_2Cl^- \longrightarrow CH_3COCH_2CH_2\overset{H_+}{N}(C_2H_5)_2Cl^-$
   (66–75%)

3[c] $CH_3CH_2CH_2CH{=}O + CH_2O + (CH_3)_2\overset{+}{N}H_2Cl^- \xrightarrow[\text{2) distill}]{\text{1) 60°, 6 hr}} CH_2{=}CCH{=}O$
   $\overset{|}{CH_2CH_3}$ (73%)

4[d] $+ PhCOCH_2CH_2N(CH_3)_2 \xrightarrow{NaOH}$ CH_2CH_2COPh (52%)

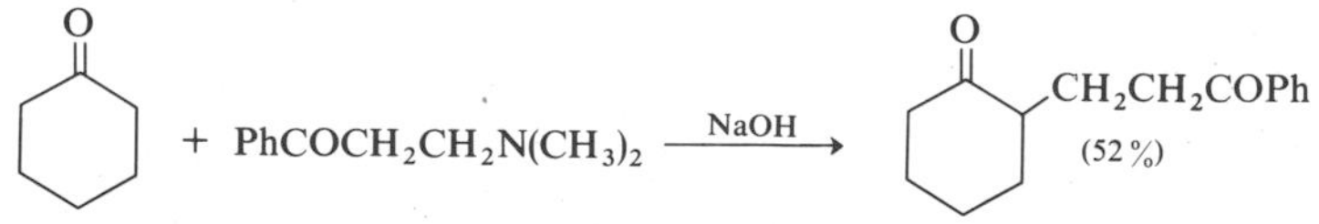

5[e] $PhCOCH_2CH_2N(CH_3)_2 + KCN \longrightarrow PhCOCH_2CH_2CN$ (67%)

a. C. E. Maxwell, *Org. Synth.* **III**, 305 (1955).
b. A. L. Wilds, R. M. Nowak, and K. E. McCaleb, *Org. Synth.* **IV**, 281 (1963).
c. C. S. Marvel, R. L. Myers, and J. H. Saunders, *J. Am. Chem. Soc.* **70**, 1694 (1948).
d. A. C. Cope and E. C. Hermann, *J. Am. Chem. Soc.* **72**, 3405 (1950).
e. E. B. Knott, *J. Chem. Soc.*, 1190 (1947).

quaternary salts leads to α-methylene ketones. The decomposition of the quaternary salts is particularly facile, and they can be used as *in situ* sources of the

$$(CH_3)_2CHCHCH{=}O \xrightarrow{\Delta} (CH_3)_2CHCCH{=}O$$
$$\underset{CH_2N(CH_3)_2}{|} \qquad\qquad \underset{CH_2}{||}$$
   Ref. 16

α,β-unsaturated carbonyl compounds. These are useful synthetic intermediates, as, for example, in Michael additions (Chapter 1) and hydroboration (Chapter 3). Entries 4 and 5 in Scheme 2.4 represent such synthetic elaboration via Michael addition to α,β-unsaturated intermediates.

The Mannich reaction, or at least a close mechanistic analog, is important in the formation of many nitrogen-containing molecules in nature. As a result, the Mannich reaction has played an important role in the total synthesis of such compounds, especially in syntheses patterned after the mode of biosynthesis, i.e., *biogenetic-type synthesis*. The earliest example of the use of the Mannich reaction in this context was the successful synthesis of tropinone, a derivative of the alkaloid tropine, by Sir

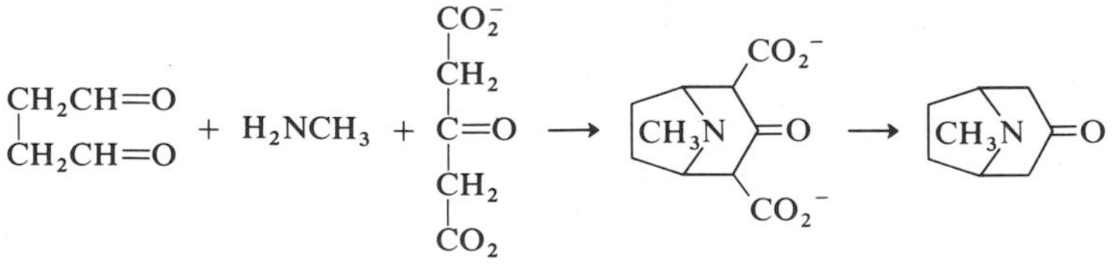

$$\underset{CH_2CH{=}O}{\overset{CH_2CH{=}O}{|}} + H_2NCH_3 + \underset{CO_2}{\overset{CO_2^-}{\underset{|}{\overset{|}{\underset{C=O}{\overset{CH_2}{|}}}}}}$$

$\longrightarrow$ Ref. 17

16. C. S. Marvel, R. L. Myers, and J. H. Saunders, *J. Am. Chem. Soc.* **70**, 1694 (1948).
17. R. Robinson, *J. Chem. Soc.*, 762 (1917).

Scheme 2.5. Mannich Reaction in the Biosynthesis of Lupine Alkaloids

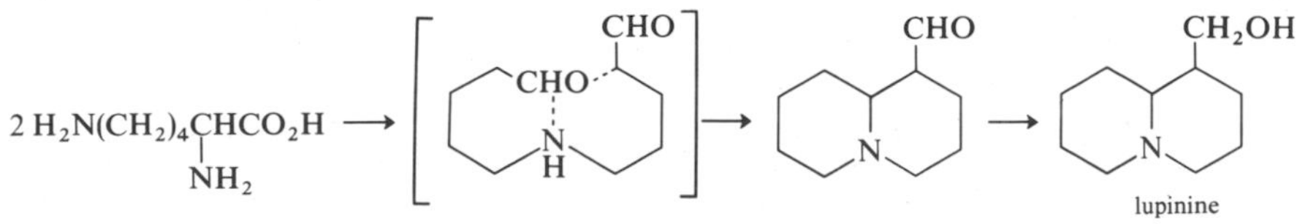

Robert Robinson in 1917. More recently, modern biosynthetic work has provided a great deal of information about alkaloid biosynthesis and many alkaloids have been efficiently synthesized by routes that generally parallel those followed in nature. The case of the lupine alkaloids can be cited as one example of the role of the Mannich reaction in biosynthesis. Scheme 2.5 provides a rough outline of the biosynthesis of this alkaloid system from the amino acid lysine.

2.4. Acylation of Nucleophilic Carbon

Enolate anions and other nucleophilic carbon species are acylated when addition to a carbonyl group is followed by elimination of one of the carbonyl substituents. A classic example of this type of reaction pattern is the base-catalyzed

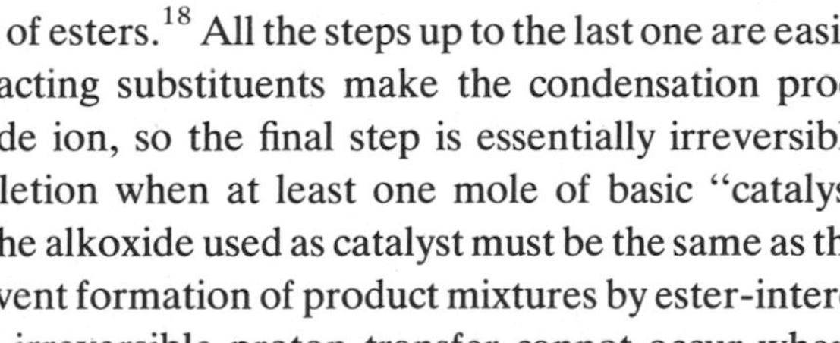

self-condensation of esters.[18] All the steps up to the last one are easily reversible. The two electron-attracting substituents make the condensation product quite acidic relative to alkoxide ion, so the final step is essentially irreversible and drives the reaction to completion when at least one mole of basic "catalyst" is used. As a practical matter, the alkoxide used as catalyst must be the same as the alcohol portion of the ester to prevent formation of product mixtures by ester-interchange reactions. Because the final irreversible proton transfer cannot occur when α-disubstituted esters are employed, these compounds do not condense when alkoxide ions are used as catalysts. This limitation can be overcome by the use of a very strong base that converts the starting ester essentially completely to its enolate. One procedure employs triphenylmethylsodium for this purpose (entry 2, Scheme 2.6).

18. C. R. Hauser and B. E. Hudson, Jr., *Org. React.* **1**, 266 (1942).

Scheme 2.6. Acylation of

A. Intermolecular Ester Condensations

1^a $CH_3(CH_2)_3CO_2C_2H_5$ $\xrightarrow{NaOEt}$ $CH_3(CH_2)_3COCHCO_2C_2H_5$ (77%)
$\qquad\qquad\qquad\qquad\qquad\qquad\qquad\qquad\qquad\quad$ |
$\qquad\qquad\qquad\qquad\qquad\qquad\qquad\qquad\qquad\quad CH_2CH_2CH_3$

2^b $CH_3CH_2CHCO_2C_2H_5$ $\xrightarrow{Ph_3C^- Na^+}$ (63%)
$\qquad\qquad$ |
$\qquad\qquad CH_3$

B. Cyclization of Diesters

3^c $C_2H_5O_2C(CH_2)_4CO_2C_2H_5$ $\xrightarrow{Na,\ toluene}$

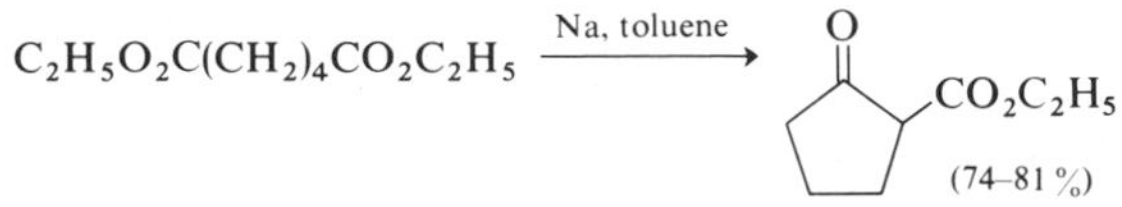

(74–81%)

4^d $CH_3-N\begin{smallmatrix}CH_2CH_2CO_2C_2H_5\\ \\CH_2CH_2CO_2C_2H_5\end{smallmatrix}$ $\xrightarrow[benzene]{NaOEt}$ $\xrightarrow{HCl}$

(71%)

5^e $C_2H_5O_2CCH_2CH_2CHCHCH_3$ $\xrightarrow{NaH}$
$\qquad\qquad\qquad\qquad\quad |\ \ \ |$
$\qquad\qquad\qquad CO_2C_2H_5\ \ CO_2C_2H_5$

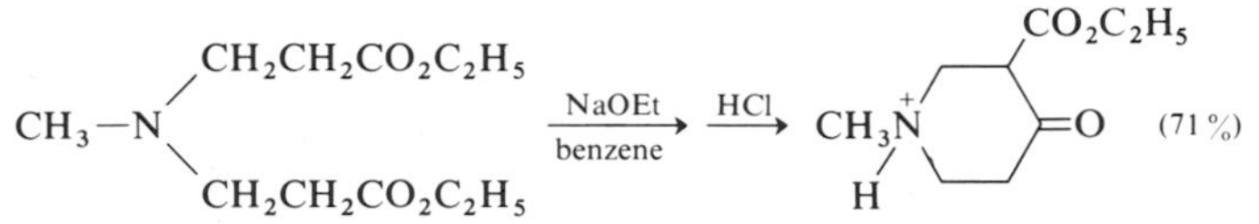

(92%)

6^f $\xrightarrow{NaH}$ (21%)

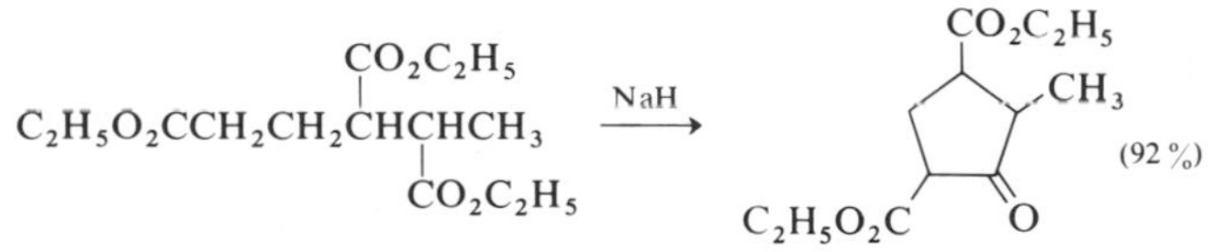

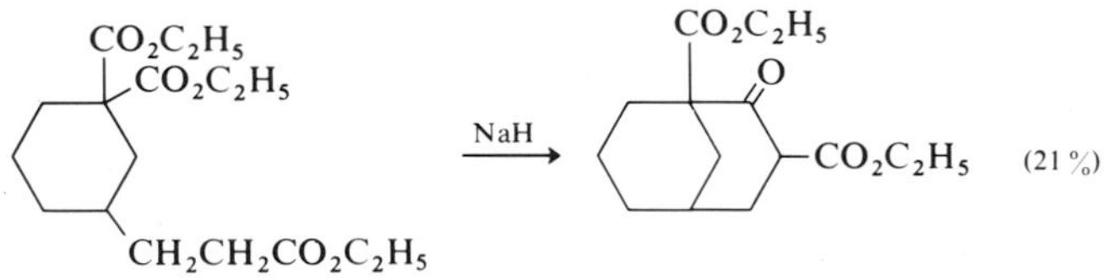

7^g $\longrightarrow$

$\xrightarrow[\text{dilute solution}]{[(CH_3)_3Si]_2NNa}$ (77%)

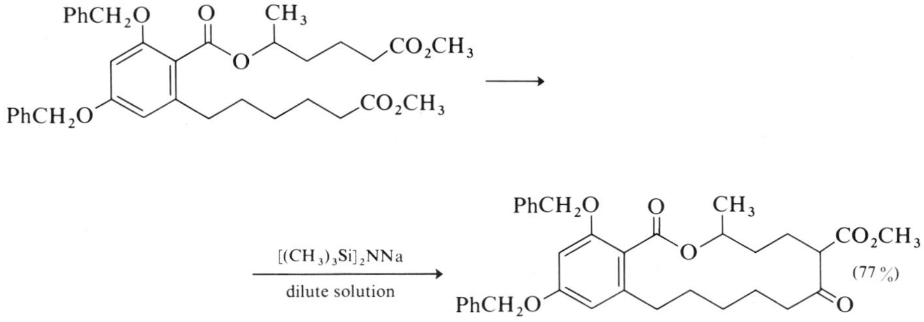

C. Mixed Ester Condensations

8^h $(CH_2CO_2C_2H_5)_2$ + $(CO_2C_2H_5)_2$ $\xrightarrow{NaOEt}$ $\begin{smallmatrix}COCO_2C_2H_5\\ |\\CHCO_2C_2H_5\\ |\\CH_2CO_2C_2H_5\end{smallmatrix}$

9^i

$+ \ CH_3(CH_2)_2CO_2C_2H_5 \ \xrightarrow{\text{NaH}}$

10^j $\quad C_{17}H_{35}CO_2C_2H_5 \ + \ (CO_2C_2H_5)_2 \ \xrightarrow{\text{NaOEt}} \ C_{16}H_{33}\underset{\underset{COCO_2C_2H_5}{|}}{CH}CO_2C_2H_5 \quad (68\text{--}71\%)$

11^k

D. Acylation with Anhydrides and Acyl Halides

12^l $\quad PhCOCOC_2H_5 \ + \ C_2H_5OMgCH(CO_2C_2H_5)_2 \ \rightarrow \ PhCOCH(CO_2C_2H_5)_2$
$\qquad\qquad\qquad\qquad\qquad\qquad\qquad\qquad\qquad\qquad\qquad (68\text{--}75\%)$

13^m $\quad CH_3\overset{O^-}{\underset{}{C}}=CHCO_2C_2H_5 \ + \ PhCOCl \ \rightarrow \ \left[PhC\overset{\overset{O}{||}}{\underset{}{-}}\overset{CCH_3}{\underset{}{C}}CO_2C_2H_5 \right] \ \rightarrow \ Ph\overset{O}{\underset{}{C}}CH_2CO_2C_2H_5$
$\qquad\qquad\qquad\qquad\qquad\qquad\qquad\qquad\qquad\qquad\qquad\qquad\qquad\qquad\qquad (68\text{--}71\%)$

14^n

15° $\quad CH_3\overset{O}{\underset{}{C}}=CHCO_2C_2H_5 \ + \ Cl\overset{O}{\underset{}{C}}(CH_2)_3CO_2C_2H_5 \ \rightarrow \ C_2H_5O_2C(CH_2)_3\overset{\overset{O}{||}}{\underset{}{C}}\text{-}\overset{CCH_3}{\underset{}{CH}}CO_2C_2H_5$
$\qquad\qquad\qquad\qquad\qquad\qquad\qquad\qquad\qquad\qquad\qquad\qquad\qquad (61\text{--}66\%)$

16^p $\quad CH_3CO_2C_2H_5 \ \xrightarrow{\text{R}_2\text{NLi}} \ LiCH_2CO_2C_2H_5 \ \xrightarrow[-78°]{\overset{O}{\overset{||}{(CH_3)_3CCCl}}} \ (CH_3)_3C\overset{O}{\underset{}{C}}CH_2CO_2C_2H_5$
$\qquad\qquad\qquad\qquad\qquad\qquad\qquad\qquad\qquad\qquad\qquad\qquad\qquad (70\%)$

a. R. R. Briese and S. M. McElvain, *J. Am. Chem. Soc.* **55**, 1697 (1933).
b. B. E. Hudson, Jr., and C. R. Hauser, *J. Am. Chem. Soc.* **63**, 3156 (1941).
c. P. S. Pinkney, *Org. Synth.* **II**, 116 (1943).
d. E. A. Prill and S. M. McElvain, *J. Am. Chem. Soc.* **55**, 1233 (1933).
e. M. S. Newman and J. L. McPherson, *J. Org. Chem.* **19**, 1717 (1954).
f. J. P. Ferris and N. C. Miller, *J. Am. Chem. Soc.* **85**, 1325 (1963).
g. R. N. Hurd and D. H. Shah, *J. Org. Chem.* **38**, 390 (1973).
h. E. M. Bottorff and L. L. Moore, *Org. Synth.* **44**, 67 (1964).
i. F. W. Swamer and C. R. Hauser, *J. Am. Chem. Soc.* **72**, 1352 (1950).
j. D. E. Floyd and S. E. Miller, *Org. Synth.* **IV**, 141 (1963).
k. E. E. Royals and D. G. Turpin, *J. Am. Chem. Soc.* **76**, 5452 (1954).
l. J. A. Price and D. S. Tarbell, *Org. Synth.* **IV**, 285 (1963).
m. J. M. Straley and A. C. Adams, *Org. Synth.* **IV**, 415 (1963).
n. G. A. Reynolds and C. R. Hauser, *Org. Synth.* **IV**, 708 (1963).
o. M. Guha and D. Nasipuri, *Org. Synth.* **42**, 41 (1962).
p. M. W. Rathke and J. Deitch, *Tetrahedron Lett.*, 2953 (1971).

The intramolecular version of ester condensation is often referred to as the *Dieckmann condensation*.[19] It is an important tool for the closure to five- and six-membered rings during synthetic sequences, and has occasionally been employed for formation of larger rings. Entries 3–7 in Scheme 2.6 are illustrative.

Modern workers have often chosen sodium hydride and a small amount of alcohol as the catalyst system. It is probable that the effective catalyst is actually the sodium alkoxide formed by reaction of the alcohol released in the condensation with sodium hydride. The sodium alkoxide is also no doubt the active catalyst in reactions

$$R'OH + NaH \rightarrow R'ONa + H_2$$

in which sodium metal is present. The alkoxide is formed by reaction between metallic sodium and the alcohol liberated as the condensation proceeds. The driving force for the reaction is the formation of a stable enolate system. Since the reaction is easily reversible, it is governed by thermodynamic control, and in situations where more than one enolate is possible, the product derived from the more stable one will be formed. An example of this effect is the cyclization of the diester **1**.[20] Only **3** is formed, because **2** cannot be converted to a stable enolate. If **2**, synthesized by another means, is subjected to conditions of the cyclization, it is isomerized to **3** by way of the reversible condensation mechanism:

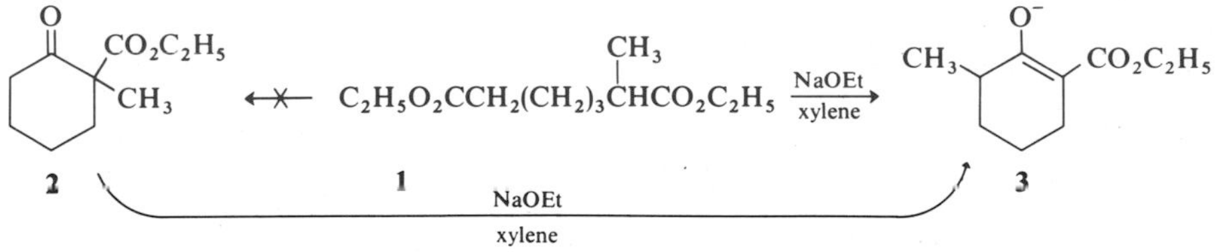

Successful mixed condensations of esters are subject to the same general restrictions as outlined in the consideration of mixed aldol condensations. One carbonyl compound must act preferentially as the acceptor and the other as the nucleophile. To compete with self-condensation of aliphatic esters, the carbonyl acceptor must be relatively electrophilic. The systems that have been commonly employed are esters of aromatic acids, formate esters, and oxalate esters. In each instance, these esters contain groups that are electron-withdrawing relative to alkyl and do not possess enolizable hydrogens. They are therefore good electrophiles, but cannot function as the nucleophile. Some examples are shown in Section C of Scheme 2.6.

The preparation of diethyl benzoylmalonate (entry 12) represents the use of an acid anhydride, a function in which it is much more reactive than an ester, as the acylating agent. The reaction must be carried out in nonnucleophilic solvents to prevent solvolysis of the anhydride from competing with the desired reaction. Other limitations on the use of highly reactive acylating agents, such as acid anhydrides and acid chlorides, in reactions with enolates derive from the fact that O-acylation may be the dominant reaction. The magnesium salt of diethyl malonate (entries 12 and

19. J. P. Schaefer and J. J. Bloomfield, *Org. React.* **15**, 1 (1967).
20. N. S. Vul'fson and V. I. Zaretskii, *J. Gen. Chem. USSR* **29**, 2704 (1959).

Scheme 2.7. Acylation of Ketones with Esters 51

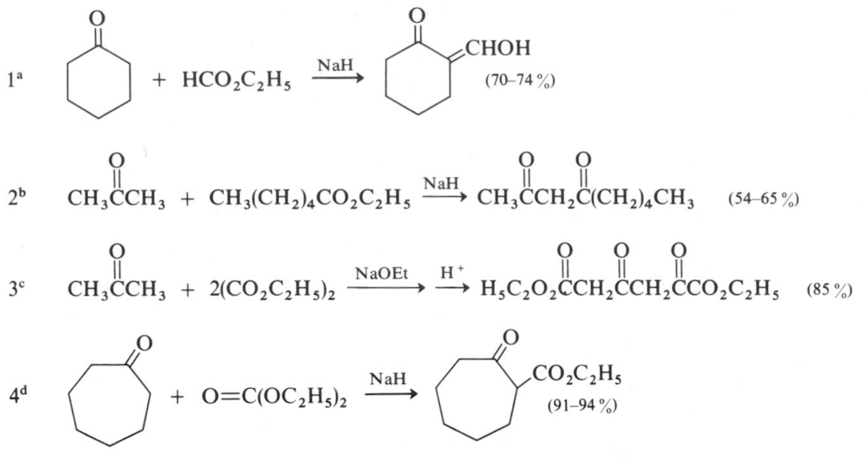

a. C. Ainsworth, *Org. Synth.* **IV**, 536 (1963).
b. N. Green and F. B. La Forge, *J. Am. Chem. Soc.* **70**, 2287 (1948); F. W. Swamer and C. R. Hauser, *J. Am. Chem. Soc.* **72**, 1352 (1950).
c. E. R. Riegel and F. Zwilgmeyer, *Org. Synth.* **II**, 126 (1943).
d. A. P. Krapcho, J. Diamanti, C. Cayen, and R. Bingham, *Org. Synth.* **47**, 20 (1967).

14) has proved to be a satisfactory substrate in these reactions, in part because it is soluble in nonnucleophilic solvents such as ether. Low temperatures permit successful acylation of lithium enolates with acid chlorides (entry 16).

The acylation of enolates derived from ketones with esters is an important tool for enhancing reactivity and selectivity in synthetic modification of ketones. Some representative examples are given in Scheme 2.7. The most common example of this is the formylation of ketone enolates by formate esters:

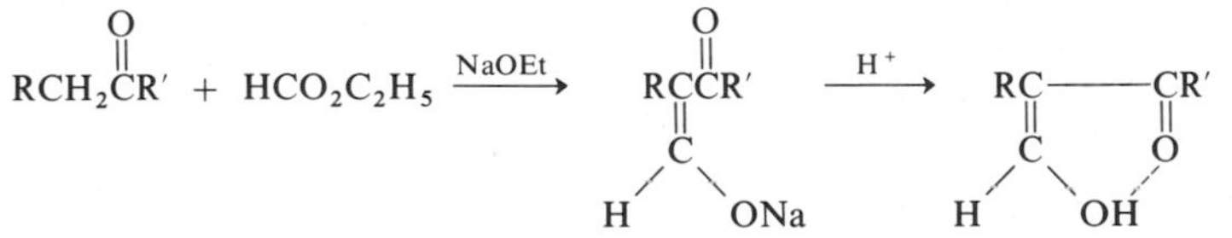

Since the β-ketoaldhydes that result from acidification exist with the formyl group extensively enolized, the compounds are often referred to as *hydroxymethylene derivatives*. The formation of the product is governed by thermodynamic control; therefore, the dominant product expected from unsymmetrical ketones can be predicted on the basis of considerations of relative stability. Once formed, hydroxymethylene compounds have several synthetic uses. A hydroxymethylene group can be converted to methyl via reaction with a mercaptan followed by reduction.[21]

21a. R. E. Ireland and J. A. Marshall, *J. Org. Chem.* **27**, 1615 (1962).
 b. J. D. Metzger, M. W. Baker, and R. J. Morris, *J. Org. Chem.* **37**, 789 (1972).

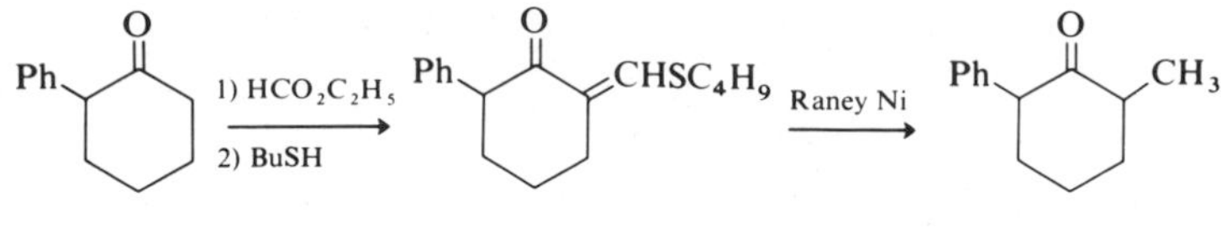

Aluminum hydride reduction of sodium enolates of hydroxymethylene ketones gives hydroxymethyl ketones.[22]

A sequence for directing an alkylation to a monosubstituted site has been developed; the sequence involves a dianion intermediate[23]:

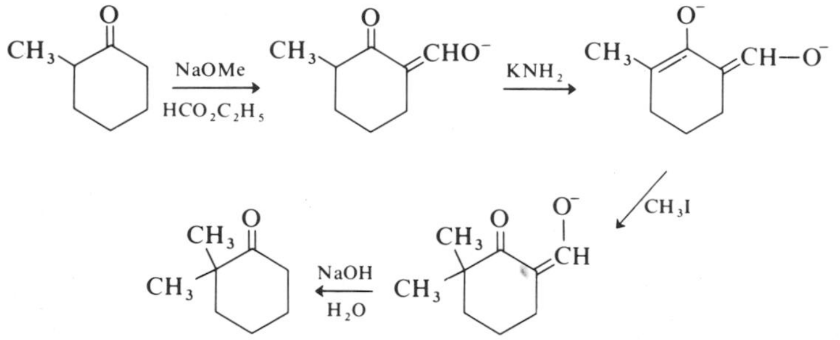

On the other hand, if an unsymmetrical ketone is to be alkylated at the site where acylation would be favored, the hydroxymethylene derivative can be directly subjected to alkylation. Condensation with ethyl carbonate or diethyl oxalate before alkylation can also serve the purpose of increasing reactivity at the α-position of a ketone. The resulting β-ketoesters are readily alkylated by the procedures described in the previous chapter.

Another class of synthetic intermediates are the β-ketosulfoxides, which are prepared by acylation of the dimethyl sulfoxide anion with esters[24]:

$$\underset{\text{RCOR}'}{\overset{\overset{\text{O}}{\|}}{}} + {}^-\text{CH}_2\text{SCH}_3 \rightleftharpoons \underset{\text{RCCHSCH}_3}{\overset{\overset{\text{O}\quad\text{O}}{\|\_\|}}{}} + \text{R}'\text{OH}$$

Mechanistically, this reaction is similar to an ester condensation and results in the

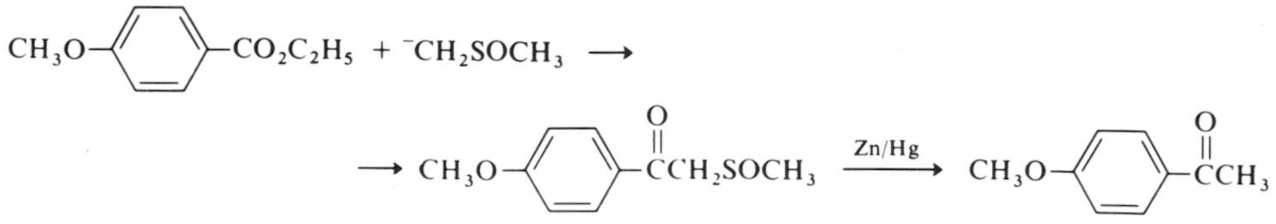

Ref. 25

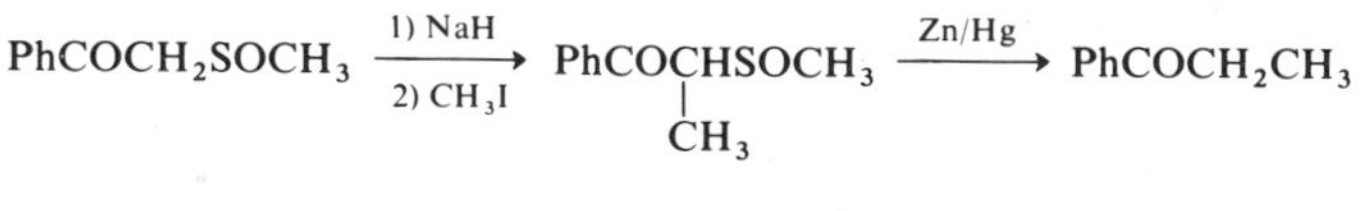

$$\text{PhCOCH}_2\text{SOCH}_3 \xrightarrow[\text{2) CH}_3\text{I}]{\text{1) NaH}} \underset{\text{CH}_3}{\overset{}{\text{PhCOCHSOCH}_3}} \xrightarrow{\text{Zn/Hg}} \text{PhCOCH}_2\text{CH}_3$$

Ref. 26

22. E. J. Corey and D. E. Cane, *J. Org. Chem.* **36**, 3070 (1971).
23. S. Boatman, T. M. Harris, and C. R. Hauser, *Org. Synth.* **48**, 40 (1968).
24a. E. J. Corey and M. Chaykovsky, *J. Am. Chem. Soc.* **87**, 1345 (1965).
   b. H. D. Becker, G. J. Mikol, and G. A. Russell, *J. Am. Chem. Soc.* **85**, 3410 (1963).
25. G. A. Russell and G. J. Mikol, *J. Am. Chem. Soc.* **88**, 5498 (1966).
26. P. G. Gassman and G. D. Richmond, *J. Org. Chem.* **31**, 2355 (1966).

formation of a quite stable carbanion. These substances are of general synthetic utility because the molecule can be reductively cleaved with removal of the methyl-sulfinyl substituent. Thus, a two-step sequence involving condensation and reduction converts esters to methyl ketones. If an alkylation step is included, higher ketones can be prepared. Dimethyl sulfone can be subjected to a similar sequence of reactions.[27] Russell and co-workers[25] have developed a number of other synthetic

$$CH_3(CH_2)_4CO_2C_2H_5 + {}^-CH_2SO_2CH_3 \longrightarrow CH_3(CH_2)_4COCH_2SO_2CH_3$$

$$\downarrow \begin{array}{l} 1) \ NaH \\ 2) \ CH_3I \end{array}$$

$$CH_3(CH_2)_4\overset{\overset{\displaystyle O}{\|}}{C}CH_2CH_3 \xleftarrow{\ Zn/Hg\ } CH_3(CH_2)_4\overset{\overset{\displaystyle O}{\|}}{C}\underset{\underset{\displaystyle CH_3}{|}}{C}HSO_2CH_3$$

procedures that are based on the tendency of β-ketosulfoxides to rearrange on exposure to acid. By using one of a variety of reaction conditions, it is possible to

$$R\overset{\overset{\displaystyle O}{\|}}{C}CH_2SOCH_3 \xrightarrow{\ H^+\ } R\overset{\overset{\displaystyle O}{\|}}{C}\underset{\underset{\displaystyle OH}{|}}{C}HSCH_3$$

convert the rearrangement products to compounds with a variety of functional groups on the terminal two carbon atoms. These include combinations of hydroxyl and carbonyl oxidation states, such as ketoaldehydes, hydroxyaldehydes, glycols, and ketoesters.[25,28]

2.5. The Wittig Reaction

Phosphorus and sulfur ylides are important groups of nucleophilic carbon species. An ylide is a molecule that has a contributing Lewis structure with opposite charges on adjacent atoms when these atoms have octets of electrons. While this definition is broad enough to include a number of other classes of compounds, we will restrict the following discussion to ylides in which a negative charge is on carbon. The synthetic applications of ylides are particularly significant when the positively charged atom is a second-row element, such as phosphorus or sulfur.

Phosphorus ylides are usually stable, albeit highly reactive, compounds and can be represented by two limiting Lewis structures. These are sometimes referred to as the *ylide* and *ylene* form. Using $(CH_3)_3PCH_2$ (trimethylphosphonium methylide) as an example, these two forms can be illustrated as follows:

$$(CH_3)_3\overset{+}{P}{-}CH_2^- \leftrightarrow (CH_3)_3P{=}CH_2$$

ylide   ylene

27. H. O. House and J. K. Larson, *J. Org. Chem.* **33**, 61 (1968).
25. See p. 52.
28. G. A. Russell and L. A. Ochrymowycz, *J. Org. Chem.* **34**, 3618 (1969).

The stability of phosphorus ylides is ascribed to resonance between the two Lewis structures in which the ylene form implies electron donation into phosphorus $3d$-orbitals. Recent careful ^{1}H, ^{13}C, and ^{31}P nmr spectroscopic studies of trimethyl-phosphonium methylide, however, are more consistent with the dipolar ylide structure's having an sp^2-hybridized carbon and sp^3-hybridized phosphorus and suggest only a minor contribution from the ylene structure.[29]

Phosphorus ylides are most commonly prepared by deprotonation of phosphonium salts with strong base. The phosphonium salts, in turn, are prepared by the reaction of trialkyl- or triarylphosphines with alkyl halides:

$$R_3P + R'CH_2X \rightarrow R_3\overset{+}{P}CH_2R' \ X^-$$

$$X = I. Br. Cl. etc.$$

$$R_3\overset{+}{P}CH_2R' + base \rightarrow R_3\overset{+}{P}-\overset{-}{C}HR'$$

Although ylides had been known for many years, their synthetic potential was not appreciated until G. Wittig and his associates at the University of Heidelberg established their utility in olefin synthesis.[30] The reaction of a phosphorus ylide with an aldehyde or ketone provides a means of introducing a carbon–carbon double bond in place of a carbon–oxygen double bond, as shown by the equation:

$$R_3\overset{+}{P}-\overset{-}{C}R_2' + R_2''C=O \rightarrow R_2''\overset{\overset{\displaystyle O^-}{|}}{C}-\overset{\overset{\displaystyle \overset{+}{P}R_3}{|}}{C}R_2' \rightarrow R_2''C=CR_2' + R_3P=O$$

The mechanism usually proposed involves nucleophilic addition of the ylide carbon to the carbonyl group to yield a dipolar intermediate (a betaine), followed by elimination of phosphine oxide. The elimination might be concerted, or it might take place via a four-membered dihydrooxaphosphetane intermediate.[31] It has also been suggested that the dihydrooxaphosphetane intermediate might be formed directly, without the intermediacy of the betaine, by a cycloaddition process.[32]

Alkylphosphonium salts are only weakly acidic. Proton abstraction is usually done with organolithium reagents or with the anion of dimethyl sulfoxide. The resulting ylides are very reactive. β-Ketophosphonium salts are considerably more acidic and undergo ylide formation with weaker bases. The resulting ylides are stabilized by enolate resonance and are substantially less reactive. Vigorous conditions are often required to bring about reaction with ketones. Modifications of the Wittig reaction to be discussed in succeeding sections are often more convenient for synthesis of olefins bearing electron-withdrawing substituents.

$$R\overset{\overset{\displaystyle O}{||}}{C}-\overset{-}{C}H-\overset{+}{P}R_3' \leftrightarrow R\overset{\overset{\displaystyle O^-}{|}}{C}=CH-\overset{+}{P}R_3' \leftrightarrow R\overset{\overset{\displaystyle O}{||}}{C}-CH=PR_3'$$

The Wittig reaction in its original form is not highly stereoselective. Instances of

29. H. Schmidbaur, W. Buchner, and D. Schentzow, *Chem. Ber.* **106**, 1251 (1973).
30. General review of the Wittig reaction: A. Maercker, *Org. React.* **14**, 270 (1965).
31. C. Trindle, J.-T. Hwang, and F. A. Carey, *J. Org. Chem.* **38**, 2664 (1973).
32. E. Vedejs and K. A. J. Snoble, *J. Am. Chem. Soc.* **95**, 5778 (1973).

1^a [cyclohexanone]=O + CH$_2$=P(Ph)$_3$ → [cyclohexane]=CH$_2$ (35–40%)

2^b PhCH=O + PhCH=P(Ph)$_3$ → PhCH=CHPh   (70% *trans*, 30% *cis*)

3^c PhCH=O + CH$_3$CH=P(Ph)$_3$ → PhCH=CHCH$_3$ (13% *trans*, 87% *cis*)

4^d PhCH=O + PhCH=CH—CH$_2\overset{+}{P}$(Ph)$_3$ $\xrightarrow{\text{LiOEt}}$ PhCH=CH—CH=CHPh   (60–67%)

5^e (CH$_3$)$_2$N—[benzene ring]—CH=O + Cl$_2$C=P(Ph)$_3$ → (CH$_3$)$_2$N—[benzene ring]—CH=CCl$_2$ (39–56%)

6^f [cyclohexanone]=O (C$_2$H$_5$O)$_2\overset{O}{\overset{\|}{P}}CH_2CO_2C_2H_5$ $\xrightarrow{\text{NaH}}$ [cyclohexane]=CHCO$_2$C$_2$H$_5$ (67–77%)

7^g CH$_2$=C$\overset{\text{C}_2\text{H}_5}{\underset{\text{CH}=\text{O}}{}}$ + (C$_2$H$_5$O)$_2\overset{O}{\overset{\|}{P}}CH_2CO_2C_2H_5$ $\xrightarrow{\text{NaOEt}}$ CH$_2$=C$\overset{\text{C}_2\text{H}_5}{}$—C=C$\overset{\text{H}}{\underset{\text{CO}_2\text{C}_2\text{H}_5}{}}$ with H   (66%)

8^h (Ph)$_2$C=O + (CH$_3$)$_2\overset{O}{\overset{\|}{C}}\underset{\text{Li}}{P}$[N(CH$_3$)$_2$]$_2$ → (Ph)$_2$C=C(CH$_3$)$_2$ (94%)

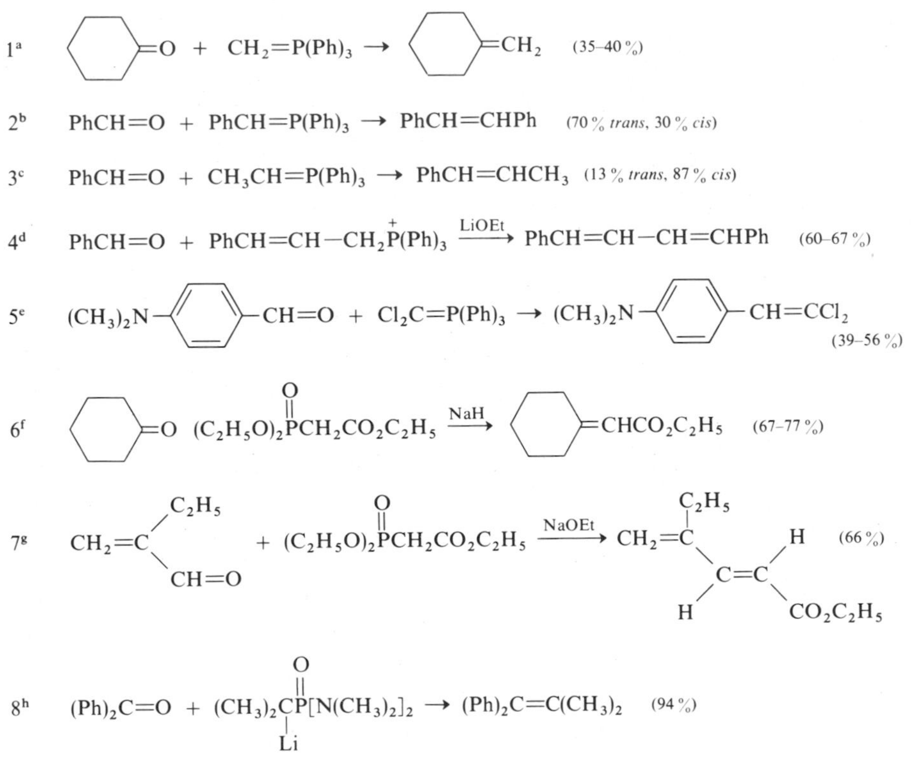

a. G. Wittig and U. Schöllkopf, *Org. Synth.* **40**, 66 (1960).
b. G. Wittig and U. Schöllkopf, *Chem. Ber.* **87**, 1318 (1954).
c. M. Schlosser and K. F. Christmann, *Justus Liebigs Ann. Chem.* **708**, 1 (1967).
d. R. N. McDonald and T. W. Campbell, *Org. Synth.* **40**, 36 (1960).
e. A. J. Speziale, K. W. Ratts, and D. E. Bissing, *Org. Synth.* **45**, 33 (1965).
f. W. S. Wadsworth, Jr., and W. D. Emmons, *Org. Synth.* **45**, 44 (1965).
g. R. J. Sundberg, P. A. Bukowick, and F. O. Holcombe, *J. Org. Chem.* **32**, 2938 (1967).
h. E. J. Corey and G. T. Kwiatkowski, *J. Am. Chem. Soc.* **90**, 6816 (1968).

domination of both *cis* and *trans* isomers are known (see, for example, entries 2 and 3 in Scheme 2.8).[33] There has been considerable research on the effect of reaction conditions on the isomer ratio, but conditions that assure high stereoselectivity have not been developed.[34] The factors that determine the stereochemistry of the product olefin can be considered with reference to the mechanism shown earlier. Since the final elimination involves a four-center transition state, it is necessarily a *syn* elimination. Thus, the stereochemistry present in the betaine intermediate as it attains the elimination transition state governs olefin geometry. The *erythro* betaine

33. M. Schlosser, *Top. Stereochem.* **5**, 1 (1970).
34. L. D. Bergelson and M. M. Shemyakin, *Tetrahedron* **19**, 149 (1963); L. D. Bergelson, V. A. Vaver, L. I. Barsukov, and M. M. Shemyakin, *Tetrahedron Lett.*, 2669 (1964); L. D. Bergelson, L. I. Barsulov, and M. M. Shemyakin, *Tetrahedron* **23**, 2709 (1967).

(**A**) must give *cis* olefin and the *threo* isomer (**B**) gives *trans*. It follows that when

elimination of triphenylphosphine oxide occurs rapidly with respect to the reverse of addition, the olefin stereochemistry will be governed by the relative rates of formation of the betaines. Usually, this step is not highly stereoselective, and a mixture of olefins therefore results. If betaine formation is rapidly reversible, the two betaines are equilibrated. When this occurs, olefin stereochemistry will be governed by the relative rates of decomposition of the betaines.

The approximate reaction-energy profiles for the two limiting situations are shown in Figure 2.2. When the betaines are in equilibrium, *trans* olefin ordinarily is dominant. This dominance is attributable to the relative stabilities of the *cis* and *trans* olefins. The steric repulsions that make the *cis* olefin less stable begin to appear as the elimination transition state is approached; therefore, the transition state leading to *cis* olefin is of higher energy than that leading to *trans* olefin. Very high *trans* : *cis* ratios are characteristic of the phosphorus ylides that contain electron-attracting substituents such as carboalkoxy and acyl. This must be because the ylide, being more stable, can undergo reversal of addition relatively easily; it may also be due in part to a lower electronic charge at oxygen, which results in a relatively slow rate of ylide decomposition. An additional factor may contribute to the preference for *trans*

olefin when carbonyl substituents are present. In the transition state leading to *cis* olefin, there is an unfavorable steric interaction that interferes with the coplanarity of the carbonyl group and olefin. The planar array possible in the *trans* olefin is favorable because of the delocalization it permits.

An important modification of the Wittig reaction that involves phosphonate esters has the advantage of being highly stereoselective. The use of phosphonate esters rather than phosphonium salts is illustrated in the accompanying equations:

$$RCH_2X + P(OC_2H_5)_3 \longrightarrow RCH_2\overset{\overset{\displaystyle O}{\|}}{P}(OC_2H_5)_2 + C_2H_5X$$

$$RCH_2\overset{\overset{\displaystyle O}{\|}}{P}(OC_2H_5)_2 \xrightarrow{base} R\bar{C}H\overset{\overset{\displaystyle O}{\|}}{P}(OC_2H_5)_2$$

$$R\bar{C}H\overset{\overset{\displaystyle O}{\|}}{P}(OC_2H_5)_2 + R_2'C=O \longrightarrow R_2'\underset{\underset{\displaystyle R}{|}}{\overset{\overset{\displaystyle ^-O}{|}}{C}}-\underset{}{\overset{\overset{\displaystyle P(OC_2H_5)_2}{|}}{CH}} \longrightarrow R_2'C=CHR + (C_2H_5O)_2\overset{\overset{\displaystyle O}{\|}}{P}-O^-$$

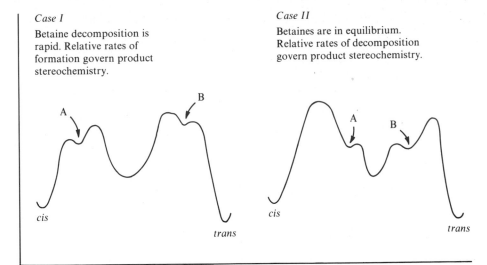

Case I

Betaine decomposition is
rapid. Relative rates of
formation govern product
stereochemistry.

Case II

Betaines are in equilibrium.
Relative rates of decomposition
govern product stereochemistry.

Figure 2.2. Stereochemistry of the Wittig reaction as a function of the reaction-energy profile.

This modification has been used primarily for cases in which R is a group, such as phenyl, acyl, or alkoxycarbonyl, that can assist in stabilization of the carbanionic center.[35] When R is a simple alkyl group, the addition reaction occurs, but the adduct is stable toward elimination.[36] In those cases where R is an electron-withdrawing group, the phosphonate modification of the Wittig reaction shows a high preference for *trans* olefin.[37] The factors responsible are the same as those considered in the case of the standard Wittig procedure. The stabilized phosphonate carbanions add reversibly to the carbonyl compound, and the ease of elimination favors the *trans* product. Entries 6 and 7 in Scheme 2.8 are examples of the phosphonate modification of the Wittig reaction.

Wittig reactions employing alkylphosphonium salts can be made stereoselective by utilizing a modified procedure. The betaine is formed at low temperature and in the presence of lithium halide. Under these conditions, betaine does not rapidly eliminate triphenylphosphine oxide. The intermediate is then treated with a second mole of an organolithium compound to form a new ylide:

Finally, addition of an equivalent amount of a proton donor regenerates the original

35. J. Boutagy and R. Thomas, *Chem. Rev.* **74**, 87 (1974).
36. E. J. Corey and G. T. Kwiatkowski, *J. Am. Chem. Soc.* **88**, 5654 (1966).
37. D. H. Wadsworth, O. E. Schupp, III, E. J. Seus, and J. A. Ford, Jr., *J. Org. Chem.* **30**, 680 (1965).

ylide, but as the more stable *threo* isomer. Subsequent decomposition of this stereoisomer leads to olefin formation with *trans : cis* ratios of higher than 95 : 5.[38] The anion obtained by reaction of the betaine intermediate with a second mole of alkyllithium reagent can also be subjected to other synthetic transformations. The most useful procedure developed to date involves treatment of the anion with formaldehyde. On warming to promote elimination of the triphenylphosphine oxide, an allylic alcohol is produced. The reaction is highly stereoselective and has proved to be of value in synthesis of natural products.[39]

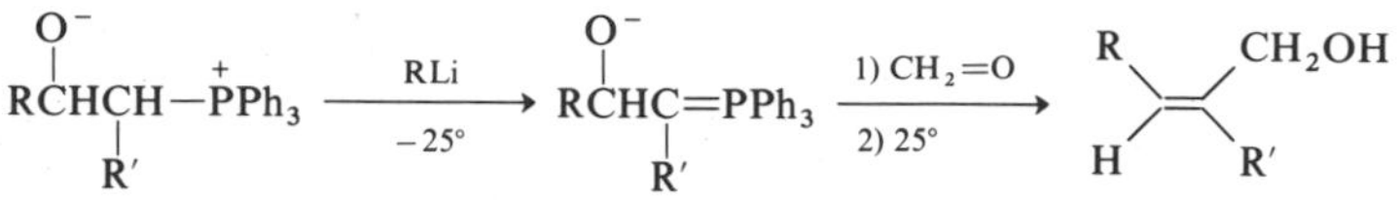

The synthetic utility of the Wittig reaction can be expanded beyond its ability to form double bonds by the use of certain functionalized ylides. Thus, methoxymethylene[40] and phenoxymethylene[41] derivatives lead to vinyl ethers, which can subsequently be hydrolyzed to aldehydes:

A similar principle is involved in the use of 1-(methylthio)alkylphosphonates in the synthesis of ketones.[42] The vinyl sulfides that are the products of the modified Wittig reactions are hydrolyzed to aldehydes in the presence of mercuric salts.

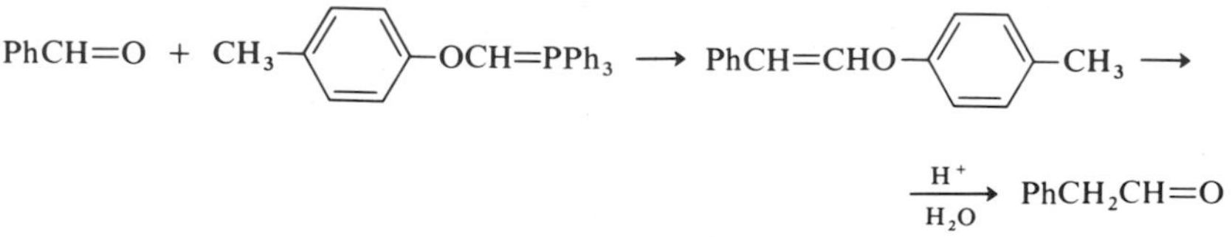

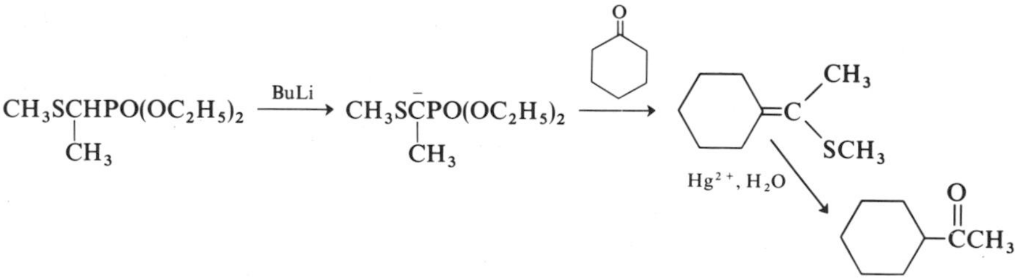

38. M. Schlosser and K. F. Christmann, *Justus Liebigs Ann. Chem.* **708**, 1 (1967).
39. E. J. Corey and H. Yamamoto, *J. Am. Chem. Soc.* **92**, 226 (1970); E. J. Corey, H. Yamamoto, D. K. Herron, and K. Achiwa, *J. Am. Chem. Soc.* **92**, 6635 (1970; E. J. Corey and H. Yamamoto, *J. Am. Chem. Soc.* **92**, 6636 (1970); E. J. Corey and H. Yamamoto, *J. Am. Chem. Soc.* **92**, 6637 (1970); E. J. Corey, J. I. Shulman, and H. Yamamoto, *Tetrahedron Lett.*, 447 (1970).
40. S. G. Levine, *J. Am. Chem. Soc.* **80**, 6150 (1958).
41. G. Wittig, W. Böll, and K. H. Krück, *Chem. Ber.* **95**, 2514 (1962).
42. E. J. Corey and J. I. Shulman, *J. Org. Chem.* **35**, 777 (1970).

Organolithium reagents bearing α-trimethylsilyl substituents react with aldehydes and ketones to produce olefins in a manner similar to the Wittig reaction.[43] The presence of the silicon substituent may facilitate formation of the anionic carbon

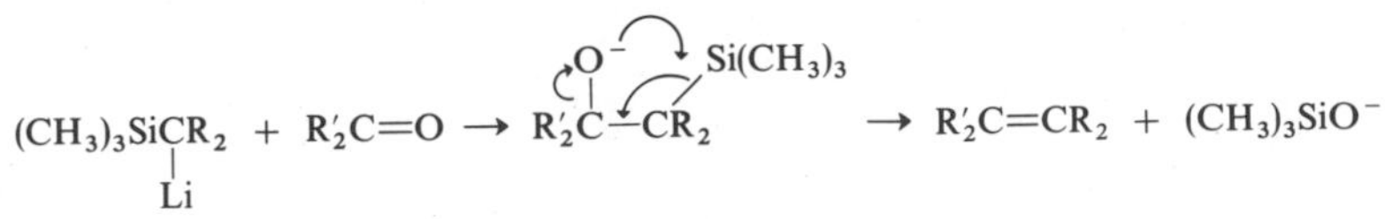

$$(CH_3)_3SiCR_2 + R'_2C=O \rightarrow R'_2C-CR_2 \rightarrow R'_2C=CR_2 + (CH_3)_3SiO^-$$
$$\underset{Li}{|}$$

in the deprotonation step. Organolithium compounds are used as the base. The primary function of the silicon, however, is to provide a low-energy path for decomposition of the intermediate that is formed in the addition step. The reaction involving organosilicon reagents has been applied to a variety of substituted olefins, although it is not as widely used as the Wittig reaction. α,β-Unsaturated esters[44a] and vinyl sulfoxides[44b] are two examples of the types of compounds that have been prepared using organosilicon reagents. The carboxylate and sulfur substituents help to stabilize the carbanion intermediate:

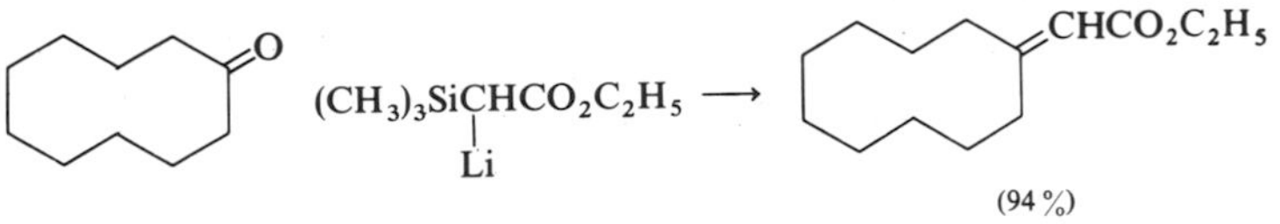

Ref. 44a

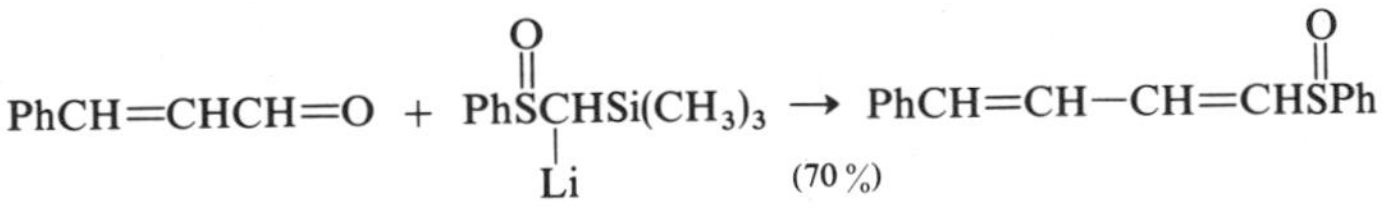

(94%)

$$PhCH=CHCH=O + \underset{Li}{PhSCHSi(CH_3)_3} \rightarrow PhCH=CH-CH=CHSPh$$
(70%) Ref. 44b

2.6. Sulfur Ylides as Nucleophiles

Ylides derived from sulfur rank next in importance to those derived from phosphorus in synthetic utility. The first widely used sulfur ylides were dimethyloxo-sulfonium methylide (4) and dimethylsulfonium methylide (5).[45] The preparations of these compounds are outlined below:

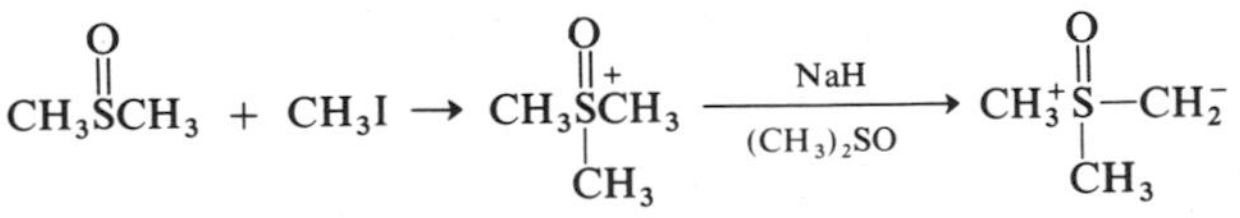

43. D. J. Peterson, *J. Org. Chem.* **33**, 780 (1968).
44a. K. Shimoji, H. Taguchi, K. Oshima, H. Yamamoto, and H. Nozaki, *J. Am. Chem. Soc.* **96**, 1620 (1974).
  b. F. A. Carey and O. Hernandez, *J. Org. Chem.* **38**, 2670 (1973).
45. E. J. Corey and M. Chaykovsky, *J. Am. Chem. Soc.* **87**, 1353 (1965).

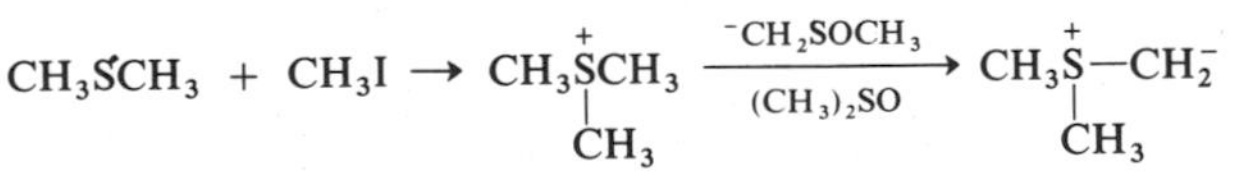

60

CHAPTER 2
REACTIONS OF
NUCLEOPHILIC
CARBON
SPECIES

Scheme 2.9. Syntheses

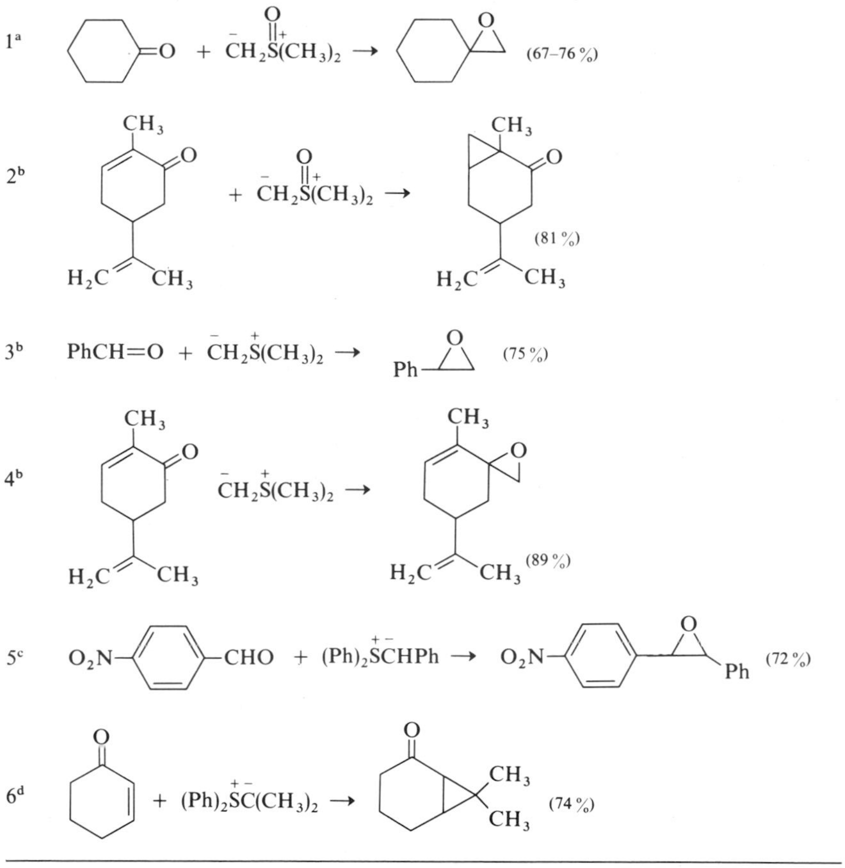

As in the case of the phosphorus ylides, it is believed that *d*-orbitals on the sulfur participate in bonding, resulting in stabilization of the carbanion site. On reaction with unconjugated ketones and aldehydes, dimethyloxosulfonium methylide gives epoxides. The reaction course involves carbonyl addition, followed by an intramolecular nucleophilic displacement. Several specific examples are given in

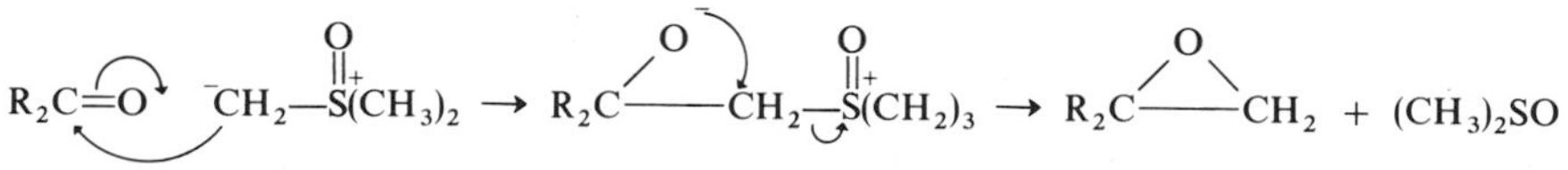

Scheme 2.9. In cyclohexane derivatives, equatorial attack dominates. With α,β-unsaturated ketones, the reaction occurs by conjugate addition. The addition step is

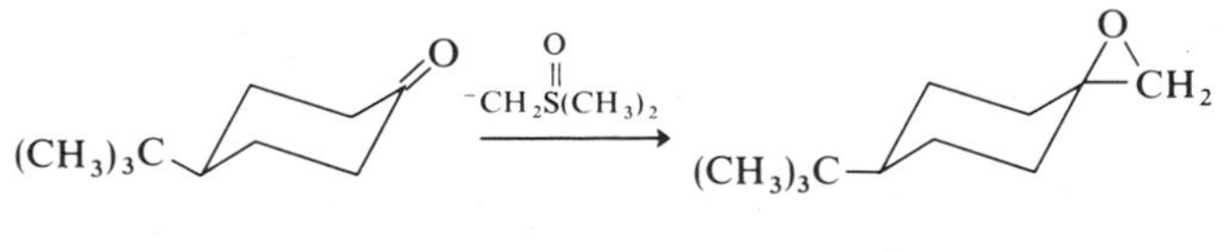

followed by intramolecular nucleophilic displacement, resulting in closure to a

7[e] (63%)

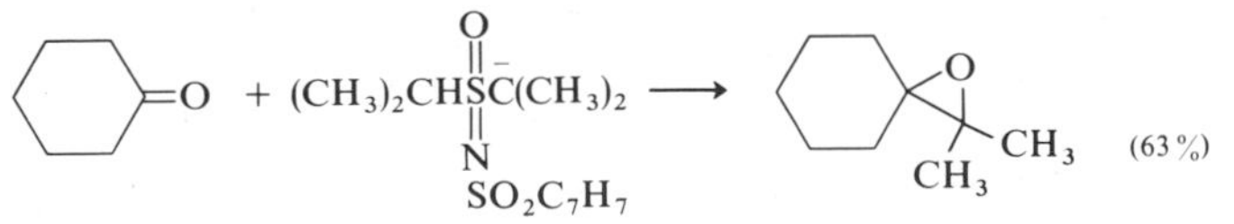

8[e] (77%)

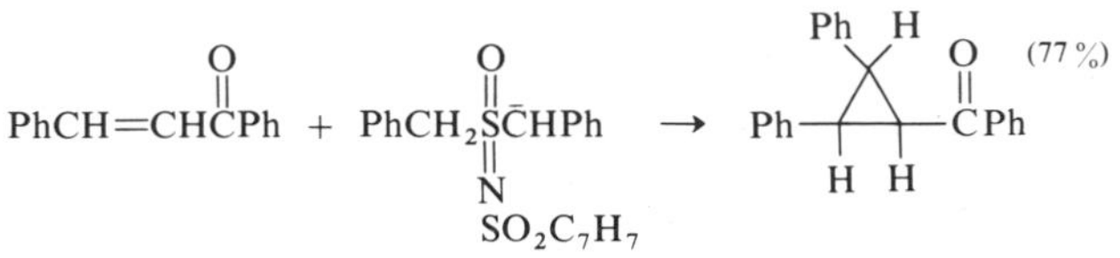

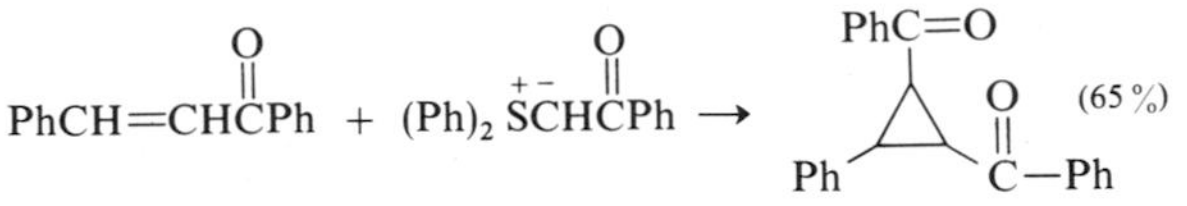

9[f] (65%)

10[g] (39%)

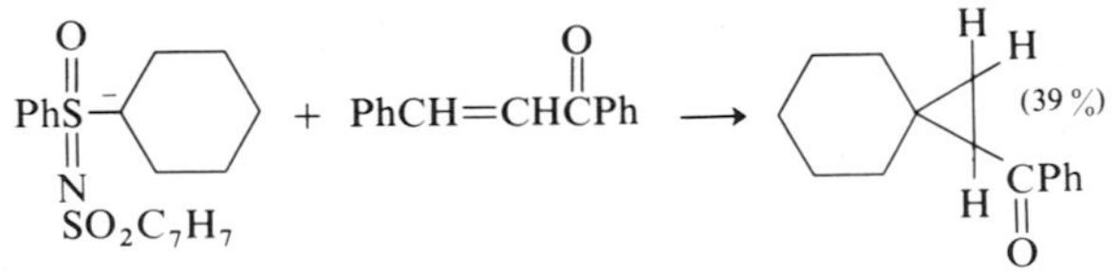

a. E. J. Corey and M. Chaykovsky, *Org. Synth.* **49**, 78 (1969).
b. E. J. Corey and M. Chaykovsky, *J. Am. Chem. Soc.* **87**, 1353 (1965).
c. A. W. Johnson, V. J. Hruby, and J. L. Williams, *J. Am. Chem. Soc.* **86**, 918 (1964).
d. E. J. Corey and M. Jautelat, *J. Am. Chem. Soc.* **89**, 3912 (1967).
e. C. R. Johnson and G. F. Katekar, *J. Am. Chem. Soc.* **92**, 5753 (1970).
f. B. M. Trost, *J. Am. Chem. Soc.* **89**, 138 (1967).
g. C. R. Johnson, G. F. Katekar, R. F. Huxol, and E. R. Janiga, *J. Am. Chem. Soc.* **93**, 3771 (1971).

cyclopropane ring. Dimethylsulfonium methylide provides epoxides from both

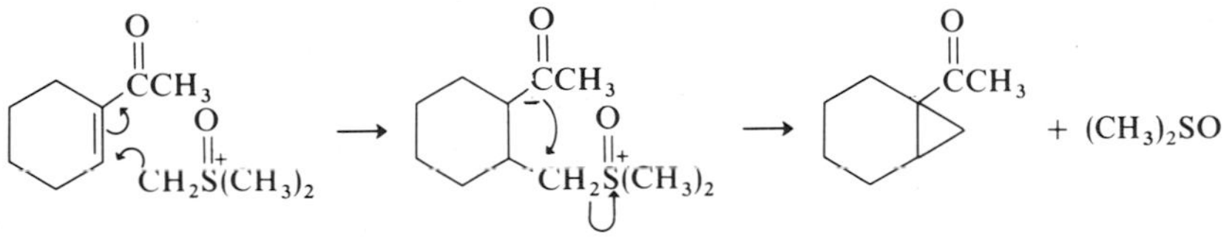

conjugated and unconjugated ketones. This ylide is the more reactive of the two, but is also less stable and must be treated with the carbonyl compound immediately after

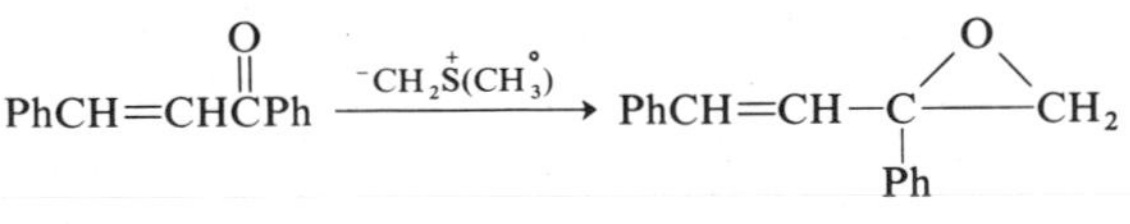

formation if decomposition is to be avoided. Dimethylsulfonium methylide also shows contrasting behavior to dimethyloxosulfonium methylide with regard to stereochemistry of attack at six-membered rings. In simple cyclohexanones, axial attack is preferred.[45,46] The contrasting stereoselectivity of the two reagents may reflect their considerably different stability. It has been suggested that dimethylsulfonium methylide is sufficiently unstable that the addition step is irreversible. Under these conditions, the direction of addition to the carbonyl group would determine the epoxide stereochemistry. With the more stable dimethylsulfoxonium methylide, the addition may be reversible, in which case the more stable epoxide could be favored in the cyclization step.[47] This argument closely parallels that presented in discussing the stereochemistry of the Wittig reaction with stable and unstable phosphonium ylides.

Extension of the synthetic versatility of sulfur ylides so that substituted methylene groups, as well as methylene itself, can be transferred has been accomplished by modification of the sulfur substituents. Entry 6 in Scheme 2.9 illustrates a transfer of an isopropylidene unit. Even greater structural flexibility has been achieved by use of nucleophiles derived from sulfoximines.[48] The N-p-toluenesulfinyl sulfoximines appear to be particularly promising reagents. They are prepared from sulfoxides by reaction with p-toluenesulfonyl azide. Strong base

$$
\underset{\substack{\parallel \\ \text{O}}}{\text{PhSCHR}_2} + \text{ArSO}_2\text{N}_3 \rightarrow \underset{\substack{\parallel \\ \text{ArSO}_2\text{N}}}{\overset{\substack{\text{O} \\ \parallel}}{\text{PhSCHR}_2}}
$$

effects conversion to the anion, and the mechanistic pattern established in the sulfoxonium ylide reactions is followed. Epoxides result from saturated ketones,

$$
\underset{\substack{\parallel \\ \text{ArSO}_2\text{N}}}{\overset{\substack{\text{O} \\ \parallel}}{\text{PhSCHR}_2}} \xrightarrow{\text{NaH}} \underset{\substack{\parallel \\ \text{ArSO}_2\text{N}}}{\overset{\substack{\text{O} \\ \parallel}}{\text{PhS}-\overset{-}{\text{CR}}_2}} \quad \text{R}_2'\text{C}=\text{O} \rightarrow \underset{\substack{\parallel \\ \text{ArSO}_2\text{N}}}{\overset{\substack{\text{O} \\ \parallel}}{\text{PhS}}}\underset{\text{R}}{\overset{\text{R}}{-\text{C}-}}\underset{}{\overset{\text{O}^-}{\text{CR}_2'}}
$$

whereas conjugated carbonyl compounds give cyclopropanes (entries 7 and 8, Scheme 2.9).

Sulfur ylides provide a useful route to the spiro[2,2]pentane system. A cyclopropyl substituent in the sulfur compound is transferred to an α,β-unsaturated system, forming the second cyclopropane ring.[49] The ring system incorporated may

45. See p. 59.
46. C. E. Cook, R. C. Corley, and M. E. Wall, *J. Org. Chem.* **33**, 2789 (1968).
47. C. R. Johnson and C. W. Schroeck, *J. Am. Chem. Soc.* **93**, 5303 (1971); C. R. Johnson, C. W. Schroeck, and J. R. Shanklin, *J. Am. Chem. Soc.* **95**, 7424 (1973).
48a. C. R. Johnson, E. R. Janiga, and M. Haake, *J. Am. Chem. Soc.* **90**, 3890 (1968).
  b. C. R. Johnson and G. F. Katekar, *J. Am. Chem. Soc.* **92**, 5753, (1970).
  c. C. R. Johnson, R. A. Kirchhoff, R. J. Reischer, and G. F. Katekar, *J. Am. Chem. Soc.* **95**, 4287 (1973).
49a. B. M. Trost and M. J. Bogdanowicz, *J. Am. Chem. Soc.* **95**, 5307 (1973).
  b. C. R. Johnson, G. F. Katekar, R. F. Huxol, and E. R. Janiga, *J. Am. Chem. Soc.* **93**, 3771 (1971); C. R. Johnson and E. R. Janiga, *J. Am. Chem. Soc.* **95**, 7692 (1973).
  c. B. M. Trost and M. J. Bogdanowicz, *J. Am. Chem. Soc.* **95**, 5311, 5321 (1973).

$$Ph_2\overset{+}{S}-\triangleleft \;\underset{}{\overset{NaOH}{\rightleftharpoons}}\; Ph_2S=\triangleleft \;\overset{CH_2=CHCO_2CH_3}{\longrightarrow}\; \triangleleft\!\!\triangleright\!\!-CO_2CH_3$$

(83%)   Ref. 49a

$$(CH_3)_2N-\overset{Ph}{\underset{O}{\overset{|}{S}}}-\triangleleft \;\overset{NaH}{\longrightarrow}\; (CH_3)_2N-\overset{Ph}{\underset{O}{\overset{|}{S}}}=\triangleleft \;\overset{(CH_3)_2C=CHCCH_3}{\longrightarrow}\;$$

(61%)   Ref. 49b

also be larger than three-membered. These compounds have proved to be versatile synthetic intermediates.[49c]

2.7. Nucleophilic Addition–Cyclization

The pattern of nucleophilic addition at a carbonyl group followed by intramolecular nucleophilic displacement of a leaving group present in the nucleophile can also be recognized in a much older synthetic technique, the *Darzens reaction*.[50] The first step is an addition of the enolate derived from an α-haloester to

$$R_2C=O + ClCH_2COR' \longrightarrow \overset{R}{\underset{R}{\diagdown}}\overset{O}{\underset{}{\overset{}{C}}}-\overset{}{\underset{H}{\overset{CO_2R'}{C}}}$$

the carbonyl compound. The alkoxide oxygen then effects nucleophilic attack on the carbon–halogen bond, forming an epoxide. The reaction is not very stereoselective,

$$R_2\overset{O}{\overset{||}{C}}\overset{}{\underset{Cl}{\overset{}{C}}}HCO_2C_2H_5 \rightarrow R_2C-\overset{O^-}{\underset{Cl}{\overset{}{C}}}HCO_2C_2H_5 \rightarrow R_2C-CHCO_2C_2H_5$$

so a mixture of isomers is usually formed from unsymmetrical ketones.[51] Scheme 2.10 gives some typical examples of the Darzens reaction.

The epoxy esters formed in the Darzens condensation have been used in the synthesis of ketones and aldehydes. The acids formed by saponification of the esters

$$\overset{R}{\underset{R}{\diagdown}}\overset{O}{\underset{}{\overset{}{C-C}}}\overset{\overset{O}{\overset{||}{C}}-O-H}{\underset{H}{}} \rightarrow \overset{R}{\underset{R}{\diagdown}}C=C\overset{OH}{\underset{H}{\diagup}} \rightarrow R_2CHCH=O$$

49a–c. See p. 62.
50. M. S. Newman and B. J. Magerlein, *Org. React.* **5**, 413 (1951).
51. F. W. Bachelor and R. K. Bansal, *J. Org. Chem.* **34**, 3600 (1969).

Scheme 2.10. Darzens Condensation Reactions

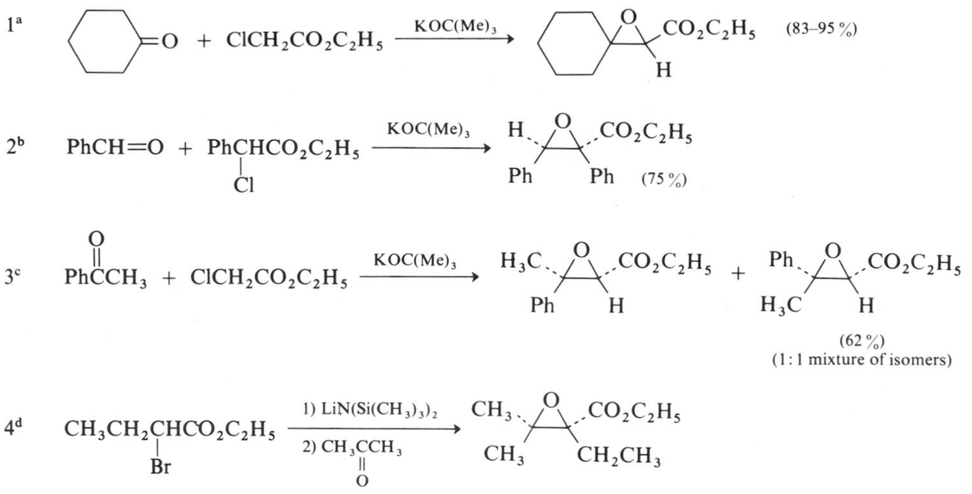

1^a (cyclohexanone) $=O$ + $ClCH_2CO_2C_2H_5$ $\xrightarrow{\text{KOC(Me)}_3}$ (spiro epoxide) $CO_2C_2H_5$ (83–95%)

2^b $PhCH{=}O$ + $PhCHCO_2C_2H_5$ (with Cl) $\xrightarrow{\text{KOC(Me)}_3}$ (epoxide) H, Ph, Ph, $CO_2C_2H_5$ (75%)

3^c $Ph\overset{O}{\overset{\|}{C}}CH_3$ + $ClCH_2CO_2C_2H_5$ $\xrightarrow{\text{KOC(Me)}_3}$ H_3C, Ph, $CO_2C_2H_5$, H + Ph, H_3C, $CO_2C_2H_5$, H

     (62%)
     (1:1 mixture of isomers)

4^d $CH_3CH_2CHCO_2C_2H_5$ (with Br) $\xrightarrow[\text{2) } CH_3CCH_3 \text{ (O)}]{\text{1) LiN(Si(CH}_3)_3)_2}$ CH_3, CH_3, $CO_2C_2H_5$, CH_2CH_3

a. R. H. Hunt, L. J. Chinn, and W. S. Johnson, *Org. Synth.* **IV**, 459 (1963).
b. H. E. Zimmerman and L. Ahramjian, *J. Am. Chem. Soc.* **82**, 5459 (1960).
c. F. W. Bachelor and R. K. Bansal, *J. Org. Chem.* **34**, 3600 (1969).
d. R. F. Borch, *Tetrahedron Lett.*, 3761 (1972).

are thermally decarboxylated to give the carbonyl compound. This is usually viewed as a concerted process via the bicyclic transition state indicated below:

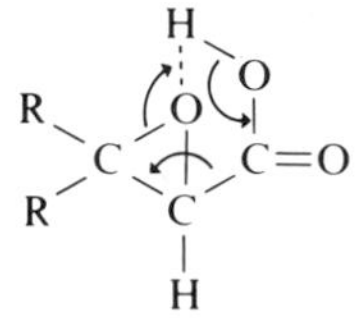

General References

Aldol Condensation

A. T. Nielsen and W. J. Houlihan, *Org. React.* **16**, 1 (1968).

Mannich Reaction

F. F. Blicke, *Org. React.* **1**, 303 (1942).

Wittig Reaction

A. Maercker, *Org. React.* **14**, 270 (1965); A. W. Johnson, *Ylide Chemistry*, Academic Press, New York, NY, 1966.

(*References for these problems will be found on page 498.*)

1. Predict the structure and stereochemistry of the product or products expected from each of the following reactions:

   (a) $(CH_3)_2CHCO_2C_2H_5$ $\xrightarrow[\text{2) PhCHO}]{\text{1) LiNR}_2}$

   (b) $CH_3CH_2NO_2 + CH_2O$ $\xrightarrow{\text{NaOH}}$

   (c)
   $+ PhCHO$ $\xrightarrow{\text{NaOH}}$

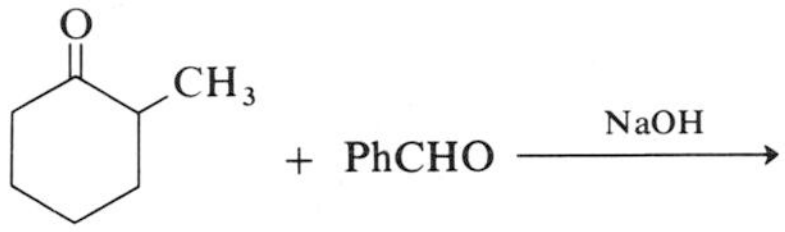

   (d)
   $CHO + PhCH_2\overset{O}{\overset{\|}{C}}CH_3$ $\xrightarrow{\text{NaOH, H}_2\text{O}}$

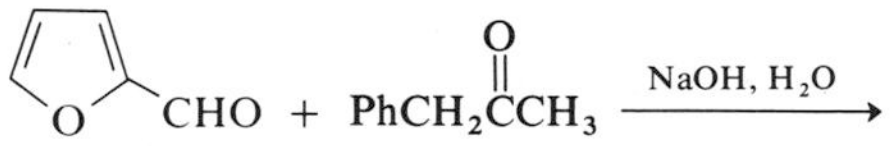

   (e) $Ph\overset{O}{\overset{\|}{C}}CH_2CH_3 + CH_2O + (CH_3)_2NH$ $\longrightarrow$

2. When heated with *m*-chlorobenzaldehyde and triphenylphosphine in ethanol, the epoxides **A** and **B** give rise to products **C**, **D**, **E**, and **F** in the yields indicated. Much more **D** is formed from the *trans* epoxide than from the *cis*. Explain these observations and point out their relevance to the mechanism of the Wittig reaction.

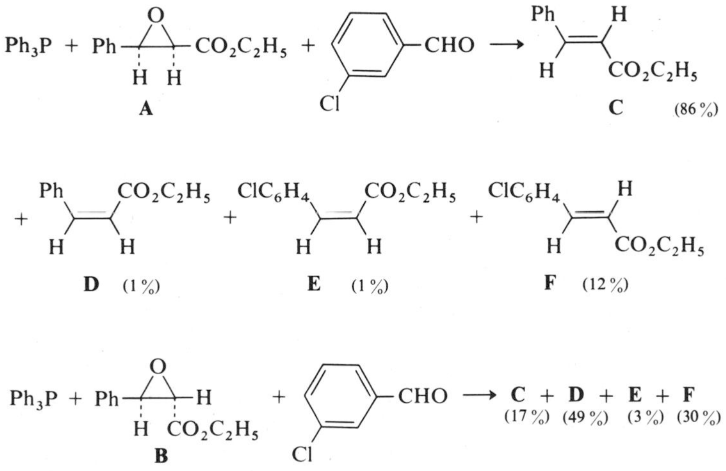

3. Compounds **A** and **B** are key intermediates in one total synthesis of cholesterol. Rationalize their formation by the routes shown:

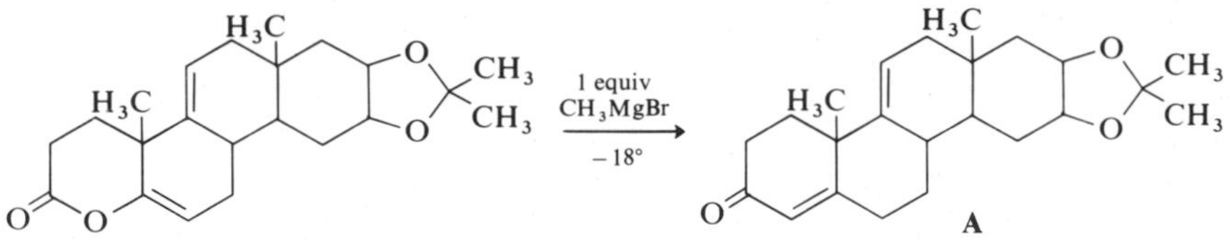

A

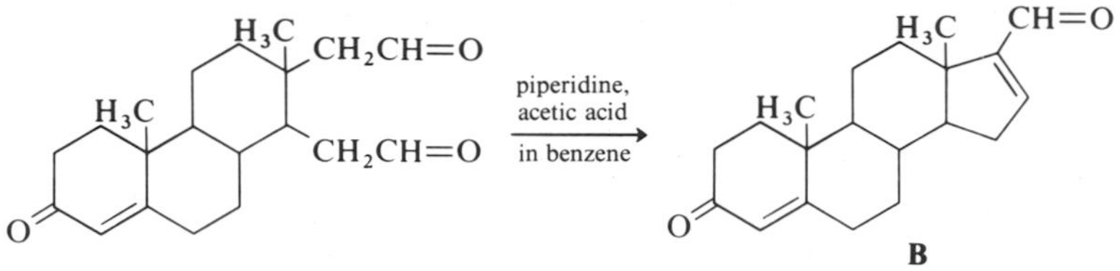

B

4. Offer a mechanism for each of the following transformations:

(a)

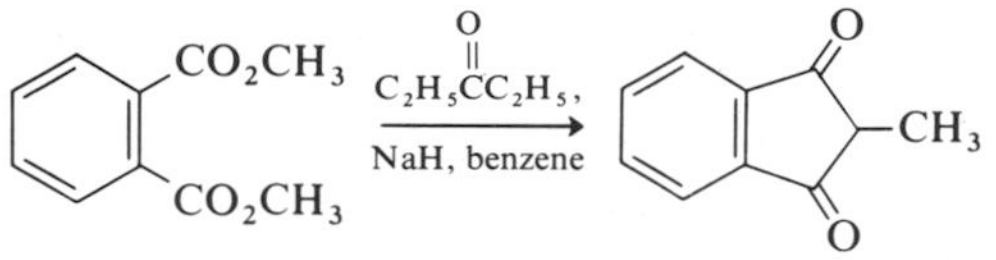

(b)

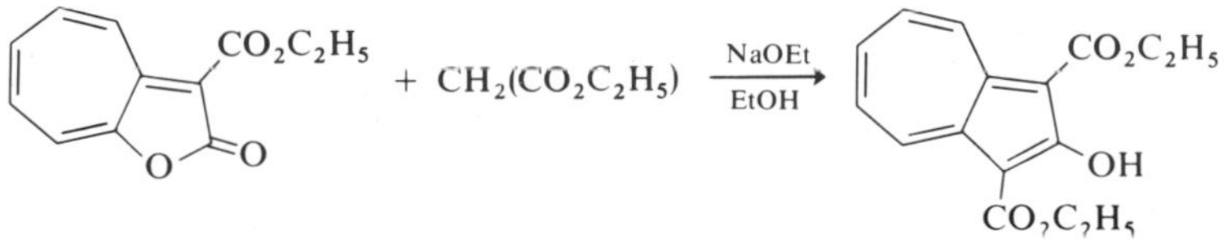

(c)

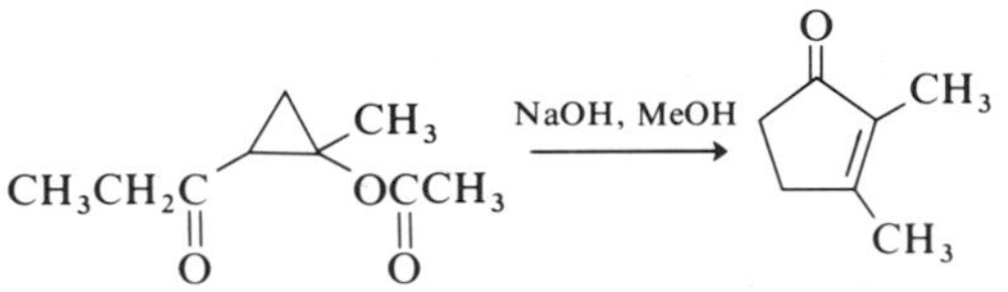

(d)

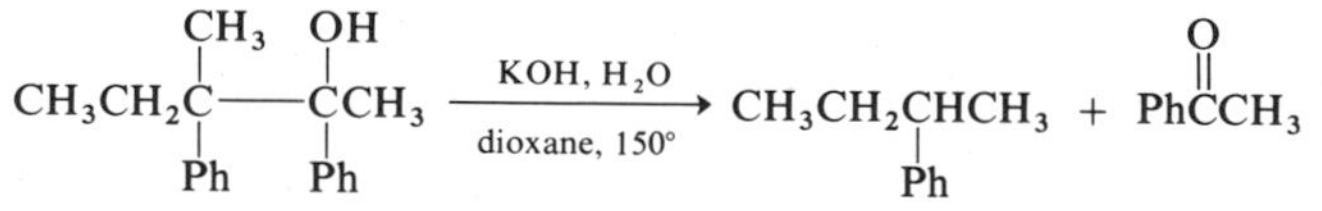

5. Tetraacetic acid (or a biological equivalent) has been suggested as an intermediate in the biosynthesis of phenolic natural products. Its synthesis has been
described, as has its ready conversion to orsellinic acid. Suggest a mechanism for
formation of orsellinic acid under the conditions specified:

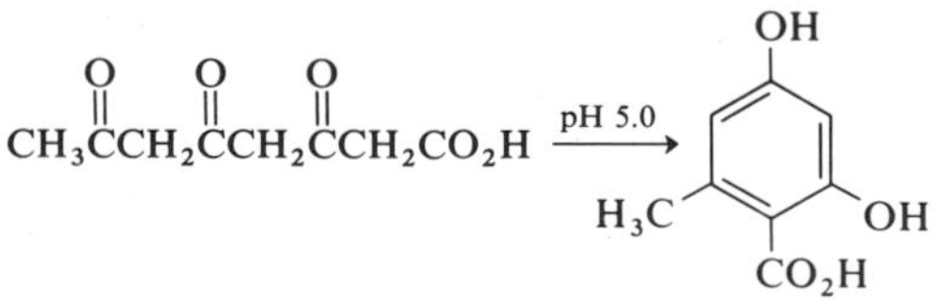

orsellinic acid

6. Indicate reaction conditions or a series of reactions that could effect each of the following synthetic conversions:

(a) $CH_3CO_2C(CH_3)_3 \rightarrow (CH_3)_2\overset{\overset{\displaystyle OH}{|}}{C}CH_2CO_2C(CH_3)_3$

(b)

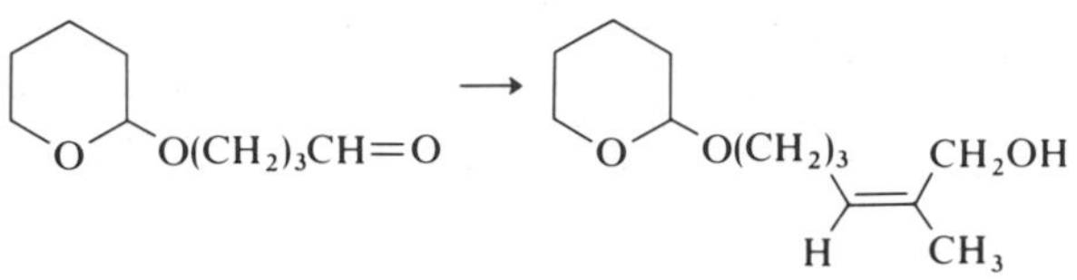

(c)

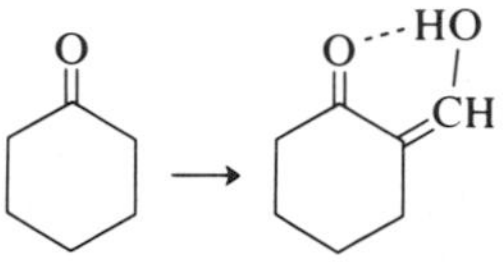

(d) $Ph_2C=O \rightarrow$

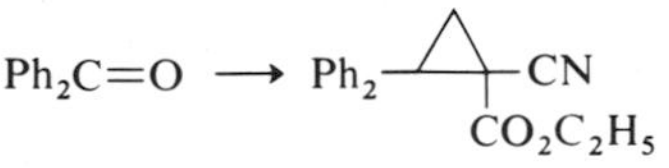

(e)

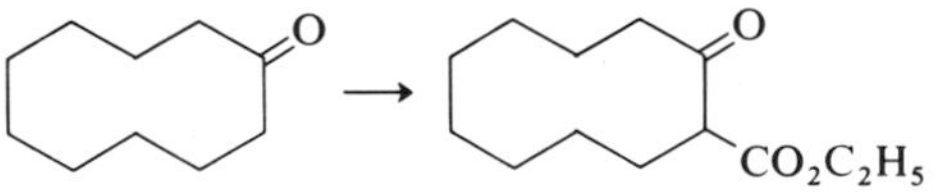

(f)

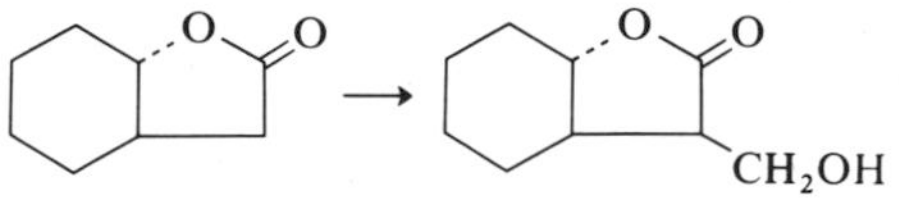

(g)

(h)

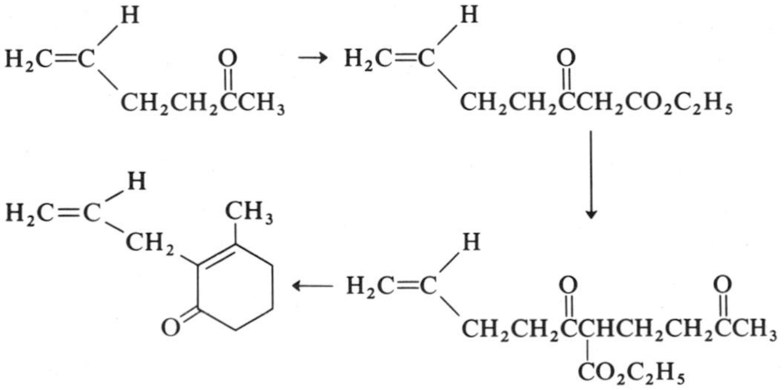

(i) $H_5C_2O_2CCH_2CH_2CO_2C_2H_5 \rightarrow$

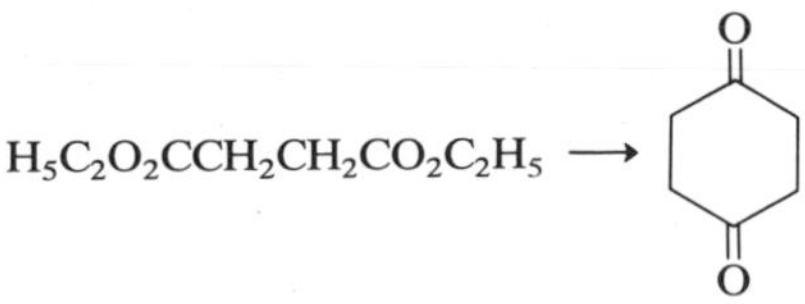

(j)

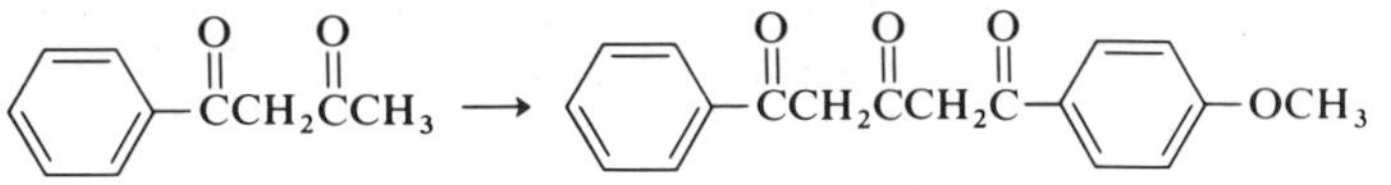

7. The first few steps of the synthesis of the alkaloid conessine produce **B** from **A**. Suggest a sequence of reactions for effecting this conversion:

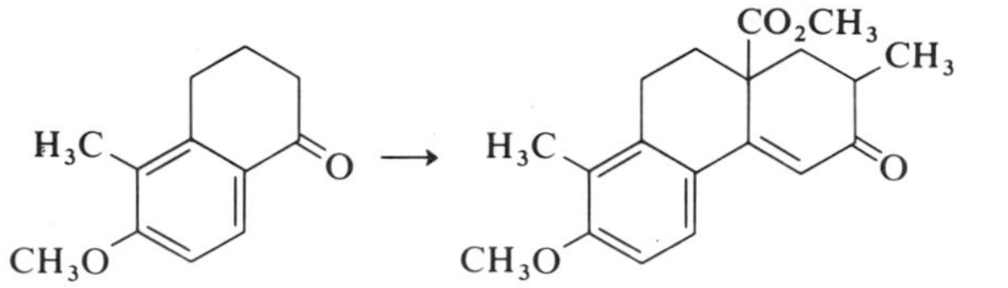

A   B

8. Predict the stereochemistry of the products expected from each of the following reactions:

(a)

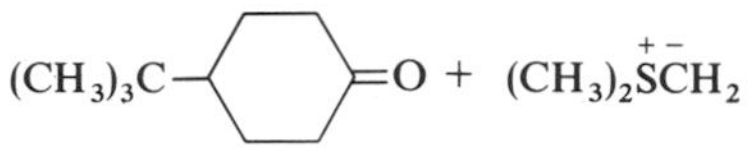

$(CH_3)_3C$—⬡=O + $(CH_3)_2\overset{+}{S}\overset{-}{C}H_2$

(c)

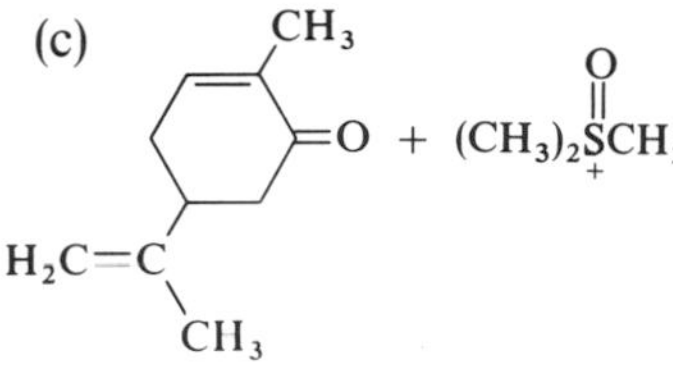

=O + $(CH_3)_2\overset{+}{S}CH_2^-$

(b) $(CH_3)_3C$—⬡=O + $(CH_3)_2\overset{+}{S}CH_2^-$

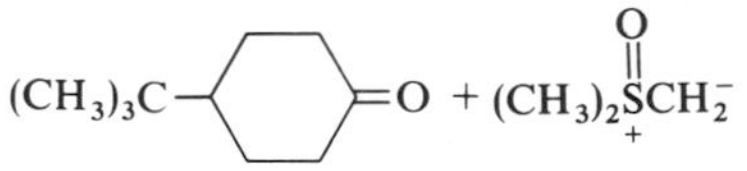

(d)

+ $(CH_3)_2\overset{+}{S}CH_2^-$

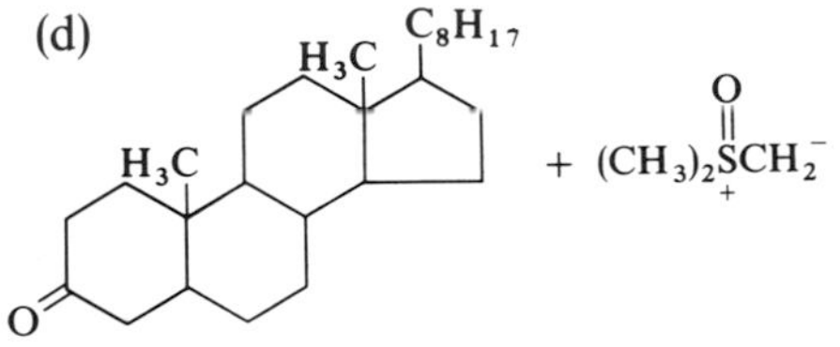

9. Suggest a mechanism for each of the following reactions:

(a)

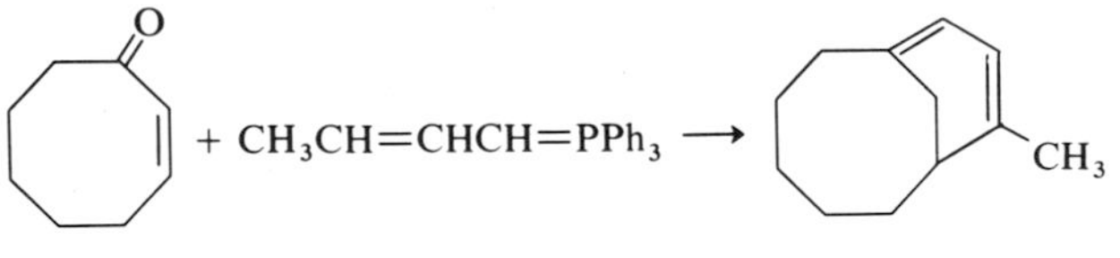

+ $CH_3CH{=}CHCH{=}PPh_3$ ⟶

(b)

+ $H_2C{=}CHCCH_3$ ⟶

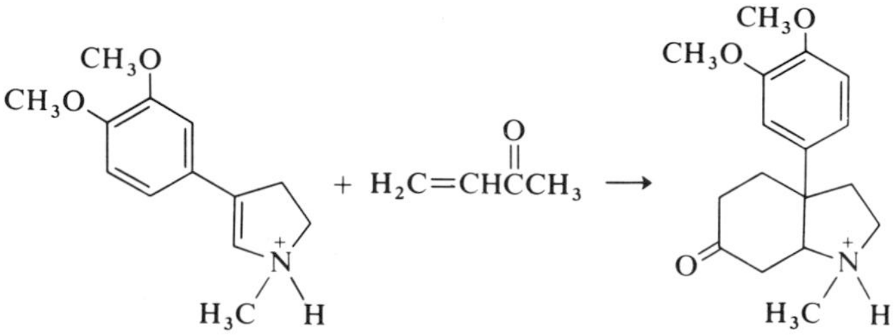

10. A stereospecific method for deoxygenating epoxides to alkenes involves reaction of the epoxide with the diphenylphosphide ion, followed by methyl iodide. The method results in overall inversion of the alkene stereochemistry. Thus,

cis-cyclooctene epoxide gives *trans*-cyclooctene. Propose a mechanism for this process and discuss the relationship of the reaction to the Wittig reaction.

11. Propose a mechanism for the following reaction:

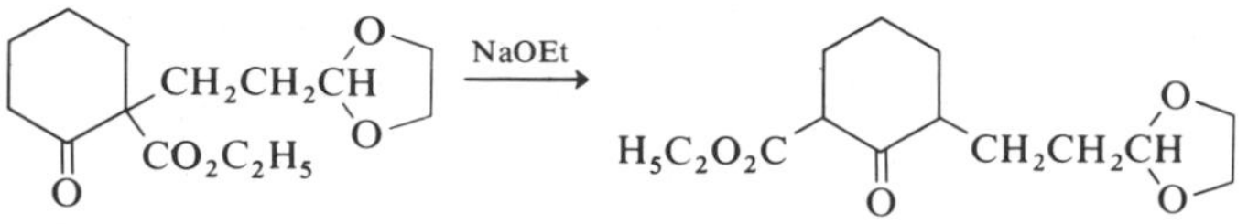

12. Indicate how you might prepare suitable reagents for synthesis of each of the following types of functionalized olefins, using the Wittig reaction with a readily available carbonyl compound:

(a) $H_2C=CHCO_2CH_3$

(b) $H_2C=CHCl$

(c)

(d) $Ph_2C=C=C=CH-$$-NO_2$

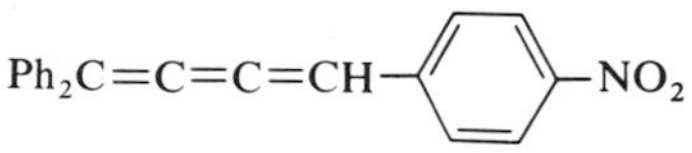

13. A fairly general method for ring closure that involves vinyltriphenylphosphonium halides has been developed. Two examples are shown. Comment on the mechanism of the reaction and suggest two additional types of rings that could be synthesized, using vinyltriphenylphosphonium salts.

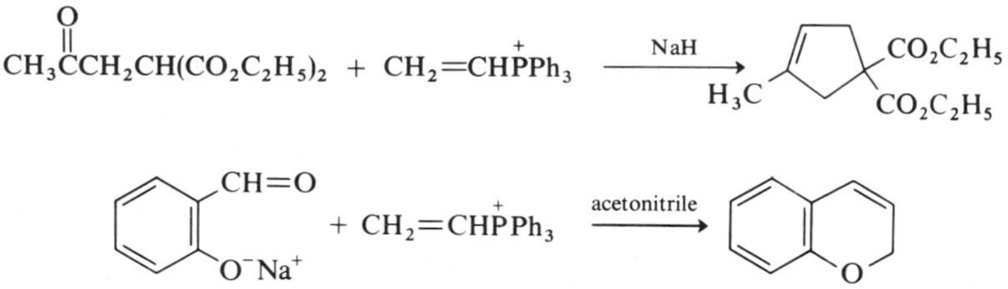

14. Although the sulfur ylide **A** does not react with simple ketones, the carboxylate analog **B** does, leading to epoxides. Explain the difference in reactivity.

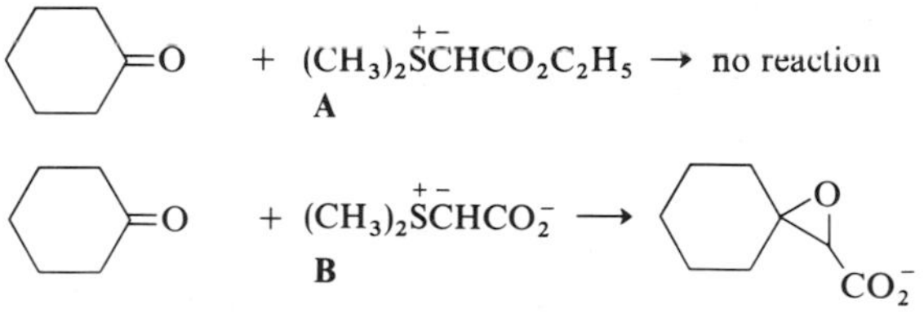

15. Suggest routes for each of the following synthetic transformations. More than one step may be required. Other reagents may be necessary.

(a) $Ph_2C=O$, $CH_3CH=O$, $\rightarrow$ $Ph_2C=CHCH=O$

(b) $CH_3CH_2CH_2CH=O$, $ClCH_2CO_2C_2H_5$ $\rightarrow$ $CH_3CH_2\underset{\underset{CH_2}{\|}}{C}CH=CHCO_2C_2H_5$

(c) $\quad CH_3NH_2, \ CH_2=CHCO_2C_2H_5 \longrightarrow$

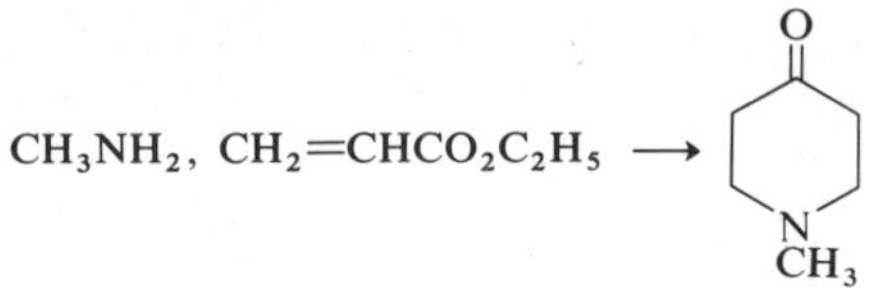

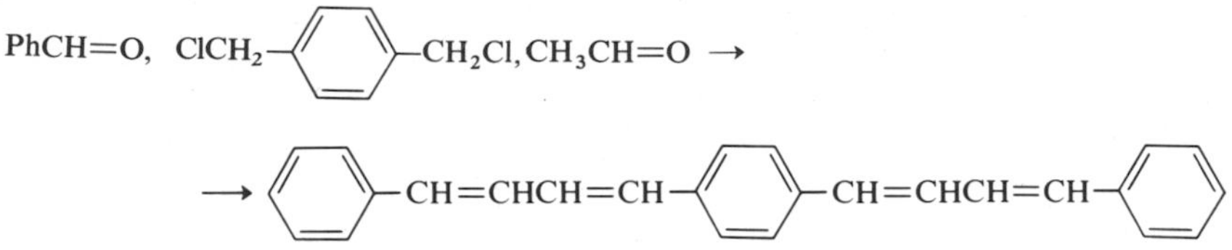

(d) $\quad PhCH=O, \ ClCH_2-$$-CH_2Cl, CH_3CH=O \ \rightarrow$

(e) $\quad (CH_3)_2CHCH=O, \ CH_2=CHCCH_3 \longrightarrow$

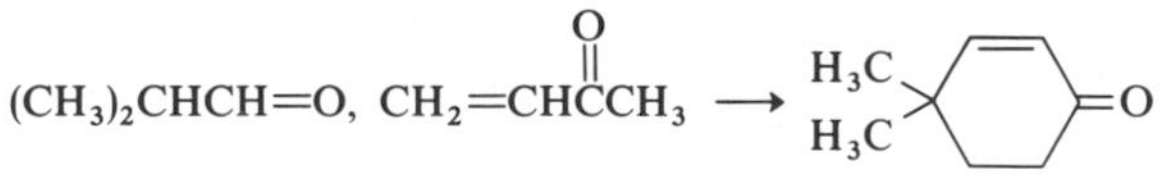

16. When the ylide shown below is treated with 1 equiv of lithium diisopropylamide in THF at $-78°$, a red solution is formed. Reaction with n-propyl bromide is slow and produces only a 10% yield of alkylated product after 1 hr at 25°. If the red solution is treated with an additional equivalent of lithium diisopropylamide, a black solution forms. This reacts rapidly with n-propyl bromide at $-78°$ to give the product in good yield. Explain.

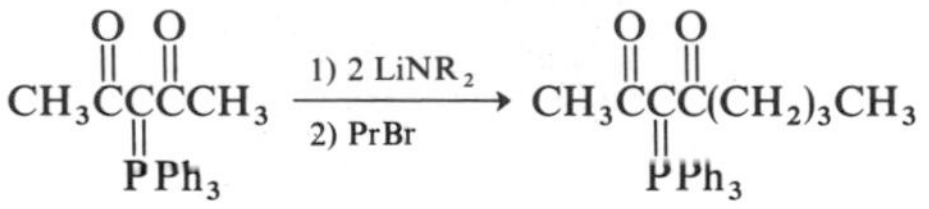

17. The cyclization of **A** could conceivably lead to **B** or **C**, but **B** is the predominant product. In order to define the mechanism that determines the product structure, the following experiments were done. What conclusion is possible about the identity of the product-determining step?

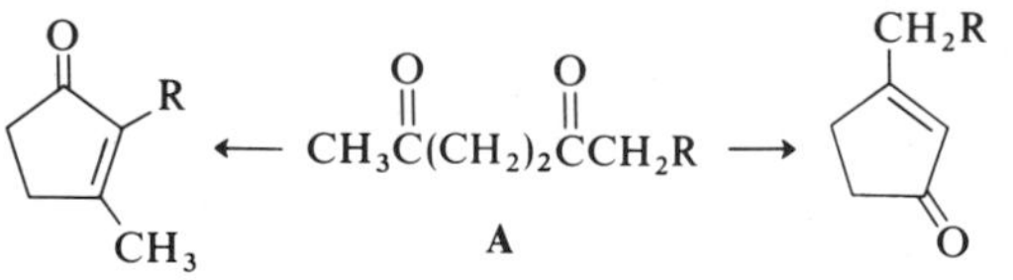

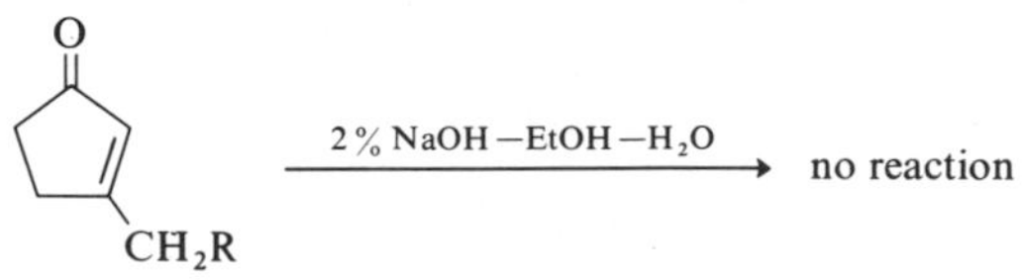

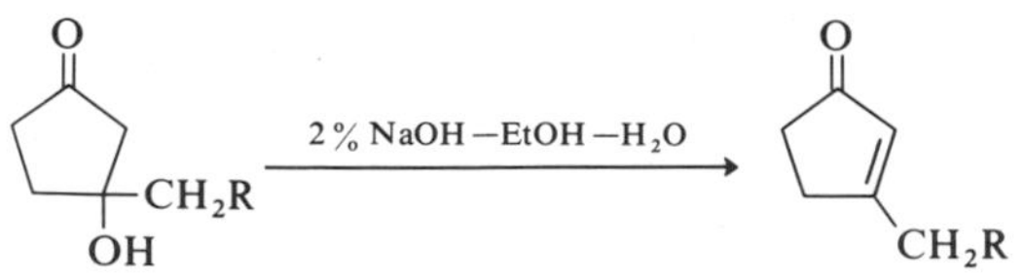

$$CH_3\overset{O}{\underset{}{C}}CH_2CH_2\overset{O}{\underset{}{C}}CH_2R \xrightarrow[\substack{\text{(shorter time than} \\ \text{for other runs)}}]{2\% \text{ NaOH}-\text{EtOH}-D_2O} CD_3\overset{O}{\underset{}{C}}CD_2CD_2\overset{O}{\underset{}{C}}CD_2R$$

18. Treatment of catechin (**A**) with 0.5% NaOH for 45 min at reflux gives compound **B**. Formulate a mechanism.

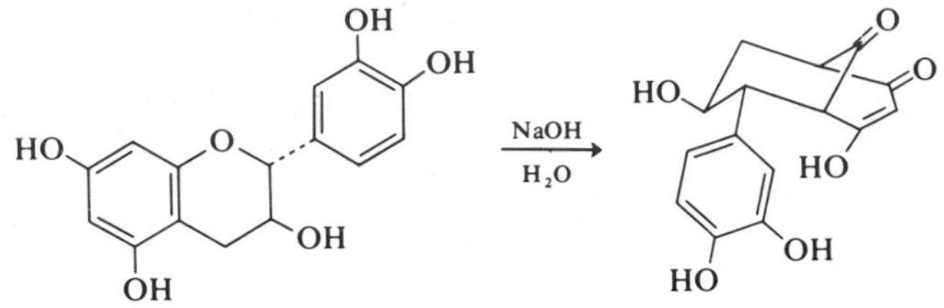

19. Suggest routes for synthesis of each of the following compounds from readily available starting materials:

(a)

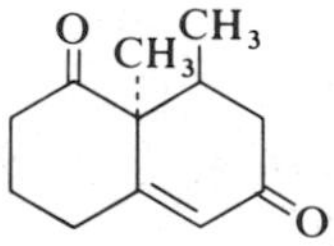

(b)

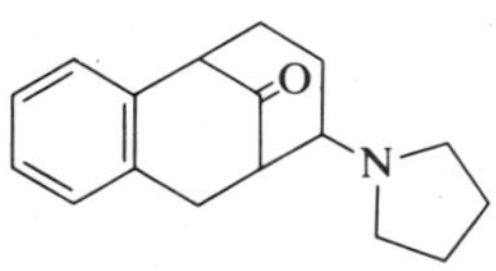

20. The two phosphonium salts **A** and **B** have both been used in syntheses of cyclohexadienes. Suggest appropriate coreactants and catalysts that would be expected to lead to cyclohexadienes.

$$H_2C{=}CHCH_2\overset{+}{P}Ph_3 \qquad H_2C{=}CHCH{=}CH\overset{+}{P}Ph_3$$
$$\textbf{A} \qquad\qquad\qquad \textbf{B}$$

Addition Reactions of Carbon–Carbon Multiple Bonds

In this chapter, the general topic of additions to carbon–carbon double bonds is discussed. Several classes of additions are excluded, since they seem to be more appropriate for other sections. These include nucleophilic additions to electrophilic olefins (Chapter 1), cycloadditions (Chapter 6), additions that are primarily oxidative processes (Chapter 9), and additions that occur by free-radical mechanisms (Part A, Chapter 12).

3.1. Addition of Hydrogen

The most widely used method of adding the elements of hydrogen to a double bond is catalytic hydrogenation. Except for sterically hindered olefins, this reaction

$$RCH{=}CHR \ + \ H_2 \xrightarrow{\text{catalyst}} RCH_2CH_2R$$

usually proceeds rapidly and cleanly. The common catalysts are various finely divided forms of transition metals, particularly platinum, palladium, rhodium, nickel, and copper. Since, from a synthetic point of view, it is important to know the response of other functional groups toward catalytic hydrogenation conditions, the catalytic hydrogenation of other functional groups, as well as the carbon–carbon double bond, will be discussed in this section.

Although many details remain uncertain, a rudimentary understanding of the mechanism and stereochemistry of catalytic hydrogenation has been developed. It is

74

CHAPTER 3
ADDITION
REACTIONS OF
CARBON–CARBON
MULTIPLE
BONDS

known that hydrogen is adsorbed onto the surface of materials that function as hydrogenation catalysts. Carbon–carbon multiple bonds and other functional groups interact with the metal surface, forming intermediates in which the organic molecule is strongly adsorbed.

For simple alkenes, at least three types of intermediates have been implicated in processes that occur during hydrogenation. The intermediate initially formed is pictured as adsorbed at both of the carbon atoms of the alkene bond, as shown in **A**. The π-orbitals of the alkene are used for bonding to the metal surface. Hydrogen

adsorbed on the surface can be added to the organic residue leading to **B**, a monoadsorbed species attached to the metal surface by what approximates a sigma bond. This intermediate can react with adsorbed hydrogen to give the saturated product. A third species that can account for the double-bond migration and exchange of hydrogen that sometimes competes with simple addition of hydrogen is the species **C**. The organic molecule in **C** approximates the allyl radical derived from the alkene by abstraction of a hydrogen atom. The coordination of this intermediate probably involves π- rather than σ-orbitals. In Chapter 5, the structure of discrete organometallic compounds is discussed. We will encounter examples of each of these three types of bonding of organic molecules to metal atoms in small discrete complexes. This lends some plausibility to the existence of similar entities on metal surfaces.

The metal surface should not be regarded as uniform. There are apparently sites that show preference for specific types of interaction with organic molecules. Also, the surface is presumably not smooth but consists of irregular features that would have variable steric interactions with the adsorbed molecule.

In most cases, both hydrogen atoms are added to the same side of the substrate (*syn* addition). This can result from nearly simultaneous addition of both hydrogen atoms. If hydrogenation occurs in two steps, the intermediate must remain bonded to the metal surface in such a way that single-bond rotations do not obscure the original stereochemical configuration. Adsorption to the catalyst surface normally involves the less sterically hindered side of the molecule. Scheme 3.1 illustrates some hydrogenations in which the *syn* addition from the less hindered side is observed. There are sufficient examples of alternate modes of addition, as is noted in Scheme 3.1, that independent corroboration of the stereochemistry is normally desirable.

The stereochemistry of hydrogenation is affected by the presence of polar functional groups that can govern the mode of adsorption of the molecule to the catalyst surface. For instance, there are a number of examples where the presence of a carboxylic acid or hydroxyl group results in hydrogen being introduced from the

Scheme 3.1. Stereochemistry of Hydrogenation of Some Olefins

A. Examples of *syn* Addition from Less Hindered Side

1[a]

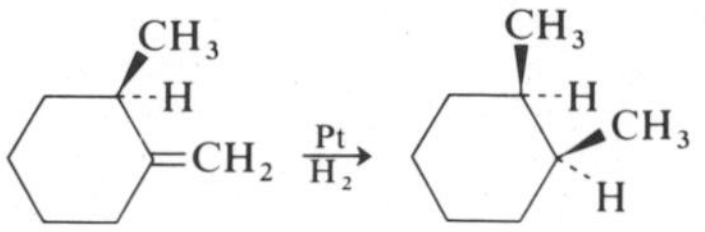

2[b]

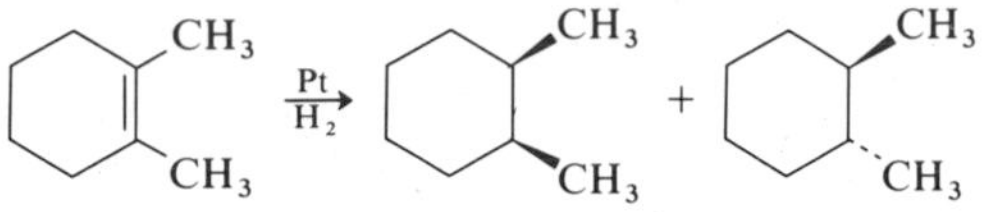

(70–85 %)   (30–15 %)

3[c]

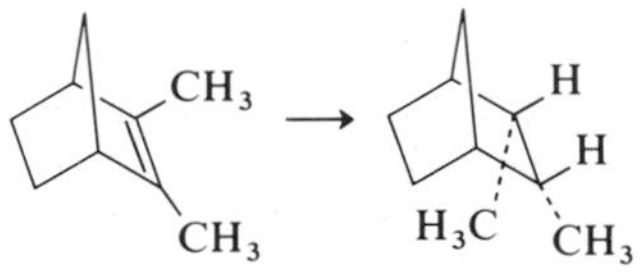

4[b]

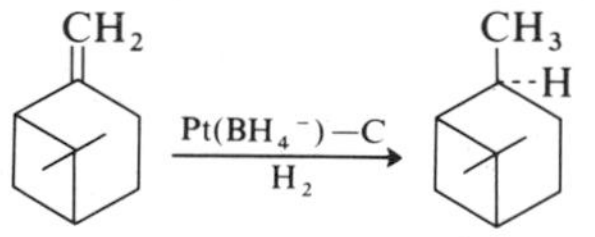

B. Exceptions

5[a]

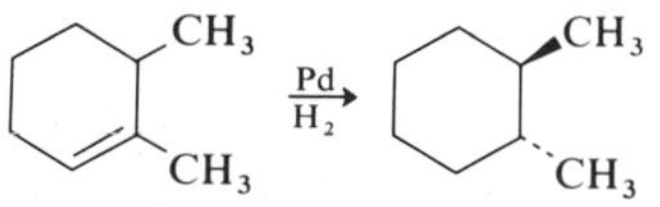

6[d]

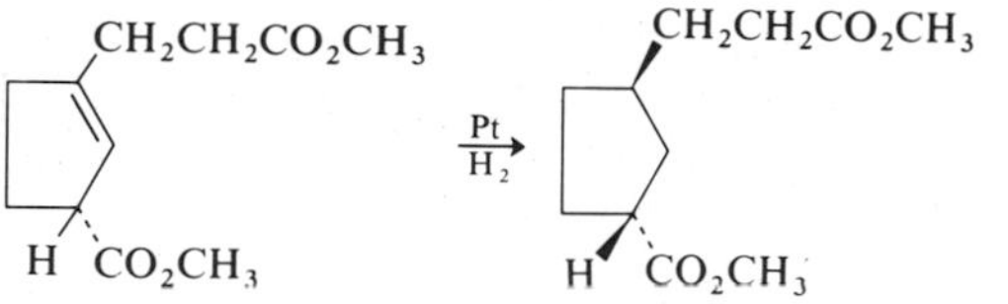

7[e]

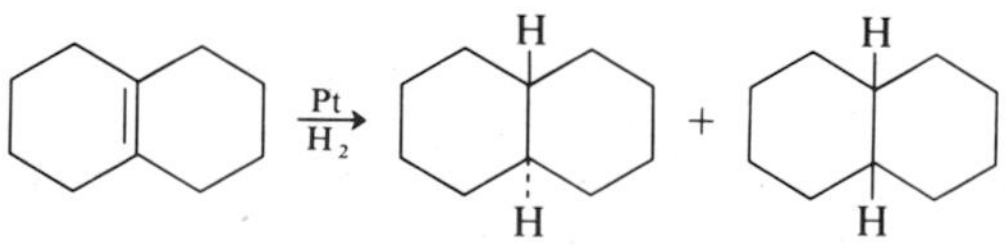

(79 %)   (21 %)

a. S. Siegel and G. V. Smith, *J. Am. Chem. Soc.* **82**, 6082, 6087 (1960).
b. C. A. Brown, *J. Am. Chem. Soc.* **91**, 5901 (1969).
c. K. Alder and W. Roth, *Chem. Ber.* **87**, 161 (1954).
d. J. P. Ferris and N. C. Miller, *J. Am. Chem. Soc.* **88**, 3522 (1966).
e. J. F. Sauvage, R. H. Baker, and A. S. Hussey, *J. Am. Chem. Soc.* **83**, 3874 (1961).

76

CHAPTER 3
ADDITION
REACTIONS OF
CARBON–CARBON
MULTIPLE
BONDS

side of the molecule carrying the polar group.[1] This implies that the molecule is adsorbed preferentially in such a way that the hydroxyl group can interact strongly with the catalyst surface.

If desorption of the partially hydrogenated olefin is competitive with complete reduction, olefin isomerization can result:

$$
\begin{array}{ccc}
\underset{\displaystyle \underset{/\!/\!/\!/\!/\!/\!/\!/\!/}{\overset{\text{H}\quad\quad\text{H--H}}{}}}{\overset{\displaystyle \overset{\text{H}\ \text{H}\ \text{H}}{R-\overset{|}{\underset{|}{C}}-\overset{|}{C}=\overset{|}{C}-R'}}{}}
& \rightleftarrows &
R-\overset{\text{H}\ \text{H}\ \text{H}}{\overset{|}{\underset{|}{C}}-\overset{|}{C}-\overset{|}{C}-R'}
\rightleftarrows
R\overset{\text{H}\ \text{H}\ \text{H}}{C=\overset{|}{C}-\overset{|}{C}-R'}
\end{array}
$$

For example, hydrogenation of 1-pentene over Raney nickel is accompanied by isomerization to *cis*- and *trans*-2-pentene.[2] The isomerized products are converted to pentane, but at a slower rate than 1-pentene. When the stereochemical course of hydrogenation is to be interpreted, the prior isomerization of the olefin can be a complicating factor, since the steric requirements for hydrogenation of the isomer may differ substantially from those of the original substrate. Exchange of hydrogen atoms from the substrate for hydrogen from the catalyst surface has often been detected by using deuterium-labeled substrates or by carrying out the catalytic reduction in a deuterium atmosphere. Allylic positions undergo such exchange particularly rapidly.[3]

The conditions required for the hydrogenation of various functional groups differ with the catalyst. Table 3.1 summarizes some general information about catalytic reduction of various functional groups.

Table 3.1 shows that not all catalytic hydrogenations result in simple addition of hydrogen. Aromatic ketones, alcohols, and amines undergo cleavage of C–O or C–N bonds adjacent to the benzene ring. Alkyl and aryl halides undergo replacement of halogen by hydrogen. The term *hydrogenolysis* is used to describe such reactions, which involve cleavage of bonds and replacement of a substituent with hydrogen.

Reductions that involve removal of oxygen from carbonyl groups usually occur by stepwise processes in which the oxygen atom is lost as a water molecule:

$$RCO_2H \rightarrow R-\overset{\text{OH}}{\underset{\text{H}}{\overset{|}{\underset{|}{C}}}}-OH \rightarrow RCH{=}O \rightarrow RCH_2OH$$

$$RCONH_2 \rightarrow R-\overset{\text{OH}}{\underset{\text{H}}{\overset{|}{\underset{|}{C}}}}-NH_2 \rightarrow RCH{=}NH \rightarrow RCH_2NH_2$$

$$RNO_2 \rightarrow R-\underset{\text{OH}}{\overset{|}{\underset{|}{N}}}-OH \rightarrow RNHOH \rightarrow RNH_2$$

1. S. Mitsui, Y. Senda, and H. Saito, *Bull. Chem. Soc. Jpn.* **39**, 694 (1966).
2. H. C. Brown and C. A. Brown, *J. Am. Chem. Soc.* **85**, 1005 (1963).
3. G. V. Smith and J. R. Swoap, *J. Org. Chem.* **31**, 3904 (1966).

The reductive cleavage of carbon–oxygen bonds occurs readily when the carbon atom is substituted by a phenyl ring:

$$\text{Ph–CH}_2\text{OR} \xrightarrow{\text{H}_2,\ \text{Pd}} \text{Ph–CH}_3 + \text{HOR}$$

This facile cleavage of the benzyl oxygen bond has made the benzyl group a useful "protecting group" during multistep synthetic schemes. A particularly important example is the use of the carbobenzyloxy group in peptide synthesis:

$$\underset{\substack{\|\\ \text{O}}}{\text{PhCH}_2\text{OC}}\text{NH}\underset{\substack{|\\ \text{R}}}{\text{CH}}\text{CO}_2\text{H} + \text{NH}_2\underset{\substack{|\\ \text{R}'}}{\text{CH}}\text{CO}_2\text{R}'' \rightarrow \text{PhCH}_2\overset{\substack{\text{O}\\ \|}}{\text{OC}}\text{NH}\overset{\substack{\text{R}\\ |}}{\text{CH}}\text{CONH}\overset{\substack{\text{R}'\\ |}}{\text{CH}}\text{CO}_2\text{R}'' \rightarrow$$

$$\xrightarrow[\text{H}_2]{\text{Pd}} \text{H}_2\text{N}\overset{\substack{\text{R}\\ |}}{\text{CH}}\text{CONH}\overset{\substack{\text{R}'\\ |}}{\text{CH}}\text{CO}_2\text{R}''$$

The protecting group is removed by hydrogenolysis after the coupling reaction. The substituted carbamic acid generated by benzyl hydrogenolysis decarboxylates spontaneously:

$$\text{PhCH}_2\overset{\substack{\text{O}\\ \|}}{\text{OC}}\text{NHR} \rightarrow \text{PhCH}_3 + \text{HO}\overset{\substack{\text{O}\\ \|}}{\text{C}}\text{NHR} \rightarrow \text{CO}_2 + \text{H}_2\text{NR}$$

Nearly all carbon–halogen bonds are susceptible to cleavage under the conditions of catalytic hydrogenation. Aromatic halogen substituents are readily removed, and only the most reactive functional groups can be reduced without competitive loss of halogen. Alkyl halides are less reactive, although dehalogenation is strongly promoted by bases.

Most catalytic reductions are carried out in systems in which gaseous hydrogen is introduced from commercial high-pressure cylinders. There are also techniques available for generating hydrogen chemically.[4] This method is particularly convenient for hydrogenation that must be carried out at low temperature.[5] A solution of sodium borohydride is introduced into an acidic solution and reaction occurs, generating hydrogen. A simple apparatus has been constructed that automatically adds sodium borohydride to maintain the pressure of hydrogen at 1 atm. The progress of the reaction can be monitored by the volume of solution that has been added at any point.

It has also been found that highly active catalysts can be prepared *in situ* by reducing a platinum, palladium, or rhodium salt with sodium borohydride in the presence of a carbon support.[6] Similar reductions of nickel salts produce colloidal nickel–boron, which is highly selective toward olefins of different structural types.[2] The normal order of reactivity—i.e., terminal > disubstituted > trisubstituted—is observed, but the reactivity spread is sufficiently large that selective hydrogenation of polyenes is possible, as illustrated by a step in a synthesis of the natural product

4. C. A. Brown and H. C. Brown, *J. Org. Chem.* **31**, 3989 (1966).
5. C. A. Brown, *J. Am. Chem. Soc.* **91**, 5901 (1969).
6. H. C. Brown and C. A. Brown, *J. Am. Chem. Soc.* **84**, 2827 (1962).
2. See p. 76.

Table 3.1. Conditions for Catalytic

CHAPTER 3
ADDITION
REACTIONS OF
CARBON–CARBON
MULTIPLE
BONDS

Functional group	Reduction product	Common catalysts	Typical reaction conditions
$\diagdown$C=C$\diagup$	$-\overset{\mid}{\underset{H}{C}}-\overset{\mid}{\underset{H}{C}}-$	Pd, Pt, Ni, Ru, Rh	Rapid at R.T. and 1 atm except for highly substituted or hindered cases
$-C\equiv C-$	$\underset{H}{\diagdown}$C=C$\underset{H}{\diagup}$	Pd	R.T. and low pressure, quinoline or lead added to deactivate catalyst
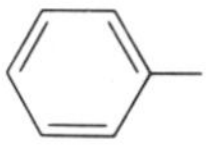 ⬡—	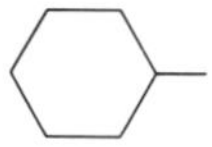 ⬡—	Rh, Pt	Moderate pressure (5–10 atm), 50–100°
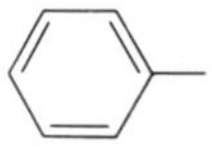 ⬡—	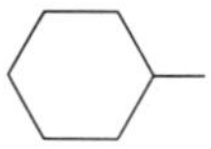 ⬡—	Ni, Pd	High pressure (100–200 atm), 100–200°
$\overset{O}{\overset{\|}{RCR}}$	$\overset{RCHR}{\underset{\mid}{OH}}$	Pt, Ru	Moderate rate at R.T. and 1–4 atm, acid-catalyzed
$\overset{O}{\overset{\|}{RCR}}$	$\overset{RCHR}{\underset{\mid}{OH}}$	Cu–Cr, Ni	High pressure, 50–100°
⬡$-\overset{O}{\overset{\|}{CR}}$ or ⬡$-\overset{OR}{\underset{\mid}{CHR}}$	⬡$-CH_2R$	Pd	R.T., 1–4 atm, acid-catalyzed
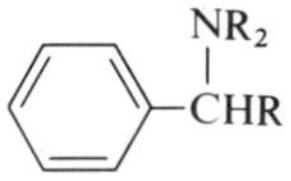 ⬡$-\overset{NR_2}{\underset{\mid}{CHR}}$	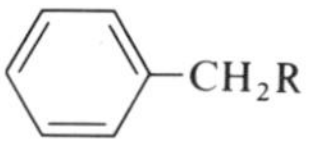 ⬡$-CH_2R$	Pd, Ni	50–100°, 1–4 atm

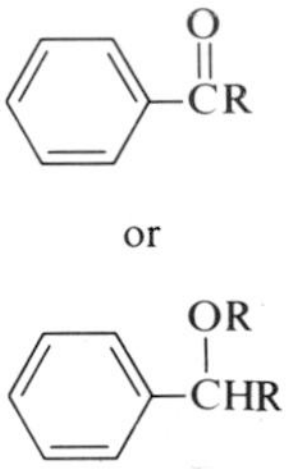

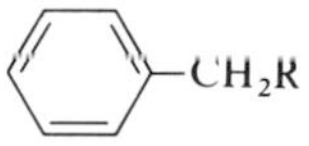

sirenin.[7] The acetylinic linkage is reduced to a disubstituted olefin, which, in turn, is

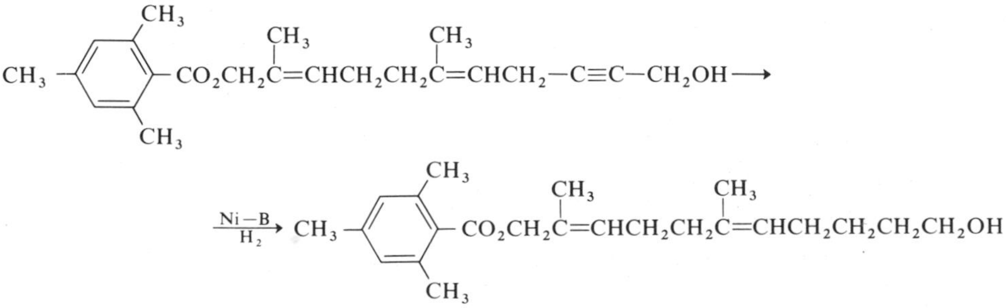

7. E. J. Corey, K. Achiwa, and J. A. Katzenellenbogen, *J. Am. Chem. Soc.* **91**, 4318 (1969).

Reduction of Various Groups[a]

Functional group	Reduction product	Common catalysts	Typical reaction conditions
$\underset{RCCl}{\overset{O}{\parallel}}$	$\underset{RCH}{\overset{O}{\parallel}}$	Pd	R.T., 1 atm, quinoline or other catalyst moderator used
$\underset{RCOH}{\overset{O}{\parallel}}$	RCH_2OH	Pd, Ni, Ru	Very strenuous conditions required
$\underset{RCOR}{\overset{O}{\parallel}}$	RCH_2OH	Cu–Cr, Ni	200°, high pressure
$RC\equiv N$	RCH_2NH_2	Ni, Rh	50–100°, usually high pressure, NH_3 added to increase yield of primary amine
$\underset{RCNH_2}{\overset{O}{\parallel}}$	RCH_2NH_2	Cu–Cr	Very strenuous conditions required
RNO_2	RNH_2	Pd, Ni, Pt	R.T., 1–4 atm
$\underset{RCR}{\overset{NR}{\parallel}}$	R_2CHNHR	Pd, Pt	R.T., 4–100 atm
R—Cl R—Br R—I	R—H	Pd	Order of reactivity: I > Br > Cl > F, bases promote reaction for R = alkyl
epoxide	$-\overset{\overset{H}{\vert}}{C}-\overset{\overset{OH}{\vert}}{C}-$	Pt, Pd	Proceeds slowly at R.T., 1–4 atm, acid-catalyzed

a. General references: R. L. Augustine, *Catalytic Hydrogenation*, Marcel Dekker, New York, NY, 1965; P. N. Rylander, *Catalytic Hydrogenation over Platinum Metals*, Academic Press, New York, NY, 1973; M. Freifelder, *Practical Catalytic Hydrogenation*, Wiley–Interscience, New York, NY, 1971.

reduced more rapidly than the two trisubstituted alkene groups present in the molecule.

A recent development in catalytic reduction has been the discovery of soluble metal complexes that promote hydrogenation in homogeneous solution. Most of these catalytic species are complex ions of metals of the platinum family containing various ligand species. The ligands serve to enhance the solubility of the complexes in organic media. Some of the active species are shown in Table 3.2. Many of these catalysts are highly selective in their reducing abilities. Homogeneous catalysts are also advantageous for specific introduction of deuterium, since the exchange processes that sometimes occur with heterogeneous catalysts are minimized.

An interesting application of these catalysts is for asymmetric addition of the elements of hydrogen. Chiral catalysts are synthesized by use of optically active phosphine ligands. This constitutes a very efficient way of generating centers of

80

CHAPTER 3
ADDITION
REACTIONS OF
CARBON–CARBON
MULTIPLE
BONDS

Table 3.2. Homogeneous Hydrogenation Catalysts

Composition	Selectivity	Ref.
$(Ph_3P)_3RhCl$	Highly selective for olefinic bonds in preference to other easily reduced groups; less hindered olefins reduced in preference to more hindered	a
MeO_2C—⟨benzene⟩—$Cr(CO)_3$	Selective for reduction of dienes to *cis*-olefins	b
$(Ph_3P)_2PtCl_2$—$SnCl_2$	Selective for terminal olefin groups	c
$(Ph_3P)_2IrCOCl$	Electron-attracting groups in conjugation with olefin accelerate reduction	d

a. J. A. Osborn, F. H. Jardine, J. F. Young, and G. Wilkinson, *J. Chem. Soc. A*, 1711 (1966).
b. E. N. Frankel and R. O. Butterfield, *J. Org. Chem.* **34**, 3930 (1969).
c. R. W. Adams, G. E. Batley, and J. C. Bailar, *J. Am. Chem. Soc.* **90**, 6051 (1968).
d. L. Vaska and R. E. Rhodes, *J. Am. Chem. Soc.* **87**, 4970 (1965); W. Strohmeier and R. Fleischman, *Z. Naturforsch.* **24B**, 1217 (1969).

optical activity because the catalyst is not destroyed. The optically active diphos-

1

phine **1**, when incorporated as a ligand on a Rh atom, gives a catalyst that permits synthesis of α-amino acids from achiral precursors in optical yields of around 70%:

Ref. 8

(95% yield, 70% optical purity)

Another important method for addition of hydrogen to olefinic bonds involves the unstable molecule diimide, $HN\!=\!NH$. Several methods for generation of diimide have been developed and are summarized in Scheme 3.2. Simple olefins are effectively reduced by the reagent, but most of the common easily reduced functional groups, particularly nitro and carbonyl groups, are unaffected by diimide. This permits reduction of carbon–carbon double bonds in the presence of other easily reduced groups. The mechanism of the reaction has been pictured as a transfer of hydrogen to the multiple bond via a nonpolar transition state with loss of nitrogen:

8. H. B. Kagan and T.-P. Dang, *J. Am. Chem. Soc.* **94**, 6429 (1972).

Scheme 3.2. Reductions with Diimide 81

1^a $CH_2=CHCH_2OH$ $\xrightarrow[\text{RCO}_2\text{H, 25°}]{\overset{+}{\text{Na}}\,^-O_2C-N=N-CO_2^-\overset{+}{\text{Na}}}$ $CH_3CH_2CH_2OH$ (78%)

2^b $\xrightarrow{\text{C}_7\text{H}_7\text{SO}_2\text{NHNH}_2,\,\text{heat}}$ (98%)

3^c $(CH_2=CHCH_2)_2S$ $\xrightarrow{\text{C}_7\text{H}_7\text{SO}_2\text{NHNH}_2,\,\text{heat}}$ $(CH_3CH_2CH_2)_2S$ (93–100%)

4^d $\xrightarrow{\text{NH}_2\text{NH}_2,\,\text{O}_2,\,\text{Cu(II)}}$

5^e $O_2N-\langle\ \rangle-CH=CHCO_2H$ $\xrightarrow[\text{NH}_2\text{OH}]{\text{NH}_2\text{OSO}_3^-}$ $O_2N-\langle\ \rangle-CH_2CH_2CO_2H$ (87%)

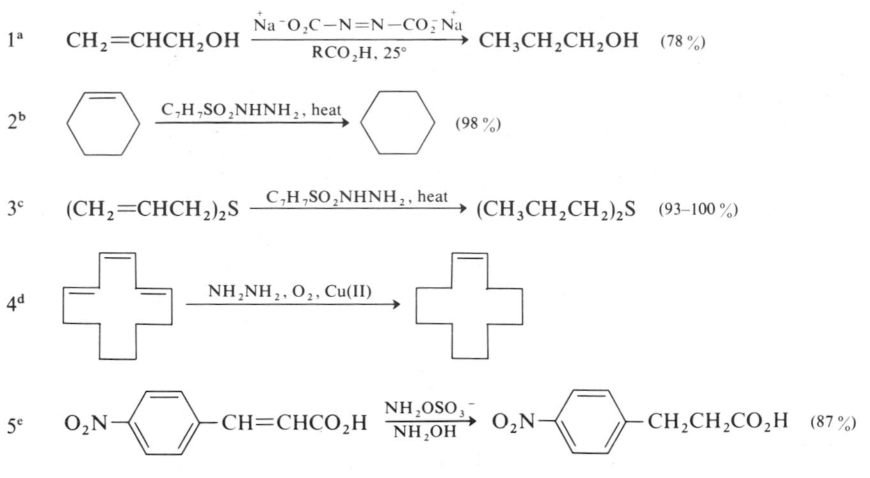

a. E. E. van Tamelen, R. S. Dewey, and R. J. Timmons, *J. Am. Chem. Soc.* **83**, 3725 (1961).
b. R. S. Dewey and E. E. van Tamelen, *J. Am. Chem. Soc.* **83**, 3729 (1961).
c. E. E. van Tamelen, R. S. Dewey, M. F. Lease, and W. H. Pirkle, *J. Am. Chem. Soc.* **83**, 4302 (1961).
d. M. Ohno and M. Okamoto, *Org. Synth.* **49**, 30 (1969).
e. W. Dürckheimer, *Justus Liebigs Ann. Chem.* **721**, 240 (1969).

In agreement with this mechanism is the fact that the addition has been demonstrated to be *syn* for several typical alkenes.[9] The rate of diimide reductions has been shown to be affected by torsional and angle strain in the alkene.[10] More strained double bonds react at accelerated rates. For example, the more strained *trans* double bond is selectively reduced in *cis, trans*-1,5-cyclodecadiene.[11]

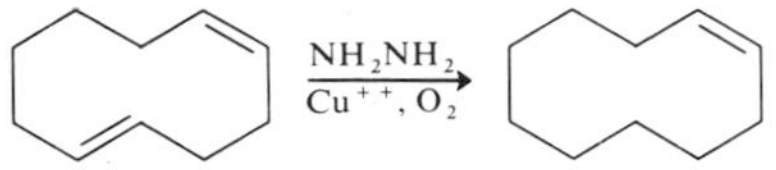

$\xrightarrow[\text{Cu}^{++},\,\text{O}_2]{\text{NH}_2\text{NH}_2}$

3.2. Addition of Hydrogen Halides

Hydrogen chloride and hydrogen bromide add to olefins to give addition products. Many years ago, it was noted that additions usually take place to give the product in which the halogen atom is attached to the more substituted end of the olefin. This type of behavior was sufficiently general that the name *Markownikoff's*

9. E. J. Corey, D. J. Pasto and W. L. Mock, *J. Am. Chem. Soc.* **83**, 2957 (1961).
10. E. W. Garbisch, Jr., S. M. Schildcrout, D. B. Patterson, and C. M. Sprecher, *J. Am. Chem. Soc.* **87**, 2932 (1965).
11. J. G. Traynham, G. R. Franzen, G. A. Knesel, and D. J. Northington, Jr., *J. Org. Chem.* **32**, 3285 (1967).

82

CHAPTER 3
ADDITION
REACTIONS OF
CARBON–CARBON
MULTIPLE
BONDS

$$R_2C{=}CH_2 \ + \ HX \ \rightarrow \ R_2\underset{\underset{X}{|}}{C}CH_3$$

$$R_2C{=}CHR' \ + \ HX \ \rightarrow \ R_2\underset{\underset{X}{|}}{C}CH_2R'$$

rule was given to the statement generalizing this observed mode of addition. A very rudimentary picture of the mechanism of addition reveals the mechanistic foundation of Markownikoff's rule. The first step in the addition involves protonation of the

$$R_2C{=}CH_2 \ + \ HX \ \rightarrow \ \underset{R}{\overset{R}{\diagdown}}\!\!\overset{+}{C}{-}CH_3 \ + \ X^-$$

$$\underset{R}{\overset{R}{\diagdown}}\!\!\overset{+}{C}{-}CH_3 \ + \ X^- \ \rightarrow \ R_2\underset{\underset{X}{|}}{C}CH_3$$

olefin or the formation of a transition state involving partial protonation of the carbon–carbon double bond. In the first olefin above, addition of a proton can lead to either a tertiary or a primary carbonium ion (see Figure 3.1). The reaction follows the lower energy course, and the more substituted carbonium ion results. Addition is completed when the carbonium ion and halide ion react to give product. A more complete discussion of the mechanism of ionic addition of hydrogen halides to alkene was given in Part A, Chapter 6. In particular, the question of whether or not a discrete carbonium ion is always involved is considered there.

The general terms *regioselective* and *regiospecific* have been introduced to describe addition reactions that proceed selectively or exclusively in one direction with unsymmetrical olefins.[12] Thus, addition of HBr to styrene is described as Br-phenyl-regiospecific. Markownikoff's rule then describes a general case of regioselectivity that operates because of the stabilizing effect of alkyl and aryl groups on carbonium ion centers.

In nucleophilic solvents, products that arise from reaction of the solvent with the intermediate may be encountered. For example, reaction of cyclohexene with hydrogen bromide in acetic acid gives cyclohexyl acetate as well as cyclohexyl

12. A. Hassner, *J. Org. Chem.* **33**, 2684 (1968).

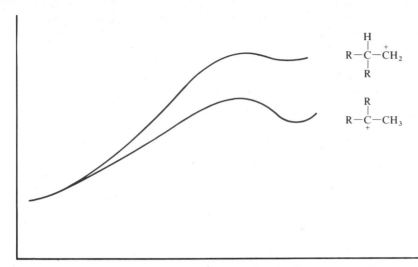

Figure 3.1. Relative reaction energy in partial protonation of primary versus tertiary carbon atoms.

bromide. This result is readily understood as resulting from competition between acetic acid and bromide ion acting as nucleophiles in the addition reaction.

Because of the involvement of carbonium ion intermediates, rearrangement is a possibility. Reaction of *t*-butylethylene with hydrogen chloride in acetic acid gives both rearranged and unrearranged product.[13] The rearranged acetate may also be

$$(CH_3)_3CCH\text{=}CH_2 \xrightarrow[\text{HCl}]{\text{AcOH}} (CH_3)_3CCHCH_3 + (CH_3)_2CCH(CH_3)_2 +$$
$$\qquad\qquad\qquad\qquad\qquad\qquad | \qquad\qquad\qquad |$$
$$\qquad\qquad\qquad\qquad\qquad\quad Cl \qquad\qquad\qquad Cl$$
$$\qquad\qquad\qquad\qquad\qquad (35\text{--}40\%) \qquad\quad (40\text{--}50\%)$$

$$(CH_3)_3CCHCH_3 + \left[(CH_3)_2CCH(CH_3)_2 \right]$$
$$\qquad\qquad\quad | \qquad\qquad\qquad\qquad\qquad |$$
$$\qquad\qquad OAc \qquad\qquad\qquad\qquad\quad OAc$$
$$\qquad (15\text{--}20\%)$$

formed, but it is unstable under the reaction conditions, being converted to the rearranged chloride.

The stereochemistry of the addition of hydrogen halides to a variety of alkenes has been investigated. The addition of hydrogen chloride to 1-methylcyclopentene is $96 \pm 4\%$ *anti*[14]:

Addition of hydrogen bromide to cyclohexene and *cis*- and *trans*-2-butene also takes place by *anti* addition.[15] In other instances, additions of hydrogen halides to

13. R. C. Fahey and C. A. McPherson, *J. Am. Chem. Soc.* **91**, 3865 (1969).
14. R. C. Fahey and R. A. Smith, *J. Am. Chem. Soc.* **86**, 5035 (1964).
15. D. J. Pasto, G. R. Meyer, and S. Kang, *J. Am. Chem. Soc.* **91**, 2163 (1969).

84

CHAPTER 3
ADDITION
REACTIONS OF
CARBON–CARBON
MULTIPLE
BONDS

Table 3.3. Stereochemistry of Addition of Hydrogen Halides to Olefins

Olefin	Hydrogen halide	Stereochemistry	Ref.
1,2-Dimethylcyclohexene	HBr	*anti*	a
1,2-Dimethylcyclohexene	HCl	solvent- and temperature-dependent	a
Cyclohexene	HBr	*anti*	b
cis-2-Butene	DBr	*anti*	c
trans-2-Butene	DBr	*anti*	c
1,2-Dimethylcyclopentene	HBr	*anti*	d
1-Methylcyclopentene	HCl	*anti*	e
Norbornene	HBr	*syn* and rearrangement	f
Norbornene	HCl	*syn* and rearrangement	g
trans-1-Phenylpropene	HBr	*syn* (9 : 1)	h
cis-1-Phenylpropene	HBr	*syn* (8 : 1)	h
Bicyclo[3.1.0]hex-2-ene	DCl	*syn*	i
1-Phenyl-4-*t*-butylcyclohexene	DCl	*syn*	j

a. G. S. Hammond and T. D. Nevitt, *J. Am. Chem. Soc.* **76**, 4121 (1954); R. C. Fahey and C. A. McPherson, *J. Am. Chem. Soc.* **93**, 2445 (1971); K. B. Becker and C. A. Grob, *Synthesis*, 789 (1973).
b. R. C. Fahey and R. A. Smith, *J. Am. Chem. Soc.* **86**, 5035 (1964).
c. D. J. Pasto, G. R. Meyer, and B. Lepeska, *J. Am. Chem. Soc.* **96**, 1858 (1974).
d. G. S. Hammond and C. H. Collins, *J. Am. Chem. Soc.* **82**, 4323 (1960).
e. Y. Pocker and K. D. Stevens, *J. Am. Chem. Soc.* **91**, 4205 (1969).
f. H. Kwart and J. L. Nyce, *J. Am. Chem. Soc.* **86**, 2601 (1964).
g. J. K. Stille, F. M. Sonnenberg, and T. H. Kinstle, *J. Am. Chem. Soc.* **88**, 4922 (1966).
h. M. J. S. Dewar and R. C. Fahey, *J. Am. Chem. Soc.* **85**, 3645 (1963).
i. P. K. Freeman, F. A. Raymond, and M. F. Grostic, *J. Org. Chem.* **32**, 24 (1967).
j. K. D. Berlin, R. O. Lyerla, D. E. Gibbs, and J. P. Devlin, *Chem. Commun.*, 1246 (1970).

olefins have been shown to be *syn* additions.[16] The stereochemistry of addition of some olefins, such as 1,2-dimethylcyclohexene, is solvent- and temperature-dependent.[17] *Syn* addition is particularly common with olefins having a phenyl substituent. Table 3.3 lists examples of several olefins for which the stereochemistry of addition of hydrogen chloride or hydrogen bromide has been studied.

The stereochemistry of the addition depends upon the details of the addition mechanism. Two general mechanisms have been encountered for alkenes. The addition can proceed through an ion pair that involves a discrete carbonium ion:

$$RCH{=}CH_2 + HCl \rightarrow \underset{Cl^-}{\overset{+}{RCHCH_3}} \rightarrow \underset{Cl}{RCHCH_3}$$

Many alkenes, however, react via a transition state that involves the alkene, hydrogen chloride, and a third species, either solvent or halide ion. This termolecular mechanism is generally pictured as involving nucleophilic attack on an alkene–hydrogen chloride complex and bypassing a discrete carbonium ion:

16. M. J. S. Dewar and R. C. Fahey, *J. Am. Chem. Soc.* **85**, 3645 (1963).
17. R. C. Fahey and C. A. McPherson, *J. Am. Chem. Soc* **93**, 2445 (1971); K. B. Becker and C. A. Grob, *Synthesis*, 789 (1973).

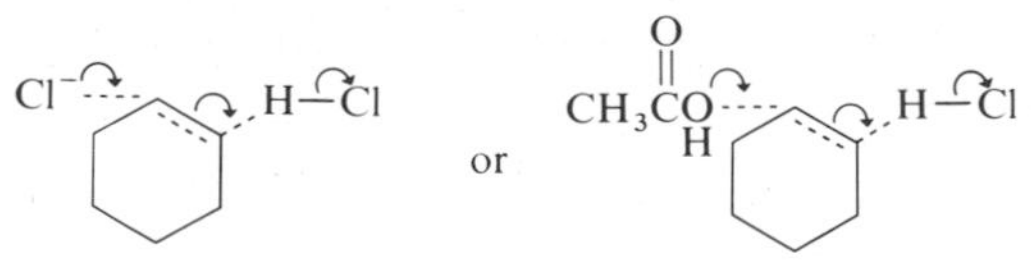

The ion-pair mechanism would not be expected to be stereospecific, since it involves an intermediate that permits loss of the stereochemistry present in the initial olefin. It might be expected that the ion-pair mechanism could lead to a preference for *syn* addition, since at the moment of formation of the ion pair, the chloride ion is necessarily on the same side of the multiple bond from which hydrogen was added. On the other hand, the termolecular mechanism could be expected to give *anti* addition. Concerted attack by the nucleophile would occur from the side opposite from proton addition:

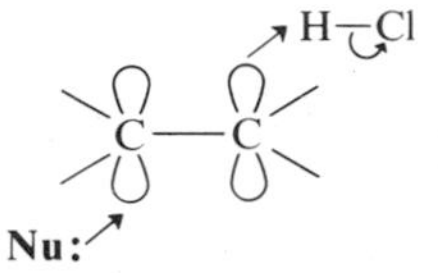

What structural features can determine whether a molecule will be involved in an ion-pair mechanism or the termolecular *anti* addition? Olefins that form highly stable carbonium ions are likely candidates for the ion-pair mechanism. Structural features that lead to a stable carbonium ion will lower the energy required to reach the transition state leading to a discrete carbonium ion and result in formation of the ion pair. Steric hindrance should also favor the ion-pair mechanism, since approach of the nucleophile is resisted and may be delayed to a point on the reaction coordinate at which C–H bond formation is essentially complete. The fact that the only olefins that give much *syn* addition are those capable of forming relatively stable carbonium ions is in agreement with this mechanistic description.

These basic mechanistic ideas set some limits to the synthetic utility of the ionic addition of hydrogen halides to alkenes. Because of the involvement of intermediates with carbonium-ion character, potential reactants must be examined for the likelihood of rearrangements. The dependence of the stereochemistry of hydrogen halide addition on the stability of the potential carbonium-ion intermediates gives some basis for predicting stereoselectivity. Disubstituted alkenes generally give *anti* addition. For trisubstituted alkenes or styrene derivatives, however, the degree of stereoselectivity is a sensitive function of reactant structure and reaction conditions, and no single generalization can adequately predict the stereochemistry of the hydrogen halide addition.

Besides the ionic mechanism for hydrogen bromide addition, there is an efficient free-radical mechanism. As emphasized in Part A, Section 12.4, this mechanism results in regioselectivity which is opposite to that of the ionic mechanism.

3.3. Hydration and Other Acid-Catalyzed Additions

In addition to halide ions, a variety of nucleophilic species can be added to

86

CHAPTER 3
ADDITION
REACTIONS OF
CARBON–CARBON
MULTIPLE
BONDS

olefinic bonds under acidic conditions. A fundamental example is the hydration of olefins in aqueous systems:

$$R_2C=CH_2 \ + \ H^+ \ \longrightarrow \ R_2\overset{+}{C}CH_3 \ \xrightarrow{H_2O} \ R_2\underset{\underset{+}{O}H_2}{C}CH_3 \ \xrightarrow{-H^+} \ R_2\underset{\underset{}{O}H}{C}CH_3$$

Addition of a proton occurs to give the more substituted carbonium ion, so that addition follows the orientation predicted by Markownikoff's rule. A more detailed discussion of the reaction mechanism is given in Part A, Section 6.2. The cationic intermediates are prone to rearrangement when a more stable carbonium ion can be formed by alkyl, aryl, or hydrogen migration. The reaction is most easily applied to the synthesis of tertiary alcohols:

$$(CH_3)_2C=CHCH_2CH_2\overset{O}{\overset{\|}{C}}CH_3 \ \xrightarrow[H_2O]{H_2SO_4} \ (CH_3)_2\underset{\underset{}{O}H}{C}CH_2CH_2\overset{O}{\overset{\|}{C}}CH_3 \qquad \text{Ref. 18}$$

Additions of nucleophilic solvents such as alcohols and carboxylic acids can be effected by use of strong acid catalysts[19]:

$$(CH_3)_2C=CH_2 \ + \ MeOH \ \xrightarrow{HBF_4} \ (CH_3)_3COCH_3$$

$$CH_3CH=CH_2 \ + \ CH_3CO_2H \ \xrightarrow{HBF_4} \ (CH_3)_2CHOCOCH_3$$

Because of the strongly acidic and rather vigorous reaction conditions required to effect hydration of most olefins, the reaction is applicable only to molecules that have no other very sensitive functional groups. A much milder procedure for alkene hydration is discussed in the next section.

Trifluoroacetic acid is a sufficiently strong acid to react with olefins under relatively mild conditions.[20] The addition is regiospecific in the direction predicted by Markownikoff's rule. The reaction is catalyzed by sulfuric acid, and ring strain

$$ClCH_2CH_2CH_2CH=CH_2 \ \xrightarrow[\Delta]{CF_3CO_2H} \ ClCH_2CH_2CH_2\underset{\underset{}{O_2CCF_3}}{C}HCH_3$$

enhances olefin reactivity. Norbornene undergoes rapid addition at 0°.[21] Again,

$$(CH_3)_3CCH=CH_2 \ \xrightarrow{CF_3CO_2H} \ (CH_3)_2\overset{+}{C}-\underset{\underset{}{H}}{C}(CH_3)_2 \ \longrightarrow \ (CH_3)_2\underset{\underset{}{O_2CCF_3}}{C}-CH(CH_3)_2$$

18. J. Meinwald, *J. Am. Chem. Soc.* **77**, 1617 (1955).
19. R. D. Morin and A. E. Bearse, *Ind. Eng. Chem.* **43**, 1596 (1951); D. T. Dalgleish, D. C. Nonhebel, and P. L. Pauson, *J. Chem. Soc. C*, 1174 (1971).
20. P. E. Peterson, R. J. Bopp, D. M. Chevli, E. L. Curran, D. E. Dillard, and R. J. Kamat, *J. Am. Chem. Soc.* **89**, 5902 (1967).
21. H. C. Brown, J. H. Kawakami, and K.-T. Liu, *J. Am. Chem. Soc.* **92**, 5536 (1970).

rearrangement is important in substrates that lead to a carbonium ion that can be stabilized by migration.[22]

3.4. Oxymercuration

Electrophilic attack on a carbon–carbon double bond by a metal ion can be followed by addition of a nucleophilic species, often a molecule of solvent. By far the

$$RCH{=}CHR \ + \ M^{+n} \ + \ \overset{..}{N} \ \rightarrow \ R{-}\underset{\underset{N}{|}}{\overset{\overset{H}{|}}{C}}{-}\underset{\underset{H}{|}}{\overset{\overset{M}{|}}{C}}{-}R$$

most completely studied examples of this type of reaction involve mercuric acetate.[23] It is not entirely certain whether the initial intermediate is a bridged mercurinium ion or an open carbonium ion[24]:

$$RCH{=}CH_2 \ + \ Hg(II) \ \rightarrow \ RCH\overset{\overset{2+}{Hg}}{=\!=\!=}CH_2 \quad or \quad R\overset{+}{C}H{-}\overset{\overset{+}{Hg}}{\underset{|}{}}CH_2$$

It has been possible to detect mercurinium ions when they are generated in a nonnucleophilic solvent medium[25]:

$$CH_3CH{=}CHCH_3 \ \xrightarrow[FSO_3H{-}SbF_5]{Hg(O_2CCF_3)_2} \ CH_3CH\overset{\overset{2+}{Hg}}{=\!=\!=}CHCH_3$$

The high preference for *anti* addition also suggests the involvement of a bridged intermediate. With 4-*t*-butylcyclohexene, for example, the two possible diaxial products are formed. This stereochemical behavior is characteristic of nucleophilic opening of 3-membered rings in the cyclohexane series. In molecules where the two

Ref. 26

possible directions of approach to the π-system are nonidentical, there is a preference for addition of mercury to the less hindered side of the molecule. Entry 2 in Scheme 3.4 (p. 90) is an example of this.

The reactivity of alkenes toward mercuration appears to be governed by a combination of steric and electronic factors.[27] Terminal double bonds are more

22. V. J. Shiner, Jr., R. D. Fisher, and W. Dowd, *J. Am. Chem. Soc.* **91**, 7748 (1969).
23. W. Kitching, *Organomet. Chem. Rev.* **3**, 61 (1968).
24. H. C. Brown and J. H. Kawakami, *J. Am. Chem. Soc.* **95**, 8665 (1973).
25. G. A. Olah and P. R. Clifford, *J. Am. Chem. Soc.* **95**, 6067 (1973).
26. D. J. Pasto and J. A. Gontarz, *J. Am. Chem. Soc.* **92**, 7480 (1970); W. L. Waters, T. G. Traylor, and A. Factor, *J. Org. Chem.* **38**, 2306 (1973).
27. H. C. Brown and P. J. Geoghegan, Jr., *J. Org. Chem.* **37**, 1937 (1972).

88

CHAPTER 3
ADDITION
REACTIONS OF
CARBON–CARBON
MULTIPLE
BONDS

Table 3.4. Relative Reactivity of Some Alkenes in Oxymercuration

1-Pentene	6.6[a]
2-Methyl-1-pentene	48
cis-2-Pentene	0.56
trans-2-Pentene	0.17
2-Methyl-2-pentene	1.24

a. Relative to cyclohexene; Ref. 27.

reactive than internal ones. Disubstituted terminal olefins, however, are more reactive than monosubstituted cases, as would be expected for electrophilic attack. The differences in relative reactivities are large enough that selectivity can be achieved in certain dienes. Relative reactivity data for some pentene derivatives are given in Table 3.4.

Other nucleophilic solvents besides water can be captured by the mercurinium ion.

$$CH_2{=}CH_2 \xrightarrow[\text{Hg(OAc)}_2]{\text{HOAc}} CH_3CO_2CH_2CH_2HgO_2CCH_3$$

The synthetic utility of the reaction for hydration of carbon–carbon double bonds depends on reactions that can replace the mercury atom by hydrogen. This is accomplished by various reducing agents, in particular sodium borohydride.[28]

The mechanism of the reductive replacement of mercury by hydrogen involves a free-radical intermediate generated by decomposition of an intermediate alkylmercury hydride[29]:

$$RHgX + NaBH_4 \rightarrow RHgH$$

$$RHgH \rightarrow R{\cdot} + Hg^{(I)}H$$

$$R{\cdot} + RHg^{(I)}H \rightarrow RH + Hg^{(0)} + R{\cdot}$$

The evidence for this mechanism includes the fact that the course of the reaction can be diverted by oxygen, an efficient radical scavenger. Also, the stereochemistry of the reduction, as studied by using NaBD$_4$ as the reducing agent, is consistent with a radical intermediate.[30] For example, a 50 : 50 mixture of *erythro*- and *threo*-3-deuterio-2-butanol is obtained by oxymercuration of either *cis*- or *trans*-2-butene, followed by reduction with NaBD$_4$:

28. F. G. Bordwell and M. L. Douglass, *J. Am. Chem. Soc.* **88**, 993 (1966); F. R. Jensen, J. J. Miller, S. J. Cristol, and R. S. Beckley, *J. Org. Chem.* **37**, 4341 (1972).
29. C. L. Hill and G. M. Whitesides, *J. Am. Chem. Soc.* **96**, 870 (1974).
30 D. J. Pasto and J. A. Gontarz, *J. Am. Chem. Soc.* **91**, 719 (1969); G. A. Gray and W. R. Jackson, *J. Am. Chem. Soc.* **91**, 6205 (1969).

Scheme 3.3. Regioselectivity in Formation of Alcohols via Oxymercuration[a]

Olefin	Alcohol composition[b]

$CH_3CH_2CH_2CH_2CH{=}CH_2 \rightarrow CH_3CH_2CH_2CH_2\underset{\underset{OH}{|}}{C}HCH_3 + CH_3CH_2CH_2CH_2CH_2OH$

(99.5 %)   (0.5 %)

$(CH_3)_3CCH{=}CH_2 \rightleftharpoons (CH_3)_3C\underset{\underset{OH}{|}}{C}HCH_3 + (CH_3)_3CCH_2CH_2OH$

(97 %)   (3 %)

$(CH_3)_3CCH{=}CHCH_3 \rightarrow (CH_3)_3C\underset{\underset{OH}{|}}{C}HCH_2CH_3 + (CH_3)_3CCH_2\underset{\underset{OH}{|}}{C}HCH_3$

(5 %)   (95 %)

$CH_3CH_2CH{=}CHCH_3 \rightarrow CH_3CH_2\underset{\underset{OH}{|}}{C}HCH_2CH_3 + CH_3CH_2CH_2\underset{\underset{OH}{|}}{C}HCH_3$

(44 %)   (56 %)

$Ph\underset{\underset{CH_3}{}}{\overset{\overset{CH_3}{|}}{C}}{=}CH_2 \rightarrow Ph\underset{\underset{OH}{|}}{C}(CH_3)_2$ (100 %)

a. All data from H. C. Brown and P. J. Geoghegan, Jr., *J. Org. Chem.* **35**, 1844 (1970).
b. All yields of alcohols exceeded 90%.

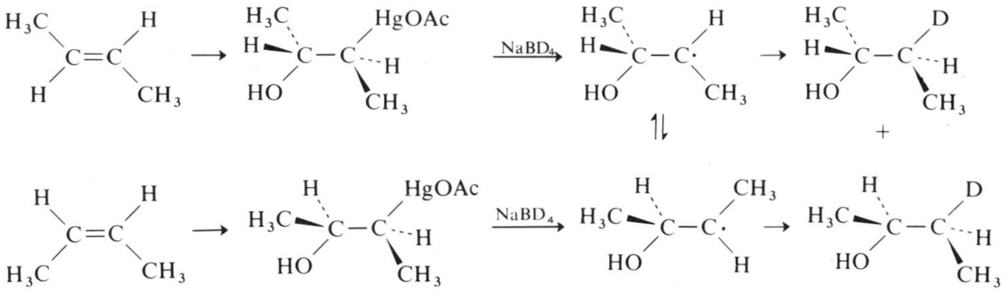

The overall oxymercuration–demercuration procedure is highly regioselective, with the nucleophile being introduced at the most highly substituted carbon:

$CH_3CH_2CH_2CH{=}CH_2 \xrightarrow[\text{THF}]{\text{1) Hg(OAc)}_2,\ H_2O} CH_3CH_2CH_2\underset{\underset{OH}{|}}{C}HCH_2HgOAc$

$CH_3CH_2CH_2\underset{\underset{OH}{|}}{C}HCH_2HgOAc \xrightarrow{\text{2) NaBH}_4} CH_3CH_2CH_2\underset{\underset{OH}{|}}{C}HCH_3$ Ref. 31

(97 %)

Scheme 3.3 reports the isomer ratios for some representative olefins. Synthetic

31. H. C. Brown and P. J. Geoghegan, Jr., *J. Org. Chem.* **35**, 1844 (1970).

90

CHAPTER 3
ADDITION
REACTIONS OF
CARBON–CARBON
MULTIPLE
BONDS

Scheme 3.4. Synthesis of Alcohols, Ethers, and Amides via Mercuration

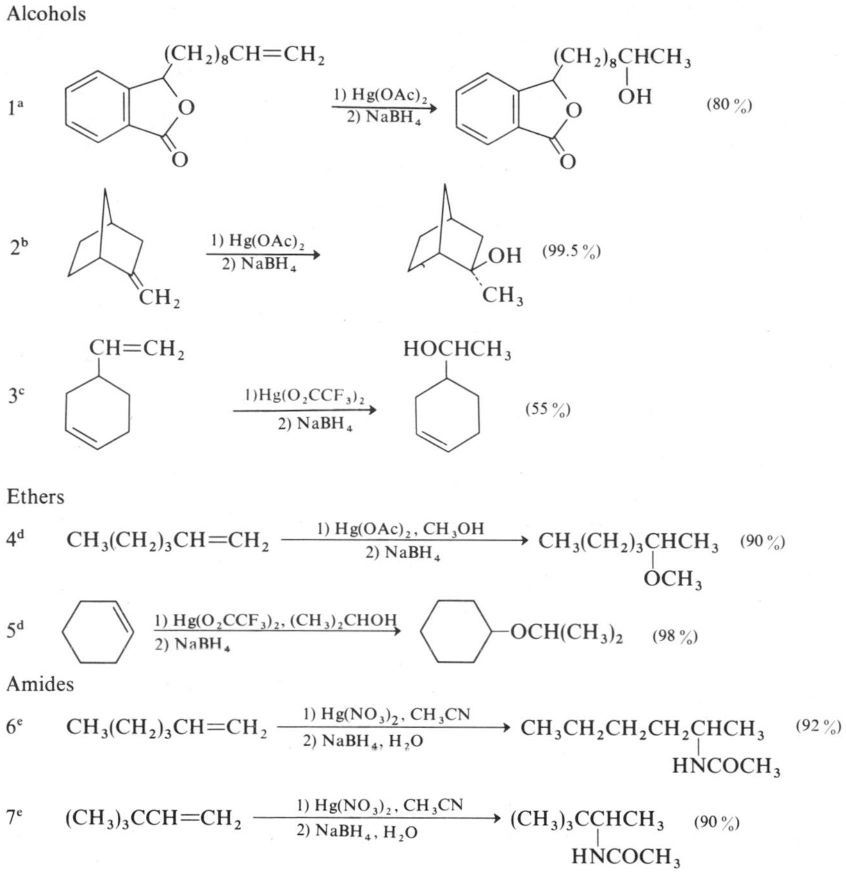

Alcohols

1^a (ring structure, $(CH_2)_8CH=CH_2$) $\xrightarrow[\text{2) NaBH}_4]{\text{1) Hg(OAc)}_2}$ (ring structure, $(CH_2)_8CHCH_3$, OH)   (80 %)

2^b (bicyclic structure, CH_2) $\xrightarrow[\text{2) NaBH}_4]{\text{1) Hg(OAc)}_2}$ (bicyclic structure, OH, CH_3)   (99.5 %)

3^c (cyclohexene with $CH=CH_2$) $\xrightarrow[\text{2) NaBH}_4]{\text{1)Hg(O}_2\text{CCF}_3)_2}$ (cyclohexene with $HOCHCH_3$)   (55 %)

Ethers

4^d $CH_3(CH_2)_3CH=CH_2$ $\xrightarrow[\text{2) NaBH}_4]{\text{1) Hg(OAc)}_2, \text{ CH}_3\text{OH}}$ $CH_3(CH_2)_3CHCH_3$, OCH_3 (90 %)

5^d (cyclohexene) $\xrightarrow[\text{2) NaBH}_4]{\text{1) Hg(O}_2\text{CCF}_3)_2, (CH_3)_2\text{CHOH}}$ (cyclohexane)–$OCH(CH_3)_2$ (98 %)

Amides

6^e $CH_3(CH_2)_3CH=CH_2$ $\xrightarrow[\text{2) NaBH}_4, \text{ H}_2\text{O}]{\text{1) Hg(NO}_3)_2, \text{ CH}_3\text{CN}}$ $CH_3CH_2CH_2CH_2CHCH_3$, $HNCOCH_3$ (92 %)

7^e $(CH_3)_3CCH=CH_2$ $\xrightarrow[\text{2) NaBH}_4, \text{ H}_2\text{O}]{\text{1) Hg(NO}_3)_2, \text{ CH}_3\text{CN}}$ $(CH_3)_3CCHCH_3$, $HNCOCH_3$ (90 %)

a. H. L. Wehrmeister and D. E. Robertson, *J. Org. Chem.* **33**, 4173 (1968).
b. H. C. Brown and W. J. Hammar, *J. Am. Chem. Soc.* **89**, 1524 (1967).
c. H. C. Brown, P. J. Geoghegan, Jr., G. J. Lynch, and J. T. Kurek, *J. Org. Chem.* **37**, 1941 (1972).
d. H. C. Brown and M.-H. Rei, *J. Am. Chem. Soc.* **91**, 5646 (1969).
e. H. C. Brown and J. T. Kurek, *J. Am. Chem. Soc.* **91**, 5648 (1969).

use of solvomercuration for synthesis of alcohols, ethers, and amides is illustrated by the reactions shown in Scheme 3.4.

3.5. Addition of Halogens to Olefins

The addition of chlorine and bromine to olefins is a very general reaction.

$$RCH=CHR + X_2 \rightarrow RCH-CHR$$
$$\qquad\qquad\qquad\quad X \quad\; X$$

Considerable insight has been gained into the mechanism of halogen-addition reactions by studies on the stereochemistry of the reaction. A number of classes of olefins are known to add bromine in a highly stereospecific manner, giving the products of *anti* addition. Among the olefins that are known to undergo *anti* addition are maleic and fumaric acid, *cis*-2-butene, *trans*-2-butene, and a number of cyclic olefins.[32] Cyclic, positively charged bromonium-ion intermediates offer an attractive

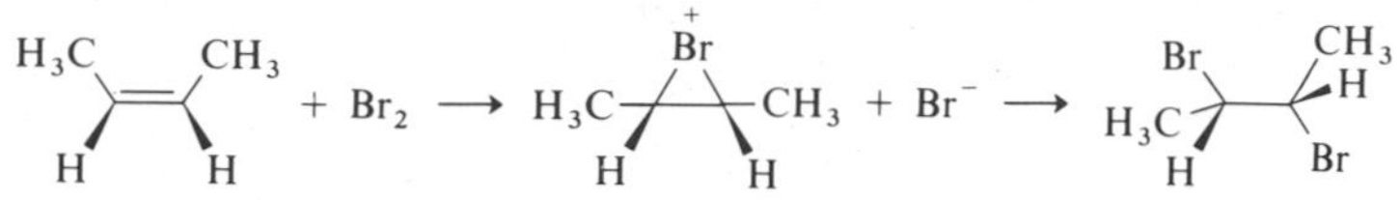

explanation for the observed stereospecificity. The bridging bromine atom prevents rotation around the bond linking the olefinic carbon atoms, and normal backside nucleophilic opening of the ring by bromide can lead to the observed *anti* addition. Physical evidence for the existence of bromonium ions has been obtained from nmr measurements.[33] A bromonium-ion salt (counterion, Br_3^-) has been isolated from the reaction of the olefin adamantylideneadamantane.[34]

Substantial amounts of *syn* addition have been observed for *cis*-1-phenylpropene (27–80% *syn* addition), *trans*-1-phenylpropene (17–29% *syn* addition), and *cis*-stilbene (up to 90% *syn* addition in polar solvents). A common feature

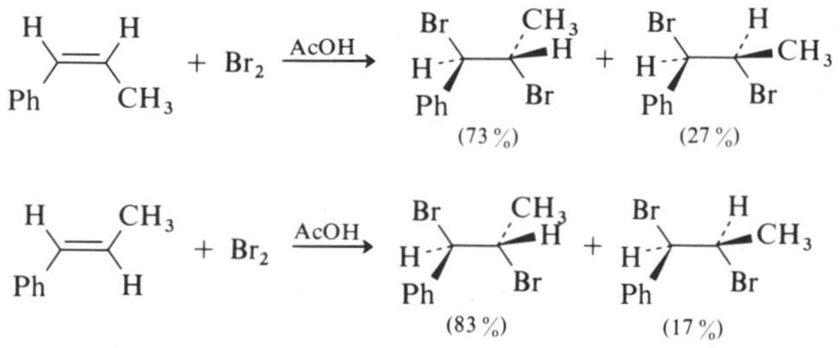

of the olefins that give extensive *syn* addition is the presence of a phenyl ring on at least one of the olefinic carbons. The phenyl substituent is believed to be responsible for the loss of stereospecificity. The presence of a phenyl substituent diminishes the necessity for strong bromonium-ion bridging, since it can effectively stabilize the positive charge. Weakened, unsymmetrical bridging can be depicted as follows:

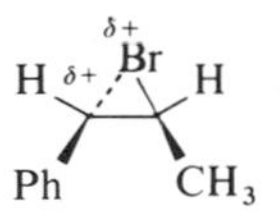

The intermediate has carbonium character at the phenyl-substituted carbon. The effect of this diminished bridging is to increase the rate of rotation around the central C–C bond. If such rotation takes place, *anti* stereospecificity is, of course, lost.

32. J. H. Rolston and K. Yates, *J. Am. Chem. Soc.* **91**, 1469, 1477 (1969).
33. G. A. Olah, J. M. Bollinger, and J. Brinich, *J. Am. Chem. Soc.* **90**, 2587 (1968); G. A. Olah, P. Schilling, P. W. Westerman, and H. C. Lin, *J. Am. Chem. Soc.* **96**, 3581 (1974).
34. J. Strating, J. H. Wieringa, and H. Wynberg, *Chem Commun.*, 907 (1969).

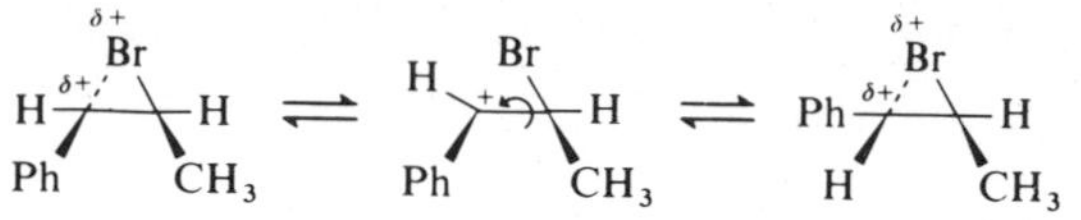

92

CHAPTER 3
ADDITION
REACTIONS OF
CARBON–CARBON
MULTIPLE
BONDS

Chlorination shows relatively less tendency to give *anti* addition products than does bromination. Although chlorination of aliphatic olefins gives largely *anti* addition, *syn* addition is often dominant for phenyl-substituted olefins[35,36]:

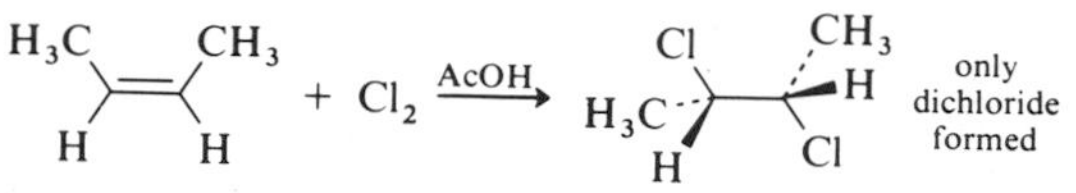

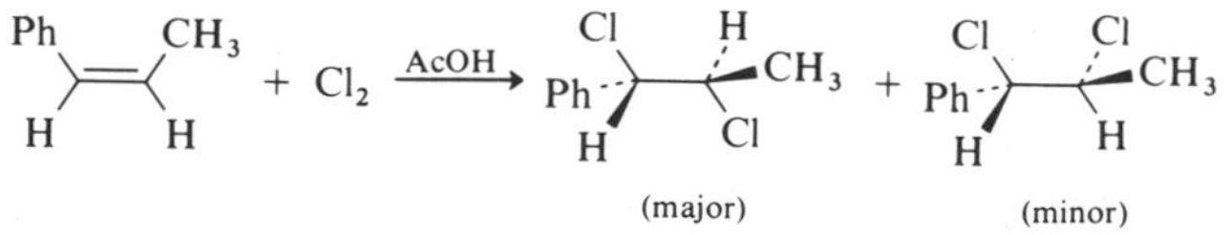

These results again reflect a difference in the extent of bridging in the intermediates. With aliphatic olefins, there is strong bridging with resulting high *anti* stereospecificity. Phenyl substitution leads to increased carbonium-ion character at the benzylic site, and the addition is far less stereospecific, showing a preference for *syn* addition. Because of its smaller polarizability, the chlorine atom is less effective than bromine at bridging in any particular olefin, and bromination therefore generally shows a higher degree of stereospecific addition than chlorination.[37]

Chlorination can be accompanied by reactions characteristic of carbonium-ion intermediates. Branched olefins can give products that are the result of elimination of a proton from a cationic intermediate[38]:

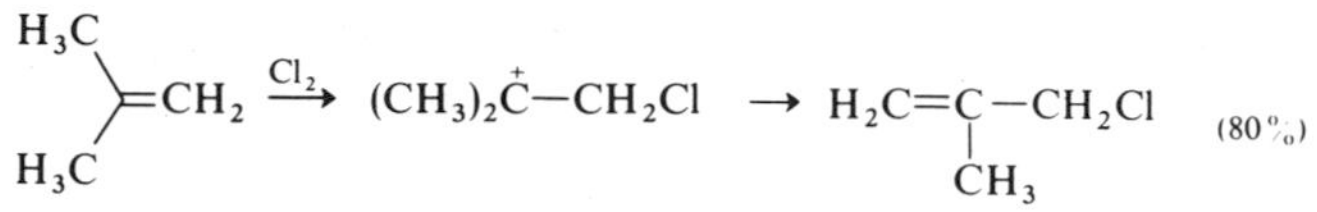

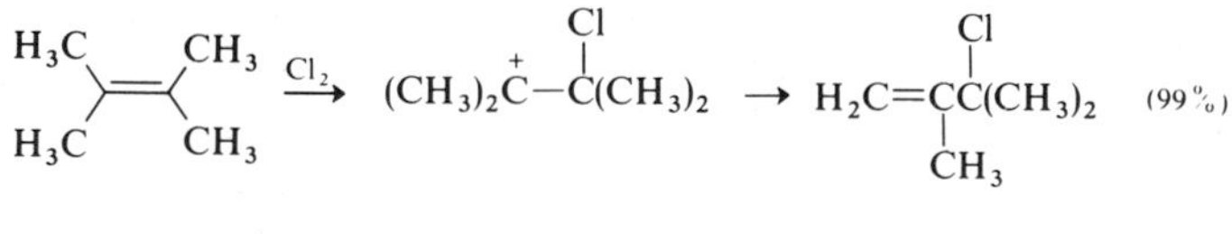

Skeletal rearrangement has also been observed, but is important only in systems that are very prone toward migration:

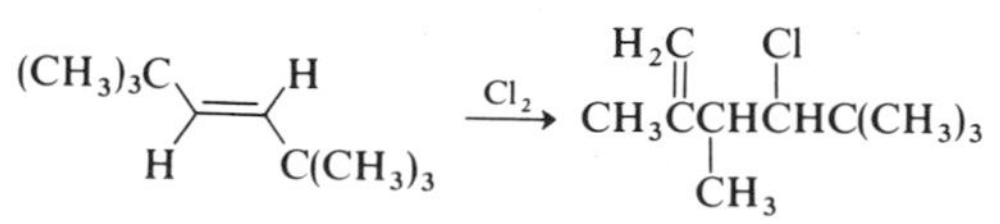

Ref. 36

35. M. L. Poutsma, *J. Am. Chem. Soc.* **87**, 2161, 2172 (1965).
36. R. C. Fahey, *J. Am. Chem. Soc.* **88**, 4681 (1966); R. C. Fahey and C. Schubert, *J. Am. Chem. Soc.* **87**, 5172 (1965).
37. R. J. Abraham and J. R. Monasterios, *J. Chem. Soc. Perkin Trans. I*, 1446 (1973).
38. M. L. Poutsma, *J. Am. Chem. Soc.* **87**, 4285 (1965).

$$Ph_3CCH=CH_2 \xrightarrow{Br_2} Ph_3CCHCH_2Br + Ph_2C=CCH_2Br \qquad \text{Ref. 39}$$
$$\underset{Br}{|} \qquad \underset{Ph}{|}$$

Since halogenation involves electrophilic attack on the olefin, substituent groups that increase the electron density increase the rate of reaction, whereas electron-withdrawing substituents have the opposite effect. Bromination of simple alkenes is extremely fast, and a trend toward increasing reactivity with increasing alkyl substitution is evident. Specific rate data are tabulated and discussed in Part A, Section 6.3.

In nucleophilic solvents, the solvent can compete with halide ion for the positively-charged intermediate. For example, the bromination of styrene in acetic acid leads to substantial amounts of the acetoxybromo derivative:

$$PhCH=CH_2 + Br_2 \xrightarrow{AcOH} PhCHCH_2Br + PhCHCH_2Br$$
$$\underset{\substack{Br \\ (80\%)}}{|} \qquad \underset{\substack{OAc \\ (20\%)}}{|}$$

The acetoxy group is introduced exclusively at the benzylic carbon. This is in accord with the picture of the intermediate bromonium ion as a weakly bridged species.

There is a preference for attack by the nucleophilic acetic acid at the benzylic end of the bromonium ion because the C–Br bond here is weaker, and the extent of positive charge greater, than at the other carbon of the bromonium-ion species. Addition of bromide salts to the reaction mixture diminishes the amount of acetoxy compound formed by tipping the competition between acetic acid and bromide ion for the electrophilic site in favor of the bromide ion by a concentration effect.

From a synthetic point of view, the participation of water in brominations, leading to bromohydrins, is probably the most important example of nucleophilic participation by solvent. In the case of an unsymmetrical bromonium ion, the water molecule will react at the carbon atom with the most pronounced carbonium-ion character. If it is desired to favor introduction of water, it is clear that the concentration of the competing bromide ion should be kept as low as possible. One method for

accomplishing this is to use N-bromosuccinimide as the bromine source. High yields of bromohydrins are realized under such conditions. Dimethyl sulfoxide is a

39. R. O. C. Norman and C. B. Thomas, *J. Chem. Soc. B*, 598 (1967).

94

CHAPTER 3
ADDITION
REACTIONS OF
CARBON–CARBON
MULTIPLE
BONDS

particularly effective solvent for the reaction.[40] The reaction exhibits a high degree of stereospecificity with predominant *anti* addition, ruling out a discrete carbonium-ion intermediate. As in normal brominations, a bridged bromonium ion is invoked to explain the *anti* stereospecificity. For the reaction in dimethyl sulfoxide, it has been shown that the initial nucleophilic attack is by the solvent DMSO, which subsequently reacts with water to give the bromohydrin. In accord with the Markownikoff rule, the hydroxyl group is introduced at the carbon best able to support positive charge.

$$
(CH_3)_3CC{=}CH_2 \xrightarrow[\substack{DMSO \\ H_2O}]{NBS} (CH_3)_3C\overset{\overset{\displaystyle CH_3}{|}}{C}{-}CH_2Br \quad (60\%)
$$
with OH below the C

$$
PhCH{=}CH_2 \xrightarrow[\substack{DMSO \\ H_2O}]{NBS} Ph\overset{}{C}H{-}CH_2Br \quad (76\%)
$$
with OH below CH

$$
PhCH{=}CHCH_3 \xrightarrow[\substack{DMSO \\ H_2O}]{NBS} PhCH{-}CHCH_3 \quad (92\%)
$$
with OH Br below

Chlorinations in nucleophilic solvents can also lead to solvent incorporation, as for example in the chlorination of phenylpropene in methanol[36,41]:

$$
PhCH{=}CHCH_3 + Cl_2 \xrightarrow{CH_3OH} PhCHCHCH_3 + PhCH{-}CHCH_3
$$
with CH$_3$O Cl (82%) and Cl Cl (18%) below

Addition of iodine to olefins can be accomplished by a photochemically initiated reaction. Elimination of iodine is catalyzed by excess iodine radicals, but the diiodo

$$
RCH{=}CHR + I_2 \rightleftharpoons RCH{-}CHR
$$
with I I below

compounds can be obtained if unreacted iodine is removed.[42] The diiodo compounds are very sensitive to light and have not been used very often in synthesis.

Conjugated dienes often give mixtures of halogen-addition products. The case of isoprene can serve as an example.[43] The first two products are examples of "1,2-addition"—i.e., both halogens are introduced at adjacent carbons—whereas the other two are "1,4-addition products," the bromine atoms having been added at the extremities of the conjugate system. Electrophilic attack by bromine on a conjugated diene generates an intermediate with the character of an allylic cation. The stability of the allylic system presumably diminishes the requirement for strong bromine

40. D. R. Dalton, V. P. Dutta, and D. C. Jones, *J. Am. Chem. Soc.* **90**, 5498 (1968); G. O. Guss and R. Rosenthal, *J. Am. Chem. Soc.* **77**, 2549 (1955).
36. See p. 92.
41. M. L. Poutsma and J. L. Kartch, *J. Am. Chem. Soc.* **89**, 6595 (1967).
42. P. S. Skell and R. R. Pavlis, *J. Am. Chem. Soc.* **86**, 2956 (1964); R. L. Ayres, C. J. Michejda, and E. P. Rack, *J. Am. Chem. Soc.* **93**, 1389 (1971).
43. V. L. Heasley, C. L. Frye, R. T. Gore, Jr., and P. S. Wilday, *J. Org. Chem.* **33**, 2342 (1968).

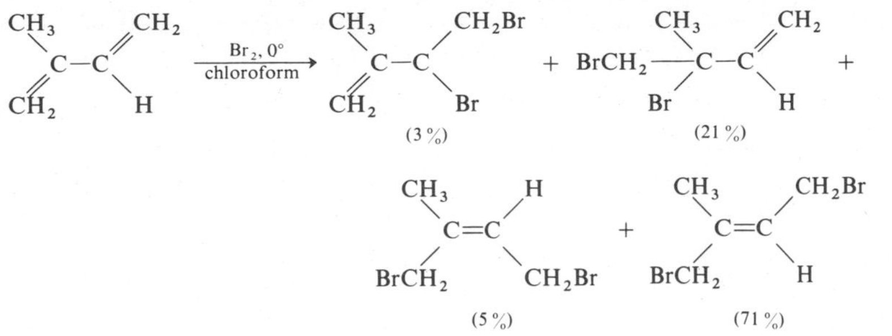

(3%)   (21%)

(5%)   (71%)

bridging. Nucleophilic attack by bromide ion can occur at either of the positive centers in the allylic system, leading to the mixture of 1,2- and 1,4-addition products.

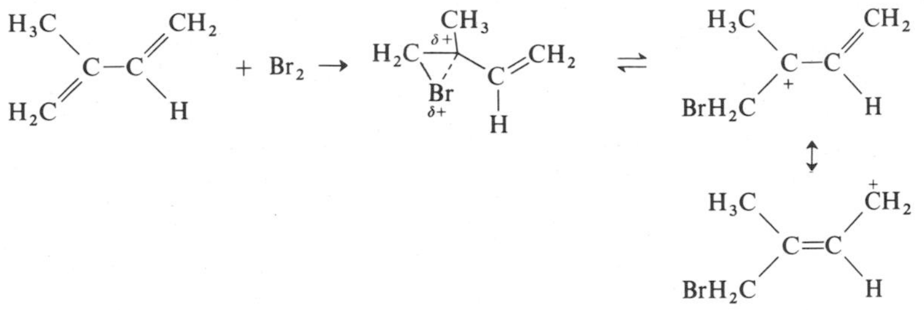

Related reactions that have been studied include the chlorination[44] and bromination[45] of butadiene. Both reactions give mixtures containing slightly more of the 1,2-addition product than the 1,4-addition product.

3.6. Addition of Other Electrophilic Reagents

Many other small molecules react with olefins to give addition products. Those that are believed to occur by mechanisms similar to halogenation are discussed here. Table 3.5 lists some of the more important electrophilic reagents. The different reagents show varying behavior with respect to both stereochemistry and regioselectivity. Attack on the olefin is initiated by the positive portion of the reagent, and for all of the examples in Table 3.5, bridged intermediates are believed to be involved. Bridging may be symmetrical or unsymmetrical, depending on the ability of the carbon atoms to accommodate positive charge. As noted in Table 3.5, the direction of opening of the bridged intermediate is usually governed by electronic factors. That is, the addition is completed by attack of the nucleophile at the most positive carbon atom of the bridged intermediate. When, however, the bridging is very strong and most of the positive charge is centered on the heteroatom, as when X = S, steric factors outweigh the small electronic differences between the two carbon atoms, and

44. M. L. Poutsma, *J. Org. Chem.* **31**, 4167 (1966).
45. V. L. Heasley and S. K. Taylor, *J. Org. Chem.* **34**, 2779 (1969).

96

CHAPTER 3
ADDITION
REACTIONS OF
CARBON–CARBON
MULTIPLE
BONDS

Table 3.5. Addition of Electrophilic Reagents to Olefins

Reagent	Control of regioselectivity	Stereospecificity in aliphatic systems	Product	Ref.
$\delta^+ \mid \delta^-$ I┼N=C=O	electronic (moderate)	*anti* (high)	RCH—CHR $\mid$ $\quad$ $\mid$ I $\quad$ NCO	a
Br┼N=N=N	electronic (moderate)	*anti*	RCH—CHR $\mid$ $\quad$ $\mid$ Br $\quad$ N$_3$	b
I┼N=N=N	electronic (high)	*anti* (high)	RCH—CHR $\mid$ $\quad$ $\mid$ I $\quad$ N$_3$	c
O=N┼Cl	electronic	?	RC—CHR $\parallel$ $\quad$ $\mid$ HON $\quad$ Cl	d
O=N┼O$_2$CH	electronic	*anti*	RC—CHR $\parallel$ $\quad$ $\mid$ HON $\quad$ O$_2$CH	e
RS┼Cl	steric (high)	*anti* (high)	RCH—CHR $\mid$ $\quad$ $\mid$ RS $\quad$ Cl	f
NCS┼SCN	electronic (high)	*anti*	RCH—CHR $\mid$ $\quad$ $\mid$ NCS $\quad$ SCN	g

a. A. Hassner, R. P. Hoblitt, C. Heathcock, J. E. Kropp, and M. Lorber, *J. Am. Chem. Soc.* **92**, 1326 (1970); A. Hassner, M. E. Lorber, and C. Heathcock, *J. Org. Chem.* **32**, 540 (1967).
b. A. Hassner, F. P. Boerwinkle, and A. B. Levy, *J. Am. Chem. Soc.* **92**, 4879 (1970).
c. F. W. Fowler, A. Hassner, and L. A. Levy, *J. Am. Chem. Soc.* **89**, 2077 (1967).
d. J. Meinwald, Y. C. Meinwald, and T. N. Baker, III, *J. Am. Chem. Soc.* **86**, 4074 (1967); P. P. Kadzyauskas and N. S. Zefirov, *Russ. Chem. Rev.* **37**, 543 (1968).
e. H. C. Hamann and D. Swern, *J. Am. Chem. Soc.* **90**, 6481 (1968).
f. W. H. Mueller and P. E. Butler, *J. Am. Chem. Soc.* **90**, 2075 (1968); W. A. Thaler, *J. Org. Chem.* **34**, 871 (1969).
g. E. J. Corey, F. A. Carey, and R. A. E. Winter, *J. Am. Chem. Soc.* **87**, 934 (1965); D. J. Pettit and G. K. Helmkamp, *J. Org. Chem.* **29**, 2702 (1964).

ease of approach by the nucleophile (steric control) becomes dominant. Put another way, the case of "electronic control" corresponds to the direction of addition being governed by ease of bond-breaking; i.e., at the transition state, bond-breaking is more complete than bond-making. In the case of "steric control," bond-making is more complete in the transition state and opening takes place in the direction in which bond-making is more easily accomplished. Figure 3.2 illustrates electronic versus steric control of ring-opening. The general analogy between factors governing electronic versus steric control of ring-opening and the factors governing nucleophilic substitution by the S_N1 and S_N2 mechanisms should be recognized.

Addition via bridged intermediates is characterized by *anti* stereochemistry, since the nucleophile opens the ring by backside attack:

Electronic control of ring-opening—nucleophilic attack at more positive carbon

Steric control of ring-opening—nucleophilic attack at less substituted carbon

Figure 3.2. Electronic versus steric control of ring-opening.

In cyclic olefins, addition occurs by diaxial addition:

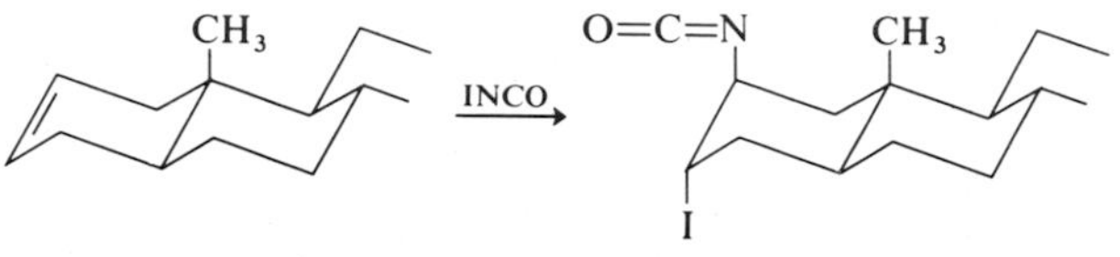

Ref. 46

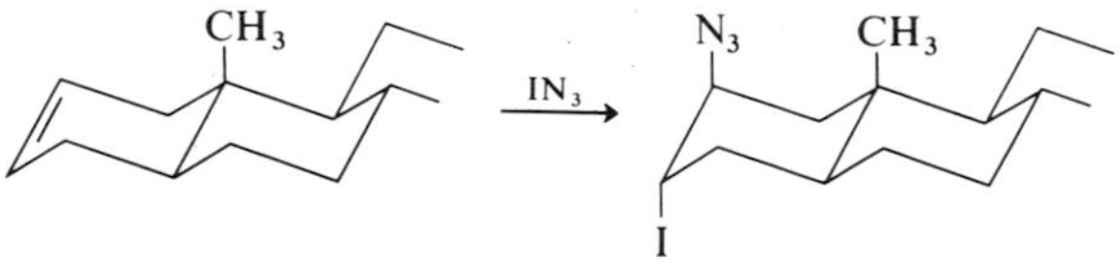

Ref. 47

The addition of nitrosyl chloride to olefins is accompanied by subsequent reactions if the nitroso group is not tertiary. The nitroso compound may dimerize or rearrange to the more stable oxime tautomer.

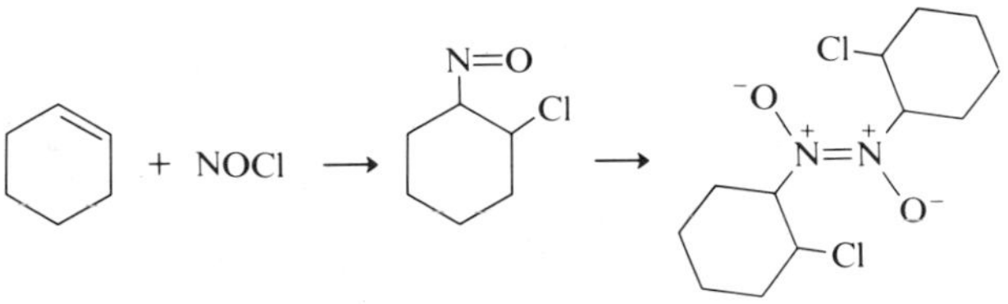

Ref. 48

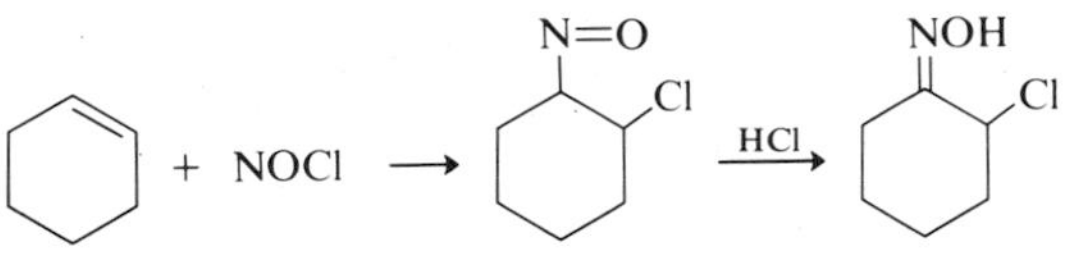

Ref. 49

Hydrolysis of the nitrosyl chloride adducts can provide a convenient route to

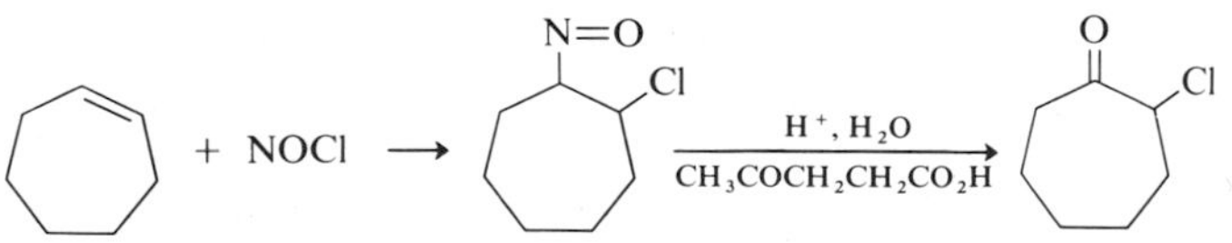

Ref. 50

46. A. Hassner and C. Heathcock, *J. Org. Chem.* **30**, 1748 (1965).
47. A. Hassner and F. Boerwinkle, *J. Am. Chem. Soc.* **90**, 216 (1968).
48. B. W. Ponder, T. E. Walton, and W. J. Pollock, *J. Org. Chem.* **33**, 3957 (1968).
49. M. Ohno, N. Naruse, S. Torimitsu, and M. Okamoto, *Bull. Chem. Soc. Jpn.* **39**, 1119 (1966).
50. J. Meinwald, Y. C. Meinwald, and T. N. Baker, III, *J. Am. Chem. Soc.* **86**, 4074 (1964); B. W. Ponder and D. R. Walker, *J. Org. Chem.* **32**, 4136 (1967).

98

CHAPTER 3
ADDITION
REACTIONS OF
CARBON–CARBON
MULTIPLE
BONDS

α-chloroketones. Hydrolysis presumably proceeds via the oxime tautomer of the monomeric nitroso compound. Levulinic acid is used to react with the hydroxylamine formed in the oxime hydrolysis. The stereochemistry of nitrosyl chloride addition has not been investigated for a wide variety of olefins. Both *syn* and *anti* addition have been reported in systems studied to date.

Phenylselenyl halides, phenylselenyl trifluoroacetate, and phenylselenyl acetate add to alkenes[51,52]:

$$\text{RCH=CHR} \xrightarrow{\text{PhSeBr}} \underset{\overset{|}{\text{Br}}}{\overset{\overset{\text{SePh}}{|}}{\text{RCH—CHR}}} \xrightarrow{\text{RCO}_2\text{H}}$$

$$\underset{\overset{|}{\text{RCO}_2}}{\overset{\overset{\text{SePh}}{|}}{\text{RCHCHR}}}$$

$$\text{RCH=CHR} \xrightarrow{\text{PhSeO}_2\text{CR}}$$

These adducts are synthetically useful because they readily undergo elimination when the selenium is oxidized to the selenoxide state:

$$\underset{\overset{|}{\text{CH}_3\text{CO}_2}}{\overset{\overset{\text{SePh}}{|}}{\text{CH}_3(\text{CH}_2)_2\text{CHCH}(\text{CH}_2)_2\text{CH}_3}} \xrightarrow{\text{H}_2\text{O}_2} \underset{\overset{|}{\text{CH}_3\text{CO}_2}}{\text{CH}_3(\text{CH}_2)_2\text{CHCH=CHCH}_2\text{CH}_3} \qquad \text{Ref. 51}$$

$$(85\%)$$

The overall process therefore accomplishes oxidation with a double-bond shift.

3.7. Electrophilic Substitution Alpha to Carbonyl Groups

Although the reaction of ketones and other carbonyl compounds with electrophiles such as bromine is formally a substitution process, it is mechanistically closely related to electrophilic additions to alkenes. The enol or enolate derived from the carbonyl compound is the reactive species, and the initial attack is similar to the electrophilic attack on alkenes. The reaction is completed by restoration of the carbonyl bond, rather than by addition. The acid- and base-catalyzed halogenations of ketones, which were discussed briefly in Part A, Chapter 7, are the best-studied examples of the reaction. The most common preparative procedures for such

$$\underset{\overset{\text{O}}{\overset{||}{\text{R}_2\text{CHCR}'}}}{} \underset{\xrightarrow{\text{H}^+}}{\rightleftharpoons} \underset{\overset{\text{OH}}{\overset{|}{\text{R}_2\text{C=CR}'}}}{} \xrightarrow{\text{Br}_2} \underset{\underset{\text{Br—Br}}{\overset{\text{OH}}{\overset{|}{\text{R}_2\text{C=CR}'}}}}{} \rightarrow \underset{\overset{|}{\text{Br}}}{\overset{\overset{\text{O}}{\overset{||}{\text{R}_2\text{CCR}'}}}}{}$$

$$\underset{\overset{\text{O}}{\overset{||}{\text{R}_2\text{CHCR}'}}}{} \underset{\xrightarrow{-\text{OH}}}{\rightleftharpoons} \underset{\overset{\text{O}^-}{\overset{|}{\text{R}_2\text{C=CR}'}}}{} \xrightarrow{\text{Br}_2} \underset{\underset{\text{Br—Br}}{\overset{\text{O}^-}{\overset{|}{\text{R}_2\text{C=CR}'}}}}{} \rightarrow \underset{\overset{|}{\text{Br}}}{\overset{\overset{\text{O}}{\overset{||}{\text{R}_2\text{CCR}'}}}}{}$$

51. K. B. Sharpless and R. F. Lauer, *J. Org. Chem.* **39**, 429 (1974).
52. H. J. Reich, *J. Org. Chem.* **39**, 428 (1974).

reactions involve use of the halogen in acetic acid. Other available halogenating agents include *N*-bromosuccinimide and sulfuryl chloride:

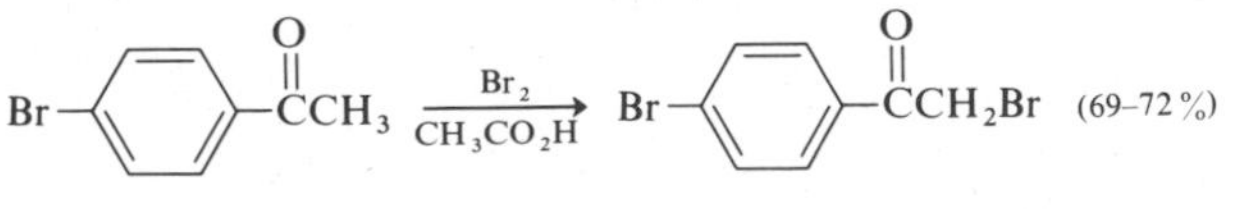

99

SECTION 3.7.
ELECTROPHILIC
SUBSTITUTION
ALPHA TO
CARBONYL
GROUPS

Ref. 53

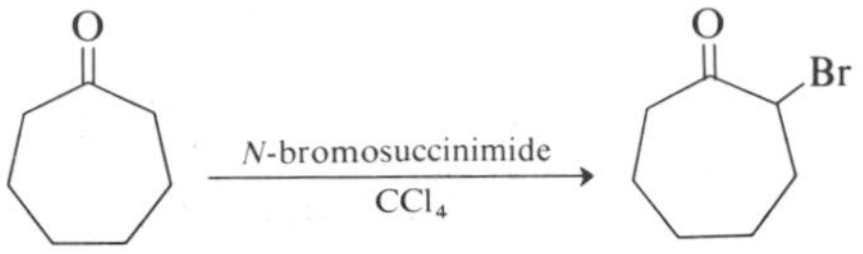

Ref. 54

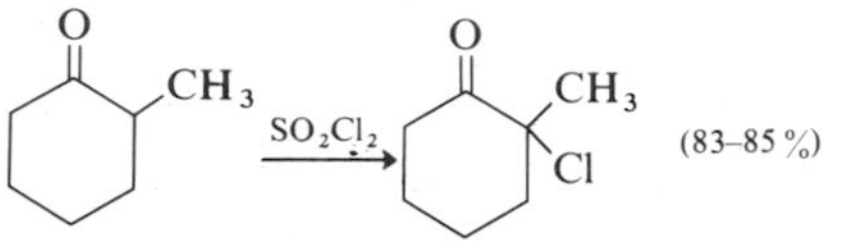

Ref. 55

Since the reactions involving bromine or chlorine evolve the hydrogen halide, they are autocatalytic. In reactions with *N*-bromosuccinimide, no hydrogen bromide is formed, so this method may be preferable in the case of acid-sensitive compounds.

As was pointed out in Part A, Section 7.3, under many conditions halogenation is fast, relative to enolization. When this is true, the position of substitution in unsymmetrical ketones is governed by the relative rates of formation of the isomeric enols. In general, mixtures are formed with unsymmetrical ketones. The presence of a halogen substituent decreases the rate of enolization and retards the rate of introduction of a second halogen on a carbon. Monohalogenation can therefore usually be carried out satisfactorily in acidic solution. Base-catalyzed halogenations tend to proceed to polyhalogenated products. The use of cupric chloride[56] and cupric bromide[57] in organic solvents such as chloroform is also an efficient method for monohalogenation of ketones:

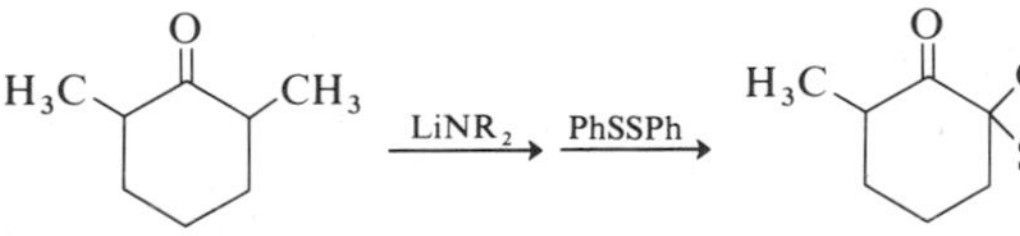

Sulfur substituents can be introduced α to carbonyl groups by reaction of the enolate with a disulfide. This sulfenylation has been carried out successfully for esters,[59] ketones,[59] and the dianions of carboxylic acids.[60] These α-substituted

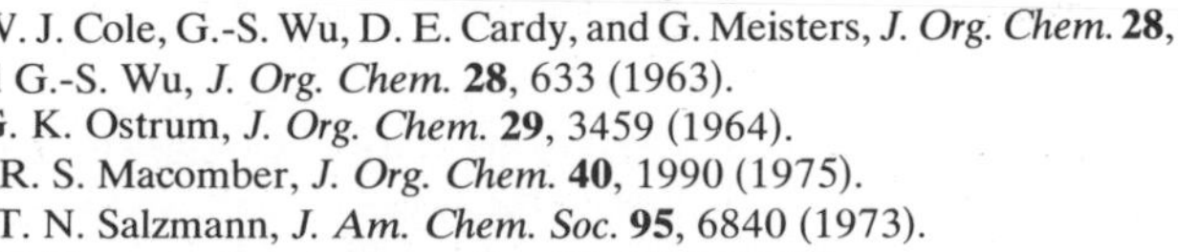

Ref. 59

53. W. D. Langley, *Org. Synth.* **1**, 127 (1944).
54. E. J. Corey, *J. Am. Chem. Soc.* **75**, 2301 (1954).
55. E. W. Warnhoff, D. G. Martin, and W. S. Johnson, *Org. Synth.* **IV**, 162 (1963).
56. E. M. Kosower, W. J. Cole, G.-S. Wu, D. E. Cardy, and G. Meisters, *J. Org. Chem.* **28**, 630 (1963); E. M. Kowower and G.-S. Wu, *J. Org. Chem.* **28**, 633 (1963).
57. L. C. King and G. K. Ostrum, *J. Org. Chem.* **29**, 3459 (1964).
58. D. P. Bauer and R. S. Macomber, *J. Org. Chem.* **40**, 1990 (1975).
59. B. M. Trost and T. N. Salzmann, *J. Am. Chem. Soc.* **95**, 6840 (1973).
60. B. M. Trost and Y. Tamaru, *J. Am. Chem. Soc.* **97**, 3528 (1975).

compounds have been used to introduce unsaturation in conjugation with the carbonyl group. The sulfides are oxidized to sulfoxides, which undergo thermolysis to the α,β-unsaturated carbonyl compounds:

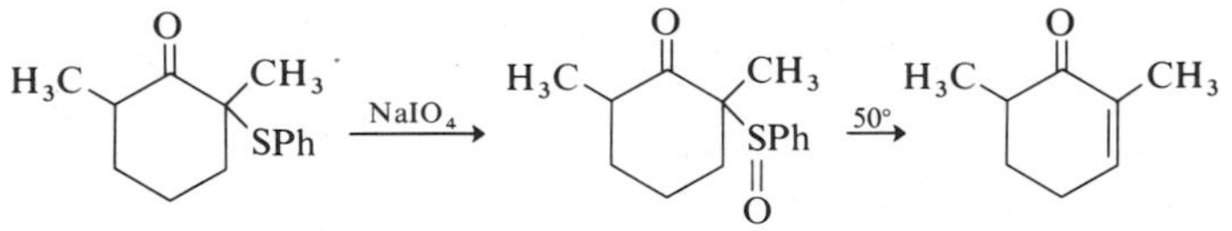

The procedure has been applied to both ketones and esters.

There are similar procedures involving organic selenides. Reaction of ketones and aldehydes with phenylselenyl chloride occurs readily to give the α-seleno derivative, presumably by a mechanism similar to α-halogenation. Alternatively, ketones and esters can be converted to the α-seleno derivatives by reaction of the enolates with phenylselenyl chloride. The α-seleno compounds serve as precursors to α,β-unsaturated carbonyl compounds because oxidation with hydrogen peroxide leads to spontaneous elimination of PhSeOH.

$$CH_3CH_2CH_2\overset{O}{\overset{\|}{C}}CH_2CH_2CH_3 \xrightarrow{PhSeCl} CH_3CH_2\underset{PhSe}{\underset{|}{C}H}\overset{O}{\overset{\|}{C}}CH_2CH_2CH_3 \longrightarrow$$

$$\xrightarrow{H_2O_2} CH_3CH{=}CH\overset{O}{\overset{\|}{C}}CH_2CH_2CH_3 \quad (64\%) \qquad \text{Ref. 61}$$

$$PhCH_2CH_2CO_2C_2H_5 \xrightarrow[2)\ PhSeCl]{1)\ LiNR_2} PhCH_2\underset{PhSe}{\underset{|}{C}H}CO_2C_2H_5 \xrightarrow{H_2O_2} PhCH{=}CHCO_2C_2H_5 \quad (80\%)$$

$$Ph\overset{O}{\overset{\|}{C}}CH_2CH_2CH_2Ph \xrightarrow[2)\ PhSeBr]{1)\ LiNR_2} Ph\overset{O}{\overset{\|}{C}}\underset{SePh}{\underset{|}{C}H}CH_2CH_2Ph \xrightarrow{H_2O_2} Ph\overset{O}{\overset{\|}{C}}CH{=}CHCH_2Ph \quad (84\%) \qquad \text{Ref. 62}$$

3.8. Hydroboration

The addition of diborane and substituted boranes to carbon–carbon double bonds is the basis of a very important method of synthesis of alcohols. The addition process is called *hydroboration*. This reaction was discovered relatively recently (during the 1950's) by H. C. Brown. Subsequently, other procedures for the utilization of organoborane intermediates have been developed by Brown. The hydroboration reaction involves addition in a single step. Diborane or substituted boranes can provide the B–H bonds. Except in the case of very sterically hindered

61. K. B. Sharpless, R. F. Lauer, and A. Y. Teranishi, *J. Am. Chem. Soc.* **95**, 6137 (1973).
62. H. J. Reich, I. L. Reich, and J. M. Renga, *J. Am. Chem. Soc.* **95**, 5813 (1973).

olefins, the reaction proceeds rapidly until all B–H bonds have reacted, providing trialkylboranes. Diborane can be generated *in situ* from sodium borohydride and boron trifluoride. Solutions in tetrahydrofuran are commercially available. An

$$3NaBH_4 + 4BF_3 \rightarrow 2B_2H_6 + 3NaBF_4$$

alternative commercially available reagent is the borane–dimethyl sulfide complex.

$$H_3\bar{B}-\overset{+}{S}(CH_3)_2$$

It is more amenable to storage over extended periods than diborane and exhibits comparable reactivity toward alkenes[63]:

$$CH_3CH_2CH_2CH{=}CH_2 + B_2H_6 \rightarrow (CH_3CH_2CH_2CH_2CH_2)_3B$$

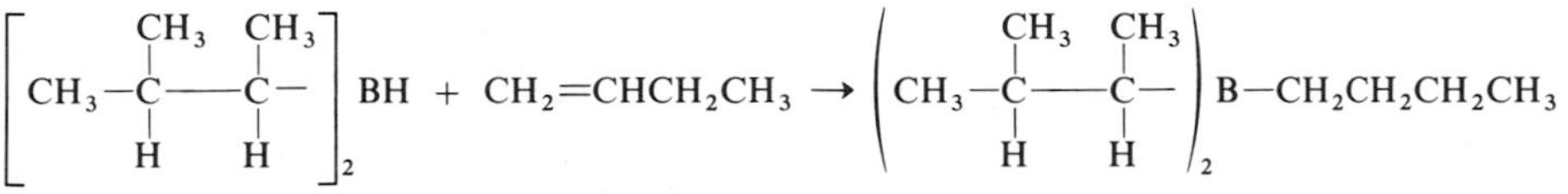

Studies with many olefins have shown that the addition of the B–H bond is highly selective both with respect to orientation (regioselectivity) and with respect to the stereochemistry of addition. The boron becomes bonded primarily to the *less substituted* carbon atom of the olefin. A combination of steric and electronic effects works together in favoring this orientation. Boron is less electronegative than hydrogen. Any positive charge that develops in the transition state is best accommo-

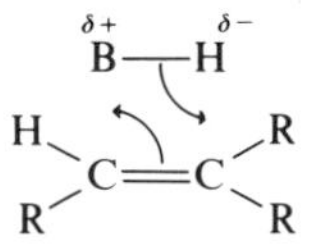

dated by the more highly substituted end of the multiple bond. Markownikoff's rule is thus followed in the addition, but, in contrast to the common examples of addition of acids to olefins, the hydrogen is *not* the most positive position in the attacking reagent. Steric factors also favor addition of the boron atom to the less substituted end of the multiple bond. In the final trialkylborane, considerable steric interferences develop if the alkyl chains are highly branched. These unfavorable steric effects are minimized by addition of the boron to the less substituted carbon. With electronic

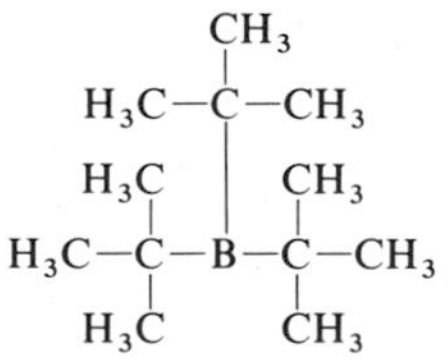

severe nonbonded   nonbonded repulsions
repulsions   reduced

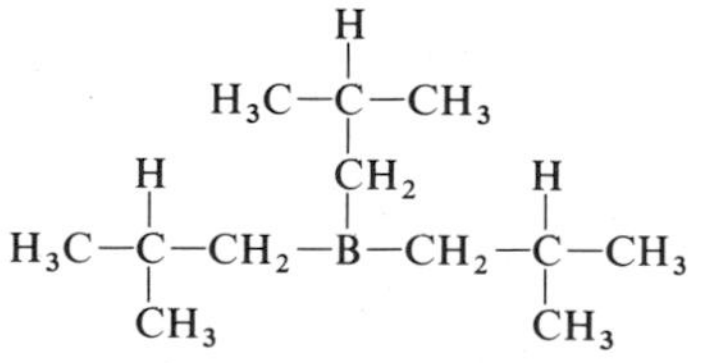

63. L. M. Braun, R. A. Braun, H. R. Crissman, M. Opperman, and R. M. Adams, *J. Org. Chem.* **36**, 2388 (1971); C. F. Lane, *J. Org. Chem.* **39**, 1437 (1974).

102

CHAPTER 3
ADDITION
REACTIONS OF
CARBON–CARBON
MULTIPLE
BONDS

and steric effects operating in the same direction in the addition reaction, high and predictable selectivity is consistently observed. Table 3.6 provides several examples of systems in which the regioselectivity of the addition has been measured.

It would be expected that the steric directing effect would become successively more important as the boron atom becomes substituted and therefore of higher steric requirement. It is possible to use this effect to enhance the selectivity of hydroborations. The hydroborations of certain hindered olefins can be terminated at intermediate stages by using the proper ratio of reactants. If these substituted boranes are

$$2\,(CH_3)_2C=CHCH_3 \ + \ BH_3 \ \rightarrow \ \left(\overset{\overset{\displaystyle CH_3}{|}}{(CH_3)_2CHCH-}\right)_2 BH$$

$$(CH_3)_2C=C(CH_3)_2 \ + \ BH_3 \ \rightarrow \ (CH_3)_2CH\overset{\overset{\displaystyle CH_3}{|}}{\underset{\underset{\displaystyle CH_3}{|}}{C}}-BH_2$$

then used to hydroborate other olefins, enhanced selectivity is observed. For example, when 1-hexene is hydroborated by diborane, oxidation of the borane gives 94% 1-hexanol and 6% 2-hexanol. Hydroboration with bis(3-methyl-2-butyl)borane followed by oxidation gives 99% 1-hexanol and only 1% 2-hexanol. Similar trends are observed with other olefins. Styrene, for example, gives 20% 1-phenylethanol and 80% 2-phenylethanol when treated with diborane and then oxidized. Use of bis(3-methyl-2-butyl)borane changes the ratio to 98% 2-phenylethanol and only 2% 1-phenylethanol. A bicyclic dialkylborane available

$$\left(\overset{\overset{\displaystyle CH_3}{|}}{(CH_3)_2CHCH-}\right)_2 BH \ + \ PhCH=CH_2 \ \longrightarrow \ \underset{OH^-}{\overset{H_2O_2}{\longrightarrow}} \ PhCH_2CH_2OH$$
$$(98\%)$$

from hydroboration of 1,5-cyclooctadiene shows a similarly enhanced selectivity in hydroboration of other olefins.[64] This reagent is commonly referred to as 9-BBN, an abbreviation derived from the systematic name 9-borabicyclo[3.3.1]nonane:

$$CH_3(CH_2)_4CH_2OH \ + \ CH_3(CH_2)_3\overset{\overset{\displaystyle }{}}{\underset{\underset{\displaystyle OH}{|}}{C}}HCH_3 \qquad\qquad (CH_3)_2CHCH_2CHCH_3 \ + \ (CH_3)_2CHCHCH_2CH_3$$

CH₃(CH₂)₄CH₂OH + CH₃(CH₂)₃CHCH₃   (CH₃)₂CHCH₂CHCH₃ + (CH₃)₂CHCHCH₂CH₃
|OH |OH |OH
(99%)   (1%)   (97%)   (3%)

64. E. F. Knights and H. C. Brown, *J. Am. Chem. Soc.* **90**, 5281 (1968); C. G. Scouten and H. C. Brown, *J. Org. Chem.* **38**, 4092 (1973).

Table 3.6 Orientation in Hydroboration Reactions[a]

7%
93%
$CH_3CH_2CH{=}CH_2$
43%
57%
$(CH_3)_2CHCH{=}CHCH_3$
1%
99%
$CH_3CH_2C{=}CH_2$
CH_3
0%
100%
$C{=}CH_2$
CH_3
2%
98%
$(CH_3)_2C{=}CHCH(CH_3)_2$

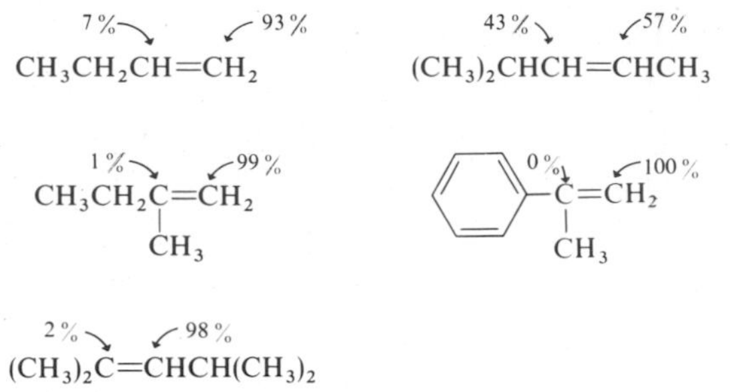

a. From G. Zweifel and H. C. Brown, *Org. React.* **13**, 1 (1963).

The hydroboration reaction is also very predictable with regard to the stereochemistry of addition. The addition occurs stereospecifically *syn* through a four-center transition state with essentially simultaneous bonding to boron and hydrogen. Both the new C–B and C–H bonds are therefore formed from the same side of the multiple bond. In molecular orbital terms, the addition reaction is viewed as taking place by interaction of the olefin π-orbital with the empty p-orbital on trivalent boron. Formation of the carbon–boron bond is accompanied by concerted rupture of a B–H bond[65]:

H—B
B
H
H—B
H
B

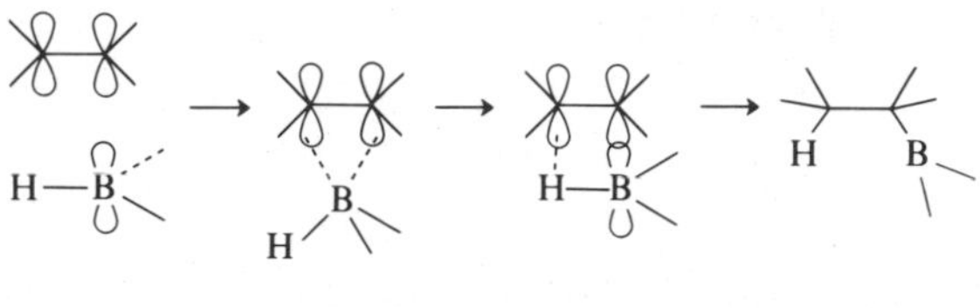

Scheme 3.5 records some examples that illustrate the strong preference for *syn* addition from the less hindered side of the molecule.

Hydroboration is thermally reversible. At 160° and above, B–H moieties are eliminated from alkylboranes, but in this temperature range the equilibrium is still in favor of the addition. These reversible additions lead to migration of boron to the

R
R—C—CH—CH$_3$
H
B
R
R—C=CH—CH$_3$
+
H—B
R H
R—C—C=CH$_2$
H
H—B
R
R—C—CH—CH$_2$—B
H

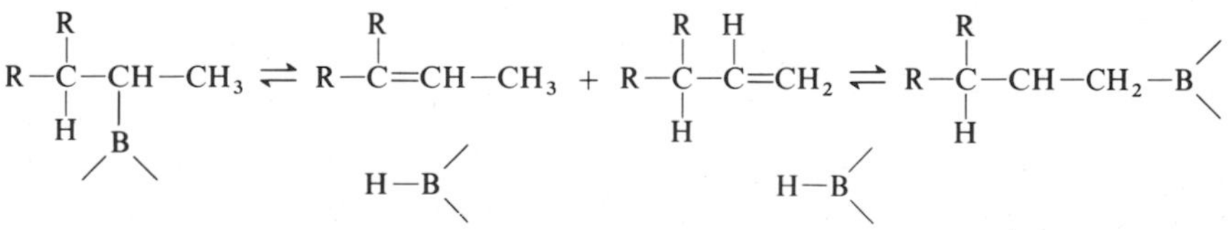

least substituted carbon atom by a series of eliminations and additions. Migration

65. D. J. Pasto, B. Lepeska, and T.-C. Cheng, *J. Am. Chem. Soc.* **94**, 6083 (1972); P. R. Jones, *J. Org. Chem.* **37**, 1886 (1972).

104

CHAPTER 3
ADDITION
REACTIONS OF
CARBON–CARBON
MULTIPLE
BONDS

Scheme 3.5. Stereochemistry of Hydroboration–Oxidation

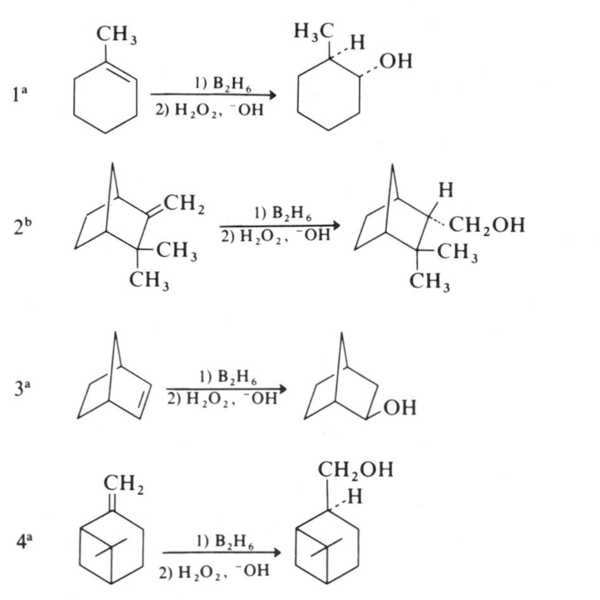

a. H. C. Brown and G. Zweifel, *J. Am. Chem. Soc.* **83**, 2544 (1961).
b. R. Dulou and Y. Chrétien-Bessière, *Bull. Soc. Chim. Fr.*, 1362 (1959).

cannot occur past a fully substituted carbon, however, since the required elimination is blocked. Some examples of thermal isomerizations of boranes are shown below:

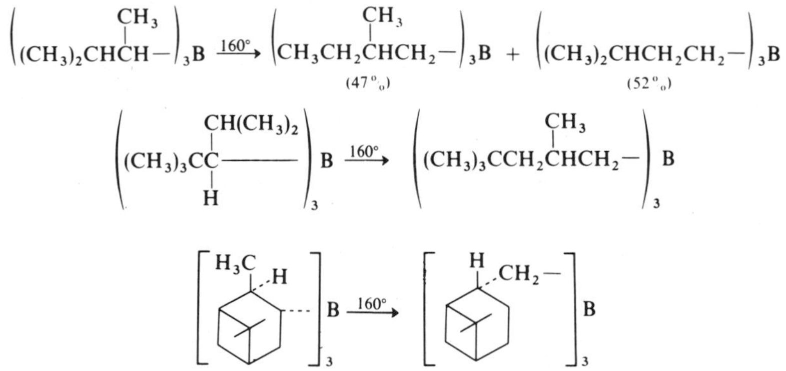

The organoboranes formed by hydroboration or isomerization are of interest in organic synthesis because of the subsequent reactions they undergo. Early work showed that the boron atom could be replaced by hydroxyl groups, amino groups, and halogen atoms. The addition–replacement sequence can thereby accomplish conversion of alkenes to a variety of other types of organic compounds. The most widely used reaction of organoboranes is the oxidation to alcohols. Alkaline aqueous hydrogen peroxide is the reagent used to effect the oxidation. The mechanism is as outlined:

$$R_3B + HOO^- \rightarrow R-\overset{\overset{\displaystyle R}{|}}{\underset{\underset{\displaystyle R}{|}}{B}}-O \overset{\frown}{-}OH \rightarrow R-\overset{\overset{\displaystyle R}{|}}{B}-OR + \overset{-}{O}H$$

$$R_2BOR + HOO^- \rightarrow R-\overset{\overset{\displaystyle R-O}{|}}{\underset{\underset{\displaystyle R}{|}}{B}}-O \overset{\frown}{-}O-H \rightarrow R-\overset{\overset{\displaystyle RO}{|}}{\underset{\underset{\displaystyle RO}{|}}{B}} + \overset{-}{O}H$$

$$(RO)_2BR + HOO^- \rightarrow (RO)_2\overset{-}{\underset{\underset{\displaystyle R}{|}}{B}}-O \overset{\frown}{-}O-H \rightarrow (RO)_3B$$

The R–O–B bonds are hydrolyzed during the oxidation, resulting in generation of the alcohol. It will be noted that the oxidation mechanism involves a series of B-to-O migrations of the alkyl groups, which migrate with their bonding electrons. The overall stereochemical outcome of the oxidation involves replacement of a C–B bond with retention of configuration. In combination with the orientation effects previously described, this result allows the structure and stereochemistry of alcohols produced by the hydroboration–oxidation sequence to be predicted with confidence. Several examples are shown in Scheme 3.5. Recently, conditions that permit oxidation of organoboranes to alcohols using molecular oxygen as oxidant have been discovered.[66] More vigorous oxidizing agents effect replacement of boron and oxidation of the substituted carbon atom, permitting the synthesis of ketones.[67]

The boron atom can also be replaced by an NH_2 group.[68] The reagents that allow this conversion are chloramine or hydroxylamine-O-sulfonic acid. The mechanisms of these reactions are very similar to that of the hydrogen peroxide oxidation of organoboranes. The nitrogen-containing reagents react as nucleophiles by adding to the boron, and rearrangement with expulsion of chloride ion or sulfate ion follows. The amine is freed by hydrolysis of the B–N bonds. Secondary amines

$$R_3B + NH_2X \rightarrow R_3\overset{-}{B}-\overset{\overset{\displaystyle H}{|}}{\underset{\underset{\displaystyle H}{|}}{N}}\overset{\frown}{-}X \rightarrow R_2B-\overset{\overset{\displaystyle NH}{|}}{R}$$

$$X = Cl^- \text{ or } OSO_3^-$$

are formed by reaction of trisubstituted boranes with azides. The most efficient boranes to use for this purpose are monoalkyldichloroboranes, which are generated

66. H. C. Brown, M. M. Midland, and G. W. Kabalka, *J. Am. Chem. Soc.* **93**, 1024 (1971).
67. H. C. Brown and C. P. Garg, *J. Am. Chem. Soc.* **83**, 2951 (1961).
68. M. W. Rathke, N. Inoue, K. R. Varma, and H. C. Brown, *J. Am. Chem. Soc.* **88**, 2870 (1966).

106

CHAPTER 3
ADDITION
REACTIONS OF
CARBON–CARBON
MULTIPLE
BONDS

by reaction of an alkene with $BHCl_2 \cdot Et_2O$.[69] The entire sequence of steps and the mechanism of the final stages are summarized by the equations below:

$$LiBH_4 + BCl_3 \xrightarrow{ether} BHCl_2 \cdot Et_2O + LiCl$$

$$BHCl_2 \cdot Et_2O + RCH = CH_2 \longrightarrow RCH_2CH_2BCl_2$$

$$RCH_2CH_2BCl_2 + R'-N_3 \longrightarrow \underset{RCH_2\overset{|}{C}H_2}{Cl_2\overset{\overset{R'}{|}}{B}-N-N\equiv N}$$

$$\downarrow$$

$$\underset{Cl_2B\overset{\overset{R'}{|}}{N}CH_2CH_2R}{}$$

$$\downarrow H_2O$$

$$R'NHCH_2CH_2R$$

Organoborane intermediates can also be used to synthesize alkyl halides. Replacement of boron by iodine is rapid in the presence of base.[70] Only two of the alkyl groups are efficiently used for primary alkyl groups, and only one with secondary alkyl groups. A similar process using bromine and sodium hydroxide

$$R_3B + 2I_2 + 2NaOH \rightarrow 2RI + RB(OH)_2 + 2NaI$$

affords bromides in good yields.[71] It should be noted that since the halogen atom replaces the boron atom, the regioselectivity of these reactions is opposite to direct addition of the hydrogen halide. Terminal olefins give primary halides.

$$RCH = CH_2 \xrightarrow[\text{2) Br}_2,\text{ NaOH}]{\text{1) B}_2\text{H}_6} RCH_2CH_2Br$$

Although each of the preceding reactions is an important method for the specific introduction of functional groups into organic molecules, none is a direct method for building larger carbon frameworks. Other work, again largely from the laboratories of H. C. Brown, has shown that the organoboranes available from hydroboration reactions can undergo numerous reactions that permit efficient elaboration of carbon skeletons. One of the first reactions of this type to be discovered involves treatment of organoboranes with silver nitrate.[72] An electron transfer is believed to occur, which leads to radical intermediates. These couple, so that the net course of the reaction involves forming a new carbon–carbon bond between two alkyl groups. This

$$2R_3B + AgNO_3 \rightarrow 3R-R$$

reaction suffers from a clear limit to its versatility, in that it can be applied only to the synthesis of symmetrical compounds.

The discovery that carbon monoxide reacts with organoboranes under mild

69. H. C. Brown, M. M. Midland, and A. B. Levy, *J. Am. Chem. Soc.* **95**, 2394 (1973).
70. H. C. Brown, M. W. Rathke, and M. M. Rogić, *J. Am. Chem. Soc.* **90**, 5038 (1968).
71. H. C. Brown and C. F. Lane, *J. Am. Chem. Soc.* **92**, 6660 (1970).
72. H. C. Brown and C. H. Snyder, *J. Am. Chem. Soc.* **83**, 1002 (1961).

conditions has led to the development of procedures that permit synthesis of primary alcohols, tertiary alcohols, and ketones from organoboranes.[73] The type of product is determined by controlling reaction conditions under which a boron-to-carbon migration of the alkyl groups occurs. If the organoborane is heated with carbon monoxide at 100–125°, all groups migrate and a tertiary alcohol is obtained after

$$R_3B + CO \rightarrow [R_3\overset{-}{B}-C\equiv\overset{+}{O}] \rightarrow [O=B-CR_3] \xrightarrow{H_2O_2, \ ^-OH} HOCR_3$$

oxidation. The addition of water to the carbonylation reaction mixture causes the reaction to cease after migration of two alkyl groups from boron to carbon. Oxidation of the reaction mixture at this stage gives dialkyl ketones.[74] Primary alcohols are obtained when the carbonylation stage of the reaction is carried out in

$$R_3B + CO \xrightarrow{H_2O} [RB\underset{OH}{\overset{|}{\longrightarrow}}\underset{OH}{\overset{|}{C}R_2}] \xrightarrow[NaOH]{H_2O_2} R_2CO$$

the presence of sodium borohydride or lithium borohydride.[75] The hydride reducing agent reduces the product of the first migration step:

$$R_3\overset{-}{B}-\overset{+}{C}\equiv O \rightarrow R_2B-\overset{\overset{\textstyle R}{|}}{C}=O \xrightarrow{BH_4^-} RCH_2OH + 2ROH$$

It should be noted that the latter synthesis utilizes only one third of the alkyl groups in the starting organoborane, a considerable limitation in the case of valuable olefins. The remaining alkyl groups are converted to the unhomologated alcohol and present a separation problem.

The versatility of the ketone synthesis has been extended to permit formation of unsymmetrical ketones. 2,3-Dimethylbutene can be hydroborated under controlled conditions to yield a monoalkylborane called "thexylborane." This compound can then be successively alkylated with two olefins having the structures desired for the two groups of the ketone (see entry 5, Scheme 3.6). The "thexyl" group migrates less readily than less branched alkyl groups. When carbonylation and migration are carried out so that only two groups migrate, oxidation affords the desired ketone.[76]

A route to aldehydes from olefins via carbonylation has also been developed.[77] The 9-BBN reagent formed on hydroboration of 1,5-cyclooctadiene is used to effect hydroboration of the olefin. Carbonylation of the resulting trialkylborane proceeds

73. H. C. Brown and M. W. Rathke, *J. Am. Chem. Soc.* **89**, 2737 (1967).
74. H. C. Brown and M. W. Rathke, *J. Am. Chem. Soc.* **89**, 2738 (1967).
75. M. W. Rathke and H. C. Brown, *J. Am. Chem. Soc.* **89**, 2740 (1967).
76. H. C. Brown and E. Negishi, *J. Am. Chem. Soc.* **89**, 5285 (1967).
77. H. C. Brown, E. F. Knights, and R. A. Coleman, *J. Am. Chem. Soc.* **91**, 2144 (1969); H. C. Brown and R. A. Coleman, *J. Am. Chem Soc.* **91**, 4606 (1969); H. C. Brown, E. F. Knights, and C. G. Scouten, *J. Am. Chem. Soc.* **96**, 7765 (1974).

CHAPTER 3
ADDITION
REACTIONS OF
CARBON–CARBON
MULTIPLE
BONDS

Scheme 3.6. Syntheses via Carbonylation of Organoboranes

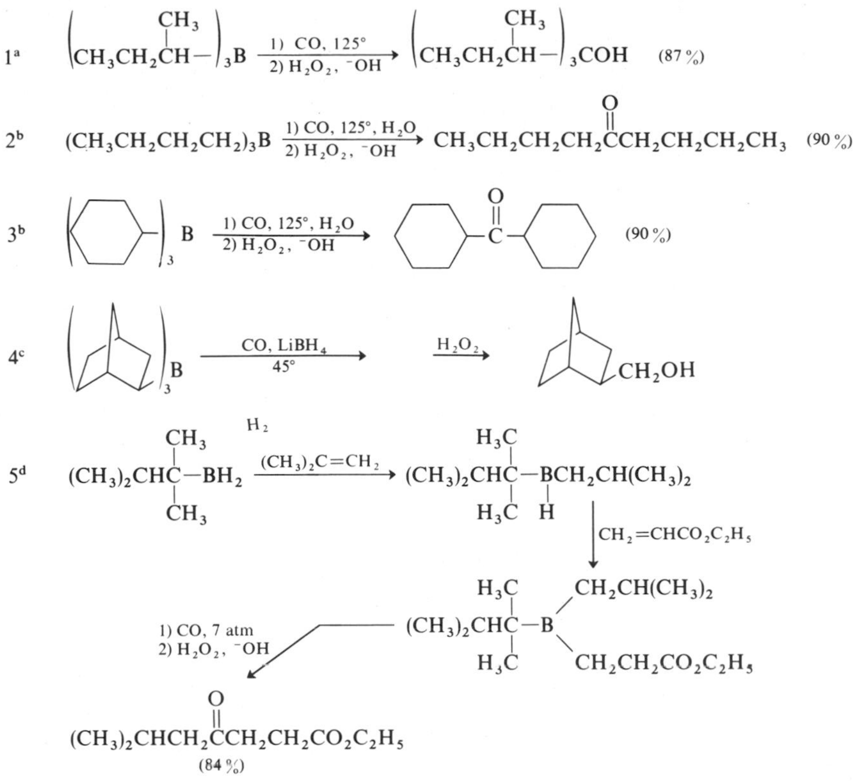

1[a] $\left(\text{CH}_3\text{CH}_2\overset{\overset{\text{CH}_3}{|}}{\text{CH}}-\right)_3\text{B}$ $\xrightarrow[\text{2) H}_2\text{O}_2,\ ^-\text{OH}]{\text{1) CO, 125°}}$ $\left(\text{CH}_3\text{CH}_2\overset{\overset{\text{CH}_3}{|}}{\text{CH}}-\right)_3\text{COH}$ (87%)

2[b] $(\text{CH}_3\text{CH}_2\text{CH}_2\text{CH}_2)_3\text{B}$ $\xrightarrow[\text{2) H}_2\text{O}_2,\ ^-\text{OH}]{\text{1) CO, 125°, H}_2\text{O}}$ $\text{CH}_3\text{CH}_2\text{CH}_2\text{CH}_2\overset{\overset{\text{O}}{\|}}{\text{C}}\text{CH}_2\text{CH}_2\text{CH}_2\text{CH}_3$ (90%)

3[b] $\xrightarrow[\text{2) H}_2\text{O}_2,\ ^-\text{OH}]{\text{1) CO, 125°, H}_2\text{O}}$ (90%)

4[c] $\xrightarrow[45°]{\text{CO, LiBH}_4}$ $\xrightarrow{\text{H}_2\text{O}_2}$

5[d] $(\text{CH}_3)_2\text{CH}\overset{\overset{\text{CH}_3}{|}}{\underset{\underset{\text{CH}_3}{|}}{\text{C}}}-\text{BH}_2$ $\xrightarrow[\text{(CH}_3)_2\text{C=CH}_2]{\text{H}_2}$ $(\text{CH}_3)_2\text{CH}\overset{\overset{\text{H}_3\text{C}}{|}}{\underset{\underset{\text{H}_3\text{C}}{|}}{\text{C}}}-\text{B}\overset{\overset{\text{H}}{|}}{\text{CH}_2}\text{CH}(\text{CH}_3)_2$

$\Big\downarrow \text{CH}_2\text{=CHCO}_2\text{C}_2\text{H}_5$

$(\text{CH}_3)_2\text{CH}\overset{\overset{\text{H}_3\text{C}}{|}}{\underset{\underset{\text{H}_3\text{C}}{|}}{\text{C}}}-\text{B}\overset{\nearrow\text{CH}_2\text{CH}(\text{CH}_3)_2}{\searrow\text{CH}_2\text{CH}_2\text{CO}_2\text{C}_2\text{H}_5}$

$\xrightarrow[\text{2) H}_2\text{O}_2,\ ^-\text{OH}]{\text{1) CO, 7 atm}}$

$(\text{CH}_3)_2\text{CHCH}_2\overset{\overset{\text{O}}{\|}}{\text{C}}\text{CH}_2\text{CH}_2\text{CO}_2\text{C}_2\text{H}_5$
(84%)

a. H. C. Brown and M. W. Rathke, *J. Am. Chem. Soc.* **89**, 2737 (1967).
b. H. C. Brown and M. W. Rathke, *J. Am. Chem. Soc.* **89**, 2738 (1967).
c. M. W. Rathke and H. C. Brown, *J. Am. Chem. Soc.* **89**, 2740 (1967).
d. H. C. Brown and E. Negishi, *J. Am. Chem. Soc.* **89**, 5285 (1967).

with selective rearrangement of the exocyclic alkyl group. Oxidation gives the desired aldehyde. The olefin can contain such functional groups as esters and nitriles without interfering with the aldehyde preparation.

Conditions have been developed that allow the alkyl groups in organoboranes to be alkylated by certain conjugated olefins and reactive halides. Typical α,β-unsaturated carbonyl compounds such as acrolein and methyl vinyl ketone (see Scheme 3.7 for other examples) alkylate organoboranes. Evidence has been developed that indicates that the alkylation proceeds via free-radical intermediates.[78] It therefore cannot be assumed that such alkylations will always proceed with the retention of configuration that is characteristic of oxidations and carbonylations of organoboranes.[79]

78. G. W. Kabalka, H. C. Brown, A. Suzuki, S. Honma, A. Arase, and M. Itoh, *J. Am. Chem. Soc.* **92**, 710 (1970).
79. H. C. Brown, M. M. Rogić, M. W. Rathke, and G. W. Kabalka, *J. Am. Chem. Soc.* **91**, 2151 (1969).

1^a $\left(\vphantom{}\right)_3 B$ + $CH_2{=}CHCOCH_3$ → $-CH_2CH_2COCH_3$ (86 %)

2^b $\left(\vphantom{}\right)_3 B$ + (structure) → (structure) (90 %)

3^c $\left(CH_3CH_2\overset{CH_3}{\underset{|}{CH}}-\right)_3 B$ + $CH_2{=}CHCHO$ → $CH_3CH_2\overset{CH_3}{\underset{|}{C}}HCH_2CH_2CHO$ (96 %)

4^c $\left(\vphantom{}\right)_3 B$ + $CH_2{=}CHCHO$ → CH_2CH_2CHO

5^d $(C_2H_5)_3B$ + $CH_3CH{=}CHCOCH_3$ → $CH_3CH_2\overset{CH_3}{\underset{|}{C}}HCH_2COCH_3$ (70 %)

6^e (structure) $B-\overset{CH_3}{\underset{CH_3}{|\;C\;|}}-CH_3$ + (cyclohexenone) → (structure) $C(CH_3)_3$ (73 %)

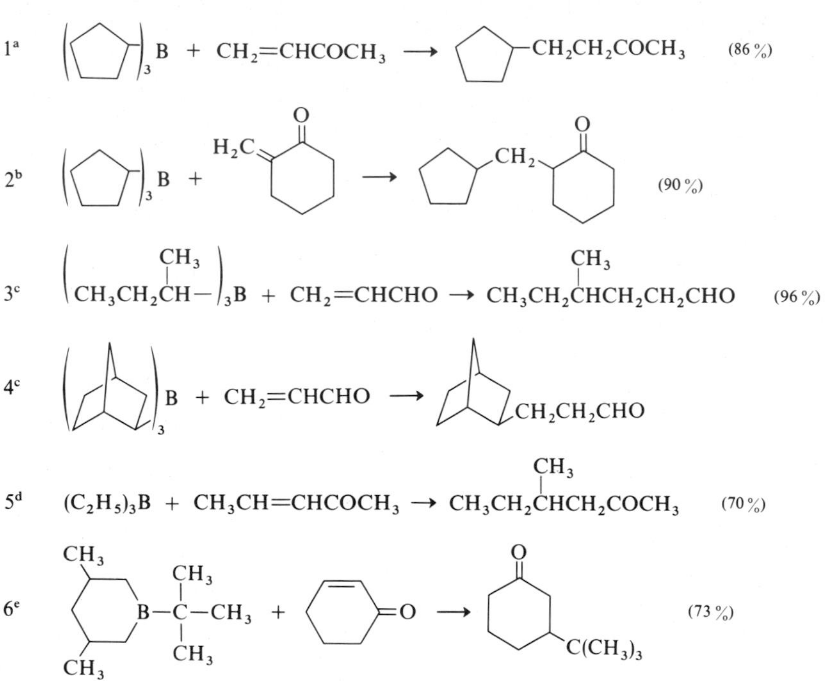

a. A. Suzuki, A. Arase, H. Matsumoto, M. Itoh, H. C. Brown, M. M. Rogić, and M. W. Rathke, *J. Am. Chem. Soc.* **89**, 5708 (1967).
b. H. C. Brown, M. W. Rathke, G. W. Kabalka, and M. M. Rogić, *J. Am. Chem. Soc.* **90**, 4166 (1968).
c. H. C. Brown, M. M. Rogić, M. W. Rathke, and G. W. Kabalka, *J. Am. Chem. Soc.* **89**, 5709 (1967).
d. H. C. Brown and G. W. Kabalka, *J. Am. Chem. Soc.* **92**, 714 (1970).
e. E. Negishi and H. C. Brown, *J. Am. Chem. Soc.* **95**, 6757 (1973).

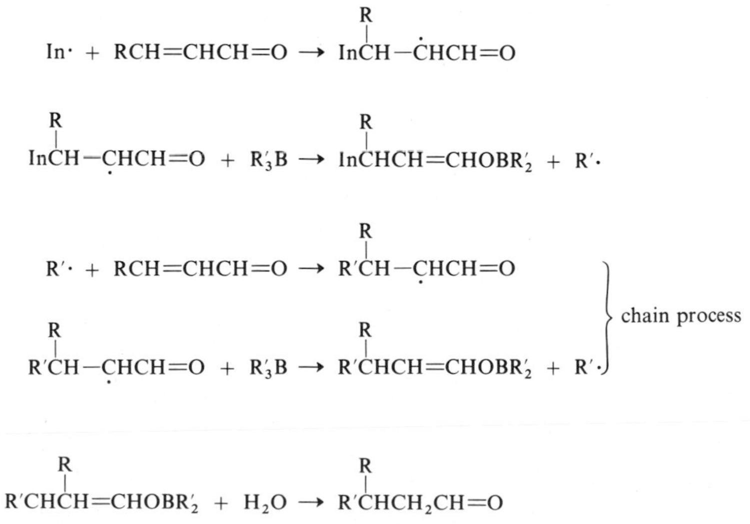

$In{\cdot}$ + $RCH{=}CHCH{=}O$ → $In\overset{R}{\underset{|}{C}}H-\overset{\cdot}{C}HCH{=}O$

$In\overset{R}{\underset{|}{C}}H-\underset{\cdot}{C}HCH{=}O$ + $R_3'B$ → $In\overset{R}{\underset{|}{C}}HCH{=}CHOBR_2'$ + $R'{\cdot}$

$R'{\cdot}$ + $RCH{=}CHCH{=}O$ → $R'\overset{R}{\underset{|}{C}}H-\underset{\cdot}{C}HCH{=}O$

$R'\overset{R}{\underset{|}{C}}H-\underset{\cdot}{C}HCH{=}O$ + $R_3'B$ → $R'\overset{R}{\underset{|}{C}}HCH{=}CHOBR_2'$ + $R'{\cdot}$

} chain process

$R'\overset{R}{\underset{|}{C}}HCH{=}CHOBR_2'$ + H_2O → $R'\overset{R}{\underset{|}{C}}HCH_2CH{=}O$

110

CHAPTER 3
ADDITION
REACTIONS OF
CARBON–CARBON
MULTIPLE
BONDS

A modified version of this reaction utilizes cyclic trialkylboron compounds (borinanes if the ring is six-membered). These are made by reacting diborane first with a diene and then with the alkyl group that is to be synthetically elaborated[80]:

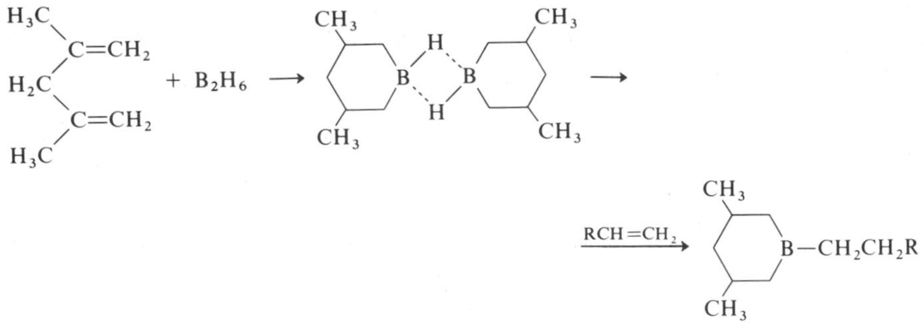

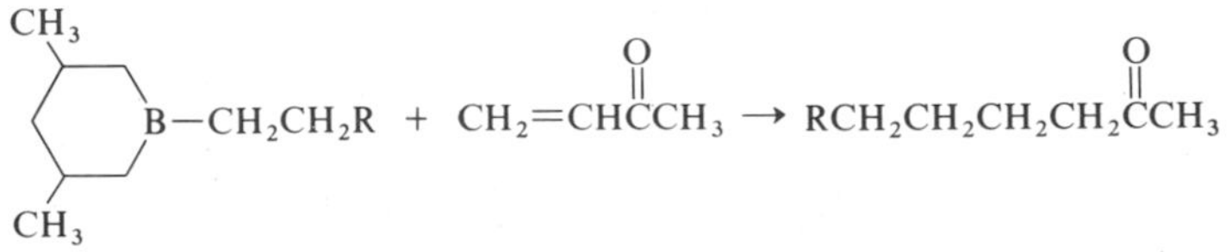

A second means of effecting alkylations of organoboranes involves reactions with highly reactive alkyl halides, especially α-halocarbonyl compounds.[81] For example, ethyl bromoacetate has been found to alkylate a number of trialkylboranes in excellent yield. This synthetic transformation is more efficiently carried out using a

$$R_3B \; + \; BrCH_2CO_2Et \; \xrightarrow{\;^-OCMe_3\;} \; RCH_2CO_2Et$$

trialkylborane prepared from the olefin to be alkylated and the dialkylborane 9-BBN. This procedure has the advantage of utilizing all of the starting olefin. Direct hydroboration of the olefin to the corresponding trialkylborane results in only one of the three alkyl groups undergoing reaction. α-Haloketones and α-halonitriles are also capable of alkylating organoboranes. A number of examples of this reaction are summarized in Scheme 3.8.

The mechanism by which these alkylations occur is fundamentally similar to the oxidation of organoboranes to alcohols. It is believed that the enolate of the haloester or haloketone reacts with the borane. Subsequently, elimination of halide, followed by migration of one of the boron substituents, occurs. In agreement with this mechanism, retention of configuration of the migrating group is observed.[82]

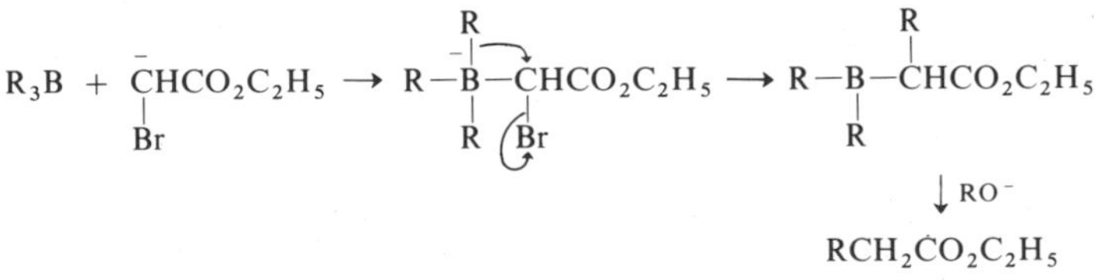

80. E. Negishi and H. C. Brown, *J. Am. Chem. Soc.* **95**, 6757 (1973).
81. H. C. Brown, M. M. Rogić, M. W. Rathke, and G. W. Kabalka, *J. Am. Chem. Soc.* **90**, 818 (1968).
82. H. C. Brown, M. M. Rogić, M. W. Rathke, and G. W. Kabalka, *J. Am. Chem. Soc.* **91**, 2151 (1969).

Scheme 3.8. Alkylation of Trialkylboranes with Haloesters, Haloketones, and Halonitriles

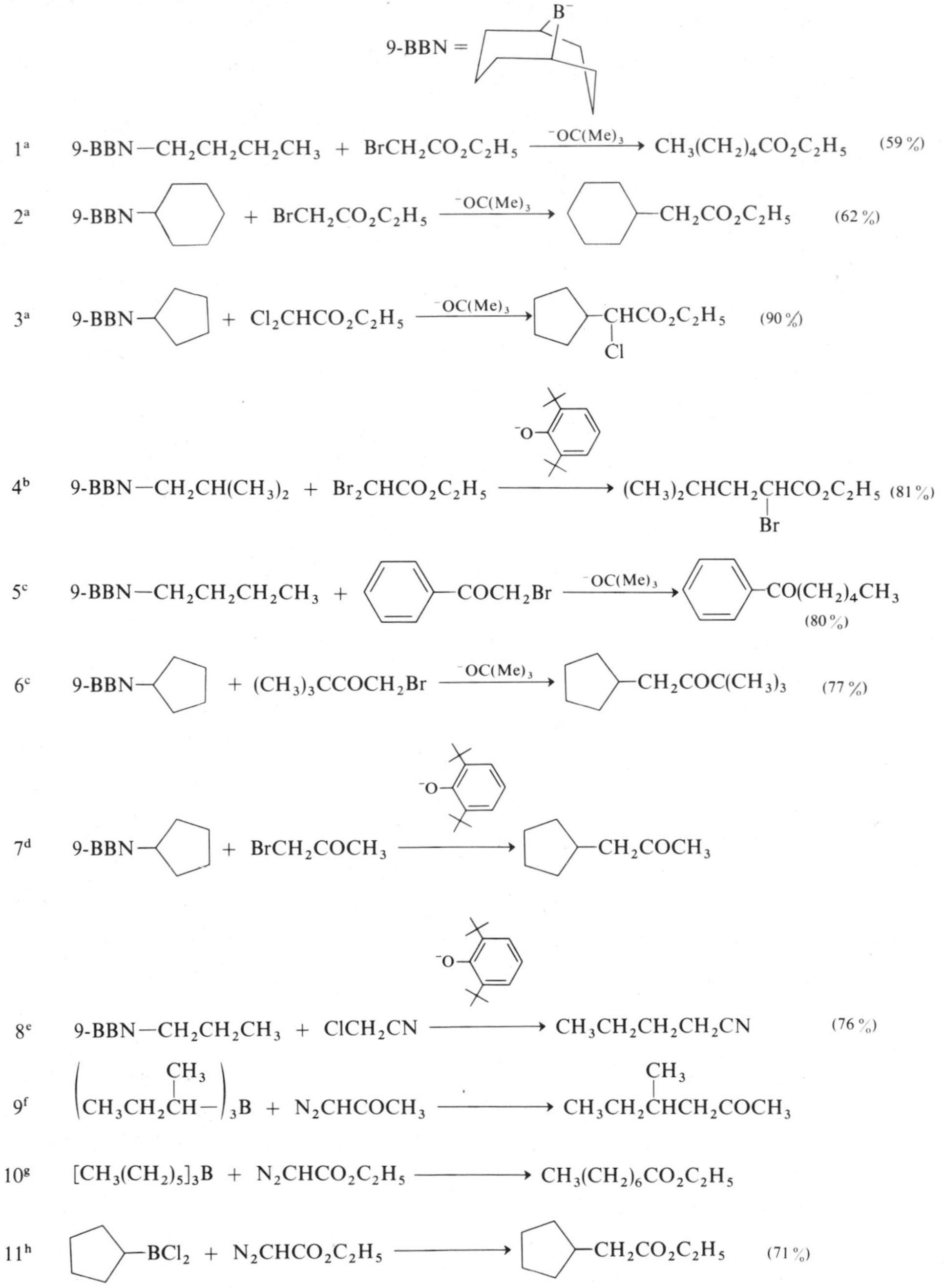

$$9\text{-BBN} =$$

1[a] $9\text{-BBN}-CH_2CH_2CH_2CH_3$ + $BrCH_2CO_2C_2H_5$ $\xrightarrow{^-OC(Me)_3}$ $CH_3(CH_2)_4CO_2C_2H_5$ (59 %)

2[a] $9\text{-BBN}-$⬡ + $BrCH_2CO_2C_2H_5$ $\xrightarrow{^-OC(Me)_3}$ ⬡$-CH_2CO_2C_2H_5$ (62 %)

3[a] $9\text{-BBN}-$⬠ + $Cl_2CHCO_2C_2H_5$ $\xrightarrow{^-OC(Me)_3}$ ⬠$-\underset{\underset{Cl}{|}}{C}HCO_2C_2H_5$ (90 %)

4[b] $9\text{-BBN}-CH_2CH(CH_3)_2$ + $Br_2CHCO_2C_2H_5$ $\xrightarrow{}$ $(CH_3)_2CHCH_2\underset{\underset{Br}{|}}{C}HCO_2C_2H_5$ (81 %)

5[c] $9\text{-BBN}-CH_2CH_2CH_2CH_3$ + ⬡$-COCH_2Br$ $\xrightarrow{^-OC(Me)_3}$ ⬡$-CO(CH_2)_4CH_3$
   (80 %)

6[c] $9\text{-BBN}-$⬠ + $(CH_3)_3CCOCH_2Br$ $\xrightarrow{^-OC(Me)_3}$ ⬠$-CH_2COC(CH_3)_3$ (77 %)

7[d] $9\text{-BBN}-$⬠ + $BrCH_2COCH_3$ $\xrightarrow{}$ ⬠$-CH_2COCH_3$

8[e] $9\text{-BBN}-CH_2CH_2CH_3$ + $ClCH_2CN$ $\xrightarrow{}$ $CH_3CH_2CH_2CH_2CN$ (76 %)

9[f] $\left(CH_3CH_2\underset{\underset{CH_3}{|}}{C}H-\right)_3B$ + $N_2CHCOCH_3$ $\xrightarrow{}$ $CH_3CH_2\underset{\underset{CH_3}{|}}{C}HCH_2COCH_3$

10[g] $[CH_3(CH_2)_5]_3B$ + $N_2CHCO_2C_2H_5$ $\xrightarrow{}$ $CH_3(CH_2)_6CO_2C_2H_5$

11[h]   ⬠$-BCl_2$ + $N_2CHCO_2C_2H_5$ $\xrightarrow{}$ ⬠$-CH_2CO_2C_2H_5$ (71 %)

a. H. C. Brown and M. M. Rogić, *J. Am. Chem. Soc.* **91**, 2146 (1969).
b. H. C. Brown, H. Nambu, and M. M. Rogić, *J. Am. Chem. Soc.* **91**, 6855 (1969).
c. H. C. Brown, M. M. Rogić, H. Nambu, and M. W. Rathke, *J. Am. Chem. Soc.* **91**, 2147 (1968).
d. H. C. Brown, H. Nambu, and M. M. Rogić, *J. Am. Chem. Soc.* **91**, 6853 (1969).
e. H. C. Brown, H. Nambu, and M. M. Rogić, *J. Am. Chem. Soc.* **91**, 6855 (1969).
f. J. Hooz and S. Linke, *J. Am. Chem. Soc.* **90**, 5936 (1968).
g. J. Hooz and S. Linke, *J. Am. Chem. Soc.* **90**, 6891 (1968).
h. J. Hooz, J. N. Bridson, J. G. Caldaza, H. C. Brown, M. M. Midland, and A. B. Levy, *J. Org. Chem.* **38**, 2574 (1973).

112

CHAPTER 3
ADDITION
REACTIONS OF
CARBON–CARBON
MULTIPLE
BONDS

A closely related reaction employs α-diazoesters or α-diazoketones.[83,84]

$$RBCl_2 + N_2CHCO_2CH_3 \rightarrow RCH_2CO_2CH_3$$

Molecular nitrogen then acts as the leaving group in the migration step. The best results are achieved with dialkylchloroboranes or alkyldichloroboranes.[84]

As can be readily judged from the preceding section, the organoboranes are versatile intermediates. The hydroboration–oxidation sequence has become an important means of alcohol synthesis, and the carbonylation- and alkylation-type reactions also seem likely to become widely used synthetic procedures. Although diborane is a reducing agent, the reductions are sufficiently slow that ester, cyano, and nitro groups do not interfere with hydroboration. On the other hand, ketone, aldehyde, carboxylic acid, and amide groups are reduced rapidly with diborane, and such reductions can be competitive with hydroboration.[85]

3.9. Additions to Allenes and Alkynes

Both allenes and alkynes require special consideration with regard to mechanisms of electrophilic addition. The attack by a proton on allene can conceivably lead to the allyl cation or the 2-propenyl cation:

$$CH_2=C=CH_2$$

$$^+CH_2-\overset{\overset{\displaystyle H}{|}}{C}=CH_2 \qquad CH_3-\overset{+}{C}=CH_2$$

An immediate presumption that the more stable allyl ion will be formed ignores the stereoelectronic facets of the reaction. Protonation at the center carbon without

rotation of the terminal methylene groups leads to a primary carbonium ion unstabilized by resonance, since the remaining π-bond is orthogonal to the empty p-orbital. Direct formation of an allyl cation therefore involves a more complex process than protonation.

The addition of HCl, HBr, and HI to allene has been studied in some detail.[86] In each case, the halogen is found at the center carbon in the product, so that protonation occurs at the terminal carbon. The initial product also undergoes some

83. H. C. Brown, M. M. Midland, and A. B. Levy, *J. Am. Chem. Soc.* **94**, 3662 (1972).
84. J. Hooz, J. N. Bridson, J. G. Calzada, H. C. Brown, M. M. Midland, and A. B. Levy, *J. Org. Chem.* **38**, 2574 (1973).
85. H. C. Brown, P. Heim, and N. M. Yoon, *J. Am. Chem. Soc.* **92**, 1637 (1970).
86. K. Griesbaum, W. Naegele, and G. G. Wanless, *J. Am. Chem. Soc.* **87**, 3151 (1965).

addition, giving rise to 2,2-dihalopropanes. Dimers are also formed, but we will not consider these.

$$CH_2\!\!=\!\!C\!\!=\!\!CH_2 + HX \rightarrow \underset{X}{\overset{X}{CH_3\overset{|}{C}\!\!=\!\!CH_2}} + \underset{X}{\overset{X}{H_3C\overset{|}{C}CH_3}}$$

The presence of a phenyl group results in the formation of products from protonation at the center carbon[87]:

$$\text{C}_6\text{H}_5\text{—CH}\!\!=\!\!\text{C}\!\!=\!\!\text{CH}_2 \xrightarrow[\text{HOAc}]{\text{HCl}} \text{C}_6\text{H}_5\text{—}\underset{H}{\overset{|}{C}}\!\!=\!\!\text{CHCH}_2\text{Cl}$$

Two alkyl groups, as in 1,1-dimethylallene, have the same effect[88]:

$$(CH_3)_2C\!\!=\!\!C\!\!=\!\!CH_2 \rightarrow (CH_3)_2C\!\!=\!\!CHCH_2Cl$$

These substituent effects are presumably due to stabilization of the cation that is generated by protonation at the center carbon. Even if the allylic conjugation is not effective in the transition state, the aryl and alkyl substituents can stabilize the charge that develops.

Mercuration with mercuric acetate reveals a regioselectivity pattern similar to that observed for the hydrogen halides. The electrophilic mercury is found on the terminal carbon for allene, but on the 2-carbon for 1,1-dimethylallene.[89] Products

$$CH_2\!\!=\!\!C\!\!=\!\!CH_2 + Hg(OAc)_2 \xrightarrow{CH_3OH} \underset{OMe}{\overset{OMe}{AcOHgCH_2\overset{|}{\underset{|}{C}}CH_2HgOAc}}$$

$$CH_2\!\!=\!\!C\!\!=\!\!C(CH_3)_2 + Hg(OAc)_2 \xrightarrow{CH_3OH} CH_2\!\!=\!\!C\overset{\diagup HgOAc}{\underset{\diagdown \underset{OCH_3}{\overset{|}{C(CH_3)_2}}}{}}$$

exhibiting both types of orientation were isolated from methylallene.

Alkynes are capable of undergoing addition reactions with the typical electrophilic reagents discussed in detail for alkenes. In general, the alkynes are less reactive. A major contribution to this difference in reactivity is the substantially higher energy of the vinyl-cation intermediate formed by electrophilic attack on an alkyne as opposed to an alkyl carbonium ion generated from an alkene. This energy difference is roughly 10 kcal/mol, depending specifically on the electrophile X^+ and

$$RC\!\!\equiv\!\!CH \xrightarrow{X^+} R\overset{+}{C}\!\!=\!\!CH\!\!-\!\!X$$

$$RCH\!\!=\!\!CH_2 \xrightarrow{X^+} R\overset{+}{C}H\!\!-\!\!CH_2\!\!-\!\!X$$

87. T. Okuyama, K. Izawa, and T. Fueno, *J. Am. Chem. Soc.* **95**, 6749 (1973).
88. T. L. Jacobs and R. N. Johnson, *J. Am. Chem. Soc.* **82**, 6397 (1960).
89. W. L. Waters and E. F. Kiefer, *J. Am. Chem. Soc.* **89**, 6261 (1967).

114

CHAPTER 3
ADDITION
REACTIONS OF
CARBON–CARBON
MULTIPLE
BONDS

Table 3.7. Relative Reactivity of Alkenes and Alkynes[a]

	Ratio of second-order rate constants (alkene/alkyne)		
	Bromination, acetic acid	Chlorination, acetic acid	Acid-catalyzed hydration, water
$CH_3CH_2CH_2CH_2CH=CH_2$ $CH_3CH_2CH_2CH_2C\equiv CH$	1.8×10^5	5.3×10^5	3.6
trans-$CH_3CH_2CH=CHCH_2CH_3$ $CH_3CH_2C\equiv CCH_2CH_3$	3.4×10^5	$\sim 1 \times 10^5$	16.6
$PhCH=CH_2$ $PhC\equiv CH$	2.6×10^3	7.2×10^2	0.65

a. From data tabulated in Ref. 90.

the particular system under study.[90] Table 3.7 summarizes some data that provide an insight into the relative reactivity of alkenes versus alkynes.

A major factor in determining the magnitude of relative rates is the solvent used. Thus, high alkene : alkyne reactivity ratios are found in organic solvents of low dielectric constant (acetic acid), but the rates of reaction are comparable when water is the solvent. This is true not only for the hydration but also for bromination. Polar solvents, water in particular, are evidently capable of minimizing the energy difference between the two transition states. Whether this is accomplished by very strong solvation or by some other mechanism is not entirely clear. Addition reactions of alkenes and alkynes with trifluoroacetic acid also take place at comparable rates.[91]

Acid-catalyzed additions to alkynes follow the Markownikoff rule. The initial addition products are not always stable, however. Addition of acetic acid, for example, results in the formation of enol acetates, which are easily converted to the corresponding carbonyl compound[92]:

$$
\begin{array}{ccc}
 & RC\equiv CH & \\
HCl \swarrow HOAc & & \searrow HCl \\
RC=CH_2 & O & RC=CH_2 \\
| & \| & | \\
CH_3CO & \rightarrow RCCH_3 & Cl \\
\| & & \\
O & & \\
\end{array}
$$

In aqueous solution, enols are formed and rapidly converted to the carbonyl compound.

The mercuric-ion-catalyzed hydration of terminal acetylenes has been a significant method for synthesis of methyl ketones. The reaction can also be used with internal acetylenes if they are symmetrical. Unsymmetrical alkynes would be expected to give a mixture of the two possible ketones.[93] Some examples of acetylene hydrations are given in Scheme 3.9.

90. K. Yates, G. H. Schmid, T. W. Regulski, D. G. Garratt, H.-W. Leung, and R. McDonald, *J. Am. Chem. Soc.* **95**, 160 (1973).
91. P. E. Peterson and J. E. Duddey, *J. Am. Chem. Soc.* **88**, 4990 (1966).
92. R. C. Fahey and D.-J. Lee, *J. Am. Chem. Soc.* **90**, 2124 (1968).
93. G. F. Hennion and C. J. Pillar, *J. Am. Chem. Soc.* **72**, 5317 (1950).

Scheme 3.9. Ketones from Acid-Catalyzed and Mercuric-Ion-Catalyzed Hydrations of Terminal Alkynes

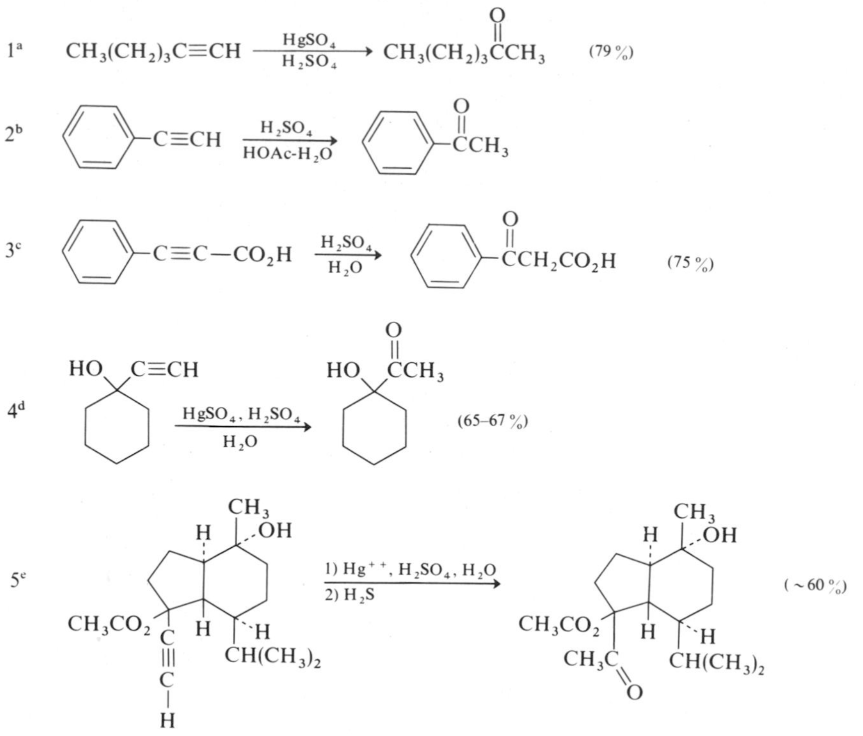

115

SECTION 3.9.
ADDITIONS TO
ALLENES AND
ALKYNES

1[a] $CH_3(CH_2)_3C{\equiv}CH \xrightarrow[H_2SO_4]{HgSO_4} CH_3(CH_2)_3\overset{\displaystyle O}{\overset{\|}{C}}CH_3$ (79%)

2[b] $-C{\equiv}CH \xrightarrow[HOAc-H_2O]{H_2SO_4}$ $-\overset{\displaystyle O}{\overset{\|}{C}}CH_3$

3[c] $-C{\equiv}C-CO_2H \xrightarrow[H_2O]{H_2SO_4}$ $-\overset{\displaystyle O}{\overset{\|}{C}}CH_2CO_2H$ (75%)

4[d] $\xrightarrow[H_2O]{HgSO_4,\ H_2SO_4}$ (65–67%)

5[e] $\xrightarrow[2)\ H_2S]{1)\ Hg^{++},\ H_2SO_4,\ H_2O}$ (~60%)

a. R. J. Thomas, K. N. Campbell, and G. F. Hennion, *J. Am. Chem. Soc.* **60**, 718 (1938).
b. R. W. Bott, C. Eaborn, and D. R. M. Walton, *J. Chem. Soc.*, 384 (1965).
c. D. S. Noyce, M. A. Matesich, and P. E. Peterson, *J. Am. Chem. Soc.* **89**, 6225 (1967).
d. G. N. Stacy and R. A. Mikulec, *Org. Synth.* **IV**, 13 (1963).
e. D. Caine and F. N. Tuller, *J. Org. Chem.* **38**, 3663 (1973).

Addition of 1 mol of hydrogen to the carbon–carbon triple bond can be accomplished stereospecifically. Catalytic reduction leads to the *cis* isomer. This is most often carried out using "Lindlar catalyst," a lead-poisoned palladium-on-calcium carbonate preparation.[94] Palladium on BaSO$_4$ is an alternative.[95] Some examples are recorded in Scheme 3.10. Numerous other catalyst systems have been employed to effect the same reduction. Many specific cases are cited in reviews of catalytic hydrogenations.[96] If the *trans* alkene is desired, the usual method is a dissolving-metal reduction in ammonia. This reaction is believed to involve two successive series of reduction by sodium and protonation:

94. H. Lindlar and R. Dubuis, *Org. Synth.* **46**, 89 (1966).
95. D. J. Cram and N. L. Allinger, *J. Am. Chem. Soc.* **78**, 2518 (1956).
96a. M. Freifelder, *Practical Catalytic Hydrogenation*, Wiley–Interscience, New York, NY, 1971, pp. 84–110.
   b. R. L. Augustine, *Catalytic Hydrogenation*, Marcel Dekker, New York, NY, 1965.

116

CHAPTER 3
ADDITION
REACTIONS OF
CARBON–CARBON
MULTIPLE
BONDS

$$R-C\equiv C-R + Na\cdot \rightarrow \underset{Na}{\overset{R}{\diagdown}}C=C\overset{R}{\cdot} \xrightarrow{H^+} \underset{H}{\overset{R}{\diagdown}}C=C\overset{R}{\cdot}$$

$$\underset{H}{\overset{R}{\diagdown}}C=C\cdot + Na\cdot \rightarrow \underset{H}{\overset{R}{\diagdown}}C=C\overset{Na}{\underset{R}{\diagup}} \xrightarrow{H^+} \underset{H}{\overset{R}{\diagdown}}C=C\overset{H}{\underset{R}{\diagup}}$$

The preference for the *trans* isomer is the result of the protonation and reduction steps being faster than isomerism of the vinyl-radical center.[97] Alternatively, lithium aluminum hydride can be used to convert acetylenes to *trans* olefins.[98]

Addition of chlorine to alkynes is slow in the absence of light. For example, 1-butyne is at least a factor of 10^2 less reactive than 1-butene. The addition reaction is readily initiated by light. The major product when butyne is in large excess is *trans*-1,2-dichlorobutene[99]:

$$CH_3CH_2C\equiv CH + Cl_2 \rightarrow \underset{Cl}{\overset{CH_3CH_2}{\diagdown}}C=C\overset{Cl}{\underset{H}{\diagup}}$$

The requirement for photoinitiation indicates that a radical-chain mechanism must be involved. Chlorination of 1-pentyne carried out in a gas phase reactor at higher temperatures and with higher chlorine:alkyne ratios gives both the *trans*-dichloroalkene and the saturated tetrachloro compound derived from addition of a second mole of chlorine[100]:

$$CH_3CH_2CH_2C\equiv CH \xrightarrow{50°} \underset{Cl}{\overset{CH_3CH_2CH_2}{\diagdown}}C=C\overset{Cl}{\underset{H}{\diagup}} + CH_3CH_2CH_2CCl_2CCl_2H$$
$$(19\%) \qquad\qquad\qquad (15\%)$$

The mechanism operative under these conditons has not been determined. Bromination of 1-phenylpropyne in acetic acid gives the *trans* dibromo adduct as the major product, but significant amounts of the *cis* isomer and products derived from incorporation of acetic acid are also observed. Addition of LiBr, however, makes the *trans* dibromo compound the overwhelming product[101]:

$$Ph-C\equiv C-CH_3 + Br_2 \xrightarrow{LiBr} \underset{Br}{\overset{Ph}{\diagdown}}C=C\overset{Br}{\underset{CH_3}{\diagup}}$$

97. H. O. House and E. F. Kinloch, *J. Org. Chem.* **39**, 747 (1974).
98. J. D. Chanley and H. Sobotka, *J. Am. Chem. Soc.* **71**, 4140 (1949); E. F. Magoon and L. H. Slaugh, *Tetrahedron* **23**, 4509 (1967).
99. M. L. Poutsma and J. L. Kartch, *Tetrahedron* **22**, 2167 (1966).
100. A. T. Morse and L. C. Leitch, *Can. J. Chem.* **33**, 6 (1955).
101. J. A. Pincock and K. Yates, *J. Am. Chem. Soc.* **90**, 5643 (1968).

Scheme 3.10. Reductions of Alkynes to Alkenes

117

SECTION 3.9.
ADDITIONS TO
ALLENES AND
ALKYNES

A. Catalytic Reduction

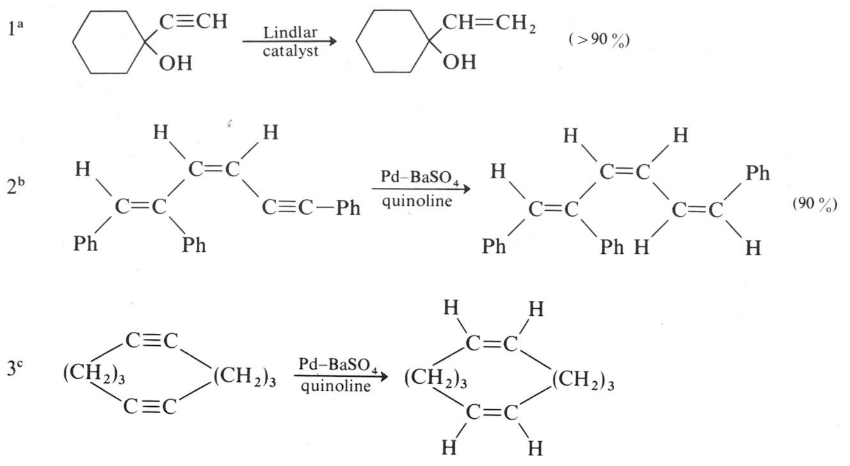

1[a]

(>90 %)

2[b]

(90 %)

3[c]

B. Dissolving-Metal Reduction

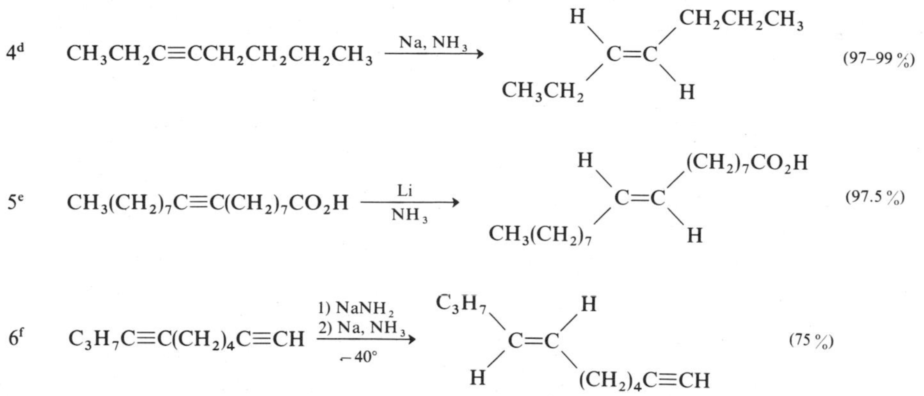

4[d]

$CH_3CH_2C\equiv CCH_2CH_2CH_2CH_3 \xrightarrow{Na, NH_3}$

(97–99 %)

5[e]

$CH_3(CH_2)_7C\equiv C(CH_2)_7CO_2H \xrightarrow[NH_3]{Li}$

(97.5 %)

6[f]

$C_3H_7C\equiv C(CH_2)_4C\equiv CH \xrightarrow[-40°]{\substack{1) NaNH_2 \\ 2) Na, NH_3}}$

(75 %)

a. W. Kimel, J. D. Surmatis, J. Weber, G. O. Chase, N. W. Sax, and A. Ofner, *J. Org. Chem.* **22**, 1611 (1957).
b. A. Padwa, L. Brodsky, and S. Clough, *J. Am. Chem. Soc.* **94**, 6767 (1972).
c. D. J. Cram and N. L. Allinger, *J. Am. Chem. Soc.* **78**, 2518 (1956).
d. A. L. Henne and K. W. Greenlee, *J. Am. Chem. Soc.* **65**, 2020 (1943).
e. R. E. A. Dear and F. L. M. Pattison, *J. Am. Chem. Soc.* **85**, 622 (1963).
f. N. A. Dobson and R. A. Raphael, *J. Chem. Soc.*, 3558 (1955).

3-Hexyne gives high conversions to the corresponding *trans* dibromo derivatives on reaction with an equimolar amount of bromine.[102] The stereospecificity of this reaction can be interpreted in terms of a bromonium-ion intermediate, as is the case with alkenes. The mechanistic studies that have been reported to date, however, are not so detailed as in the case of alkenes.

102. J. A. Pincock and K. Yates, *Can. J. Chem.* **48**, 3332 (1970).

118

CHAPTER 3
ADDITION
REACTIONS OF
CARBON–CARBON
MULTIPLE
BONDS

Alkynes are reactive toward hydroboration reagents. The most useful procedures begin with addition of a disubstituted borane to the alkyne. Treatment of the

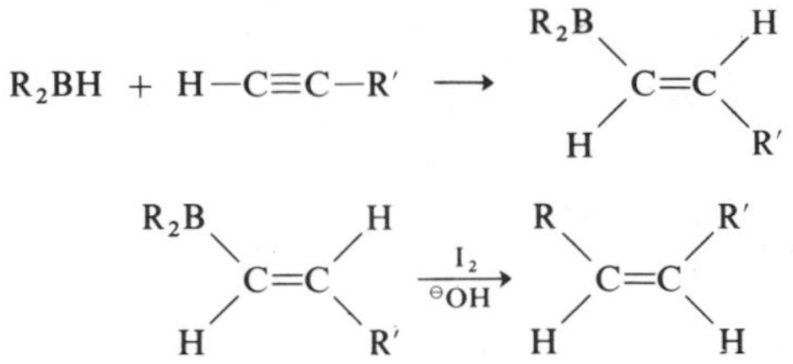

$$R_2BH + H-C\equiv C-R' \longrightarrow \underset{H}{\overset{R_2B}{\diagdown}}C=C\underset{R'}{\overset{H}{\diagup}}$$

$$\underset{H}{\overset{R_2B}{\diagdown}}C=C\underset{R'}{\overset{H}{\diagup}} \xrightarrow[\ominus OH]{I_2} \underset{H}{\overset{R}{\diagdown}}C=C\underset{H}{\overset{R'}{\diagup}}$$

resulting dialkyl vinylborane with iodine and base results in the formation of the *cis* alkene.[103] The mechanism suggested for this process involves alkyl migration initiated by iodination of the double bond, followed by stereospecific elimination of RBI_2:

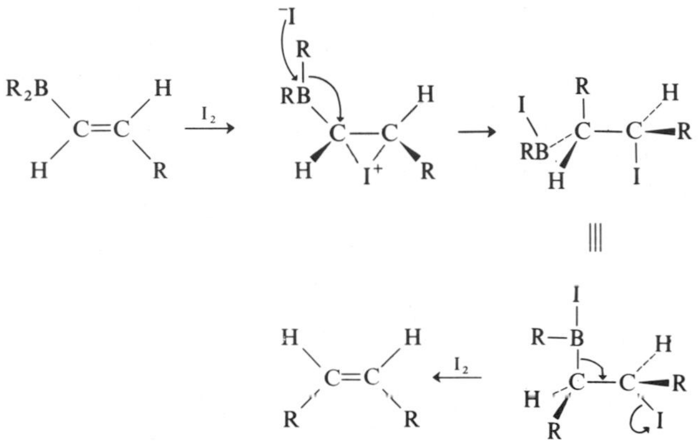

Vinylboronic esters derived from acetylenes and 1,3,2-benzodioxaborole

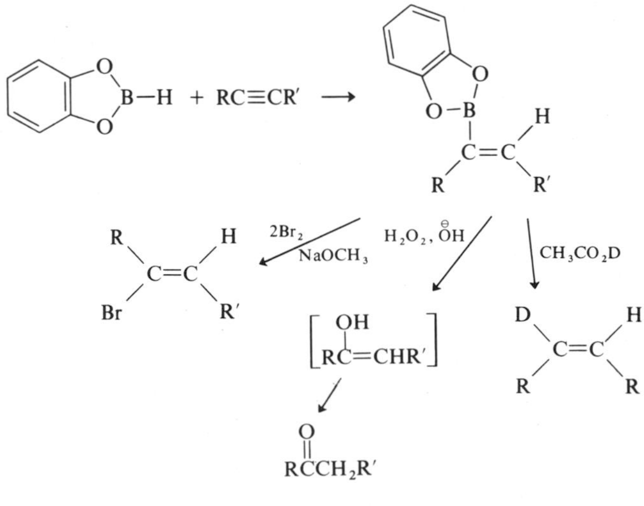

103. G. Zweifel, R. P. Fisher, J. T. Snow, and C. C. Whitney, *J. Am. Chem. Soc.* **93**, 6309 (1971).

exhibit typical reactions of boranes. Bromine in the presence of base converts the alkenylboronic esters to vinyl halides. Protonolysis with acetic acid results in net reduction of the alkyne to the *cis* alkene, whereas oxidation leads via the enol to a ketone.[104] Hydroboration takes place regioselectively with unsymmetrical alkynes so as to place the boron at the less hindered end of the triple bond. The addition is a stereospecific *cis* addition.[105]

Organoaluminum reagents also undergo a variety of synthetically useful reactions with alkynes. The addition of dialkylaluminum hydrides to alkynes is a stereospecific *cis* addition. Vinylalanes that are capable of being converted to substituted alkenes by reaction with electrophilic reagents are formed.

$$RC{\equiv}CH + (i\text{-Bu})_2AlH \rightarrow \begin{array}{c} H \quad\quad Al(i\text{-Bu})_2 \\ C{=}C \\ R \quad\quad\quad H \end{array}$$

$$RC{\equiv}CR + (i\text{-Bu})_2AlH \rightarrow \begin{array}{c} H \quad\quad Al(i\text{-Bu})_2 \\ C{=}C \\ R \quad\quad\quad R \end{array}$$

Treatment of the vinylalanes with bromine or iodine under mild conditions (tetrahydrofuran, −50°) produces the corresponding vinyl halide. Cleavage of the carbon–aluminum bond proceeds with complete retention of configuration.[106]

$$\begin{array}{c} CH_3(CH_2)_3 \quad\quad H \\ C{=}C \\ H \quad\quad Al(i\text{-Bu})_2 \end{array} \xrightarrow{I_2} \begin{array}{c} CH_3(CH_2)_3 \quad\quad H \\ C{=}C \\ H \quad\quad I \end{array} \quad (74\%)$$

Direct carbonation of vinylalanes as a route to α,β-unsaturated carboxylic acids is not efficient, but high yields can be obtained by a simple modification.[107] Addition of methyllithium to a vinylalane produces an *ate* complex that reacts with carbon dioxide either directly, or, more likely, after decomposing to a vinyllithium and trialkylalane, to afford α,β-unsaturated acids in high yield and stereospecificity:

$$\begin{array}{c} CH_3(CH_2)_3 \quad\quad H \\ C{=}C \\ H \quad\quad Al(i\text{-Bu})_2 \end{array} \xrightarrow{CH_3Li} \begin{array}{c} CH_3(CH_2)_3 \quad\quad H \\ C{=}C \\ H \quad\quad \overset{-}{Al}(i\text{-Bu})_2 \quad Li^+ \\ \quad\quad\quad CH_3 \end{array}$$

$$\downarrow \begin{array}{l} 1)\ CO_2 \\ 2)\ H^+ \end{array}$$

$$\begin{array}{c} CH_3(CH_2)_3 \quad\quad H \\ C{=}C \\ (78\%) \quad H \quad\quad CO_2H \end{array}$$

104. H. C. Brown, T. Hamaoka, and N. Ravidran, *J. Am. Chem. Soc.* **95**, 6456 (1973).
105. H. C. Brown and S. K. Gupta, *J. Am. Chem. Soc.* **94**, 4370 (1972).
106. G. Zweifel and C. C. Whitney, *J. Am. Chem. Soc.* **89**, 2753 (1967).
107. G. Zweifel and R. B. Steele, *J. Am. Chem. Soc.* **89**, 2754 (1967).

CHAPTER 3
ADDITION
REACTIONS OF
CARBON–CARBON
MULTIPLE
BONDS

Scheme 3.11. Syntheses Using Aluminum Alkyls

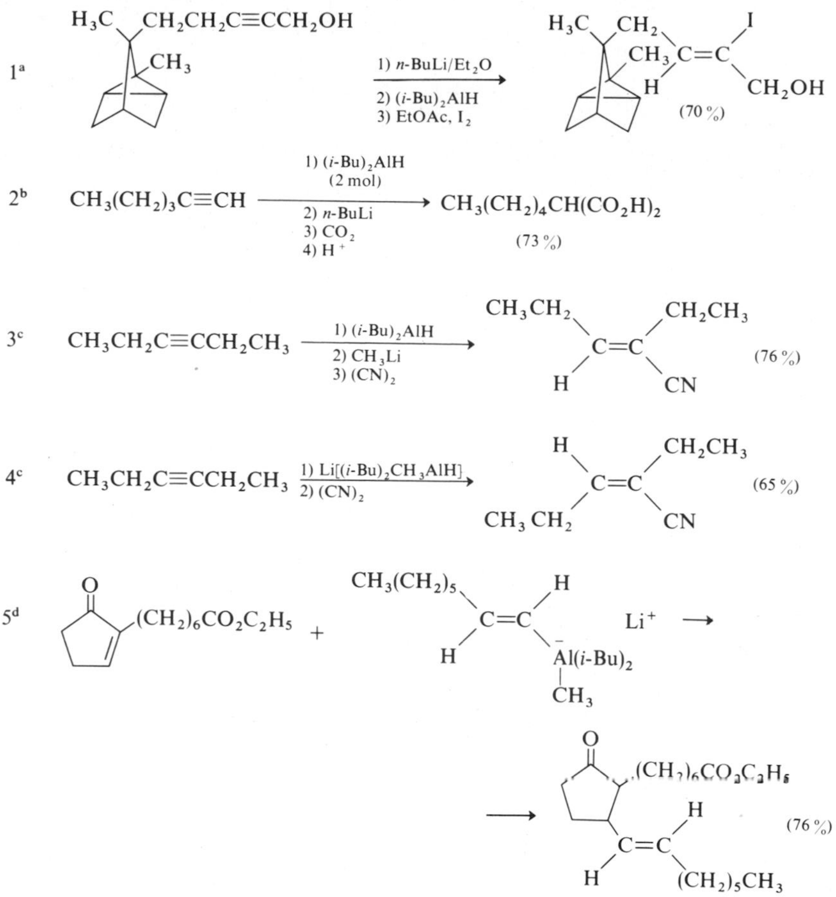

a. E. J. Corey, H. A. Kirst, and J. A. Katzenellenbogen, *J. Am. Chem. Soc.* **92**, 6314 (1970).
b. G. Zweifel and R. B. Steele, *Tetrahedron Lett.*, 6021 (1966).
c. G. Zweifel, J. T. Snow, and C. C. Whitney, *J. Am. Chem. Soc.* **90**, 7139 (1968).
d. K. F. Bernady and M. J. Weiss, *Tetrahedron Lett.* 4083 (1972).

A useful complement to the reactions above is based on the observation that the addition of lithium diisobutylmethylaluminum hydride (formed *in situ* from methyllithium and diisobutylaluminum hydride) to alkynes is a stereospecific *trans* addition[108]:

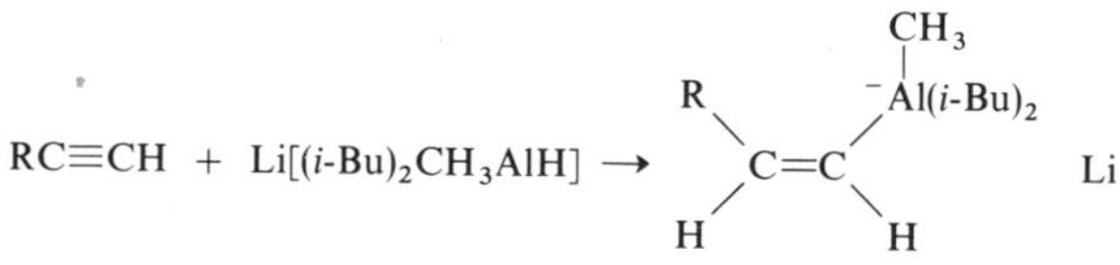

108. G. Zweifel and R. B. Steele, *J. Am. Chem. Soc.* **89**, 5085 (1967).

$$RC\equiv CR \;+\; Li[(i\text{-Bu})_2CH_3AlH] \;\rightarrow\; \underset{H}{\overset{R}{\diagdown}}C{=}C\underset{R}{\overset{\overset{\displaystyle CH_3}{|}}{\overset{-Al(i\text{-Bu})_2}{\diagup}}} \qquad Li^+$$

The *ate* complexes that result may be converted to vinyl halides or α,β-unsaturated carboxylic acids by procedures analogous to those employed in the previous examples. Organoaluminum reagents possess the characteristic of versatility, in that an alkyne may be converted stereospecifically to a *cis*- or *trans*-substituted alkene by appropriate choice of reagents. In the case of α,β-unsaturated acids, the flexibility is even more pronounced, in that it is simply the order of addition of reagents that determines the stereochemistry of the product.

Scheme 3.11 depicts some synthetic transformations that either involve vinylalanes or vinylalanates directly or in which their intermediacy is reasonable.

General References

Catalytic Reduction and Other Hydrogen-Addition Reactions

R. L. Burwell, *Acc. Chem. Res.* **2**, 289 (1969).
S. Mitzui and A. Kasahara, in *Chemistry of Alkenes*, Vol. 2, J. Zabicky (ed.), Interscience, New York, NY, 1970, Chap. 4.
M. Freifelder, *Practical Catalytic Hydrogenation*, Wiley–Interscience, New York, NY, 1971.
R. L. Augustine, *Catalytic Hydrogenation*, Marcel Dekker, New York, NY, 1965.
B. R. James, *Homogeneous Hydrogenations*, John Wiley, New York, NY, 1973.

Hydrogen Halide Additions

R. C. Fahey, *Top. Stereochem.* **2**, 237 (1968).

Solvomercuration

W. Kitching, *Organomet. Chem. Rev.* **3**, 61 (1968).

Additions of Pseudohalogens

A. Hassner, *Acc. Chem. Res.* **4**, 9 (1971).

Hydroboration

H. C. Brown and M. M. Rogić, *Organomet. Chem. Synth.* **1**, 305 (1972).
H. C. Brown and M. M. Midland, *Angew. Chem. Int. Ed. Engl.* **11**, 692 (1972).

Additions to Acetylenes

T. F. Rutledge, *Acetylenes and Allenes*, Reinhold Book Corp., New York, NY, 1969.
G. Modena and V. Tonellato, *Adv. Phys. Org. Chem.* **9**, 185 (1971).

Additions to Allenes

K. Griesbaum, *Angew. Chem. Int. Ed. Engl.* **5**, 933 (1966).
D. R. Taylor, *Chem. Rev.* **67**, 317 (1967).

122

CHAPTER 3
ADDITION
REACTIONS OF
CARBON–CARBON
MULTIPLE
BONDS

Problems

(References for these problems will be found on page 498.)

1. Predict the direction of addition and structure of the product in each of the following reactions:

   (a) $(CH_3)_2C=CH_2$ + PhSCl $\longrightarrow$

   (b) $CH_3CH_2O-C\equiv CH$ $\xrightarrow[H_2O]{H^+}$

   (c) 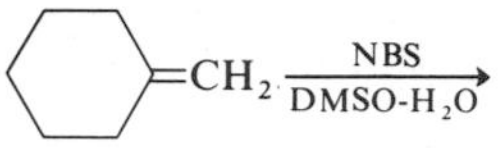   =CH_2 $\xrightarrow[DMSO-H_2O]{NBS}$

   (d) $(CH_3)_3CCH=CHCH_3$ $\xrightarrow[2)\,H_2O_2]{1)\,B_2H_6}$

   (e) 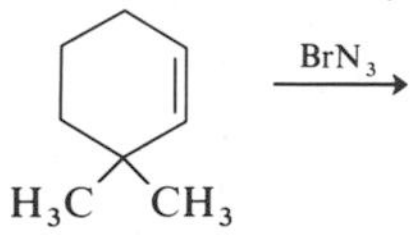 $\xrightarrow{BrN_3}$

2. Account for the fact that *trans*-decalin is the major product in the catalytic hydrogenation of Δ^9-octalin.

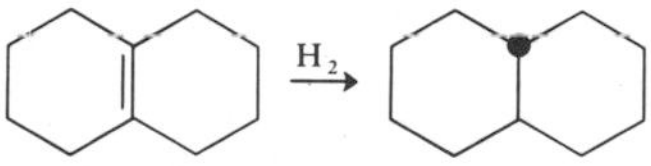

3. (a) Predict the stereochemistry of catalytic reduction of Compound **A**:

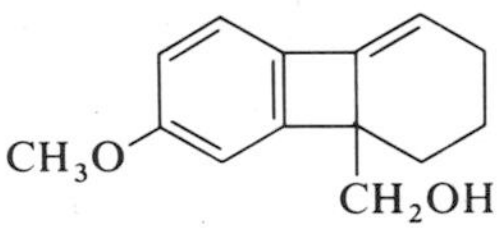

   (b) Predict the stereochemistry of reduction of **B**, using the homogeneous hydrogenation catalyst tris(triphenylphosphine)chlororhodium:

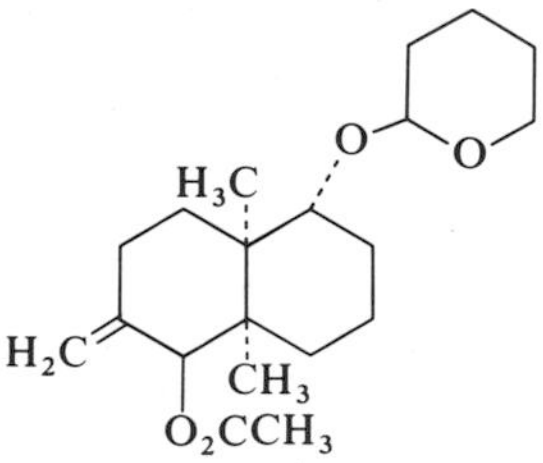

4. When 1-hexene is subjected to hydrogenation over platinum catalysts, if hydrogenation is stopped at about 95% completion, most of the unreacted alkene is a mixture of *cis*- and *trans*-2-hexene. Account for this observation.

5. Bromination of 4-*t*-butylcyclohexene in methanol gives a 45 : 55 mixture of two compounds, each of composition $C_{11}H_{21}BrO$. Predict the structure and stereochemistry of these two products. How would you confirm your prediction?

6. Suggest stereoselective syntheses of each of the isomeric epoxides derived from the steroidal olefin shown:

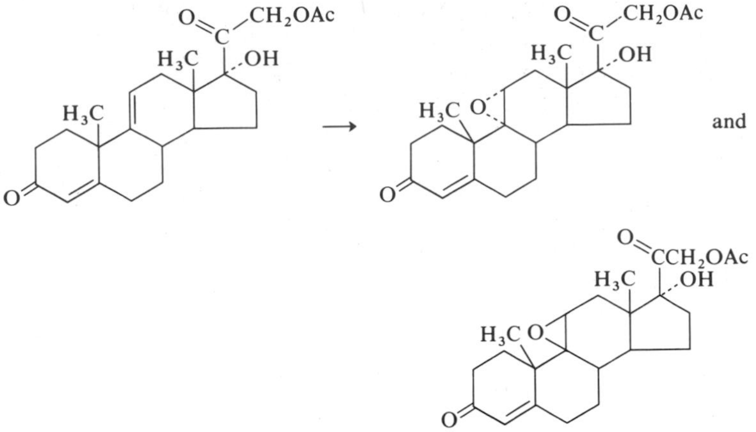

7. Starting with an alkene $RCH{=}CH_2$, indicate how organoborane intermediates could be used for each of the following synthetic transformations:

(a) $RCH{=}CH_2 \rightarrow RCH_2CH_2CH_2\overset{\overset{\displaystyle O}{\|}}{C}{-}\langle\text{Ph}\rangle$

(b) $RCH{=}CH_2 \rightarrow RCH_2CH_2CH{=}O$

(c) $RCH{=}CH_2 \rightarrow RCH_2CH_2\overset{\overset{\displaystyle CH_3}{|}}{C}HCH_2\overset{\overset{\displaystyle O}{\|}}{C}CH_3$

(d) $RCH{=}CH_2 \rightarrow$
$$\begin{array}{c}RCH_2CH_2CH_3\\ \diagdown\diagup\\ C{=}C\\ \diagup\diagdown\\ HH\end{array}$$

(e) $RCH{=}CH_2 \rightarrow RCH_2CH_2\overset{\overset{\displaystyle O}{\|}}{C}CH_2CH_2R$

(f) $RCH{=}CH_2 \rightarrow RCH_2CH_2CH_2CO_2C_2H_5$

8. In addition to carbon monoxide, sodium cyanide or chlorodifluoromethane can serve as the one carbon fragment in conversion of trialkylboranes to trialkylcarbinols. In the former case, the borane is first treated with sodium cyanide, and the resulting adduct is then treated with trifluoroacetic anhydride. With chlorodifluoromethane, the other reagent required is a potassium t-alkoxide. An oxidation step (hydrogen peroxide) then yields the product. Formulate a mechanism for each system.

9. Hydroboration–oxidation of $PhCH{=}CHOC_2H_5$ gives **A** as the major product if the hydroboration step is of short duration (7 sec), but **B** is the major product if the hydroboration is allowed to proceed for a longer time (2 hr). Explain.

$$\underset{\textbf{A}}{\overset{\displaystyle Ph\underset{\underset{\displaystyle OH}{|}}{C}HCH_2OC_2H_5}{}}\qquad\qquad\underset{\textbf{B}}{PhCH_2CH_2OH}$$

124

CHAPTER 3
ADDITION
REACTIONS OF
CARBON–CARBON
MULTIPLE
BONDS

10. Oxymercuration of 4-*t*-butylcyclohexene, followed by $NaBH_4$ reduction, gives *cis*-4-*t*-butylcyclohexanol and *trans*-4-*t*-butylcyclohexanol in approximately equal amounts. 1-Methyl-4-*t*-butylcyclohexene under similar conditions gives only *cis*-4-*t*-butyl-1-methylcyclohexanol. Formulate a mechanism for the oxymercuration–reduction process that is consistent with this stereochemical result.

11. Predict the major product of oxymercuration-reductive demercuration of 1,3-pentadiene.

12. Suggest synthetic sequences that could accomplish each of the following transformations:

(a)

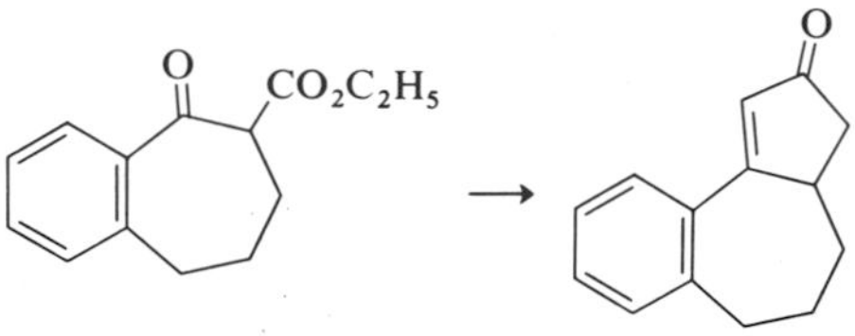

(b)

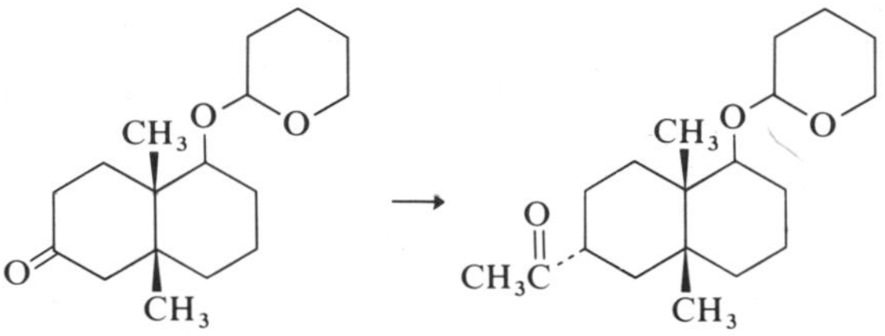

(c)

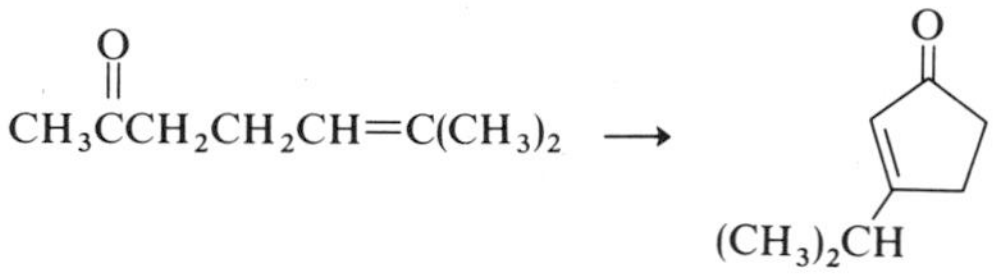

$$CH_3CCH_2CH_2CH=C(CH_3)_2 \longrightarrow$$

(d)

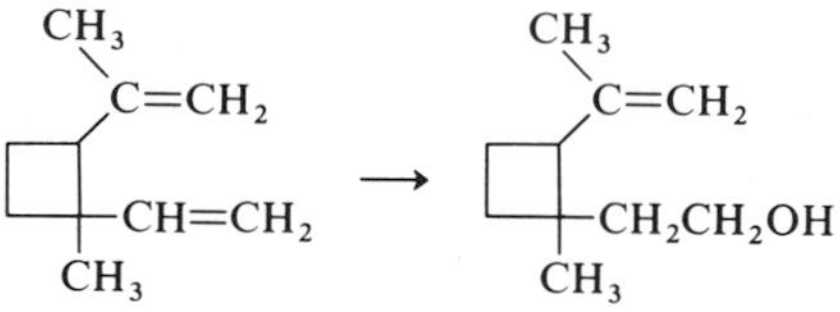

13. Each of the following compounds gives a product other than simple acid-catalyzed solvent addition under the conditions outlined. Suggest possible structures on the basis of mechanistic arguments.

(a)

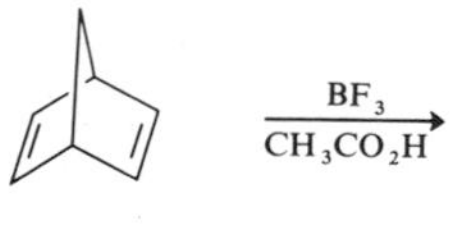

(b)

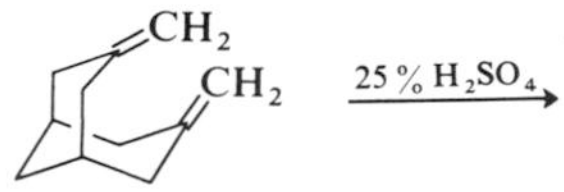

(c)

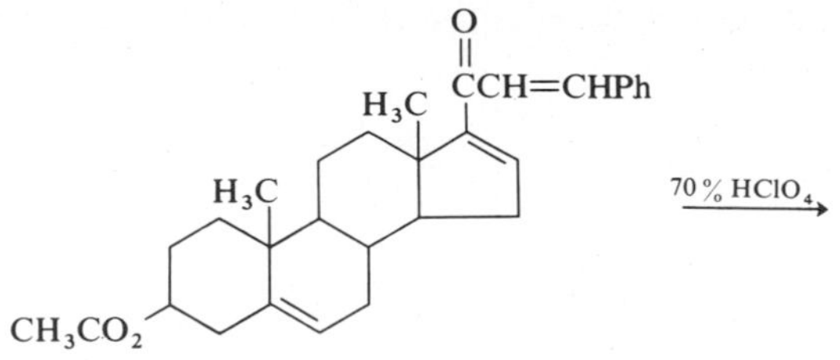

H_3C O‖CCH=CHPh

H_3C

CH_3CO_2

70 % HClO₄ →

(d)   HC≡CHCH₂CH₂CH₂Cl $\xrightarrow{CF_3CO_2H}$

(e)

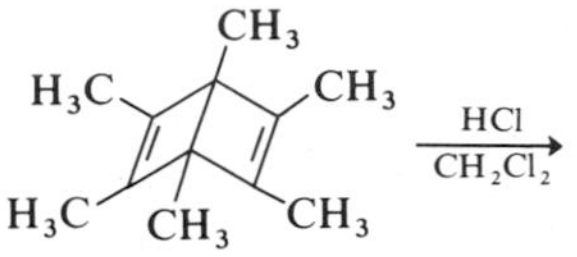

14. (a) The relative rates of addition of hydrogen chloride to *trans*-1-phenyl-1,3-butadiene, *cis*-1-phenyl-1,3-butadiene, and 1-phenyl-1,2-butadiene are 63:1.0:12, respectively. The product in each case is *trans*-3-chloro-1-phenyl-1-butene. Discuss these observations.

   (b) The hydration of 5-undecyn-2-one with mercuric sulfate and sulfuric acid in methanol is regioselective, giving 2,5-undecadione in 85% yield. Suggest an explanation for the high selectivity.

15. One approach to obtain information about the intermediates involved in catalytic hydrogenation is to examine the pattern of introduction of deuterium when saturated hydrocarbons are exposed to catalytic surfaces that are saturated with deuterium. Below are shown the initial products of several such partial exchange reactions. (Continued exposure eventually results in more random distribution of deuterium.) Discuss the pattern revealed by these exchange studies and indicate what information this provides about the absorbed intermediate in hydrogenation reactions.

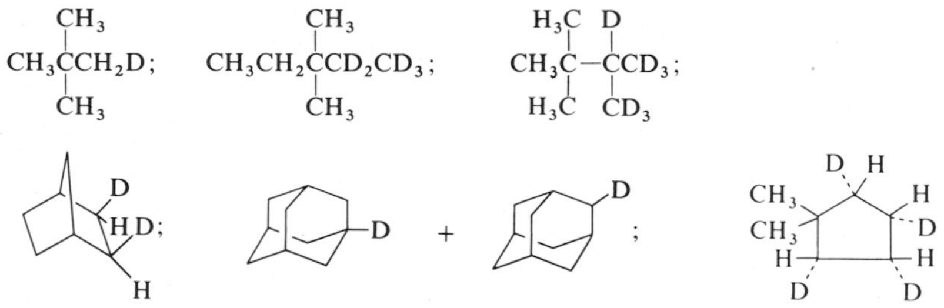

16. When trialkylboranes react with α-diazoketones or α-diazoesters in D_2O, the resulting products are monodeuterated. Formulate the reaction mechanism in sufficient detail to account for this fact.

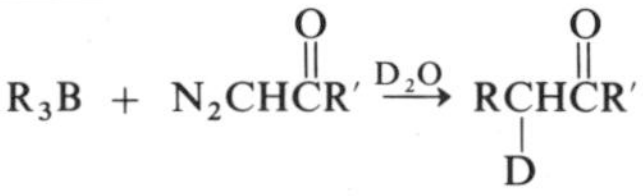

$$R_3B \ + \ N_2CHCR' \xrightarrow{D_2O} RCHCR'$$

126

CHAPTER 3
ADDITION
REACTIONS OF
CARBON–CARBON
MULTIPLE
BONDS

17. Indicate reaction conditions that would permit each of the following transformations:

(a) $CH_3CH_2C\equiv CCH_2CH_3 \rightarrow$

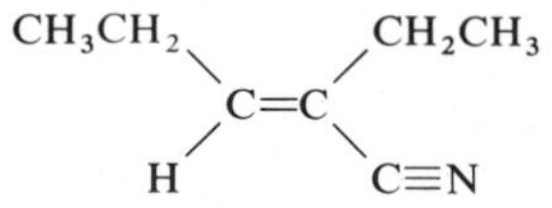

$$\underset{H}{\overset{CH_3CH_2}{\diagdown}}C=C\underset{C\equiv N}{\overset{CH_2CH_3}{\diagup}}$$

(b) $CH_3CH=CHCH_3 \rightarrow CH_3CH_2\overset{\overset{\textstyle CH_3}{|}}{CH}CH_2CH_2CH=O$

(c) $PhCH=CHCH_3 \rightarrow CH_3O\overset{\overset{\textstyle O}{\|}}{C}NH\overset{\overset{\textstyle Ph}{|}}{CH}\overset{\overset{\textstyle}{|}}{\underset{I}{CH}}CH_3$

(d) $CH_3CH_2CH_2CH_2C\equiv CH \rightarrow$

$$\underset{H}{\overset{CH_3CH_2CH_2CH_2}{\diagdown}}C=C\underset{I}{\overset{H}{\diagup}}$$

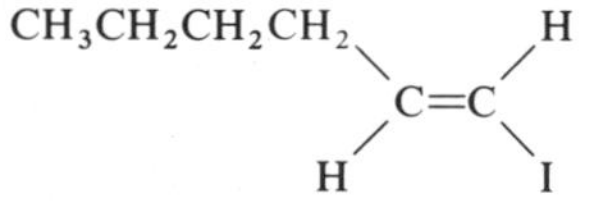

(e) $CH_3C\equiv CC(CH_3)_3 \rightarrow$

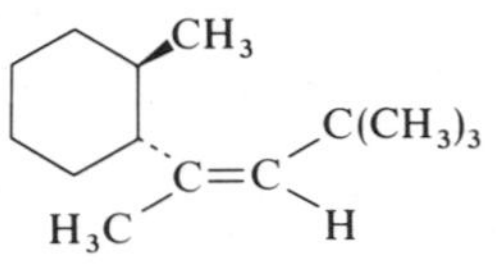

(f)

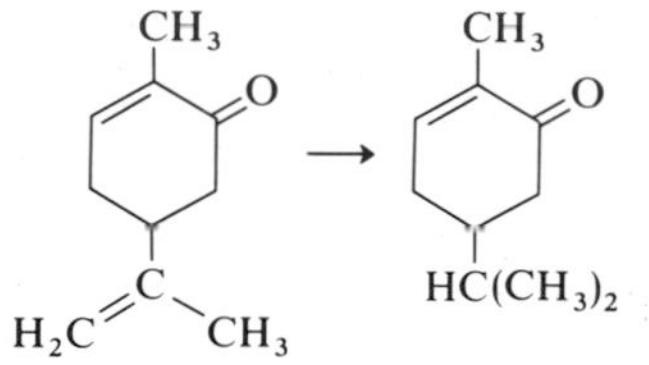

(g)

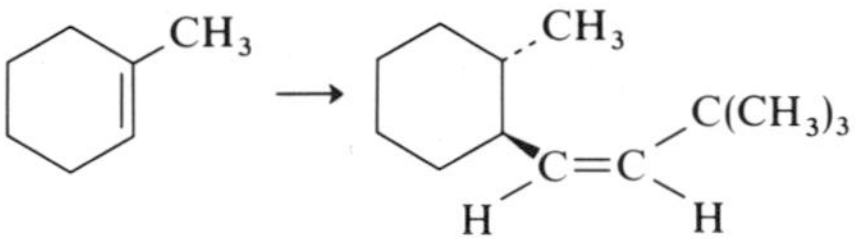

(h) $CH_3(CH_2)_3C\equiv CH \rightarrow$

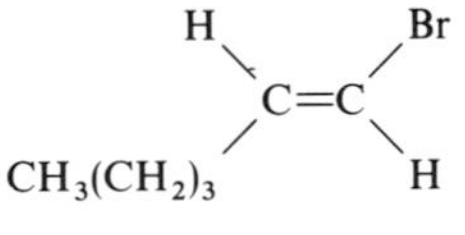

$$\underset{CH_3(CH_2)_3}{\overset{H}{\diagdown}}C=C\underset{H}{\overset{Br}{\diagup}}$$

(i)

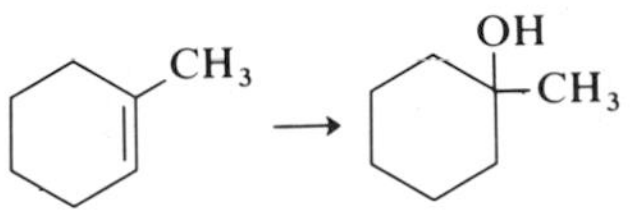

(j)

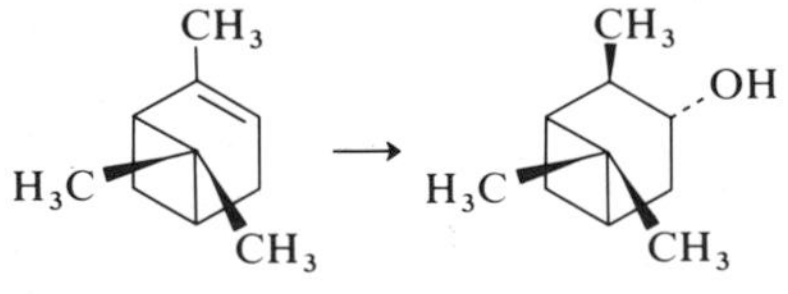

18. The bromination of vinylcyclopropane is at least 300 times faster than that of 1-hexene. In contrast, the rates of reaction of the two alkenes toward arylsulfenyl halides are similar (vinylcyclopropane is faster by a factor of 2). What explanation can you offer for this difference? What predictions about product structure do the rate data suggest?

Reduction of Carbonyl and Other Functional Groups

4.1. Hydride-Transfer Reagents

Most reductions of carbonyl and other functional groups are now done with reagents that transfer a hydride ion from a group III atom. The numerous reagents of this type that have become available provide a considerable degree of selectivity and stereochemical control. Sodium borohydride and lithium aluminum hydride are not only the most familiar examples, but also indicate the range in selectivity possible with hydride reducing agents. Sodium borohydride is a mild reducing agent, reducing only aldehydes and ketones rapidly. Lithium aluminum hydride is one of the most powerful hydride-transfer agents; it will reduce ketones, esters, acids, and even amides quite rapidly. The reactivity of these reagents, along with that of several other related reducing agents, is summarized in Table 4.1.

The mechanism by which all the group III complex hydrides effect reduction is believed to be quite similar. It involves hydride transfer accompanied by coordination with the metalloid atoms. Since, in general, each of the hydrogens can eventually

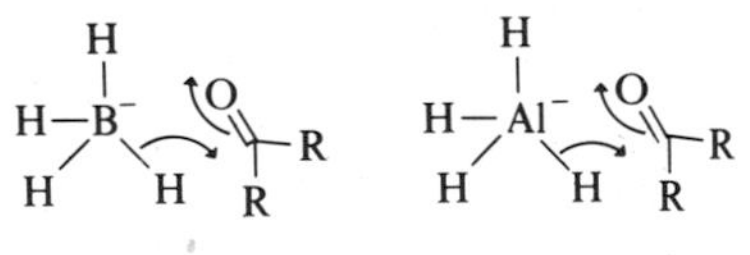

be transferred, there are actually several distinct reducing agents functioning during

$$BH_4^- + R_2CO \rightarrow R_2CHO\bar{B}H_3$$

$$R_2CHO\bar{B}H_3 + R_2CO \rightarrow [R_2CHO]_2\bar{B}H_2$$

$$[R_2CHO]_2\bar{B}H_2 + R_2CO \rightarrow [R_2CHO]_3\bar{B}H$$

$$[R_2CHO]_3B\bar{H} + R_2CO \rightarrow [R_2CHO]_4\bar{B}$$

129

Table 4.1. Reactivity of Hydride-Transfer Reducing Agents

Reducing agent	Usual products					
	Acyl chloride	Aldehyde	Ketone	Ester	Acid	Amide
NaBH₄[a]	alcohol	alcohol	alcohol	NR		NR
LiAlH₄[a]	alcohol	alcohol	alcohol	alcohol	alcohol	amine
LiAl(t-BuO)₃H[b]	aldehyde	alcohol	alcohol	very slow		aldehyde
AlH₃[c]	alcohol	alcohol	alcohol	alcohol	alcohol	amine
B₂H₆[d]		alcohol	alcohol	NR	alcohol	amine
$\left[(CH_3)_2CHCH{\overset{\displaystyle CH_3}{|}} \right]_2 BH$[e]	NR	alcohol	alcohol	NR[g]	NR	aldehyde
[(CH₃)₂CHCH₂]₂AlH[f]		alcohol	alcohol	aldehyde		aldehyde

a. See General References at end of this chapter.
b. H. C. Brown and R. F. McFarlin, *J. Am. Chem. Soc.* **78**, 252 (1956); **80**, 5372 (1958); H. C. Brown and B. C. Subba Rao, *J. Am. Chem. Soc.* **80**, 5377 (1958); H. C. Brown and A. Tsukamoto, *J. Am. Chem. Soc.* **86**, 1089 (1964).
c. H. C. Brown and N. M. Yoon, *J. Am. Chem. Soc.* **88**, 1464 (1966).
d. H. C. Brown, P. Heim, and N. M. Yoon, *J. Am. Chem. Soc.* **92**, 1637 (1970); N. M. Yoon, C. S. Pak, H. C. Brown, S. Krishnamurthy, and T. P. Stocky, *J. Org. Chem.* **38**, 2786 (1973); H. C. Brown and P. Heim, *J. Org. Chem.* **38**, 912 (1973).
e. H. C. Brown, D. B. Bigley, S. K. Arora, and N. M. Yoon, *J. Am. Chem. Soc.* **92**, 7161 (1970); H. C. Brown and V. Varma, *J. Org. Chem.* **39**, 1631 (1974).
f. K. E. Wilson, R. T. Seidner, and S. Masamune, *Chem. Commun.*, 213 (1970); E. J. Corey and E. A. Broger, *Tetrahedron Lett.*, 1779 (1969); Z. I. Zakharkin and L. P. Sorokina, *Zh. Obshch. Khim.* **37**, 561 (1967.
g. Lactones give lactols; ref. 16 (see p.133).

the course of the reduction. This somewhat complicates the interpretation of rates and stereochemistry, but has not detracted seriously from the synthetic utility of these reducing agents.[1] The alcohol formed is liberated by hydrolysis of the alkoxyboranes. This can take place during the reaction or workup.

$$[R_2CHO]_4\bar{B} + H_2O \xrightarrow{H^+} R_2CHOH + B(OH)_3$$

Hydride reductions of carboxylic acid derivatives such as conversion of esters to alcohols involve elimination steps in addition to hydride transfer. In the case of

amides, amines are formed selectively because the nitrogen group is a poorer leaving

1. B. Rickborn and M. T. Wuesthoff, *J. Am. Chem. Soc.* **92**, 6894 (1970).

group than the oxygen group. Reductions of amides constitute an important method

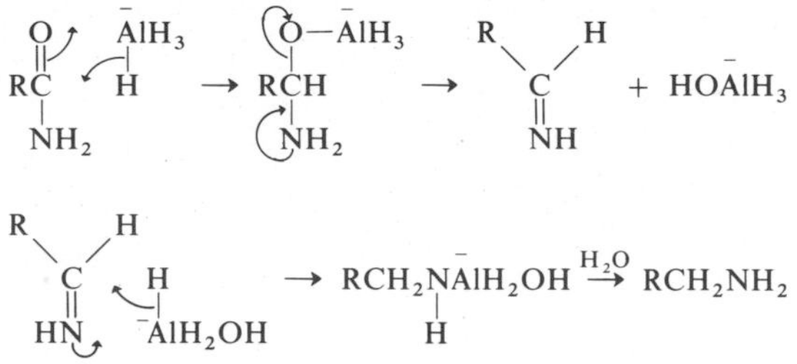

of synthesis of amines:

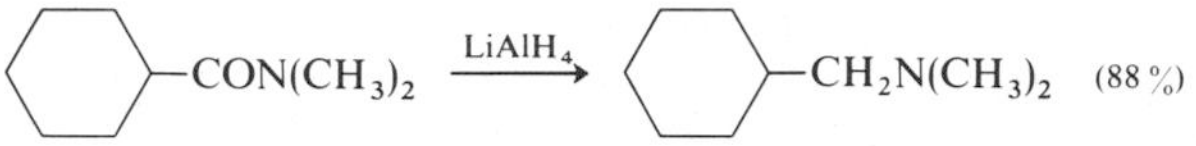

Ref. 2

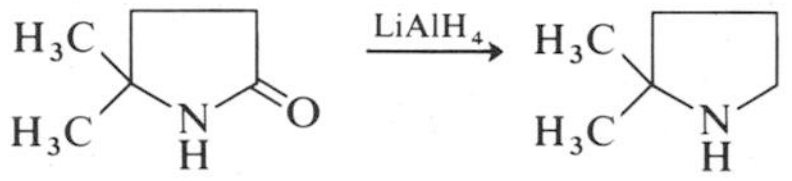

Ref. 3

The reduction of ketones and aldehydes to CH_2 groups occurs with hydride reducing agents only when some special structural feature promotes elimination of the oxygen atom from the initially formed carbinol. This sometimes occurs with aromatic ketones if a strongly electron-releasing group is present on the ring. It is also common in reduction of ketones derived from electron-rich heterocyclic rings such as the pyrrole ring.

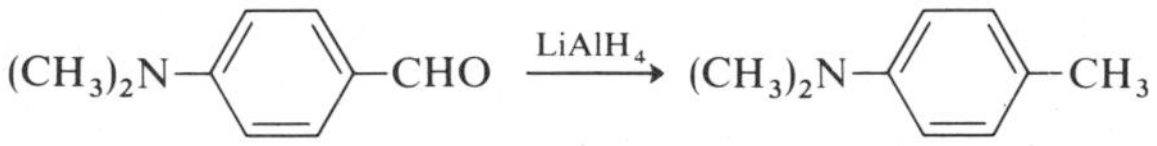

Ref. 4

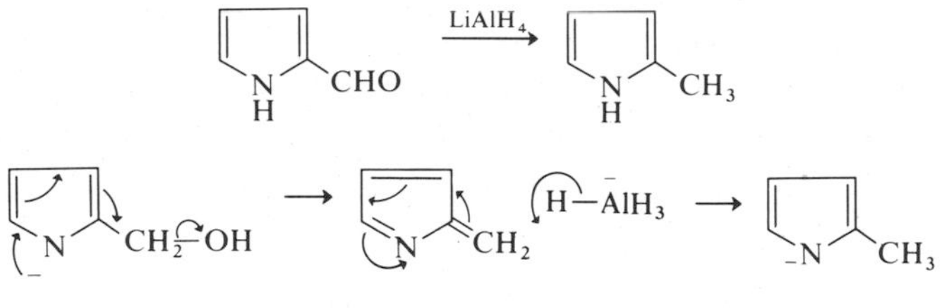

Ref. 5

One of the more difficult selective reductions to accomplish is the conversion of a carboxylic acid derivative to an aldehyde without overreduction to the alcohol. Several approaches have been used to achieve this selectivity. One is to replace some of the hydrides by bulky groups, thus modifying reactivity by steric factors. Lithium tri-t-butoxyaluminum hydride[6] is an example of this approach. This reagent is prepared by addition of an appropriate quantity of t-butanol to $LiAlH_4$. As noted in

2. A. C. Cope and E. Ciganek, *Org. Synth.* **IV**, 339 (1963).
3. R. B. Moffett, *Org. Synth.* **IV**, 354 (1963).
4. L. H. Conover and D. S. Tarbell, *J. Am. Chem. Soc.* **72**, 3586 (1950).
5. R. L. Hinman and S. Theodoropulos, *J. Org. Chem.* **28**, 3052 (1963).
6. H. C. Brown and B. C. Subba Rao, *J. Am. Chem. Soc.* **80**, 5377 (1958).

Table 4.1, lithium tri-t-butoxyaluminum hydride can be used to reduce acid chlorides to the aldehyde stage without extensive overreduction to the alcohol. Other trialkoxyaluminum hydrides prepared from less bulky primary alcohols have been prepared and their reactivity studied.[7] In many instances, they show selectivity superior to that of the simple hydrides.

Unique properties of solubility and reactivity are encountered when certain functionalized alcohols interact with lithium aluminum hydride.[8] Sodium bis(2-methoxyethoxy)aluminum hydride ("Redal"), for example, is soluble in benzene even at $-70°$. It is possible to reduce esters to aldehydes with this reagent:

$$(CH_3)_3CCO_2CH_3 \ + \ (CH_3OCH_2CH_2O)_2\bar{A}lH_2 \ \rightarrow \ (CH_3)_3CCH{=}O \ \ (81\%)$$

Lactones are reduced to cyclic hemiacetals (lactols).

Another approach to aldehydes by partial reduction of acid derivatives involves introduction of groups that stabilize the partially reduced intermediate against elimination of water. The aziridine ring has this effect because of the I-strain that is introduced if elimination occurs.[9] The partially reduced carbinolamine inter-mediates are stable to further reduction, and aldehydes are formed on subsequent

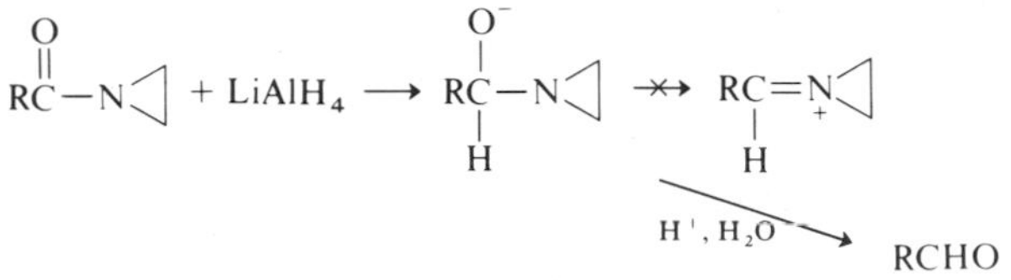

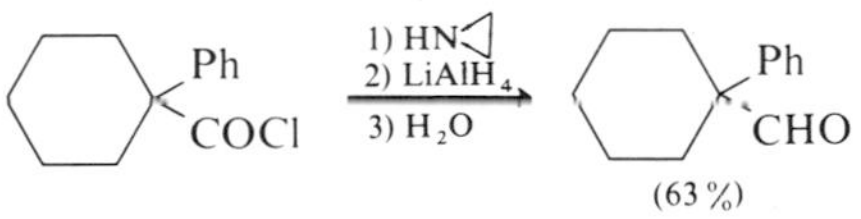

Ref. 10

hydrolysis. The acyl derivatives of certain aromatic nitrogen heterocycles are also reduced only to the aldehyde stage. Here, elimination is unfavorable because the nitrogen lone pair is delocalized into the aromatic π-system and does not effectively assist the departure of the oxygen atom[11]:

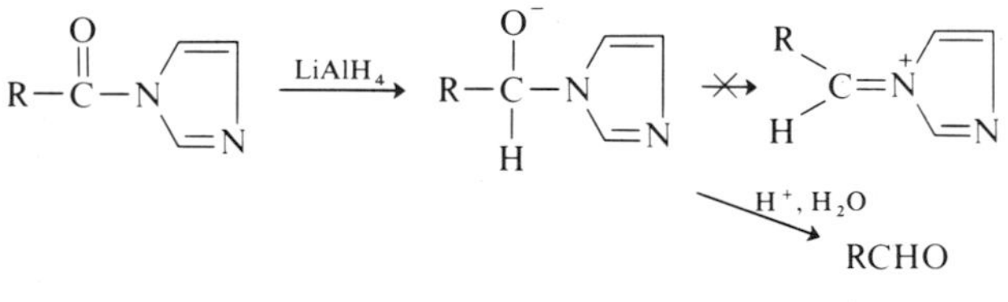

The reduction of nitriles with triethoxyaluminum hydride has been examined as a synthetic method for aldehydes, and generally good yields were reported.[12]

7. H. C. Brown and P. M. Weissman, *J. Am. Chem. Soc.* **87**, 5614 (1965).
8. J. Málek and M. Černý, *Synthesis*, 217 (1972).
9. H. C. Brown and A. Tsukamoto, *J. Am. Chem. Soc.* **83**, 4549 (1961).
10. J. W. Wilt, J. M. Kosturik, and R. C. Orlowski, *J. Org. Chem.* **30**, 1052 (1965).
11. H. A. Staab and H. Bräunling, *Justus Liebigs Ann. Chem.* **654**, 119 (1962).
12. H. C. Brown and C. P. Garg, *J. Am. Chem. Soc.* **86**, 1085 (1964).

Another procedure utilizes *N,N*-dimethylamides as the substrate and triethoxy-aluminum hydride (or diethoxyaluminum hydride) as the reducing agent.[13] The success of both procedures depends on the stability of the first reduced intermediate toward further reduction. Hydrolytic workup subsequently liberates the aldehyde:

$$RC{\equiv}N \; + \; (EtO)_3\bar{A}lH \; \longrightarrow \; RC{=}N{-}\bar{A}l(OEt)_3 \; \xrightarrow[H^+]{H_2O} \; RCH{=}O$$
$$\underset{H}{|}$$

$$\underset{\displaystyle RCN(CH_3)_2}{\overset{\displaystyle O}{\overset{\|}{}}} \; + \; (EtO)_3\bar{A}lH \; \longrightarrow \; \underset{\underset{H}{|}}{\overset{\overset{\displaystyle O{-}Al(OEt)_3}{|}}{RCN(CH_3)_2}} \; \xrightarrow[H^+]{H_2O} \; RCH{=}O$$

The dialkylboranes and dialkylaluminums are also of value when partial reduction of an ester or amide is desired. The intermediates formed by the first hydride transfer are stable under the conditions of the reduction. Subsequent hydrolysis then provides the carbonyl compound at the aldehyde reduction stage. This method using diisobutylaluminum hydride has been particularly useful for reduction of esters to aldehydes,[14,15] lactones to lactols,[16] and for cleavage of amides by partial reduc-

Ref. 14

tion.[17] Partial reduction of nitriles by diisobutylaluminum hydride to imines, which are subsequently hydrolyzed to aldehydes, has also been used preparatively[18,19]:

$$CH_3CH{=}CHCH_2CH_2CH_2C{\equiv}N \; \xrightarrow[2)\,H^+,\,H_2O]{1)\,(i\text{-}Bu)_2AlH} \; CH_3CH{=}CHCH_2CH_2CH_2CH{=}O \qquad \text{Ref. 18}$$
$$(64\%)$$

Sodium cyanoborohydride[20] is a close relative of sodium borohydride, but is considerably less reactive toward carbonyl groups at neutral pH. The reagent is useful for reducing C=N systems to amines. This reaction occurs rapidly at pH values where the C=N system is protonated. At pH 6–7, $NaBH_3CN$ is essentially unreactive toward carbonyl compounds, but readily reduces the protonated imines. This permits preparation of amines from carbonyl compounds without isolation of the intermediate imines. The carbonyl compound and ammonia or an alkylamine are

13. H. C. Brown and A. Tsukamoto, *J. Am. Chem. Soc.* **86**, 1089 (1964).
14. G. Grethe, H. S. Lee, T. Mitt, and M. R. Uskoković, *J. Am. Chem. Soc.* **93**, 5904 (1971).
15. E. J. Corey and E. A. Broger, *Tetrahedron Lett.*, 1779 (1969).
16. H. C. Brown, D. B. Bigley, S. K. Arora, and N. M. Yoon, *J. Am. Chem. Soc.* **92**, 7161 (1970).
17. J. Gutzwiller and M. Uskoković, *J. Am. Chem. Soc.* **92**, 204 (1970).
18. N. A. LeBel, M. E. Post, and J. J. Wang, *J. Am. Chem. Soc.* **86**, 3759 (1964).
19. R. V. Stevens and J. T. Lai, *J. Org. Chem.* **37**, 2138 (1972); S. Trofimenko, *J. Org. Chem.* **29**, 3046 (1964).
20. C. F. Lane, *Synthesis*, 135 (1975).

in equilibrium with the imine. Since the imine is selectively reduced as it is formed *in situ*, the reaction leads in good yield to the substituted amine[21]:

$$R_2C=O + R'NH_2 \rightleftarrows R_2C=NR'$$

$$R_2C=NR' + H^+ \rightleftarrows R_2C\overset{H}{\underset{+}{=}}NR'$$

$$R_2C\overset{H}{\underset{+}{=}}NR' + BH_3CN^- \rightarrow R_2CHNHR'$$

Diborane itself has some utility as a reducing agent. Perhaps most useful is its ability to selectively reduce carboxylic acid groups to primary alcohols under mild conditions.[22] Diborane is much less reactive toward some of the other functional groups that would be attacked by the ionic hydride reducing agents or by catalytic hydrogenation under the conditions necessary to reduce the carboxyl group. For example, ester, nitro, and cyano groups are much more slowly reduced by diborane than is the carboxyl group. Diborane is also useful for reducing amides. Tertiary and secondary amides are quite readily reduced, but primary amides react only slowly.[23] Amides require vigorous conditions for complete reduction by $LiAlH_4$. Selective reduction of amide groups in the presence of such functional groups as ester and nitro are therefore best done with diborane.

Another important aspect of the reactivity of hydride reducing agents is their stereoselectivity. The stereochemistry of hydride reduction has been studied most thoroughly with cyclohexanone derivatives. Some reagents give predominantly axial alcohols from cyclohexanones, while others give primarily equatorial alcohols. It has been observed that axial alcohols are most likely to be formed with sterically hindered hydride transfer agents. There has been general agreement that this occurs because the equatorial direction of attack is more open and therefore preferred by bulky reagents. The phrase *steric approach control* was coined to describe this behavior.[24]

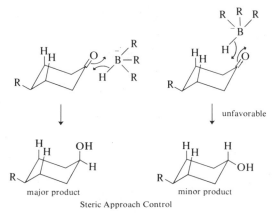

Steric Approach Control

21. R. F. Borch, M. D. Bernstein, and H. D. Durst, *J. Am. Chem. Soc.* **93**, 2897 (1971).
22. M. N. Yoon, C. S. Pak, H. C. Brown, S. Krishnamurthy, and T. P. Stocky, *J. Org. Chem.* **38**, 2786 (1973).
23. H. C. Brown and P. Heim, *J. Org. Chem.* **38**, 912 (1973).
24. W. G. Dauben, G. J. Fonken, and D. S. Noyce, *J. Am. Chem. Soc.* **78**, 2579 (1956).

With less hindered hydride donors, particularly sodium borohydride and lithium aluminum hydride, cyclohexanones give predominantly the equatorial alcohol. There has been less agreement about the factors that lead to this result. The equatorial alcohols are, of course, the more stable of the two isomers. The stereochemistry of hydride reduction is determined by kinetic control, but it was argued that the relative stability of the equatorial alcohol might be reflected in the transition state and be the dominant factor when no major steric problems intervened. The term *product development control* was introduced to indicate this explanation of the reaction stereochemistry. A number of objections were raised to this idea, primarily on the basis of the Hammond principle. The common hydride reductions are exothermic reactions with low activation energies. The transition state should resemble starting ketone and reflect little of the structural features that are present in the product, so that it is difficult to see why product stability should determine the product composition.

A more recent explanation of the preference for equatorial alcohols in the absence of overriding steric factors involves an analysis of the torsional strain that develops in the two transition states.[25,26] The preference for equatorial alcohol is

Torsional strain as oxygen passes through
an eclipsed conformation

Oxygen moves away from equatorial
hydrogens; no torsional strain

then explained on the basis of the torsional strain that develops in the transition state leading to axial alcohols.

Table 4.2 compares the course of reduction of several ketones with hydrides of increasing steric bulk. The trends in the table illustrate the increasing importance of steric approach control as both the hydride reagent and the ketone become more highly substituted. Highly hindered hydride reducing agents are therefore chosen if an axial cyclohexanol is desired.

When the ketones are relatively hindered—as, for example, in the bicyclo[2.2.1]heptan-2-one system—steric effects govern the stereochemistry of

25. M. Chérest, H. Felkin, and N. Prudent, *Tetrahedron Lett.*, 2199 (1968); M. Chérest and H. Felkin, *Tetrahedron Lett.*, 2205 (1968).
26. For an alternative explanation stressing electronic factors, see J. Klein, *Tetrahedron Lett.*, 4307 (1973).

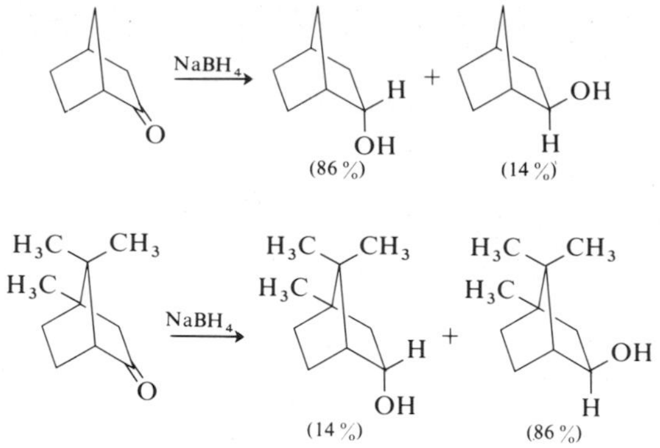

reduction even for small hydride reagents such as sodium borohydride and lithium aluminum hydride.

Among the most selective of the hydride reducing agents are those containing several large alkyl substituents. These reagents can be prepared from organoboranes

Table 4.2. Stereochemistry of Hydride Reductions[a]

	Percentage alcohol favored by steric approach control				
Reducing agent	 % axial	 % axial	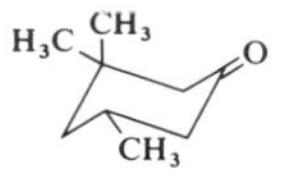 % axial	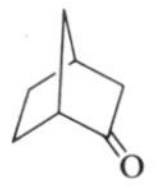 % endo	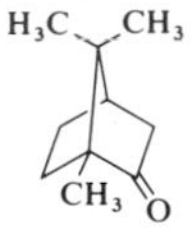 % exo
NaBH$_4$	20[b]	25[c]	58[c]	86[d]	86[d]
LiAlH$_4$	8	24	83	89	92
LiAl(OMe)$_3$H	9	69		98	99
LiAl(t-BuO)$_3$H	9[e]	36[f]	95	94[f]	94[f]
$\left[(CH_3)_2CHCH(CH_3) \right]_2BH$	37	94		94	100
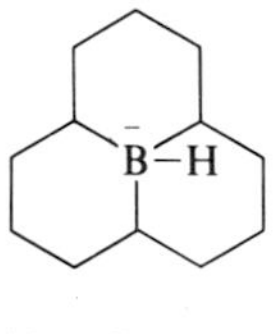 (decalylborane) Li$^+$	54	94	99	99	99
(CH$_3$CH$_2$CH(CH$_3$)—)$_3$BH Li$^+$	93[g]	98[g]	99.8[g]	99.6[g]	99.6[g]

a. Except where otherwise noted, data are those given by H. C. Brown and W. C. Dickason, *J. Am. Chem. Soc.* **92**, 709 (1970). Data for many other cyclic ketones and reducing agents are given by A. V. Kamernitzky and A. A. Akhrem, *Tetrahedron* **18**, 705 (1962).
b. P. T. Lansbury and R. E. MacLeay, *J. Org. Chem.* **28**, 1940 (1963).
c. B. Rickborn and W. T. Wuesthoff, *J. Am. Chem. Soc.* **92**, 6894 (1970).
d. H. C. Brown and J. Muzzio, *J. Am. Chem. Soc.* **88**, 2811 (1966).
e. J. Klein, E. Dunkelblum, E. L. Eliel, and Y. Senda, *Tetrahedron Lett.*, 6127 (1968).
f. E. C. Ashby, J. P. Sevenair, and F. R. Dobbs, *J. Org. Chem.* **36**, 197 (1971).
g. H. C. Brown and S. Krishnamurthy, *J. Am. Chem. Soc.* **94**, 7159 (1972).

or aluminum alkyls. For example, the cyclic borane **1** reacts with lithium hydride to give a trialkylborohydride.[27] This hydride-transfer agent is characterized by a very

B + LiH ⟶ B⁻—H Li⁺

1　　　**2**

high stereoselectivity. Corey and co-workers observed superior stereoselectivity in a ketone reduction with the highly hindered trialkylborohydride **3**[28]:

$$H_3C \quad \overset{H}{\underset{CH_3}{B^-}} \overset{CH_3}{\underset{|}{C}} -CH(CH_3)_2$$

3

The highest stereoselectivity reported to date apparently is that of lithium tris(*s*-butyl)borohydride.[29] This reagent is sufficiently bulky that even 4-alkyl-cyclohexanones are reduced by equatorial approach, giving the axial alcohol.

The stereochemistry of reduction of acyclic ketones can be predicted on the basis of steric factors. An asymmetric carbon atom adjacent to the carbonyl center undergoing reduction can control the direction of approach of the reducing agent:

minor

H_5C_2 — Ph
O = — CH_3 ⟶
H

major

H_5C_2 — OH — Ph
H — CH_3
H
(75%)

H_5C_2 — H — Ph
HO — CH_3
H
(25%)

Ref. 30

A prediction on the basis of the steric factors is made by assuming a conformation in which the carbonyl group is staggered between the smallest and medium groups on the α-carbon. Addition of hydride from the less hindered side in this conformation leads to a correct prediction of stereochemistry. The reason for choosing this conformation is that it places the carbonyl oxygen in the least hindered environment. Although this is not necessarily the most stable conformation of the ground-state ketone, the transition state for hydride reduction places the rather large aluminum or boron group at the carbonyl oxygen, greatly increasing its steric requirement.

Steric factors arising from groups more remote from the center undergoing reduction can also influence the stereochemical course of reduction. These steric

27. H. C. Brown and W. C. Dickason, *J. Am. Chem. Soc.* **92**, 709 (1970).
28. E. J. Corey, S. M. Albonico, U. Koelliker, T. K. Schaaf, and R. K. Varma, *J. Am. Chem. Soc.* **93**, 1491 (1971).
29. H. C. Brown and S. Krishnamurthy, *J. Am. Chem. Soc.* **94**, 7159 (1972). This reagent is available commercially and is referred to as "Selectride."
30. D. J. Cram and F. A. Abd Elhafez, *J. Am. Chem. Soc.* **74**, 5828 (1952).

factors are magnified by using bulky reducing agents. For example, a 4.5 : 1 preference for **5** over **6** is achieved by using the very bulky trialkylborohydride **3** in reduction of the side-chain of a prostaglandin intermediate[28]:

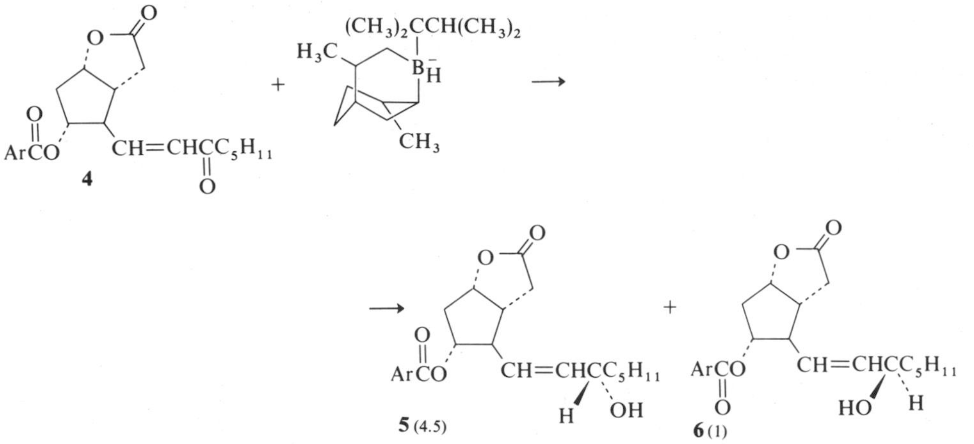

The hydride reduction of α,β-unsaturated carbonyl compounds can take one of two courses. Initial reaction at the carbonyl group gives allylic alcohols. Usually, no further reduction takes place, since the unconjugated carbon–carbon double bond is inert to nucleophilic species. If initial attack is at the double bond, an enolate is produced. In protic solvents, this can lead to the carbonyl compounds, which can, in turn, be reduced, resulting ultimately in the saturated alcohol. For this reason, it is

$$R_2C{=}CH\overset{O}{\overset{\|}{C}}R' + [H^-] \rightarrow R_2C{=}CH\overset{O^-}{\underset{H}{C}}R' \xrightarrow{H^+} R_2C{=}CH\overset{OH}{C}HR'$$

1,2-reduction

$$R_2C{=}CH\overset{O}{\overset{\|}{C}}R' + [H^-] \rightarrow R_2CH{-}CH{=}\overset{O^-}{C}R' \xrightarrow{H^+} R_2CHCH_2\overset{O}{\overset{\|}{C}}R'$$

$$R_2CHCH_2\overset{O}{\overset{\|}{C}}R' + [H^-] \rightarrow R_2CHCH_2\overset{O^-}{C}HR' \xrightarrow{H^+} R_2CHCH_2\overset{OH}{C}HR'$$

1,4-reduction leading to saturated alcohol

fairly common to find both saturated and unsaturated alcohols from $NaBH_4$ or $LiAlH_4$ reduction of conjugated unsaturated ketones.[31] The extent of reduction to the saturated alcohol is usually greater with $NaBH_4$ than with $LiAlH_4$.

Diisobutylaluminum hydride has been recommended for reduction of enones to allylic alcohols, since this reagent has not been observed to cause reduction of the adjacent double bond.[32] If it is desired to reduce the double bond without involving

28. See p. 137.
31. M. R. Johnson and B. Rickborn, *J. Org. Chem.* **35**, 1041 (1970); W. R. Jackson and A. Zurqiyah, *J. Chem. Soc.*, 5280 (1965).
32. K. E. Wilson, R. T. Seidner, and S. Masamune, *Chem. Commun.*, 213 (1970).

the carbonyl group, this reduction can usually be accomplished by catalytic hydrogenation. A reagent prepared from copper hydride and an alkyllithium compound will also selectively reduce the carbon–carbon double bond.[33]

Although reductions of the common carbonyl and carboxylic acid derivatives are the most widely used cases of complex metal-hydride reductions, there are certain other systems that have been of sufficient synthetic utility to discuss at this point. Scheme 4.1 summarizes some of these systems.

The enol ethers of β-dicarbonyl compounds are reduced to α,β-unsaturated ketones by LiAlH₄ followed by hydrolysis.[34] Reduction proceeds only to the allylic alcohol, but subsequent treatment with acid effects hydrolysis of the enol ether and dehydration.

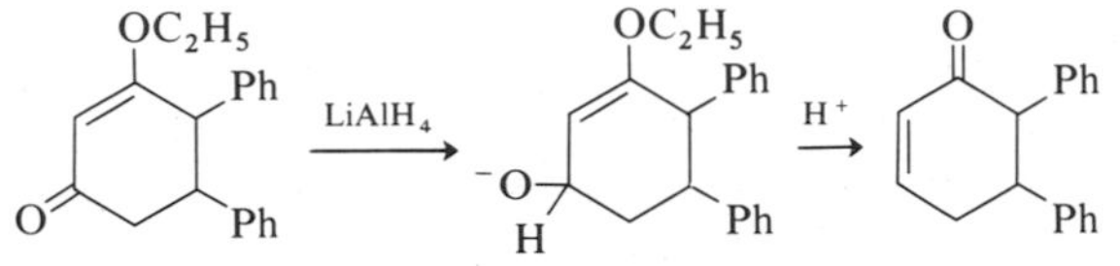

Epoxides are converted to alcohols by LiAlH₄. The reaction occurs by nucleophilic attack, which takes place preferentially at the less hindered carbon of the

epoxide. Cyclohexene epoxides are reduced through transition states involving diaxial opening[35]:

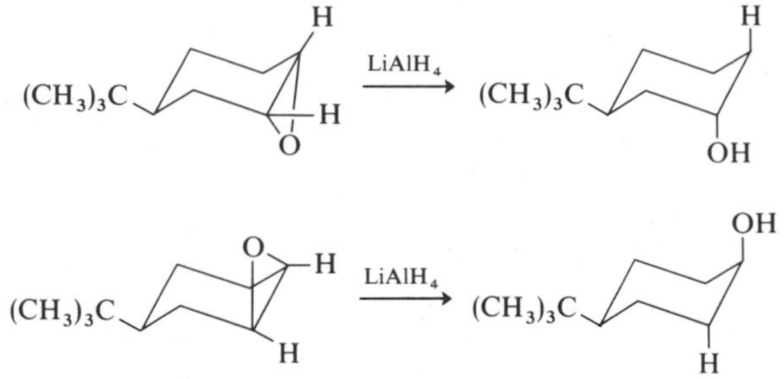

Lithium triethylborohydride is a superior reagent for reduction of epoxides that are relatively unreactive or prone to rearrangement.[36]

The complex metal hydrides can effect replacement of halogen or sulfonate groups under reaction conditions that favor S_N2 processes. Sodium borohydride in solvents such as DMSO or sulfolane that enhance nucleophilic reactivity has been

33. S. Masamune, G. S. Bates and P. E. Georghiou, *J. Am. Chem. Soc.* **96**, 3686 (1974).
34. H. E. Zimmerman and D. I. Schuster, *J. Am. Chem. Soc.* **84**, 4527 (1962).
35. B. Rickborn and J. Quartucci, *J. Org. Chem.* **29**, 3185 (1964); B. Rickborn and W. E. Lamke, II, *J. Org. Chem.* **32**, 537 (1967); D. K. Murphy, R. L. Alumbaugh, and B. Rickborn, *J. Am. Chem. Soc.* **91**, 2649 (1969).
36. S. Krishnamurthy, R. M. Schubert, and H. C. Brown, *J. Am. Chem.Soc.* **95**, 8486 (1973).

**Scheme 4.1. Reduction of Other Functional Groups by Complex
Metal Hydrides**

3-Alkoxycyclohexenones

1[a]

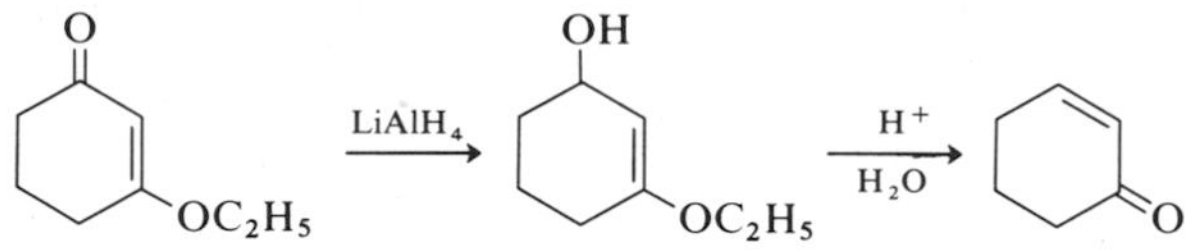

Epoxides

2[b]

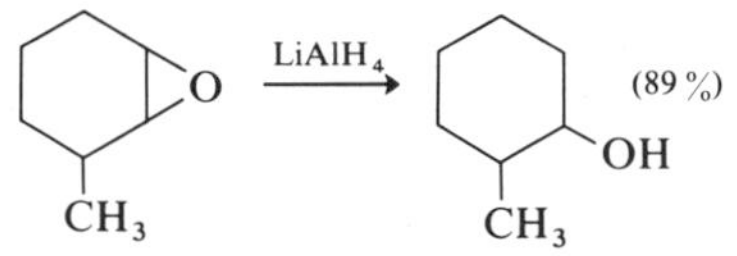

(89 %)

Halides

3[c] $CH_3(CH_2)_5\underset{\underset{Cl}{|}}{CH}CH_3 \xrightarrow[DMSO]{NaBH_4} CH_3(CH_2)_6CH_3$ (67 %)

4[d] $CH_3(CH_2)_8CH_2I \xrightarrow[HMPA]{NaBH_3CN} CH_3(CH_2)_8CH_3$ (88–90 %)

5[e]

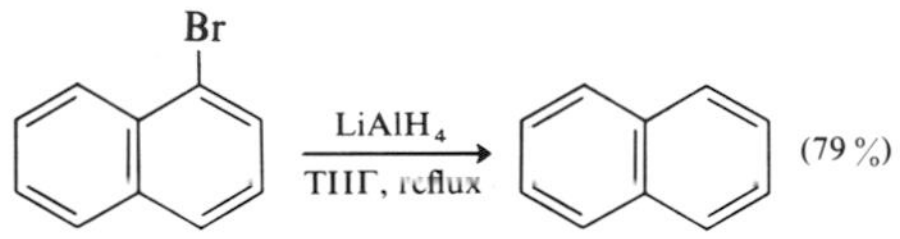

$\xrightarrow[THF, reflux]{LiAlH_4}$ (79 %)

Sulfonate Esters

6[f]

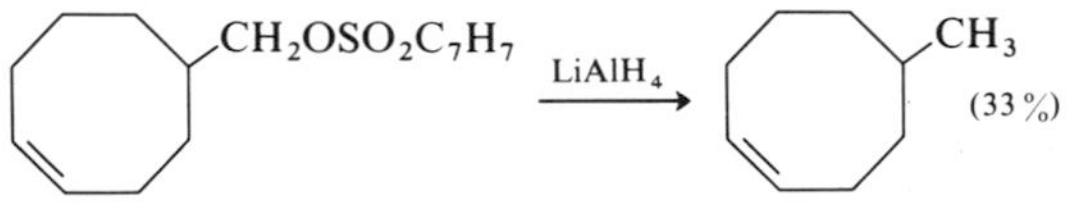

$\xrightarrow{LiAlH_4}$ (33 %)

7[g]

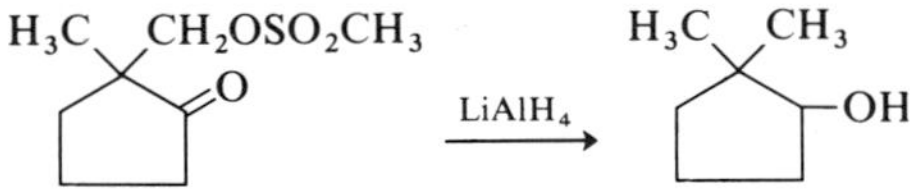

$\xrightarrow{LiAlH_4}$

8[h]

$\xrightarrow{LiCuHC_4H_9}$

(75 %)

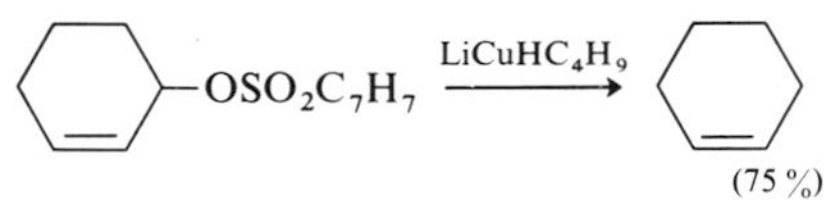

a. W. F. Gannon and H. O. House *Org. Synth.* **40**, 14 (1960).
b. B. Rickborn and W. E. Lamke, II, *J. Org. Chem.* **32**, 537 (1967).
c. R. O. Hutchins, D. Hoke, J. Keogh, and D. Koharski, *Tetrahedron Lett.*, 3495 (1969); H. M. Bell, C. W. Vanderslice, and A. Spehar, *J. Org. Chem.* **34**, 3923 (1969).
d. R. O. Hutchins, C. A. Milewski, and B. E. Maryanoff, *Org. Synth.* **53**, 107 (1973).
e. H. C. Brown and S. Krishnamurthy, *J. Org. Chem.* **34**, 3918 (1969).
f. A. C. Cope and G. L. Woo, *J. Am. Chem. Soc.* **85**, 3601 (1963).
g. A. Eschenmoser and A. Frey, *Helv. Chim. Acta* **35**, 1660 (1952).
h. S. Masamune, G. S. Bates, and P. E. Geoghiou, *J. Am. Chem. Soc.* **96**, 3686 (1974).

found to reduce halides and sulfonates to the corresponding hydrocarbons.[37] Lithium aluminum hydride is also an effective reducing agent for sulfonates. Sulfonation followed by lithium aluminum hydride reduction therefore constitutes an important method for reductive removal of hydroxy groups. Lithium triethyl-

$$ROH + ArSO_2Cl \rightarrow ROSO_2Ar$$

$$ROSO_2Ar + LiAlH_4 \rightarrow R-H$$

borohydride is the most reactive reagent known for reduction of alkyl halides. It gives high yields of the corresponding hydrocarbon.[38]

Although some of these reductions proceed with inversion of configuration, indicating an S_N2 mechanism,[39] the S_N2 process is evidently not the only mechanism by which $LiAlH_4$ and related hydrides can reduce halides. Such compounds as vinyl halides, bridgehead halides, and cyclopropyl halides, all of which are extremely resistant to S_N2 processes, can be reduced to hydrocarbons in good yield by $LiAlH_4$.[40] Aryl halides are also reduced by $LiAlH_4$ in refluxing THF.[41] Bromides and iodides are reduced substantially more rapidly than chlorides. The mechanisms by which these reductions occur have not yet been studied in detail, but clearly, simple nucleophilic displacement on carbon does not apply.

There is also a group of reactions in which hydride is transferred from carbon. The carbon–hydrogen bond has little intrinsic polarity or tendency to break in the way required for hydride transfer. These reactions usually proceed via cyclic transition states in which new C–H bonds are formed simultaneously with the cleavage. Hydride transfer is also facilitated by high charge density on the donor carbon atom. The Cannizzaro reaction, the base-catalyzed disproportionation of aldehydes, is one example of a hydride-transfer reaction. A general mechanism is outlined below. The hydride transfer is believed to occur from a species bearing two

$$PhCH{=}O + \overset{-}{O}H \rightleftharpoons PhC\overset{\overset{\displaystyle H}{|}}{\underset{\underset{\displaystyle OH}{|}}{}}{-}O^-$$

$$PhC\overset{\overset{\displaystyle H}{|}}{\underset{\underset{\displaystyle OH}{|}}{}}{-}O^- + \overset{-}{O}H \rightleftharpoons PhC\overset{\overset{\displaystyle H}{|}}{\underset{\underset{\displaystyle O^-}{|}}{}}{-}O^- + H_2O$$

$$PhC{-}H \quad CPh \rightleftharpoons PhC{-}O^- + PhCH_2O^-$$

37. R. O. Hutchins, D. Hoke, J. Keogh, and D. Koharski, *Tetrahedron Lett.*, 3495 (1969); H. M. Bell, C. W. Vanderslice, and A. Spehar, *J. Org. Chem.* **34**, 3923 (1969).
38. H. C. Brown and S. Krishnamurthy, *J. Am. Chem. Soc.* **95**, 1669 (1973).
39. G. K. Helmкamp and B. F. Rickborn, *J. Org. Chem.* **22**, 479 (1957).
40. C. W. Jefford, D. Kirkpatrick, and F. Delay, *J. Am. Chem. Soc.* **94**, 8905 (1972).
41. H. C. Brown and S. Krishnamurthy, *J. Org. Chem.* **34**, 3918 (1969); P. Olavi, I. Virtanen, and P. Jaakkola, *Tetrahedron Lett.*, 1223 (1969).

negative charges, and presumably this high charge density is responsible for the ease of hydride transfer. The reaction is not adaptable to aldehydes with enolizable hydrogen because the aldol condensation and subsequent transformations intervene. The reaction has limited modern synthetic utility. The role of reducing agent and substrate can be assumed by two different aldehydes. If this is desired, formaldehyde is usually used as the reducing agent. The synthesis of polylhydroxymethyl compounds makes use of fomaldehyde both for aldol condensation and reduction:

$$RCH_2CH{=}O \; + \; 2\,CH_2{=}O \; \underset{}{\overset{NaOH}{\rightleftharpoons}} \; \underset{\underset{CH_2OH}{|}}{\overset{\overset{CH_2OH}{|}}{R\overset{}{C}CH{=}O}}$$

$$\underset{\underset{CH_2OH}{|}}{\overset{\overset{CH_2OH}{|}}{R\overset{}{C}CH{=}O}} \; + \; CH_2{=}O \; \xrightarrow{NaOH} \; \underset{\underset{CH_2OH}{|}}{\overset{\overset{CH_2OH}{|}}{R\overset{}{C}CH_2OH}} \; + \; HCO_2H$$

Aluminum alkoxides catalyze transfer of hydride from alcohols to ketones. This reaction can be driven to completion if one of the ketones is removed from the reaction system—by distillation, for example. This reaction, usually carried out with aluminum isopropoxide, is known as the *Meerwein–Pondorff–Verley reaction.*[42] The

$$3\,R_2C{=}O \; + \; Al[OCH(CH_3)_2]_3 \; \rightarrow \; [R_2CHO]_3Al \; + \; 3\,CH_3\underset{\underset{O}{\|}}{C}CH_3$$

reaction is thought to proceed via a cyclic transition state involving coordination of the aluminum atom with the carbonyl group. Hydride donation usually takes place

from the less hindered side, so that in reductions of cyclohexanones, for example, the axial alcohol is a major product.[43]

Similar hydride transfers occur with sodium alkoxides on heating. This constitutes an important means of effecting stereochemical equilibration of a hydroxyl group. The normal procedure is to add a small amount of a carbonyl compound, frequently benzophenone, to act as the initial hydride acceptor and thus catalyze the

reaction. Repeated formation and oxidation of the alcohol leads to the establishment of the equilibrium composition.

42. A. L. Wilds, *Org. React.* **2**, 178 (1944).
43. F. Nerdel, D. Frank, and G. Barth, *Chem. Ber.* **102**, 395 (1969).

When normal addition of Grignard reagents to ketones is severely hindered by steric factors, a hydride-transfer reaction becomes the dominant reaction.[44] This process will be discussed further in Chapter 5.

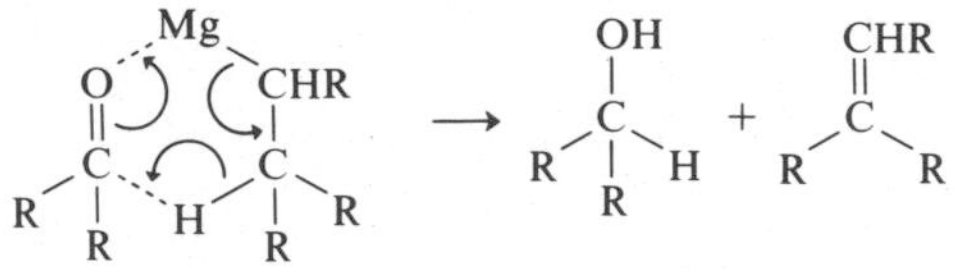

A novel reduction process, which is characterized by very high axial-to-equatorial product ratios, apparently involves hydride transfer from the isopropanol used in the reaction, although the detailed mechanism of the reaction is not established.[45]

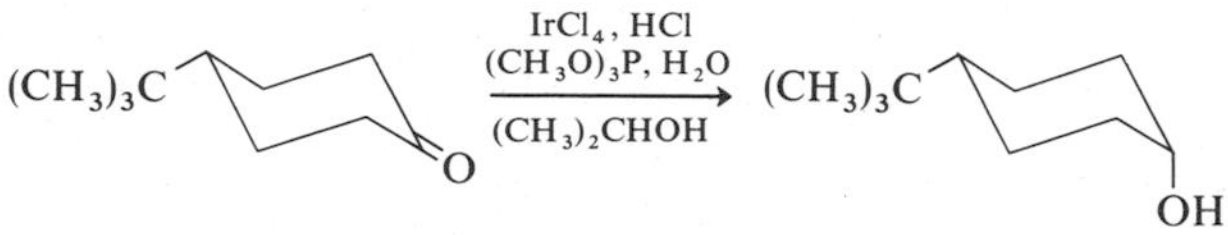

Carbonium ions can abstract hydride from potential hydride donors. The silicon–hydrogen bond is very reactive toward carbonium ions, resulting in reduction of the carbonium ion to the hydrocarbon.[46] This reaction is preparatively useful for reduction of alcohols that can be converted to carbonium ions in trifluoroacetic acid.

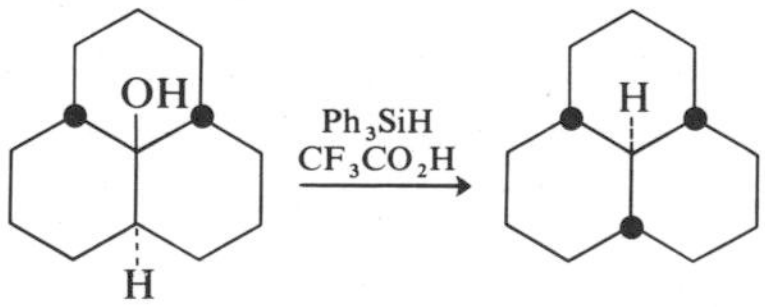

Ref. 46

4.2. Hydrogen-Atom Donors

Reduction by hydrogen-atom donors necessarily involves intermediates with unpaired electrons. Tri-n-butyltin hydride is the most important example of this type of reducing agent. It is able to reductively replace halogen by hydrogen in many types of halogen compounds. Mechanistic studies have indicated a free-radical chain mechanism.[47] The ability of trisubstituted stannanes to function effectively in these

$$In\cdot \ + \ Bu_3SnH \ \rightarrow \ In-H \ + \ Bu_3Sn\cdot$$

$$Bu_3Sn\cdot \ + \ R-X \ \rightarrow \ R\cdot \ + \ Bu_3SnX$$

$$R\cdot \ + \ Bu_3SnH \ \rightarrow \ RH \ + \ Bu_3Sn\cdot$$

44. J. S. Birtwistle, K. Lee, J. D. Morrison, W. A. Sanderson, and H. S. Mosher, *J. Org. Chem.* **29**, 37 (1964).
45. E. L. Eliel, T. W. Doyle, R. O. Hutchins, and E. C. Gilbert, *Org. Synth.* **50**, 13 (1970).
46. F. A. Carey and H. S. Tremper, *J. Org. Chem.* **36**, 758 (1971).
47. L. W. Menapace and H. G. Kuivila, *J. Am. Chem. Soc.* **86**, 3047 (1964).

Scheme 4.2. Dehalogenations with Stannanes

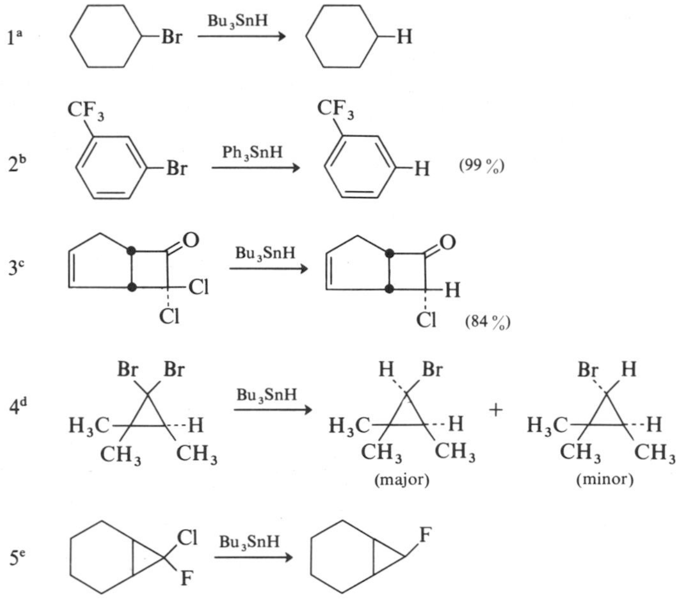

a. H. G. Kuivila, L. W. Menapace, and C. R. Warner, *J. Am. Chem. Soc.* **84**, 3584 (1962).
b. D. H. Lorenz, P. Shapiro, A. Stern, and E. I. Becker, *J. Org. Chem.* **28**, 2332 (1963).
c. W. T. Brady and E. F. Hoff, Jr., *J. Org. Chem.* **35**, 3733 (1970).
d. D. Seyferth, H. Yamazaki, and D. L. Alleston, *J. Org. Chem.* **28**, 703 (1963).
e. T. Ando, F. Namigata, H. Yamanaka, and W. Funasaka, *J. Am. Chem. Soc.* **89**, 5719 (1967).

reactions depends on the facility with which they serve as hydrogen-atom donors. The order of reactivity for the halides is $RI > RBr > RCl > RF$,[48] which reflects the relative ease of the halogen-atom abstraction.

Tri-*n*-butyltin hydride shows substantial selectivity toward polyhalogenated compounds, permitting partial dehalogenation. The reason for the higher reactivity of more highly halogenated carbons toward reduction lies in the stabilizing effect that the remaining halogen has on the radical intermediate. This selectivity has been used, for example, to reduce dihalocyclopropanes to monohalocyclopropanes (entries 4 and 5, Scheme 4.2).

The reductive dehalogenation can also be applied to the conversion of acid chlorides to aldehydes.[49] A competing reaction leading to conversion of some of the

$$RCOCl \xrightarrow{Bu_3SnH} RCHO$$

aldehyde to an ester occurs under some conditions. Reductions by processes

48. H. G. Kuivila and L. W. Menapace, *J. Org. Chem.* **28**, 2165 (1963).
49. H. G. Kuivila and E. J. Walsh, Jr., *J. Am. Chem. Soc.* **88**, 571 (1966); E. J. Walsh, Jr., and H. G. Kuivila, *J. Am. Chem. Soc.* **88**, 576 (1966).

$$\underset{\substack{\| \\ O}}{RCCl} + Bu_3Sn\cdot \rightarrow \underset{\substack{\| \\ O}}{RC}\cdot + Bu_3SnCl$$

$$\underset{\substack{\| \\ O}}{RC}\cdot + RCH{=}O \rightarrow \underset{\substack{\| \\ O}}{RC}-O-\overset{\cdot}{C}HR$$

$$\underset{\substack{\| \\ O}}{RC}-O-\overset{\cdot}{C}HR + Bu_3SnH \rightarrow \underset{\substack{\| \\ O}}{RC}OCH_2R + Bu_3Sn\cdot$$

involving hydrogen-atom donors also occur in a number of photochemical processes and in several types of free-radical reactions. The discussion of these processes can be found in Part A, Chapters 11 and 12.

4.3. Dissolving-Metal Reductions

A number of useful reduction processes involve metals as the reducing agent. Scheme 4.3 lists some of the more important reactions that fall into this group. Before the advent of the metal hydrides, the reduction of ketones to alcohols with active metals, primarily sodium, using alcohols such as ethanol and propanol as solvent was an important synthetic process. This procedure, sometimes called the *Bouvault–Blanc reduction*, is seldom the method of choice nowadays. The mechanism involves electron transfer from the metal to the carbonyl in two separate stages.

$$Na\cdot + R_2CO \rightarrow Na^+ + R_2CO\cdot \xrightarrow{R'OH} R_2CHO\cdot$$

$$R_2CHO\cdot + Na\cdot \rightarrow R_2CHO^- + Na^+$$

Usually, the reaction leads predominantly to the more stable alcohol of a stereoisomeric pair because the reaction conditions promote equilibration by the hydride-transfer process discussed in Section 4.1.

Ketones can be reduced to symmetrical diols when coupling of the one-electron reduction intermediates is faster than subsequent reduction to the alkoxide. The conditions required to favor the coupling involve a diminished rate of protonation of the initial radical anion, so that it couples in preference to protonation. For this

$$R_2C{=}O + e^- \rightarrow R_2\overset{\cdot}{C}O^-$$

$$2\,R_2\overset{\cdot}{C}O^- \rightarrow R_2\overset{\substack{O^- \\ |}}{C}\!-\!-\!\overset{\substack{O^- \\ |}}{C}R_2$$

reason, reductive dimerizations are usually carried out in hydrocarbon solvent. Magnesium, zinc, and aluminum amalgam are the most commonly used metals. The reduction of acetone to 2,3-dimethylbutane-2,3-diol (pinacol) is given as an example in Scheme 4.3.

Reductive dimerization of esters proceeds with sodium metal in inert solvents. The reaction is sometimes referred to as the *acyloin condensation*, after the class

Scheme 4.3. Dissolving-

Ketones

1^a $CH_3CO(CH_2)_4CH_3 \xrightarrow[\text{EtOH}]{\text{Na}} CH_3\underset{\underset{OH}{|}}{C}H(CH_2)_4CH_3$

2^b $2(CH_3)_2CO \xrightarrow{\text{Mg(Hg)}} (CH_3)_2\underset{\underset{HO}{|}}{C}-\underset{\underset{OH}{|}}{C}(CH_3)_2$

Esters

3^c $C_2H_5O_2C(CH_2)_8CO_2C_2H_5 \xrightarrow[\text{EtOH}]{\text{Na}} HOCH_2(CH_2)_8CH_2OH$

4^d $CH_3O_2C(CH_2)_8CO_2CH_3 \xrightarrow[\text{xylene}]{\text{Na}}$ $\xrightarrow{\text{H}^+}$

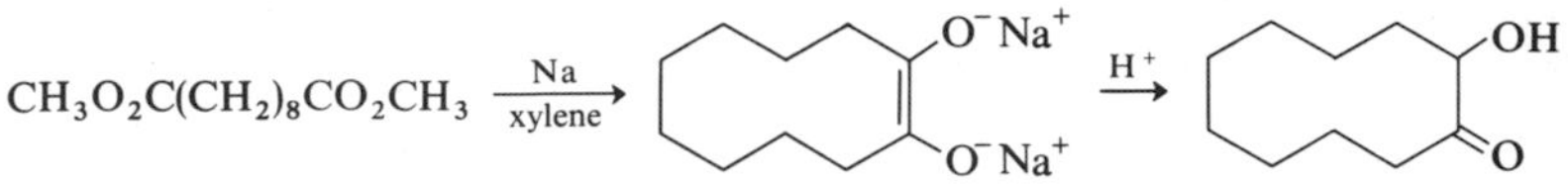

5^e $\xrightarrow[\text{xylene}]{\text{Na}}$

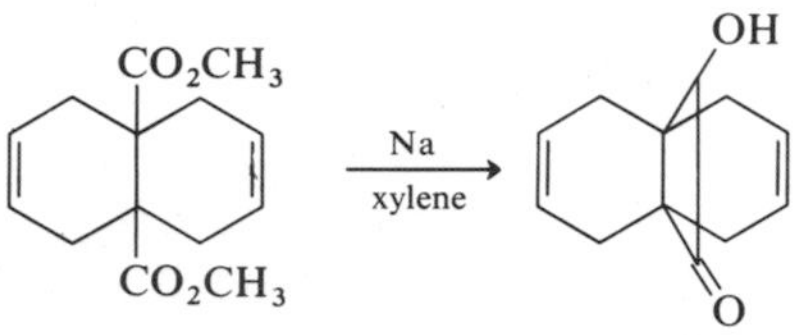

Nitrogen-Containing Groups

6^f $\xrightarrow[\text{H}^+]{\text{Fe}}$ (74 %)

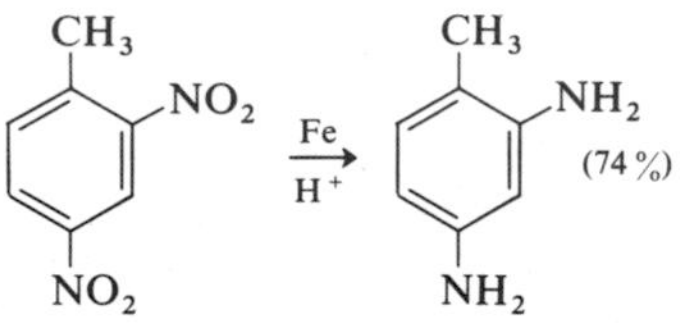

7^g $CH_3(CH_2)_5CH{=}NOH \xrightarrow[\text{EtOH}]{\text{Na}} CH_3(CH_2)_6NH_2$ (60–73 %)

α,β-Unsaturated Ketones

8^h $\xrightarrow[\text{NH}_3]{\text{Li}}$ $\xrightarrow{\text{H}^+}$

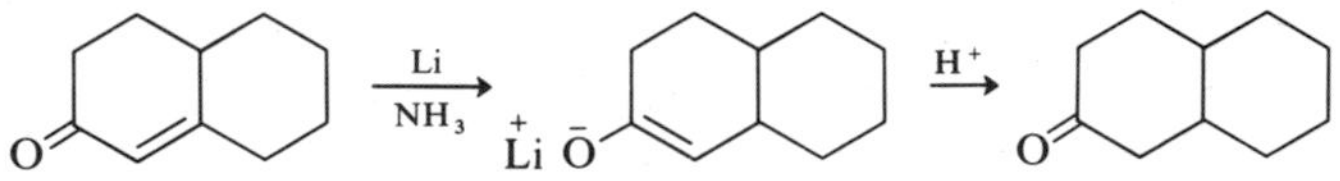

9^i $\xrightarrow[\text{NH}_3]{\text{Li, }t\text{-BuOH}}$

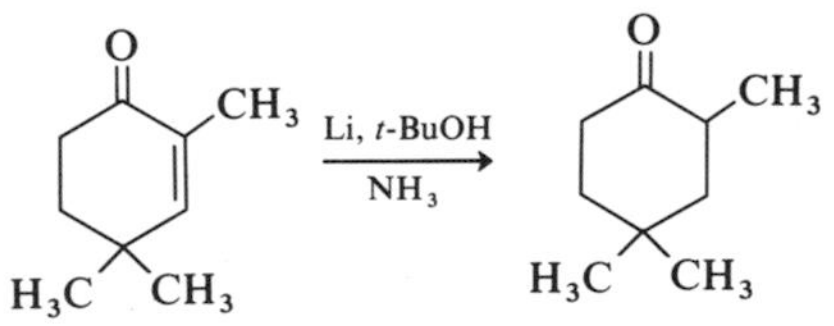

Epoxides

10^j $\xrightarrow[\text{H}_2NCH_2CH_2NH_2]{\text{Li}}$

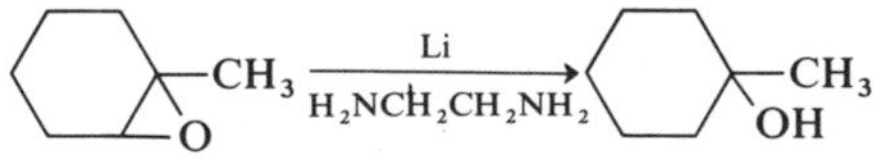

Metal Reductions

147

SECTION 4.3.
DISSOLVING-
METAL
REDUCTIONS

Dehalogenations

11[k]

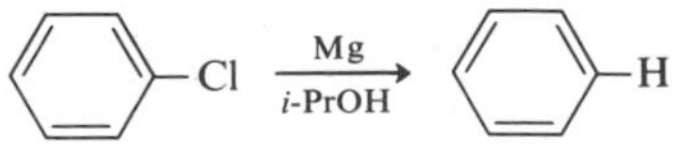

12[l]

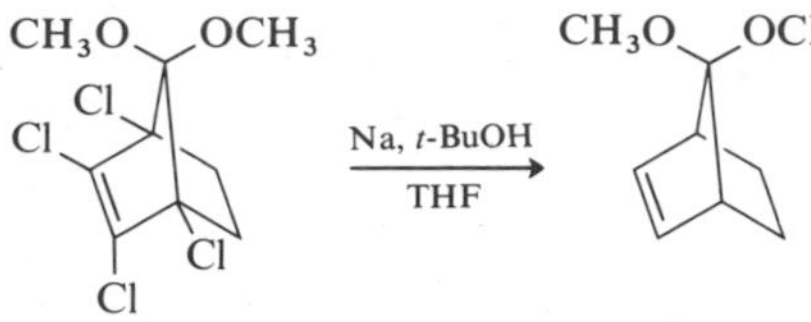

(40%)

13[m]

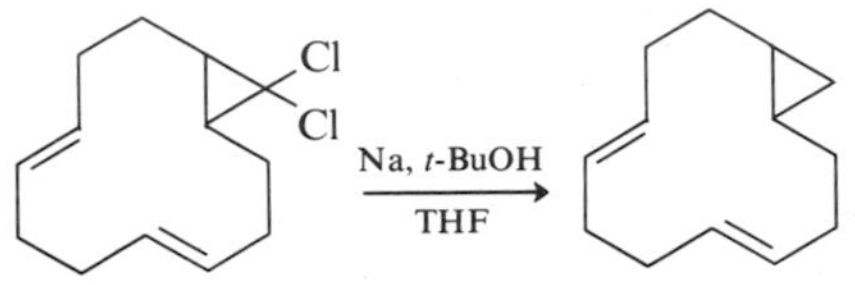

Reductions of Aromatic Rings

14[n]

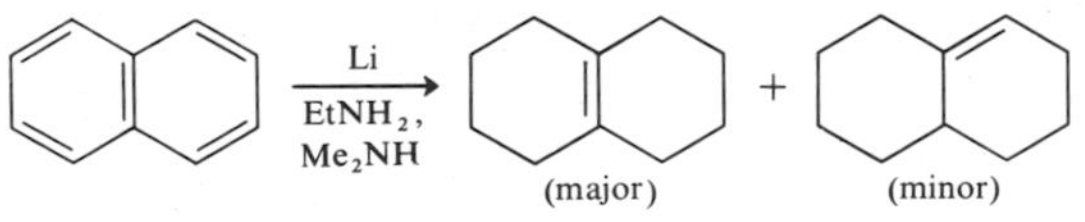

(major) + (minor)

15[o]

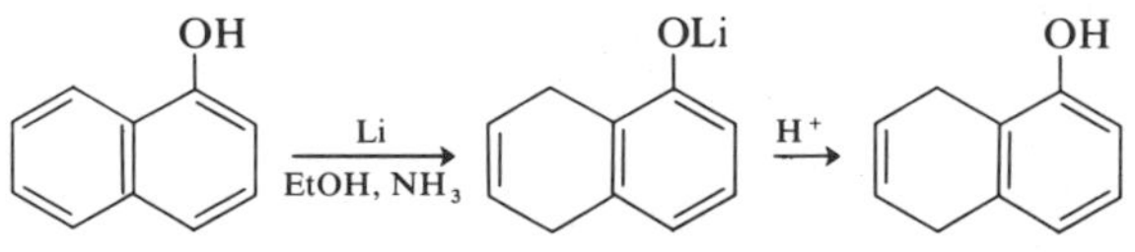

(97–99%)

16[p]

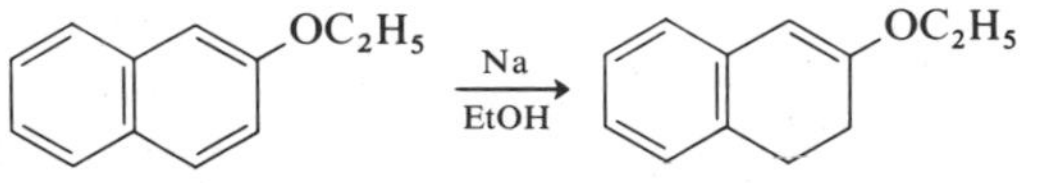

17[q]

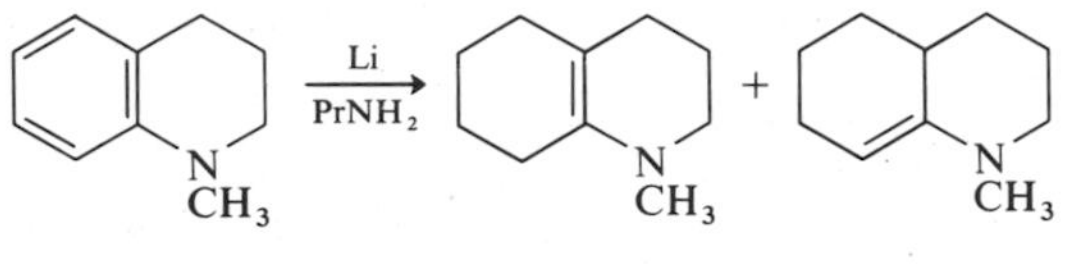

a. F. C. Whitmore and T. Otterbacher, *Org. Synth.* **II**, 317 (1943).
b. R. Adams and E. W. Adams, *Org. Synth.* **I**, 448 (1932).
c. R. H. Manske, *Org. Synth.* **II**, 154 (1943).
d. N. L. Allinger, *Org. Synth.* **IV**, 840 (1963).
e. J. J. Bloomfield and J. R. S. Irelan, *J. Org. Chem.* **31**, 2017 (1966).
f. S. A. Mahood and P. V. L. Schaffner, *Org. Synth.* **II**, 160 (1943).
g. W. H. Lycan, S. V. Puntambeker, and C. S. Marvel, *Org. Synth.* **II**, 318 (1943).
h. G. Stork, P. Rosen, N. Goldman, R. V. Coombs, and J. Tsuji, *J. Am. Chem. Soc.* **87**, 275 (1965).
i. H. A. Smith, B. J. L. Huff, W. J. Powers, III, and D. Caine, *J. Org. Chem.* **32**, 2851 (1967).
j. H. C. Brown, S. Ikegami, and J. H. Kawakami, *J. Org. Chem.* **35**, 3243 (1970).
k. D. Bryce-Smith and B. J. Wakefield, *Org. Synth.* **47**, 103 (1967).
l. P. G. Gassman and P. G. Pape, *J. Org. Chem.* **29**, 160 (1964); P. G. Gassman and J. L. Marshall, *Org. Synth.* **48**, 68 (1968).
m. P. G. Gassman, J. Seter, and F. J. Williams, *J. Am. Chem. Soc.* **93**, 1673 (1971).
n. E. M. Kaiser and R. A. Benkeser, *Org. Synth.* **50**, 88 (1970).
o. C. D. Gutsche and H. H. Peter, *Org. Synth.* **IV**, 887 (1963).
p. M. D. Soffer, M. P. Bellis, H. E. Gellerson, and R. A. Stewart, *Org. Synth.* **IV**, 903 (1963).
q. N. J. Leonard, C. K. Steinhardt, and C. Lee, *J. Org. Chem.* **27**, 4027 (1963).

name of the reaction products. Diesters react intramolecularly under these conditions to give cyclic acyloins.[50] This reaction is one of the important methods for preparation of medium- and large-sized rings, as illustrated by entry 4 in Scheme 4.3.

$$RCOR' + Na\cdot \rightarrow R\overset{\cdot}{C}OR'$$

$$2\,RCOR' \rightarrow R\overset{|}{C}\!\!-\!\!\overset{|}{C}R \rightarrow R\overset{\parallel}{C}\!\!-\!\!\overset{\parallel}{C}R$$

$$R\overset{\parallel}{C}\!\!-\!\!\overset{\parallel}{C}R + 2\,Na\cdot \rightarrow R\overset{-}{C}\!\!=\!\!\overset{-}{C}R \overset{H^+}{\rightarrow} RCCHR$$

An alternative mechanism has been suggested. It bypasses the α-dicarbonyl intermediate, since this is a species with questionable involvement in the reaction[51]:

$$RCO_2R' + Na\cdot \rightarrow R\overset{\cdot}{C}OR' \qquad R\overset{\cdot}{C}OR' + RCO_2R' \rightarrow R\overset{|}{C}\!\!-\!\!O\!\!-\!\!\overset{|}{C}R \rightarrow$$

$$\overset{Na\cdot}{\rightarrow} R\overset{|}{C}\!\!-\!\!O\!\!-\!\!\overset{|}{C}R \rightarrow R\overset{|}{C}\!\!-\!\!\overset{|}{C}R \overset{Na}{\rightarrow} R\overset{|}{C}\!\!-\!\!\overset{|}{C}R \rightarrow R\overset{-}{C}\!\!=\!\!\overset{-}{C}R \overset{Na\cdot}{\rightarrow} R\overset{-}{C}\!\!=\!\!\overset{-}{C}R$$

Zinc and hydrochloric acid can accomplish reduction of carbonyl groups to methylene. The oxygen is believed to be eliminated as water from a partially reduced intermediate bound to zinc. This reaction is known as the *Clemmensen reduction.*[52] The corresponding alcohols are not reduced to hydrocarbons under the conditions of the reactions, so they obviously cannot be intermediates in the overall process. The

$$R\overset{O}{\overset{\parallel}{C}}R + H^+ \rightleftharpoons R\overset{+OH}{\overset{\parallel}{C}}R \overset{Zn}{\rightarrow} R\underset{Zn\!-\!Zn\!-\!Zn}{\overset{OH}{\overset{|}{C}}}R$$

$$R\underset{Zn\!-\!Zn\!-\!Zn}{\overset{OH}{\overset{|}{C}}}R + H^+ \rightarrow R\overset{O}{\underset{Zn\!-\!Zn\!-\!Zn}{\overset{\parallel}{C}}}R + H_2O \overset{H^+}{\rightarrow} RCH_2R$$

Clemmensen reduction is usually quite efficient for aryl ketones, but less reliable when the carbonyl group is not conjugated. Several examples of the reaction are included in Scheme 4.4.

50. K. T. Finley, *Chem. Rev.* **64**, 573 (1964); K. T. Finley and N. A. Sasaki, *J. Am. Chem. Soc.* **88**, 4267 (1966).
51. J. J. Bloomfield, D. C. Owsley, C. Ainsworth, and R. E. Robertson, *J. Org. Chem.* **40**, 393 (1975).
52. E. Vedejs, *Org. React.* **22**, 401 (1975).

Scheme 4.4. Carbonyl-to-Methylene Reductions

Clemmensen

1[a]

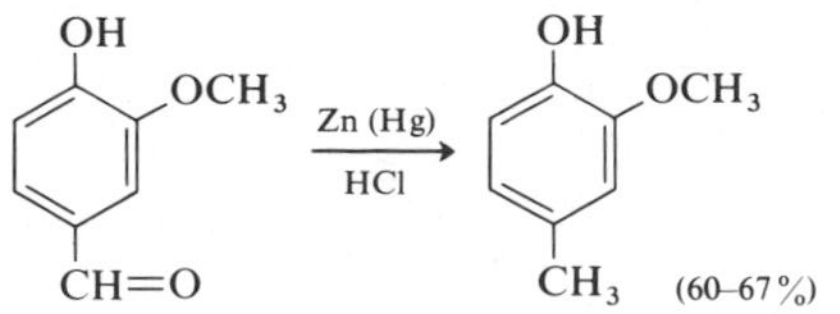

(60–67%)

2[b]

$CH_2(CH_2)_5CH_3$ (81–86%)

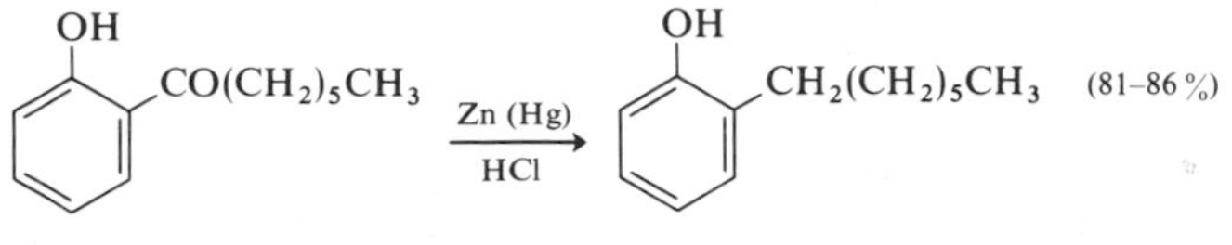

Wolff–Kishner

3[c] $HO_2C(CH_2)_4CO(CH_2)_4CO_2H \xrightarrow[\text{KOH}]{\text{NH}_2\text{NH}_2} HO_2C(CH_2)_9CO_2H$ (87–93%)

4[d] $Ph\underset{\underset{NNH_2}{\|}}{-C-}Ph \xrightarrow[\text{DMSO}]{\text{KOC(CH}_3)_3} PhCH_2Ph$ (90%)

Caglioti

5[e]

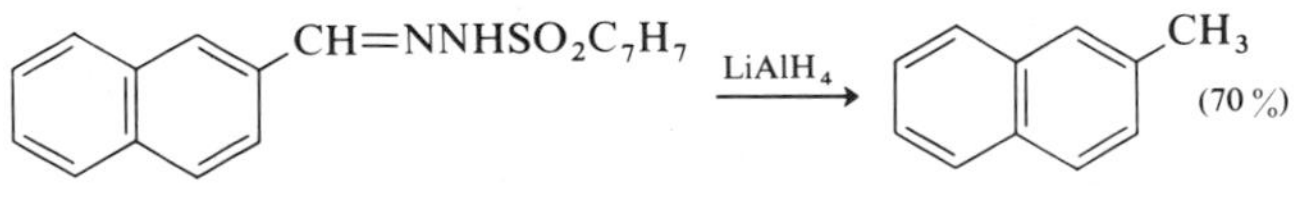

(70%)

6[f] $(CH_3)_3C-$⬡$=O \xrightarrow[\text{NaBH}_3\text{CN}]{\text{C}_7\text{H}_7\text{SO}_2\text{NHNH}_2} (CH_3)_3C-$⬡   (77%)

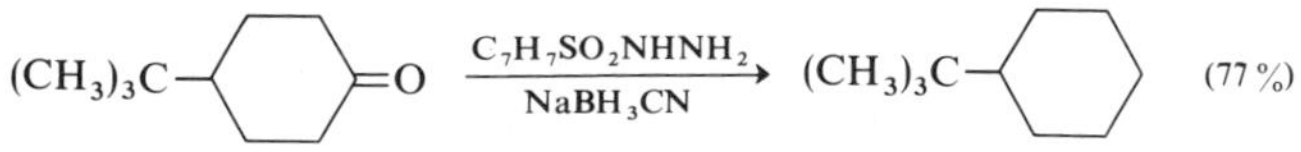

Thioketal Desulfurization

7[g]

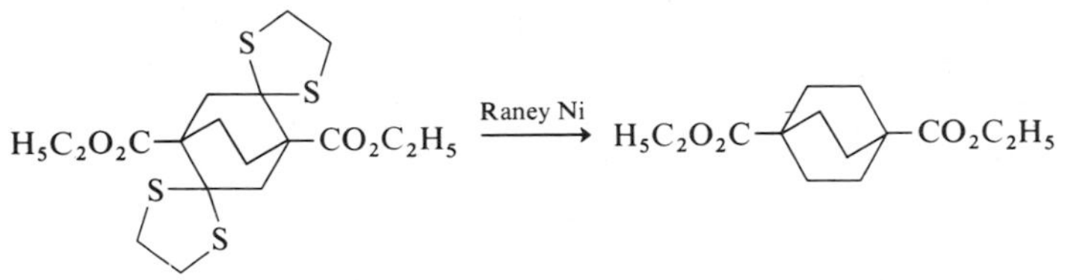

8[h]

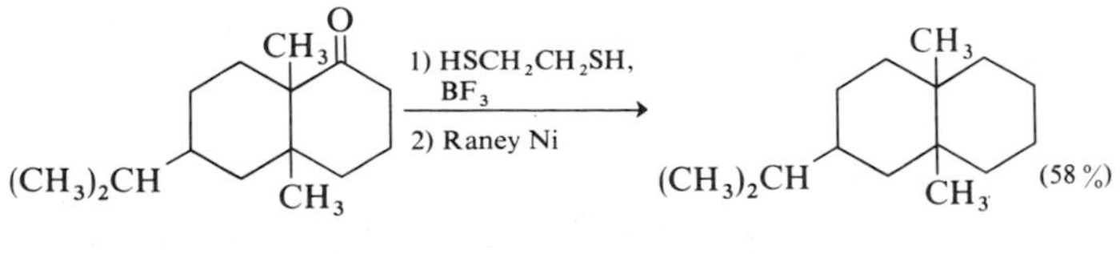

(58%)

---

a. R. Schwarz and H. Hering, *Org. Synth.* **IV**, 203 (1963).
b. R. R. Read and J. Wood, Jr., *Org. Synth.* **III**, 444 (1955).
c. L. J. Durham, D. J. McLeod, and J. Cason, *Org. Synth.* **IV**, 510 (1963).
d. D. J. Cram, M. R. V. Sahyun, and G. R. Knox, *J. Am. Chem. Soc.* **84**, 1734 (1962).
e. L. Caglioti and M. Magi, *Tetrahedron* **19**, 1127 (1963).
f. R. O. Hutchins, B. E. Maryanoff, and C. A. Milewski, *J. Am. Chem. Soc.* **93**, 1793 (1971).
g. J. D. Roberts and W. T. Moreland, Jr., *J. Am. Chem. Soc.* **75**, 2167 (1953).
h. P. N. Rao, *J. Org. Chem.* **36**, 2426 (1971).

Although they are not dissolving-metal processes, several other important means for effecting reduction of a carbonyl group to methylene can be conveniently discussed at this point. The Wolff–Kishner reaction involves the base-catalyzed decomposition of hydrazones.[53] Alkyl diimides are believed to be formed and then to collapse with loss of nitrogen:

$$R_2C{=}N{-}NH_2 + \overset{-}{O}H \rightleftharpoons R_2C{=}N{-}\overset{-}{N}H + H_2O$$

$$R_2CH_2 \xleftarrow{-N_2} R_2\underset{H}{C}{-}N{=}N{-}H$$

The reduction of tosylhydrazones by LiAlH$_4$ or NaBH$_4$ also converts carbonyl groups to methylene.[54] It is believed that a diimide intermediate is involved, as in the

$$R_2C{=}NNHSO_2Ar \xrightarrow{NaBH_4} R_2CHN{-}\overset{H}{\underset{}{N}}{-}SO_2Ar \rightarrow R_2CHN{=}NH \rightarrow R_2CH_2$$

Wolff–Kishner reaction. Excellent yields of carbonyl-to-methylene reduction products have been reported using the mild reducing agent sodium cyanoborohydride.[55] This reagent is added to a mixture of the carbonyl compound to be reduced and p-toluenesulfonylhydrazide. Hydrazone formation is faster than reduction of the carbonyl group by cyanoborohydride. As the hydrazone is formed, it is reduced to the hydrocarbon by NaBH$_3$CN. Carbonyl groups can also be reduced to methylene via thioketal intermediates. The preparation of the cyclic thioketals derived from ethanedithiol is common. Heating the thioketal with excess Raney nickel causes hydrogenolysis of the C–S bonds.

(81%)   Ref. 56

Lithium in ammonia is also an important reduction system. The utility of this method for preparation of specific enolates by reduction of α,β-unsaturated ketones was discussed in Section 1.3. The enolate is not sufficiently basic to abstract a proton from ammonia, so it is stable and can subsequently undergo the various reactions characteristic of enolates:

$$\underset{RCCH=CR_2'}{\overset{O}{\parallel}} + Li \rightarrow \underset{R\overset{}{C}{=}CH\bar{C}R_2'}{\overset{O}{\mid}} \xrightarrow{NH_3} \underset{R\overset{}{C}{=}CHCHR_2'}{\overset{O\cdot}{\mid}} \xrightarrow{Li} \underset{R\overset{}{C}{=}CHCHR_2'}{\overset{O^-}{\mid}}$$

53. D. Todd, *Org. React.*, **4**, 378 (1948); Huang-Minlon, *J. Am. Chem. Soc.* **68**, 2487 (1946).
54. L. Caglioti, *Tetrahedron* **22**, 487 (1966).
55. R. O. Hutchins, B. E. Maryanoff, and C. E. Milewski, *J. Am. Chem. Soc.* **93**, 1793 (1971); R. O. Hutchins, C. A. Milewski, and B. E. Maryanoff, *J. Am. Chem. Soc.* **95**, 3662 (1973).
56. F. Sondheimer and S. Wolfe, *Can. J. Chem.* **37**, 1870 (1959).

If reduction is desired, a proton source such as an alcohol is added to the reaction mixture so that protonation of the enolate occurs.

Dissolving-metal systems constitute the most general method for partial reduction of aromatic rings.[57] The usual reducing medium is lithium or sodium and an alcohol in liquid ammonia. The reaction is initiated by electron transfer, and the radical anion is then protonated by the alcohol:

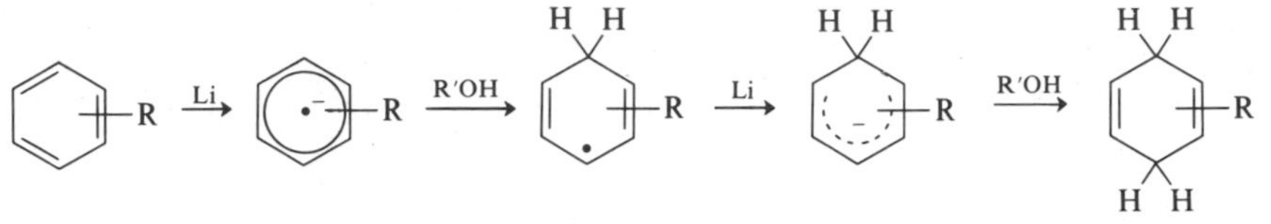

The isolated double bonds in the dihydro system are much less easily reduced than the conjugated ring system, so the reduction stops at the dihydro stage. In the absence of the added alcohol, reduction can proceed to the tetrahydro stage via the

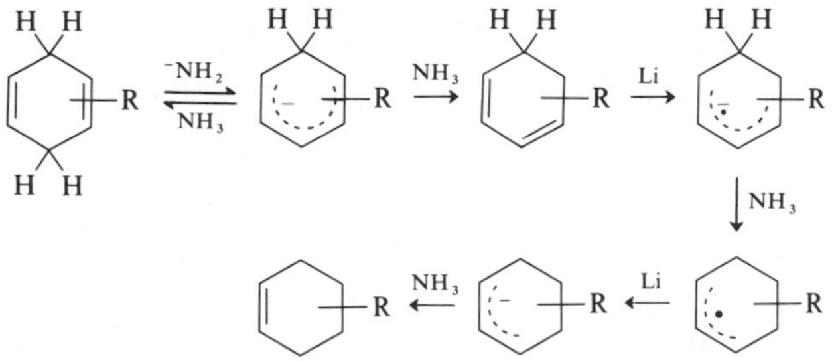

conjugated diene. The rate of reduction is affected in a predictable way by substituent groups. Electron-releasing groups retard the electron transfer, whereas electron-withdrawing groups facilitate reduction. The substituents also govern the position of protonation. Electron-withdrawing substituents lead to formation of 1-substituted-1,4-dihydrobenzenes, while electron-releasing substituents give 1-substituted-2,5-dihydrobenzenes.

The reduction of methoxybenzenes is of importance in the synthesis of cyclohexanones via hydrolysis of the intermediate enol ethers:

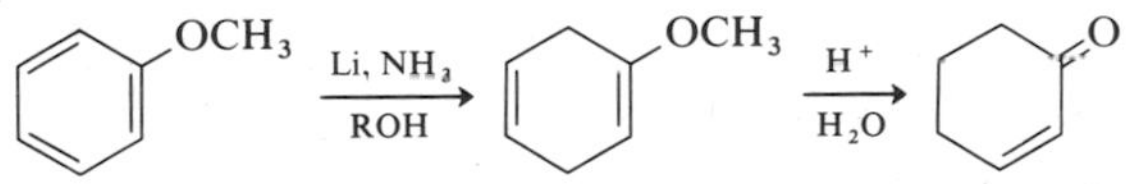

Lithium or sodium dissolving in tetrahydrofuran-t-butanol can accomplish efficient reductive dechlorination of chlorine compounds. This synthetic method has been important in obtaining hydrocarbons via halogenated intermediates. Important examples are conversion of dichlorocyclopropanes (prepared by dichlorocarbene

57. A. J. Birch, Q. Rev. (London) 4, 69 (1950); R. G. Harvey, Synthesis, 161 (1970); A. J. Birch and H. Smith, Q. Rev. (London) 12, 17 (1958).

additions; see Section 8.2) and chlorinated Diels–Alder adducts (from hexa-chlorocyclopentadiene; see Section 6.1) to the corresponding hydrocarbons (entries 11–13 in Scheme 4.3. In solvents containing oxygen-bound deuterium, the halogen atom is replaced by deuterium:

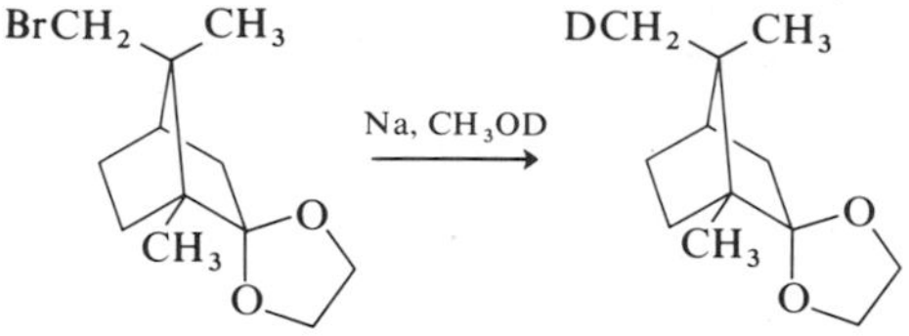

Ref. 58

Epoxides are opened reductively by lithium dissolving in ethylenediamine.[59] This method is distinctly preferable to $LiAlH_4$ reduction for hindered epoxides. Unsymmetrical epoxides give secondary alcohols to the exclusion of primary alcohols, but trisubstituted epoxides give both possible alcohols.

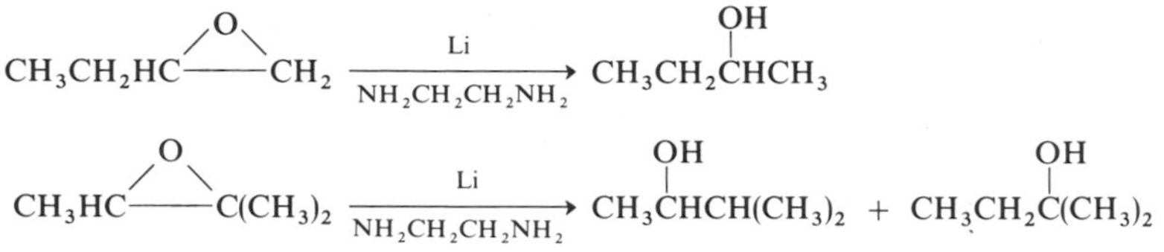

General References

N. C. Gaylord, *Reduction with Complex Metal Hydrides*, Interscience, New York, NY, 1956.
E. Schenker, in *Newer Methods of Preparative Organic Chemistry*, Vol. IV, W. Foerst (ed.), Verlag Chemie, Weinheim, Ger., 1968.
R. L. Augustine, *Reduction; Techniques and Applications to Organic Synthesis*, Marcel Dekker, New York, NY, 1968.
H. C. Brown, *Organic Synthesis via Boranes*, Interscience, New York, NY, 1975.
H. C. Brown, *Boranes in Organic Chemistry*, Cornell University Press, Ithaca, NY, 1972.

Problems

(*References for these problems will be found on page 499.*)

1. Indicate the product to be expected from each of the following reductions:

(a) $CH_3CH_2\overset{\displaystyle O}{\overset{\displaystyle \|}{C}}NH_2 \xrightarrow{LiAlH_4}$

58. M. R. Willcott, III, and C. J. Boriack, *J. Am. Chem. Soc.* **93**, 2354 (1971).
59. H. C. Brown, S. Ikegami, and J. H. Kawakami, *J. Org. Chem.* **35**, 3243 (1970).

(b)

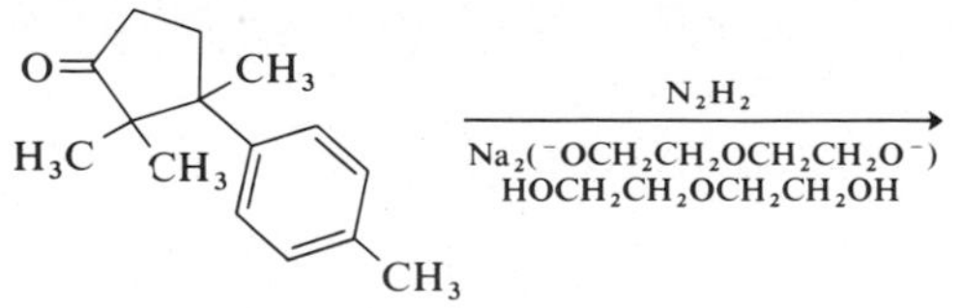

$$\xrightarrow[\substack{Na_2(^-OCH_2CH_2OCH_2CH_2O^-)\\HOCH_2CH_2OCH_2CH_2OH}]{N_2H_2}}$$

(c)

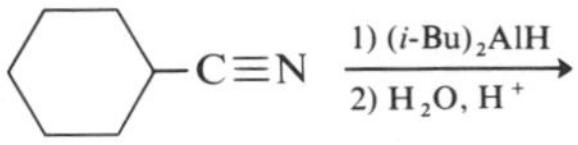

$$\xrightarrow[\text{2) } H_2O,\ H^+]{\text{1) } (i\text{-Bu})_2AlH}$$

(d)

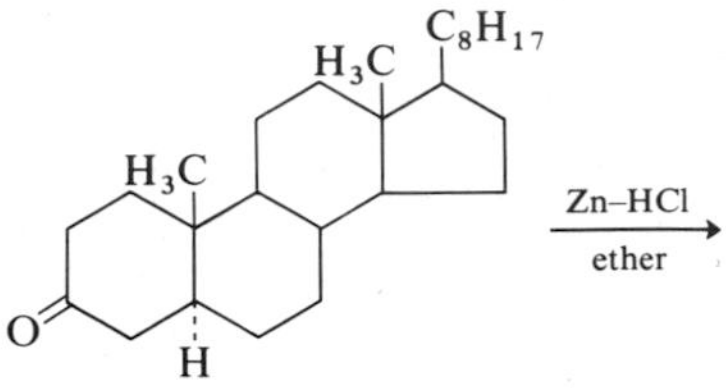

$$\xrightarrow[\text{ether}]{\text{Zn–HCl}}$$

(e)

$(CH_3)_3C-$ $\xrightarrow{\text{LiAlH}_4}$

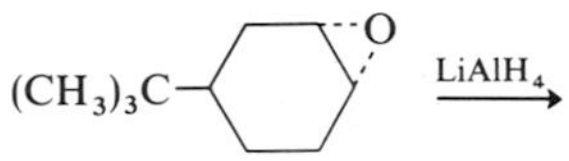

2. Indicate the stereochemistry of the major alcohol that would be formed by sodium borohydride reduction of each of the cyclohexanone derivatives shown:

(a)

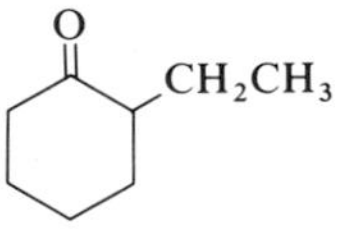

(c)

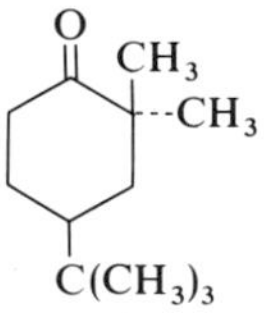

(b)

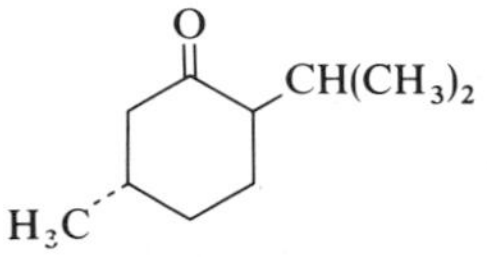

(d)

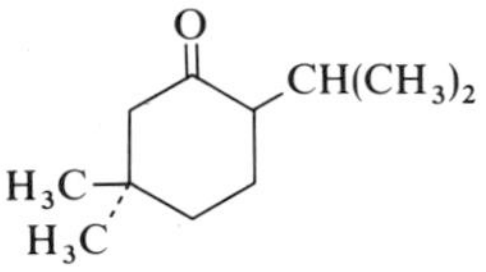

3. Indicate reaction conditions that would accomplish each of the following transformations in one step:

(a)

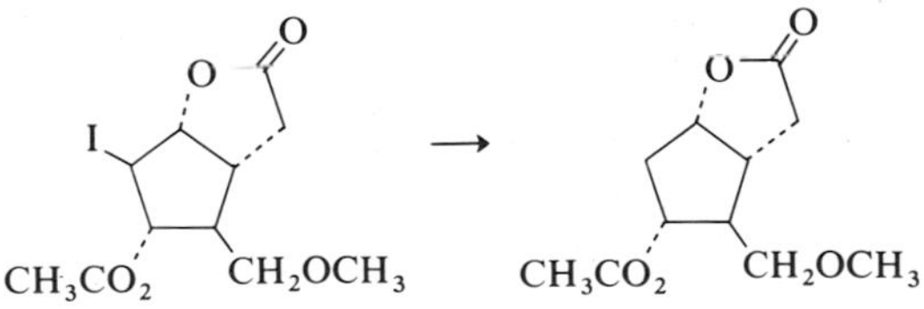

$\longrightarrow$

(b)

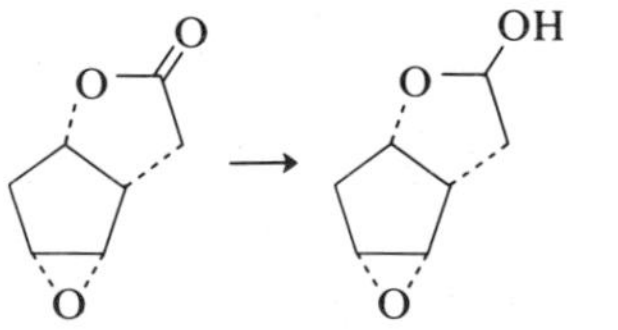

$\longrightarrow$

(c)

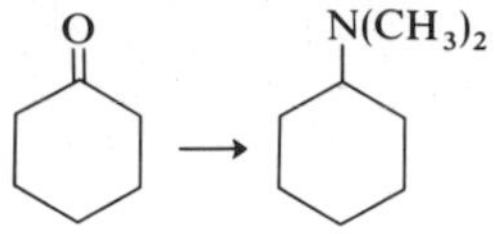

(d)

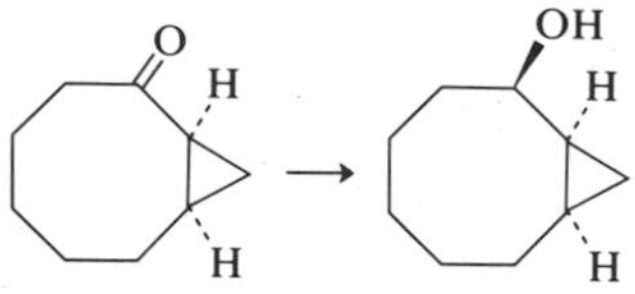

4. From the list of commercially available hydride reducing agents given below, choose one or more that would be appropriate for one-step reduction of each compound to the product shown. If more than one of the reagents seem appropriate, make a choice between them and explain the basis of your choice.

Reagents: $LiAlH_4$ $LiAlH(OC(CH_3)_3)_3$ $LiAlH_2(OCH_2CH_2OCH_3)_2$
$NaBH_4$ $AlH[CH_2CH(CH_3)_2]_2$ $LiBH(CHC_2H_5)_3$
$NaBH_3CN$ CH_3
B_2H_6 $LiBH(C_2H_5)_3$

(a)

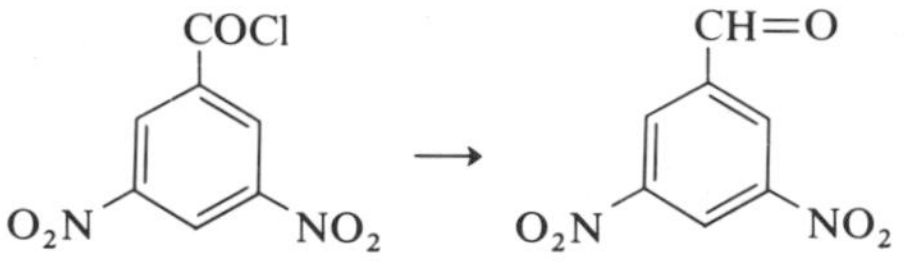

(b) $CH_3CO_2CH_2\overset{CH_3}{\underset{|}{C}}=CHCH=CHCH=O \rightarrow CH_3CO_2CH_2\overset{CH_3}{\underset{|}{C}}=CHCH=CHCH_2OH$

(c)

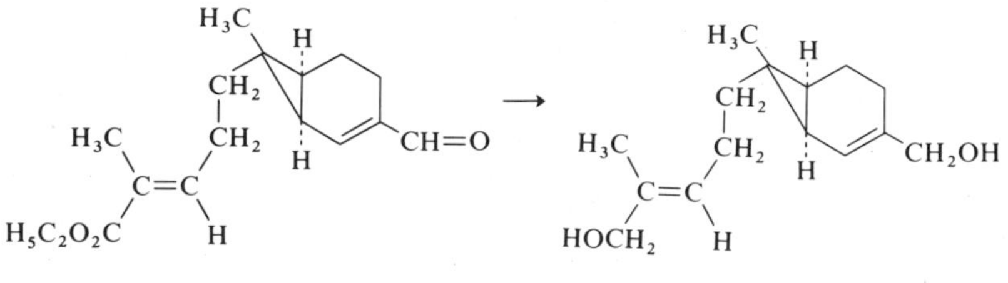

(d)

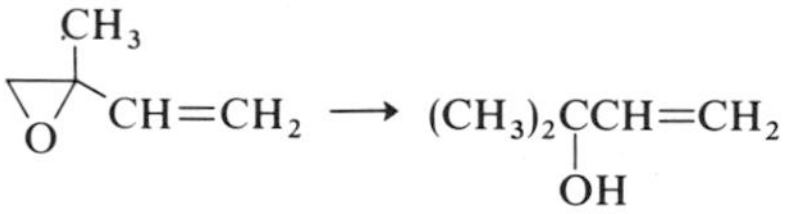

$\rightarrow (CH_3)_2CCH=CH_2$ with $\overset{}{\underset{OH}{|}}$

(e)

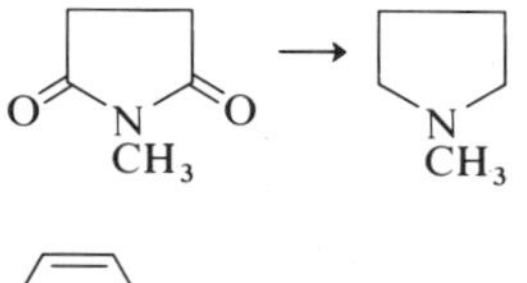

(f)

$-CH_2CO_2C_3H_7 \rightarrow$ $-CH_2CHO$

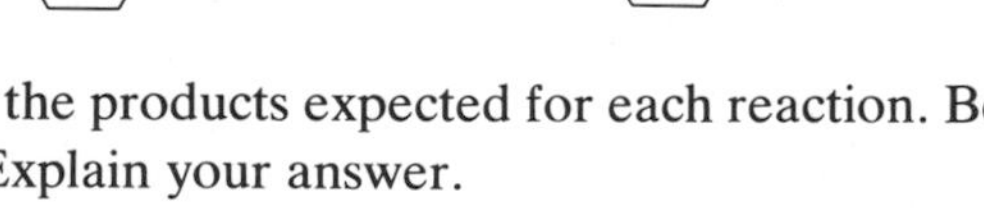

5. Give the products expected for each reaction. Be sure to specify stereochemistry. Explain your answer.

(a) 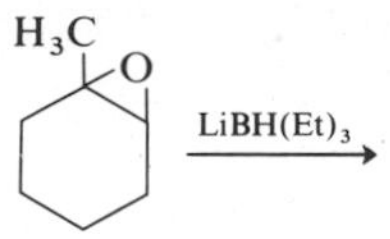
H3C
O
LiBH(Et)3

(b) 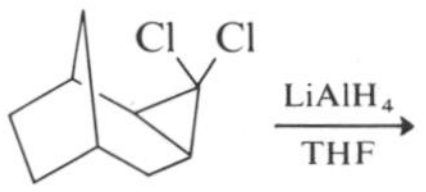
Cl Cl
LiAlH4
THF

(c) 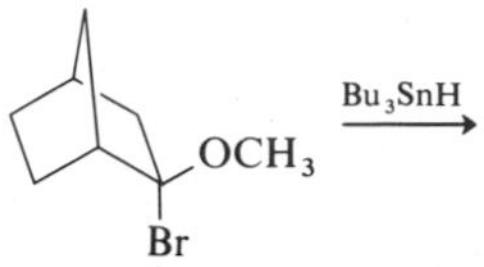
OCH3
Br
Bu3SnH

(d) 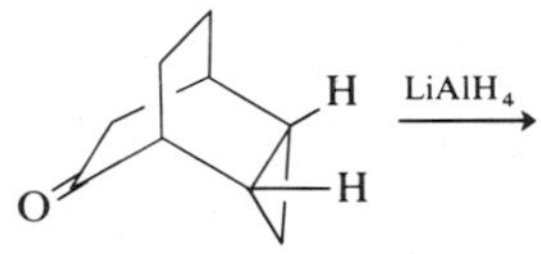
H
H
O
LiAlH4

(e) 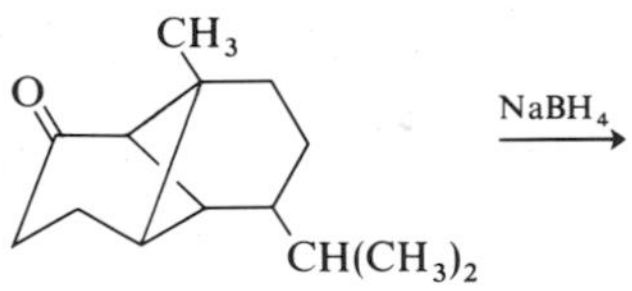
CH3
O
CH(CH3)2
NaBH4

6. For a series of substituted cyclohexanones and norbornanones, it was found that lithium trimethoxyaluminum hydride had a higher tendency than lithium tri-*t*-butoxyaluminum hydride to add to the less hindered side of the molecule. Suggest possible explanations for this behavior.

7. Each of the following reactions has been shown to proceed with high stereoselectivity. Discuss the principal factors that you feel govern the stereochemistry in each case.

(a) 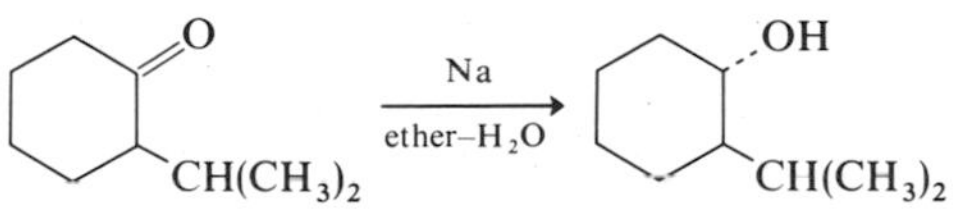
O
CH(CH3)2
Na
ether–H2O
OH
CH(CH3)2

(b) 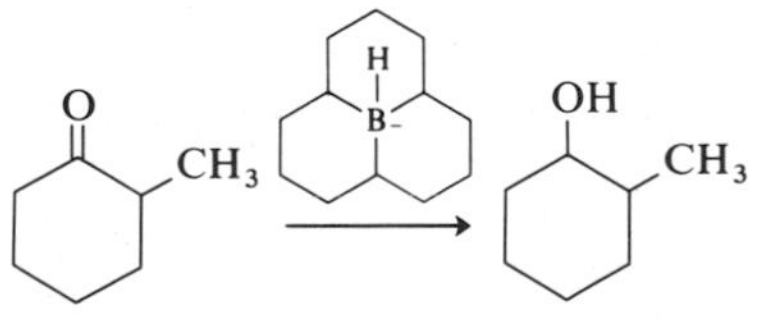
O
CH3
H
B
OH
CH3

(c) 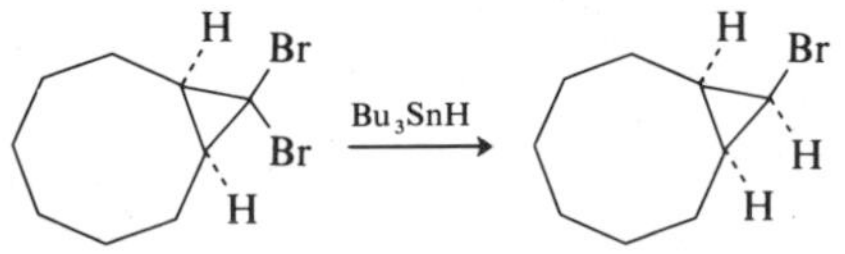
H
Br
Br
H
Bu3SnH
H
Br
H
H

(d)

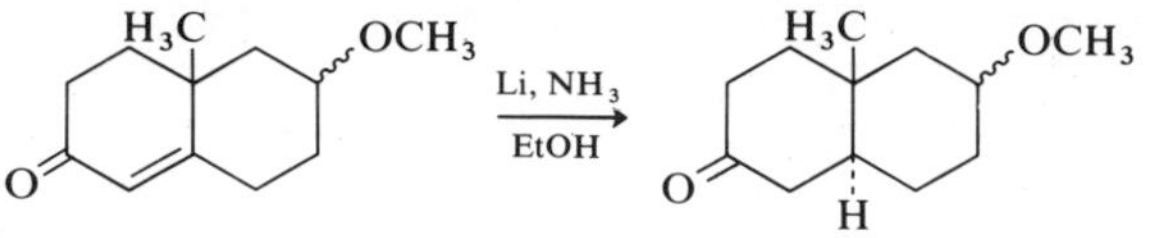

8. When the sodium cyanoborohydride reduction of tosylhydrazones discussed on page 150 is applied to α,β-unsaturated carbonyl compounds, double-bond migration invariably occurs. Suggest a mechanism for this process.

$$PhCH{=}CHCH{=}O \xrightarrow[\text{NaBH}_3\text{CN}]{\text{ArSO}_2\text{NHNH}_2} PhCH_2CH{=}CH_2$$

9. The reduction shown below was accomplished with Li metal in liquid ammonia. Formulate a mechanism. Be sure to account for the observed stereochemistry.

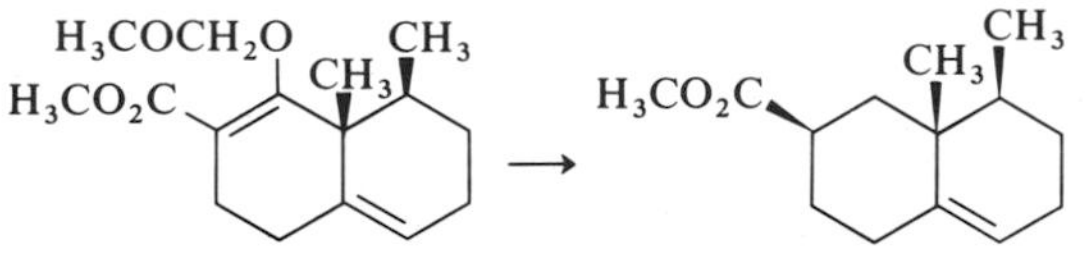

10. Explain the basis of the selectivity observed in each reduction reactions:

(a)

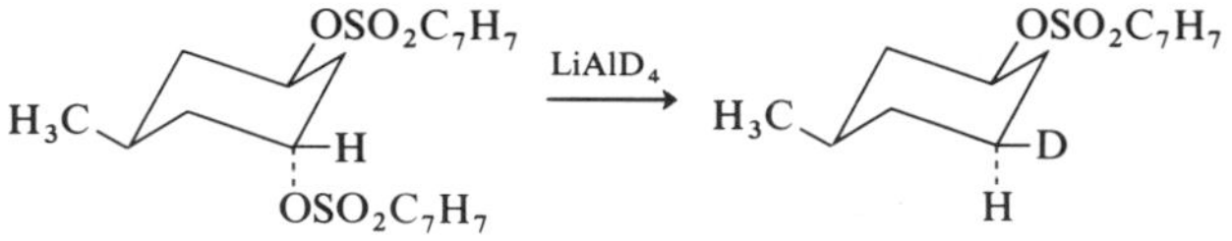

(b)

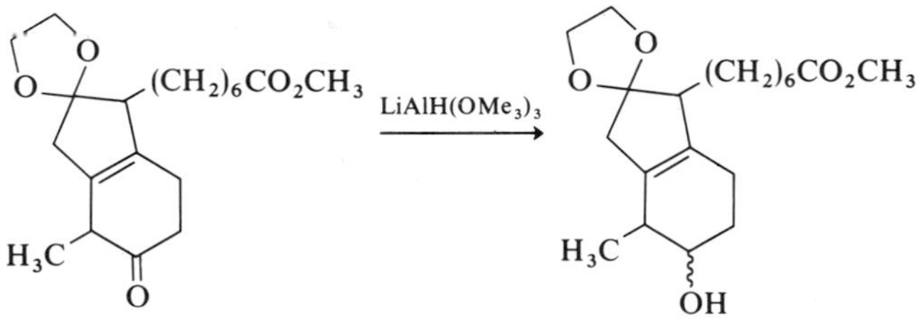

(c)

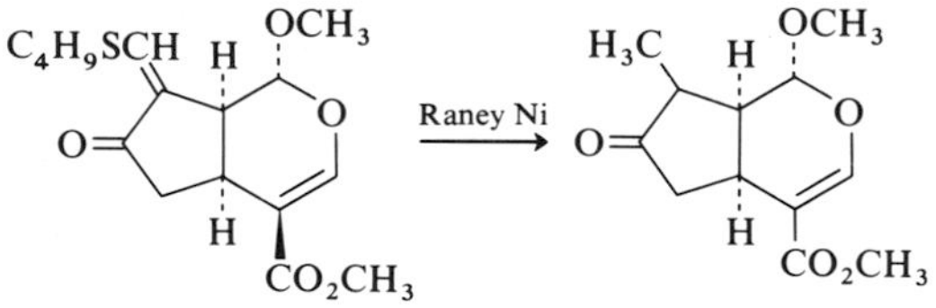

11. Attempted acyloin condensation of **A** does not take the usual course. Instead, **B** is formed. Suggest a mechanism.

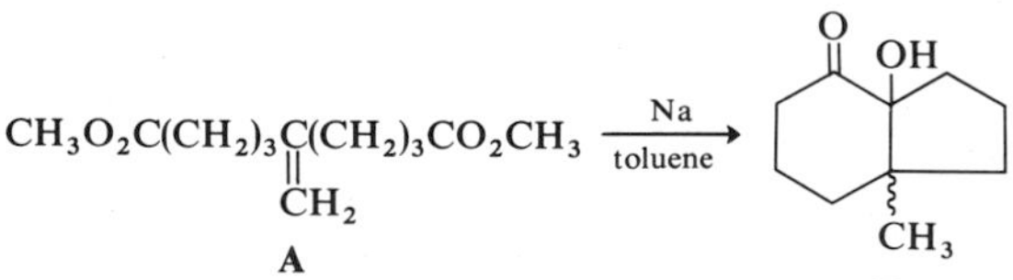

12. Suggest reagents and reaction conditions that would be suitable for each of the following selective or partial reductions:

(a) $HO_2C(CH_2)_4CO_2C_2H_5 \rightarrow HOCH_2(CH_2)_4CO_2C_2H_5$

(b)

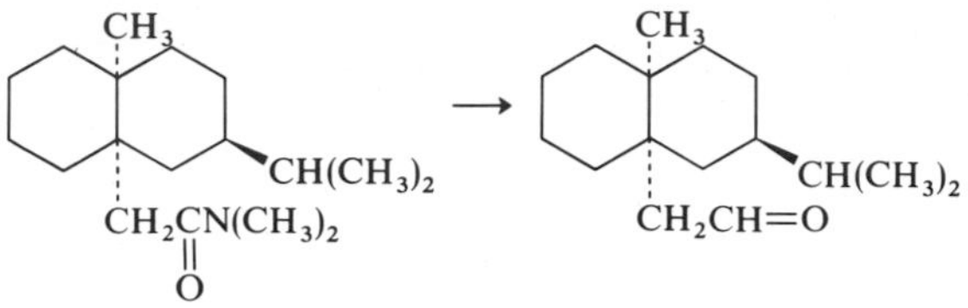

(c)

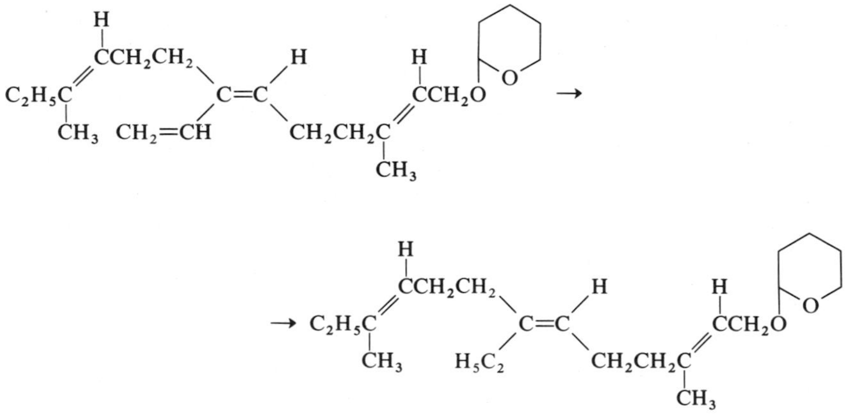

(d)

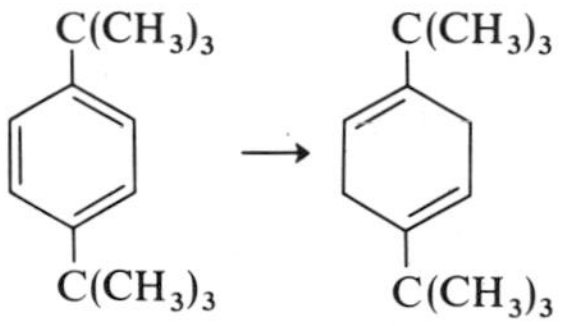

(e)

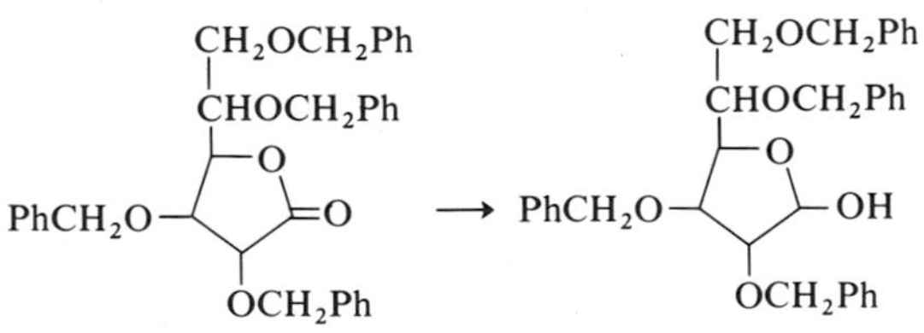

(f)

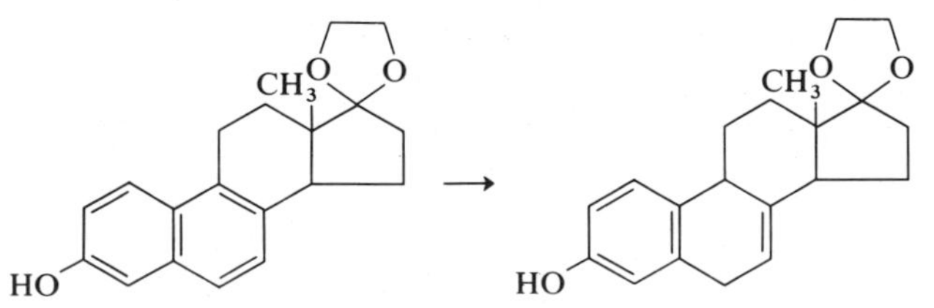

(g)

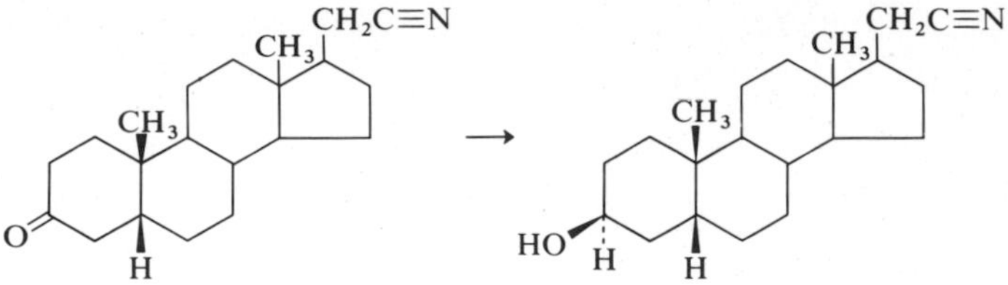

(h) 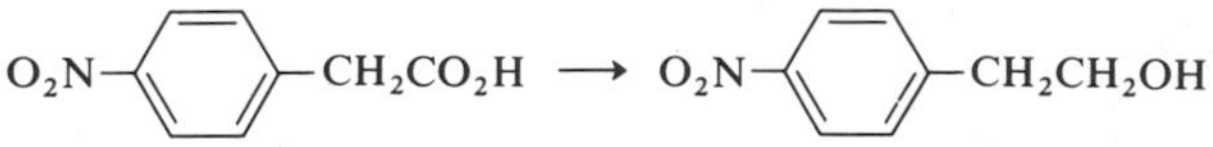 O_2N—⟨⟩—CH_2CO_2H ⟶ O_2N—⟨⟩—CH_2CH_2OH

(i)

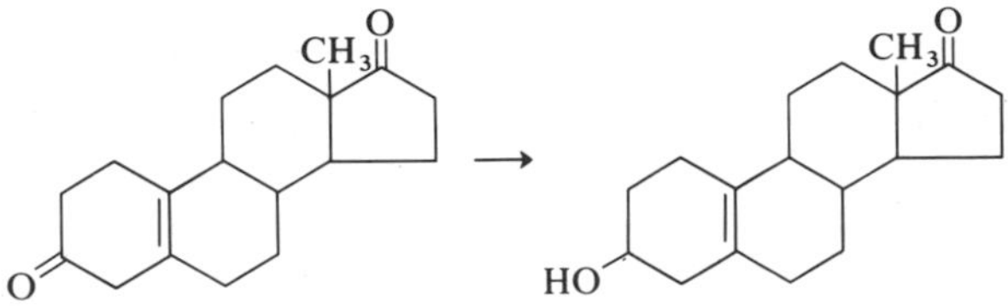

(j) $CH_3\overset{O}{\overset{\|}{C}}(CH_2)_2CO_2C_8H_{17}$ ⟶ $CH_3(CH_2)_3CO_2C_8H_{17}$

(k) 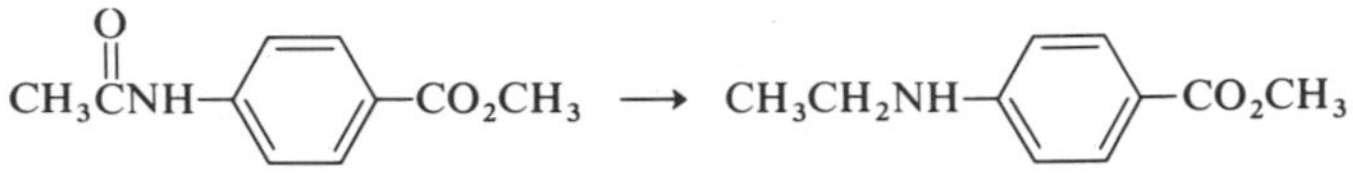 $CH_3\overset{O}{\overset{\|}{C}}NH$—⟨⟩—$CO_2CH_3$ ⟶ CH_3CH_2NH—⟨⟩—CO_2CH_3

(l) 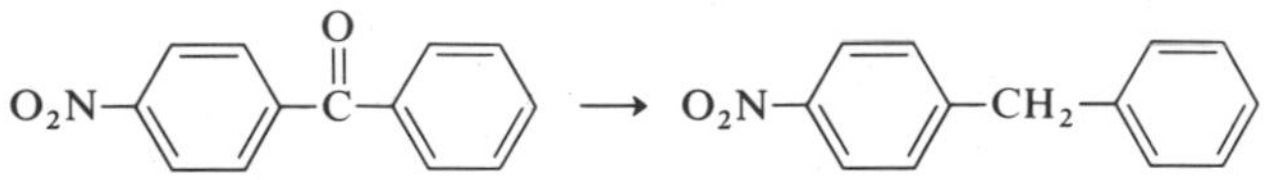 O_2N—⟨⟩—$\overset{O}{\overset{\|}{C}}$—⟨⟩ ⟶ O_2N—⟨⟩—CH_2—⟨⟩

(m)

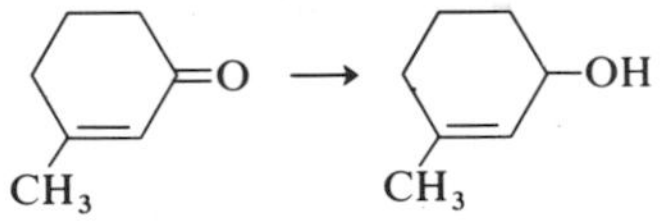

13. In each of the following syntheses, one or more reduction steps can be employed advantageously. Suggest a sequence of reactions, including at least one reduction, that could achieve each transformation.

(a)

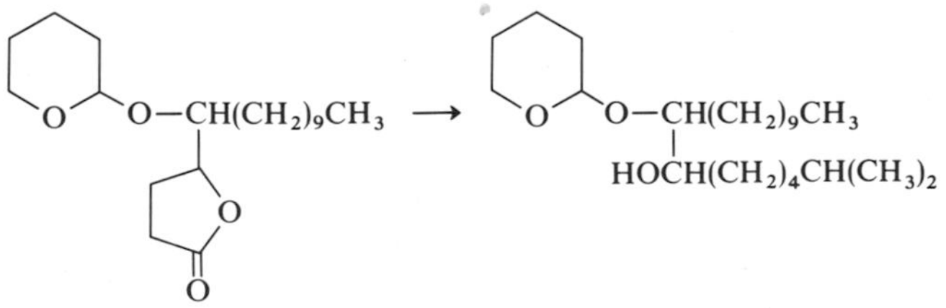

(b)

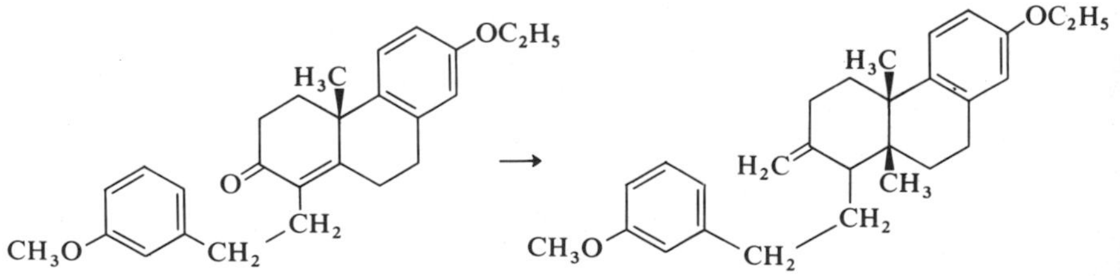

(c)

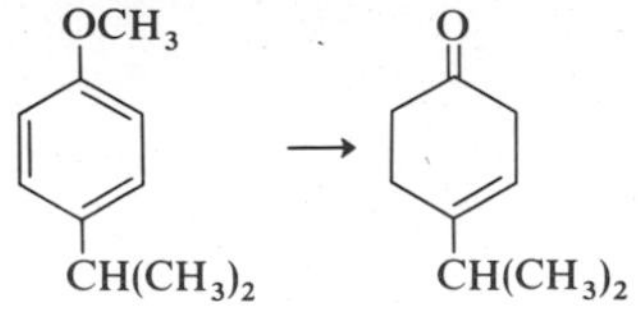

(d)

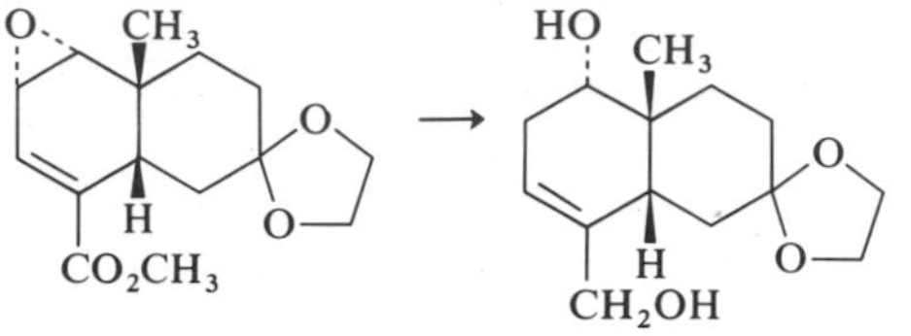

(e)

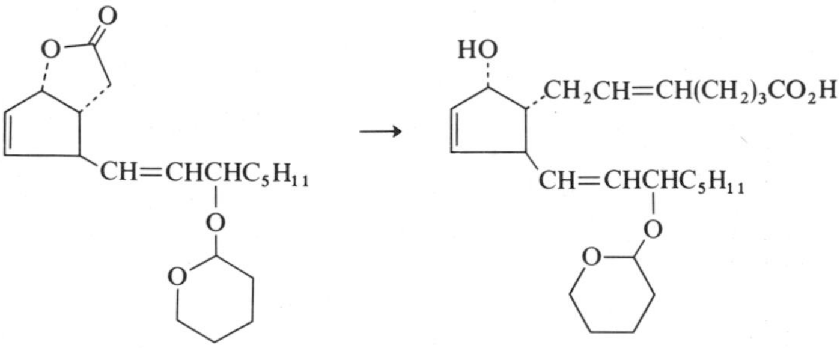

14. Many aromatic ketones and aldehydes can be reduced to the corresponding methylene compound by reaction with triethylsilane in trifluoroacetic acid. The reaction fails for ketones, such as *p*-cyanoacetophenone, which have strong electron-attracting substituents. In this case, the trifluoroacetate ester of the corresponding alcohol is produced. Suggest a mechanism for these reactions.

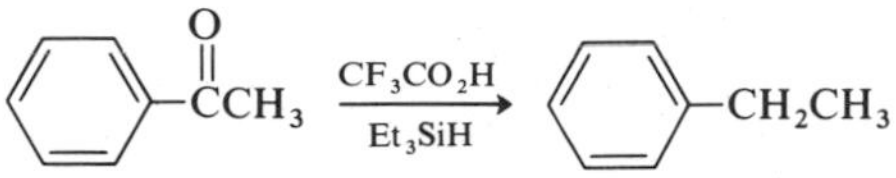

15. Normally, ester groups are unaffected by $NaBH_4$ under mild conditions. Two exceptions are shown. Explain the basis for the unusually facile reduction in each case.

(a) 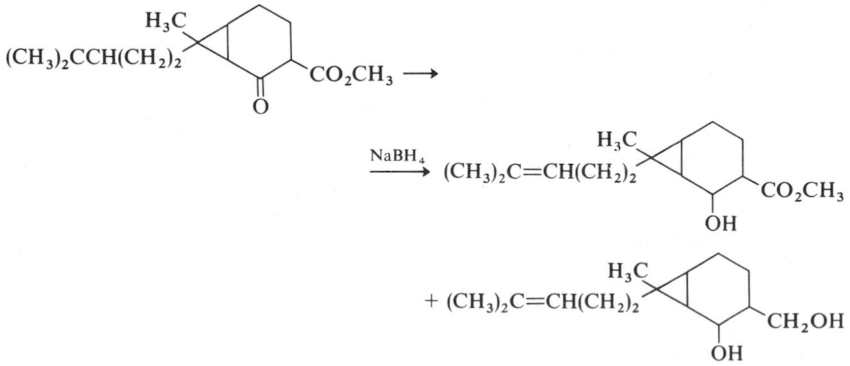 $(CH_3)_2CCH(CH_2)_2$

(b) $PhCH_2CH_2CO_2CH_2CF_3 \xrightarrow{NaBH_4} PhCH_2CH_2CH_2OH$

16. Predict the stereoselectivity for each of the following reductions. Explain the basis of your prediction.

(a) 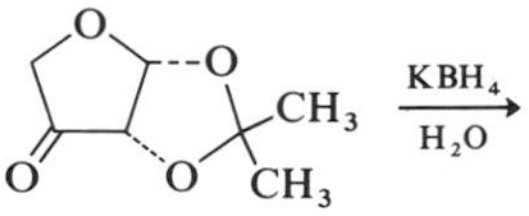 $\xrightarrow[H_2O]{KBH_4}$

(b) 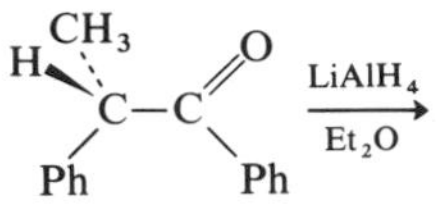 $\xrightarrow[Et_2O]{LiAlH_4}$

(c) 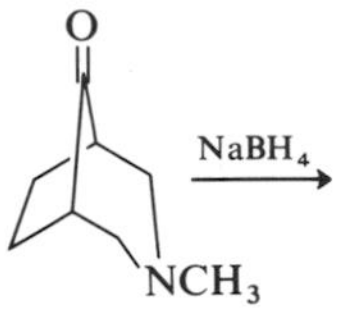 $\xrightarrow{NaBH_4}$

(d) 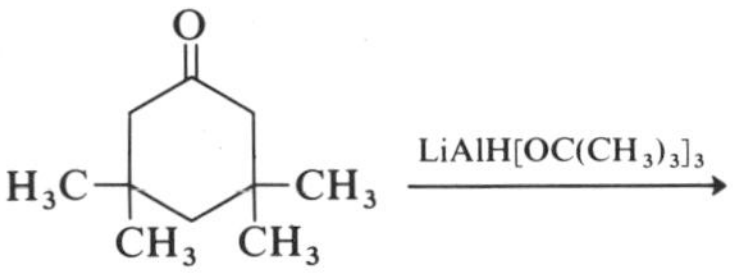 H_3C $\xrightarrow{LiAlH[OC(CH_3)_3]_3}$

(e) 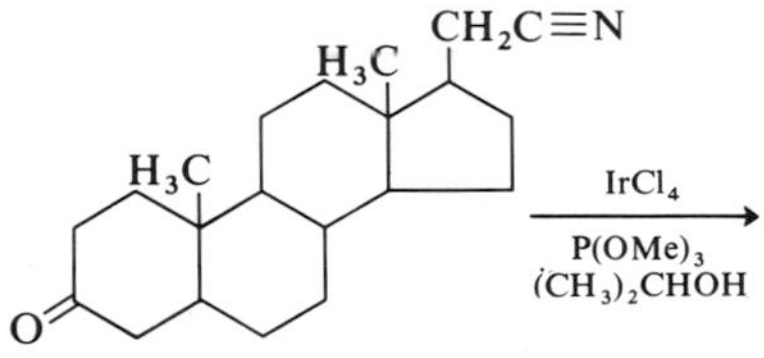 $\xrightarrow[\substack{P(OMe)_3 \\ (CH_3)_2CHOH}]{IrCl_4}$

17. Wolff–Kishner reduction of ketones that bear other functional groups sometimes gives products other than the corresponding methylene compound. Some examples are given. Indicate a mechanism for each of the reactions.

(a) $(CH_3)_3CCCH_2OPh \longrightarrow (CH_3)_3CCH=CH_2$

(b) 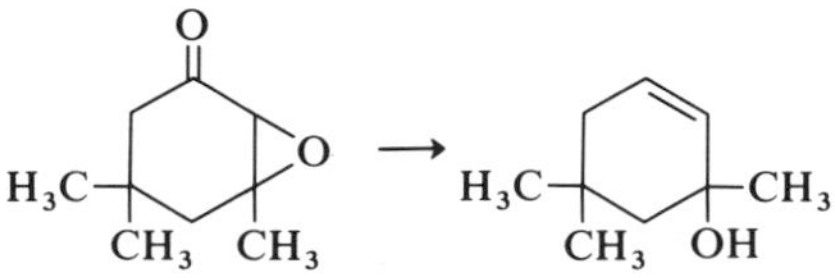 $\longrightarrow$

(c) 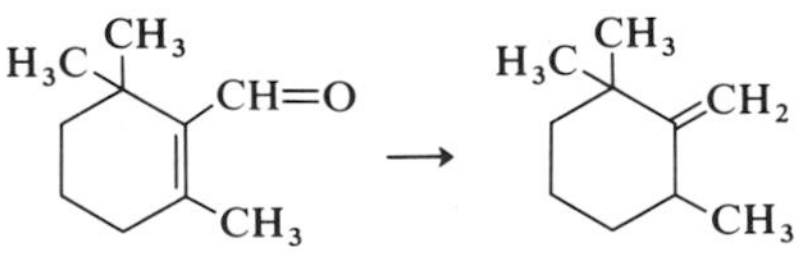 $\longrightarrow$

(d) 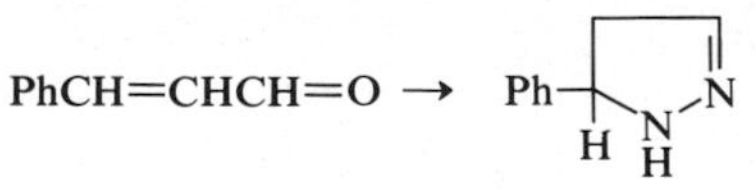   PhCH=CHCH=O → Ph—

18. Birch reductions are important in each of the following synthetic transformations. Give a sequence of steps that would accomplish each transformation.

(a)

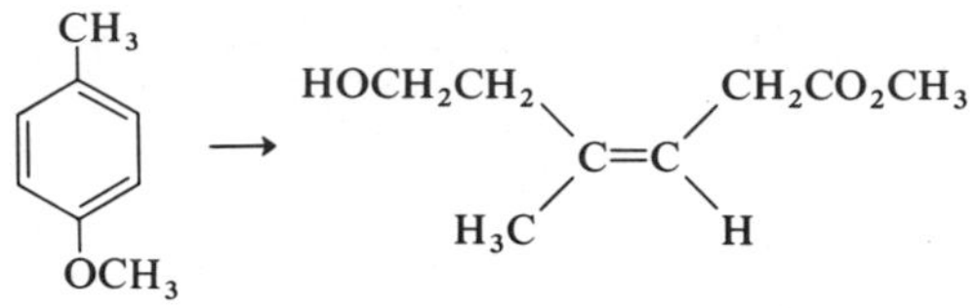

(b)

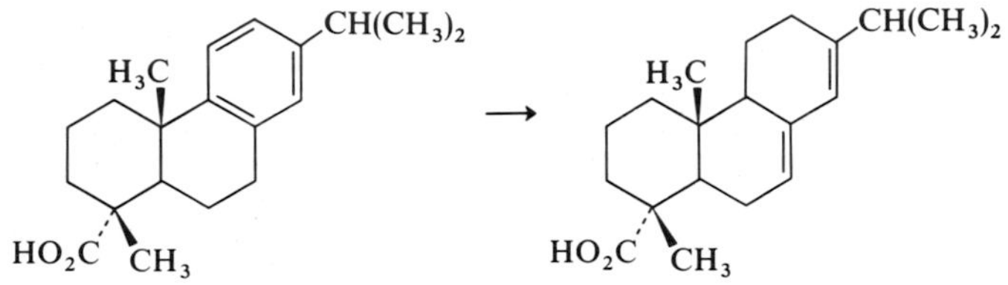

Organometallic Compounds

5.1. Organic Derivatives of Group I and II Metals

5.1.1. Preparation and Properties

The organic derivatives of lithium, magnesium, and sodium have been extensively investigated and are the most important of the Group I and II organometallics.[1] The metals in these two groups are the most electropositive of the elements. The polarity of their bonds to carbon is such as to place high charge density on carbon. This gross electronic distribution is responsible for the strong nucleophilicity and basicity that characterize this group of compounds. As will be discussed in detail below, these

$$\overset{\delta-}{R}-\overset{\delta+}{M}$$

compounds react rapidly with normal carbonyl groups to give addition products.

$$R-M + R_2'C=O \rightarrow R-\overset{\overset{\textstyle O^-M^+}{|}}{\underset{\underset{\textstyle R'}{|}}{C}}-R'$$

They also abstract protons rapidly from all OH and NH groups to generate the hydrocarbon. From the point of view of wide synthetic utility, the most useful of the

$$R-M + ROH \rightarrow R-H + \overset{+}{M}\ \overset{-}{OR}$$

organometallics are the magnesium and lithium compounds.

The discovery by Victor Grignard that organic halides react with metallic magnesium to give nucleophilic organomagnesium compounds was a landmark in organic synthesis. These compounds continue to occupy an important place in

---

1. General Reference: G. E. Coates and K. Wade, *Organometallic Compounds,* Vol. I, Methuen and Co., London, 1967, pp. 1–176.

organic chemistry, and reaction of the halide with metallic magnesium in diethyl

$$RX + Mg \rightarrow RMgX$$

ether remains the principal method of synthesis. The order of reactivity is $RI > RBr > RCl > RF$.

Solutions of some of the common Grignard reagents such as methylmagnesium bromide, ethylmagnesium bromide, and phenylmagnesium bromide are now available commercially. Some Grignard reagents are formed in tetrahydrofuran more rapidly than in ether. This is true of vinyl Grignard reagents, for example.[2] The preference for ether solvents is a result of the excellent solubility of Grignard reagents in ethers. This solubility is the result of strong Lewis acid–base complex formation between the ether molecules and the magnesium atom. The ether

$$
\begin{array}{c}
Br \\
| \\
R-Mg{\leftarrow}OR'_2 \\
\uparrow \\
OR'_2
\end{array}
$$

molecules are held quite tightly in the magnesium coordination sphere. For example, phenylmagnesium bromide has been shown by X-ray methods to retain the coordinated ether molecules in the crystalline state.[3]

A special technique that has been used to prepare Grignard reagents from unreactive halides is known as *entrainment.* The halide of interest, along with a more reactive halide, is added to the magnesium and solvent. Ethylene dibromide is useful as the reactive halide, since it decomposes to ethylene rather than giving a second organomagnesium reagent, which would compete in subsequent reactions[4]:

$$BrCH_2CH_2Br + Mg \rightarrow [BrMg{-}CH_2CH_2{-}Br] \rightarrow CH_2{=}CH_2 + MgBr_2$$

A technique that has occasionally been used to carry out Grignard reactions with unstable reagents involves simultaneous equimolar addition of the halide and carbonyl or other coreactant to magnesium. In this way, the organomagnesium compound reacts as it is formed.[5] Another promising technique is to prepare an extremely reactive form of magnesium as a black powder by reducing magnesium salts with sodium or potassium metal. This material reacts much more rapidly with organic halides than does the metal in the form of small shavings.[6] With the highly reactive powder, it is even possible to prepare Grignard reagents from fluorides, which are usually inert to magnesium metal. Vinyl halides are also easily converted to Grignard reagents with this form of magnesium.

Acetylenes are sufficiently acidic to act as proton donors toward alkyl Grignard

2. D. Seyferth and F. G. A. Stone, *J. Am. Chem. Soc.* **79**, 515 (1957); H. Normant, *Adv. Org. Chem.* **2**, 1 (1960).
3. G. D. Stucky and R. E. Rundle, *J. Am. Chem. Soc.* **85**, 1002 (1963).
4. D. E. Pearson, D. Cowan, and J. D. Beckler, *J. Org. Chem.* **24**, 504 (1959); D. E. Pearson and D. Cowan, *Org. Synth.* **44**, 78 (1964).
5. M. P. Dreyfuss, *J. Org. Chem.* **28**, 3269 (1963).
6. R. D. Rieke and S. E. Bales, *J. Am. Chem. Soc.* **96**, 1775 (1974).

reagents. This reaction is the usual method of preparing alkynylmagnesium compounds.

$$HC{\equiv}CH \; + \; C_2H_5MgBr \; \rightarrow \; HC{\equiv}CMgBr \; + \; C_2H_6$$

165

SECTION 5.1.
ORGANIC
DERIVATIVES
OF GROUP I
AND II
METALS

Although ethers are the usual solvents, Grignard reagents can also be formed in hydrocarbons.[7] Alkylmagnesium halides are insoluble in alkanes, but homogeneous solutions are obtained in aromatic hydrocarbons if one mole of a tertiary amine such as triethylamine is present.[8] The amine exerts its solubilizing effect by complexing with the magnesium atom in a manner similar to that described for ethers.

The familiar designation RMgX is the correct structural representation for most of the alkyl groups in ether solution, but it is well established that an equilibrium exists with magnesium bromide and the dialkylmagnesium. The position of the

$$2 \, RMgX \; \rightleftharpoons \; R_2Mg + MgX_2$$

equilibrium depends upon the solvent and the identity of the specific organic group, but lies far to the left in ether solvents for simple aryl-, alkyl-, and alkenylmagnesium halides.

Solutions of dialkylmagnesium that are free of halide ion can be prepared from alkylmercury compounds.[9] This is an example of a general class of reactions in

$$R_2Hg \; + \; Mg \; \rightarrow \; R_2Mg \; + \; Hg$$

organometallic chemistry known as *metal–metal exchange*. Exchange between an organometallic and metal takes place if the elemental metal is more electropositive than the metal present in the organometallic compound. Exchange in this direction is thermodynamically favorable, largely because the metal atom bound to carbon assumes a partial positive charge resulting from the ionic character of the bond to carbon.

Exchange between organometallics and metal salts takes place in the direction that forms the salt of the more electropositive metal. This is the favored direction,

$$2 \, RMgCl \; + \; CdCl_2 \; \rightarrow \; R_2Cd \; + \; 2 \, MgCl_2$$

since it places the more electropositive metal in the state in which it assumes the largest positive charge. In the case of an exchange involving a salt, the highest positive charge on the metal is in the ionic salt.

Solutions of organomagnesium compounds in diethyl ether contain aggregated species.[10] Dimers are the dominant form in ether solutions of alkylmagnesium

$$2 \, RMgCl \; \rightleftharpoons \; R{-}Mg \underset{Cl}{\overset{Cl}{\diagup\diagdown}} Mg{-}R$$

7. D. Bryce-Smith and E. T. Blues, *Org. Synth.* **47**, 113 (1967).
8. E. C. Ashby and R. Reed, *J. Org. Chem.* **31**, 971 (1966).
9. H. O. House, D. D. Traficante, and R. A. Evans, *J. Org. Chem.* **28**, 348 (1963).
10. E. C. Ashby and M. B. Smith, *J. Am. Chem. Soc.* **86**, 4363 (1964); F. W. Walker and E. C. Ashby, *J. Am. Chem. Soc.* **91**, 3845 (1969).

chlorides. The corresponding bromides and iodides show concentration-dependent behavior; in very dilute solutions, they exist as monomers. In tetrahydrofuran, there is less tendency to aggregate, and several alkyl and aryl Grignard reagents have been found to be monomeric in this solvent.

The aryl- and alkyllithiums can be prepared from the corresponding halides by reaction with lithium metal. A mole of lithium halide is produced as a by-product

$$2 \, Li + RX \rightarrow RLi + LiX$$

Techniques for the preparation of alkenyllithiums are also available.[11] Lithium compounds free of halide can be prepared from mercurials by metal–metal

$$2 \, Li + R_2Hg \rightarrow 2 \, RLi + Hg$$

exchange. Transmetalation involving exchange with tin derivatives is of value in the preparation of several lithium compounds, for example, vinyllithium and allyllithium[12]:

$$4 \, PhLi + (CH_2{=}CH)_4Sn \rightarrow 4 \, CH_2{=}CHLi + Ph_4Sn$$

This technique has been applied to more complex vinyllithium compounds as well.[13] Metal–metal exchanges are occasionally the method of choice in other cases as well, for instance, in the preparation of o-dilithiobenzene from o-phenylenemercury[14]:

Halogen–metal exchange and hydrogen–metal exchange are important processes for preparation of certain lithium compounds. The exchange reactions are particularly favorable when the new organometallic is stabilized by resonance. Metalation often occurs selectively at positions near heteroatoms. Nitrogen and oxygen can stabilize the negative charge on carbon by inductive effects, but even more important is stabilization by chelation involving the substituent group as a ligand at the metal atom.[15] Such exchange processes are accelerated by use of tetramethylethylenediamine to give alkyllithiums of greater reactivity. The effect of the added amine is believed to be the result of its coordination with lithium, which leads to the breakdown of organolithium aggregates and formation of smaller, more reactive systems. Scheme 5.1 shows some examples of the preparation of organolithium compounds by hydrogen–metal exchange.

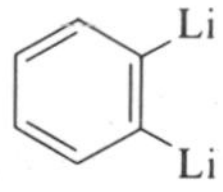

A variety of data indicate that the simple alkyllithium compounds exist as

11. E. A. Braude, *Prog. Org. Chem.* **3**, 172 (1955).
12. D. Seyferth and M. A. Weiner, *J. Am. Chem. Soc.* **83**, 3583 (1961); *J. Org. Chem.* **26**, 4797 (1961); *Org. Synth.* **41**, 30 (1961).
13. E. J. Corey and R. H. Wollenberg, *J. Org. Chem.* **40**, 2265 (1975).
14. H. J. S. Winkler and G. Wittig, *J. Org. Chem.* **28**, 1733 (1963).
15. R. G. Jones and H. Gilman, *Org. React.* **6**, 339 (1951); H. Gilman and J. W. Morton, Jr., *Org. React.* **8**, 258 (1954); R. E. Dessy, W. Kitching, T. Psarras, R. Salinger, A. Chen, and T. Chivers, *J. Am. Chem. Soc.* **88**, 460 (1966); H. J. S. Winkler and H. Winkler, *J. Am. Chem. Soc.* **88**, 964 (1966).

167

SECTION 5.1.
ORGANIC
DERIVATIVES
OF GROUP I
AND II
METALS

Scheme 5.1. Organolithium Compounds by Metalation

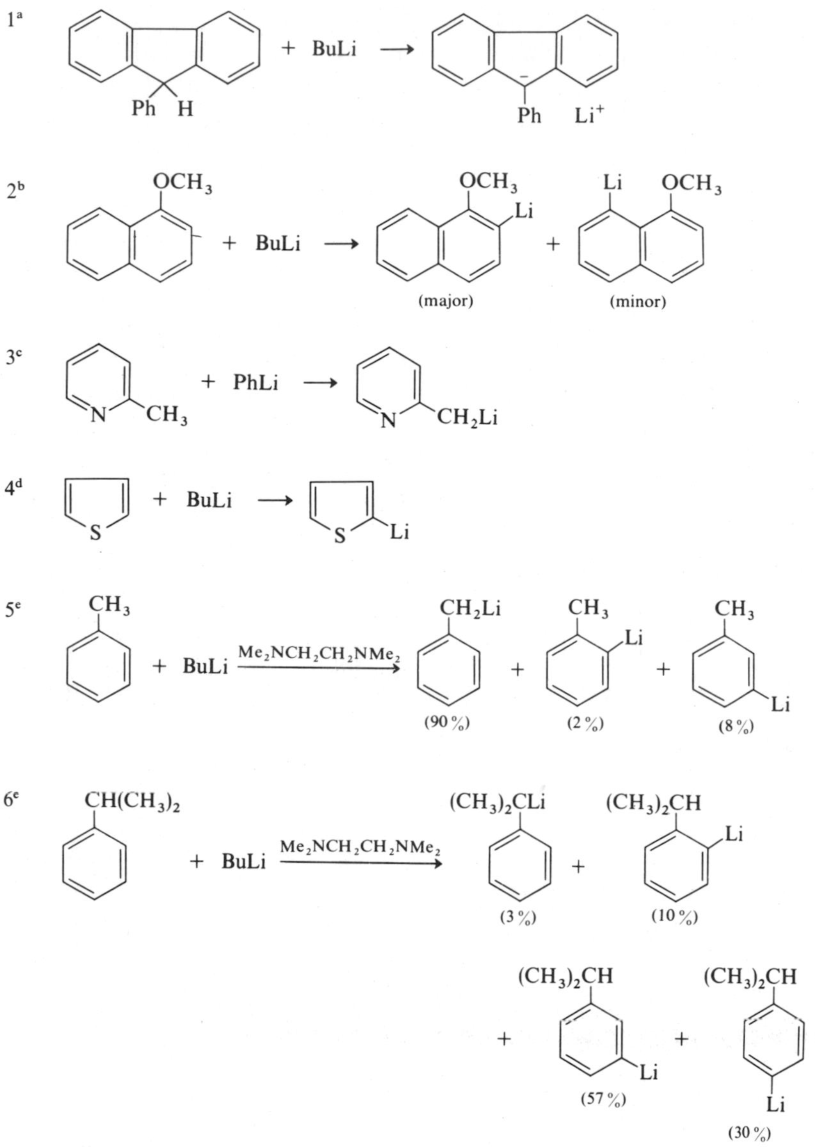

1[a]

2[b]

3[c]

4[d]

5[e]

6[e]

---

a. H. Gilman and J. W. Morton, Jr., *Org. React.* **8**, 258 (1954).
b. B. M. Graybill and D. A. Shirley, *J. Org. Chem.* **31**, 1221 (1966).
c. R. B. Woodward and E. C. Kornfeld, *Org. Synth.* **III**, 413 (1955).
d. E. Jones and I. M. Moodie, *Org. Synth.* **50**, 104 (1970).
e. C. D. Broaddus, *J. Org. Chem.* **35**, 10 (1970).

hexamers in hydrocarbon solvents. In the common ether solvents, the evidence indicates that tetramers are dominant.[16] The tetramers, in turn, are solvated with ether molecules.[17] Certain highly hindered alkyllithiums have been observed to be more reactive than simpler alkyl systems. This has been attributed to steric hindrance, which prevents the formation of the tetramer. Higher reactivity is generally associated with the less aggregated species.[18]

The preparation of alkyllithiums and alkylmagnesium bromides[19] from the alkyl halides by reaction with the metals occurs with loss of stereochemical integrity at the site of the reaction. Stereoisomeric halides give rise to organometallics of identical

$$\text{H}_3\text{C} \overset{\text{Cl}}{\underset{\text{CH(CH}_3)_2}{}} \longrightarrow \text{H}_3\text{C} \overset{\text{Li}}{\underset{\text{CH(CH}_3)_2}{}} \longleftarrow \text{H}_3\text{C} \overset{\text{Cl}}{\underset{\text{CH(CH}_3)_2}{}}$$

composition. Cyclopropyl halides and vinyl halides are converted to Grignard reagents with partial retention of configuration, and the reagents, once formed, are configurationally stable.[20]

Once formed, secondary organomagnesium compounds undergo stereochemical inversion only slowly. *Endo-* and *exo-*norbornylmagnesium bromide require one day at room temperature to equilibrate.[21] Nmr studies have also demonstrated that inversion of configuration at the carbon–magnesium bond is slow on the nmr time scale even up to 170°.[22] In contrast, the inversion of configuration in primary alkylmagnesium halides is fast.[23] The half-life for inversion of 2-methylbutylmagnesium bromide, for example, is less than a second at 25° in ether. This difference in the primary and secondary systems may be the result of a mechanism in which inversion accompanies exchange of alkyl groups between magnesium atoms:

$$\text{XMg}-\overset{}{\text{C}} \quad \overset{+}{\text{MgX}} \longrightarrow \overset{\delta+}{\text{XMg}}\cdots\text{C}\cdots\overset{\delta+}{\text{MgX}} \longrightarrow \overset{+}{\text{XMg}} \quad \text{C}-\text{MgX}$$

If such a bridged intermediate is required, the larger steric bulk of the secondary systems could greatly retard the reaction.

16. H. L. Lewis and T. L. Brown, *J. Am. Chem. Soc.* **92**, 4664 (1970); P. West and R. Waack, *J. Am. Chem. Soc.* **89**, 4395 (1967); T. L. Brown, *Adv. Organometal. Chem.* **3**, 365 (1965).
17. P. D. Bartlett, C. V. Goebel, and W. P. Weber, *J. Am. Chem. Soc.* **91**, 7425 (1969).
18. W. H. Glaze and C. H. Freeman, *J. Am. Chem. Soc.* **91**, 7198 (1969).
19. W. H. Glaze and C. M. Selman, *J. Org. Chem.* **33**, 1987 (1968); W. H. Glaze and C. M. Selman, *J. Organometal. Chem.* **11**, P3 (1968).
20. T. Yoshino and Y. Manabe, *J. Am. Chem. Soc.* **85**, 2860 (1963); H. M. Walborsky and A. E. Young, *J. Am. Chem. Soc.* **86**, 3288 (1964).
21. F. R. Jensen and K. L. Nakamaye, *J. Am. Chem. Soc.* **88**, 3437 (1966); N. G. Krieghoff and D. O. Cowan, *J. Am. Chem. Soc.* **88**, 1322 (1966).
22. E. Pechhold, D. G. Adams, and G. Fraenkel, *J. Org. Chem.* **36**, 1368 (1971).
23. G. M. Whitesides, M. Witanowski, and J. D. Roberts, *J. Am. Chem. Soc.* **87**, 2854 (1965); G. M. Whitesides and J. D. Roberts, *J. Am. Chem. Soc.* **87**, 4878 (1965); G. Fraenkel and D. T. Dix, *J. Am. Chem. Soc.* **88**, 979 (1966).

169

SECTION 5.1.
ORGANIC
DERIVATIVES
OF GROUP I
AND II
METALS

Significant retention of configuration occurs when organolithium compounds are prepared by metal–halogen exchange or metal–metal exchange. Net retention is low in the preparation of alkyllithium compounds by these methods,[24] but high for cyclopropyl and vinyl systems.[25] With alkyl systems, the extent of racemization is increased in the presence of ethers. Hydrocarbons are the preferred solvent if racemization is to be avoided.

The alkylsodiums have not been nearly so useful in organic synthesis as the lithium and magnesium compounds. They are highly insoluble in hydrocarbon solvents, and are so strongly basic that they react rapidly with other solvents, including ethers. Direct preparation of alkylsodium reagents from halides is not feasible because the Wurtz coupling reaction (see subsection 5.1.2) and the competing processes associated with this reaction take place. The alkylsodiums are prepared

$$R_2Hg + \underset{\text{(excess)}}{Na} \rightarrow RNa + Na\text{-}Hg$$

by metal–metal exchange using alkylmercury compounds. Hydrogen–metal exchange is used to prepare phenyl- or benzylsodium:

Procedures for preparation of phenylsodium and vinylsodium from the halides have been described. These procedures involve slow addition of the halide to finely divided dispersions of sodium metal. In this way, the halide concentration is kept low. Since aryl and vinyl halides are relatively unreactive toward displacement, halogen–metal exchange is faster than Wurtz coupling. The organic derivatives of the higher

Ref. 26

$$CH_2{=}CHCl + 2\,Na \rightarrow CH_2{=}\overset{-}{C}H\overset{+}{N}a + NaCl \qquad \text{Ref. 27}$$

alkali metals resemble organosodium compounds in their properties insofar as they have been studied. Increasing reactivity with increasing atomic number of the metal is to be expected from simple electronegativity relationships.

24. R. L. Letsinger, *J. Am. Chem. Soc.* **72**, 4842 (1950); D. Y. Curtin and W. J. Koehl, Jr., *J. Am. Chem. Soc.* **84**, 1967 (1962).
25. H. M. Walborsky, F. J. Impastato, and A. E. Young, *J. Am. Chem. Soc.* **86**, 3283 (1964); D. Seyferth and L. G. Vaughan, *J. Am. Chem. Soc.* **86**, 883 (1964); M. J. S. Dewar and J. M. Harris, *J. Am. Chem. Soc.* **91**, 3652 (1969).
26. H. Ruschig, R. Fugmann, and W. Meixner, *Angew. Chem.* **70**, 71 (1958).
27. R. G. Anderson, M. B. Silverman, and D. M. Ritter, *J. Org. Chem.* **23**, 750 (1958).

5.1.2. Reactions

The organometallic compounds of the Group I and II metals are strongly basic and nucleophilic. Although the alkylation of organometallics by alkyl halides might seem to offer a general synthetic method for unsymmetrical hydrocarbons, this reaction is seldom synthetically useful. A very old reaction of this type is the Wurtz coupling of halides by sodium metal, in which an organosodium compound has been supposed to be an intermediate. Recent studies have indicated that these couplings

$$RX + Na \rightarrow RNa$$

$$RNa + RX \rightarrow R-R$$

involve radicals generated by redox processes, rather than S_N2 substitutions.[28] The sodium coupling has been difficult to study in detail because of the extreme reactivity

$$RX + Na \rightarrow [RX]^{\cdot-} + Na^+$$

$$[RX]^{\cdot-} \rightarrow R\cdot + X^-$$

$$2 R\cdot \rightarrow R-R$$

of organosodium compounds. More extensive studies have been carried out on the organolithium–alkyl halide reaction. Several results from simple alkyllithium–alkyl halide coupling reactions lead to the conclusion that radicals are involved. Intermediate radicals have been detected spectroscopically by ESR and CIDNP techniques.[29] The simple alkylation product is accompanied by products formed by disproportionation and symmetrical coupling of the postulated radical intermediates.[30]

$$n\text{-}C_4H_9Br + n\text{-}C_4H_9Li \xrightarrow{HO(CH_3)_3} CH_3(CH_2)_6CH_3 + CH_3CH_2CH=CH_2 +$$
$$\text{(43\%)} \qquad\qquad \text{(3\%)}$$

$$CH_3CH_2CH_2CH_3 +$$
$$\text{(19\%)}$$

With allyl halides, a cyclic concerted mechanism has been proposed.[31] Support

$$\rightarrow PhCH_2CH=CH_2$$

28. J. F. Garst and R. H. Cox, *J. Am. Chem. Soc.* **92**, 6389 (1970).
29. G. A. Russell and D. W. Lamson, *J. Am. Chem. Soc.* **91**, 3967 (1969); H. R. Ward and R. G. Lawler, *J. Am. Chem. Soc.* **89**, 5518 (1967); A. R. Lepley and R. L. Landau, *J. Am. Chem. Soc.* **91**, 748 (1969); H. R. Ward, R. G. Lawler, and R. A. Cooper, *J. Am. Chem. Soc.* **91**, 746 (1969).
30. D. Bryce-Smith, *J. Chem. Soc.*, 1603 (1956).
31. R. M. Magid and J. G. Welch, *J. Am. Chem. Soc.* **90**, 5211 (1968); R. M. Magid, E. C. Nieh, and R. D. Gandour, *J. Org. Chem.* **36**, 2099 (1971); R. M. Magid and E. C. Nieh, *J. Org. Chem.* **36**, 2105 (1971).

171

SECTION 5.1.
ORGANIC
DERIVATIVES
OF GROUP I
AND II
METALS

for this mechanism includes an isotopic labeling study that showed that when allyl chloride-1-^{14}C reacts with phenyllithium, about three fourths of the product has the labeled carbon at the terminal methylene group.

Organolithium reagents in which the carbanion is a relatively delocalized one also appear to constitute an exception to the radical-coupling mechanism. Allyllithium and benzyllithium both react with secondary alkyl bromides in fair yield, with a high degree of inversion of configuration at the site of bromide displacement[32]:

$$PhCH_2Li + \underset{H_3C}{\overset{CH_3CH_2}{C}}\!\!-Br \longrightarrow PhCH_2-\underset{H}{\overset{CH_2CH_3}{C}}\cdots CH_3 \quad \text{(58\% yield, 100\% inversion)}$$

Vinyllithium reagents can be alkylated in good yields by alkyl iodides.[33]

The alkylation of Grignard reagents is of some synthetic value when methyl, allyl, or benzyl halides are involved:

$$Ph\underset{(CH_2)_4CH_3}{\overset{CH_3}{C}}\!\!-Cl + CH_3(CH_2)_9MgBr \longrightarrow Ph\underset{(CH_2)_4CH_3}{\overset{CH_3}{C}}\!\!-(CH_2)_9CH_3 \quad \text{Ref. 34}$$

(30%)

Synthetically useful alkylations of Grignard reagents also occur with alkyl sulfates and sulfonates:

$$PhCH_2MgCl + CH_3CH_2CH_2CH_2OSO_2C_7H_7 \longrightarrow PhCH_2CH_2CH_2CH_2CH_3 \quad \text{Ref. 35}$$

(50–59%)

(52–60%)

By far the most important type of reaction of Grignard reagents is the family of reactions involving addition to carbonyl groups. The transition state for addition of Grignard reagents is often represented as a cyclic array containing the carbonyl and

32. L. H. Sommer and W. D. Korte, *J. Org. Chem.* **35**, 22 (1970).
33. J. Millon, R. Lorne, and G. Linstrumelle, *Synthesis,* 434 (1975).
34. C. F. Hobbs and W. C. Hammann, *J. Org. Chem.* **35**, 4188 (1970); R. G. Gough and J. A. Dixon, *J. Org. Chem.* **33**, 2148 (1968).
35. H. Gilman and J. Robinson, *Org. Synth.* **II**, 47 (1943).
36. L. I. Smith, *Org. Synth.* **II**, 360 (1943).

two molecules of Grignard reagent. There is in fact considerable evidence for a termolecular transition state arising by attack of a Grignard reagent on a carbonyl–Grignard complex.[37] Dialkylmagnesiums are more reactive than alkylmagnesium halides toward ketones. This greater reactivity, and the presence of the dialkylmagnesiums in equilibria with alkylmagnesium halides, allows for the possibility that some of the typical Grignard additions might involve dialkylmagnesiums.[38]

$$2\,RMgX \rightleftharpoons R_2Mg + MgX_2$$

$$R_2Mg + R_2'CO \xrightarrow{fast} R_2'\underset{\underset{R}{|}}{C}OMgR$$

When the initial Grignard carbonyl adduct contains a potential leaving group, this adduct breaks down and a second or third Grignard reagent can add. It is

$$RMgX + R'\overset{O}{\overset{\|}{C}}OR'' \rightarrow R-\underset{\underset{R'}{|}}{\overset{\overset{OMgX}{|}}{C}}-OR''$$

$$R-\underset{\underset{R'}{|}}{\overset{\overset{OMgX}{|}}{C}}OR'' \rightarrow R\overset{O}{\overset{\|}{C}}R' + R''OMgX$$

$$R\overset{O}{\overset{\|}{C}}R' + RMgX \xrightarrow{fast} R_2\overset{OMgX}{\overset{|}{C}}R'$$

relatively difficult to control Grignard additions to esters in such a way as to provide ketones, because the ketonic products are more reactive than the esters. The addition of Grignard reagents to ketones, aldehydes, and esters forms the basis for synthetic methods that can provide alcohols having a variety of structures, as summarized in Scheme 5.2.

In addition, Grignard reagents are useful for the synthesis of ketones via reactions with nitriles. Aldehydes can be obtained by reaction with triethyl orthofor-

$$RMgX + R'C\equiv N \rightarrow R\overset{NMgX}{\overset{\|}{C}}R' \xrightarrow{H_2O} R\overset{O}{\overset{\|}{C}}R'$$

mate. The addition in this case must be preceded by elimination of one of the alkoxy groups. The elimination is presumably catalyzed by magnesium acting as a Lewis

$$\underset{C_2H_5O}{\overset{C_2H_5O}{}}\overset{\overset{R}{\underset{}{\downarrow}}}{Mg-X} \rightarrow HC\overset{OC_2H_5}{\underset{OC_2H_5}{(+}} + C_2H_5OMgR + X^-$$

$$RMgX + HC\overset{OC_2H_5}{\underset{OC_2H_5}{(+}} \rightarrow R\overset{OC_2H_5}{\underset{OC_2H_5}{CH}}$$

37. E. C. Ashby, R. B. Duke, and H. M. Neumann, *J. Am. Chem. Soc.* **89**, 1964 (1967).
38. H. O. House and J. E. Oliver, *J. Org. Chem.* **33**, 929 (1968); E. C. Ashby, J. Laemmle, and H. M. Neumann, *J. Am. Chem. Soc.* **93**, 4601 (1971).

acid.[39] The acetals formed by addition are stable to the reaction conditions, but are hydrolyzed to aldehydes by aqueous acid. Carboxylic acids are obtained easily from Grignard reagents by reaction with carbon dioxide. Scheme 5.2 summarizes some of

173

SECTION 5.1.
ORGANIC
DERIVATIVES
OF GROUP I
AND II
METALS

$$RMgX + CO_2 \rightarrow RCOMgX \xrightarrow[H_2O]{H^+} RCO_2H$$

the most widely applied synthetic procedures involving Grignard reagents, using examples of procedures described in *Organic Syntheses* for illustration.

Grignard reagents, it must be remembered, are severely restricted in the functional groups that can be present in either the organometallic or carbonyl compound. Alkene, ketal, and acetal groups ordinarily cause no difficulty, but unprotected OH, NH, or SH groups and carbonyl groups cannot be present.

Grignard additions are sensitive to steric effects, and with hindered ketones, a competing process involving reduction of the carbonyl group is observed. A cyclic transition state similar to that proposed for Meerwein–Pondorff–Verley reduction can account for this transformation. The extent of this reaction increases with

increasing steric bulk in both the ketone and the Grignard reagent. For example, no addition occurs between diisopropyl ketone and isopropylmagnesium bromide, and the reduction product diisopropylcarbinol is formed in 70% yield.[40]

The reduction process has not been of great synthetic importance, although its potential for asymmetric reductions of ketones has been recognized and investigated.[41] For example, the reduction of isopropyl phenyl ketone to the corresponding alcohol occurs in 82% optical yield with the optically active Grignard reagent 1:

Enolization of the ketone is also occasionally a competing reaction. Since the enolate is unreactive toward Grignard addition, the ketone is recovered after hydrolysis. Enolization has been shown to be especially important when a considerable portion of the Grignard reagent is present as an alkoxide.[42] Alkoxides are formed as the addition reaction proceeds. They are also formed by oxidation processes if

39. E. L. Eliel and F. W. Nader, *J. Am. Chem. Soc.* **92**, 584 (1970).
40. D. O. Cowan and H. S. Mosher, *J. Org. Chem.* **27**, 1 (1962).
41. J. S. Birtwistle, K. Lee, J. D. Morrison, W. A. Sanderson, and H. S. Mosher, *J. Org. Chem.* **29**, 37 (1964).
42. H. O. House and D. D. Traficante, *J. Org. Chem.* **28**, 355 (1963).

A. Primary Alcohols from Formaldehyde

1^a [cyclohexyl]—$MgCl$ + CH_2O $\longrightarrow$ $\xrightarrow[H^+]{H_2O}$ [cyclohexyl]—CH_2OH (64–69%)

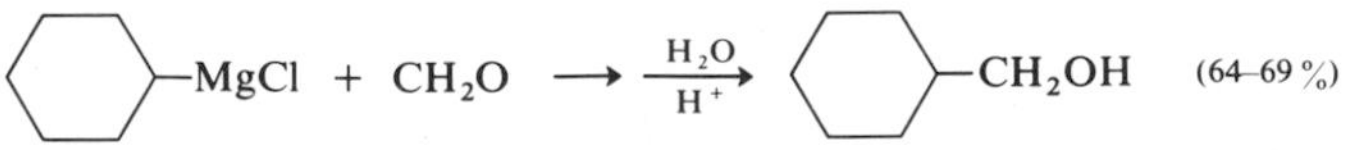

B. Primary Alcohols from Ethylene Oxide

2^b $CH_3(CH_2)_3MgBr$ + $H_2C\overset{\displaystyle O}{\overline{\diagdown\diagup}}CH_2$ $\longrightarrow$ $CH_3(CH_2)_5OH$ (60–62%)

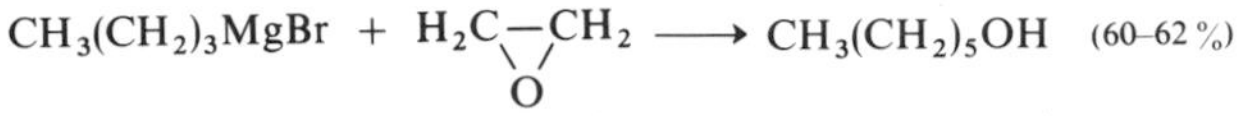

C. Secondary Alcohols from Aldehydes

3^c $PhCH{=}CHCH{=}O$ + $HC{\equiv}CMgBr$ $\longrightarrow$ $\xrightarrow[H^+]{H_2O}$ $HC{\equiv}C\overset{OH}{\underset{|}{C}}HCH{=}CHPh$ (58–69%)

4^d [3-chlorophenyl]—$MgBr$ + $CH_3CH{=}O$ $\longrightarrow$ $\xrightarrow{H_2O}$ [3-chlorophenyl]—$CHOHCH_3$ (82–85%)

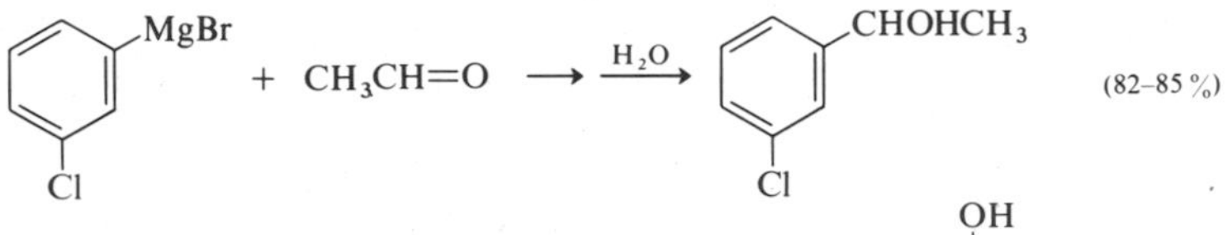

5^e $CH_3CH{=}CHCH{=}O$ + CH_3MgCl $\longrightarrow$ $\xrightarrow{H_2O}$ $CH_3CH{=}CH\overset{OH}{\underset{|}{C}}HCH_3$ (81–86%)

6^f $(CH_3)_2CHMgBr$ + $CH_3CH{=}O$ $\longrightarrow$ $(CH_3)_2CH\overset{OH}{\underset{|}{C}}HCH_3$ (53–54%)

D. Secondary Alcohols from Formate Esters

7^g $2\,CH_3(CH_2)_3MgBr$ + $HCO_2C_2H_5$ $\longrightarrow$ $\xrightarrow[H^+]{H_2O}$ $(CH_3CH_2CH_2CH_2)_2CHOH$ (83–85%)

E. Tertiary Alcohols from Esters and Lactones

8^h $3\,C_2H_5MgBr$ + $(C_2H_5O)_2CO$ $\longrightarrow$ $\xrightarrow{H_2O}{NH_4Cl}$ $(C_2H_5)_3COH$ (82–88%)

9^i $2\,PhMgBr$ + $PhCO_2C_2H_5$ $\longrightarrow$ $\xrightarrow{H_2O}$ Ph_3COH (89–93%)

10^j [lactone: $CH_3(CH_2)_4$ ring with O and $=O$] + $2\,CH_3MgBr$ $\longrightarrow$ $\xrightarrow[H^+]{H_2O}$ $CH_3(CH_2)_4\overset{}{\underset{OH}{C}}H(CH_2)_2\overset{}{\underset{OH}{C}}(CH_3)_2$ (57%)

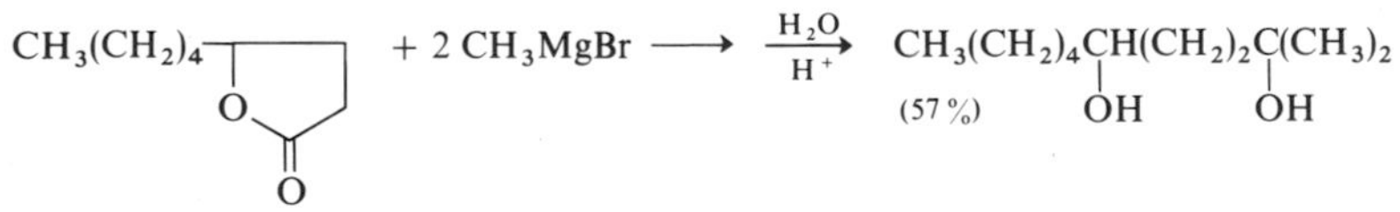

F. Aldehydes from Ethyl Orthoformate

11^k [phenanthrenyl]—$MgBr$ + $HC(OC_2H_5)_3$ $\longrightarrow$ $\xrightarrow[H^+]{H_2O}$ [phenanthrenyl]—$CH{=}O$ (40–42%)

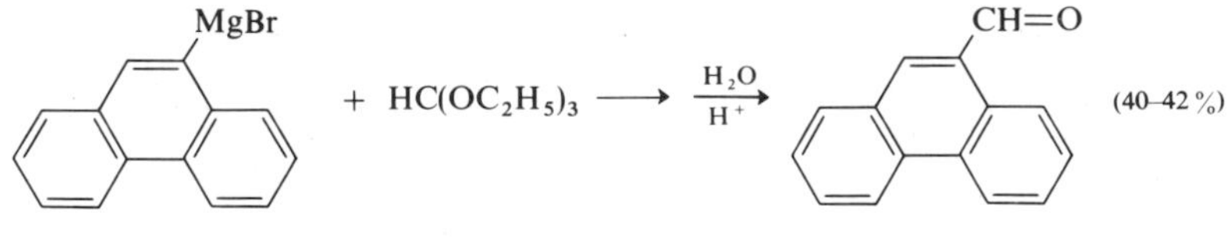

12^l $CH_3(CH_2)_4MgBr$ + $HC(OC_2H_5)_3$ $\longrightarrow$ $\xrightarrow[H^+]{H_2O}$ $CH_3(CH_2)_4CH{=}O$ (45–50%)

Involving Grignard Reagents

175

SECTION 5.1.
ORGANIC
DERIVATIVES
OF GROUP I
AND II
METALS

G. Ketones from Nitriles

13[m]

$+ CH_3MgI \longrightarrow \xrightarrow[HCl]{H_2O}$ (52–59%)

14[n] $CH_3OCH_2C{\equiv}N + PhMgBr \longrightarrow \xrightarrow[HCl]{H_2O} PhCCH_2OCH_3$ (71–78%)

H. Carboxylic Acids by Carbonation

15[o]

$+ CO_2 \longrightarrow \xrightarrow[H^+]{H_2O}$ (86–87%)

16[p] $CH_3CH_2\underset{\underset{MgBr}{|}}{C}HCH_3 + CO_2 \longrightarrow \xrightarrow[H^+]{H_2O} CH_3CH_2\underset{\underset{CO_2H}{|}}{C}HCH_3$ (76–86%)

17[q]

$+ CO_2 \longrightarrow \xrightarrow[H^+]{H_2O}$
(65%)

I. Amines from Imines

18[r] $PhCH{=}NCH_3 + PhCH_2MgCl \longrightarrow \xrightarrow{H_2O} Ph\underset{\underset{CH_3NH}{|}}{C}HCH_2Ph$ (96%)

J. Olefins after Dehydration of Intermediate Alcohols

19[s] $PhCH{=}CHCH{=}O + CH_3MgBr \longrightarrow \xrightarrow{H_2SO_4} PhCH{=}CHCH{=}CH_2$ (75%)

20[t] $2\ PhMgBr + CH_3CO_2C_2H_5 \longrightarrow \xrightarrow[H_2O]{H^+} Ph_2C{=}CH_2$ (67–70%)

a. H. Gilman and W. E. Catlin, *Org. Synth.* **I**, 182 (1932).
b. E. E. Dreger, *Org. Synth.* **I**, 299 (1932).
c. L. Skattebøl, E. R. H. Jones, and M. C. Whiting, *Org. Synth.* **IV**, 792 (1963).
d. C. G. Overberger, J. H. Saunders, R. E. Allen, and R. Gander, *Org. Synth.* **III**, 200 (1955).
e. E. R. Coburn, *Org. Synth.* **III**, 696 (1955).
f. N. L. Drake and G. B. Cooke, *Org. Synth.* **II**, 406 (1943).
g. G. H. Coleman and D. Craig, *Org. Synth.* **II**, 179 (1943).
h. W. W. Moyer and C. S. Marvel, *Org. Synth.* **II**, 602 (1943).
i. W. E. Bachman and H. P. Hetzner, *Org. Synth.* **III**, 839 (1955).
j. J. Colonge and R. Marey, *Org. Synth.* **IV**, 601 (1963).
k. C. A. Dornfeld and G. H. Coleman, *Org. Synth.* **III**, 701 (1955).
l. G. B. Bachman, *Org. Synth.* **II**, 323 (1943).
m. J. E. Callen, C. A. Dornfeld, and G. H. Coleman, *Org. Synth.* **III**, 26 (1955).
n. R. B. Moffett and R. L. Shriner, *Org. Synth.* **III**, 562 (1955).
o. D. M. Bowen, *Org. Synth.* **III**, 553 (1955).
p. H. Gilman and R. H. Kirby, *Org. Synth.* **I**, 353 (1932).
q. D. E. Pearson and D. Cowan, *Org. Synth.* **44**, 78 (1964).
r. R. B. Moffett, *Org. Synth.* **IV**, 605 (1963).
s. O. Grummitt and E. I. Becker, *Org. Synth.* **IV**, 771 (1963).
t. C. F. H. Allen and S. Converse, *Org. Synth.* **I**, 221 (1932).

$$ROMgX + R'\overset{\overset{O}{\|}}{\underset{\underset{H}{|}}{C}}CR''_2 \xrightarrow[RMgX]{} ROH + R'\overset{\overset{-O}{\|}}{C}=CR''_2 \xrightarrow{H^+} R'\overset{\overset{O}{\|}}{\underset{\underset{H}{|}}{C}}CR''_2$$

$$RH$$

oxygen is not excluded. As with reduction, enolization is most seriously competitive in cases where addition is retarded by steric factors.

Addition of Grignard reagents to electrophilic carbon–carbon double bonds has been observed. Two factors have been demonstrated to favor conjugate addition in α,β-unsaturated ketones and esters. Steric hindrance in the vicinity of the carbonyl group favors conjugate addition. For example, conjugate addition is efficient for the 2-butyl ester of crotonic acid, but less hindered esters react at the carbonyl group[43]:

$$CH_3CH=CHCO_2\overset{\overset{CH_3}{|}}{C}HCH_2CH_3 + CH_3CH_2CH_2CH_2MgBr \rightarrow$$

$$\xrightarrow[H_2O]{H^+} CH_3CH_2CH_2CH_2\overset{\overset{CH_3}{|}}{C}HCH_2CO_2\overset{\overset{CH_3}{|}}{C}HCH_2CH_3$$

The presence of catalytic amounts of cuprous salts exerts a profound influence on the ratio of carbonyl to conjugate addition.[44] Mechanistic studies of this phenomenon have indicated that organocopper intermediates are involved. Copper-catalyzed

$$CH_3CH=CH\overset{\overset{O}{\|}}{C}CH_3 + CH_3MgBr \longrightarrow CH_3CH=CH\overset{\overset{}{\underset{\underset{OH}{|}}{C}}}{C}(CH_3)_2 + \quad (90\%)$$

$$(CH_3)_2CHCH_2\overset{\overset{}{\underset{\underset{OH}{|}}{C}}}{C}(CH_3)_2 \quad (3\%)$$

$$CH_3CH=CH\overset{\overset{O}{\|}}{C}CH_3 + CH_3MgBr \xrightarrow{n\text{-}Bu_3P\text{-}CuI} CH_3CH=CH\overset{\overset{}{\underset{\underset{OH}{|}}{C}}}{C}(CH_3)_2 + \quad (1\%)$$

$$(CH_3)_2CHCH_2\overset{\overset{}{\underset{\underset{OH}{|}}{C}}}{C}(CH_3)_2 \quad (95\%)$$

conjugate additions will be further discussed in the section on organometallic derivatives of transition metals. Grignard reagents give substantial amounts of conjugate-addition products when reacted with alkylidene derivatives of dialkyl malonates or alkyl cyanoacetates. The extent of conjugate addition is also increased by addition of cuprous salts in these systems.[45] This reaction has been a useful method for generation of highly branched alkyl chains (entry 3, Scheme 5.3).

Besides their versatility in the construction of carbon frameworks, the Grignard reagents have also been occasionally used to introduce oxygen functions. Grignard reagents react rapidly with oxygen to give a hydroperoxide salt, which then oxidizes a second molecule of Grignard reagent. The reactivity of Grignard reagents toward

43. J. Munch-Petersen, *Org. Synth.* **41**, 60 (1961).
44. H. O. House, W. L. Respess, and G. M. Whitesides, *J. Org. Chem.* **31**, 3128 (1966).
45. R. M. Schisla and W. C. Hammann, *J. Org. Chem.* **35**, 3224 (1970).

Scheme 5.3. Conjugate Additions of Grignard Reagents 177

SECTION 5.1.
ORGANIC
DERIVATIVES
OF GROUP I
AND II
METALS

1[a]
$$\underset{\underset{C_2H_5}{}}{\overset{CH_3}{}}C=C\underset{\underset{C{\equiv}N}{}}{\overset{CO_2C_2H_5}{}} + PhCH_2MgCl \longrightarrow \xrightarrow{H_2O} PhCH_2\underset{\underset{C_2H_5}{}}{\overset{CH_3}{C}}\text{—}\underset{\underset{CO_2C_2H_5}{}}{CHC{\equiv}N} \quad (95\%)$$

2[b]
$$(CH_3)_2C=C(CO_2C_2H_5)_2 + CH_3MgI \xrightarrow[Et_2O]{CuCl} \xrightarrow[H^+]{H_2O} (CH_3)_3CCH(CO_2C_2H_5)_2 \quad (84\text{–}94\%)$$

3[c]
$$(CH_3)_2C=C(CO_2C_2H_5)_2 + CH_3(CH_2)_4MgBr \xrightarrow{CuI} CH_3(CH_2)_4\underset{\underset{CH_3}{}}{\overset{CH_3}{C}}CH(CO_2C_2H_5)_2 \quad (65\%)$$

4[d]
$$(CH_3)_2C=CHCOPh + (CH_3)_2CHMgBr \xrightarrow{CuI} (CH_3)_2CH\underset{\underset{CH_3}{}}{\overset{CH_3}{C}}CH_2COPh$$

5[e]

(68 %)

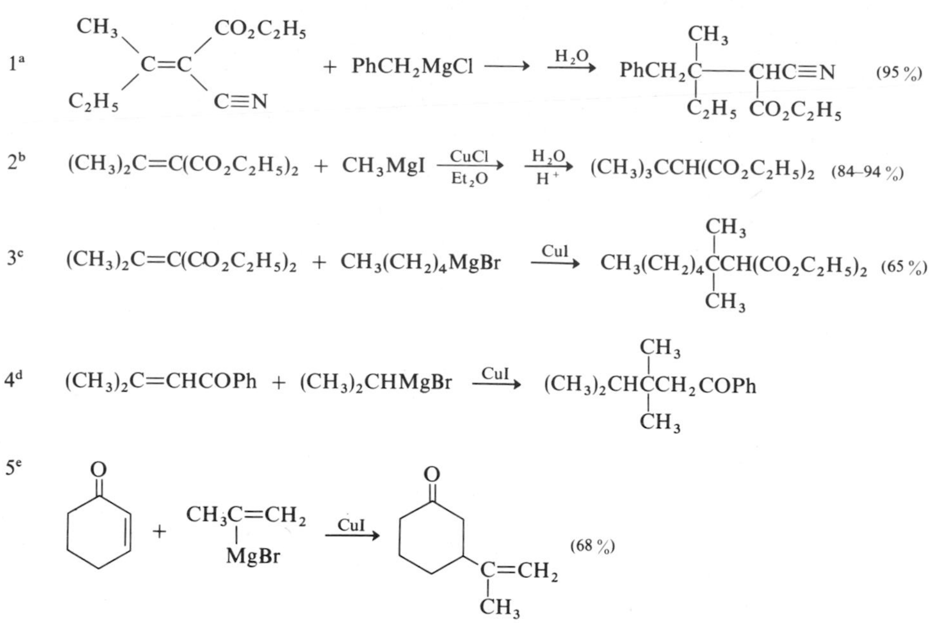

a. F. S. Prout, R. J. Hartman, E. P.-Y. Huang, C. J. Korpics, and G. R. Tichelaar, *Org. Synth.* **IV**, 93 (1963).
b. E. L. Eliel, R. O. Hutchins, and M. Knoeber, *Org. Synth.* **50**, 38 (1970).
c. R. M. Schisla and W. C. Hammann, *J. Org. Chem.* **35**, 3224 (1970).
d. F. D. Lewis and T. A. Hilliard; *J. Am. Chem. Soc.* **92**, 6672 (1970).
e. H. O. House, R. A. Latham, and C. D. Slater, *J. Org. Chem.* **31**, 2667 (1966).

$$RMgX + O_2 \rightarrow R\text{—}O\text{—}O\text{—}MgX$$

$$R\text{—}O\text{—}O\text{—}MgX + RMgX \rightarrow 2\,ROMgX$$

$$ROMgX \xrightarrow{H^+} ROH$$

oxygen–oxygen single bonds is the basis of a synthetic method for *t*-butyl ethers:

$$PhMgBr + (CH_3)_3CO\overset{\overset{O}{\|}}{\text{—}O}CPh \rightarrow PhOC(CH_3)_3 \quad (35\text{–}38\%) \qquad \text{Ref. 46}$$

The reactivity of the alkyl- and aryllithium compounds can conveniently be discussed in terms of comparison with the Grignard reagents. The alkyllithiums are more reactive than the Grignard reagents, and have less tendency to undergo reductions or conjugate additions. Organolithium reagents show higher reactivity toward addition to carbonyl groups. This is illustrated by the comparative reactivity of ethyllithium and ethylmagnesium bromide toward adamantanone; and 83% yield of the desired tertiary alcohol is obtained with ethyllithium, whereas the Grignard reagent gives mainly the reduction product 2-adamantanol. Even *t*-butyllithium

---

46. C. Frisell and S.-O. Lawesson, *Org. Synth.* **41**, 91 (1961).

Scheme 5.4. Synthesis of Ketones from Carboxylate Salts

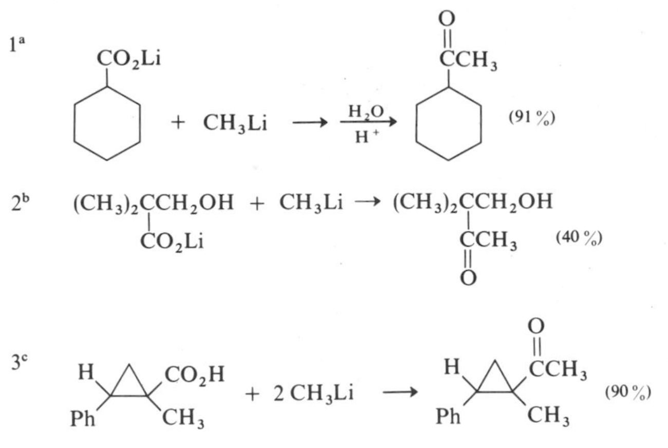

a. T. M. Bare and H. O. House, *Org. Synth.* **49**, 81 (1969).
b. R. A. Schneider and J. Meinwald, *J. Am. Chem. Soc.* **89**, 2023 (1966).
c. C. H. DePuy, F. W. Breitbeil, and K. R. DeBruin, *J. Am. Chem. Soc.* **88**, 3347 (1966).

adds smoothly to adamantanone, giving an 80% yield of the highly crowded 2-*t*-butyl-2-adamantanol.[47] Reactions involving organolithium reagents are subject to the same limitations on the types of functional groups that can be present in the reacting molecules. One reaction that is quite efficient for alkyllithium reagents but poor for Grignard reagents is the synthesis of ketones from carboxylic acids.[48] The success of this reaction depends upon the stability of the dilithio adduct that is formed. This intermediate does not break down until hydrolysis, at which time the ketone is liberated. Some examples of this reaction are shown in Scheme 5.4.

In contrast to the reactions involved in their formation from halides, electrophilic substitution on magnesium and lithium reagents often occurs with high stereospecificity. Retention of configuration has been demonstrated several times with carbon dioxide and ethyl chloroformate as the electrophiles.[25,49] Reaction of alkyl organometallics with bromine has been observed to give both net inversion and retention.[49,50] There is at this time no comprehensive theory of electrophilic substitution reactions of organometallic compounds capable of explaining these features of stereochemistry.

47. J. L. Fry, E. M. Engler, and P. von R. Schleyer, *J. Am. Chem. Soc.* **94**, 4628 (1972).
48. M. J. Jorgenson, *Org. React.* **18**, 1 (1971).
25. See p. 169.
49. D. E. Applequist and G. N. Chmurny, *J. Am. Chem. Soc.* **89**, 875 (1967).
50. W. H. Glaze, C. M. Selman, A. L. Ball, Jr., and L. E. Bray, *J. Org. Chem.* **34**, 641 (1969).

179

SECTION 5.1.
ORGANIC
DERIVATIVES
OF GROUP I
AND II
METALS

The stereochemistry of the addition of methylmagnesium bromide, dimethyl-magnesium, and methyllithium, as well as of other organometallics, has been studied.[51] Although the stereoselectivity is usually not high, there is generally a preference for attack from the equatorial direction to give the axial alcohol; this preference for equatorial attack increases as the size of the alkyl group increases. Generalization from these results suggests that steric factors govern the direction of

addition of organometallic compounds to carbonyl groups. Bicyclic ketones also react with organolithium and organomagnesium reagents to give the products of addition from the less hindered side of the carbonyl group.

Structural rearrangements are not usually encountered with saturated Grignard reagents. Allyl and homoallyl systems can give products resulting from structural isomerization. 2-Butenylmagnesium bromide and 1-methylpropenylmagnesium bromide are in equilibrium in solution. Addition products are derived from the latter compound, although it is the minor component at equlibrium.[52] Addition is believed

$$CH_3CH=CHCH_2MgBr \rightleftharpoons CH_2=CHCHMgBr \overset{CH_3}{\underset{}{|}} \rightleftharpoons CH_3CH=CHCH_2MgBr$$
$$\textit{trans} \qquad\qquad\qquad \textit{cis}$$

to occur through a cyclic transition state that leads to allylic rearrangement. This

mode of addition is supplanted by reaction at the primary carbon when highly hindered ketones are involved. This is undoubtedly a steric effect.

3-Butenylmagnesium bromide is in mobile equilibrium with a small amount of cyclopropylmethylmagnesium bromide. The existence of the equilibrium has been established by deuterium-labeling techniques.[53] Cyclopropylmethylmagnesium bromide[54] and cyclopropylmethyllithium[55] have been prepared by working at low

51. E. C. Ashby and J. T. Laemmle, *Chem. Rev.* **75**, 521 (1975).
52. R. A. Benkeser, W. G. Young, W. E. Broxterman, D. A. Jones, Jr., and S. J. Piaseczynski, *J. Am. Chem. Soc.* **91**, 132 (1969).
53. M. E. H. Howden, A. Maercker, J. Burdon, and J. D. Roberts, *J. Am. Chem. Soc.* **88**, 1732 (1966).
54. D. J. Patel, C. L. Hamilton, and J. D. Roberts, *J. Am. Chem. Soc.* **87**, 5144 (1965).
55. P. T. Lansbury, V. A. Pattison, W. A. Clement, and J. D. Sidler, *J. Am. Chem. Soc.* **86**, 2247 (1964).

$$CH_2=CHCH_2CD_2MgBr \rightleftharpoons \underset{CD_2}{\overset{CH_2}{|}}\!\!\!\diagdown CHCH_2MgBr \rightleftharpoons BrMgCH_2CD_2CH=CH_2$$

temperatures. At room temperature, the ring-opened 3-butenyl reagents are formed. When the olefinic bond is further removed, as in 5-hexenylmagnesium bromide, there is no evidence that equilibration with a cyclic form occurs.[56]

$$CH_2=CHCH_2CH_2CH_2CH_2MgBr \;\;\not\rightleftharpoons\;\; BrMgCH_2-\!\!\!\!\triangleleft$$

5.2. Organic Derivatives of Group IIb Metals

In this section, we will discuss organometallics in which the metal is cadmium, mercury, or zinc. Grignard reagents and organolithium compounds can be converted to heavy-metal derivatives by reaction with salts of these metals. The reaction is driven forward by the tendency for formation of the ionic salt of the more electropositive metal. This reaction is well documented for organocadmium compounds and organomercurials:

$$2\,RMgX \;+\; HgX_2 \;\rightarrow\; R_2Hg \;+\; 2\,MgX_2$$

$$2\,RMgX \;+\; CdX_2 \;\rightarrow\; R_2Cd \;+\; 2\,MgX_2$$

Organomercury compounds can also be prepared from trialkylboranes[57]:

$$R_3B \;+\; 3\,Hg(O_2CCH_3)_2 \;\rightarrow\; 3\,RHgO_2CCH_3$$

The mercury and cadmium compounds are much less reactive than the lithium or magnesium systems, and are therefore useful for certain reactions where selectivity is important. The most widely applied example of this strategy is the preparation of ketones from acid chlorides and cadmium reagents. The cadmium reagent is useful because it is too unreactive to add to the product ketone, and the reaction therefore stops at the ketone stage. Scheme 5.5 lists some examples of this reaction. At

$$RCOCl \;+\; (CH_3)_2Cd \;\rightarrow\; RCOCH_3$$

present, organomercury compounds are not often used directly in organic synthesis. One promising procedure, based on the availability of the mercury compounds from alkenes via organoboranes, is the synthesis of primary halides from alkenes. This procedure has the opposite regioselectivity to the direct ionic addition of hydrogen bromide to terminal alkenes.[58]

$$CH_3(CH_2)_6CH=CH_2 \xrightarrow[\substack{2)\ Hg(O_2CCH_3)_2 \\ 3)\ Br_2}]{1)\ B_2H_6} CH_3(CH_2)_8Br \;\;(69\%)$$

56. R. C. Lamb, P. W. Ayers, M. K. Toney, and J. F. Garst, *J. Am. Chem. Soc.* **88**, 4261 (1966).
57. R. C. Larock and H. C. Brown, *J. Am. Chem. Soc.* **92**, 2467 (1970).
58. J. J. Tufariello and M. M. Hovey, *J. Am. Chem. Soc.* **92**, 3221 (1970).

Scheme 5.5. Preparation of Ketones via Organocadmium Reagents   181

SECTION 5.2.
ORGANIC
DERIVATIVES
OF GROUP IIb
METALS

1ᵃ $[(CH_3)_2CHCH_2CH_2]_2Cd + Cl\overset{O}{\overset{\|}{C}}CH_2CH_2CO_2CH_3 \rightarrow (CH_3)_2CHCH_2CH_2COCH_2CH_2COCH_3$

(73–75%)

2ᵇ $(CH_3)_2C=CHCH_2CH_2\underset{CH_3}{C}=CHCH_2CH_2CH_2COCl + (CH_3CH_2)_2Cd \rightarrow$

$\rightarrow (CH_3)_2C=CHCH_2CH_2\underset{CH_3}{C}=CHCH_2CH_2CH_2COCH_2CH_3$

(80%)

3ᶜ $\left(C_5H_{11}\underset{CH_3}{\overset{CH_3}{C}}-CH_2CH_2\right)_2Cd + Cl\overset{O}{\overset{\|}{C}}CH_2\underset{CH_3}{\overset{CH_3}{C}}-(CH_2)_6-\underset{CH_3}{\overset{CH_3}{C}}-CH_2\overset{O}{\overset{\|}{C}}Cl \rightarrow$ expected diketone   (90%)

4ᵈ

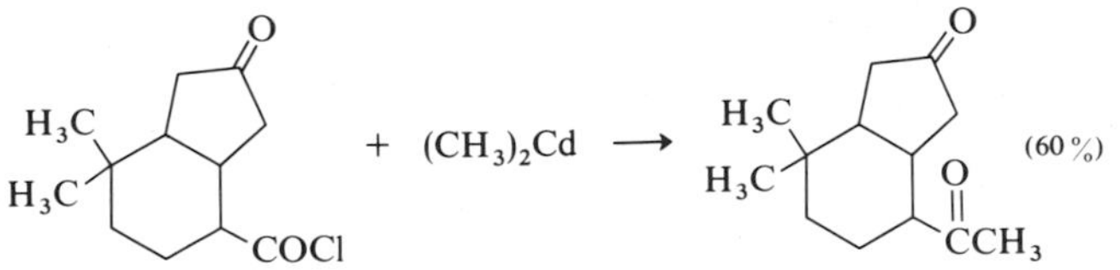

(60%)

a. J. Cason and F. S. Prout, *Org. Synth.* **III**, 601 (1955).
b. J. W. Ralls and B. Riegel, *J. Am. Chem. Soc.* **77**, 6073 (1955).
c. R. M. Schisla and W. C. Hammann, *J. Org. Chem.* **35**, 3224 (1970).
d. M. Miyano and C. R. Dorn, *J. Org. Chem.* **37**, 268 (1972).

Zinc has an important use in synthetic organic chemistry. The reaction of zinc, ethyl bromoacetate, and carbonyl compounds gives α-hydroxyesters. This reaction is known as the *Reformatsky reaction*.[59] The reaction by which the organometallic

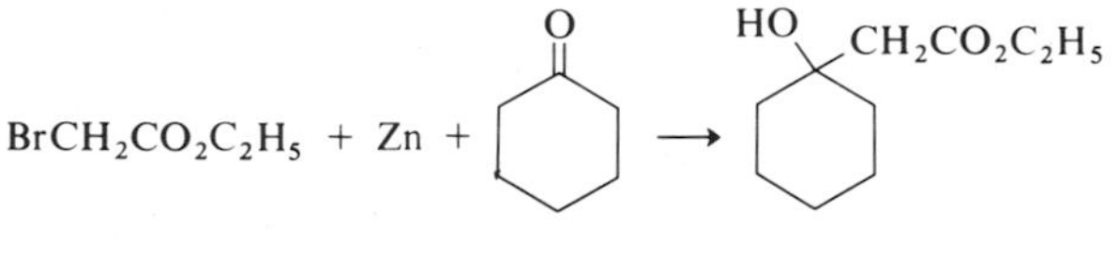

$$BrCH_2CO_2C_2H_5 + Zn + \text{(cyclohexanone)} \rightarrow \text{(1-hydroxy-1-cyclohexyl acetate ester)}$$

compound is formed is presumably similar to that in Grignard-reagent formation. The adjacent carbonyl group can delocalize the negative charge on the carbon, however, so that the nucleophile is best described as a zinc enolate.[60] The enolate can

$$H_5C_2O_2CCH_2Br + Zn \rightarrow H_5C_2O\overset{O^-Zn^{2+}}{\underset{\|}{C}}=CH_2 + Br^-$$

then carry out nucleophilic attack on the carbonyl group, in the same way as an

59. R. L. Shriner, *Org. React.* **1**, 1 (1942); M. W. Rathke, *Org. React.* **22**, 423 (1975).
60. W. R. Vaughan and H. P. Knoess, *J. Org. Chem.* **35**, 2394 (1970).

**Scheme 5.6. Condensation of α-Halocarbonyl Compounds Using Zinc—
The Reformatsky Reaction**

1[a] $CH_3(CH_2)_3\underset{\underset{C_2H_5}{|}}{C}HCH=O$ + $Br\underset{\underset{CH_3}{|}}{C}HCO_2C_2H_5$ $\xrightarrow[\text{2) H}^+]{\text{1) Zn}}$ $CH_3(CH_2)_3\underset{\underset{C_2H_5}{|}}{C}H\underset{\underset{OH}{|}}{C}H\underset{\underset{CH_3}{|}}{C}HCO_2C_2H_5$ (87%)

2[b] $PhCH=O$ + $BrCH_2CO_2C_2H_5$ $\xrightarrow[\text{2) H}^+]{\text{1) Zn}}$ $Ph\underset{\underset{OH}{|}}{C}HCH_2CO_2C_2H_5$ (61–64%)

3[c] $CH_3(CH_2)_4CH=O$ + $BrCH_2CO_2C_2H_5$ $\xrightarrow[\text{2) H}^+]{\text{1) Zn}}$ $CH_3(CH_3)_4\underset{\underset{OH}{|}}{C}HCH_2CO_2C_2H_5$

4[d] $PhCH_2CH=O$ + $BrCH_2CO_2C_2H_5$ $\xrightarrow[\text{THF}]{\text{Zn, (MeO)}_3\text{B}}$ $PhCH_2\underset{\underset{OH}{|}}{C}HCH_2CO_2C_2H_5$ (90%)

5[e]   + $BrCH_2CO_2C_2H_5$ $\xrightarrow[\text{2) H}^+]{\text{1) Zn, benzene}}$

6[f]   + $CH_3CH=O$ $\xrightarrow[\text{DMSO}]{\text{Zn, benzene}}$

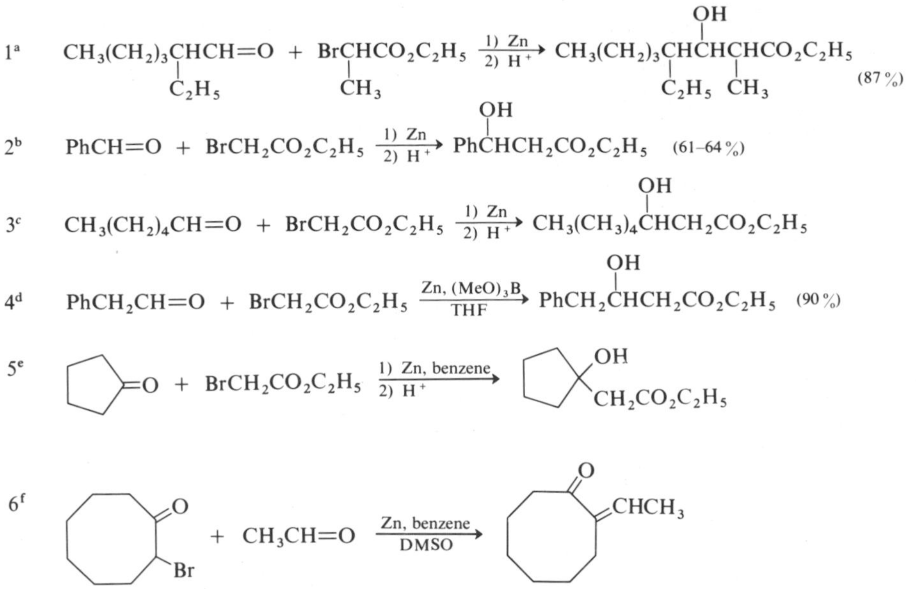

a. K. L. Rinehart, Jr., and E. G. Perkins, *Org. Synth.* **IV**, 444 (1963).
b. C. R. Hauser and D. S. Breslow, *Org. Synth.* **III**, 408 (1955).
c. J. W. Frankenfeld and J. J. Werner, *J. Org. Chem.* **34**, 3689 (1969).
d. M. W. Rathke and A. Lindert, *J. Org. Chem.* **35**, 3966 (1970).
e. J. F. Ruppert and J. D. White, *J. Org. Chem.* **39**, 269 (1974).
f. T. A. Spencer, R. W. Britton, and D. S. Watt, *J. Am. Chem. Soc.* **89**, 5727 (1967).

enolate formed by a deprotonation process. Substituted α-bromoesters and α-bromoketones also undergo addition to carbonyl groups in the presence of metallic zinc. Scheme 5.6 gives some examples of the Reformatsky reaction.

5.3. Organic Derivatives of Transition Metals

Until recently, the magnesium, lithium, and, to a lesser extent, cadmium compounds were the only organometallic systems of much use in organic synthesis. This situation has changed drastically in the last few years. Compounds of copper have proven to be especially valuable.[61] One of the first examples of these compounds is a reagent prepared from methyllithium and cuprous iodide;[44] this reagent is referred to as *lithium dimethylcuprate*. The reagent adds selectively at the β-carbon

61. G. H. Posner, *Org. React.* **19**, 1 (1972); *Org. React.* **22**, 253 (1975).
44. See p. 176.

$$2\,CH_3Li \;+\; CuI \;\rightarrow\; (CH_3)_2CuLi \;+\; LiI$$

183

SECTION 5.3.
ORGANIC
DERIVATIVES
OF TRANSITION
METALS

atom of an α,β-unsaturated carbonyl system, and also has a unique ability to effect replacement of most types of halogen atoms by methyl. Lithium dimethylcuprate reacts with allylic acetates by an S_N2' mechanism. It has been found to react with epoxides to give ring-opened alkylated alcohols. It also adds to acetylenic esters. In addition, there are examples of similar reactions involving branched alkyl, phenyl and vinyl copper reagents. Several specific examples are included in Scheme 5.7.

It has also been established that protected functional groups can be present in the organocopper species.[62-64] For example, Fried and co-workers[63] synthesized the alcohol **1**, converted it to the methoxyisopropyl ether to prevent interference by the acidic hydroxyl group, and then formed the organocopper reagent. This intermediate

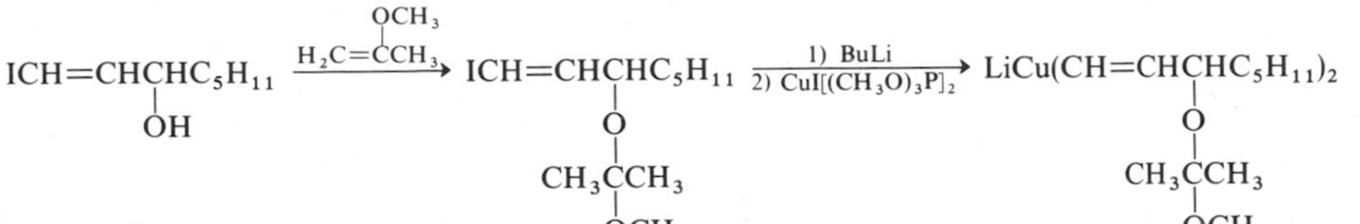

was subsequently shown to react satisfactorily with α,β-unsaturated ketones to give compounds that were utilized in the synthesis of prostaglandins.

It has also been demonstrated that mixed cuprate reagents can be prepared that selectively transfer only one of the carbon groups.[64,65] This procedure is important if relatively complex alkyl groups are to be introduced via cuprate reagents, since earlier procedures utilized only one of the two alkyl groups on copper. The procedure involves reaction of a copper acetylide with the lithium derivative of the group to be utilized in synthesis. Because of the strong binding of copper to the acetylenic ligand, the other group is transferred selectively:

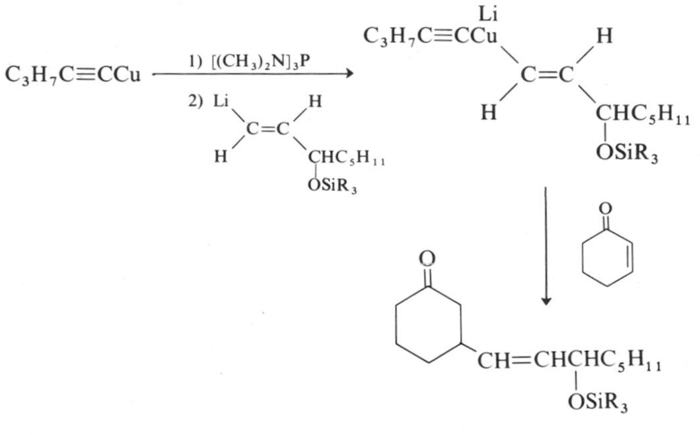

62. E. J. Corey and J. Katzenellenbogen, *J. Am. Chem. Soc.* **91**, 1851 (1969); J. B. Siddall, M. Biskup, and J. H. Fried, *J. Am. Chem. Soc.* **91**, 1853 (1969); E. J. Corey, C. U. Kim, R. H. K. Chen, and M. Takeda, *J. Am. Chem. Soc.* **94**, 4395 (1972).
63. A. F. Kluge, K. G. Untch, and J. H. Fried, *J. Am. Chem. Soc.* **94**, 7827 (1972).
64. E. J. Corey and D. J. Beames, *J. Am. Chem. Soc.* **94**, 7210 (1972).
65. H. O. House and M. J. Umen, *J. Org. Chem.* **38**, 3893 (1973).

A. Conjugate Additions

1[a] $CH_3CH{=}CHCCH_3$ + Me_2CuLi → $(CH_3)_2CHCH_2CCH_3$ (94%)

2[a]

   + Me_2CuLi →   (98%)

3[b]

   + Me_2CuLi →   (55%)

4[c]

   + $LiCu(CH{=}CH_2)_2$ →   (66%)

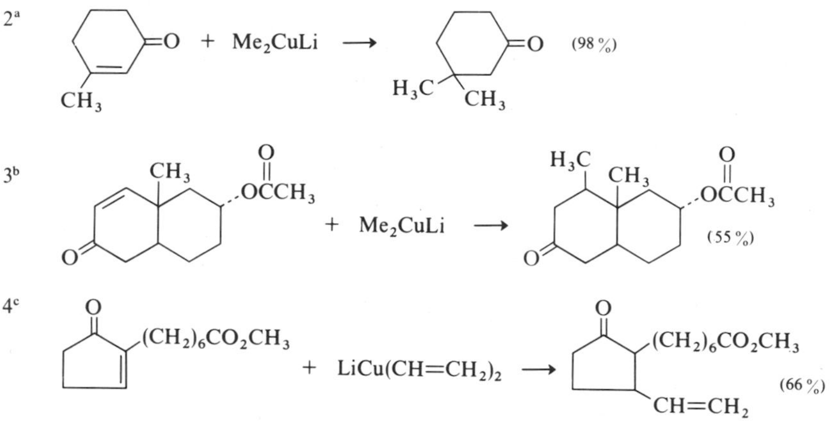

B. Halogen Substitution

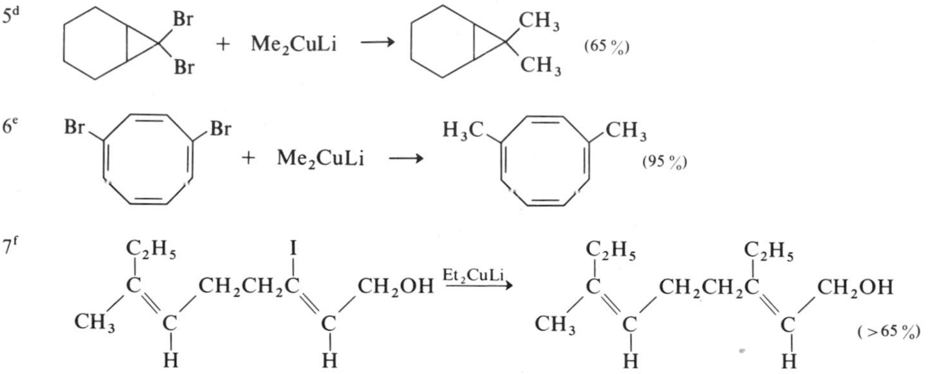

5[d]

   + Me_2CuLi →   (65%)

6[e]

   + Me_2CuLi →   (95%)

7[f]

   $\xrightarrow{Et_2CuLi}$ (>65%)

8[f] $C_5H_{11}Br$ + $[(CH_3)_3C]_2CuLi$ → $(CH_3)_3C(CH_2)_4CH_3$

Corey and Posner discovered that lithium dimethylcuprate could replace iodine or bromine by methyl in a variety of compounds, including aryl and vinyl halides.[66] This method of replacement of halide by alkyl is much more satisfactory and general

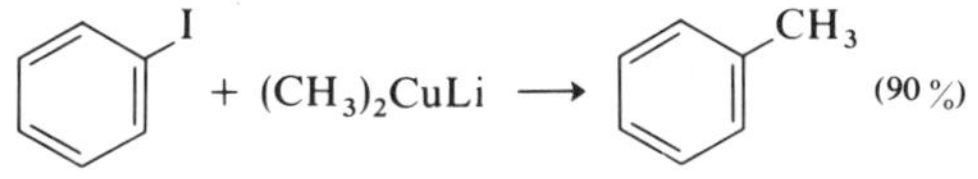

   + $(CH_3)_2CuLi$ →   (90%)

$PhCH{=}CHBr$ + $(CH_3)_2CuLi$ → $PhCH{=}CHCH_3$ (81%)

66. E. J. Corey and G. H. Posner, *J. Am. Chem. Soc.* **89**, 3911 (1967).

Lithium Copper Reagents

185

SECTION 5.3.
ORGANIC
DERIVATIVES
OF TRANSITION
METALS

9^g

$$H_3C-C(=CH_2)-CH(Cl)-CH_2CH_2-C(=CH_2)-CH(Cl)-CH_2CH_2-C(CH_3)=CH-CH_2OCPh_3 \rightarrow$$

$$\xrightarrow{Me_2CuLi} H_3C-C(C_2H_5)=CH-CH_2CH_2-C(C_2H_5)=CH-CH_2CH_2-C(CH_3)=CH-CH_2OCPh_3$$

C. Alkylation of Epoxides

10^h

$$H_3C-\overset{O}{\overset{\triangle}{\quad}}-CO_2C_2H_5 + Me_2CuLi \rightarrow CH_3\overset{CH_3}{\underset{HO}{CH}}CHCO_2C_2H_5$$

D. Alkylation of Allyl Acetates

11^i

$$(CH_3)_2C=CHCHCH_2C(CH_3)_3 + Me_2CuLi \rightarrow (CH_3)_3CCH=CHCH_2C(CH_3)_3$$
$$\underset{O_2CCH_3}{\quad}$$

E. Ketones from Acid Chlorides

12^j

$$N\equiv C(CH_2)_{10}\overset{O}{\overset{\parallel}{C}}Cl + Me_2CuLi \rightarrow N\equiv C(CH_2)_{10}\overset{O}{\overset{\parallel}{C}}CH_3$$

13^k

$$(CH_3)_3\overset{O}{\overset{\parallel}{C}}Cl + (CH_3)_2CHLi \xrightarrow{CuI} (CH_3)_3\overset{O}{\overset{\parallel}{C}}CH(CH_3)_2 \quad (94\%)$$

a. H. O. House, W. L. Respess, and G. M. Whitesides, *J. Org. Chem.* **31**, 3128 (1966).
b. J. A. Marshall and G. M. Cohen, *J. Org. Chem.* **36**, 877 (1971).
c. F. S. Alvarez, D. Wren, and A. Prince, *J. Am. Chem. Soc.* **94**, 7823 (1972).
d. E. J. Corey and G. H. Posner, *J. Am. Chem. Soc.* **89**, 3911 (1967).
e. W. E. Konz, W. Hechtl, and R. Huisgen, *J. Am. Chem. Soc.* **92**, 4104 (1970).
f. E. J. Corey, J. A. Katzenellenbogen, N. W. Gilman, S. A. Roman, and B. W. Erickson, *J. Am. Chem. Soc.* **90**, 5618 (1968).
g. E. E. van Tamelen and J. P. McCormick, *J. Am. Chem. Soc.* **92**, 737 (1970).
h. C. R. Johnson, R. W. Herr, and D. M. Wieland, *J. Org. Chem.* **38**, 4263 (1973).
i. R. J. Anderson, C. A. Henrick, and J. B. Siddall, *J. Am. Chem. Soc.* **92**, 735 (1970).
j. G. H. Posner, C. E. Whitten, and P. E. McFarland, *J. Am. Chem. Soc.* **94**, 5106 (1972).
k. J.-E. Dubois, M. Boussu, and C. Lion, *Tetrahedron Lett.*, 829 (1971).

than displacement by the Grignard or lithium reagent.[67] Allyl halides and acetates

$$H_2C=\overset{CH_3}{\underset{\underset{O}{\overset{\parallel}{CH_3C-O}}}{\overset{|}{C}}}-CH(CH_2)_3CH_3 + (CH_3)_2CuLi \rightarrow \underset{CH_3CH_2}{\overset{CH_3}{\underset{CH(CH_2)_3CH_3}{C}}} \quad (80\%)$$

67. G. M. Whitesides, W. F. Fischer, Jr., J. San Filippo, Jr., R. W. Bashe, and H. O. House, *J. Am. Chem. Soc.* **91**, 4871 (1969).

react with methylation at the olefinic carbon, accompanied by an allylic shift.[68] Acetylenic acetates also react with the shift of an electron pair, resulting in the formation of allenes[69]:

$$CH_3CO_2 \quad C{\equiv}CH + (CH_3)_2CuLi \longrightarrow =C=CHCH_3 \quad (85\%)$$

Saturated epoxides are opened in good yield by lithium dimethylcuprate.[70] The methyl group is introduced at the less hindered carbon of the epoxide:

$$CH_3CH_2 \overset{O}{\triangle} + (CH_3)_2CuLi \rightarrow CH_3CH_2\underset{OH}{CH}CH_2CH_3 \quad (88\%)$$

Epoxides having vinyl substituents undergo attack by the reagent at the double bond, with a concomitant shift of the double bond and ring-opening[71]:

$$(CH_3)_2CuLi + H_2C{=}\underset{CH_3}{C} \quad CH_3 \rightarrow CH_3CH_2\underset{OH}{C}{=}CH\underset{}{CH}CH_3$$

The cuprate reagents can also convert acid chlorides to ketones, and are alternatives to alkylcadmium reagents in this regard.[72] The acid chlorides can contain such functional groups as cyano, ester, and ketone without interfering with successful reaction.[73]

The chemical reactivity displayed by the cuprate reagents is powerful nucleophilicity toward carbon, but with a strong preference for reaction at alkene or halide sites over carbonyl groups. The mechanism(s) of the reactions have not been established with certainty. A leading proposal[74] for the conjugate addition is that reaction is initiated by a one-electron transfer from the copper–lithium species. The

$$[R_2LiCu]_2 + R'CH{=}CH\overset{O}{\overset{\|}{C}}R' \rightarrow [R_2LiCu]_2^+ + R'\overset{O^-}{\overset{|}{C}}HCH{=}CR'$$

$$R_2LiCu + Li^+ + RCu + R\overset{R'}{\underset{|}{C}}H CH{=}\overset{O^-}{\underset{|}{C}}R' \leftarrow [R_2LiCu]_2\overset{R'}{\underset{|}{C}}HCH{=}\overset{O^-}{\underset{|}{C}}R'$$

$$\downarrow H^+$$

$$R\overset{R'}{\underset{|}{C}}HCH_2\overset{O}{\overset{\|}{C}}R'$$

68. R. J. Anderson, C. A. Henrick, and J. B. Siddall, *J. Am. Chem. Soc.* **92**, 735 (1970); E. E. van Tamelen and J. P. McCormick, *J. Am. Chem. Soc.* **92**, 737 (1970).
69. P. Rona and P. Crabbé, *J. Am. Chem. Soc.* **90**, 4733 (1968).
70. C. R. Johnson, R. W. Herr, and D. M. Wieland, *J. Org. Chem.* **38**, 4263 (1973).
71. R. J. Anderson, *J. Am. Chem. Soc.* **92**, 4978 (1970); R. W. Herr and C. R. Johnson, *J. Am. Chem. Soc.* **92**, 4979 (1970).
72. G. H. Posner and C. E. Whitten, *Tetrahedron Lett.*, 4647 (1970).
73. G. H. Posner, C. E. Whitten, and P. E. McFarland, *J. Am. Chem. Soc.* **94**, 5106 (1972).
74. H. O. House and M. J. Umen, *J. Am. Chem. Soc.* **94**, 5495 (1972).

transfer of the alkyl group is pictured as occurring in an intermediate formed by combination of the two products that result from the electron-transfer step. It has been shown that the stereochemistry of the transferred alkyl group is retained in the

case of the norbornyl group.[75] This and related evidence rules out the possibility that the alkyl group might be transferred as a free radical.

A study of the relative rates of reaction of lithium di-n-butylcuprate with a series of organic halides has shown that the relative reactivity of the halides is very similar to that exhibited in S_N2 reactions.[76] Also, inversion of configuration occurs at the site of substitution.[67,77] This evidence suggests that the transition state involves nucleophilic displacement of halide, but the details of the mechanism are still open to question.

Next in importance among the transition metals in terms of utility for organic synthesis is nickel. In general, the most important reactions involving nickel species result in the coupling of two organic molecules. Allyl halides react with nickel carbonyl to give a complex in which allyl groups are bonded to the nickel. The nature of the bonding is different from that in the organometallic compounds that have been discussed previously in this chapter. The bond to nickel involves the π-orbitals; these organometallics are therefore referred to as π-allyl complexes. The details of the bonding will be discussed more fully in Section 5.5.

The important feature of these compounds for synthesis is their ability to react with a large variety of organic halides, replacing the halogen atom with an allyl group[78]:

The mechanism of the reaction has not been studied in great detail; in contrast to the reactions of the cuprate reagents, however, it is not considered likely that a nucleophilic substitution in involved. In very general terms, the reaction mechanism

75. G. M. Whitesides and P. E. Kendall, *J. Org. Chem.* **37**, 3718 (1972).
76. W. H. Mandeville and G. N. Whitesides, *J. Org. Chem.* **39**, 400 (1974).
67. See p. 185.
77. C. R. Johnson and G. A. Dutra, *J. Am. Chem. Soc.* **95**, 7783 (1973).
78. M. F. Semmelhack, *Org. React.* **19**, 115 (1972).
79. E. J. Corey and M. F. Semmelhack, *J. Am. Chem. Soc.* **89**, 2755 (1967).

187

SECTION 5.3.
ORGANIC
DERIVATIVES
OF TRANSITION
METALS

below suggests that the halide adds to the nickel reagent, and the two organic groups then couple with elimination of nickel (II):

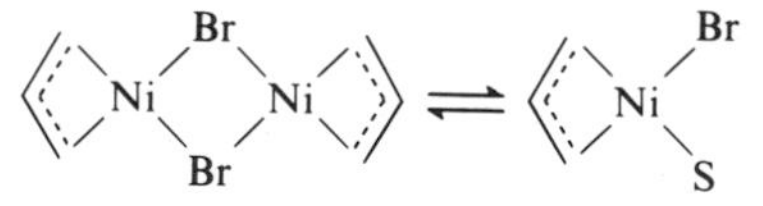

$$S = \text{solvent}$$

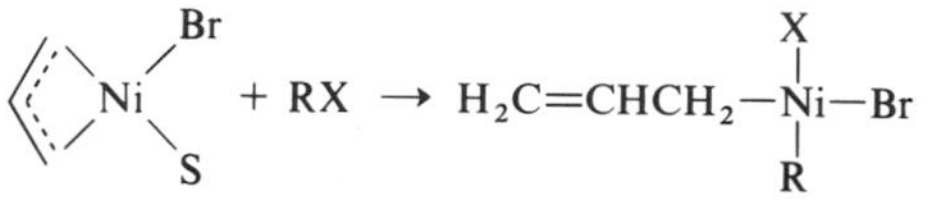

$$H_2C=CHCH_2R + NiBrX$$

Organonickel intermediates are probably involved in a group of reactions in which nickel carbonyl, $Ni(CO)_4$, is the reagent. Nickel carbonyl causes the coupling of allyl halides when the reaction is carried out in very polar solvents such as dimethylformamide or dimethyl sulfoxide. This coupling reaction has been used intramolecularly to bring about the cyclization of bis-allylic halides, and has been found useful in the preparation of rings as large as 18 atoms.

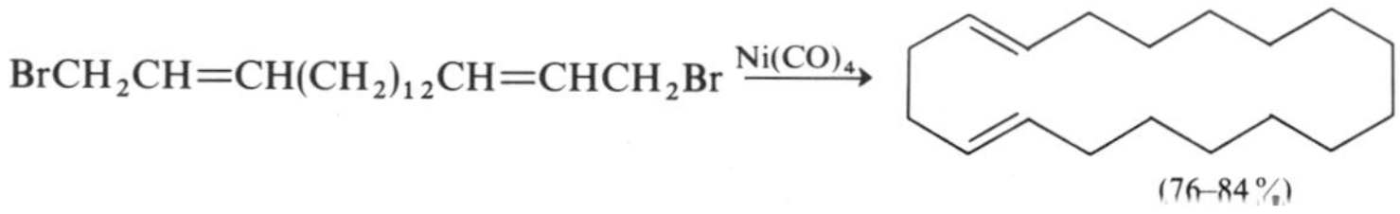

$$BrCH_2CH=CH(CH_2)_{12}CH=CHCH_2Br \xrightarrow{Ni(CO)_4}$$

(76–84%)   Ref. 80

(70–75%)   Ref. 81

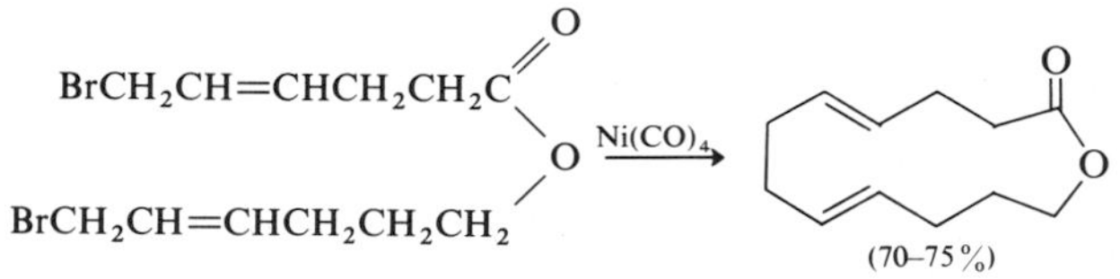

Another nickel complex, bis(1,5-cyclooctadiene)nickel(0), has been found to bring about coupling of other halides, in addition to the allylic types that are reactive toward nickel carbonyl:

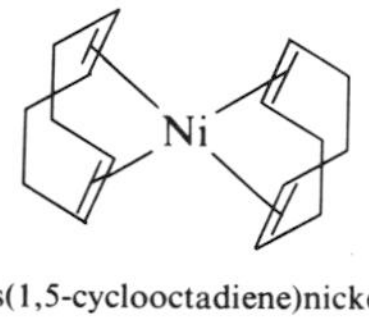

bis(1,5-cyclooctadiene)nickel(0)

$$N\equiv C-\!\!\langle\rangle\!\!-Br \xrightarrow{(COD)_2Ni} N\equiv C-\!\!\langle\rangle\!\!-\!\!\langle\rangle\!\!-C\equiv N \quad (81\%)$$

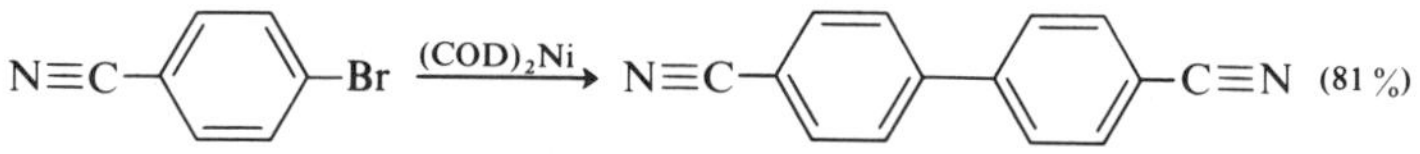

Ref. 82

80. E. J. Corey and E. K. W. Wat, *J. Am. Chem. Soc.* **89**, 2757 (1967).
81. E. J. Corey and H. A. Kirst, *J. Am. Chem. Soc.* **94**, 667 (1972).
82. M. F. Semmelhack, P. M. Helquist, and L. D. Jones, *J. Am. Chem. Soc.* **93**, 5908 (1971).

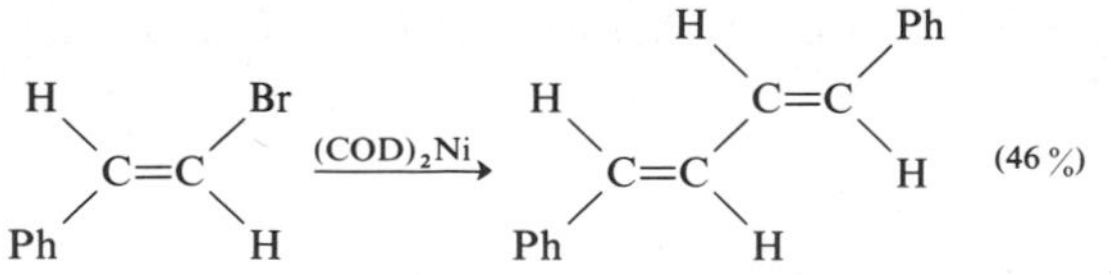

189

SECTION 5.3.
ORGANIC
DERIVATIVES
OF TRANSITION
METALS

(46%)   Ref. 83

Nickel (II) salts effectively catalyze the reaction of Grignard reagents with vinyl and aryl halides. A soluble phosphine complex, $Ni(Ph_2PCH_2CH_2PPh_2)_2Cl_2$, has been shown to be an effective catalyst.[84] With secondary Grignard reagents under

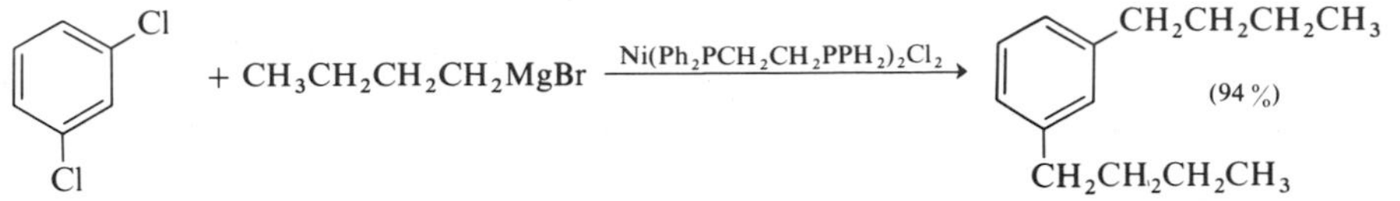

some conditions, the final product contains the corresponding primary alkyl substituent.[85] The most likely mechanism for this rearrangement involves an organonickel intermediate, which is converted to a nickel–alkene complex. Recombination of the alkene with nickel can then occur in the opposite direction from elimination:

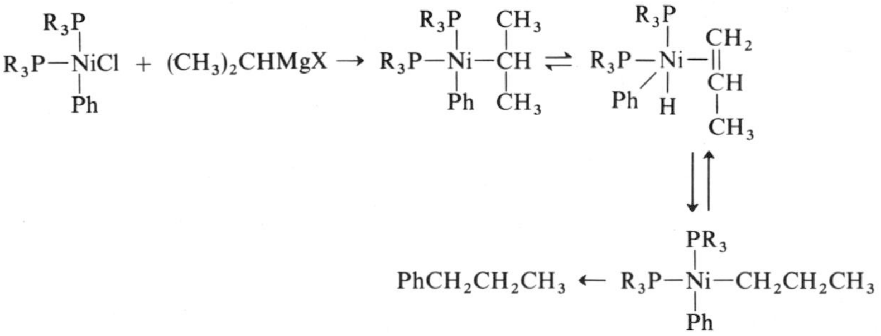

Grignard reagents react to give coupling products in the presence of Co(II) ions. Disproportionation products attributable to radical intermediates are major products from alkyl Grignards.[86] Thallium(I) bromide can effect coupling of aryl and secondary alkyl Grignard reagents, but other processes compete with primary Grignards[87]:

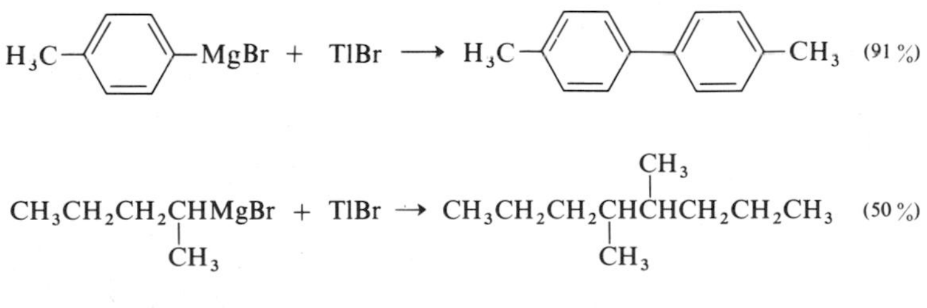

83. M. F. Semmelhack, P. M. Helquist, and J. D. Gorzynski, *J. Am. Chem. Soc.* **94**, 9234 (1972).
84. K. Tamao, K. Sumitani, and M. Kumada, *J. Am. Chem. Soc.* **94**, 4374 (1972).
85. K. Tamao, Y. Kiso, K. Sumitani, and M. Kumada, *J. Am. Chem. Soc.* **94**, 9268 (1972).
86. W. B. Smith, *J. Org. Chem.* **26**, 4206 (1961).
87. A. McKillop, L. F. Elsom, and E. C. Taylor, *J. Am. Chem. Soc.* **90**, 2423 (1968).

Scheme 5.8. Preparation of Biaryls by the Ullmann Coupling Reaction

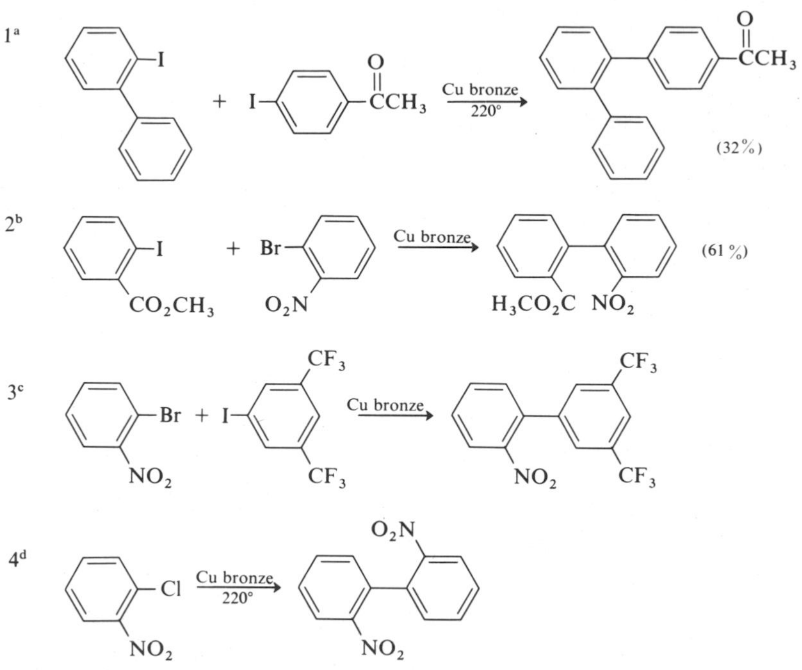

a. C. F. H. Allen and F. P. Pingert, *J. Am. Chem. Soc.* **64**, 2639 (1942).
b. M. R. Pettit and J. C. Tatlow, *J. Chem. Soc.* 1071 (1954).
c. R. J. Sundberg and R. W. Heintzelman, *J. Org. Chem.* **39**, 2546 (1974).
d. R. C. Fuson and E. A. Cleveland, *Org. Synth.* **III**, 339 (1955).

The two alkyl groups bonded to copper in dialkyllithium cuprate complexes are coupled by oxidizing agents such as oxygen or nitrobenzene[88]:

$$R_2CuLi \xrightarrow{O_2} R-R$$

Vinyl copper and vinyl silver systems, which can be prepared by metal–metal exchange from the corresponding lithium compounds, undergo thermal decomposition, resulting in coupling to butadiene, and deposition of the metal occurs near room temperature[89]:

$$CH_3CH=CHLi + [CuI(PBu_3)]_4 \rightarrow [CH_3CH=CHCu(PBu_3)]_n$$
$$\downarrow \Delta$$
$$CH_3CH=CHCH=CHCH_3$$

The mechanism of these reactions is not known in detail. They may involve one-electron oxidation of the alkyl moiety to the equivalent of a radical, followed by combination of two such groups. Both the oxidation and coupling probably take

88. G. M. Whitesides, J. San Filippo, Jr., C. P. Casey, and E. J. Panek, *J. Am. Chem. Soc.* **89**, 5302 (1967).
89. G. M. Whitesides, C. P. Casey, and J. K. Krieger, *J. Am. Chem. Soc.* **93**, 1379 (1971).

place while the alkyl groups remain in the metal coordination sphere, so that free-radical intermediates are not necessarily involved.

Organocopper compounds are probably involved[90] in a long-known procedure for coupling aromatic halides, the *Ullmann reaction*.[91] This reaction involves heating two aromatic halides in the presence of copper bronze. The order of reactivity is iodides > bromides ≫ chlorides. Reactions with iodides and bromides usually occur around 200°. This reaction can also be used satisfactorily for the preparation of unsymmetrical biaryls. Many examples of the Ullmann reaction are given in reference 91, and few selected examples are shown in Scheme 5.8. The best yields in this reaction are obtained when one of the aromatic halides also contains a nitro group. In the absence of such "activating" groups, yields are seldom good—often around 20–30%.

5.4. Catalysis of Rearrangements by Metal Ions and Complexes

Organo derivatives of transition metals are also believed to be involved in a fascinating series of reactions aptly described as *bond-switching reactions*. These reactions are typical of highly strained hydrocarbons. Scheme 5.9 lists a number of such reactions.

The view of the mechanism by which transition metals catalyze such reactions that has gained widest acceptance pictures those reactions as being initiated by electrophilic (oxidative) attack on one of the strained σ-bonds.[92] Such oxidative attack is then followed by carbonium-ion-like migrations and bond cleavages, which eventually are concluded by the expulsion of the metal ion. Some evidence for carbonium-ion intermediates has been obtained from results of reactions in nucleophilic solvents in which the intermediate "carbonium ion" can be trapped.

Ref. 93

R = t-Bu

90. T. Cohen and T. Poeth, *J. Am. Chem. Soc.* **94**, 4363 (1972); A. H. Lewin and T. Cohen, *Tetrahedron Lett.*, 4531 (1965); T. Cohen and J. G. Tirpak, *Tetrahedron Lett.*, 143 (1975).
91. P. E. Fanta, *Chem. Rev.* **64**, 613 (1964); *Synthesis*, 9 (1974).
92. T. J. Katz and S. A. Cerefice, *J. Am. Chem. Soc.* **91**, 2405 (1969), **91**, 6519 (1969); L. Cassar, P. E. Eaton, and J. Halpern, *J. Am. Chem. Soc.* **92**, 3515 (1970); P. G. Gassman and F. J. Williams, *J. Am. Chem. Soc.* **94**, 7733 (1972); L. A. Paquette, S. E. Wilson, R. P. Henzel, and G. R. Allen, Jr., *J. Am. Chem. Soc.* **94**, 7761 (1972).
93. K. L. Kaiser, R. F. Childs, and P. M. Maitlis, *J. Am. Chem. Soc.* **93**, 1270 (1971).

Scheme 5.9. Metal-Ion-Catalyzed Isomerizations

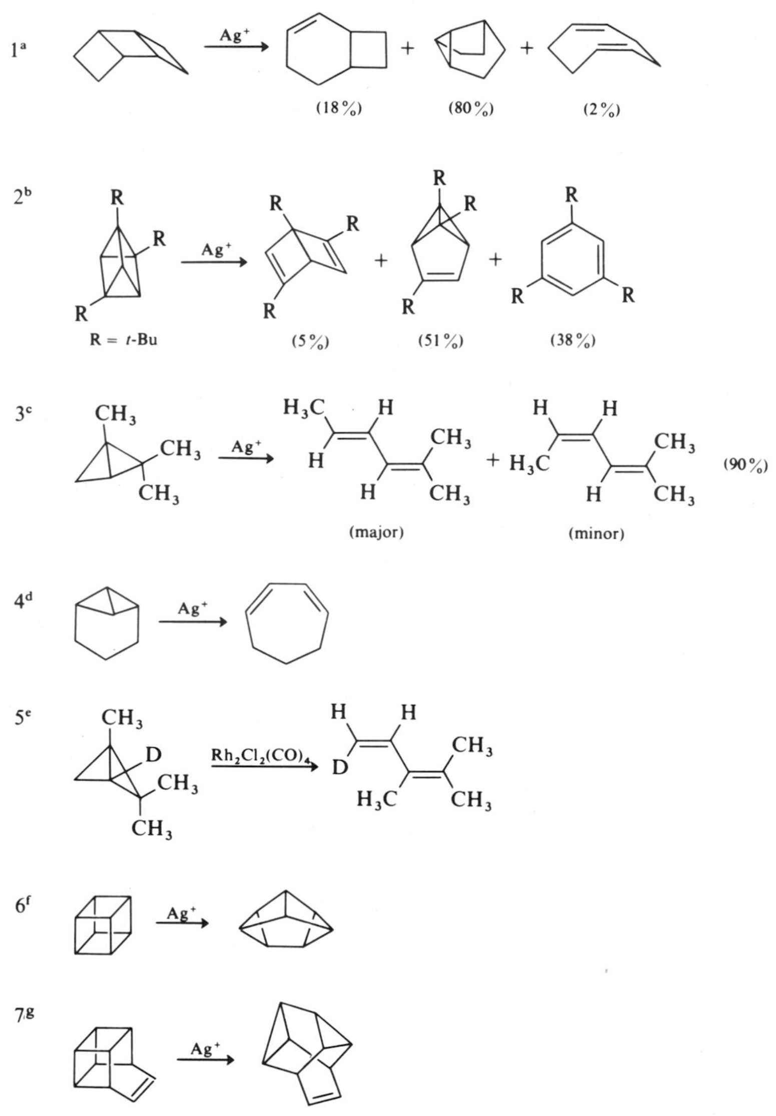

a. J. Wristers, L. Brener, and R. Pettit, *J. Am. Chem. Soc.* **92**, 7499 (1970).
b. K. L. Kaiser, R. F. Childs, and P. M. Maitlis, *J. Am. Chem. Soc.* **93**, 1270 (1971).
c. L. A. Paquette, R. P. Henzel, and S. E. Wilson, *J. Am. Chem. Soc.* **93**, 2335 (1971).
d. L. A. Paquette, G. R. Allen, Jr., and R. P. Henzel, *J. Am. Chem. Soc.* **92**, 7002 (1970).
e. P. G. Gassman and F. J. Williams, *J. Am. Chem. Soc.* **92**, 7631 (1970).
f. L. Cassar, P. E. Eaton, and J. Halpern, *J. Am. Chem. Soc.* **92**, 6366 (1970).
g. W. G. Dauben, C. H. Schallhorn, and D. L. Whalen, *J. Am. Chem. Soc.* **93**, 1446 (1971).

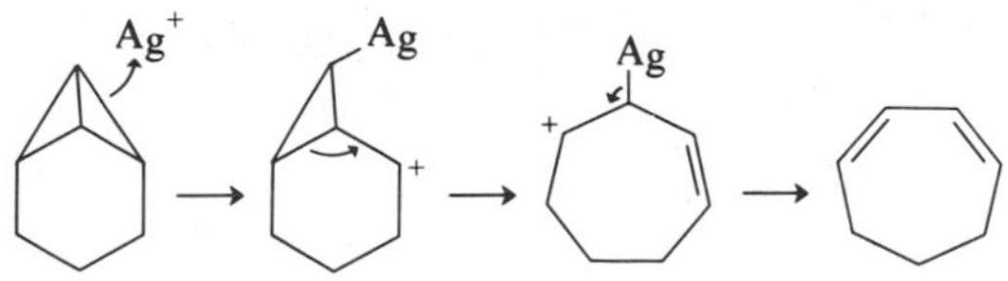

Ref. 94

Many details concerning these reactions have been established, but there are also unanswered questions regarding relationships between the catalytic metal ion and the product composition. For example, the course of isomerization of tricyclo[4.1.0.0^{2,7}]heptane, **2**, with more than a dozen transition-metal derivatives has been studied.[95] The range of products is illustrated by the results for AgBF$_4$, [Rh(CO)$_2$Cl]$_2$, and SnCl$_2$:

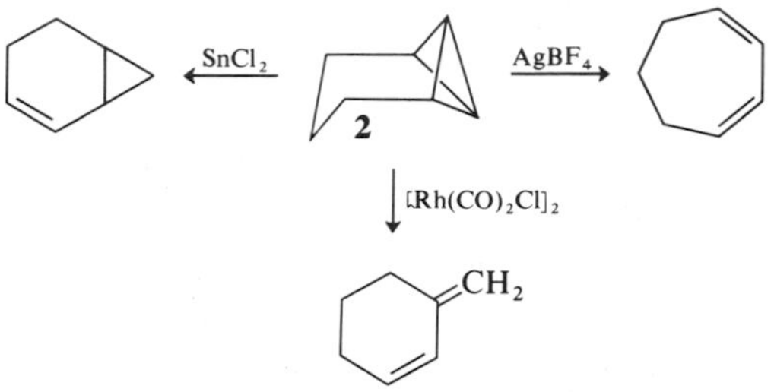

The strain energy associated with the small-ring systems in the substrates provides the driving force for these rearrangement reactions. Unstrained hydrocarbons do not react with the various transition-metal species that cause bond-switching in strained hydrocarbons.

5.5. Organometallic Compounds with π-Bonding

The compounds discussed in this section are distinguished by the nature of their carbon–metal bonds. The organometallics in the previous sections have in most cases involved carbon–metal bonds in which a single carbon atom could be recognized as being the atom to which the metal was attached. The organometallics discussed in this section, however, are bound to metals by π-molecular orbitals that are delocalized over more than one carbon atom. Among the classes of organic molecules that have been observed to form this type of bond are the simple olefins. Olefin complexes were first observed with platinum; subsequently, however, complexes derived from many of the other transition metals have been isolated. The bonding in such complexes is generally considered to be the result of two major contributions. The olefinic ligand acts as an electron donor toward the metal atom, transferring electrons from a filled π-orbital to an unfilled metal orbital. There is also considered to be a contribution called "back-bonding." This involves interaction of a

94. L. A. Paquette, S. E. Wilson, and R. P. Henzel, *J. Am. Chem. Soc.* **94**, 7771 (1972).
95. P. G. Gassman and T. J. Atkins, *J. Am. Chem. Soc.* **94**, 7748 (1972).

Figure 5.1. Representation of π-bonding in olefin–transition-metal complexes.

filled metal orbital with the empty π-antibonding orbital of the olefin. These two types of bonding are represented in Figure 5.1. This general concept of the bonding arrangement will pertain throughout this section, although details vary as the nature of the organic π-ligand and the identity of the transition metal change.

Olefin complexes have most often been prepared by displacement of some other ligand by the olefin. Both thermal and photochemical reactions have been observed. Olefins can form complexes with platinum and palladium by displacing solvent or halide molecules from the coordination sphere of the metal[96]:

$$(C_6H_5CN)_2PdCl_2 \ + \ 2\,RCH{=}CH_2 \ \rightarrow$$

Displacement of carbon monoxide from metal carbonyls is a common preparative method[97]:

The π-molecular orbital of the organic ligand can be delocalized over more than two carbon atoms, of course. There are well-characterized compounds in which the π-orbital is associated with three, four, five, six, and seven carbon atoms. Some examples of structures with allyl ligands are illustrated in Figure 5.2.

The dimeric π-allylnickel(I) bromides are prepared readily from allyl bromides and nickel carbonyl.[79] Allylmagnesium bromide reacts with nickel bromide to give bis(π-allyl)nickel[98]:

$$2\,CH_2{=}CHCH_2MgBr \ \longrightarrow \ NiBr_2 \ \longrightarrow \ \text{Ni} \ + \ 2\,MgBr_2$$

Organometallics involving organic ligands having a cyclic array of four carbon

96. M. S. Kharasch, R. C. Seyler, and F. R. Mayo, *J. Am. Chem. Soc.* **60**, 882 (1938).
97. J. Chatt and L. M. Venanzi, *J. Chem. Soc.*, 4735 (1957).
79. See p. 187.
98. D. Walter and G. Wilke, *Angew. Chem. Int. Ed. Engl.* **5**, 897 (1966).

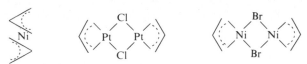

Figure 5.2. Structures of some π-organometallic compounds containing allyl groups as ligands.

atoms have been of particular interest because the ligand in this case is a cyclobutadiene derivative. Derivatives containing this ligand were first prepared in the late 1950's, and a complex of the parent cyclobutadiene was reported in 1965.[99]

$$\text{[diagram: cyclobutadiene with Cl substituents]} + Fe_2(CO)_9 \longrightarrow \text{[diagram: cyclobutadiene Fe(CO)_3 complex]}$$

Subsequent studies have provided strong evidence that oxidative decomposition of such complexes releases cyclobutadiene or its derivatives. All the evidence available indicates that these compounds have very short lifetimes, but the products derived by trapping experiments are best explained as being reaction products of cyclobutadiene. The carbocyclic ring in the cyclobutadieneiron tricarbonyl molecule behaves chemically as a reactive aromatic ring. A number of electrophilic substitutions have been carried out.[100] This type of reactivity of cyclic ligands in π-organometallics is also found with other systems that will be discussed shortly. Scheme 5.10 presents some examples of reactions in which cyclobutadiene is generated by oxidation of its iron tricarbonyl complex and then undergoes addition reactions with other reagents in solution.

The first of the π-organometallic compounds to be characterized was ferrocene. This compound is a neutral molecule that is derived from two cyclopentadienyl anions and iron(II). Two convenient methods of preparation have been described[101]; both involve reaction of the aromatic cyclopentadienide anion with Fe(II). Related

$$2 \left[\text{[cyclopentadienide anion]} \right]^- + FeCl_2 \longrightarrow \text{[ferrocene]}$$

compounds in which other metals are present have been prepared. Many of these have ligands in addition to cyclopentadienide attached to the metal.[102] The total of

99. G. F. Emerson, L. Watts, and R. Pettit, *J. Am. Chem. Soc.* **87**, 131 (1965); R. Pettit and J. Henery, *Org. Synth.* **50**, 21 (1970).
100. J. D. Fitzpatrick, L. Watts, G. F. Emerson, and R. Pettit, *J. Am. Chem. Soc.* **87**, 3254 (1965).
101. G. Wilkinson, *Org. Synth.* **IV**, 473, 476 (1963).
102. M. L. H. Green, *Organometallic Compounds*, Vol. II, Methuen and Co., London, 1968, pp. 90–139.

Scheme 5.10. Reactions of Cyclobutadiene

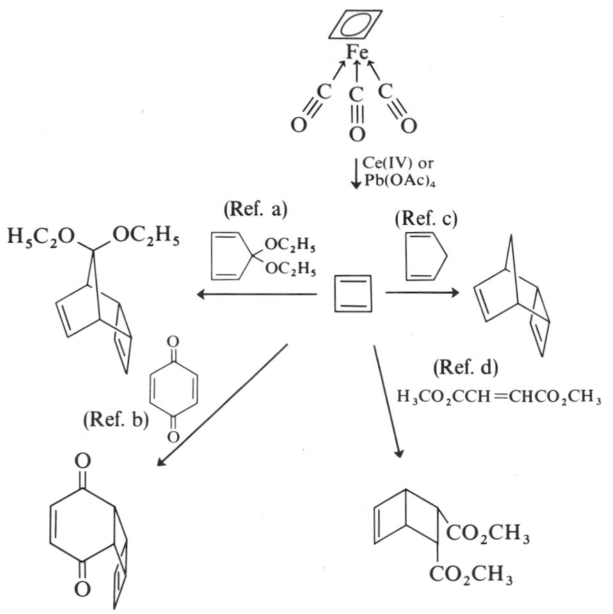

a. J. C. Barborak and R. Pettit, *J. Am. Chem. Soc.* **89**, 3080 (1967).
b. J. C. Barborak, L. Watts, and R. Pettit, *J. Am. Chem. Soc.* **88**, 1328 (1966).
c. L. Watts, J. D. Fitzpatrick, and R. Pettit, *J. Am. Chem. Soc.* **88**, 623 (1966).
d. P. Reeves, J. Henery, and R. Pettit, *J. Am. Chem. Soc.* **91**, 5889 (1969).

electrons contributed by the ligands plus the valence electrons of the metal atom (or ion) usually is 18, to satisfy the "effective atomic number rule."[103]

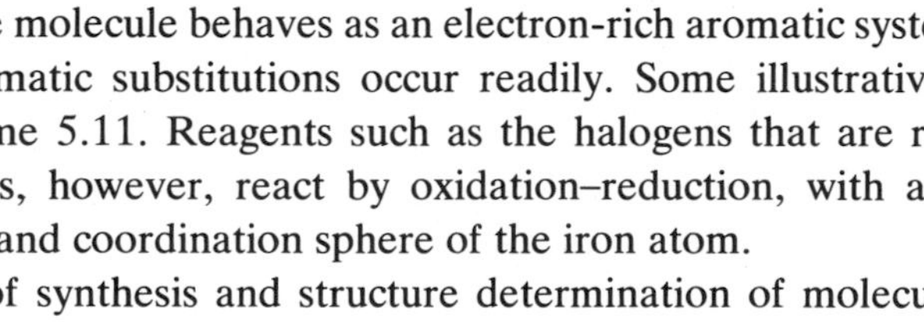

Numerous chemical reactions have been carried out on ferrocene and its derivatives. The molecule behaves as an electron-rich aromatic system, and most of the typical aromatic substitutions occur readily. Some illustrative reactions are shown in Scheme 5.11. Reagents such as the halogens that are relatively strong oxidizing agents, however, react by oxidation–reduction, with a change in the oxidation state and coordination sphere of the iron atom.

The area of synthesis and structure determination of molecules constructed

103. M. Tsutsui, M. N. Levy, A. Nakamura, M. Ichikawa, and K. Mori, *Introduction to Metal π-Complex Chemistry*, Plenum Press, New York, NY, 1970, pp. 44–45.

Scheme 5.11. Electrophilic Substitution Reactions of Ferrocene 197

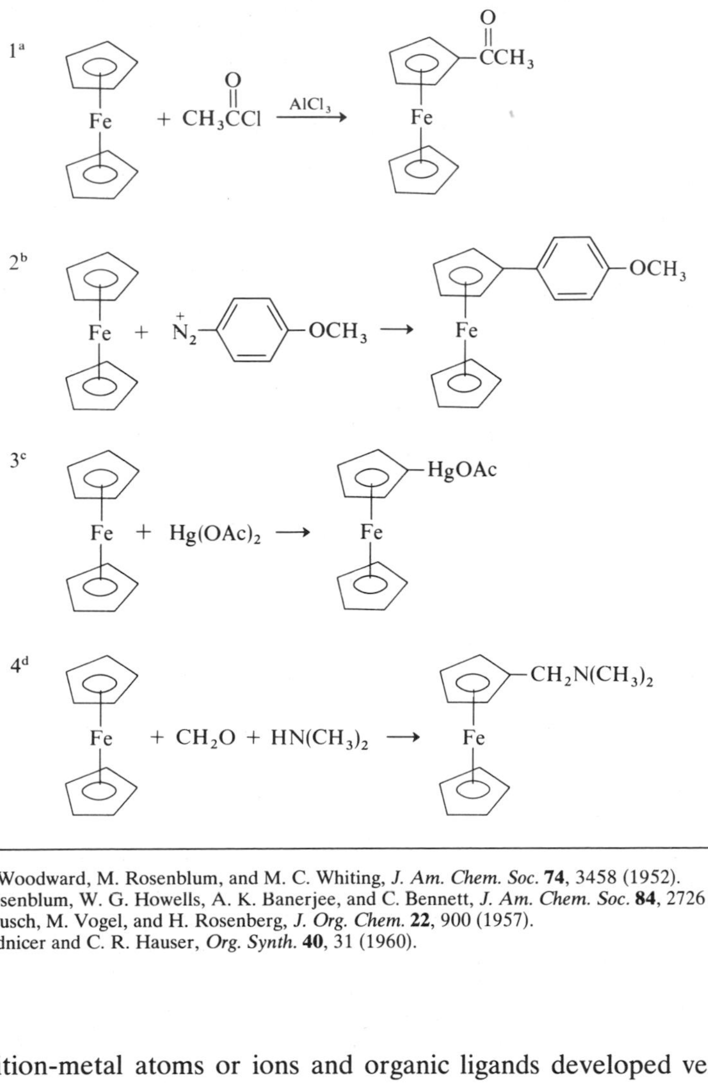

a. R. B. Woodward, M. Rosenblum, and M. C. Whiting, *J. Am. Chem. Soc.* **74**, 3458 (1952).
b. M. Rosenblum, W. G. Howells, A. K. Banerjee, and C. Bennett, *J. Am. Chem. Soc.* **84**, 2726 (1962).
c. M. Rausch, M. Vogel, and H. Rosenberg, *J. Org. Chem.* **22**, 900 (1957).
d. D. Lednicer and C. R. Hauser, *Org. Synth.* **40**, 31 (1960).

from transition-metal atoms or ions and organic ligands developed very rapidly during the 1960's, and there are now thousands of such compounds. This research field is now in itself an important area in which the merging of techniques and concepts from both organic chemistry and inorganic chemistry has resulted in the production of many interesting and potentially useful compounds.

General References

J. J. Eisch, *The Chemistry of Organometallic Compounds*, Macmillan, New York, NY, 1967.
G. E. Coates and K. Wade, *Organometallic Compounds*, Vol. I, Methuen and Co., London, 1967.
J. Wakefield, Organomagnesium compounds in organic synthesis, *Chem. Ind.* (*London*), 450 (1972).
J. M. Brown, Organolithium reagents in synthesis, *Chem. Ind.* (*London*), 454 (1972).

E. C. Ashby, Grignard reagents. Compositions and mechanisms of reaction, *Q. Rev. Chem. Soc.*, 259 (1967).

B. J. Wakefield, *The Chemistry of Organolithium Compounds*, Pergamon Press, Oxford, 1974.

M. L. H. Green, *Organometallic Compounds*, Vol. II, Methuen and Co., London, 1967.

M. Tsutsui, M. N. Levy, A. Nakamur, M. Ichikawa, and K. Mori, *Introduction to Metal π-Complex Chemistry*, Plenum Press, New York, NY, 1970.

R. B. King, *Transition-Metal Organometallic Chemistry*, Academic Press, New York, NY, 1969.

C. W. Bird, *Transition Metal Intermediates in Organic Synthesis*, Logos Press, London, 1967.

Problems

(*References for these problems will be found on page 500.*)

1. Indicate how each of the following organometallic compounds could be generated from readily available precursors:

(a)  $\text{C}_5\text{H}_5^-\ \text{Na}^+$

(b) $CH_2=\overset{|}{\underset{Li}{C}}CH_3$

(c) 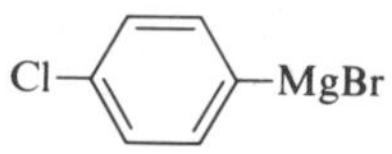 $Cl-\langle\text{C}_6\text{H}_4\rangle-MgBr$

(d) $CH_2=\overset{|}{\underset{Li}{C}}OCH_3$

(e) $[(CH_3)_3CO-Cu-C(CH_3)_3]^-\ Li^+$

(f) $(CH_3)_2NCH_2Li$

(g) 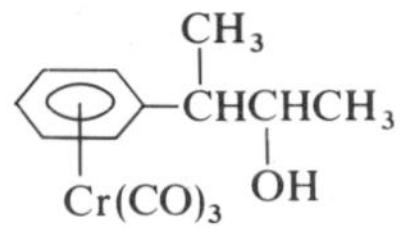 $\langle\text{C}_6\text{H}_5\rangle\text{Cr(CO)}_3-\overset{\overset{\displaystyle CH_3}{|}}{CH}CH\underset{\underset{\displaystyle OH}{|}}{CH_3}$

2. Predict the structure of the major organolithium compound that would be obtained by metalation of each compound. Explain the basis for your prediction. *n*-Butyllithium was used except for (f). In (b) and (c), tetramethylethylene-diamine was used to accelerate metalation; *t*-butyllithium was used in (f).

(a) 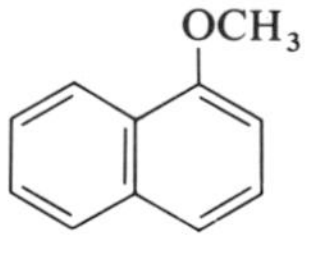 naphthalene with OCH_3 substituent

(b) $(CH_3)_2C=CH_2$

(c) 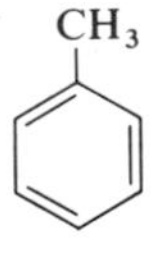 toluene (CH_3 substituent)

(d) 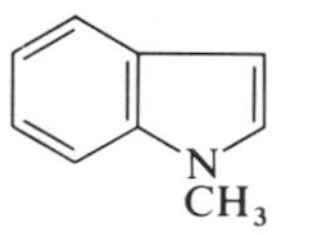 indole with N—CH_3

(e) 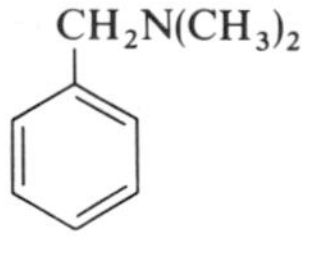 benzene with $CH_2N(CH_3)_2$ substituent

(f) $[CH_3(CH_2)_5]_2PCH_3$

3. Suggest a synthesis of each of the following compounds from readily available materials using organometallic reagents:

(a) $H_2C=CHCH_2CH_2CH_2OH$

(b) $H_2C=\overset{|}{\underset{CH_3}{C}}\overset{\overset{\displaystyle OH}{|}}{C}(CH_2CH_2CH_2CH_3)_2$

(c) $\underset{\underset{PhC(CH_2OCH_3)_2}{|}}{OH}$

(d) 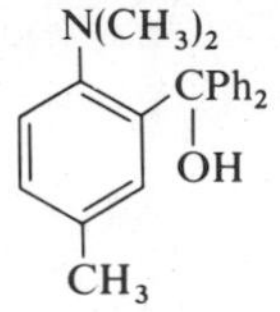
$N(CH_3)_2$
CPh_2
OH
CH_3

(e) $(CH_3)_3CCH(CO_2C_2H_5)_2$

(f) $H_2C=CHCH=CHCH=CH_2$

4. Predict the product of each of the following reactions:

(a) 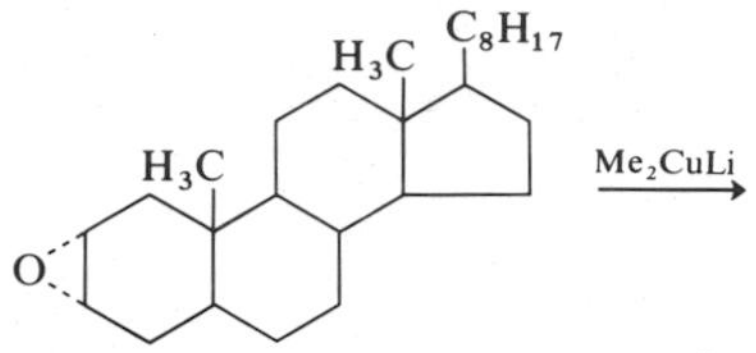
H_3C C_8H_{17}
H_3C
O
$\xrightarrow{Me_2CuLi}$

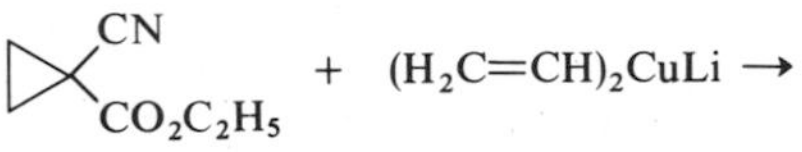

(b)
$\overset{CN}{\underset{CO_2C_2H_5}{\triangle}}$ + $(H_2C=CH)_2CuLi$ →

(c)
CO_2CH_3
$\xrightarrow{AgNO_3}$
CO_2CH_3

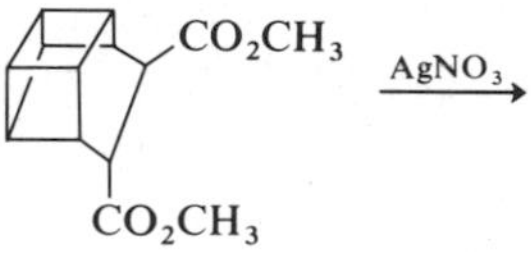

(d) 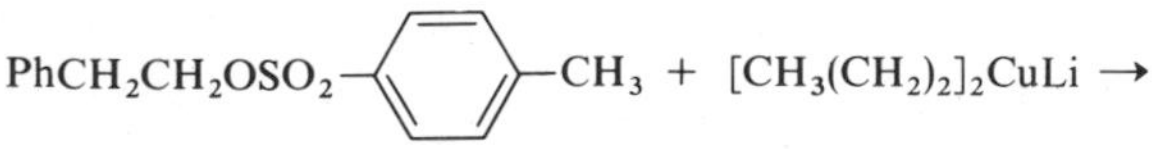 $PhCH_2CH_2OSO_2$—⟨ ⟩—CH_3 + $[CH_3(CH_2)_2]_2CuLi$ →

(e) 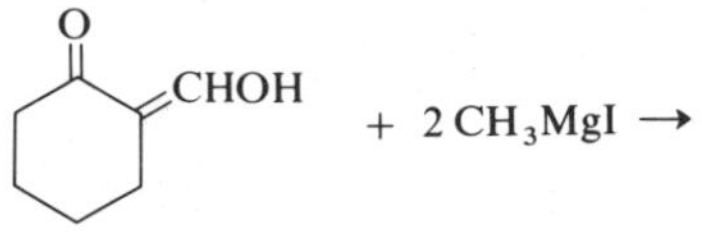
O
$CHOH$
+ 2 CH_3MgI →

(f) 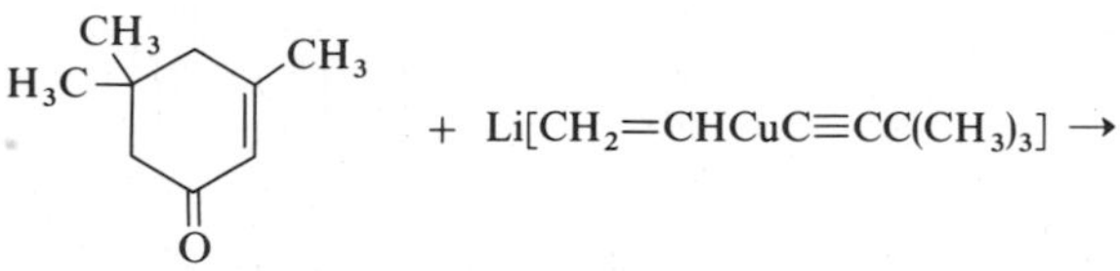
CH_3 CH_3
H_3C
O
+ $Li[CH_2=CHCuC≡CC(CH_3)_3]$ →

5. The reaction of pyrrole with ethylmagnesium bromide gives the magnesium salt of pyrrole. This, in turn, reacts with (−)-2-bromobutane to give a mixture of 2- and 3-(2-butyl)pyrrole. Oxidation of the mixture gives (+)-2-methylbutyric acid. Describe how you could determine the stereochemical course (inversion or retention) of the coupling reaction.

6. The kinetics of the reaction of CH_3MgBr with excess 2-methylbenzophenone have been studied. No simple rate law is followed. If, however, $MgBr_2$ is added in excess of the CH_3MgBr, the reaction becomes first order in CH_3MgBr, and the observed rate constant decreases. How could $MgBr_2$ be affecting the reaction kinetics? Based on your knowledge of Grignard reagents, suggest a

general kinetic expression that you would expect in the absence of added $MgBr_2$. Explain the basis for each term in you kinetic expression.

7. Metalation of **A** and **B** with lithium t-butylamide in tetrahydrofuran occurs at $-78°$. Warming to $-25°$ produces magenta-colored solutions that afford $1 : 1$ mixtures of *cis*- and *trans*-α-benzylcinnamonitrile on quenching with H_2O. The rate constant for the generation of the magenta species from **A** is 42 times greater than from **B**. Explain these results.

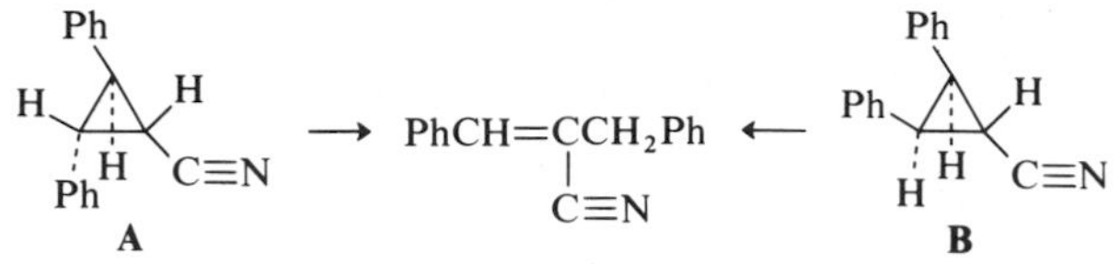

8. Reaction of the epoxide of 1-butene with methyllithium gives 3-pentanol in 90% yield. In contrast, methylmagnesium bromide under similar conditions gives the array of products shown below. Explain the difference in the reactivity of the two organometallic compounds toward this epoxide.

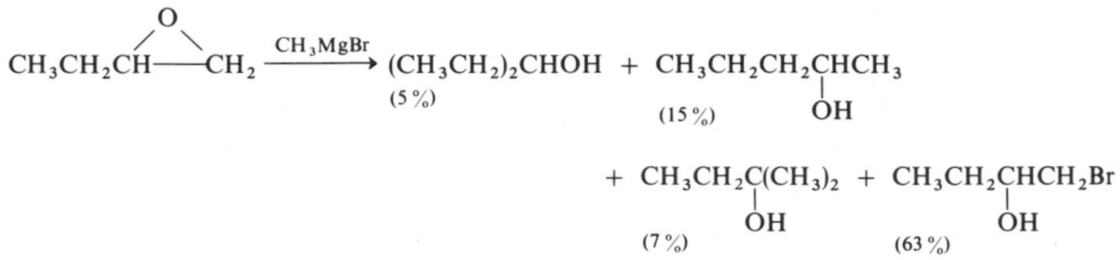

9. Formulate a mechanism for the following silver-ion-catalyzed rearrangements. Be sure that the mechanism accounts for the observed stereochemistry.

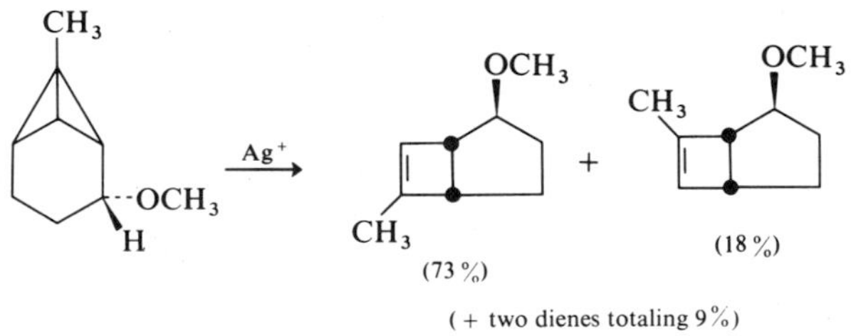

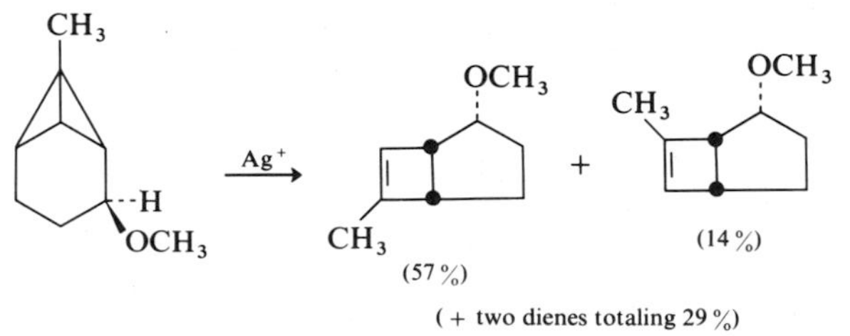

10. A series of various electrophilic transition-metal derivatives was allowed to react with the bicyclo[1.1.0]butane derivative **A**. The product consisted of varying amounts of four compounds. Show which bonds are cleaved to form

each product. In cases where there are ambiguities, suggest isotopic labeling experiments that would distinguish between the possibilities.

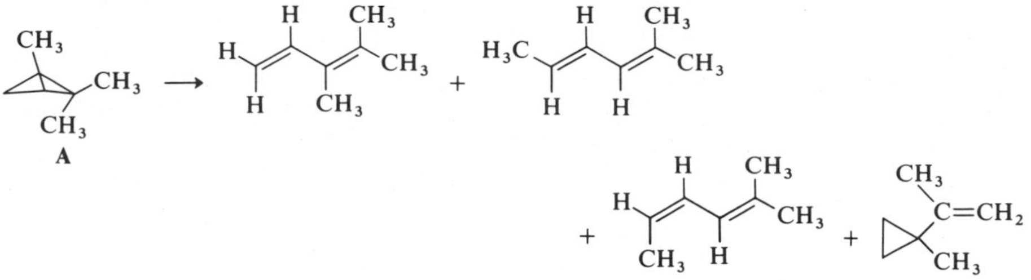

11. Formulate a mechanism for the catalysis by silver ion of the rearrangement shown below:

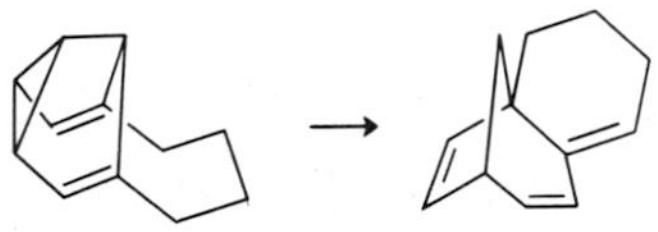

12. Suggest structures for organometallic compounds having the following compositions:

(a) $Cr(C_6H_6)(CO)_3$
(b) $Fe(norbornadiene)(CO)_3$
(c) $Ni(cyclopentadiene)(cyclopentene)(-2 H)$

(d) $Na_2Fe(CO)_4$
(e) $Co_4(CO)_{12}$

13. Suggest mechanisms for the following reactions:

(a)

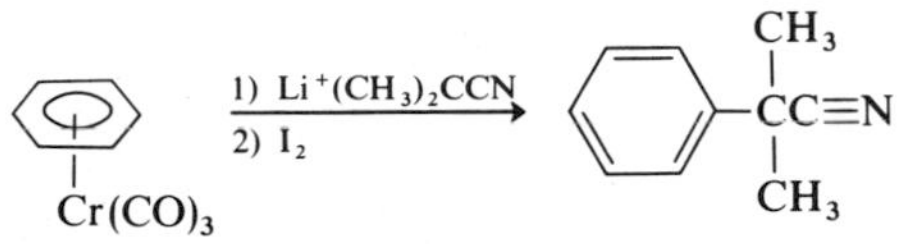

(b)

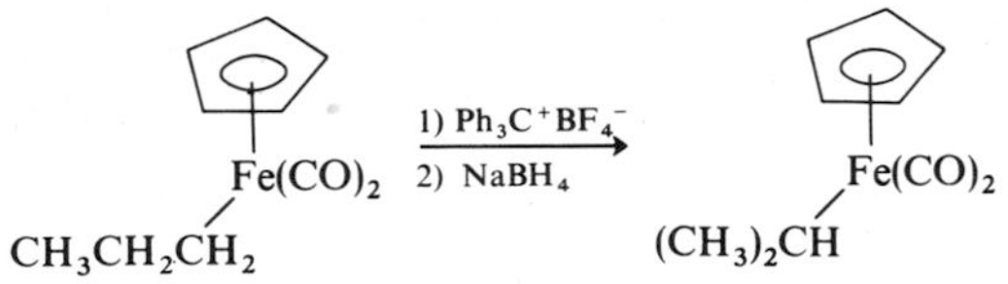

14. The following transformations require more than a single step, but in each case an organometallic reagent was used in the synthetic sequence. Suggest efficient processes for carrying out the tranformations, using the materials given as the source of all carbon atoms.

(a)

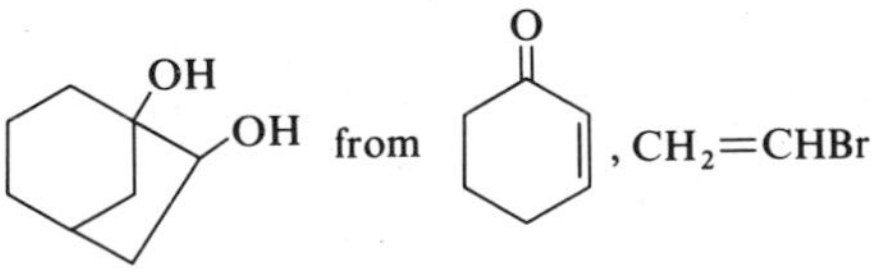

(b)

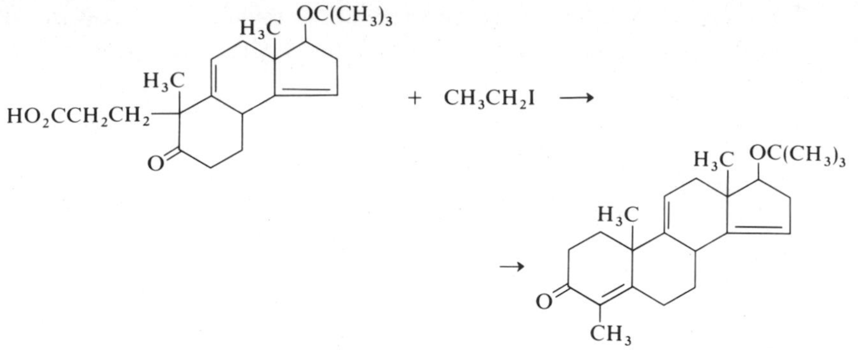

$+ \ CH_3CH_2I \ \longrightarrow$

$\longrightarrow$

(c)

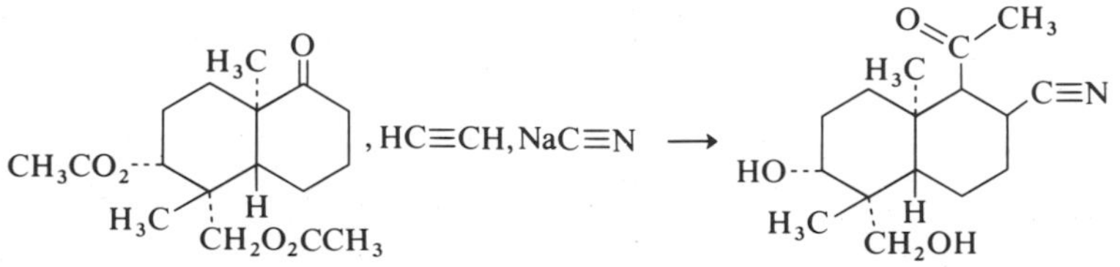

$, HC{\equiv}CH, NaC{\equiv}N \ \longrightarrow$

(d)

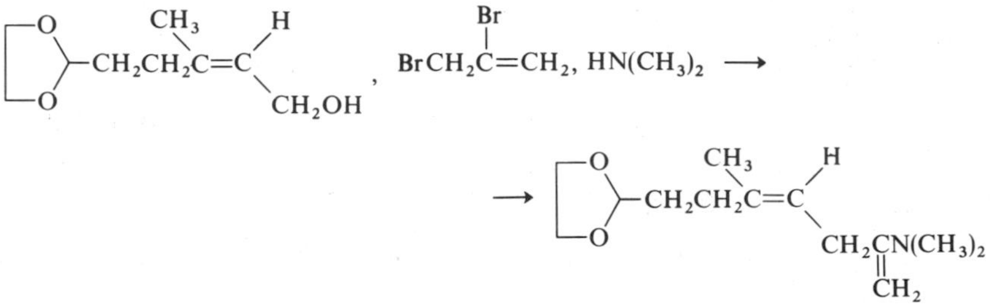

$, \ BrCH_2C{=}CH_2, \ HN(CH_3)_2 \ \longrightarrow$
with Br on the center carbon

$\longrightarrow$

(e)

from

$, HC{\equiv}CHCH_2Br, \ H_2C{=}CHBr$

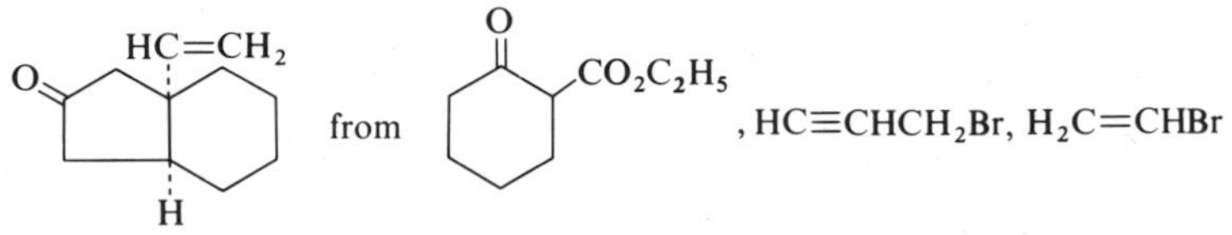

15. In each case below, an organometallic reagent permits the transformation to be accomplished in a single-stage procedure. Give the structure of the required organometallic compound.

(a) $\quad CH_3C{\equiv}CCO_2CH_3 \ \longrightarrow$

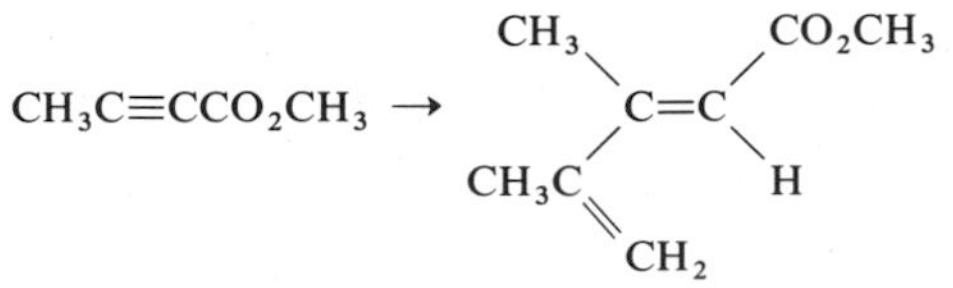

(b)

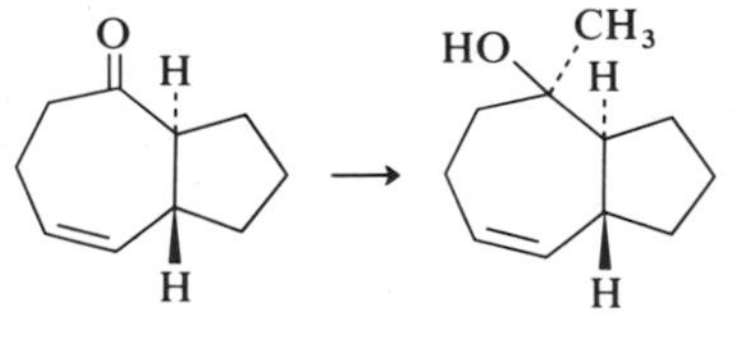

(c) 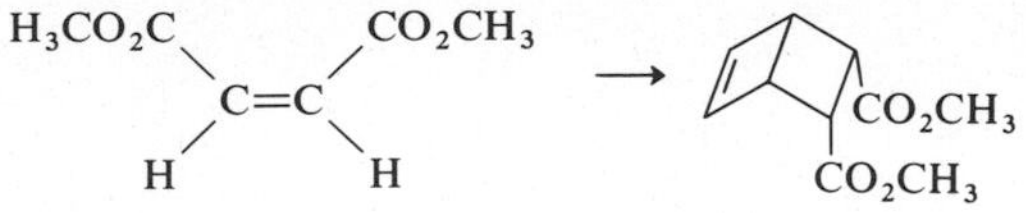 H_3CO_2C , CO_2CH_3
$\underset{H}{\overset{}{C}}=\underset{H}{\overset{}{C}}$ $\longrightarrow$ (bicyclic structure with CO_2CH_3, CO_2CH_3)

(d) $\overset{O}{\overset{\|}{ClC}}(CH_2)_6CO_2C_2H_5 \rightarrow (CH_3)_2CH(CH_2)_2\overset{O}{\overset{\|}{C}}(CH_2)_6CO_2C_2H_5$

(e) 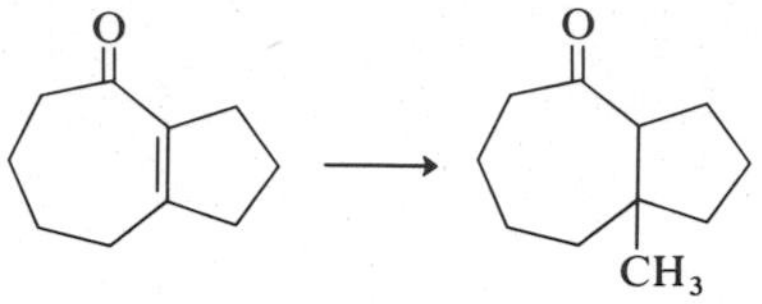   (bicyclic ketone) $\longrightarrow$ (bicyclic ketone with CH_3)

(f) 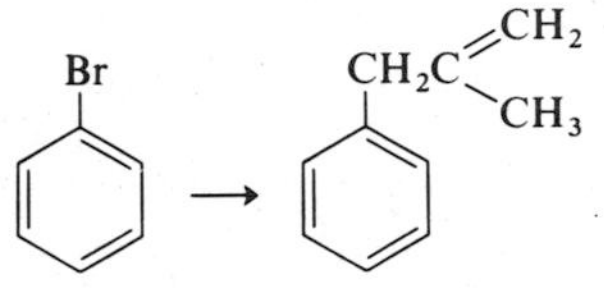   (bromobenzene, Br) $\rightarrow$ (benzene with $CH_2C\overset{CH_2}{\underset{CH_3}{}}$)

(g) 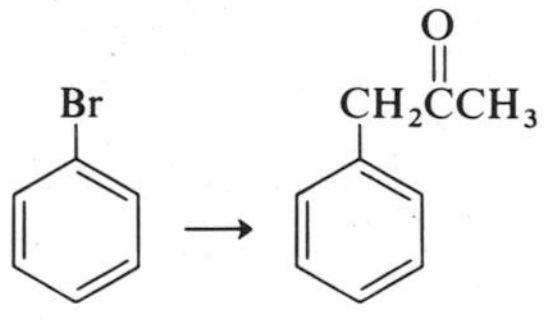   (bromobenzene, Br) $\rightarrow$ (benzene with $CH_2\overset{O}{\overset{\|}{C}}CH_3$)

(h) 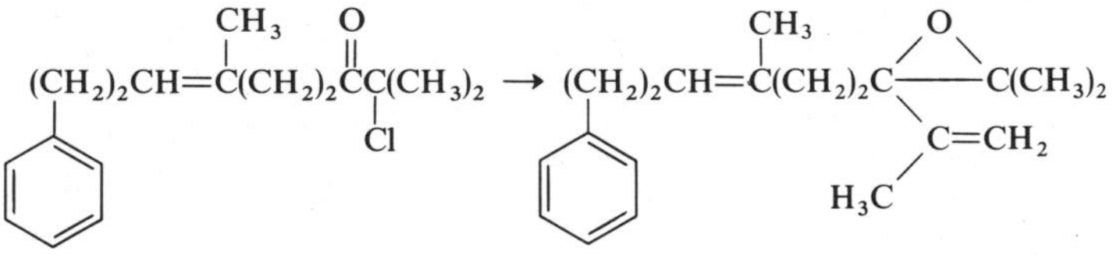 $(CH_2)_2CH{=}\overset{CH_3}{\overset{|}{C}}(CH_2)_2\overset{O}{\overset{\|}{C}}C(CH_3)_2 \rightarrow (CH_2)_2CH{=}\overset{CH_3}{\overset{|}{C}}(CH_2)_2C\overset{O}{\overbrace{\quad}}C(CH_3)_2$
$\quad\quad\quad\quad\quad\quad\quad\quad\quad\overset{|}{Cl}$
with $\overset{H_3C}{\underset{}{}}C{=}CH_2$

Cycloadditions and Unimolecular Rearrangements and Eliminations

The unifying mechanistic pattern that appears in most of the reactions to be discussed in this chapter is a reorganization of the valence electrons of the reacting molecule or molecules. Usually, these reactions proceed through a cyclic transition state containing four, five, or six atoms. Another feature these reactions have in common is that the energy to drive the reactions forward is provided by thermal or photochemical excitation of one of the reacting molecules. Other polar or radical-generating reagents having high chemical potential energy are not usually involved.

Many of the reactions to be discussed are concerted processes and may be treated mechanistically in light of the concepts of orbital-symmetry control introduced in Part A, Chapter 10. Some others are stoichiometrically similar, but on mechanistic scrutiny have been found to proceed through discrete, short-lived intermediates.

6.1. Cycloaddition Reactions

The reactions to be discussed in this section result in the formation of a new ring from two reacting molecules. A truly concerted mechanism requires that a single transition state, and therefore no intermediate, lie on the reaction path between

206

CHAPTER 6
CYCLOADDITIONS
AND
UNIMOLECULAR
REARRANGEMENTS
AND
ELIMINATIONS

reactants and adduct. Two typical cycloaddition reactions that are believed to occur by concerted mechanisms are shown below:

A firm mechanistic understanding of concerted cycloadditions had to await the formulation of the reaction mechanism within the framework of molecular orbital theory. Consideration of the molecular orbitals of reactants and products revealed that in some cases a smooth transformation of orbitals of the reactant to those of the product is possible. In other cases, reactions that appear feasible, if no consideration is given to the symmetry and spatial orientation of the orbitals, are found to require high-energy transition states when the orbital properties are considered in detail. Those considerations have permitted description of many types of cycloadditions as "allowed" or "forbidden." As has been discussed (Part A, Chapter 10), the application of orbital-symmetry relationships permits a conclusion as to whether a given concerted reaction is or is not energetically feasible. In this chapter, the synthetic application of these reactions will be emphasized. The same orbital-symmetry relationships that are informative as to the feasibility of a given reaction often are predictive of features of the stereochemistry of cycloaddition products. This predictable stereochemistry is an attractive feature for synthetic purposes.

6.1.1. Diels–Alder Reaction

The addition of alkenes to dienes is a very useful method for the formation of six-membered carbocyclic rings. The reaction is known as the *Diels–Alder reaction.*[1] The concerted nature of the mechanism was generally agreed on and the stereo-specificity of the reaction was firmly established even before the importance of orbital symmetry was recognized. In the terminology of orbital-symmetry classifica-tion, the Diels–Alder reaction is a $[_\pi 4_s + _\pi 2_s]$ cycloaddition, an allowed process. The stereochemistry of both the diene and the alkene (the alkene is often called the dienophile) is retained in the cyclization process. The transition state for addition requires the diene to adopt the *s-cis* conformation. The diene and alkene approach

1. L. W. Butz and A. W. Rytina, *Org. React.* **5**, 136 (1949); M. C. Kloetzel, *Org. React.* **4**, 1 (1948); H. L. Holmes, *Org. React.* **4**, 60 (1948); A. Wasserman, *Diels–Alder Reactions*, Elsevier, New York, NY, 1965; R. Huisgen, R. Grashey, and J. Sauer, in *Chemistry of Alkenes*, S. Patai (ed.), Interscience, New York, NY, 1964, pp. 878–928.

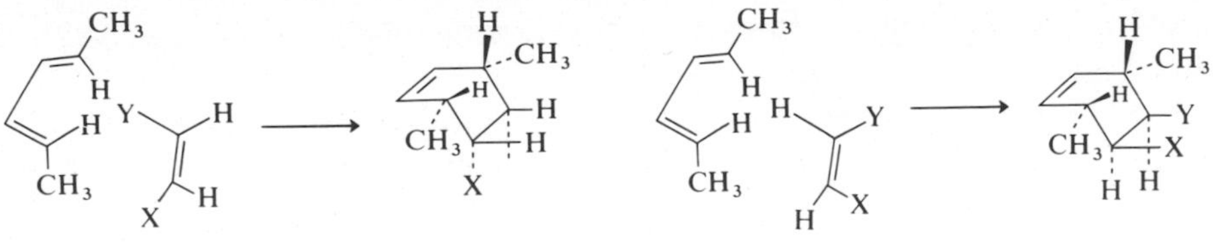

endo addition   *exo* addition

Figure 6.1. *Endo* and *exo* addition in Diels–Alder addition.

each other in parallel planes. The orbital-symmetry properties of the system permit stabilization of the transition state through bonding interactions between C-1 and C-4 of the diene and the carbon atoms of the dienophilic double bond in a six-center arrangement:

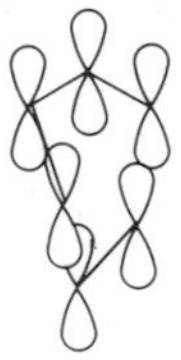

There is a further stereochemical variable in the transition state, which can lead to mixtures of products in some cases. This involves the relative orientation of the diene and dienophile in the transition state. The two possible orientations, which are referred to as *endo* and *exo* addition, are illustrated in Figure 6.1.

Whether the products of *endo* and *exo* addition will be distinguishable depends, of course, on the extent of substitution present in the diene and monoene. For example, additions involving either butadiene or a *trans*-symmetrically disubstituted alkene as one of the reactants will give the same adduct by either mode of addition. Usually, the *endo* mode of addition is preferred, especially when X or Y is an unsaturated group such as a carbonyl. The preference for this mode of addition, which is often sterically more congested, is the result of a combination of dipolar and van der Waals attractions, as well as orbital interactions involving X or Y and the diene system. The relative importance of each of these factors in determining the *exo* : *endo* ratio probably varies from system to system.[2]

There is a well-established electronic substituent effect in the Diels–Alder addition. The most favorable alkenes for reaction with most dienes are those bearing electron-attracting groups. Thus, among the most reactive dienophiles are quinones, maleic anhydride, and nitroalkenes. α,β-Unsaturated esters, ketones, and nitriles are also effective dienophiles. It is significant that if a relatively electron-deficient diene is utilized, the polarity of the transition state is apparently reversed, and electron-rich dienophiles are then preferred. For example, in the reaction of

2. Y. Kobuke, T. Sugimoto, J. Furukawa, and T. Funeo, *J. Am. Chem. Soc.* **94**, 3633 (1972); K. L. Williamson and Y.-F. L. Hsu, *J. Am. Chem. Soc.* **92**, 7385 (1970).

208

CHAPTER 6
CYCLOADDITIONS
AND
UNIMOLECULAR
REARRANGEMENTS
AND
ELIMINATIONS

hexachlorocyclopentadiene with styrenes, the reaction is facilitated by styrenes with electron-releasing groups.[3]

A question of regioselectivity arises when both the diene and the alkene are unsymmetrically substituted. Generally, there is a preference for the "*ortho*" and

"*meta*"   "*para*"
favored

"*ortho*"   "*meta*"
favored

"*para*" orientations, respectively, in the cases shown. This preference can be rationalized by qualitative molecular orbital considerations that assume that the favored transition state will be that in which there is the strongest interaction between the HOMO of the diene and the LUMO of the dienophile.[4] In a case where X is electron-releasing and Y electron-withdrawing (the most common situation), the dienophile LUMO-orbital will interact strongly with the diene HOMO. The coefficients of these molecular orbitals are largest at the 4-carbon of the diene and the 2-carbon of the dienophile. Thus, bonding to give the "*ortho*" isomer is favored, since a bond between C-4 of the diene and C-2 of the dienophile is formed in this isomer.

Diels–Alder cycloadditions are sensitive to steric effects of two major types. Bulky substituents on the dienophile or on the termini of the diene can hinder approach of the two components to each other and decrease the rate of reaction. This can be seen in the relative reactivity of 1-substituted butadienes toward maleic anhydride:[5]

R	k_{rel} (25°)
−H	1
−CH$_3$	4.2
−C(CH$_3$)$_3$	<0.05

Substitution of hydrogen by methyl results in a slight rate *increase*, probably as a result of an electronic effect, while a 1-*tert*-butyl substituent produces a significant rate *decrease*. Apparently, any steric retardation to approach of the dienophile by a methyl substituent is insignificant compared to its electronic effect. With the very large *tert*-butyl group, the steric effect is dominant.

The other steric effect has to do with intramolecular van der Waals repulsions

3. J. Sauer and H. Wiest, *Angew. Chem. Int. Ed. Engl.* **1**, 269 (1962).
4. K. N. Houk, *J. Am. Chem. Soc.* **95**, 4092 (1973).
5. D. Craig, J. J. Shipman, and R. B. Fowler, *J. Am. Chem. Soc.* **83**, 2885 (1961).

between substituents in the diene. Adoption of the *s-cis* conformation of the diene in the transition state may be accompanied by an unfavorable repulsion between substituents that do not interact strongly in the ground state. Toward tetracyanoethylene (a very reactive dienophile), *trans*-1,3-pentadiene is 10^3 times more reactive than 4-methyl-1,3-pentadiene because of the unfavorable interaction between the additional methyl substituent and the hydrogen at C-1 in the *s-cis* conformation.[6] Relatively small substituents at C-2 and C-3 of the diene exert little

R	k_{rel}
—H	1
—CH$_3$	10^{-3}

steric influence on the rate of Diels–Alder addition. 2,3-Dimethylbutadiene reacts with maleic anhydride about ten times faster than butadiene does and, again, an electronic effect is probably largely responsible. 2-*tert*-Butyl-1,3-butadiene is 27 times more reactive than butadiene because of preferential stabilization of the *s-cis* conformation relative to the *s-trans* conformation. The reaction of 2-*tert*-butyl-1,3-

butadiene with maleic anhydride has an activation energy of only 6.3 kcal/mole, and the conformational equilibrium in the ground state is reflected directly in the rate of reaction. The presence of *tert*-butyl substituents at both C-2 and C-3, however, prevents attainment of the *s-cis* conformation, and Diels–Alder reactions of 2,3-di-*tert*-butyl-1,3-butadiene are not observed.[7]

Lewis acids, particularly aluminum chloride, have been noted to catalyze Diels–Alder cycloadditions.[8-10] The catalytic effect is attributed to coordination of the Lewis acid with the dienophile. The complexed dienophile is then more electron-

deficient and therefore more reactive than the uncomplexed molecule toward the normal electron-rich dienes. The mechanism of the addition is still believed to be concerted, and high stereospecificity is observed.

6. C. A. Stewart, Jr., *J. Org. Chem.* **28**, 3320 (1963).
7. H. J. Backer, *Rec. Trav. Chim. Pays-Bas* **58**, 643 (1939).
8. P. Yates and P. Eaton, *J. Am. Chem. Soc.* **82**, 4436 (1960).
9. T. Inukai and M. Kasai, *J. Org. Chem.* **30**, 3567 (1965).
10. T. Inukai and T. Kojima, *J. Org. Chem.* **32**, 869, 872 (1967).

210

CHAPTER 6
CYCLOADDITIONS
AND
UNIMOLECULAR
REARRANGEMENTS
AND
ELIMINATIONS

Some examples of molecules that exhibit reactivity as dienophiles are given in Scheme 6.1. Some references to a few of the many known Diels–Alder additions are given in Scheme 6.2.

Polycyclic aromatics are quite reactive as the diene component in Diels–Alder reactions. Anthracene forms adducts with a number of dienophiles. The addition occurs at the center ring. There is little loss of resonance stabilization, since the

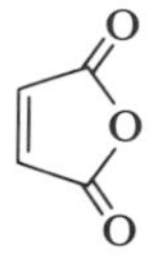

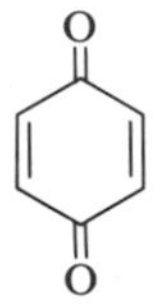

Scheme 6.1. Representative Dienophiles

A. Substituted Alkenes

B. Substituted Alkynes

7[f] $H_3CO_2CC{\equiv}CCO_2CH_3$ 8[g] $F_3CC{\equiv}CCF_3$ 9[h] $PhCC{\equiv}CCPh$ (with two C=O)

10[i] $N{\equiv}CC{\equiv}CC{\equiv}N$

C. Heteroatomic Dienophiles

11[j] $H_3CO_2CN{=}NCO_2CH_3$ 12[k]   (ring structure with N–Ph)   13[l] $RCH{=}NCO_2C_2H_5$

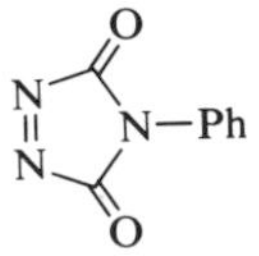

14[m] $ArSC{\equiv}N$ (with O, O)

a. M. C. Kloetzel, *Org. React.* **4**, 1 (1948).
b. L. W. Butz and A. W. Rytina, *Org. React.* **5**, 136 (1949).
c. W. M. Daniewski and C. E. Griffin, *J. Org. Chem.* **31**, 3236 (1966).
d. J. C. Philips and M. Oku, *J. Org. Chem.* **37**, 4479 (1972).
e. W. J. Middleton, R. E. Heckert, E. L. Little, and C. G. Krespan, *J. Am. Chem. Soc.* **80**, 2783 (1958).
f. H. L. Holmes, *Org. React.* **4**, 60 (1948).
g. R. E. Putnam, R. J. Harder, and J. E. Castle, *J. Am. Chem. Soc.* **83**, 391 (1961); C. G. Krespan, B. C. McKusick, and T. L. Cairns, *J. Am. Chem. Soc.* **83**, 3428 (1961).
h. J. D. White, M. E. Mann, H. D. Kirshenbaum, and A. Mitra, *J. Org. Chem.* **36**, 1048 (1971).
i. C. D. Weis, *J. Org. Chem.* **28**, 74 (1963).
j. B. T. Gillis and P. E. Beck, *J. Org. Chem.* **28**, 3177 (1963).
k. B. T. Gillis and J. D. Hagarty, *J. Org. Chem.* **32**, 330 (1967).
l. M. P. Cava, C. K. Wilkins, Jr., D. R. Dalton, and K. Bessho, *J. Org. Chem.* **30**, 3772 (1965); G. Krow, R. Rodebaugh, R. Carmosin, W. Figures, H. Pannella, G. DeVicaris, and M. Grippi, *J. Am. Chem. Soc.* **95**, 5273 (1973).
m. J. C. Jagt and A. M. van Leusen, *J. Org. Chem.* **39**, 564 (1974).

anthracene (resonance energy 1.60 eV) is replaced by two benzenoid rings (total delocalization energy $2 \times 0.87 = 1.74$ eV).[11]

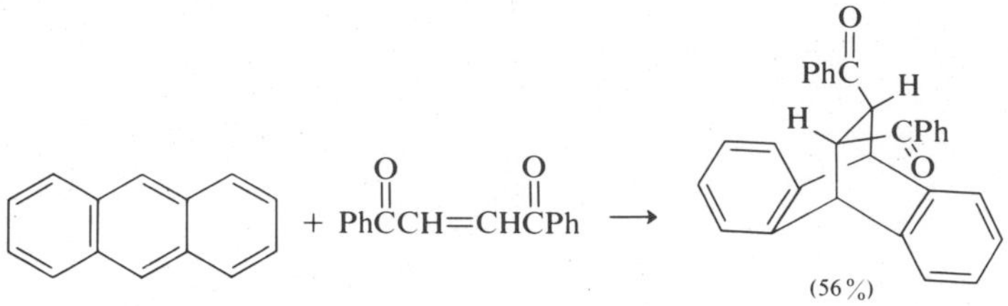

(56%)   Ref. 12

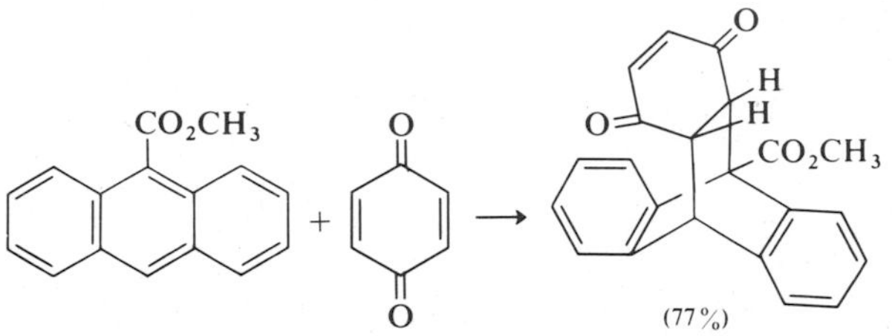

(77%)   Ref. 13

The naphthalene ring system is much less reactive. Polymethylnaphthalenes are considerably more reactive than the parent compound, and 1,2,3,4-tetramethylnaphthalene gives an adduct with maleic anhydride in 82% yield.[14] Reaction occurs exclusively at the substituted ring. It is believed that the steric repulsions between the methyl groups, which are relieved in the nonplanar adduct, exert an accelerating effect in the reaction.

With benzenoid compounds, Diels–Alder addition is rare. Formation of an adduct between benzene and dicyanoacetylene in the presence of $AlCl_3$ has been

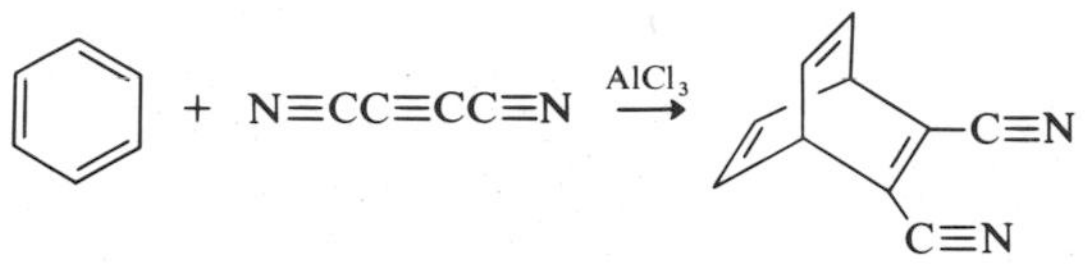

reported, however.[15] Hexafluoro-2-butyne also gives Diels–Alder adducts with benzene, toluene, and other alkylbenzenes[16]:

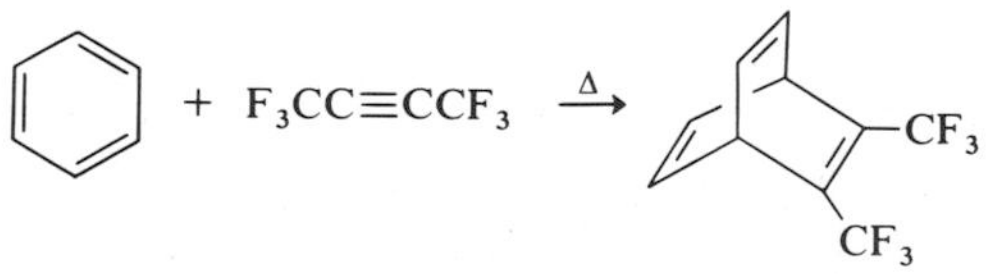

11. M. J. S. Dewar and C. de Llano, *J. Am. Chem. Soc.* **91**, 789 (1969).
12. D. M. McKinnon and J. Y. Wong, *Can. J. Chem.* **49**, 3178 (1971).
13. P. D. Bartlett and F. D. Greene, *J. Am. Chem. Soc.* **76**, 1088 (1954).
14. A. Oku, Y. Ohnishi, and F. Mashio, *J. Org. Chem.* **37**, 4264 (1972).
15. E. Ciganek, *Tetrahedron Lett.*, 3321 (1967).
16. R. S. H. Liu, *J. Am. Chem. Soc.* **90**, 215 (1968); C. G. Krespan, B. C. McKusick, and T. L. Cairns, *J. Am. Chem. Soc.* **83**, 3428 (1961).

Scheme 6.2. Some Examples

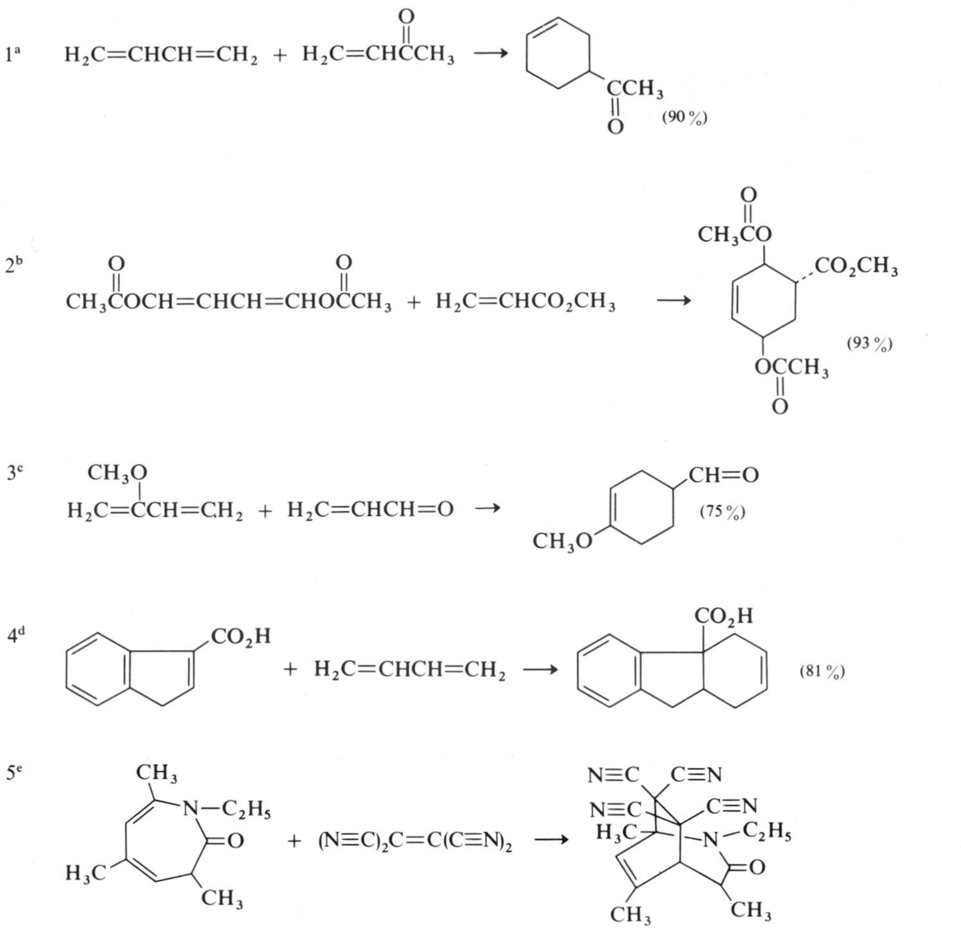

CHAPTER 6
CYCLOADDITIONS
AND
UNIMOLECULAR
REARRANGEMENTS
AND
ELIMINATIONS

a. W. K. Johnson, *J. Org. Chem.* **24**, 864 (1959).
b. E. E. Smissman, J. T. Suh, M. Oxman, and R. Daniels, *J. Am. Chem. Soc.* **84**, 1040 (1962).
c. J. I. DeGraw, L. Goodman, and B. R. Baker, *J. Org. Chem.* **26**, 1156 (1961).
d. E. F. Godefroi and L. H. Simanyi, *J. Org. Chem.* **28**, 1112 (1963).
e. L. A. Paquette, *J. Org. Chem.* **29**, 3447 (1964).

6.1.2. Dipolar Cycloaddition Reactions

In Part A, Chapter 10, the relationship of 1,3-dipolar cycloaddition reactions to the general topic of concerted cycloaddition reactions was discussed briefly. It is useful to discuss this reaction in somewhat more detail at this point, since it constitutes a quite general method for the synthesis of heterocyclic rings. Scheme 6.3 lists some of the groups of molecules that are capable of dipolar cycloaddition. These

6[f]

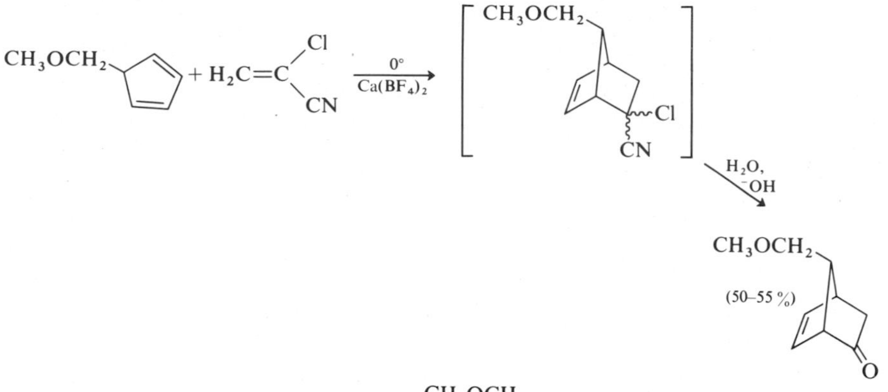

(50–55 %)

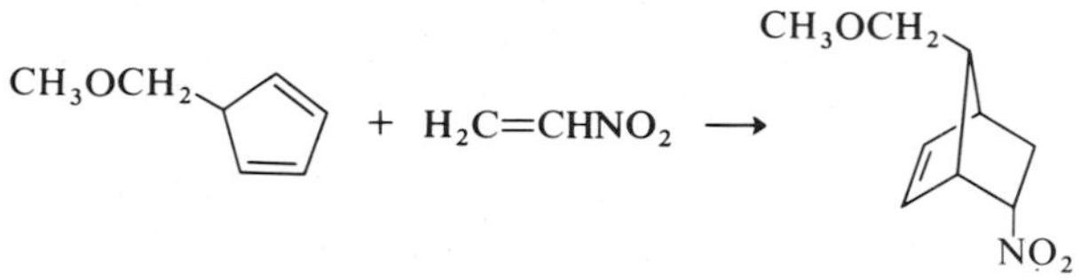

7[g]   CH$_3$OCH$_2$ + H$_2$C=CHNO$_2$ →

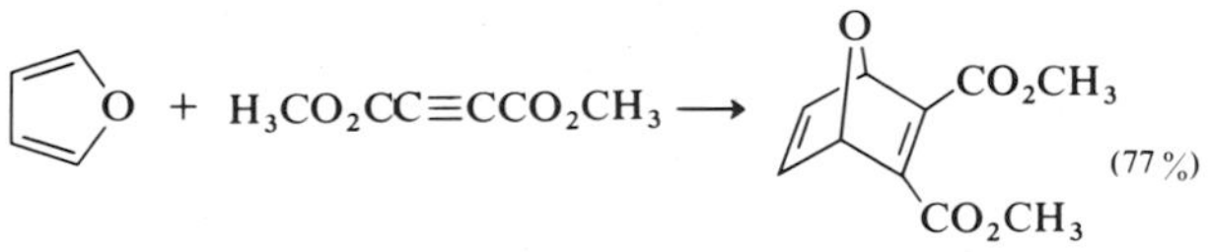

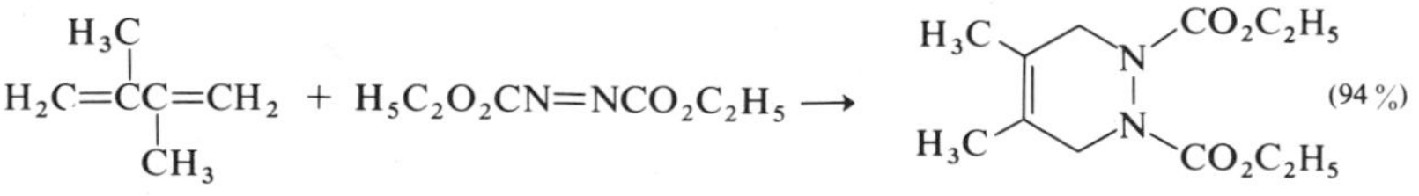

8[h]   + H$_3$CO$_2$CC≡CCO$_2$CH$_3$ →

(77 %)

9[i]   H$_2$C=CC=CH$_2$ + H$_5$C$_2$O$_2$CN=NCO$_2$C$_2$H$_5$ →

(94 %)

10[j]   + H$_2$C=NCO$_2$C$_2$H$_5$ →

(27 %)

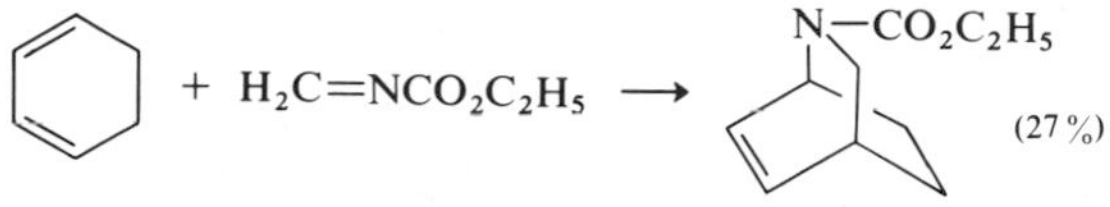

f. E. J. Corey, U. Koelliker, and J. Neuffer, *J. Am. Chem. Soc.* **93**, 1489 (1971).
g. S. Ranganathan, D. Ranganathan, and A. K. Mehrotra, *J. Am. Chem. Soc.* **96**, 5261 (1974).
h. G. Stork, E. E. van Tamelen, L. J. Friedman, and A. W. Burgstahler, *J. Am. Chem. Soc.* **75**, 384 (1953).
i. B. T. Gillis and P. E. Beck, *J. Org. Chem.* **27**, 1947 (1962).
j. M. P. Cava, C. K. Wilkins, Jr., D. R. Dalton, and K. Bessho, *J. Org. Chem.* **30**, 3772 (1965).

molecules, which are called 1,3-*dipoles*, each have more than one resonance structure. Each molecule has at least one resonance structure which indicates a separation of charge with opposite charges in a 1,3-relationship. It is from this structural
feature that the name of the reaction class comes. The other reactant in a dipolar
cycloaddition, usually an olefin or acetylene, is referred to as the *dipolarophile*. Other
multiple bonds, such as the C=N bond of imines, can also act as the dipolarophile.

Mechanistic studies have shown that the transition state for cycloaddition of
1,3-dipoles to carbon–carbon multiple bonds is not very polar. The rate of reaction is

214

CHAPTER 6
CYCLOADDITIONS
AND
UNIMOLECULAR
REARRANGEMENTS
AND
ELIMINATIONS

Scheme 6.3. 1,3-Dipolar Compounds

$$:\overset{+}{N}=\overset{..}{N}-\overset{-}{C}R_2 \leftrightarrow :N\equiv\overset{+}{N}-\overset{-}{C}R_2 \qquad \text{Diazoalkane}$$

$$:\overset{+}{N}=\overset{..}{N}-\overset{-}{N}R \leftrightarrow :N\equiv\overset{+}{N}-\overset{-}{N}R \qquad \text{Azide}$$

$$R\overset{+}{C}=\overset{..}{N}=\overset{-}{C}R_2 \leftrightarrow RC\equiv\overset{+}{N}-\overset{-}{C}R_2 \qquad \text{Nitrile ylide}$$

$$R\overset{+}{C}=\overset{..}{N}-\overset{-}{N}R \leftrightarrow RC\equiv\overset{+}{N}-\overset{-}{N}R \qquad \text{Nitrile imine}$$

$$R\overset{+}{C}=\overset{..}{N}-\overset{-}{O}: \leftrightarrow RC\equiv\overset{+}{N}-\overset{-}{O}: \qquad \text{Nitrile oxide}$$

$$R_2\overset{+}{C}-\overset{..}{N}-\overset{-}{C}R_2 \leftrightarrow R_2C=\overset{+}{N}-\overset{-}{C}R_2 \qquad \text{Azomethine ylide}$$
$$\qquad\ \ R \qquad\qquad\qquad R$$

$$R_2\overset{+}{C}-\overset{..}{N}-\overset{-}{O}: \leftrightarrow R_2C=\overset{+}{N}-\overset{-}{O}: \qquad \text{Nitrone}$$
$$\qquad\ \ R \qquad\qquad\qquad R$$

$$R_2\overset{+}{C}-\overset{..}{O}-\overset{-}{O}: \leftrightarrow R_2C=\overset{+}{O}-\overset{-}{O}: \qquad \text{Carbonyl oxide}$$

not strongly sensitive to solvent polarity, and this is in agreement with viewing the addition as a concerted process.[17] The formal destruction of charge that is indicated

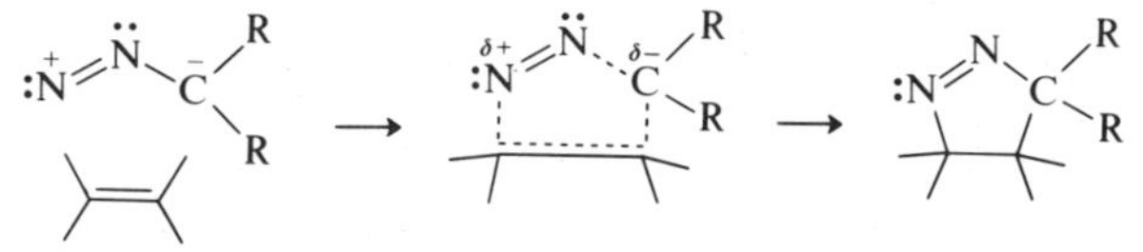

is more apparent than real, because most 1,3-dipoles are not highly polar substances. The high polarity implied by any single structure is balanced by other contributing structures.

$$:\overset{+}{N}=\overset{..}{N}-\overset{-}{C}R_2 \leftrightarrow :N\equiv\overset{+}{N}-\overset{-}{C}R_2 \leftrightarrow :\overset{-}{N}=\overset{+}{N}=CR_2 \leftrightarrow :\overset{-}{N}=\overset{..}{N}-\overset{+}{C}R_2$$

Two questions are of principal interest for predicting the structure of reaction products of 1,3-dipolar addition: (1) Is the reaction stereospecific? (2) Is the reaction regioselective? The answer to the first question is yes with respect to the dipolarophile. Many specific recorded examples demonstrate that the cyclic product results from a stereospecific *syn* addition to olefins. The stereospecific addition that is observed is exactly what would be expected on the basis of a concerted mechanism.

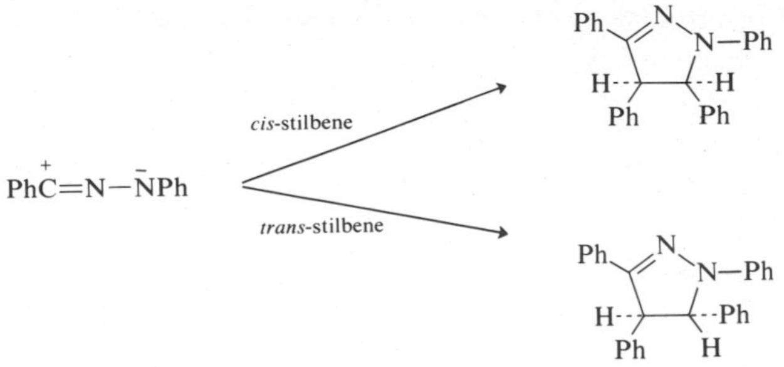

$$Ph\overset{+}{C}=N-\overset{-}{N}Ph$$

cis-stilbene

trans-stilbene

Ref. 18

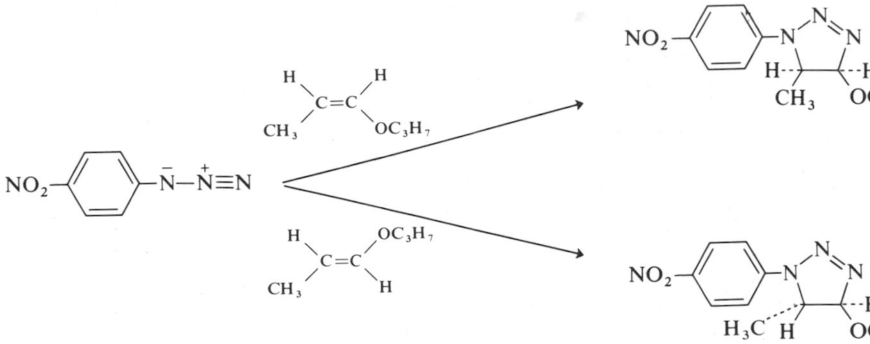

Ref. 19

$$NO_2-\underset{}{\bigcirc}-\overset{-}{N}-\overset{+}{N}\equiv N$$

With some 1,3-dipoles, two possible stereoisomers can be formed by *syn* additions differing in the relative orientation of the reacting molecules. In several cases where disubstituted diazomethanes have added to olefins, mixtures are obtained.[20] This is comparable to the situation of *endo* versus *exo* stereochemistry in the Diels–Alder reaction.

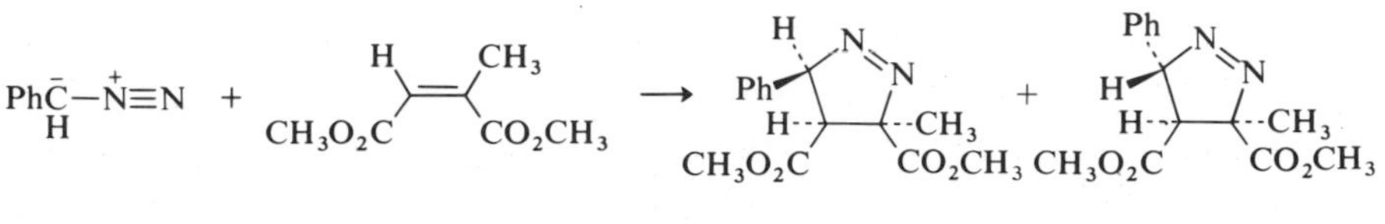

$$Ph\overset{-}{\underset{H}{C}}-\overset{+}{N}\equiv N \ + \ \underset{CH_3O_2C\quad CO_2CH_3}{\overset{H\quad CH_3}{C=C}} \ \longrightarrow$$

The question of regioselectivity is at this time a matter of some controversy, and there are no simple rules for predicting orientation in dipolar cycloadditions. In some cases, both possible cyclic adducts have been isolated from a cycloaddition reaction. There is no fundamental mechanistic requirement for regiospecificity. In particular, it is important to note that attempts to predict regioselectivity on the basis of the polarity of the 1,3-dipole and the dipolarophile are completely unreliable.

The basis for the regioselectivity that is noted in most of the cycloadditions has

17. P. K. Kadaba, *Tetrahedron* **25**, 3053 (1969); R. Huisgen, G. Szeimies, and L. Möbius, *Chem. Ber.* **100**, 2494 (1967); P. Scheiner, J. H. Schomaker, S. Deming, W. J. Libbey, and G. P. Nowack, *J. Am. Chem. Soc.* **87**, 306 (1965).
18. R. Huisgen, M. Seidel, G. Wallbillich, and H. Knupfer, *Tetrahedron* **17**, 3 (1962).
19. R. Huisgen and G. Szeimies, *Chem. Ber.* **98**, 1153 (1965).
20. R. Huisgen and P. Eberhard, *Tetrahedron Lett.*, 4343 (1971).

216

CHAPTER 6
CYCLOADDITIONS
AND
UNIMOLECULAR
REARRANGEMENTS
AND
ELIMINATIONS

been a matter of discussion. One view is that steric effects make a major contribution.[21] In many reactions, this view predicts the correct orientation for the product by assuming that the dominant product is the one in which bulky groups on the two reactant molecules are as far removed from one another as possible. There are several exceptions, however, so that steric effects cannot be the sole controlling factor.[22]

A molecular orbital analysis of several dipolar cycloadditions has suggested that regioselectivity may be governed by the distribution of electron density in the HOMO and LUMO of the reacting molecules. This analysis is based on the assumption that the preferred orientation of addition would be the one providing maximum overlap of these two frontier orbitals.[23] To carry out such an analysis, the interacting orbitals in the two addends must be identified. For electron-rich dipolarophiles, the HOMO is presumed to interact with the LUMO of the 1,3-dipole. For electron-poor dipolarophiles, the opposite obtains. The preferred orientation is then predicted by assuming that the direction of addition will be that which provides for the greatest overlap of these orbital pairs. Satisfactory rationalizations of the regioselectivity in several 1,3-dipolar cycloadditions have been made on this basis.

There have been few studies comparing the reactivity of various 1,3-dipoles. Quite a lot of work has been done, however, on comparing the reactivity of various olefins as dipolarophiles. Several generalizations have come from such studies:

1. Strain increases the reactivity of olefins. Norbornene, for example, is consistently more reactive than cyclohexene in 1,3-dipolar cycloadditions.
2. Conjugated functional groups also increase reactivity. This increased reactivity has most often been demonstrated with electron-withdrawing substituents such as carbonyl and cyano groups, but enamines and vinyl ethers, which have electron-releasing amino and alkoxy groups, are also quite reactive. That both electron-rich and electron-poor olefins are more reactive than simple alkenes again illustrates that simple considerations of ground-state electron density are not a reliable guide to reactivity in dipolar cycloadditions.

Reactivity data for a series of olefins with a few typical 1,3-dipoles are summarized in Table 6.1. Scheme 6.4 gives some examples of syntheses that employ 1,3-dipolar cycloaddition.

Dipolar cycloadditions are important as a means of synthesis of heterocyclic molecules. The cycloaddition of diazo compounds has also been used in the synthesis of cyclopropanes and other strained molecules. The cyclic adducts (pyrazolines) contain the azo linkage, and molecular nitrogen is lost on thermal or photolytic decomposition. Usually, the two carbon centers resulting from elimination of

21. R. Huisgen, *J. Org. Chem.* **33**, 2291 (1968).
22. R. A. Firestone, *J. Org. Chem.* **33**, 2285 (1968).
23. R. Sustmann, *Tetrahedron Lett.*, 2717 (1971); R. Sustmann and H. Trill, *Angew. Chem. Int. Ed. Engl.* **11**, 838 (1972); K. N. Houk, *J. Am. Chem. Soc.* **94**, 8953 (1972).

Table 6.1. Relative Reactivity of Substituted Alkenes toward 1,3-Dipoles[a,b]

Substituted alkene	Ph_2CN_2	PhN_3	$Ph\overset{+}{N}=N-\overset{-}{N}Ph$	$PhC\equiv\overset{+}{N}-\overset{-}{O}$	$Ph\underset{H}{\overset{+}{C}}=\underset{\overset{\mid}{O^-}}{N}-CH_3$
Dimethyl fumarate	100	31	283	94	18.3
Dimethyl maleate	27.8	1.25	7.9	1.61	6.25
Norbornene	1.15	700	3.1	97	0.13
Ethyl acrylate	28.8	36.5	48	66	11.1 (methyl ester)
Butyl vinyl ether	—	1.5	—	15	—
Styrene	0.57	1.5	1.6	9.3	0.32
Ethyl crotonate	1.0	1.0	1.0	1.0	1.0
Cyclopentene	—	6.9	0.13	1.04	0.022
Terminal alkene	—	0.89 (heptene)	0.15 (heptene)	2.6 (hexene)	0.072 (heptene)
Cyclohexene	—	—	0.011	0.055	—

a. Data are selected from those compiled by R. Huisgen, R. Grashey, and J. Sauer, in *Chemistry of Alkenes*, S. Patai (ed.), Interscience, New York, NY, 1964, pp. 806–877.
b. Conditions such as solvent and temperature vary for each 1,3-dipole, so comparison from dipole to dipole is not possible. Following Huisgen, Grashey, and Sauer,[a] ethyl crotonate is assigned reactivity = 1.0 for each 1,3-dipole.

nitrogen combine, forming a three-membered ring. This reaction is considered, along with eliminations of nitrogen from rings of other sizes, in Section 6.4.

An interesting variation of the 1,3-dipolar cycloadditions involves generation of 1,3-dipoles from stable strained-ring compounds. As an example, aziridines **1** and **2** give adducts apparently derived from the reaction of 1,3-dipoles **3** and **4**, respectively, with a variety of dipolarophiles.[24] The evidence for the involvement of dipoles

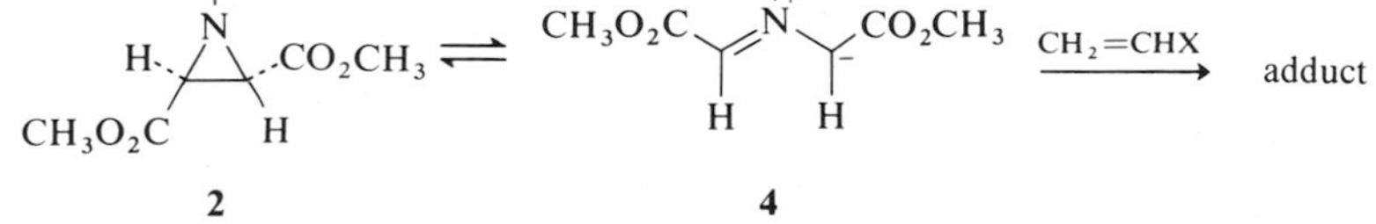

as discrete intermediates includes the observation that the reaction rates are independent of dipolarophile concentration, a result consistent with ring-opening being the rate-determining step in the reaction. This ring-opening reaction appears

24. R. Huisgen and H. Mäder, *J. Am. Chem. Soc.* **93**, 1777 (1971).

218

CHAPTER 6
CYCLOADDITIONS
AND
UNIMOLECULAR
REARRANGEMENTS
AND
ELIMINATIONS

Scheme 6.4. Typical 1,3-Dipolar Cycloaddition Reactions

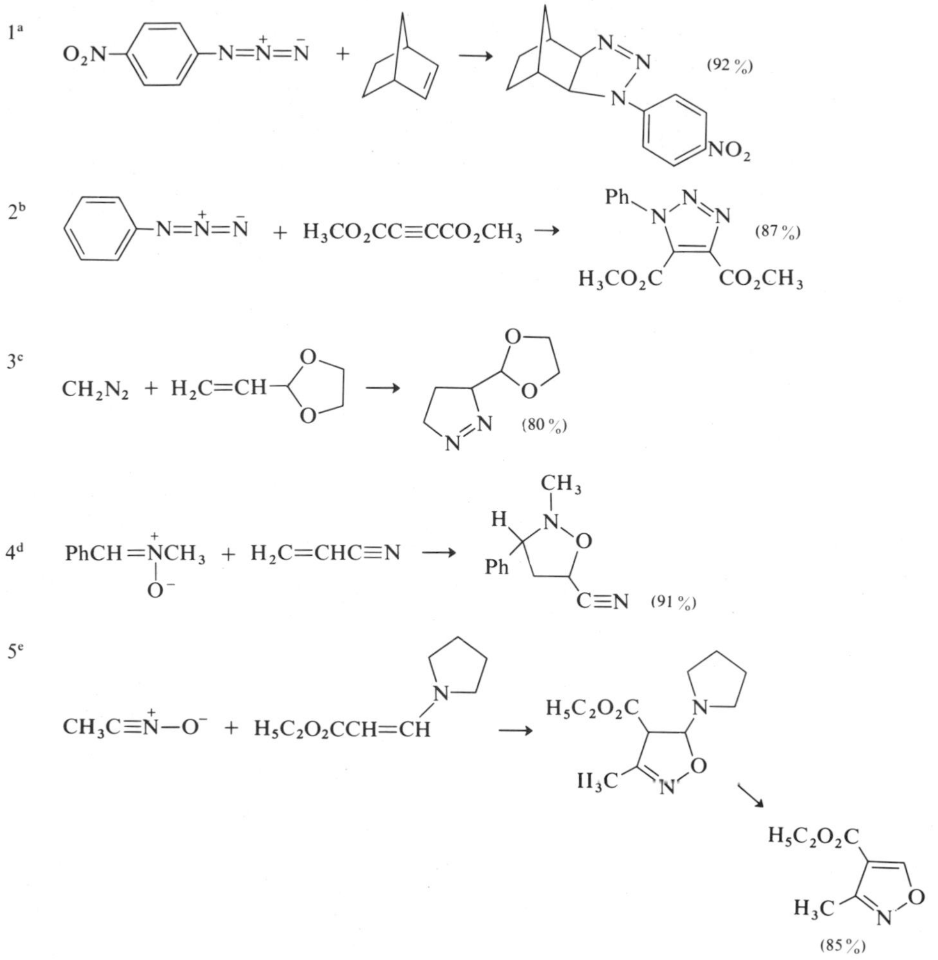

a. P. Scheiner, J. H. Schomaker, S. Deming, W. J. Libbey, and G. P. Nowack, *J. Am. Chem. Soc.* **87**, 306 (1965).
b. R. Huisgen, R. Knorr, L. Möbius, and G. Szeimies, *Chem. Ber.* **98**, 4014 (1965).
c. J. M. Stewart, C. Carlisle, K. Kem, and G. Lee, *J. Org. Chem.* **35**, 2040 (1970).
d. R. Huisgen, H. Hauck, R. Grashey, and H. Seidl, *Chem. Ber.* **101**, 2568 (1968).
e. G. Stork and J. E. McMurry, *J. Am. Chem. Soc.* **89**, 5461 (1967).

to be most facile for aziridines that have an electron-attracting substituent capable of stabilizing the carbanionic center in the dipole.

Cyclopropanones are also reactive toward cycloadditions of various types. It is suspected that a dipolar species resulting from reversible cleavage of the cyclo-

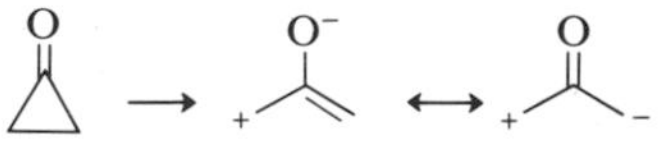

1ᵃ

2ᵇ

3ᶜ

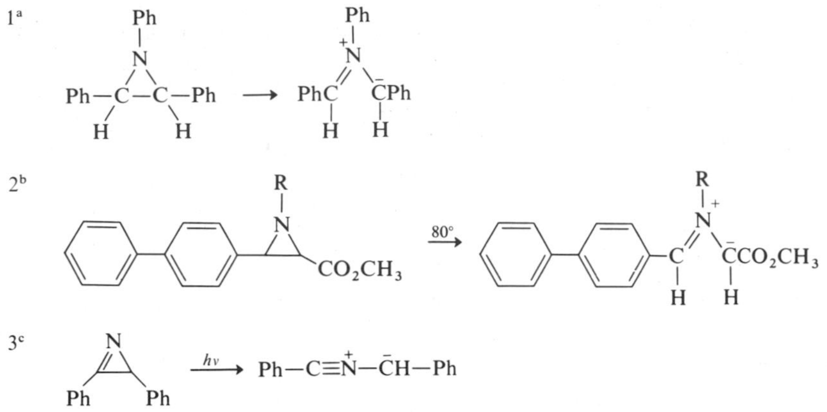

a. H. W. Heine, R. Peavy, and A. J. Durbetaki, *J. Org. Chem.* **31**, 3924 (1966).
b. P. B. Woller and N. H. Cromwell, *J. Org. Chem.* **35**, 888 (1970).
c. A. Padwa, M. Dharan, J. Smolanoff, and S. I. Wetmore, Jr., *J. Am. Chem. Soc.* **95**, 1945, 1954 (1973).

propanone ring may be the reactive species.[25] Illustrations of other 1,3-dipoles that are believed to form from ring-opening of strained rings are given in Scheme 6.5.

6.1.3. 2+2 Cycloadditions and Other Reactions Leading to Cyclobutanes

Among the cycloaddition reactions that have been shown to have some generality and synthetic utility are the 2+2 cycloadditions of ketenes with alkenes. These reactions can occur in a concerted manner by $[_\pi 2_s + _\pi 2_a]$ cycloaddition.[26] The perpendicular approach required in the transition state is more easily achieved with

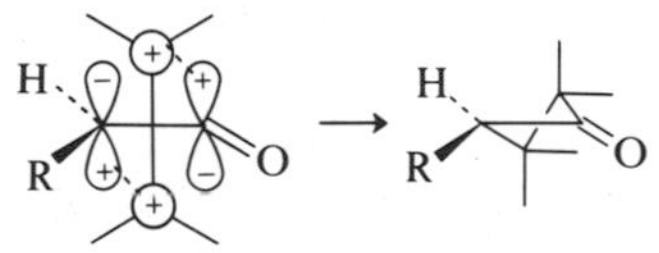

ketenes than with other alkenes because there is no interfering substituent at the carbonyl carbon of the ketene. The electron deficiency of the ketene system, which has available a low-lying C=O π^*-orbital for interaction with the approaching alkene, is also an important factor in the high reactivity of ketenes toward 2+2

25. N. J. Turro, S. S. Edelson, J. R. Williams, T. R. Darling, and W. B. Hammond, *J. Am. Chem. Soc.* **91**, 2283 (1969); S. S. Edelson and N. J. Turro, *J. Am. Chem. Soc.* **92**, 2770 (1970); N. J. Turro, *Acc. Chem. Res.* **2**, 25 (1969).
26. R. B. Woodward and R. Hoffmann, *Angew. Chem. Int. Ed. Engl.* **8**, 781 (1969).

CHAPTER 6
CYCLOADDITIONS
AND
UNIMOLECULAR
REARRANGEMENTS
AND
ELIMINATIONS

Scheme 6.6. 2+2 Cycloadditions of Ketenes

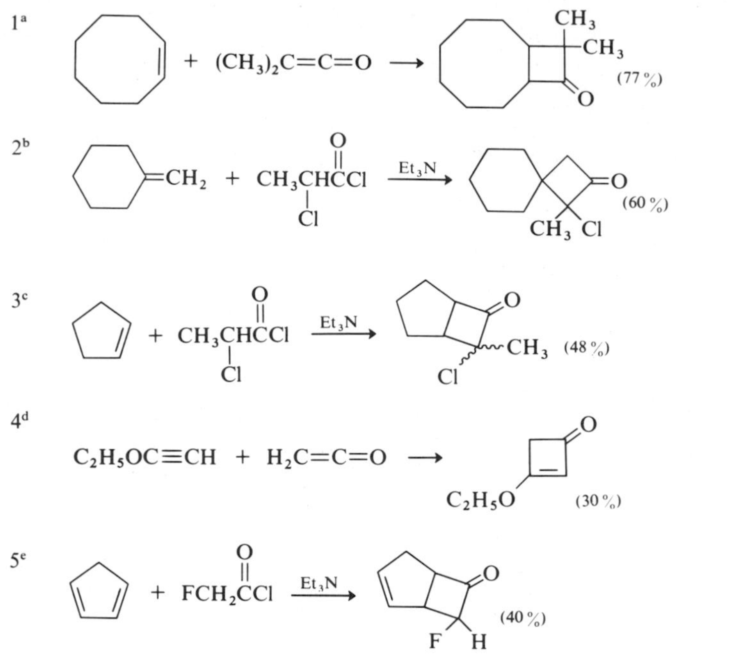

1[a]

2[b]

3[c]

4[d]

$$C_2H_5OC\equiv CH \ + \ H_2C=C=O \ \longrightarrow$$

5[e]

a. A. P. Krapcho and J. H. Lesser, *J. Org. Chem.* **31**, 2030 (1966).
b. W. T. Brady and A. D. Patel, *J. Org. Chem.* **38**, 4106 (1973).
c. W. T. Brady, R. Roe, Jr., E. F. Hoff, and F. H. Parry, III, *J. Am. Chem. Soc.* **92**, 146 (1970).
d. H. H. Wasserman, J. U. Piper, and E. V. Dehmlow, *J. Org. Chem.* **38**, 1451 (1973).
e. W. T. Brady and E. F. Hoff, Jr., *J. Org. Chem.* **35**, 3733 (1970).

cycloaddition. The reaction constitutes a fairly general synthesis of cyclobutanones. Some examples of such reactions are given in Scheme 6.6.

A few other types of compounds can react with alkenes to give cyclobutanes, but these reactions are believed to involve diradical intermediates, rather than being concerted.[27] The best substrates for such additions are fluoroalkenes and allenes. Either alkynes or alkenes can serve as the other reactive component. Dienes and other conjugated compounds are more reactive than are simple alkenes. Some examples of this class of reactions are given in Scheme 6.7.

Cyclobutanes can also be formed by noncerted reactions involving dipolar intermediates. Enamines react with certain electron-deficient alkenes to give cy-

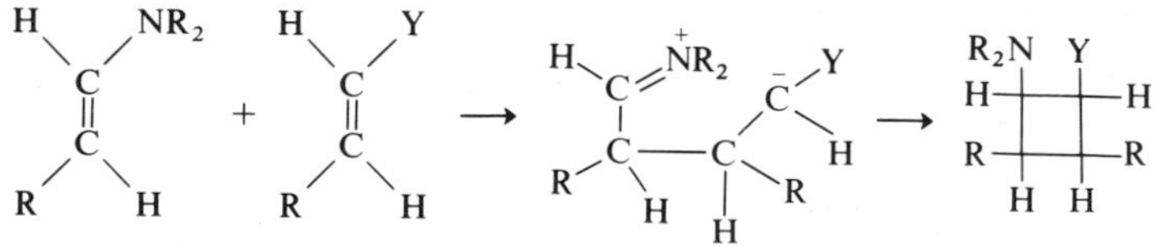

27. J. D. Roberts and C. M. Sharts, *Org. React.* **12**, 1 (1962); P. D. Bartlett, *Q. Rev. Chem. Soc.* **24**, 473 (1970).

Scheme 6.7. Formation of Cyclobutanes in Thermal Addition Reactions

1[a]
F₂C=CF₂ + CH₃CH=CH₂
200°
(45%)
2[b]
H₂C=C=CH₂ + H₂C=CHC≡N
200°
(60%)
3[c]
F₂C=CCl₂ + PhC≡CH
130°
(71%)
4[d]
+ F₂C=CF₂
475°
(58:38 ratio, 70%)

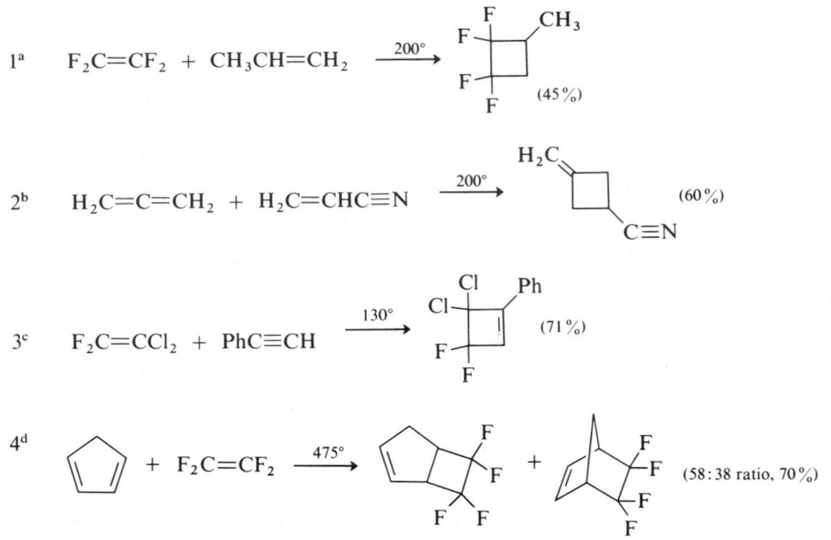

a. D. D. Coffman, P. L. Barrick, R. D. Cramer, and M. S. Raasch, *J. Am. Chem. Soc.* **71**, 490 (1949).
b. H. N. Cripps, J. K. Williams, and W. H. Sharkey, *J. Am. Chem. Soc.* **81**, 2723 (1959).
c. J. D. Roberts, G. B. Kline, and H. E. Simmons, Jr., *J. Am. Chem. Soc.* **75**, 4765 (1953).
d. J. J. Drysdale, W. W. Gilbert, H. K. Sinclair, and W. H. Sharkey, *J. Am. Chem. Soc.* **80**, 3672 (1958).

clobutane derivatives. The electron-attracting group is essential both for initiating reaction of the electron-rich enamine with the carbon–carbon double bond and for stabilizing the carbanionic center in the dipolar intermediate. Two examples of this type of reaction are cited below:

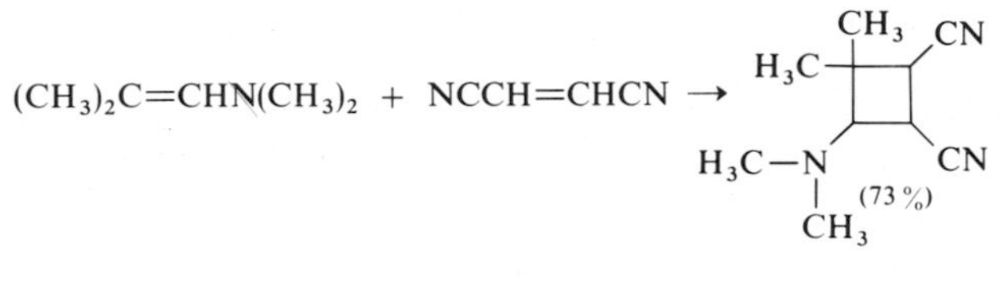

(CH₃)₂C=CHN(CH₃)₂ + NCCH=CHCN →
(73%)

Ref. 28

CH₃CH₂CH=CHN + PhCH=CHNO₂ →
(100%)

Ref. 29

Dipolar intermediates are also involved in cycloaddition of tetracyanoethylene and enol ethers. Again, the two addends are substituted so that separation of charge is energetically acceptable. This reaction strongly favors retention of the vinyl ether

28. K. C. Brannock, A. Bell, R. D. Burpitt, and C. A. Kelly, *J. Org. Chem.* **29**, 801 (1964).
29. M. E. Kuehne and L. Foley, *J. Org. Chem.* **30**, 4280 (1965).

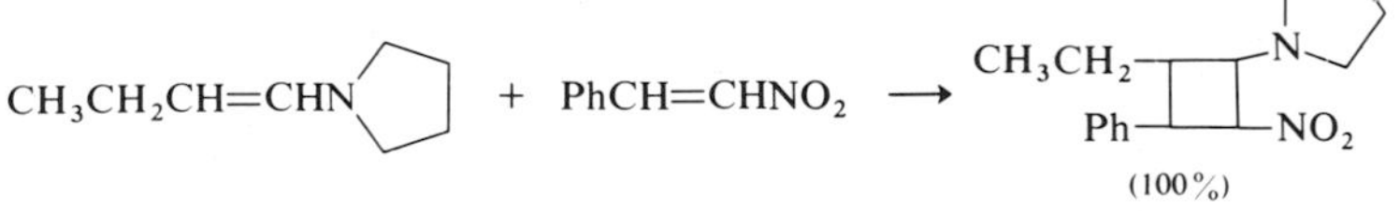

222

CHAPTER 6
CYCLOADDITIONS
AND
UNIMOLECULAR
REARRANGEMENTS
AND
ELIMINATIONS

stereochemistry in nonpolar solvents, but becomes less stereospecific in polar solvents. This change is attributed to a solvent effect on the lifetime of the zwitterion intermediate. In more polar solvents, the increased lifetime permits some rotation, leading to loss of stereospecificity. Further evidence for ionic intermediates is the fact

$$ROCH{=}CHCH_3 \rightarrow \overset{+}{RO}{=}CH{-}\underset{|}{C}HCH_3 \rightarrow ROCH{-}\underset{|}{C}HCH_3 \qquad \text{Ref. 30}$$

$$\underset{(NC)_2C=C(CN)_2}{\Big\downarrow} \qquad \underset{(NC)_2\bar{C}{-}C(CN)_2}{} \quad \underset{(NC)_2C{-\!-}C(CN)_2}{}$$

that the rate of this cycloaddition shows a strong dependence on solvent polarity, in contrast to what is observed when a concerted mechanism operates.[31]

6.2. Photochemical Cycloadditions

Photochemical cycloadditions provide a method that is often complementary to thermal cycloadditions with regard to the types of compounds that can be prepared. The theoretical basis for this complementary relationship between the thermal and photochemical modes of reaction lies in orbital-symmetry relationships and was discussed in detail in Part A, Chapter 10.

The reaction types permitted by photochemical excitation that are particularly useful for synthesis are $2+2$ additions between two carbon–carbon double bonds, and $2+2$ addition of alkenes with carbonyl compounds leading to oxetanes. Many examples of intramolecular cycloadditions involving two alkene groups have been reported, and this now constitutes an important method for constructing compounds containing four-membered rings.[32] $2+2$ Photochemical cycloadditions are not always concerted processes, the reason being that the reactive photochemical excited state is often a triplet. In this case, the intermediate adduct, which is also a triplet, must undergo spin inversion before the cycloaddition can be completed. As a result, photochemical $2+2$ cycloadditions are not always stereospecific, in contrast to concerted thermal cycloadditions such as the Diels–Alder reaction. Stereospecificity

30. R. Huisgen and G. Steiner, *J. Am. Chem. Soc.* **95**, 5054 (1973).
31. G. Steiner and R. Huisgen, *J. Am. Chem. Soc.* **95**, 5056 (1973).
32. P. de Mayo, *Acc. Chem. Res.* **4**, 41 (1971).

is lost if the intermediate 1,4-diradical undergoes bond rotation faster than ring closure.

Intermolecular addition of alkenes can be carried out by photosensitization with mercury or directly with short-wavelength sources.[33] Relatively little preparative use has been made of simple alkenes, however. Dienes can be photosensitized using such materials as benzophenone, butane-2,3-dione, and acetophenone.[34] Under these conditions, preparatively useful yields of stereoisomeric mixtures of dimers are obtained. Usually, some $4+2$ adduct accompanies the dominant $2+2$ adduct.

The photodimerizations of substituted styrenes were among the earliest photochemical reactions to be studied.[35] These compounds give good yields of dimers

$$PhCH{=}CHCO_2H \xrightarrow[\text{H}_2\text{O}]{h\nu}$$ (56%) Ref. 36

$$PhCH{=}CHNO_2 \xrightarrow[\text{H}_2\text{O}]{h\nu}$$ (77%) Ref. 37

when irradiation is carried out in the crystalline state. In solution, *cis–trans* isomerization is the dominant reaction.

Cycloaddition of carbon–carbon double bonds can also occur intramolecularly. Direct irradiation of simple dienes leads to cyclobutanes.[38] This is a singlet-state process and is concerted. The stereochemistry of the cyclobutane can be predicted on the basis of orbital-symmetry rules (Part A, Section 10.1). Nonconjugated dienes can also undergo photochemical cyclization employing mercury or carbonyl compounds as sensitizers. Cyclobutane formation is usually unfavorable with 1,4-dienes because it would result in a very strained ring system. When the alkene units are separated by at least two carbon atoms, cyclization becomes more favorable sterically:

Ref. 39

33. H. Yamazaki and R. J. Cvetanović, *J. Am. Chem. Soc.* **91**, 520 (1969).
34. G. S. Hammond, N. J. Turro, and R. S. H. Liu, *J. Org. Chem.* **28**, 3297 (1963).
35. A. Mustafa, *Chem. Rev.* **51**, 1 (1952).
36. D. G. Farnum and A. J. Mostashari, *Org. Photochem. Synth.* **1**, 103 (1971).
37. D. G. Farnum and A. J. Mostashari, *Org. Photochem. Synth.* **1**, 79 (1971).
38. R. Srinivasan, *J. Am. Chem. Soc.* **84**, 4141 (1962); **90**, 4498 (1968).
39. J. Meinwald and G. W. Smith, *J. Am. Chem. Soc.* **89**, 4923 (1967); R. Srinivasan and K. H. Carlough, *J. Am. Chem. Soc.* **89**, 4932 (1967).

CHAPTER 6
CYCLOADDITIONS
AND
UNIMOLECULAR
REARRANGEMENTS
AND
ELIMINATIONS

Scheme 6.8. Intramolecular Cyclization of Dienes

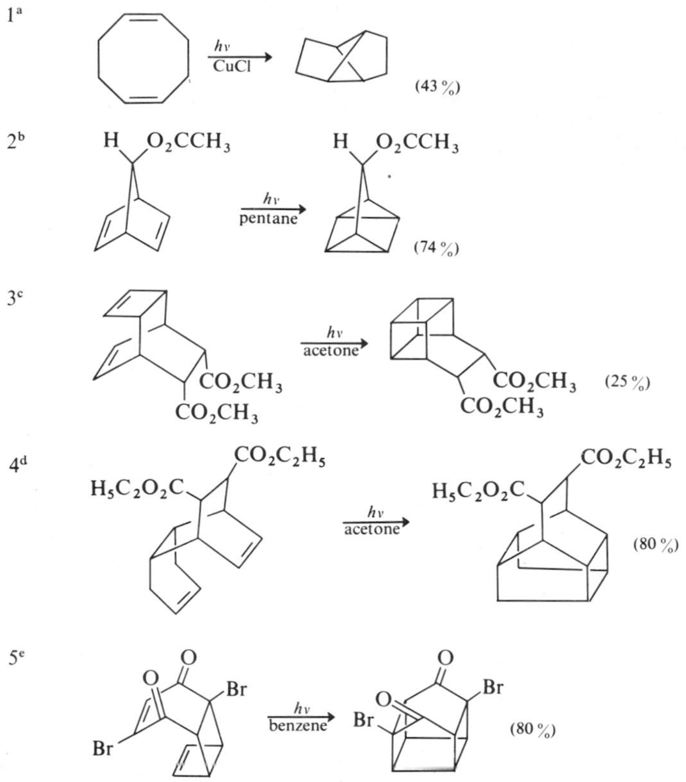

1^a

(43 %)

2^b

(74 %)

3^c

(25 %)

4^d

(80 %)

5^e

(80 %)

a. P. Srinivasan, *Org. Photochem. Synth.* **1**, 101 (1971); *J. Am. Chem. Soc.* **86**, 3318 (1964).
b. P. G. Gassman and D. S. Patton, *J. Am. Chem. Soc.* **90**, 7276 (1968).
c. W. G. Dauben, C. H. Schallhorn, and D. L. Whalen, *J. Am. Chem. Soc.* **93**, 1446 (1971).
d. B. M. Jacobson, *J. Am. Chem. Soc.* **95**, 2579 (1973).
e. J. C. Barborak, L. Watts, and R. Pettit, *J. Am. Chem. Soc.* **88**, 1328 (1966).

The most widely exploited photochemical $2+2$ cycloadditions of alkenes are intramolecular reactions that have been used to effect synthesis of a variety of complex "cage" skeletons. Cyclization is facilitated in these molecules by the proximity of the two carbon–carbon double bonds. Several examples are given in Scheme 6.8.

Another class of molecules that are quite prone to undergo photochemical cycloadditions are α,β-unsaturated carbonyl compounds. The reactive state in photochemical cycloadditions of α,β-unsaturated ketones is believed to be the $n-\pi^*$ triplet.[40] Conservation of spin then implies that the initial intermediate is a triplet diradical, and the reaction need not be stereospecific with respect to the alkene component.[41] The reaction has been most thoroughly studied in the case of cyclopen-

40. P. E. Eaton, *Acc. Chem. Res.* **1**, 50 (1968).
41. E. J. Corey, J. D. Bass, R. Le Mahieu, and R. B. Mitra, *J. Am. Chem. Soc.* **86**, 5570 (1964).

1[a]

2[b]

3[c]

4[d]

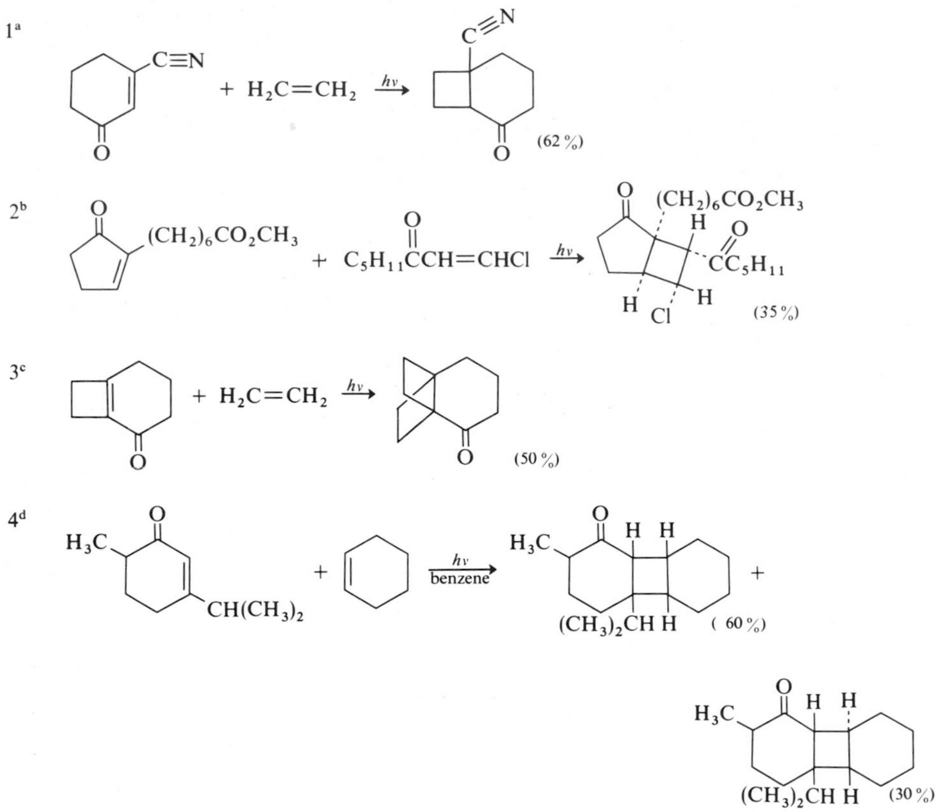

a. W. C. Agosta and W. W. Lowrance, Jr., *J. Org. Chem.* **35**, 3851 (1970).
b. J. F. Bagli and T. Bogri, *J. Org. Chem.* **37**, 2132 (1972).
c. P. E. Eaton and K. Nyi, *J. Am. Chem. Soc.* **93**, 2786 (1971).
d. P. Singh, *J. Org. Chem.* **36**, 3334 (1971).

tenones and cyclohexenones. The excited states of acyclic enones and larger ring compounds can be deactivated by *cis–trans* isomerization and do not add readily to alkenes. Isomerization is not possible in cyclopentenones or cyclohexenones because of the restraint of the ring. Alkynes can serve in the addition reaction in place of alkenes.[42,43] Unsymmetrical alkenes can undergo two modes of addition. The factors governing regioselectivity in such reactions are not entirely clear at this point.

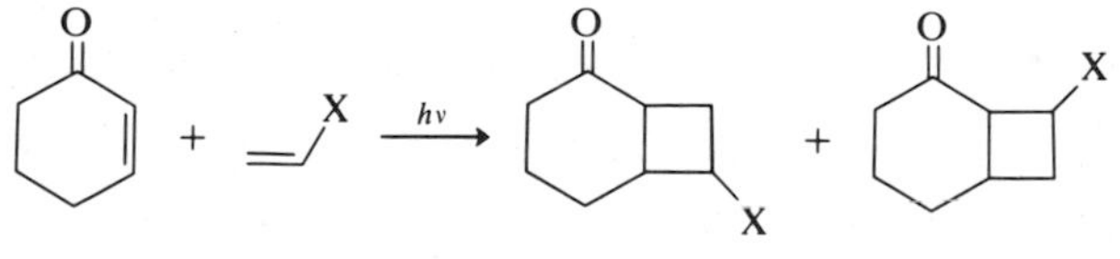

42. R. L. Cargill, T. Y. King, A. B. Sears, and M. R. Willcott, *J. Org. Chem.* **36**, 1423 (1971).
43. W. C. Agosta and W. W. Lowrance, Jr., *J. Org. Chem.* **35**, 3851 (1970).

226

CHAPTER 6
CYCLOADDITIONS
AND
UNIMOLECULAR
REARRANGEMENTS
AND
ELIMINATIONS

Scheme 6.9 records some examples of photochemical cycloaddition of enones and alkenes.

With saturated carbonyl compounds, and especially with aromatic carbonyl compounds, reaction between the photoexcited carbonyl chromophore and alkenes results in the formation of four-membered cyclic ethers (oxetane). The addition of carbonyl compounds to alkenes to give oxetanes is often referred to as the *Paterno–*

$$R_2C{=}O \ + \ R'CH{=}CHR' \ \longrightarrow$$

Büchi reaction.[44] The reaction is stereoselective for at least some aliphatic carbonyl compounds, but not for aromatic systems.[45] This result suggests that the reactive excited carbonyl compound is the singlet species for aliphatics, but a triplet for aromatics. With aromatic ketones, the predominant mode of addition can usually be predicted on the basis that the more stable of the two possible diradical intermediates will be formed by bonding of the oxygen to the less-substituted end of the carbon–carbon double bond. Many successful examples have been reported and tabulated.[44]

$$Ar_2C{=}O \ + \ CH_2{=}C(CH_3)_2 \ \xrightarrow{h\nu} \ Ar_2\overset{\cdot}{C}{-}O{-}CH_2\overset{\cdot}{C}(CH_3)_2 \ \longrightarrow$$

favored

Yields are best for aromatic ketones and aldehydes. Some examples are given in Scheme 6.10.

6.3. Sigmatropic Rearrangements

The mechanistic basis and terminology of sigmatropic rearrangements were considered in Part A, Chapter 10. The sigmatropic process that has become most firmly established as a useful tool in synthetic methodology is the [3,3]-sigmatropic rearrangement. The principles of orbital-symmetry conservation establish that this reaction can occur via a concerted mechanism. As a result, the process has the advantage of highly predictable stereospecificity. Examples are known in which

various combinations of carbon, oxygen, nitrogen, and sulfur atoms occupy places in

44. D. R. Arnold, *Adv. Photochem.* **6**, 301 (1968).
45. N. C. Yang and W. Eisenhardt, *J. Am. Chem. Soc.* **93**, 1277 (1971); D. R. Arnold, R. L. Hinman, and A. H. Glick, *Tetrahedron Lett.*, 1425 (1964); N. J. Turro and P. A. Wriede, *J. Am. Chem. Soc.* **90**, 6863 (1968); J. A. Barltrop and H. A. J. Carless, *J. Am. Chem. Soc.* **94**, 8761 (1972).

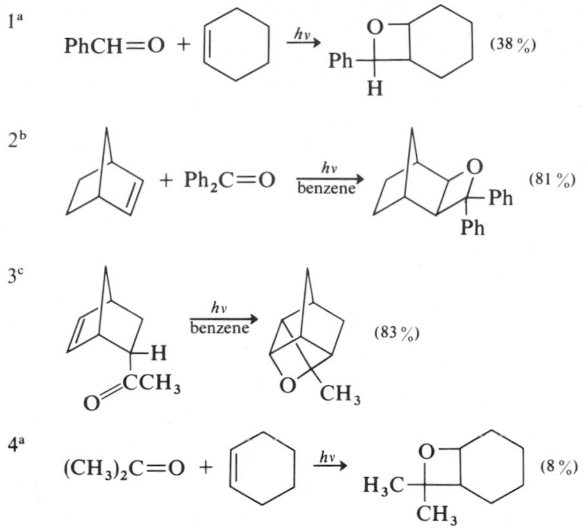

Scheme 6.10. Photocycloaddition of Carbonyl Compounds with Alkenes

227

SECTION 6.3.
SIGMATROPIC
REARRANGE-
MENTS

1[a]

$$PhCH{=}O \; + \; \text{[alkene]} \xrightarrow{h\nu} \text{[product]} \quad (38\,\%)$$

2[b]

$$\text{[norbornene]} \; + \; Ph_2C{=}O \xrightarrow[\text{benzene}]{h\nu} \text{[product]} \quad (81\,\%)$$

3[c]

$$\text{[reactant]} \xrightarrow[\text{benzene}]{h\nu} \text{[product]} \quad (83\,\%)$$

4[a]

$$(CH_3)_2C{=}O \; + \; \text{[alkene]} \xrightarrow{h\nu} \text{[product]} \quad (8\,\%)$$

a. J. S. Bradshaw, *J. Org. Chem.* **31**, 237 (1966).
b. D. R. Arnold, A. H. Glick, and V. Y. Abraitys, *Org. Photochem. Synth.* **1**, 51 (1971).
c. R. R. Sauers, W. Schinski, and B. Sickles, *Org. Photochem. Synth.* **1**, 76 (1971).

the reacting system, but from a synthetic standpoint the most important cases are those in which all atoms are carbon and the case when $X{=}O$.

The [3,3]-sigmatropic rearrangement of allyl vinyl ethers is known as the *Claisen rearrangement.* This reaction was well-studied first in the case of allyl phenyl ethers, but its potential for synthetic utility in aliphatic systems was also eventually developed. An important application of the reaction is in introducing substituent groups at the "angular" position at the junction of two six-membered rings.[46] Introduction of a substituent at such a position is frequently necessary in the synthesis of steroids and terpenes. A $-CH_2CH{=}O$ group can be introduced by thermal rearrangement of vinyl allyl ethers. The required ethers can be prepared

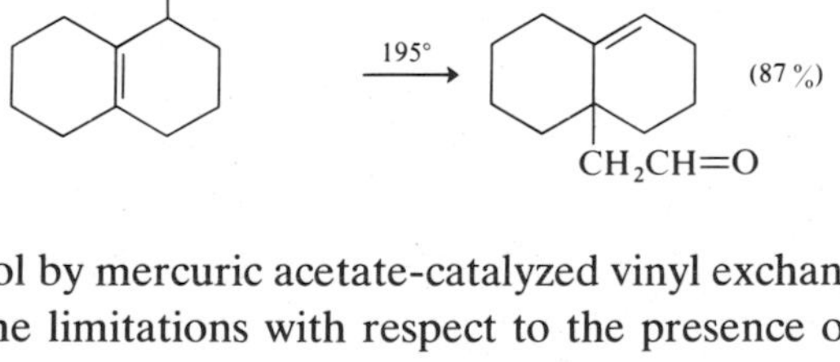

$$\text{[OCH{=}CH_2 substituted decalin]} \xrightarrow{195°} \text{[product with CH_2CH{=}O]} \quad (87\,\%) \qquad \text{Ref. 47}$$

from the allyl alcohol by mercuric acetate-catalyzed vinyl exchange.[48] This stage of the process has some limitations with respect to the presence of other functional

46. W. G. Dauben and T. J. Dietsche, *J. Org. Chem.* **37**, 1212 (1972).
47. A. W. Burgstahler and I. C. Nordin, *J. Am. Chem. Soc.* **83**, 198 (1961).
48. W. H. Watanabe and L. E. Conlon, *J. Am. Chem. Soc.* **79**, 2229 (1957).

228

CHAPTER 6
CYCLOADDITIONS
AND
UNIMOLECULAR
REARRANGEMENTS
AND
ELIMINATIONS

groups, and several other procedures have been developed that introduce the vinyl group by noncatalytic, thermal exchange reactions. In these procedures, the vinyl group is substituted so as to facilitate exchange and formation of a vinyl ether. Heating an allyl alcohol with the dimethyl acetal of N,N-dimethylacetamide results in the introduction of the allyl alcohol by a reversible thermal elimination–addition mechanism. Elimination of methanol after the allyl alcohol has been introduced

$$(CH_3)_2\overset{\overset{\displaystyle OCH_3}{|}}{\underset{\underset{\displaystyle OCH_3}{|}}{N}}CCH_3 \;\rightleftarrows\; (CH_3)_2\overset{\overset{\displaystyle OCH_3}{|}}{N}C=CH_2 \;+\; CH_3OH$$

$$RCH=CHCH_2OH \;+\; (CH_3)_2\overset{\overset{\displaystyle OCH_3}{|}}{N}C=CH_2 \;\rightleftarrows\; (CH_3)_2\overset{\overset{\displaystyle OCH_3}{|}}{\underset{\underset{\displaystyle CH_3}{|}}{N}}COCH_2CH=CHR$$

$$(CH_3)_2\overset{\overset{\displaystyle OCH_3}{|}}{\underset{\underset{\displaystyle CH_3}{|}}{N}}COCH_2CH=CHR \;\rightleftarrows\; (CH_3)_2\overset{}{\underset{\underset{\displaystyle CH_2}{\|}}{N}}COCH_2CH=CHR \;+\; CH_3OH$$

generates a vinyl allyl ether, and sigmatropic rearrangement ensues. The net effect is introduction of an N,N-dimethylacetamide side chain at carbon-3 of the allyl alcohol.[49] The reaction can be carried out by simply heating the allyl alcohol with the

$$\underset{CH_2CH}{\overset{(CH_3)_2NC=CH_2}{O}}CHR \;\rightarrow\; \underset{CH_2=CH-CHR}{\overset{\overset{\displaystyle O}{\|}}{(CH_3)_2NC-CH_2}}$$

N,N-dimethylacetamide dimethyl acetal, or with the elimination product 1-dimethylamino-1-methoxyethene, which can be prepared separately.

A carboalkoxymethyl group can be introduced in a similar way by using an ortho ester of acetic acid.[50] Again, a series of addition and elimination reactions generates a reactive allyl vinyl ether grouping:

$$RCH=CHCH_2OH \;+\; CH_3C(OC_2H_5)_3 \;\rightleftarrows\; RCH=CHCH_2O\overset{\overset{\displaystyle OCH_3}{|}}{\underset{\underset{\displaystyle OCH_3}{|}}{C}}CH_3$$

$$RCH=CHCH_2O\overset{\overset{\displaystyle OCH_3}{|}}{\underset{\underset{\displaystyle OCH_3}{|}}{C}}CH_3 \;\rightleftarrows\; RCH=CHCH_2O\overset{\overset{\displaystyle OCH_3}{|}}{C}=CH_2 \;\rightarrow\; R\overset{\overset{\displaystyle CH_2CO_2CH_3}{|}}{C}HCH=CH_2$$

The exchange is catalyzed by addition of a small amount of a weak acid, such as propionic acid.

49. A. E. Wick, D. Felix, K. Steen, and A. Eschenmoser, *Helv. Chim. Acta* **47**, 2425 (1964); D. Felix, K. Gschwend-Steen, A. E. Wick, and A. Eschenmoser, *Helv. Chim. Acta* **52**, 1030 (1969).
50. W. S. Johnson, L. Werthemann, W. R. Bartlett, T. J. Brocksom, T. Li, D. J. Faulkner, and M. R. Petersen, *J. Am. Chem. Soc.* **92**, 741 (1970).

A version of the Claisen rearrangement that utilizes 3-methoxyisoprene has proved valuable as a method for introduction of isoprene units in the synthesis of

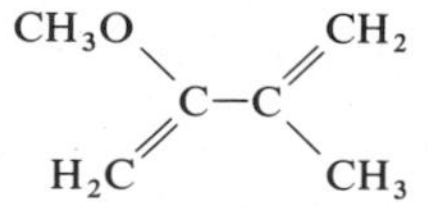

natural products. An allylic alcohol can undergo acid-catalyzed exchange with the methoxy group. The process can be repeated by reducing the α,β-unsaturated

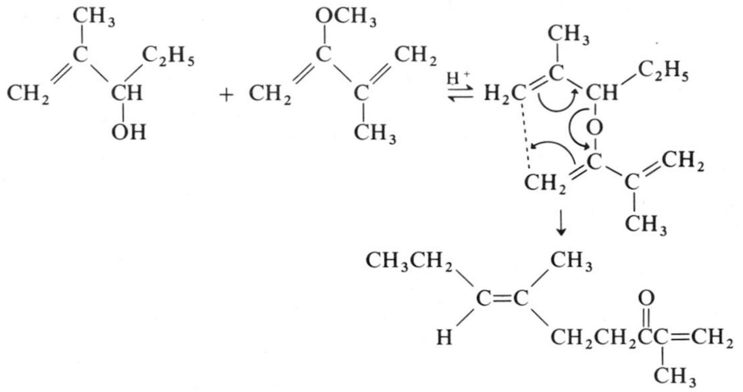

ketone that is produced to the corresponding allylic alcohol.[51] Because of the interconvertibility of vinyl ethers and ketals, the ketal **5** is the synthetic equivalent of

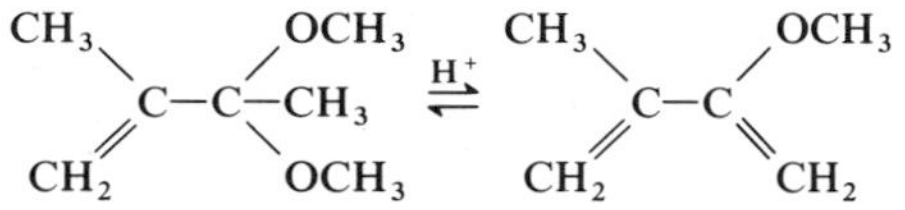

5

2-methoxyisoprene and is, in fact, advantageous to use in some syntheses.[52]

Esters of allyl alcohols can be rearranged to γ,δ-unsaturated carboxylic acids via the trimethylsilyl derivative of the ester enolate.[53] This rearrangement takes place

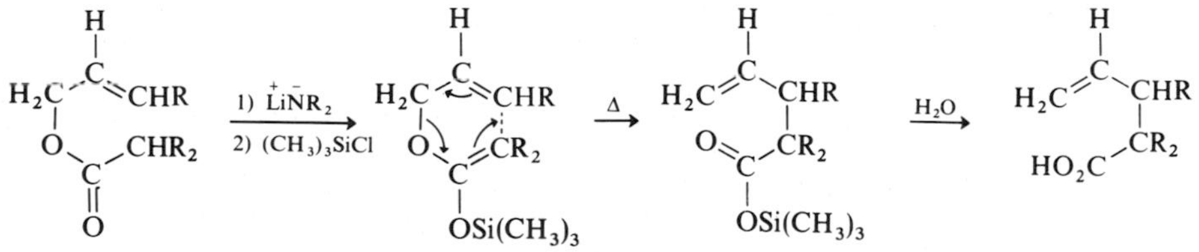

quite rapidly slightly above room temperature.

51. D. J. Faulkner and M. R. Petersen, *Tetrahedron Lett.*, 3243 (1969); *J. Am. Chem. Soc.* **95**, 553 (1973).
52. W. S. Johnson, T. J. Brocksom, P. Loew, D. H. Rich, L. Werthemann, R. A. Arnold, T. Li, and D. J. Faulkner, *J. Am. Chem. Soc.* **92**, 4463 (1970).
53. R. E. Ireland, R. H. Mueller, and A. K. Willard, *J. Am. Chem. Soc.* **98**, 2868 (1976).

230

CHAPTER 6
CYCLOADDITIONS
AND
UNIMOLECULAR
REARRANGEMENTS
AND
ELIMINATIONS

The concerted mechanism that operates in Claisen rearrangements permits predictions concerning the stereochemistry of the reaction on the basis of conformational effects in the transition state. This transition state is believed usually to have a chair-like conformation.[54] The new double bond produced in the rearrangement is

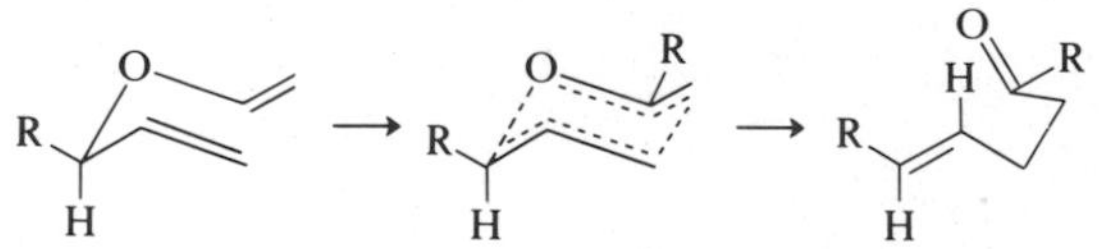

preferentially *trans* because of the tendency of the substituent alpha to the oxygen to take up an equatorial position in the transition state.

High degrees of asymmetric induction have been observed in Claisen rearrangements.[55] Treatment of optically active **6** with triethyl orthoacetate gives product **7** in 90% optical yield. The configuration of the product was established by degradation.

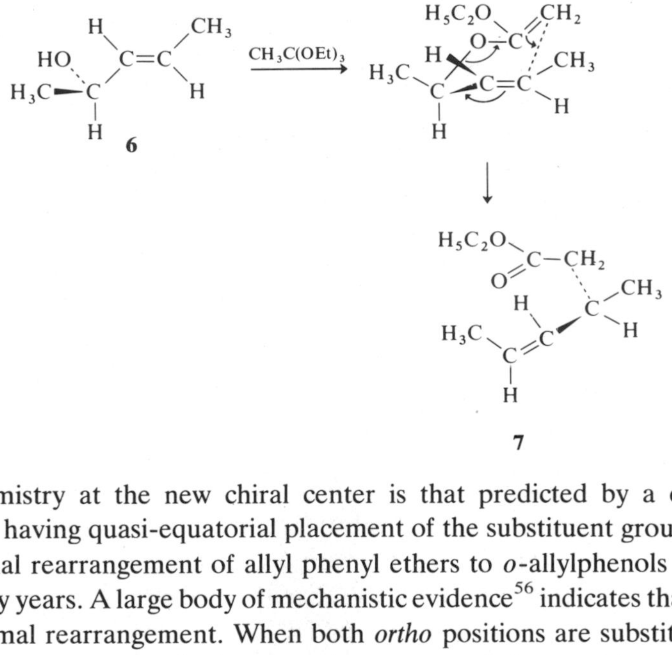

The stereochemistry at the new chiral center is that predicted by a chair-like transition state having quasi-equatorial placement of the substituent groups.

The thermal rearrangement of allyl phenyl ethers to *o*-allylphenols has been known for many years. A large body of mechanistic evidence[56] indicates that this is a concerted thermal rearrangement. When both *ortho* positions are substituted, the allyl group undergoes a second migration via a concerted sigmatropic mechanism, giving the *para*-substituted phenol:

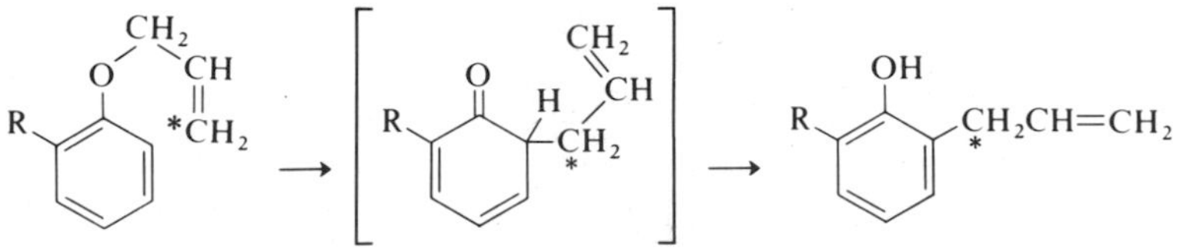

54. W. von E. Doering and W. R. Roth, *Tetrahedron* **18**, 67 (1962).
55. R. K. Hill, R. Soman, and S. Sawada, *J. Org. Chem.* **37**, 3737 (1972).
56. S. J. Rhoads, in *Molecular Rearrangements*, Vol. 1, P. de Mayo (ed.), Interscience, New York, NY, 1963, pp. 655–684.

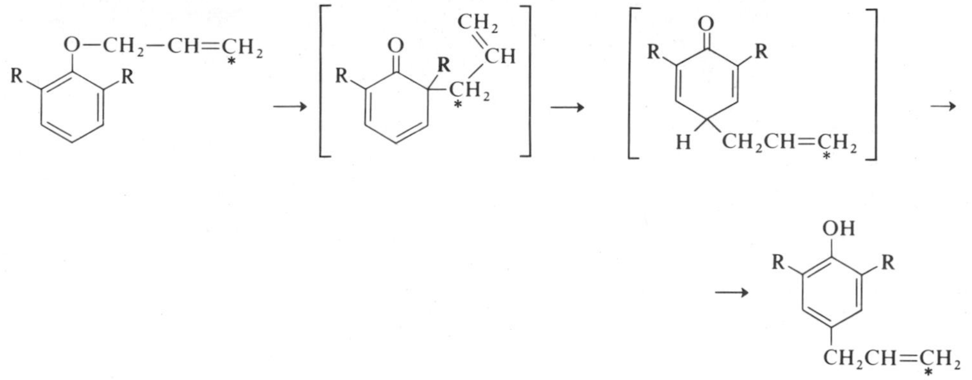

$O-CH_2-CH=CH_2$

Radioactive carbon-labeling experiments have shown that C-1 of the allyl group in the ether becomes C-3 in the product when migration terminates at the *ortho* position. When the rearrangement leads to the *para* substitution product, there are two exchanges of C-1 and C-3 of the migrating allyl group, as indicated in the outline mechanism. As a result, the carbon that is C-3 in the starting material is also C-3 in the final product.

The stereochemistry of the Claisen rearrangement of allyl phenyl ethers has been examined.[57] The new double bond is primarily *trans* whether it is formed from a *cis* or a *trans* precursor. These results are accommodated by a chair-like cyclic transition state. For the two isomeric reactants, the preferred geometry for the transition state would be **A** and **B**, respectively. Both lead to a *trans* double bond in

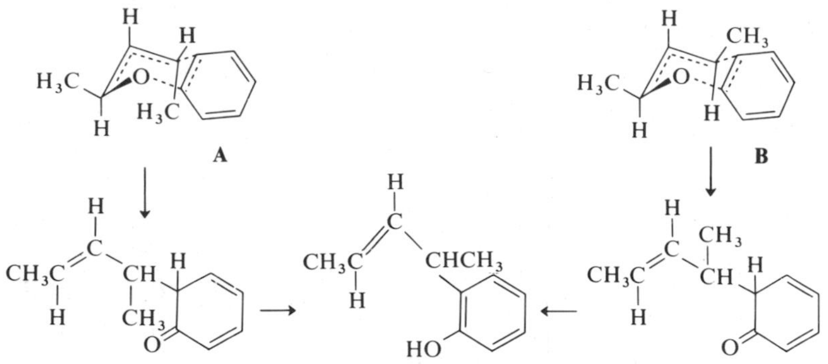

the product because of the preference for a pseudo-equatorial orientation of the methyl group.

Scheme 6.11 gives some examples of [3,3]-sigmatropic rearrangements for both aliphatic and aromatic systems.

The rearrangement of 1,5-dienes by the [3,3]-sigmatropic reaction is referred to as the *Cope rearrangement*.[58,59] This reaction was discussed in Part A, Section 10.2,

57. H. L. Goering and W. I. Kimoto, *J. Am. Chem. Soc.* **87**, 1748 (1965); E. N. Marvell, J. L. Stephenson, and J. Ong, *J. Am. Chem. Soc.* **87**, 1267 (1965).
58. A. Jefferson and F. Scheinmann, *Q. Rev. Chem. Soc.* **22**, 391 (1968).
59. S. J. Rhoads, in *Molecular Rearrangements*, Vol. 1, P. de Mayo (ed.), Interscience, New York, NY, 1963, pp. 684–696; S. J. Rhoads and N. R. Raulins, *Org. React.* **22**, 1 (1975).

232

Scheme 6.11. Claisen

CHAPTER 6
CYCLOADDITIONS
AND
UNIMOLECULAR
REARRANGEMENTS
AND
ELIMINATIONS

A. Aliphatic Compounds

1[a]

$$H_2C=CHCH_2CH_2\overset{OH}{\underset{\underset{CH_3}{|}}{\overset{|}{C}}}HC=CH_2 \xrightarrow{CH_3C(OC_2H_5)_3} H_2C=CHCH_2CH_2\overset{H}{C}$$

$$\overset{}{\underset{CH_3}{\overset{|}{C}}}CH_2CH_2CO_2C_2H_5$$

2[b]

$$\overset{H_5C_2}{\underset{H_2C}{\overset{|}{C}}}\overset{}{=}\overset{}{\underset{OH}{C}}HCH_2CH_2\overset{CH_3}{\underset{\underset{H}{|}}{\overset{|}{C}}}\overset{}{\underset{CO_2CH_3}{C}} \longrightarrow$$

$$\xrightarrow[110°]{CH_3C(OCH_3)_3} CH_3O_2CCH_2CH_2\overset{H_5C_2}{\overset{|}{C}}\overset{}{=}\overset{}{\underset{H}{\overset{|}{C}}}CH_2CH_2\overset{CH_3}{\underset{\underset{H}{|}}{\overset{|}{C}}}\overset{}{\underset{CO_2CH_3}{C}}$$

(85%)

3[c]

(74%)

4[d]

(96%)

$$\text{CH}_3\text{CH}_2\text{CHCO}_2\text{CH}_3$$

5[e]

$$(CH_3)_2\overset{CH_3}{\underset{OH}{\overset{|}{C}}}CH=CH_2 + H_2C=COCH_3 \xrightarrow[125°]{H^+} (CH_3)_2C=CHCH_2CH_2\overset{O}{\overset{||}{C}}CH_3 \quad (94\%)$$

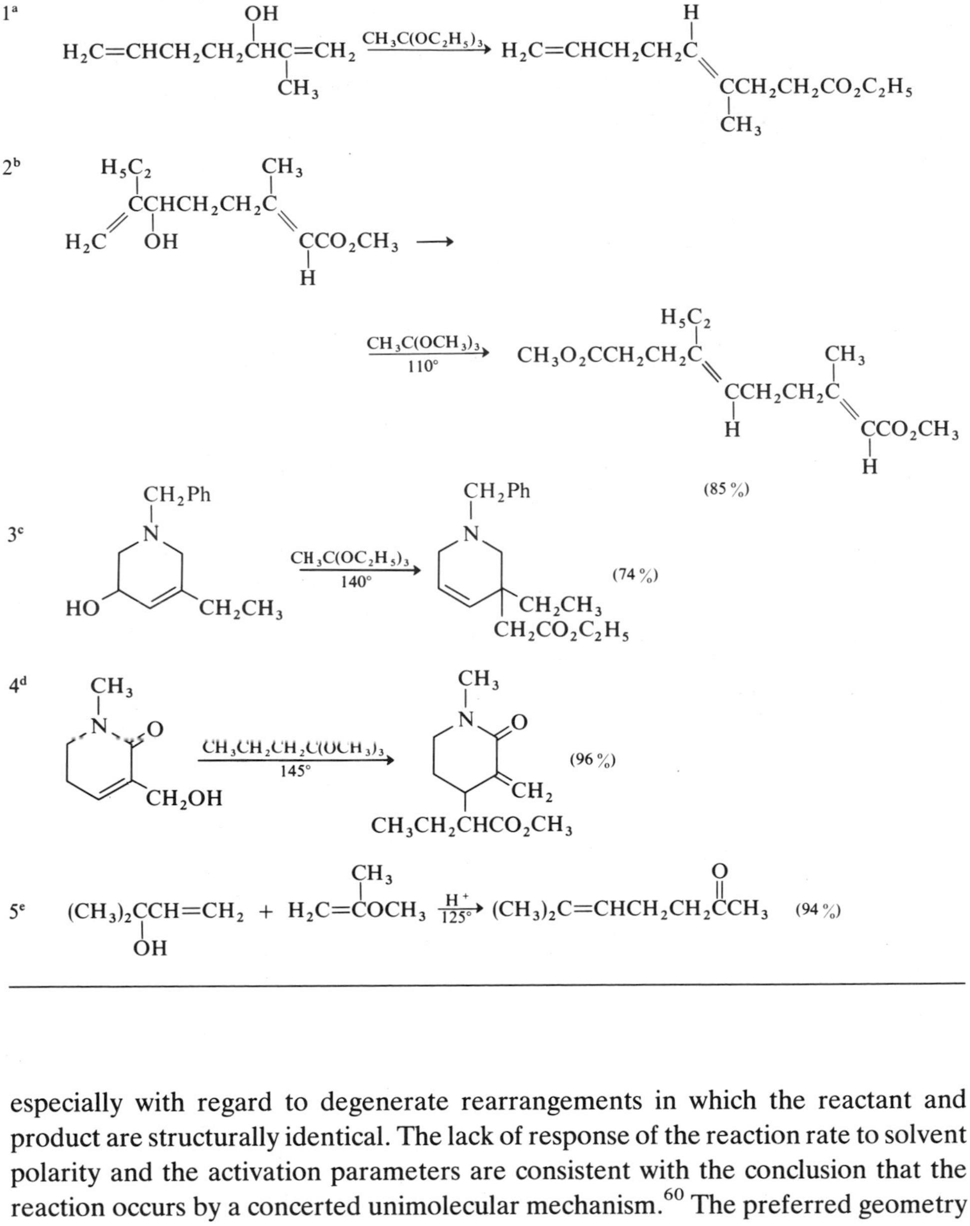

especially with regard to degenerate rearrangements in which the reactant and product are structurally identical. The lack of response of the reaction rate to solvent polarity and the activation parameters are consistent with the conclusion that the reaction occurs by a concerted unimolecular mechanism.[60] The preferred geometry

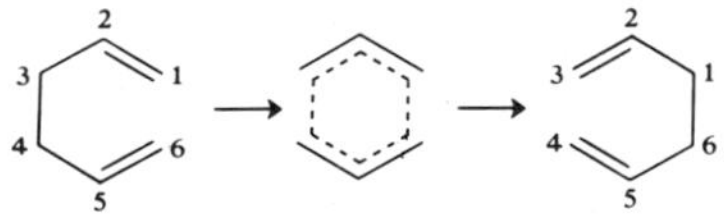

60. H. M. Frey and R. Walsh, *Chem. Rev.* **69**, 103 (1969).

Rearrangements

233

SECTION 6.3.
SIGMATROPIC
REARRANGE-
MENTS

6[f] $CH_3CH=CHCH_2OC$ (with CH_3 and $CCO_2CH_2CH=CHCH_3$, H) $\xrightarrow{140-145°}$ $CH_3\overset{O}{\overset{||}{C}}CHCO_2CH_2CH=CHCH_3$, $CH_3\overset{|}{C}HCH=CH_2$ (61%)

B. Aromatic Compounds

7[g]

$OCH_2CH=CH_2$ ... $HNCCH_3$ ($\overset{||}{O}$) $\rightarrow$ OH, $CH_2CH=CH_2$ (83%), $HNCCH_3$ ($\overset{||}{O}$)

8[h]

CH_3O, $OCH_2CH=CH_2$, OCH_3 $\rightarrow$ CH_3O, OH, OCH_3 (90%), $CH_2CH=CH_2$

9[i]

CH_2OH $\xrightarrow[H_2C=\overset{|}{C}OCH_3]{N(CH_3)_2}$ $CH_2\overset{O}{\overset{||}{C}}N(CH_3)_2$, CH_3 (94%)

a. R. I. Trust and R. E. Ireland, *Org. Synth.* **53**, 116 (1973).
b. C. A. Henrick, R. Schaub, and J. B. Siddall, *J. Am. Chem. Soc.* **94**, 5374 (1972).
c. F. E. Ziegler and G. B. Bennett, *J. Am. Chem. Soc.* **95**, 7458 (1973).
d. J. J. Plattner, R. D. Glass, and H. Rapoport, *J. Am. Chem. Soc.* **94**, 8614 (1972).
e. G. Saucy and R. Marbet, *Helv. Chim. Acta* **50**, 2091 (1967).
f. J. W. Ralls, R. E. Lundin, and G. F. Bailey, *J. Org. Chem.* **28**, 3521 (1963).
g. J. H. Burkhalter, F. H. Tendick, E. M. Jones, P. A. Jones, W. F. Holcomb, and A. L. Rawlins, *J. Am. Chem. Soc.* **70**, 1363 (1948).
h. I. A. Pearl, *J. Am. Chem. Soc.* **70**, 1746 (1948).
i. D. Felix, K. Gschwend-Steen, A. E. Wick, and A. Eschenmoser, *Helv. Chim. Acta* **52**, 1030 (1969).

of the transition state is generally considered to resemble a cyclohexane chair in conformation, although the boat conformation can be utilized where structural features restrict the chair conformation.[61]

   The rate of the reaction is influenced by the energy of the starting materials, and strained compounds react at much lower temperatures than do unstrained molecules. The 1,2-divinyl derivatives of cyclopropane, cyclobutane, and cyclopentane illustrate this point. *cis*-Divinylcyclopropane rearranges to cycloheptadiene

61. R. P. Lutz, S. Bernal, R. J. Boggio, R. O. Harris, and M. W. McNicholas, *J. Am. Chem. Soc.* **93**, 3985 (1971).

234

CHAPTER 6
CYCLOADDITIONS
AND
UNIMOLECULAR
REARRANGEMENTS
AND
ELIMINATIONS

below $-40°$.[62] *cis*-Divinylcyclobutane rearranges at 120° to 1,4-cyclooctadiene in 91% yield.[63] The *trans* isomer, in which the cyclic transition state necessary for

concerted rearrangement is impossible, does not rearrange at temperatures below 240°. At that temperature, an alternative mode of decomposition occurs. *cis*-Divinylcyclopentane is equilibrated with *cis,cis*-1,5-nonadiene at 220°, but the cyclopentane compound is the more stable of the two, and only 5% conversion occurs.[64]

Cope rearrangements are generally reversible processes, and there are no changes in the number of single and double bonds as a result of the reaction, so that to a rough approximation the total bond energy is unchanged. The position of the final equilibrium is governed by the relative stability of the starting materials and the possible products. For cyclic compounds, therefore, the relative strain of the compounds is a major factor governing the equilibrium position. For acyclic compounds, the product ratio at equilibrium will favor conjugated systems and internal over terminal carbon–carbon double bonds.

When there is a hydroxyl substituent at C-3 of the diene system, the Cope rearrangement product is an enol, which is, of course, converted to the carbonyl compound. This reaction, which is called the *oxy-Cope rearrangement*, thus has a driving force resulting from the formation of the carbonyl group.[65] This reaction has found some use in the synthesis of medium-sized rings.

(90 %)   Ref. 66

6.4. Unimolecular Thermal Elimination Reactions

The term *cycloreversion* is applied to reactions that are the reverse of cycloadditions and lead to the formation of two or more molecules from a cyclic starting material. Although some reactions of this type are concerted, others occur by a stepwise mechanism. These non-concerted processes are considered here because

62. W. von E. Doering and W. R. Roth, *Tetrahedron* **19**, 715 (1963).
63. E. Vogel, *Justus Liebigs Ann. Chem.* **615**, 1 (1958); G. S. Hammond and C. DeBoer, *J. Am. Chem. Soc.* **86**, 899 (1964).
64. E. Vogel, W. Grimme, and E. Dinné, *Angew. Chem.* **75**, 1103 (1963).
65. A. Viola, E. J. Iorio, K. K. Chen, G. M. Glover, U. Nayak, and P. J. Kocienski, *J. Am. Chem. Soc.* **89**, 3462 (1967).
66. E. N. Marvell and W. Whalley, *Tetrahedron Lett.*, 509 (1970).

235

SECTION 6.4.
UNIMOLECULAR
THERMAL
ELIMINATION
REACTIONS

the overall transformations are often closely related to those that occur by concerted mechanisms. In subsection 6.4.2, a series of synthetically important eliminations that occur by way of concerted mechanisms involving cyclic transition states are discussed.

6.4.1. Cycloreversions and Related Eliminations

Since a cycloreversion, the reverse of a cycloaddition, travels the same reaction path as the forward reaction, the considerations of stereochemistry and orbital symmetry that govern concerted cycloadditions are equally applicable to cycloreversions. The number of such reactions that have been studied in detail is not large, but there is sufficient information to establish that orbital-symmetry controls are indeed operating. The principles of orbital-symmetry conservation specify which processes can occur in concerted fashion and the stereochemical restrictions that are imposed by a concerted mechanism. We will first discuss some reactions that do occur by concerted mechanisms, and then turn to some of the elimination processes that involve discrete intermediates.

A good example of a concerted elimination is the reaction that takes place on treatment of 3-pyrrolines with N-nitrohydroxylamine. The reactive intermediate **C** is capable of elimination of molecular nitrogen. The reaction has been shown to be

completely stereospecific.[67] Furthermore, the groups at carbon atoms 2 and 5 rotate

in opposite senses in going to product (disrotatory). This stereochemistry is consistent with predictions based on conservation of orbital symmetry.

On treatment with difluoramine, aziridines undergo deamination to afford alkenes in a process that should not be concerted, according to the principles of orbital-symmetry conservation.[68] Nevertheless, the reaction has been observed to be

67. D. M. Lemal and S. D. McGregor, *J. Am. Chem. Soc.* **88**, 1335 (1966).
68. J. P. Freeman and W. H. Graham, *J. Am. Chem. Soc.* **89**, 1761 (1967).

236

CHAPTER 6
CYCLOADDITIONS
AND
UNIMOLECULAR
REARRANGEMENTS
AND
ELIMINATIONS

stereospecific for the case of formation of *cis*- and *trans*-2-butene from the corresponding *cis*- and *trans*-2,3-dimethylaziridines:

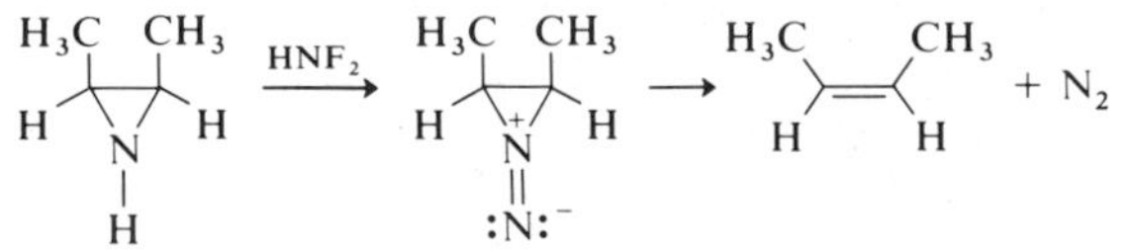

The elimination of carbon monoxide from cyclic ketones is a fairly common process. The reaction occurs photochemically with saturated systems; the mechanism of this process was discussed in Part A, Section 11.3. Facile thermal expulsion occurs only in molecules having special structural features. The elimination of carbon monoxide from bicyclo[2.2.1]heptadien-7-ones can occur by a concerted mechanism. In fact, generation of bicyclo[2.2.1]heptadien-7-ones is usually accompanied by

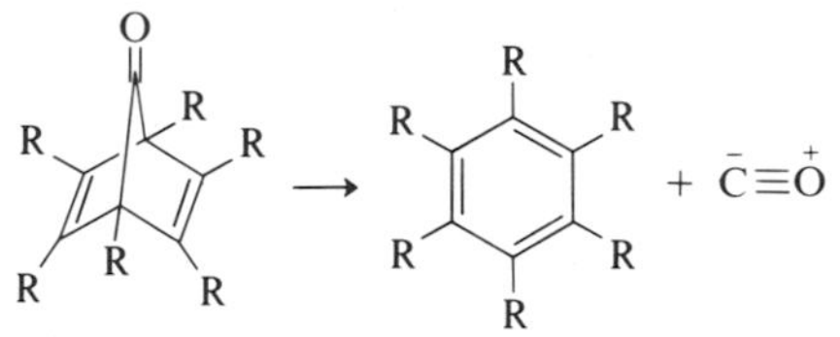

spontaneous elimination of CO. The most general route to this ring system is by Diels–Alder reaction of a substituted cyclopentadienone with an acetylene. The reaction is of some utility for the synthesis of highly substituted benzene rings.[69] Higher temperatures are required to eliminate CO from related systems that cannot lead directly to an aromatic ring. The most studied case is the conversion of bicyclo[2.2.1]hept-2-en-7-ones to cyclohexadienes.[70] The nature of the substituents on the ring system determines the ease of elimination in this case, but the temperatures required are often above 100–150°. Although this reaction can also occur by a concerted process, the transition-state energy is considerably higher. The exceptional reactivity of the bicyclo[2.2.1]heptadien-7-ones probably reflects stabilization of the transition state by virtue of the developing ring.

Exceptionally facile elimination of CO also takes place from structure **8**, in which homoaromaticity can stabilize the transition state:

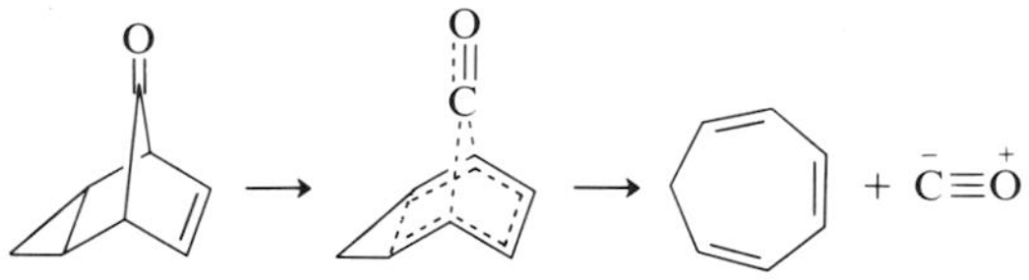

Elimination of sulfur in various oxidation states from heterocyclic molecules

69. M. A. Ogliaruso, M. G. Romanelli, and E. I. Becker, *Chem. Rev.* **65**, 261 (1965).
70. B. Halton, M. A. Battiste, R. Rehberg, C. L. Deyrup, and M. E. Brennan, *J. Am. Chem. Soc.* **89**, 5964 (1967); S. C. Clarke and B. L. Johnson, *Tetrahedron* **27**, 3555 (1971).

often occurs readily and has been the subject of mechanistic study and synthetic application. The sulfolenes **9** and **10** fragment to sulfur dioxide and *trans,trans*-2,4-hexadiene and *cis,trans*-2,4-hexadiene, respectively, at temperatures of 100–150°. The cleavages are stereospecific and involve disrotatory motion of the substituents at

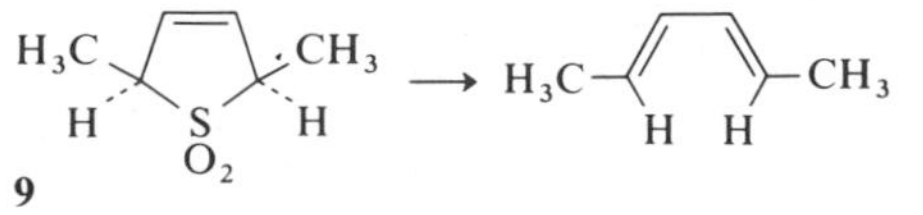

237

SECTION 6.4.
UNIMOLECULAR
THERMAL
ELIMINATION
REACTIONS

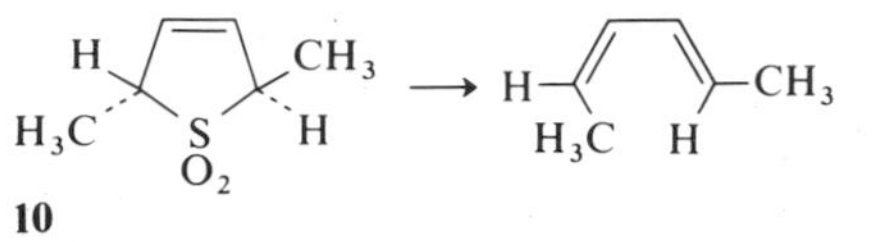

C-2 and C-5, similar to the pyrroline decomposition described previously.[71]

The loss of sulfur dioxide from thiirane 1,1-dioxides (episulfones) occurs even more readily, with *cis*-2,3-dimethylthiirane 1,1-dioxide decomposing at 60° to afford *cis*-2-butene stereospecifically.[72] Similarly, *cis*-stilbene is formed from *cis*-2,3-diphenylthiirane 1,1-dioxide[73]:

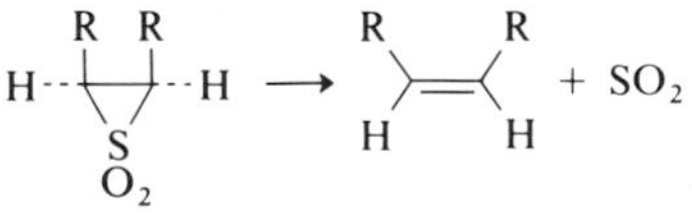

As in the case of aziridine deamination, orbital-symmetry rules forbid concerted decomposition, but the cases so far studied exhibit stereospecificity nontheless.

The most widely studied process involving elimination of small molecules from rings is the elimination of nitrogen from cyclic azo compounds. This is a useful reaction, especially for the synthesis of certain strained-ring systems. In most cases, the elimination of nitrogen from cyclic azo compounds can be carried out either photochemically or thermally. Although this reaction generally does not proceed by a concerted mechanism, but involves at least one intermediate, we will first consider some special instances where concerted thermal elimination is possible. We will then proceed to consider the mechanism and synthetic aspects of the more general case.

An interesting illustration that orbital-symmetry considerations are as applicable to cycloreversions as to cycloadditions is the contrasting stability of the azo compounds **11** and **12**. Compound **11** decomposes to norbornene and nitrogen only

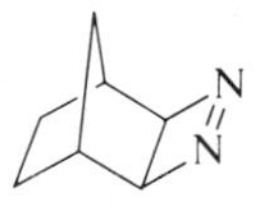

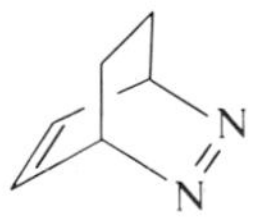

71. W. L. Mock, *J. Am. Chem. Soc.* **88**, 2857 (1966); S. D. McGregor and D. M. Lemal, *J. Am. Chem. Soc.* **88**, 2858 (1966).
72. N. P. Neureiter, *J. Am. Chem. Soc.* **88**, 558 (1966).
73. N. Tokura, T. Nagai, and S. Matsumura, *J. Org. Chem.* **31**, 349 (1966).

238

CHAPTER 6
CYCLOADDITIONS
AND
UNIMOLECULAR
REARRANGEMENTS
AND
ELIMINATIONS

above 100°. In contrast, **12** decomposes immediately on preparation even at −78°,[74] even though it is less strained than **11**. The reason for this difference is that if **11** were to undergo a concerted elimination, it would have to follow the high-energy 2+2 pathway. For **12**, the elimination can take place by a concerted 4+2 cycloreversion. The temperature range at which **11** decomposes is fairly typical of strained azo compounds, and it is presumably proceeding by a nonconcerted biradical mechanism. Because a C–N bond must be broken without concomitant compensation by

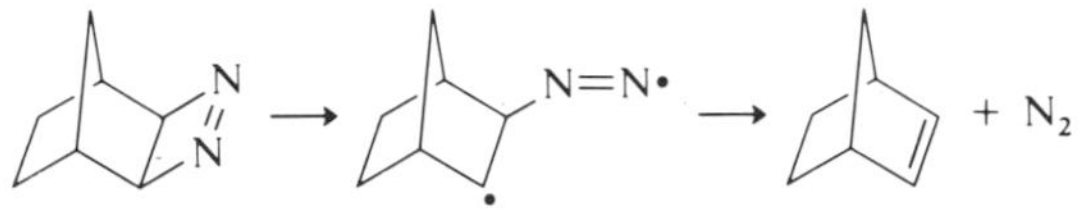

carbon–carbon double-bond formation, the activation energy is much higher than for the concerted process.

Another example of an azo compound for which a concerted mechanism is available is **13**.[75] Here, it is the electrons of the cyclopropane ring that participate in the smooth electronic reorganization associated with elimination of nitrogen. This azo compound eliminates nitrogen at room temperature:

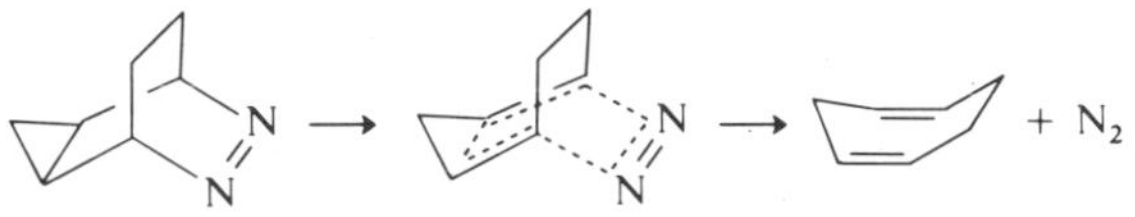

13

Further evidence that a concerted mechanism is possible when the cyclopropane ring is present comes from a study of the stereochemistry of the elimination of nitrogen from **14–16**.[76] In each case, the reaction is highly stereospecific and the diene produced is that expected from a concerted reaction. The transition state is of

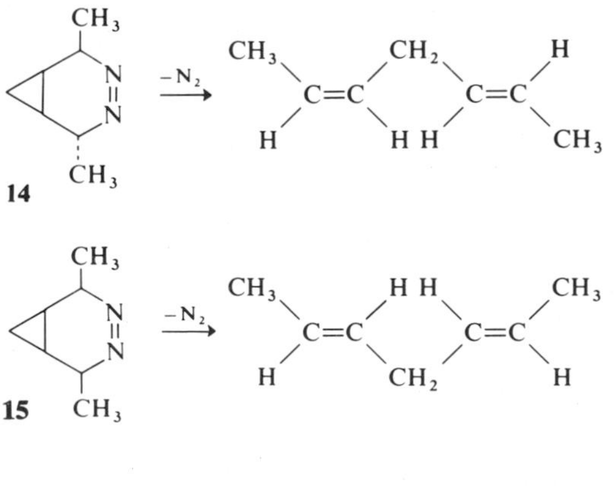

14

15

74. N. Rieber, J. Alberts, J. A. Lipsky, and D. M. Lemal, *J. Am. Chem. Soc.* **91**, 5668 (1969).
75. E. L. Allred, J. C. Hinshaw, and A. L. Johnson, *J. Am. Chem. Soc.* **91**, 3382 (1969); E. L. Allred and K. J. Voorhees, *J. Am. Chem. Soc.* **95**, 620 (1973).
76. J. A. Berson and S. S. Olin, *J. Am. Chem. Soc.* **91**, 777 (1969).

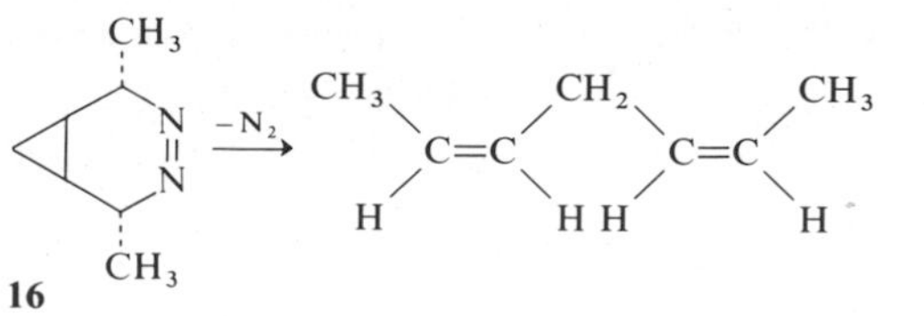

239

SECTION 6.4.
UNIMOLECULAR
THERMAL
ELIMINATION
REACTIONS

16

relatively low energy because of its homoaromatic nature. These azo compounds are unstable at room temperature.

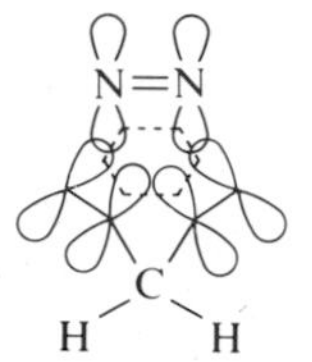

Although the concerted mechanism described in the preceding paragraphs is available only to those azo compounds with appropriate orbital arrangements, the decomposition by nonconcerted mechanisms occurs at low enough temperatures to be synthetically useful. Most often, the principal product from decomposition of a cyclic azo compound is the hydrocarbon formed by bond formation between the two

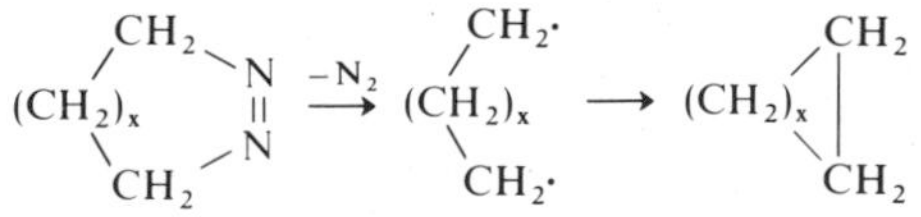

radical centers generated when nitrogen is eliminated. The decomposition can be carried out thermally or photochemically.

A mechanistic detail that is still under investigation is the question of the timing of the cleavage of the two C–N bonds. If the cleavages are simultaneous, a single intermediate, the diradical **D**, would be formed. If one bond is cleaved first, a second species, **E**, might also represent an energy minimum. At this time, there is little evidence for species **E**, but since it would be expected to be highly unstable,

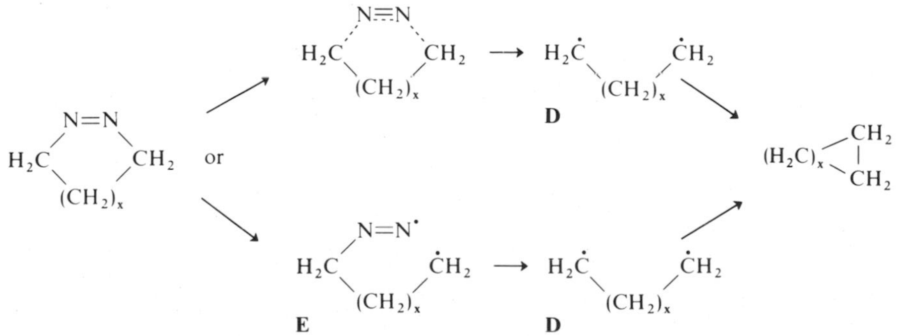

demonstrating its existence would not be expected to be easy.

The stereochemistry of decomposition of cyclic azo compounds has received

240

CHAPTER 6
CYCLOADDITIONS
AND
UNIMOLECULAR
REARRANGEMENTS
AND
ELIMINATIONS

some attention. In the decomposition of the stereoisomeric azo compounds **17** and **18** to the corresponding cyclobutanes, there is considerable stereospecificity. It is clear that a common intermediate is not reached in the reaction. The diradical

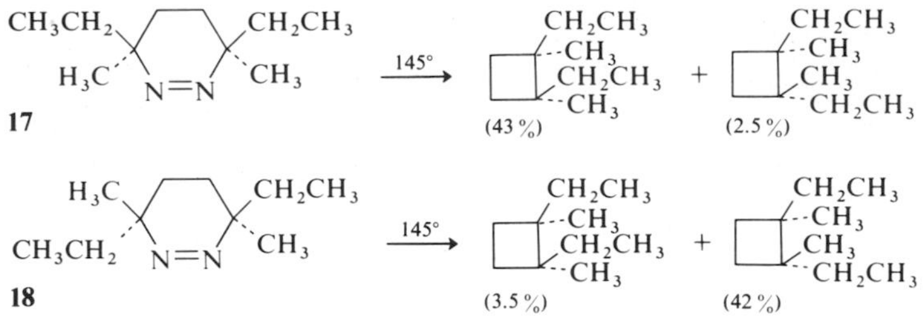

intermediate, which because of spin conservation in the loss of nitrogen is initially formed as a singlet, is apparently able to close to give the cyclobutane ring before undergoing the conformational changes that interconvert the two diradicals.[77] Ring closure seems generally to be faster than conformational interconversion of singlet

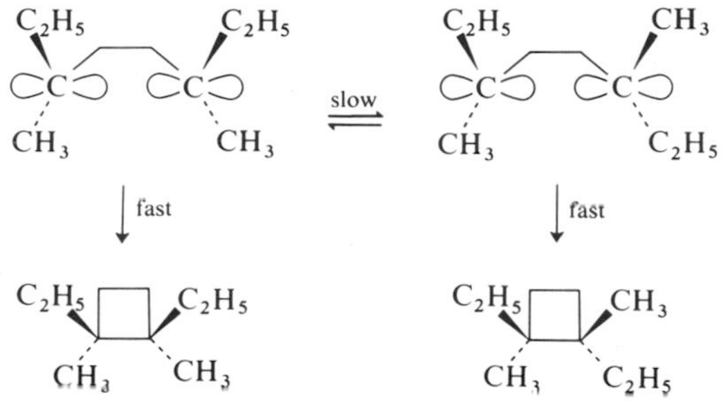

1,4-diradicals generated in an exoenergetic reaction step, but not for 1,4-diradicals at their energy minima.[78]

Decompositions of cyclic five-membered azo compounds show only modest stereospecificity. In some cases, the stereochemistry is inverted relative to that of the starting material. An explanation for this phenomenon in terms of orbital-symmetry

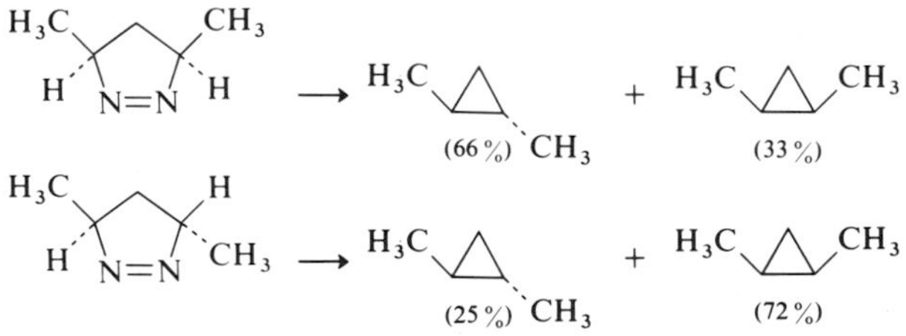

Ref. 79

principles pertaining to the formation and cyclization of the 1,3-diradical intermediate has been offered,[80] but since the stereospecificity is not very high, it is clear

77. P. D. Bartlett and N. A. Porter, *J. Am. Chem. Soc.* **90**, 5317 (1968).
78. L. M. Stephenson and J. I. Brauman, *J. Am. Chem. Soc.* **93**, 1988 (1971).
79. R. J. Crawford and A. Mishra, *J. Am. Chem. Soc.* **88**, 3963 (1966).
80. R. Hoffmann, *J. Am. Chem. Soc.* **90**, 1475 (1968).

241

SECTION 6.4.
UNIMOLECULAR
THERMAL
ELIMINATION
REACTIONS

Scheme 6.12. Photochemical and Thermal Decomposition of Cyclic Azo Compounds

1[a]

CH_2N_2 ... (major) + (minor)

2[b] $CF_3CH{=}CH_2 + CH_2N_2 \longrightarrow$ ... $\xrightarrow{260°}$ (51%)

3[c]

4[d] (74%)

5[e]

6[f]

7[g] (35%)

a. T. V. Van Auken and K. L. Rinehart, Jr., *J. Am. Chem. Soc.* **84**, 3736 (1962).
b. F. Misani, L. Speers, and A. M. Lyon, *J. Am. Chem. Soc.* **78**, 2801 (1956).
c. J. P. Freeman, *J. Org. Chem.* **29**, 1379 (1964).
d. G. L. Closs, W. A. Böll, H. Heyn, and V. Dev, *J. Am. Chem. Soc.* **90**, 173 (1968).
e. C. G. Overberger, N. R. Byrd, and R. B. Mesrobian, *J. Am. Chem. Soc.* **78**, 1961 (1956).
f. R. Anet and F. A. L. Anet, *J. Am. Chem. Soc.* **86**, 525 (1964).
g. F. M. Moriarty, *J. Org. Chem.* **28**, 2385 (1963).

242

CHAPTER 6
CYCLOADDITIONS
AND
UNIMOLECULAR
REARRANGEMENTS
AND
ELIMINATIONS

that any control by orbital symmetry cannot be so complete as is observed in most situations where it is applicable.

Cyclic azo compounds also undergo elimination of nitrogen when subjected to photolysis. The photodecomposition can be carried out directly, or photosensitization techniques can be used.[81]

An important synthetic use of pyrazolines is in the formation of cyclopropanes. This particular ring system is often accessible by addition of diazo compounds to alkenes (subsection 6.1.2). Some examples of both thermal and photochemical processes of this sort are given in Scheme 6.12.

Cyclization is sometimes accompanied by formation of alkenes in thermal decomposition. For the eight-membered azo compound **19**, for example, alkene formation is the major reaction pathway. An intramolecular disproportionation via transition state **F** may account for the large amount of alkene formed in this particular case.[82]

6.4.2. β-Eliminations Involving Cyclic Transition States

A second important family of elimination reactions have as a common feature cyclic transition states through which there occurs intramolecular transfer of hydrogen, accompanied by departure of a small molecule and formation of an olefin. Scheme 6.13 depicts the most important examples of this family of reactions. These processes also share a second common feature: they are thermal, unimolecular reactions and do not require acidic or basic catalysts. There is, however, wide variation in the temperatures at which the reactions proceed at convenient rates.

The cyclic nature of the transition states in these reactions dictates that elimination will proceed with *syn* stereochemistry. A cyclic transition state involving only five or six atoms cannot accommodate an *anti* stereochemical relationship at the site of elimination. For this reason, this family of reactions is often referred to as *thermal syn eliminations*.

Amine oxide pyrolysis occurs under milder conditions than any of the other

81. P. S. Engel, *J. Am. Chem. Soc.* **91**, 6903 (1969); P. D. Bartlett and P. S. Engel, *J. Am. Chem. Soc.* **90**, 2960 (1968).
82. C. G. Overberger and J. W. Stoddard, *J. Am. Chem. Soc.* **92**, 4922 (1970).

Scheme 6.13. Eliminations via Cyclic Transition States

243

SECTION 6.4.
UNIMOLECULAR
THERMAL
ELIMINATION
REACTIONS

Substrate	Transition state	Product	Temp. range	Ref.
$\overset{H}{\underset{R-CH-CHR}{\overset{\bar{O}-\overset{+}{N}(CH_3)_2}{|}}}$	(cyclic TS)	RCH=CHR + HON(CH$_3$)$_2$	100–150°	a
$\overset{CH_3}{\underset{R-CH-CHR}{\overset{|}{O=C-O-...H}}}$	(cyclic TS)	RCH=CHR + CH$_3$CO$_2$H	400–600°	b
$\overset{SCH_3}{\underset{R-CH-CHR}{\overset{|}{S=C-O-...H}}}$	(cyclic TS)	RCH=CHR + CH$_3$SH + SCO	150–250°	c

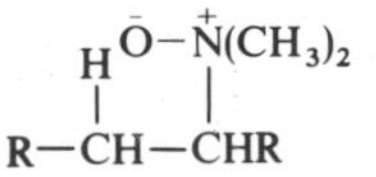

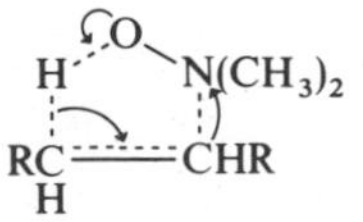

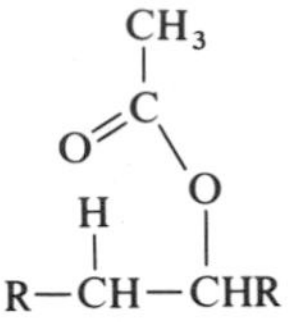

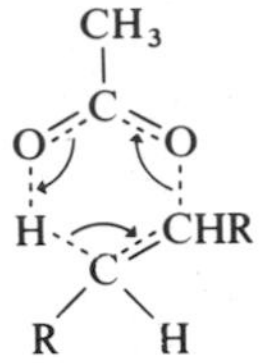

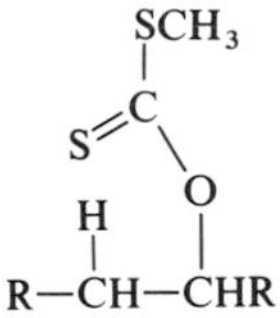

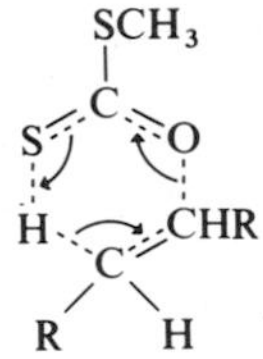

a. A. C. Cope and E. R. Trumbull, *Org. React.* **11**, 317 (1960).
b. C. H. DePuy and R. W. King, *Chem. Rev.* **60**, 431 (1961).
c. H. R. Nace, *Org. React.* **12**, 57 (1962).

common pyrolytic eliminations, and is therefore particularly valuable when a very sensitive olefin must be prepared by an elimination process. The reaction will proceed at room temperature for some amine oxides in dimethyl sulfoxide.[83] If more than one type of β-hydrogen is present in the amine oxide, a mixture of olefins will generally result. Some typical examples are shown in Scheme 6.14.

Under the mild conditions of the reaction, there is no equilibration of the alkenes, so the product composition is governed by the relative stabilities of the various transition states. Usually, more of the *trans* olefin than of its *cis* isomer is formed, presumably because steric repulsion raises the energy of the transition state leading to *cis* olefin. The selectivity is not high, however, since the ratio of *trans:cis* from some simple cases is in the range 3:1 to 2:1. In cyclic systems, conformational

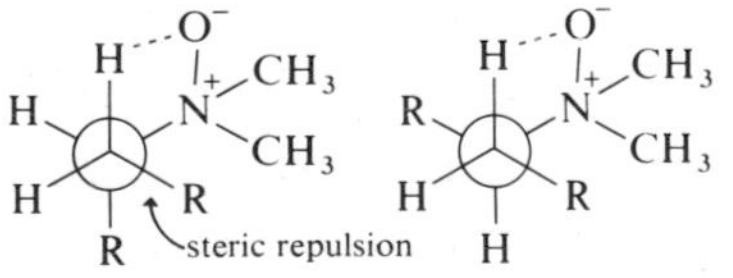

less favorable   more favorable

83. D. J. Cram, M. R. V. Sahyun, and G. R. Knox, *J. Am. Chem. Soc.* **84**, 1734 (1962).

Scheme 6.14. Thermal Eliminations

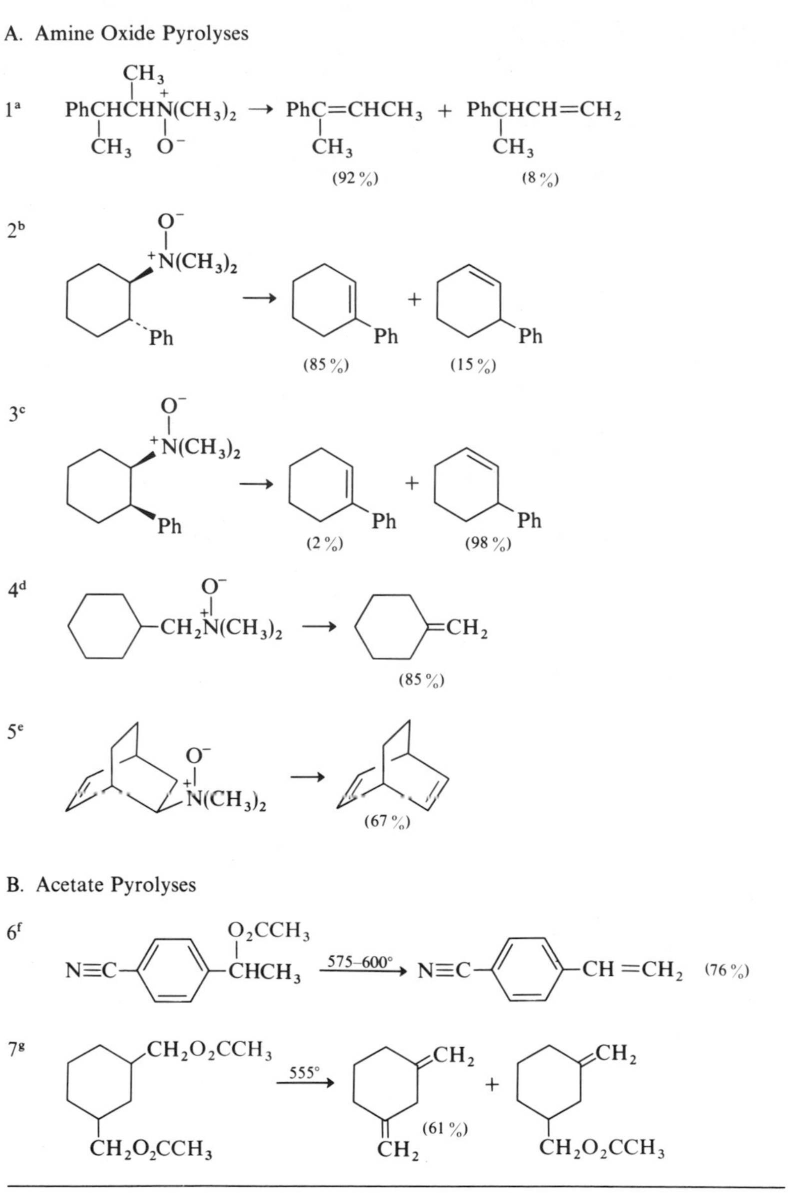

CHAPTER 6
CYCLOADDITIONS
AND
UNIMOLECULAR
REARRANGEMENTS
AND
ELIMINATIONS

A. Amine Oxide Pyrolyses

1[a]   PhCHCHN(CH$_3$)$_2$ → PhC=CHCH$_3$ + PhCHCH=CH$_2$

(92%)   (8%)

2[b]

(85%)   (15%)

3[c]

(2%)   (98%)

4[d]

(85%)

5[e]

(67%)

B. Acetate Pyrolyses

6[f]

N≡C—⟨ ⟩—CHCH$_3$ $\xrightarrow{575-600°}$ N≡C—⟨ ⟩—CH=CH$_2$ (76%)

7[g]

$\xrightarrow{555°}$

(61%)

effects and the requirement for a cyclic transition state are the most important factors in determining product composition. Elimination to give an olefin conjugated with an aromatic ring is especially favorable. The increased acidity of the hydrogen alpha to the phenyl group and the conjugation that develops in the transition state are believed to be responsible for this result. Entry 1 in Scheme 6.14 is an example of the importance of this factor in determining product composition.

via Cyclic Transition States

245

SECTION 6.4.
UNIMOLECULAR
THERMAL
ELIMINATION
REACTIONS

8[h]

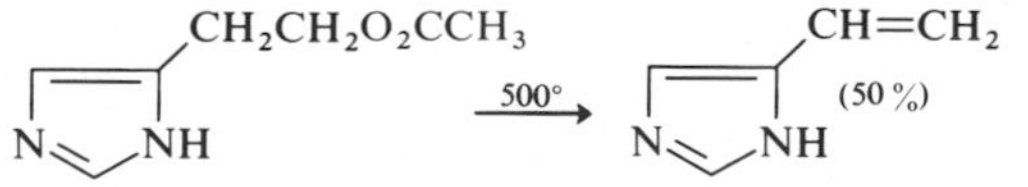

9[i]

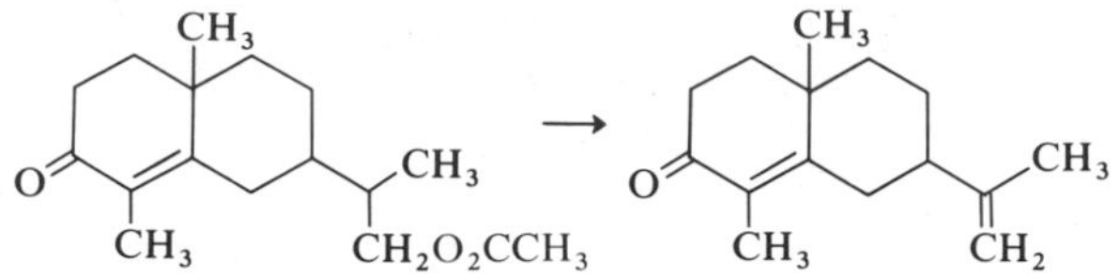

C. Xanthate Ester Pyrolyses

10[j] $PhCHCHCH_3$ $\xrightarrow[\substack{3)\ CH_3I \\ 4)\ \Delta}]{\substack{1)\ K \\ 2)\ CS_2}}$ $PhC=CHCH_3$ (91%)

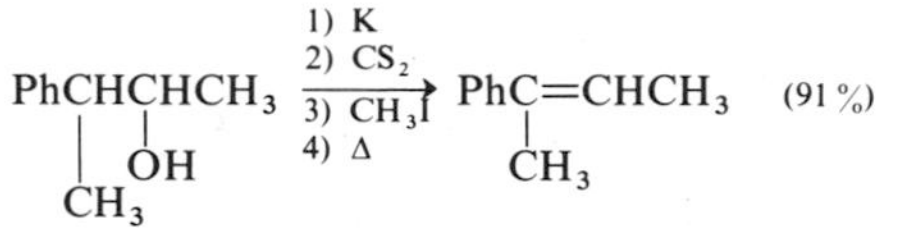

11[k]

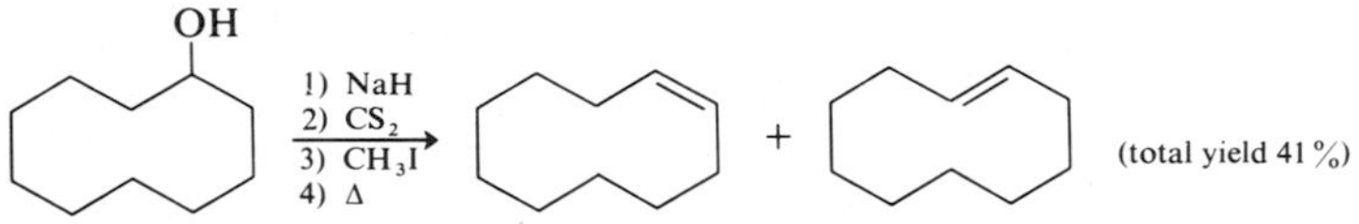

12[l]

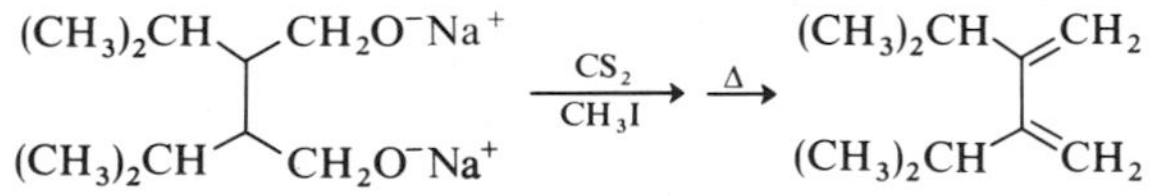

13[m]

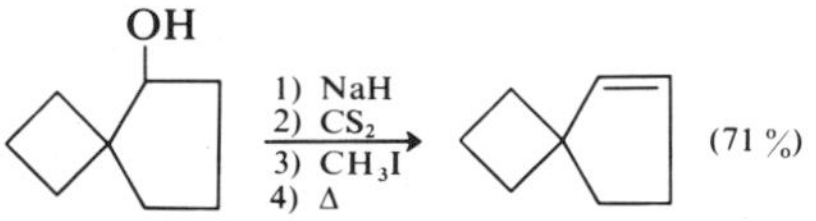

a. D. J. Cram and J. E. McCarty, *J. Am. Chem. Soc.* **76**, 5740 (1954).
b. A. C. Cope and C. L. Bumgardner, *J. Am. Chem. Soc.* **79**, 960 (1957).
c. A. C. Cope, C. L. Bumgardner, and E. C. Schweizer, *J. Am. Chem. Soc.* **79**, 4729 (1957).
d. A. C. Cope, E. Ciganek, and N. A. LeBel, *J. Am. Chem. Soc.* **81**, 2799 (1959).
e. A. C. Cope and C. L. Bumgardner, *J. Am. Chem. Soc.* **78**, 2812 (1956).
f. C. G. Overberger and R. E. Allen, *J. Am. Chem. Soc.* **68**, 722 (1946).
g. W. J. Bailey and J. Economy, *J. Org. Chem.* **23**, 1002 (1958).
h. C. G. Overberger and N. Vorchheimer, *J. Am. Chem. Soc.* **85**, 951 (1963).
i. E. Piers and K. F. Cheng, *Can. J. Chem.* **46**, 377 (1968).
j. D. J. Cram, *J. Am. Chem. Soc.* **71**, 3883 (1949).
k. A. T. Blomquist and A. Goldstein, *J. Am. Chem. Soc.* **77**, 1001 (1955).
l. A. de Groot, B. Evenhuis, and H. Wynberg, *J. Org. Chem.* **33**, 2214 (1968).
m. C. F. Wilcox, Jr., and G. C. Whitney, *J. Org. Chem.* **32**, 2933 (1967).

The pyrolysis of esters has usually been done with acetate esters. The thermal requirement for the reaction is very high, with temperatures above 500° usually required. The pyrolysis is thus a vapor-phase reaction. In the laboratory, this is usually accomplished in a packed glass tube heated with a small furnace. The reacting vapors and product are swept through the hot chamber at an appropriate rate by an inert gas such as nitrogen, and into a cold trap or other system for condensation.

246

CHAPTER 6
CYCLOADDITIONS
AND
UNIMOLECULAR
REARRANGEMENTS
AND
ELIMINATIONS

Similar reactions occur with esters derived from long-chain acids; if the boiling point of the ester is high enough, the reaction can be carried out in the liquid phase. Vapor-phase acetate pyrolysis is, however, the most generally used procedure.

That the elimination is *syn* has been established by use of deuterium labels. Deuterium was introduced stereospecifically by LiAlD$_4$ reduction of *cis*- and *trans*-stilbene oxide. The product of the subsequent ester pyrolysis is *trans*-stilbene because of eclipsing effects in the transition state. The *syn* elimination is demonstrated by retention of deuterium in the olefin from *trans*-stilbene oxide and its absence in the olefin from *cis*-stilbene oxide.[84]

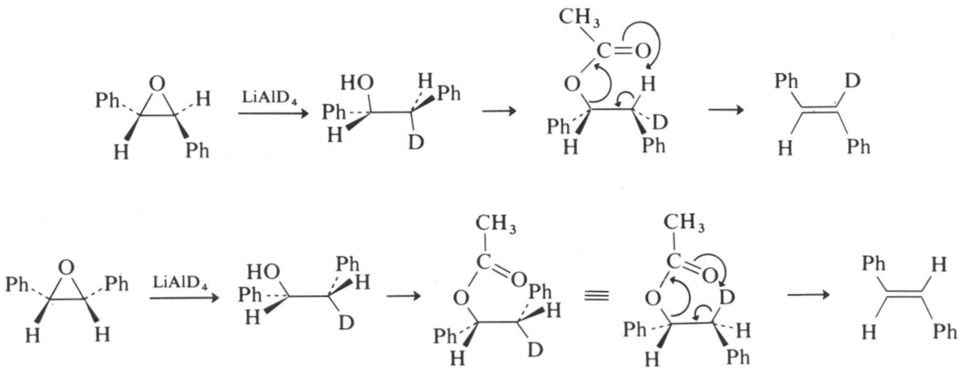

As with the amine oxides, mixtures of olefins are obtained when more than one type of β-hydrogen is present. In noncyclic compounds, the olefin composition often approaches that expected on a statistical basis from the number of each type of hydrogen. The *trans* olefin usually predominates over the *cis* for a given isomeric pair. In cyclic systems, conformational features, ring strain, and related factors usually distort the mixture from that expected on a statistical basis. Elimination in a direction in which the *syn* mechanism can operate is strongly preferred over elimination in a direction where this is impossible.

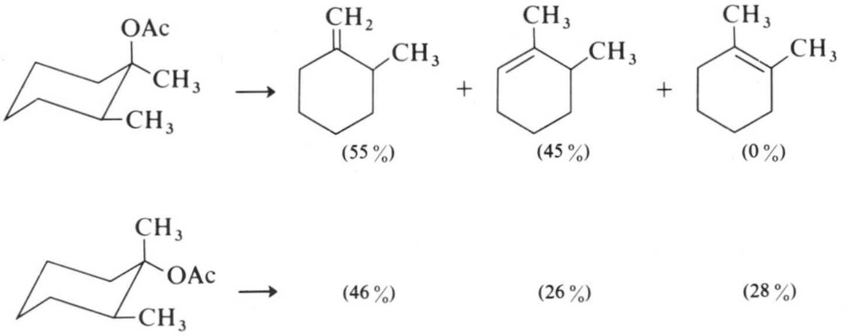

Alcohols can be dehydrated via xanthate esters at temperatures much lower

84. D. Y. Curtin and D. B. Kellom, *J. Am. Chem. Soc.* **75**, 6011 (1953).
85. D. H. Froemsdorf, C. H. Collins, G. S. Hammond, and C. H. DePuy, *J. Am. Chem. Soc.* **81**, 643 (1959).

than those required for acetate pyrolysis. The preparation of xanthate esters involves reaction of the sodium alkoxide with carbon disulfide. The resulting salt is alkylated

$$ROH + Na \rightarrow RO^-Na^+$$

$$RO^-Na^+ + CS_2 \rightarrow RO\overset{\overset{S}{\|}}{C}S^-Na^+$$

$$RO\overset{\overset{S}{\|}}{C}S^- + CH_3I \rightarrow RO\overset{\overset{S}{\|}}{C}SCH_3$$

with methyl iodide to complete the transformation. The elimination step is often effected by simple distillation:

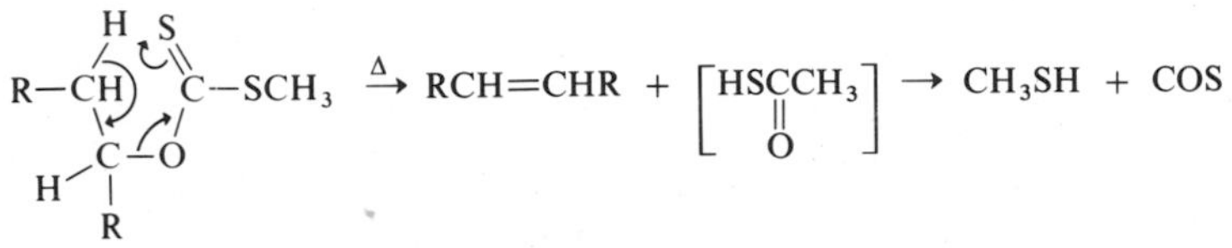

$$R-CH\begin{matrix} H \ S \\ \diagup \ \diagup \\ \end{matrix}C-SCH_3 \xrightarrow{\Delta} RCH{=}CHR + \left[HS\overset{\overset{S}{\|}}{C}CH_3 \right] \rightarrow CH_3SH + COS$$

As with the other *syn* thermal eliminations, there are no intermediates prone to undergo skeletal rearrangement. The principal application of the xanthate method has been in situations in which acid-catalyzed alcohol dehydration would be accompanied by skeletal rearrangement, or in which a very sensitive olefin is being produced.

General References

Cycloaddition Reactions

J. J. Vollmer and K. L. Servis, *J. Chem. Educ.* **47**, 491 (1970).
R. Huisgen, R. Grashey, and J. Sauer, in *The Chemistry of Alkenes*, S. Patai (ed.), Interscience, New York, NY, 1964, pp. 739–953.

Thermal Rearrangements

T. S. Stevens and W. E. Watts, *Selected Molecular Rearrangements*, Van Nostrand Reinhold, London, 1973, Chapter 8.
H.-J. Hansen, in *Mechanisms of Molecular Migrations*, Vol. 3, B. S. Thyagarajan (ed.), Wiley–Interscience, New York, NY, 1971, pp. 177–236.
S. J. Rhoads and N. R. Raulins, *Org. React.* **22**, 1 (1975).

Thermal Eliminations

W. H. Saunders, Jr., and A. F. Cockerill, *Mechanisms of Elimination Reactions*, Wiley, New York, NY, 1973, Chapter VIII.

Problems

CHAPTER 6
CYCLOADDITIONS
AND
UNIMOLECULAR
REARRANGEMENTS
AND
ELIMINATIONS

(*References for these problems will be found on page 501.*)

1. The strained alkene bicyclo[2.1.0]pent-2-ene gives a Diels–Alder adduct with cyclopentadiene. Draw the structure of the *exo* and *endo* adducts. Suggest a means of determining which stereoisomer is which.

2. Reaction of α-pyrone (**A**) with methyl acrylate at reflux for extended periods gives a mixture of **B** and other isomers. Account for the formation of this product.

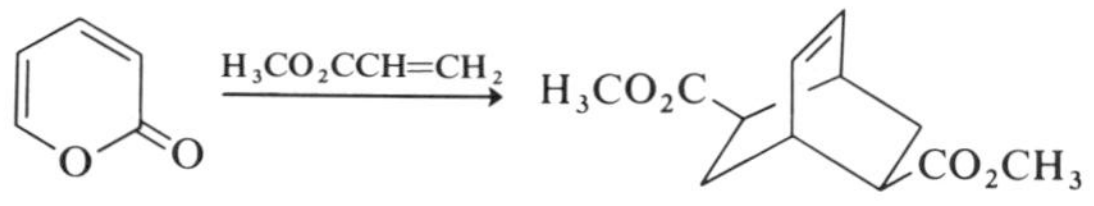

3. Predict the products of thermal rearrangement of the following compounds. Specify the stereochemistry of the expected product.

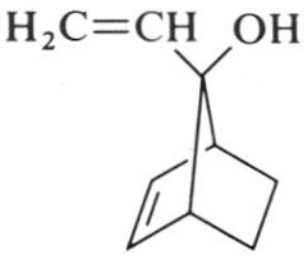

(a)

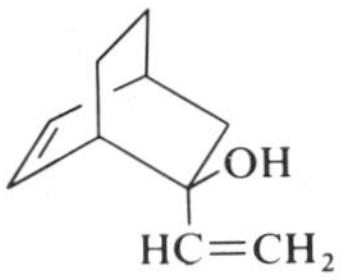

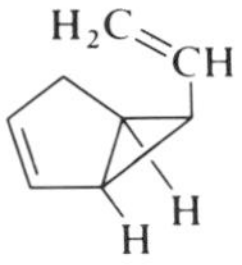

(b) H₂C=CH OH

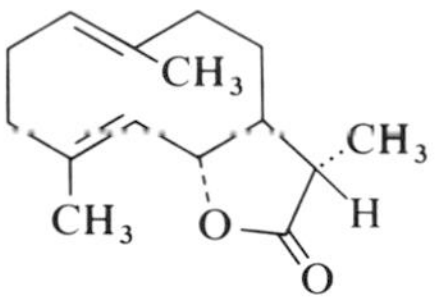

(c)

(d)

(e)

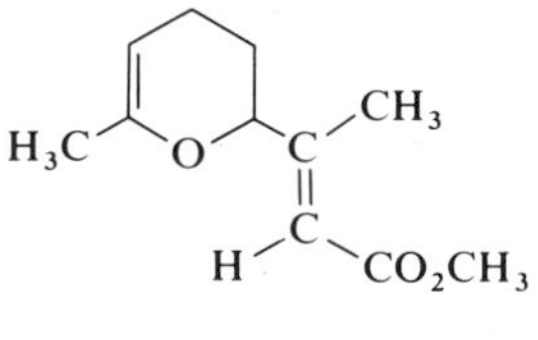

(f)

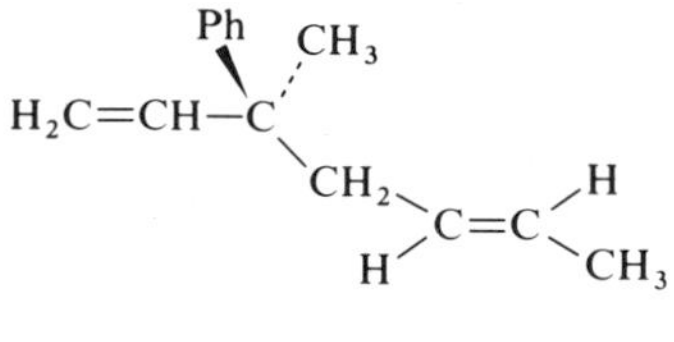

4. The dihydrofuran **A** gives 4-methyl-4-cycloheptenone on heating in the range 140–200°. Account for this transformation.

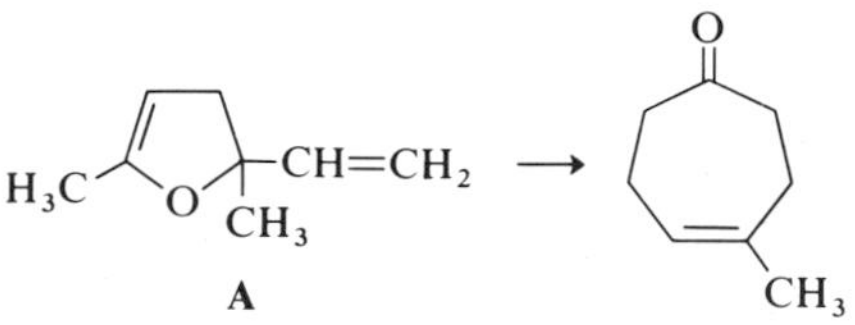

5. A route to hasubanan alkaloids has been described involving reaction of 1-butadienyl phenyl sulfoxide with the tetrahydrobenzindole **A**. Treatment of the resulting adduct with sodium sulfide in refluxing methanol gave **C**. Suggest a structure for **B**, and rationalize the formation of **C** from **B**.

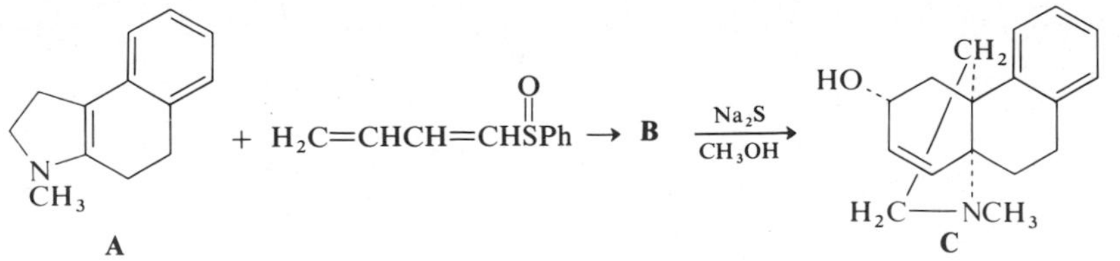

$$+ \ H_2C=CHCH=CHSPh \ \rightarrow \ B \ \xrightarrow[\text{CH}_3\text{OH}]{\text{Na}_2\text{S}}$$

6. Irradiation of the dienone **A** generates three isomeric saturated ketones, all of which contain cyclobutane rings. Postulate reasonable structures.

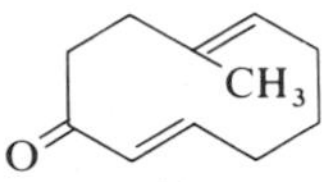

7. Show how each of the following dienophiles or dienes could be used in a Diels–Alder reaction to give a precursor for the final synthetic objective. Outline a series of reactions that could be utilized to convert the initial Diels–Alder adduct to the desired product.
   (a) 4-methylenecyclohexenes using α-bromoacrolein as the dienophile
   (b) bicyclic ketones using α-chloroacetonitrile as the dienophile
   (c) 4-acetylcyclohex-3-enone using 1-methoxy-3-trimethylsilyloxybutadiene as the diene
   (d) 2,2′,3,3′,4,4′,5,5′-octaphenyldiphenyl using tetraphenylcyclopentadienone as the diene.

8. Irradiation of a mixture of cyclohexene and 5,5-dimethylcyclohexane-1,3-dione gives **A**. Formulate a mechanism.

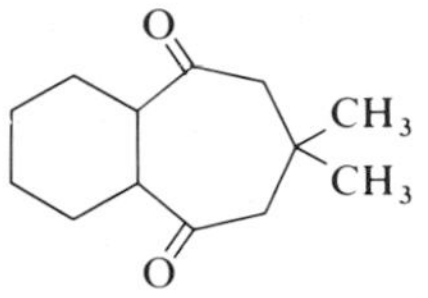

9. Assuming that the usual stereospecificity of the Diels–Alder reaction is operative, what steroisomers would be possible products for the intramolecular Diels–Alder reaction of compounds **A** and **B**?

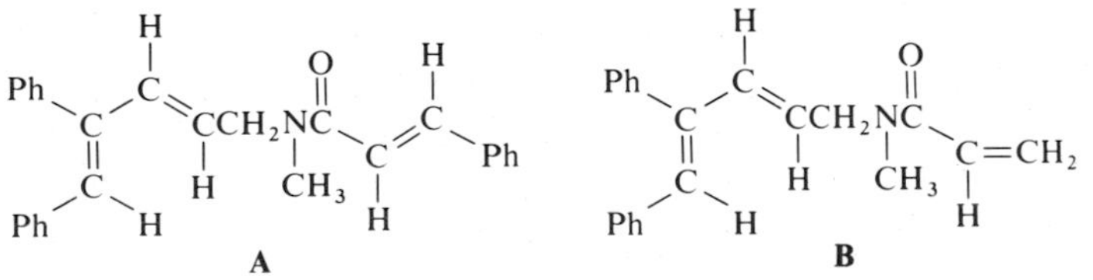

10. Irradiation of α-methylstyrene and biacetyl (butane-2,3-dione) gives the expected isomeric oxetanes in 48% yield. Also formed is compound **A**. When the methyl group of the α-methylstyrene is replaced by a CD_3 group, the

250

CHAPTER 6
CYCLOADDITIONS
AND
UNIMOLECULAR
REARRANGEMENTS
AND
ELIMINATIONS

distribution of deuterium in **A** is as shown. Account for the formation of **A** and the observed distribution of label.

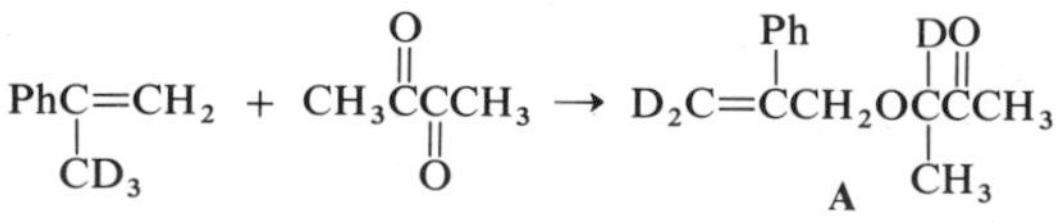

$$PhC{=}CH_2 \; + \; CH_3\overset{O}{\underset{\parallel}{C}}CCH_3 \; \rightarrow \; D_2C{=}\overset{Ph}{\underset{CH_3}{C}}CH_2O\overset{DO}{\underset{\parallel}{CC}}CH_3$$

$$\underset{CD_3}{|} \qquad\qquad \underset{O}{\parallel} \qquad\qquad\qquad \underset{A}{} \quad$$

11. Cyclopentadiene gives adducts with unsymmetrical disubstituted ketenes in which the larger ketene substituent occupies an *endo* position. Offer an explanation for this stereochemistry.

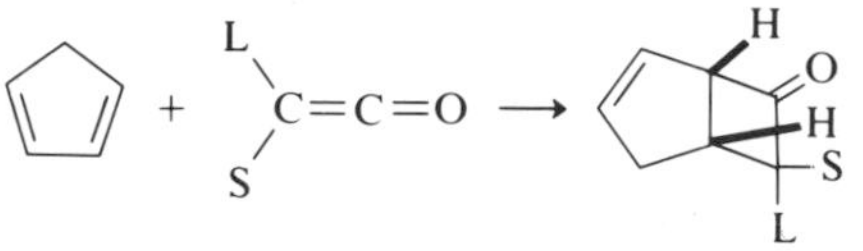

12. Explain why azo compound **C** gives diene **A** rather than **B**.

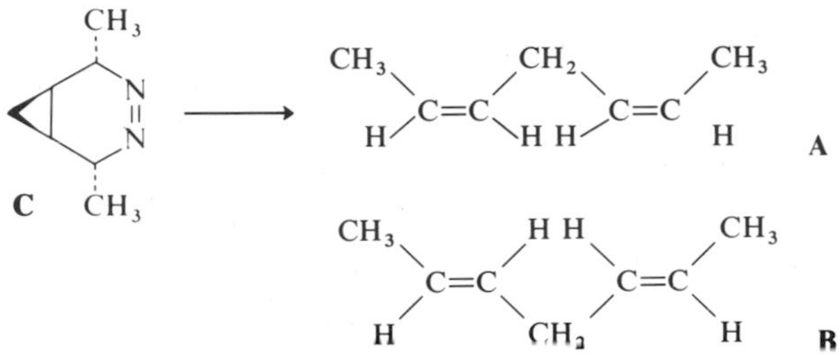

13. Irradiation of *o*-methylbenzaldehyde in the presence of maleic anhydride gives **A**. The same compound is obtained when **B** is heated with maleic anhydride. Both reactions give only the stereoisomer shown. Formulate a mechanism.

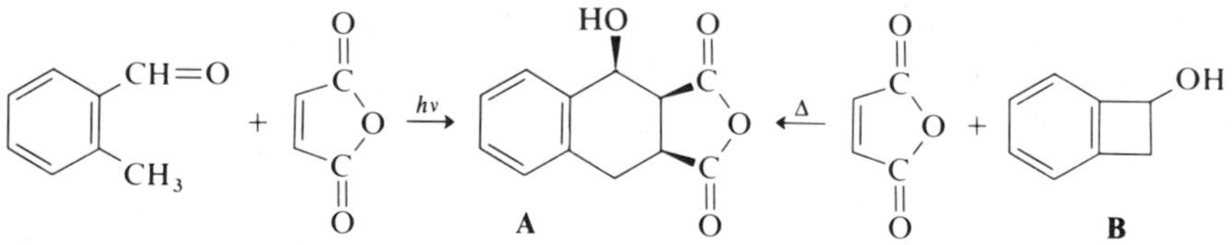

14. A solution of but-2-enal, 2-acetoxypropene, and dimethyl acetylenedicarboxylate refluxed in the presence of a small amount of an acidic catalyst gives an 80% yield of dimethyl phthalate. Explain the course of this reaction.

15. Azlactone **A** reacts with dimethyl acetylenedicarboxylate in refluxing xylene, evolving carbon dioxide and yielding pyrrole **B**. In the absence of dimethyl acetylenedicarboxylate, **A** does not evolve CO_2 under similar conditions. Compound **C** does not give an adduct with dimethyl acetylenedicarboxylate. Suggest a mechanism for the reaction that is consistent with these facts.

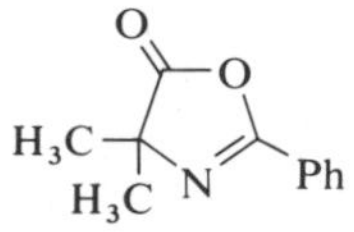

16. The terpene pregeijerene rearranges to geijerene when heated. Depict the conformation of pregeijerene that would permit this rearrangement.

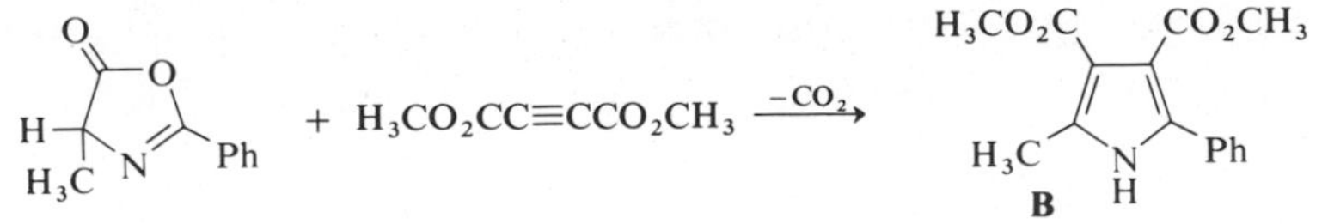

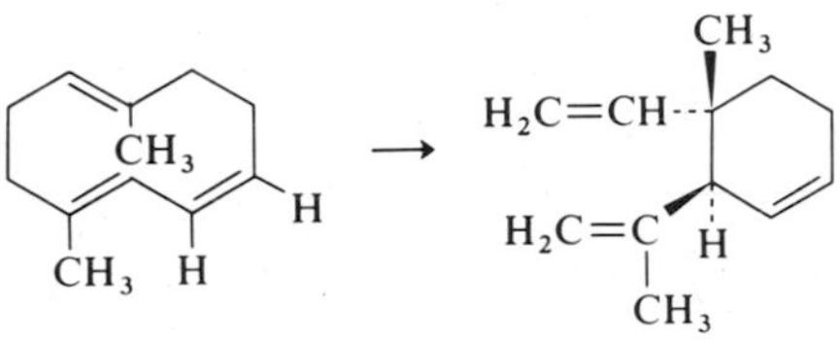

17. (a) 1,2,4,5-Tetrazines react with alkenes to give dihydropyridazines, as illustrated in the equation below. Suggest a mechanism.

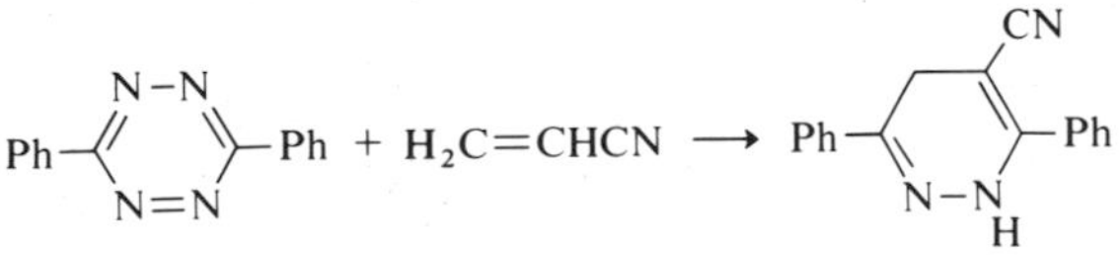

(b) Compounds **A** and **B** are both unstable toward loss of nitrogen at room temperature. Both compounds give **C** as the product of decomposition. Account for the formation of **C**.

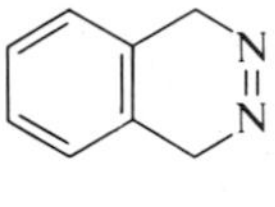

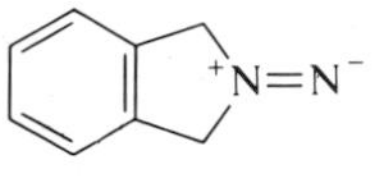

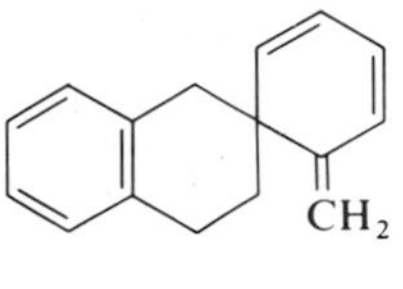

18. The product mixtures resulting from pyrolysis of the stereoisomeric pyrazolines **A** and **B** have been compared. Both compounds give a 1:1 mixture of *cis*- and *trans*-2-deuteriomethylcyclopropane. What restrictions are placed on the mechanism by this result? Formulate a mechanism that is consistent with the data.

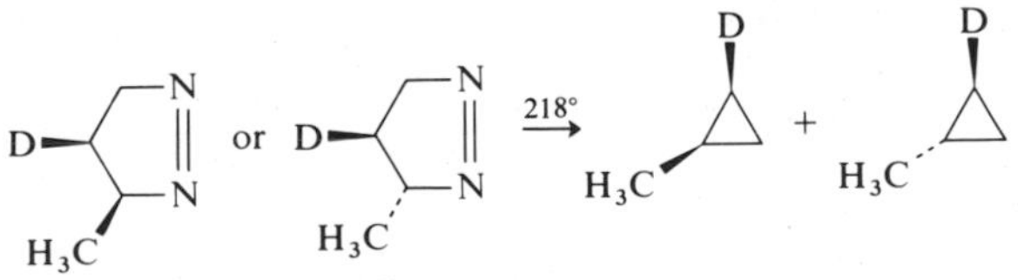

252

CHAPTER 6
CYCLOADDITIONS
AND
UNIMOLECULAR
REARRANGEMENTS
AND
ELIMINATIONS

19. Show how the indicated starting materials can be elaborated to provide substrates that would fragment thermally to produce the desired reactive species:

(a)

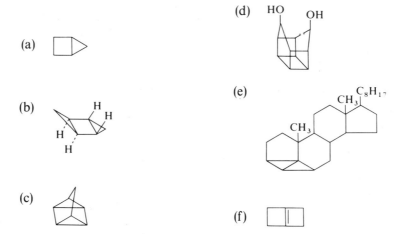

as a precursor to $(CH_3)_2Si=CH_2$

(b)

and $(CH_3O)_3P$ as precursors to CH_3OP

20. Photolysis of **A** gives an isomeric compound **B** in 83% yield. Alkaline hydrolysis of **B** affords a hydroxy carboxylic acid **C**, $C_{25}H_{32}O_4$. Treatment of **B** with silica gel in hexane yields **D**, $C_{24}H_{28}O_2$. **D** is converted by sodium periodate–potassium permanganate to a mixture of **E** and **F**. What are the structures of **B**, **C**, and **D**?

$-CO_2(CH_2)_9CH=CH_2$

A

$-CO_2H$ $HO(CH_2)_9CO_2H$

E   **F**

21. Suggest a means of synthesizing each of the following strained-ring compounds:

(a)

(b)

(c)

(d)   HO   OH

(e)

(f)

22. Suggest sequences of reactions for accomplishing each of the following synthetic transformations:

(a) squalene from succinaldehyde, isopropyl bromide, and 3-methoxy-2-methyl-1,3-butadiene

(b)

$(CH_3)_2N$ HO

from

$CH=O$

(c)

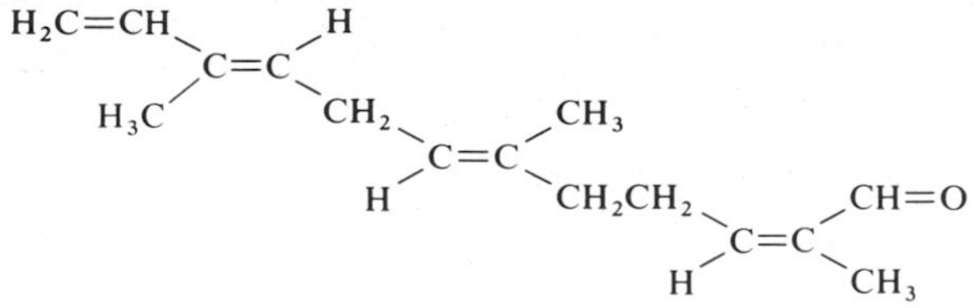

$H_2C=CH$, $C=C$, H, H_3C, CH_2, CH_3, $C=C$, H, CH_2CH_2, $C=C$, $CH=O$, H, CH_3

from

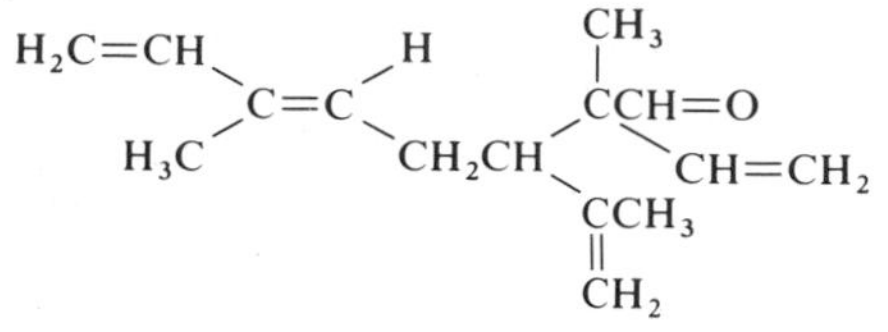

$H_2C=CH$, $C=C$, H, H_3C, CH_2CH, CH_3, $CCH=O$, $CH=CH_2$, CCH_3, CH_2

(d) $H_5C_2O_2CCH_2CH_2\overset{H_5C_2}{\underset{}{C}}=CHCH_2CH_2Cl$ from $H_2C=\overset{H_5C_2}{\underset{OH}{C}}CHCH_2CH_2Cl$

(e)

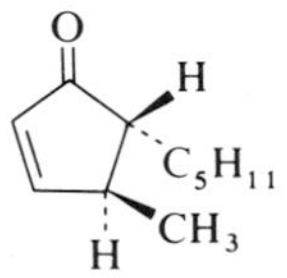

O, H, C_5H_{11}, from, $H_5C_2O_2CO_2CH_2\overset{H_5C_2O}{\underset{}{C}}=CHC_5H_{11}$, CH_3, H

(f)

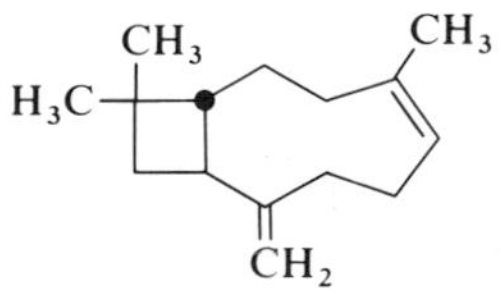

CH_3, H_3C, CH_3, CH_2 from CH_3, H_3C, CH_3, CO_2CH_3, O, O

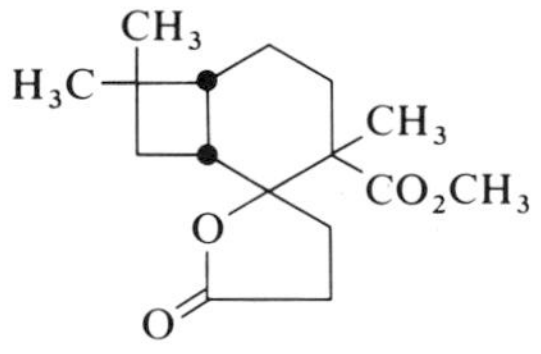

(g)

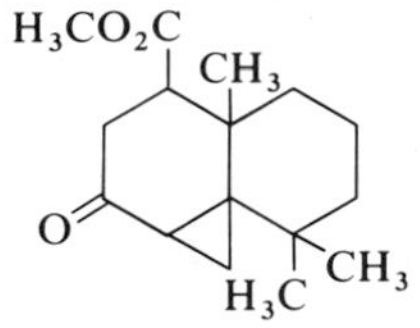

H_3CO_2C, CH_3, O, CH_3, H_3C from CO_2CH_3, HO_2CCH_2CH, CH_3, H_2C, H_3C CH_3

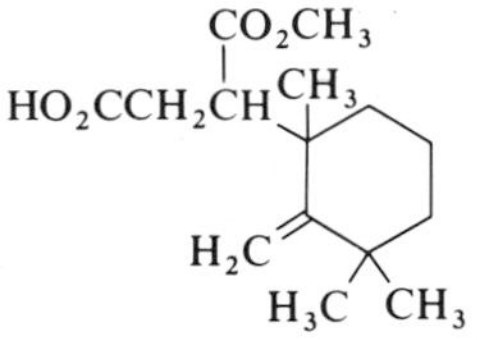

(h) $(CH_3)_2C=CHCH\overset{CH_3}{\underset{CH=CH_2}{C}}=CH_2$ from $(CH_3)_2C=CHCH=CHCH_2OH$

(i) $(CH_3)_2CHCH_2\overset{O}{\overset{\|}{C}}CH_2\overset{CH_3}{\underset{}{C}}=CH_2$ from $H_2C=CHOCH_2CH=CH_2$ and $(CH_3)_2CHBr$

(j)

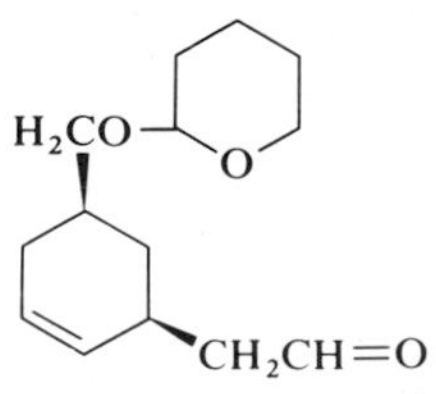

H_2CO, O, $CH_2CH=O$ from H_2CO, O, HO

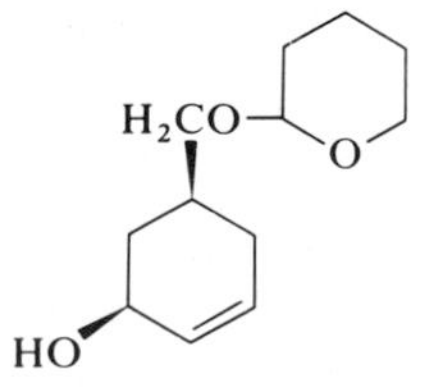

254

CHAPTER 6
CYCLOADDITIONS
AND
UNIMOLECULAR
REARRANGEMENTS
AND
ELIMINATIONS

(k)

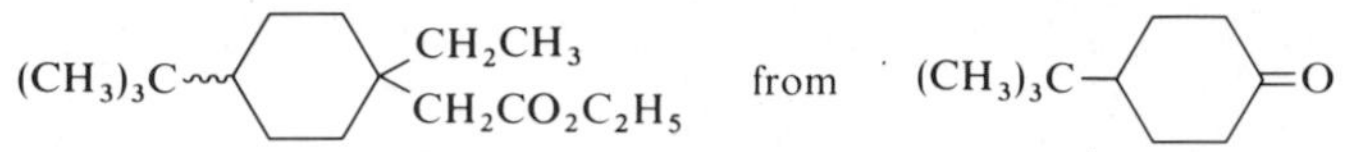

23. Vinylcyclopropane, when irradiated with benzophenone or benzaldehyde, gives a mixture of two types of products. Suggest the mechanism by which product of type **B** is formed.

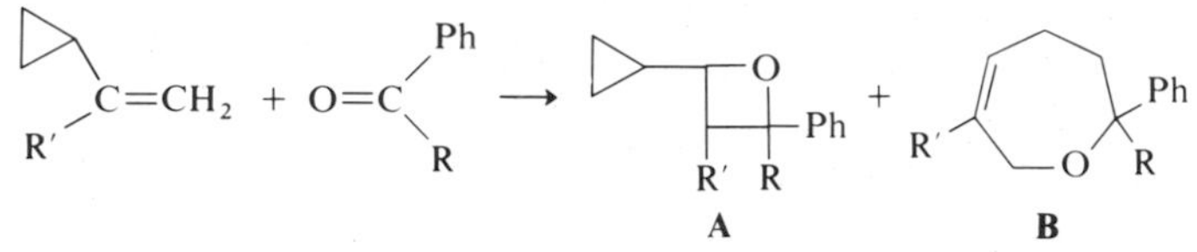

24. The addition reaction of tetracyanoethylene and ethyl vinyl ether in acetone gives 94% of the 2+2 adduct and 6% of an adduct having the composition: tetracyanoethylene + ethyl vinyl ether + acetone. If the 2+2 adduct is kept in contact with acetone for several days, it is completely converted to the minor product. Suggest a structure for this product, and indicate its mode of formation (a) in the initial reaction and (b) on standing in acetone.

25. A convenient preparation of 2-allylcyclohexanone involves simply heating the diallyl ketal of cyclohexanone in toluene containing a trace of p-toluenesulfonic acid and collecting a distillate consisting of toluene and allyl alcohol. Distillation of the residue gives a 90% yield of 2-allylcyclohexanone. Outline the mechanism of this reaction.

26. Outline syntheses that could be expected to provide the desired product from the starting materials specified:
  (a) hexaphenylbenzene from benzil, dibenzyl ketone, and diphenylacetylene

  (b)

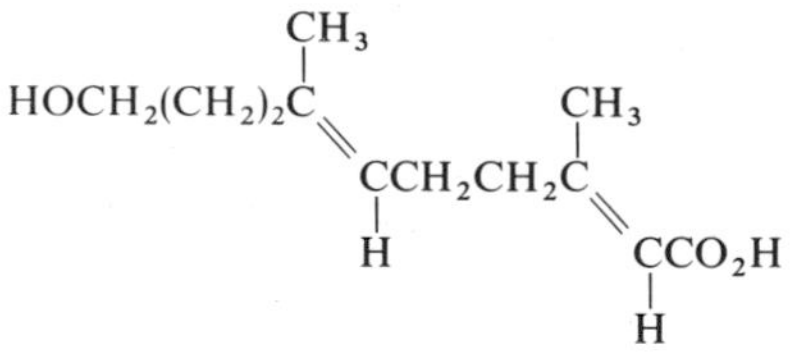

  from 1,1-dimethoxypentan-4-one, triethyl phosphonoacetate, 2-bromo-propene, and methyl orthoacetate

  (c) 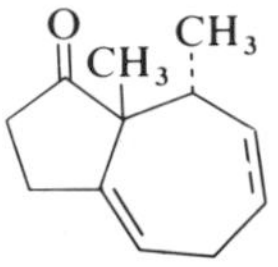 from 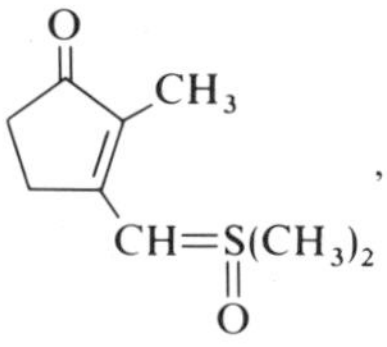 , acrolein, and ethyl iodide

27. Reaction of an allylic halide such as **A** with N-cyanomethylpyrrolidine, followed by treatment with potassium t-butoxide and then aqueous hydrolysis, is a

general synthesis of β,γ-unsaturated aldehydes. Indicate a mechanism by which this reaction might occur.

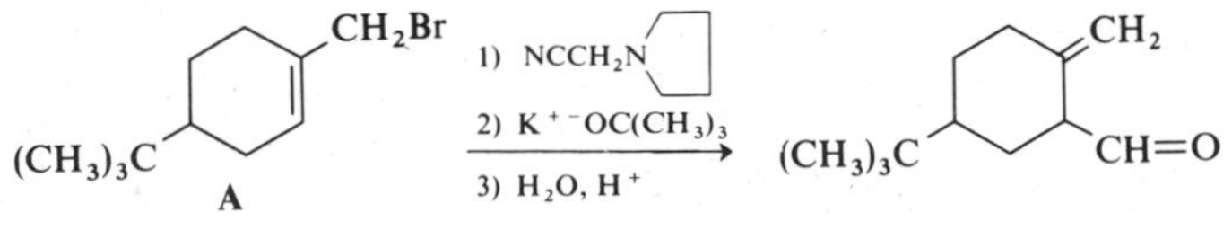

7

Aromatic Substitution Reactions

This chapter is concerned with reactions that introduce or interchange substituent groups on aromatic rings. The most important group of such reactions are the electrophilic aromatic substitutions, but there are also significant reactions that take place by nucleophilic substitution mechanisms, and still others that involve radical mechanisms. Examples of synthetically important reactions from each group will be discussed. Electrophilic aromatic substitution has also been studied in great detail from the point of view of reaction mechanism and structure–reactivity relationships; these mechanistic studies received considerable attention in Part A, Chapter 9. In this chapter, the synthetic aspects of electrophilic aromatic substitutions will be emphasized.

7.1. Electrophilic Aromatic Substitution

7.1.1. Nitration

Aromatic nitration has been an important reaction, from the standpoint both of synthesis and of understanding reaction mechanisms, since the early days of organic chemistry. Synthetically, it provides a route for introduction of amino groups onto aromatic rings, since the nitro group can be readily reduced to the amino function. As will be seen in subsection 7.2.1, the amino group, in turn, provides an entry for many other important functional groups. From the point of view of reaction mechanism, nitration was important in the early work that established the patterns of substituent-directing effects.

There are several reaction systems that are capable of effecting nitration. A major factor in the choice of reagent is the reactivity of the aromatic ring to be nitrated. Concentrated nitric acid can effect nitration, but it is not nearly so reactive as mixtures of nitric acid with sulfuric acid. In both media, the active nitrating species is the nitronium ion. A variety of physical measurements provide evidence for the existence of this species and permit estimation of its concentration under some conditions. In concentrated sulfuric acid, the dissolution of nitric acid results in the formation of nearly 4 ions per nitric acid molecule, as determined by freezing-point depression[1]:

$$HNO_3 + 2\,H_2SO_4 \rightleftarrows NO_2^+ + H_3O^+ + 2\,HSO_4^-$$

Such solutions also exhibit a Raman absorption band at 1400 cm^{-1}.[2] A similar band is observed for solutions of nitric acid in other strong acids, such as perchloric acid. The same band is present, but much weaker, in concentrated nitric acid. It is possible to prepare solid nitronium salts such as $NO_2^+ClO_4^-$, and here too this band is found.[3] This strongly supports assignment of the band to the NO_2^+ ion.

Nitration can also be carried out in organic solvents, of which acetic acid and nitromethane are perhaps the most common examples. In such solvents, the rate constants for nitration are often found to be zero-order in the aromatic substrate. The rate-controlling step is formation of the active nitrating species, the nitronium ion[4]:

$$2\,HNO_3 \rightleftarrows H_2NO_3^+ + NO_3^-$$
$$H_2NO_3^+ \xrightarrow{slow} NO_2^+ + H_2O$$
$$ArH + NO_2^+ \xrightarrow{fast} ArNO_2 + H^+$$

Another useful medium for nitration is a solution prepared by dissolving nitric acid in acetic anhydride. This is believed to generate acetyl nitrate:

$$HNO_3 + (CH_3CO)_2O \rightleftarrows CH_3\overset{O}{\overset{\|}{C}}ONO_2 + CH_3CO_2H$$

A useful feature of this reagent is that high ratios of *ortho* to *para* products are found for some substituted aromatics.[5] Finally, it possible to use nitronium-ion salts for nitration.[6] Nitronium fluoroborate has been extensively studied. The trifluoromethanesulfonate salt is also readily prepared and is an active nitrating

1. R. J. Gillespie, J. Graham, E. D. Hughes, C. K. Ingold, and E. R. A. Peeling, *J. Chem. Soc.*, 2504 (1950).
2. C. K. Ingold, D. J. Millen, and H. G. Poole, *J. Chem. Soc.*, 2576 (1950).
3. D. R. Goodard, E. D. Hughes, and C. K. Ingold, *J. Chem. Soc.*, 2559 (1950); D. J. Millen, *J. Chem. Soc.*, 2606 (1950).
4. E. D. Hughes, C. K. Ingold, and R. I. Reed, *J. Chem. Soc.*, 2400 (1950); J. G. Hoggett, R. B. Moodie, and K. Schofield, *J. Chem. Soc. B*, 1 (1969).
5. A. K. Sparks, *J. Org. Chem.* **31**, 2299 (1966).
6. S. J. Kuhn and G. A. Olah, *J. Am. Chem. Soc.* **83**, 4564 (1961); G. A. Olah and S. J. Kuhn, *J. Am. Chem. Soc.* **84**, 3684 (1962).

Scheme 7.1. Some Examples of Aromatic Nitration

1[a] CH_2CN phenyl → $\xrightarrow[H_2SO_4]{HNO_3}$ → p-NO_2-phenyl-CH_2CN (50–54%)

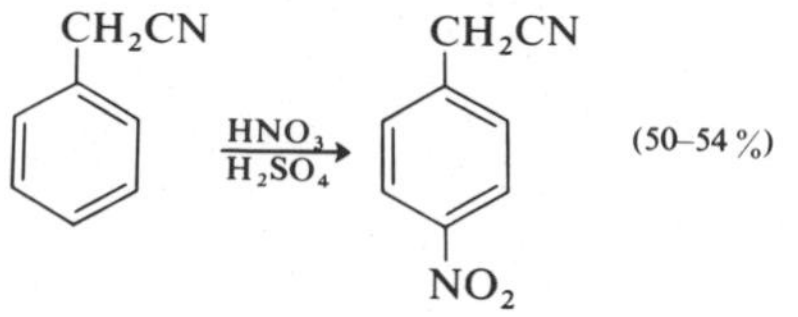

2[b] $N(CH_3)_2$ phenyl → $\xrightarrow[HNO_3]{H_2SO_4}$ → m-NO_2 (56–63%)

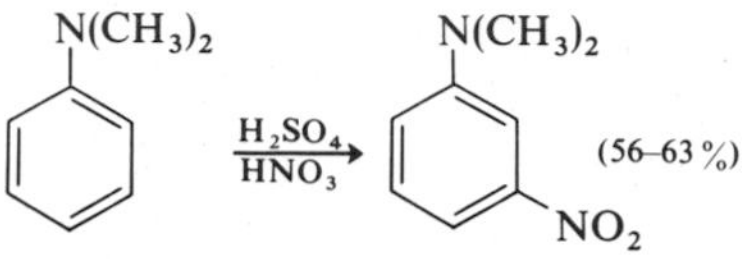

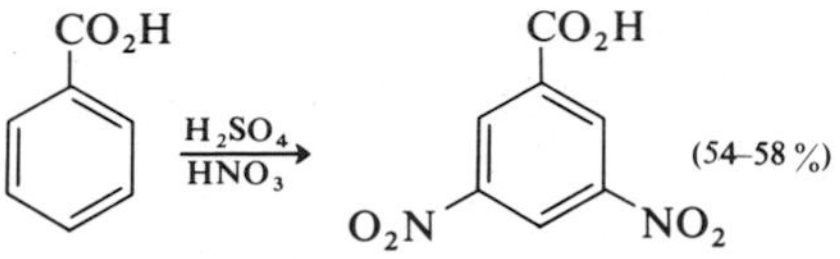

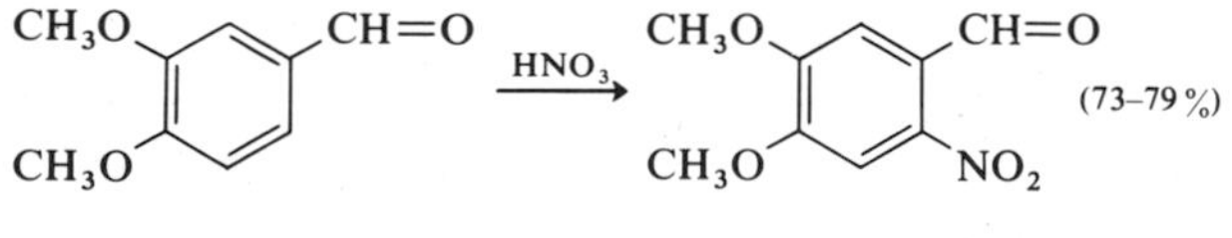

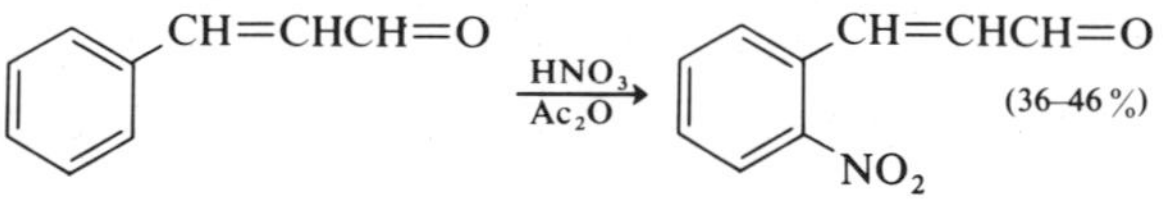

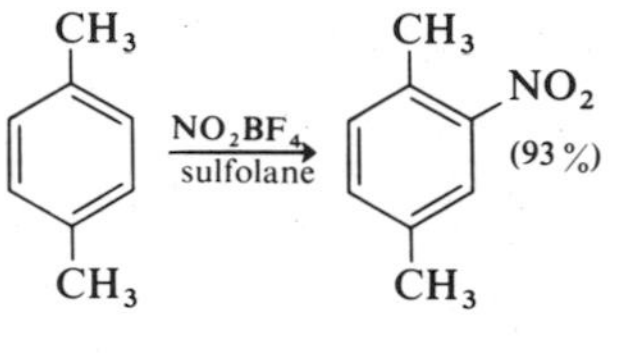

3[c] CO_2H phenyl → $\xrightarrow[HNO_3]{H_2SO_4}$ → 3,5-(O_2N)(NO_2) (54–58%)

4[d] CH_3O, CH_3O-phenyl-$CH=O$ → $\xrightarrow{HNO_3}$ → CH_3O, CH_3O-phenyl-$CH=O$, NO_2 (73–79%)

5[e] phenyl-$CH=CHCH=O$ → $\xrightarrow[Ac_2O]{HNO_3}$ → o-NO_2-phenyl-$CH=CHCH=O$ (36–46%)

6[f] 1,4-CH_3,CH_3-phenyl → $\xrightarrow[sulfolane]{NO_2BF_4}$ → CH_3, NO_2, CH_3 (93%)

7[g] CH_3-phenyl + 2-CH_3-1-NO_2-pyridinium → CH_3-phenyl-NO_2
$o:m:p$ (100%)
64:3:33

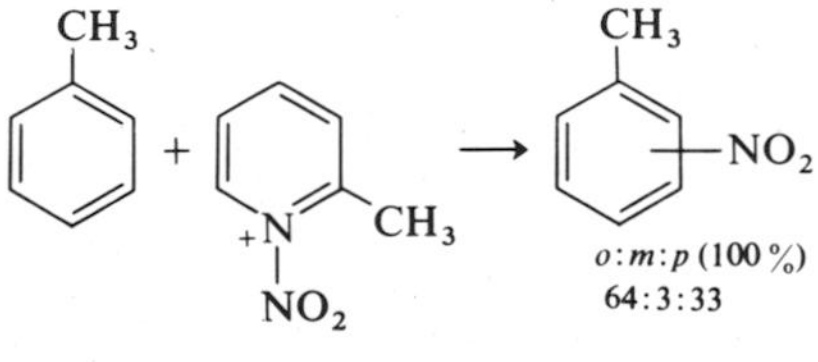

a. G. R. Robertson, *Org. Synth.* **I**, 389 (1932).
b. H. M. Fitch, *Org. Synth.* **III**, 658 (1955).
c. R. Q. Brewster, B. Williams, and R. Phillips, *Org. Synth.* **III**, 337 (1955).
d. C. A. Fetscher, *Org. Synth.* **IV**, 735 (1963).
e. R. E. Buckles and M. P. Bellis, *Org. Synth.* **IV**, 722 (1963).
f. S. J. Kuhn and G. A. Olah, *J. Am. Chem. Soc.* **83**, 4564 (1961).
g. C. A. Cupas and R. L. Pearson, *J. Am. Chem. Soc.* **90**, 4742 (1968).

agent in both organic solvents and strong acids.[7] It has been found that *N*-nitropyridinium salts can effect aromatic nitration, as illustrated by entry 7 in Scheme 7.1.[8]

7.1.2. Halogenation

The introduction of one of the halogens onto the aromatic ring is an important synthetic procedure. Chlorine and bromine are reactive toward aromatic hydrocarbons, although catalysts are often needed to achieve desirable rates. Fluorine reacts too violently to be controlled. Iodine can effect substitution of only very reactive aromatic compounds, but synthetic methods involving oxidants for direct introduction of iodine into less reactive molecules have been developed.

Chlorination often exhibits second-order kinetics in acetic acid solution.[9] In nonpolar solvents, the reaction is catalyzed by such added reagents as hydrogen chloride and trifluoroacetic acid,[10] and exhibits complex kinetics. This rate behavior has been interpreted in terms of acid-catalyzed cleavage of the Cl–Cl bond in a chlorine–substrate complex. Chlorination is much more rapid in polar than in nonpolar solvents.[11] Bromination also tends to exhibit complex kinetics in acetic acid

and other solvents, with terms ranging from first- to third-order in bromine being indicated. Second-order kinetics have been observed in aqueous acetic acid.[12]

For preparative reactions, the catalytic effect of Lewis acids is often utilized. Zinc chloride or ferric chloride are used in chlorinations, and metallic iron, which

$$MX_n + X_2 \rightleftharpoons X-X\cdots MX_n$$

7. C. L. Coon, W. G. Blucher, and M. E. Hill, *J. Org. Chem.* **38**, 4243 (1973).
8. C. A. Cupas and R. L. Pearson, *J. Am. Chem. Soc.* **90**, 4742 (1968).
9. L. M. Stock and F. W. Baker, *J. Am. Chem. Soc.* **84**, 1661 (1962).
10. L. J. Andrews and R. M. Keefer, *J. Am. Chem. Soc.* **81**, 1063 (1959); R. M. Keefer and L. J. Andrews, *J. Am. Chem. Soc.* **82**, 4547 (1960); L. J. Andrews and R. M. Keefer, *J. Am. Chem. Soc.* **79**, 5169, (1957).
11. L. M. Stock and A. Himoe, *J. Am. Chem. Soc.* **83**, 4605 (1961).
12. E. Berliner and J. C. Powers, *J. Am. Chem. Soc.* **83**, 905 (1961).

generates $FeBr_3$, is often added to bromination mixtures. The Lewis acid presumably takes over the role of protic solvent in facilitating the cleavage of the halogen–halogen bond.

It is possible to brominate a wide variety of aromatic substrates. Anilines and phenols are highly reactive, and bromination at all positions *ortho* and *para* to the activating group occurs readily. Amido, alkyl, and halobenzenes can be brominated satisfactorily, and the usual *ortho–para*-directing effect of these substituents is noted. Use of Lewis acid catalysts permits bromination of aromatic rings carrying such strongly electron-attracting substituents as nitro or cyano.

Iodination can be carried out in the presence of cupric salts.[13] The cupric salt appears to act both as a Lewis acid catalyst and as an oxidant for converting iodide to iodine. The latter function is readily evident, since iodide salts are effective as iodinating agents in the presence of cupric chloride.

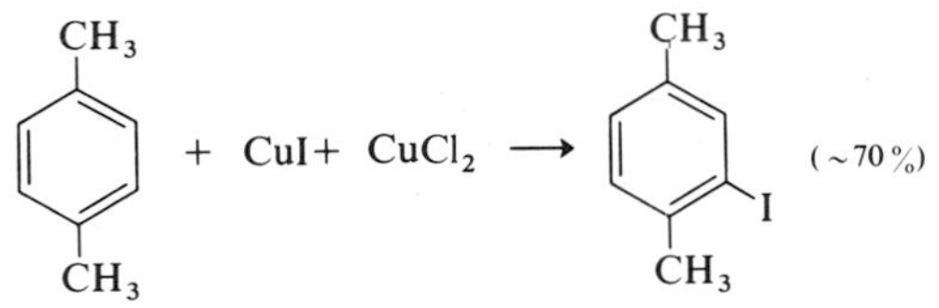

Halogenations are also strongly catalyzed by certain other metal ions. A well-studied case is catalysis by mercuric ion. In solutions of halogen and mercuric carboxylate salts, the dominant halogenating agents is the acyl hypohalite. The

$$Hg(O_2CR)_2 + X_2 \rightleftharpoons HgX(O_2CR) + RCO_2X$$

trifluoroacetyl hypohalites are extremely reactive halogenating agents. Even nitrobenzene, for example, is readily brominated by trifluoroacetyl hypobromite.[14]

Specific examples of aromatic halogenations are shown in Scheme 7.2.

7.1.3. Friedel–Crafts Alkylations and Acylations

Friedel–Crafts reactions are important methods for introducing carbon substituents on aromatic rings. The reactive intermediates are electrophilic carbon species. In some reactions, discrete carbonium ions or acylium ions are involved; in other cases, however, the electrophile no doubt consists of the alkyl or acyl group still bonded to a potential leaving group, which is displaced in the substitution step. Whether discrete carbonium ions are involved depends primarily on the stability of the potential carbonium ion.

The choice of potential alkylating agents is quite wide. Complexes of alkyl

13. W. C. Baird, Jr., and J. H. Surridge, *J. Org. Chem.* **35**, 3436 (1970).
14. J. R. Barnett, L. J. Andrews, and R. M. Keefer, *J. Am. Chem. Soc.* **94**, 6129 (1972).

A. Chlorination

1[a]

(25:72 ratio)

2[b]

3[c]

(85%)

B. Bromination

4[d]

(86–90%)

5[e]

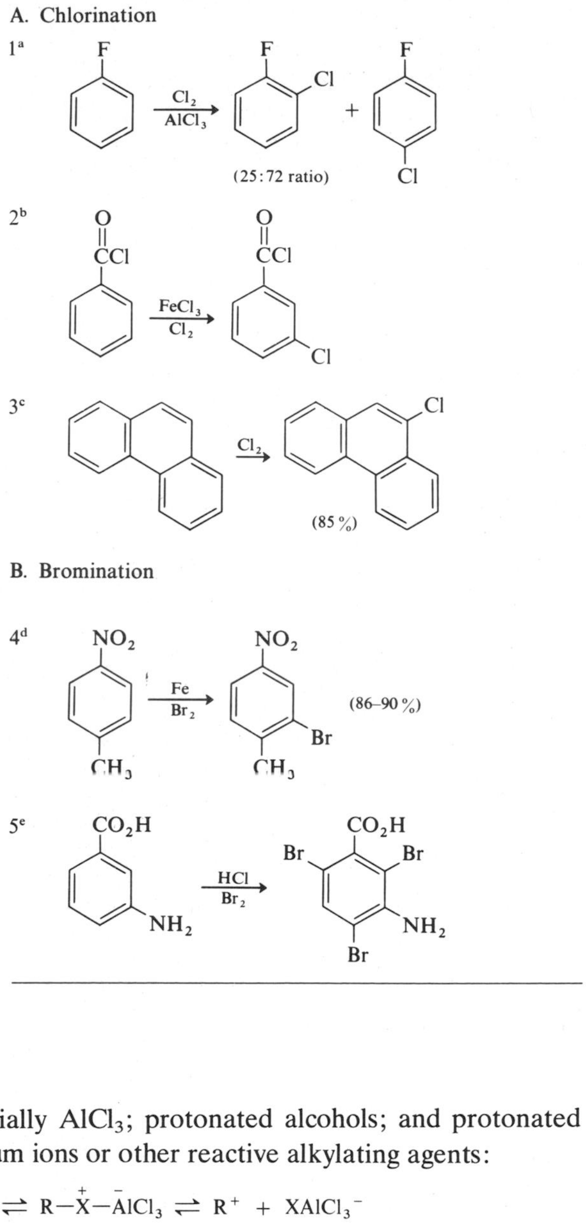

halides with Lewis acids, especially $AlCl_3$; protonated alcohols; and protonated alkenes can all provide carbonium ions or other reactive alkylating agents:

$$R{-}X \;+\; AlCl_3 \;\rightleftharpoons\; R{-}\overset{+}{X}{-}\overset{-}{A}lCl_3 \;\rightleftharpoons\; R^+ \;+\; XAlCl_3{}^-$$

$$R{-}OH \;+\; H^+ \;\rightleftharpoons\; R{-}\underset{H}{\overset{+}{O}}H \;\rightleftharpoons\; R^+ \;+\; H_2O$$

$$RCH{=}CH_2 \;+\; H^+ \;\rightleftharpoons\; R\underset{+}{C}HCH_3$$

Because of the involvement of carbonium ions and related intermediates, Friedel–Crafts reactions are often accompanied by rearrangements in the alkylating

6^f

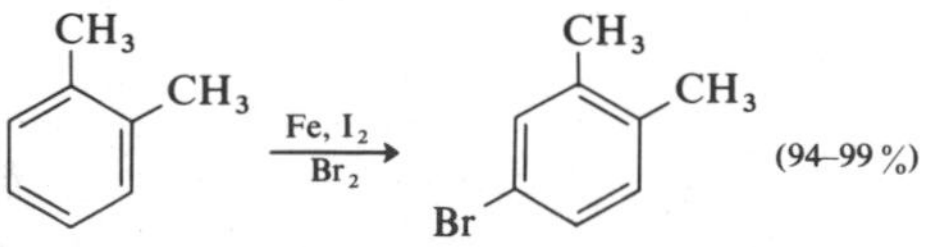

                                                                (94–99 %)

C. Iodination

7^g

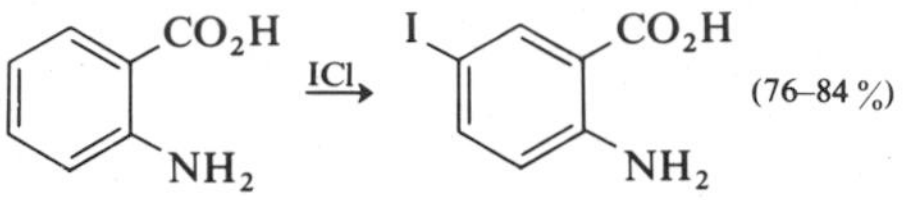

                                                                (76–84 %)

8^h

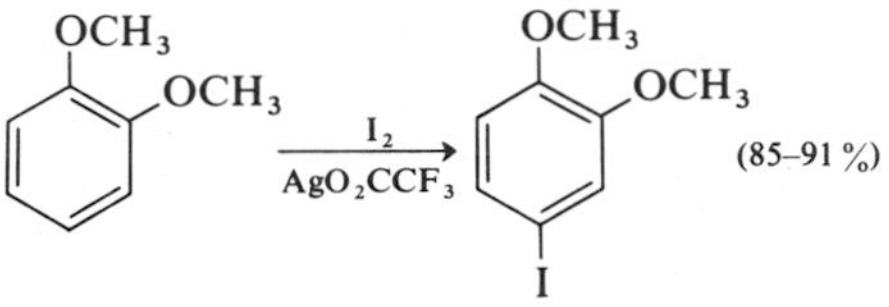

                                                                (85–91 %)

9^i

                                                                (80–81 %)

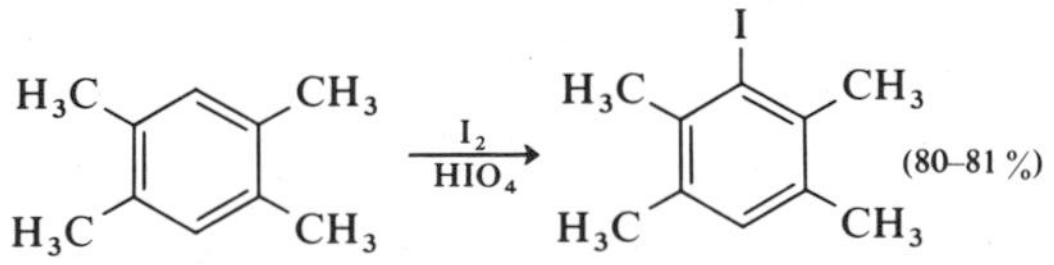

a. G. A. Olah, S. J. Kuhn, and B. A. Hardie, *J. Am. Chem. Soc.* **86**, 1055
   (1964).
b. E. Hope and G. F. Riley, *J. Chem. Soc.* **121**, 2510 (1922).
c. J. Schultz, M. A. Goldberg, E. P. Ordas, and C. Carsch, *J. Org. Chem.*
   11, 320 (1946).
e. M. M. Robison and B. L. Robison, *Org. Synth.* **IV**, 947 (1963).
f. W. A. Wisansky and S. Ansbacher, *Org. Synth.* **III**, 138 (1955).
g. V. H. Wallingford and P. A. Krueger, *Org. Synth.* **II**, 349 (1943).
h. D. E. Janssen and C. V. Wilson, *Org. Synth.* **IV**, 547 (1963).
i. H. Suzuki, *Org. Synth.* **51**, 94 (1971).

agent. For example, isopropyl groups are often introduced when *n*-propyl compounds are used as substrates[15]:

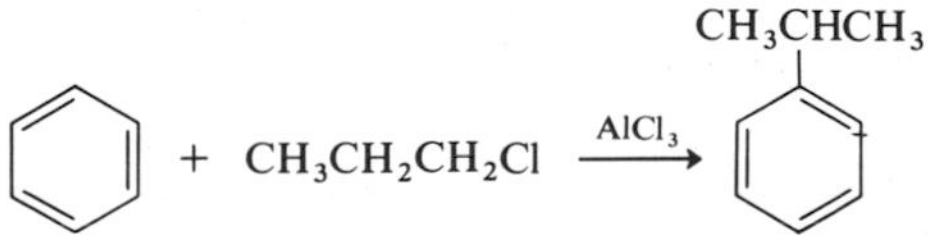

Rearrangements that might seem unusual in the sense that secondary substituent groups are derived from tertiary substrates have also been observed. This

15. S. H. Sharman, *J. Am. Chem. Soc.* **84**, 2945 (1962).

type of behavior is due to subsequent isomerization of the initial (unrearranged) reaction product under the influence of $AlCl_3$, rather than to rearrangement of a carbonium-ion intermediate.[17] In fact, milder Friedel–Crafts catalysts such as ferric chloride lead to the product expected from the unrearranged tertiary carbonium ion.

At room temperature and slightly above, migration of alkyl substituents on the benzene ring can also occur under Friedel–Crafts conditions[18]:

Such migrations usually proceed in the direction of minimizing steric interactions between the substituent groups.

Relative reactivity data show that the order of reactivity for alkyl halides is $tert > sec > pri > $ methyl.[19] This order can be accommodated by a carbonium-ion mechanism or by a process involving a complex of the halide and the catalyst. In either case, the heterolysis of the C–halide bond will be facilitated by additional alkyl groups.

The interrelationships among catalyst activity, carbonium-ion stability, and positional selectivity have been studied in detail, with the use of substituted benzyl halides and a wide variety of Friedel–Crafts catalysts.[20] These data indicate that no single mechanistic description can encompass all Friedel–Crafts alkylations. With very reactive catalysts, there is little selectivity with respect to competing aromatic substrates. In less reactive systems, substrate selectivity increases. Quantitative description of catalyst activity has not been achieved, but a number of Lewis acids have been grouped into four broad categories, based on their activity in catalyzing the benzylation of benzene. Some of the catalysts are listed in Table 7.1.

Friedel–Crafts alkylations usually give substantial amounts of both *ortho* and *para* product from alkylbenzenes. For example, reaction of toluene with isopropyl bromide and aluminum chloride–nitromethane gave 47% *ortho*, 15% *meta*, and 39% *para* product.[21] These reactions are sensitive to steric influences, and only the *meta* and *para* isomers are formed with *t*-butyl bromide.[22]

16. L. Schmerling and J. P. West, *J. Am. Chem. Soc.* **76**, 1917 (1954).
17. A. A. Khalaf and R. M. Roberts, *J. Org. Chem.* **35**, 3717 (1970).
18. R. M. Roberts and D. Shiengthong, *J. Am. Chem. Soc.* **86**, 2851 (1964).
19. H. Jungk, C. R. Smoot, and H. C. Brown, *J. Am. Chem. Soc.* **78**, 2185 (1956).
20. G. A. Olah, S. Kobayashi, and M. Tashiro, *J. Am. Chem. Soc.* **94**, 7448 (1972).
21. G. A. Olah, S. H. Flood, S. J. Kuhn, M. E. Moffatt, and N. A. Overchuck, *J. Am. Chem. Soc.* **86**, 1046 (1964).
22. G. A. Olah, S. H. Flood, and M. E. Moffatt, *J. Am. Chem. Soc.* **86**, 1060 (1964).

Table 7.1. Relative Activity of Friedel–Crafts Catalysts[a]

Very active	Moderately active	Weak
$AlCl_3$, $AlBr_3$, $GaCl_3$, $GaCl_2$, SbF_5, $MoCl_5$	$InCl_3$, $LnBr_3$, $SbCl_5$, $FeCl_3$, $AlCl_3$–CH_3NO_2, SbF_5–CH_3NO_2	BCl_3, $SnCl_4$, $TiCl_4$, $TiBr_4$, $FeCl_2$

a. G. A. Olah, S. Kobayashi, and M. Tashiro, *J. Am. Chem. Soc.* **94**, 7448 (1972).

The Friedel–Crafts alkylation reaction does not proceed successfully with aromatic substrates having electron–attracting groups. A further limitation is that the alkyl group introduced in the reaction increases the reactivity of the aromatic ring toward further substitution, so that polyalkylation can occur. Polyalkylation is usually minimized in practice by using the aromatic substrate in excess.

Besides the alkyl halide–Lewis acid combination, two other routes to carbonium ions are quite widely used in synthesis. Alcohols can serve as sources of carbonium ions in strongly acidic media such as sulfuric acid and phosphoric acid. The alkylation of aromatic rings by alcohols is also catalyzed by BF_3 and $AlCl_3$.[23]

Alkenes are also sources of carbonium ions for alkylations. Protic acids—especially sulfuric acid, phosphoric acid, and hydrogen fluoride—and Lewis acids, such as BF_3 and $AlCl_3$, are used as catalysts.[24] Insofar as comparisons have been made, such factors as position selectivity and tendency toward isomerism of the alkyl group to more stable structures are similar to those described with alkyl halides as alkylating agents.[25]

Some examples of Friedel–Crafts alkylations are given in Scheme 7.3.

Friedel–Crafts alkylation can also occur intramolecularly, in which case a new ring is formed. It is somewhat easier to form six-membered rings than five-membered rings in such reactions. Thus, while 4-phenyl-1-butanol gives a 50% yield of cyclized product in phosphoric acid, 3-phenyl-1-propanol gives mainly dehydration to alkene.[26] If a potential carbonium-ion intermediate can undergo a hydride or

23. A. Schriesheim, in *Friedel–Crafts and Related Reactions*, Vol. II, G. Olah (ed.), Interscience, New York, NY, 1964, Chapter XVIII.

24. S. H. Patinkin and B. S. Friedman, in *Friedel–Crafts and Related Reactions*, Vol. II, G. Olah (ed.), Interscience, New York, NY, 1964, Chapter XIV.

25. R. H. Allen and L. D. Yats, *J. Am. Chem. Soc.* **83**, 2799 (1961).

26. A. A. Khalaf and R. M. Roberts, *J. Org. Chem.* **34**, 3571 (1969).

Scheme 7.3. Friedel–Crafts Alkylation Reactions

A. Intermolecular Reactions

1[a]

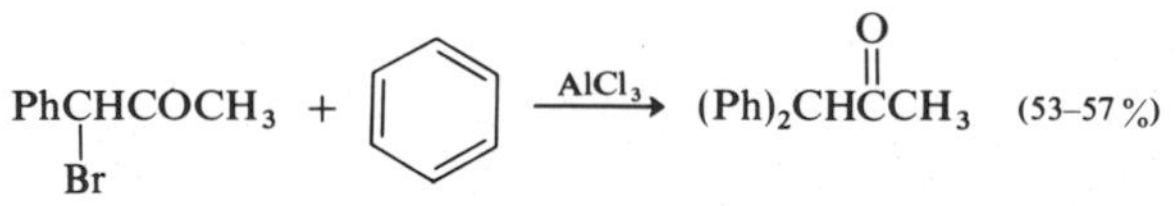
PhCHCOCH3 +
Br
AlCl3
O
(Ph)2CHCCH3
(53–57 %)

2[b]

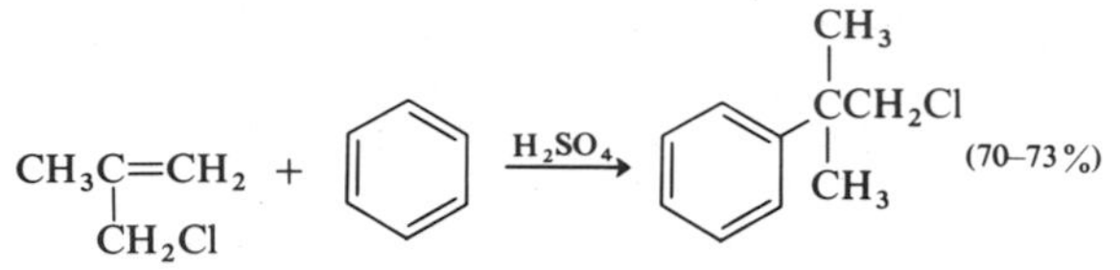
CH3C=CH2 +
CH2Cl
H2SO4
CH3
CCH2Cl
CH3
(70–73 %)

3[c]

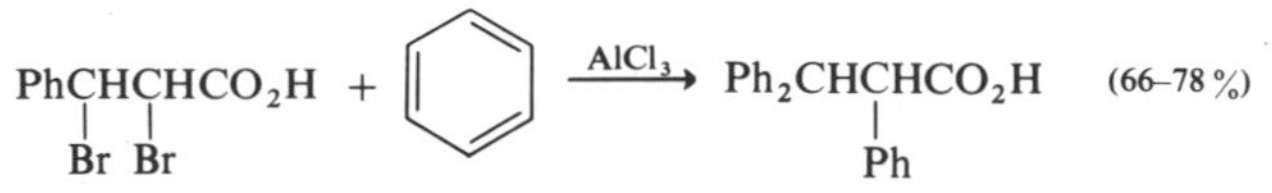
PhCHCHCO2H +
Br Br
AlCl3
Ph2CHCHCO2H
Ph
(66–78 %)

B. Intramolecular Friedel–Crafts Cyclizations

4[d]

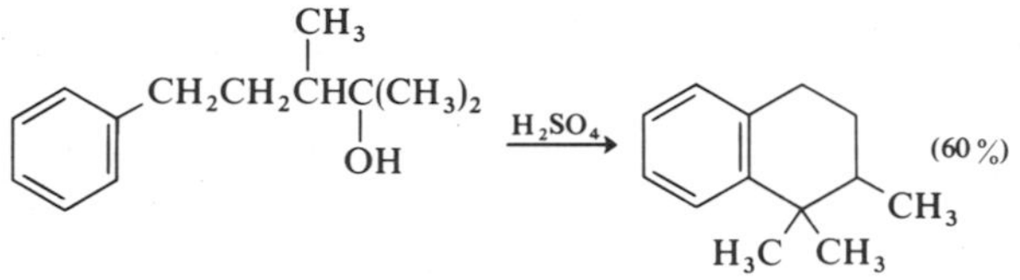
CH3
CH2CH2CHC(CH3)2
OH
H2SO4
(60 %)
CH3
H3C CH3

5[e]

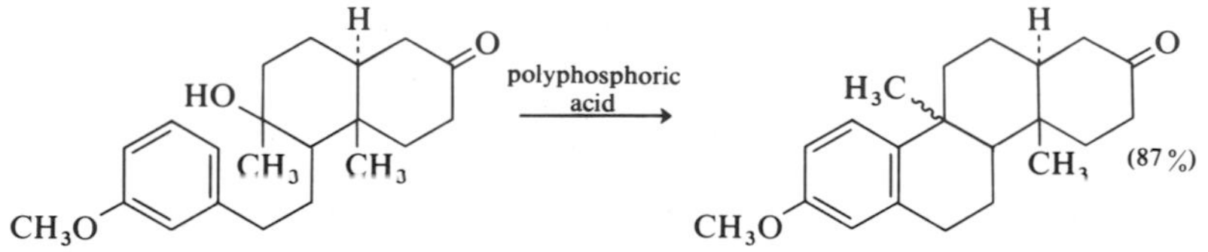
H
O
HO
CH3
CH3
CH3O
polyphosphoric
acid
H
O
H3C
CH3
(87 %)
CH3O

a. E. M. Schultz and S. Mickey, *Org. Synth.* **III**, 343 (1955).
b. W. T. Smith, Jr., and J. T. Sellas, *Org. Synth.* **IV**, 702 (1963).
c. C. P. Krimmel, L. E. Thielen, E. A. Brown, and W. J. Heidtke, *Org. Synth.* **IV**, 960 (1963).
d. A. A. Khalaf and R. M. Roberts, *J. Org. Chem.* **37**, 4227 (1972).
e. R. E. Ireland, S. W. Baldwin, and S. C. Welch, *J. Am. Chem. Soc.* **94**, 2056 (1972).

alkyl shift, this shift will occur in preference to closure of the five-membered ring:

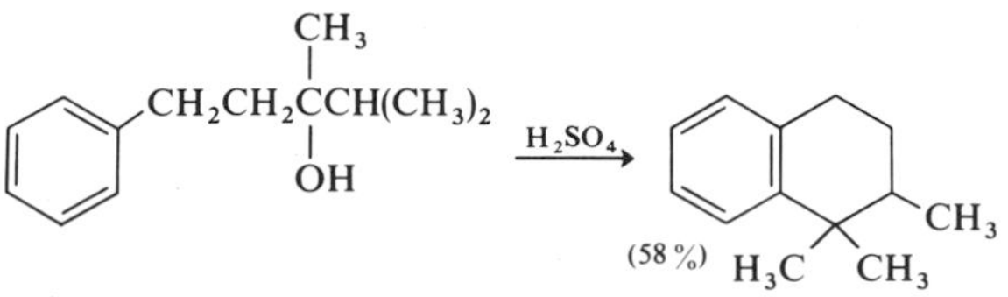
CH3
CH2CH2CCH(CH3)2
OH
H2SO4
CH3
(58 %)
H3C CH3

Ref. 27

An important use of the intramolecular Friedel–Crafts reaction is in construction of the polycyclic hydrocarbon framework of terpenes and steroids. Entry 5 in Scheme 7.3. is an example of this type of application.

27. A. A. Khalaf and R. M. Roberts, *J. Org. Chem.* **37**, 4227 (1972).

The Friedel–Crafts acylation generally involves reaction of an acid halide with a Lewis acid catalyst such as $AlCl_3$, SbF_5, or BF_3. Acid anhydrides are employed in some cases. As in alkylations, the reactive intermediate may be a dissociated organic cation (acylium ion) or a complex of the acid chloride and Lewis acid.[28]

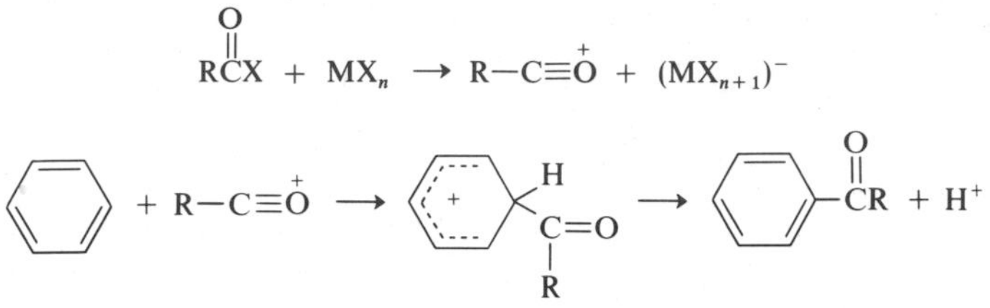

or

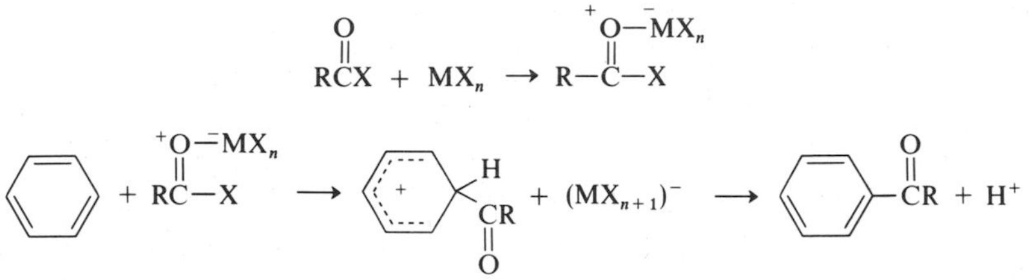

Orientation of the incoming acyl group in Friedel–Crafts acylations can be quite sensitive to the reaction solvent and other procedural variables.[29] In general, however, *para* attack predominates for alkylbenzenes. The percentage of *ortho* attack increases with the electrophilicity of the acylium ion and approaches 0.8 : 1 with such unstable, and therefore less selective, species as formyl and 2,4-dinitrobenzoyl ions.[30] For simple acyl and benzoyl derivatives, the $o:p$ ratio is usually 1 : 20 or higher.[31]

Intramolecular acylations are quite common. The normal Friedel–Crafts procedure involving an acid halide and Lewis acid is frequently used, but there are alternatives. One useful method for inducing intramolecular acylations is to dissolve the carboxylic acid in polyphosphoric acid and heat to effect cyclization. The mechanism probably involves formation of a mixed carboxylic–phosphoric anhydride.[32] Cyclizations of this type can also be carried out using "polyphosphate ester,"

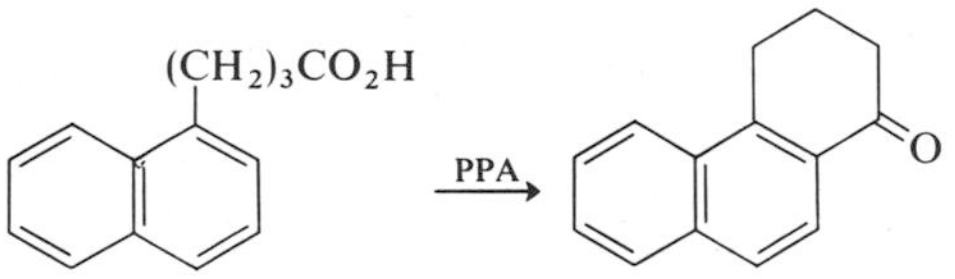

28. F. R. Jensen and G. Goldman, in *Friedel–Crafts and Related Reactions,* Vol. III, G. Olah (ed.), Interscience, New York, NY, 1964, Chap. XXXVI.
29. For example, see L. Friedman and R. J. Honour, *J. Am. Chem. Soc.* **91**, 6344 (1969).
30. G. A. Olah and S. Kobayashi, *J. Am. Chem. Soc.* **93**, 6964 (1971).
31. H. C. Brown, G. Marino, and L. M. Stock, *J. Am. Chem. Soc.* **81**, 3310 (1959); H. C. Brown and G. Marino, *J. Am. Chem. Soc.* **81**, 5611 (1959); G. A. Olah, M. E. Moffatt, S. J. Kuhn, and B. A. Hardie, *J. Am. Chem. Soc.* **86**, 2198 (1964).
32. W. E. Bachmann and W. J. Horton, *J. Am. Chem. Soc.* **69**, 58 (1947).

an esterified oligomer of phosphoric acid that is soluble in solvents such as chloroform.[33]

A procedure of long standing for introduction of a fused ring onto a benzene skeleton utilizes succinic anhydride or a derivative. An intermolecular acylation is followed by reduction and then an intramolecular acylation. The reduction is

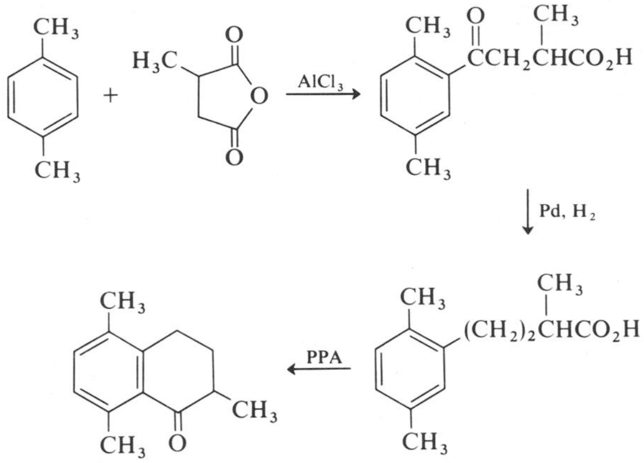

necessary to convert the deactivating acyl substituent to a more favorable alkyl substituent.

Scheme 7.4 records some typical Friedel–Crafts acylation reactions.

Certain other reactions of aromatic molecules are closely related to the Friedel–Crafts reaction. The introduction of chloromethyl groups is brought about by formaldehyde in concentrated hydrochloric acid in the presence of halide salts, especially zinc chloride.[35] The reaction is restricted in scope to benzene and derivatives with electron-releasing substituents. Several mechanistic pathways could be operative in the chloromethylation reaction, but the active electrophile is probably protonated chloromethyl alcohol.

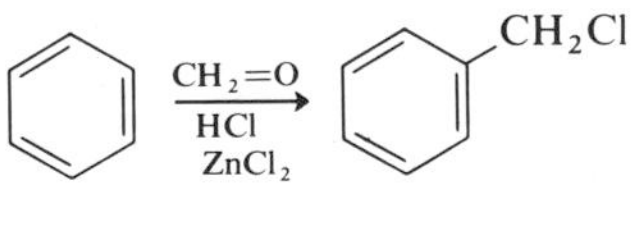

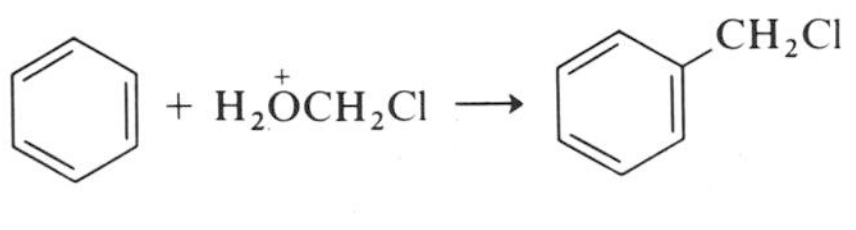

33. Y. Kanaoka, O. Yonemitsu, K. Tanizawa, and Y. Ban, *Chem. Pharm. Bull.* (Japan) **12**, 773 (1964); T. Kametani, S. Takano, S. Hibino, and T. Terui, *J. Heterocycl. Chem.* **6**, 49 (1969).

34. E. J. Eisenbraun, C. W. Hinman, J. M. Springer, J. W. Burnham, T. S. Chou, P. W. Flanagan, amd M. C. Hamming, *J. Org. Chem.* **36**, 2480 (1971).

35. R. C. Fuson and C. H. McKeever, *Org. React.* **1**, 63 (1942); G. A. Olah and S. H. Yu, *J. Am. Chem. Soc.* **97**, 2293 (1975).

Hydroxycarbonium ions are generated by protonation of ketones and aldehydes, and can effect alkylation of the aromatic ring. Usually, the products of such reactions, being substituted benzyl alcohols, are also subject to further transformations. The formation of disubstituted anthracenes from benzene and an aromatic aldehyde is illustrative:

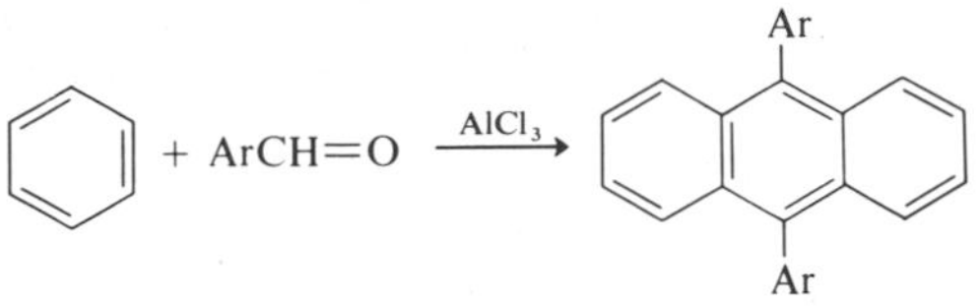

Ref. 36

A further example is the synthesis of DDT from trichloroacetaldehyde and chlorobenzene:

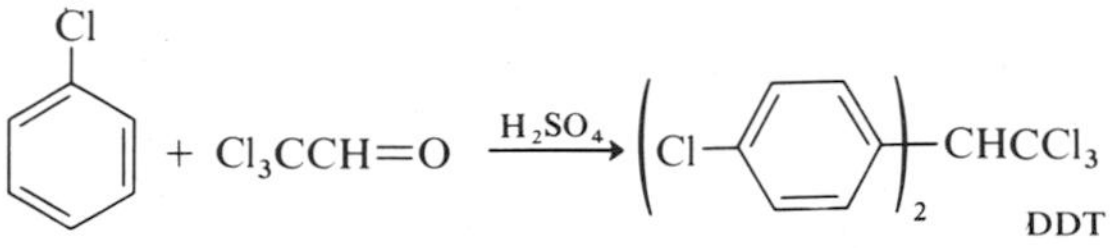

The initial carbinol product serves to alkylate a second chlorobenzene molecule:

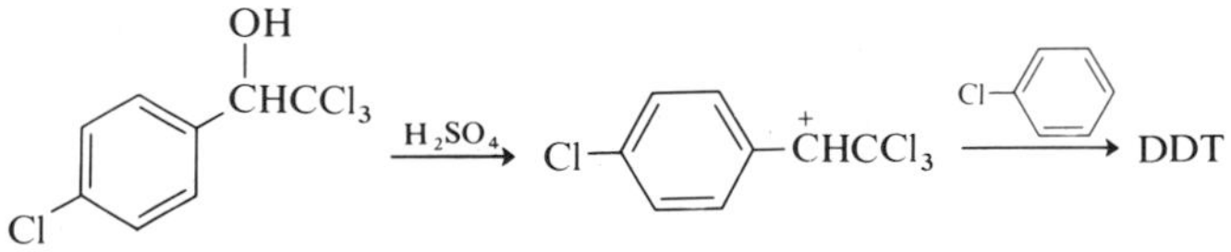

This type of reaction, which results in the formation of diphenylmethane derivatives, is especially characteristic of reactive aromatics such as mesitylene and phenol.

Carbon monoxide, hydrogen cyanide, and nitriles also react with aromatic compounds in the presence of strong acids or other Friedel–Crafts catalysts. These reactions are quite useful for synthetic purposes, since the products are formyl- or acyl-substituted aromatics. These electron-withdrawing groups retard any further electrophilic substitutions. Detailed mechanistic studies have not been carried out, but the general outlines of the mechanism of these reactions are given below:

$$\overset{-}{C}\!\equiv\!\overset{+}{O} + H^+ \rightleftharpoons H\!-\!C\!\equiv\!\overset{+}{O}$$

$$ArH + HC\!=\!\overset{+}{O} \rightarrow ArCH\!=\!O + H^+$$

$$R\!-\!C\!\equiv\!N + H^+ \rightleftharpoons R\!-\!C\!\equiv\!\overset{+}{N}\!-\!H$$

$$ArH + R\!-\!C\!\equiv\!\overset{+}{N}\!-\!H \rightarrow R\!-\!\underset{\underset{Ar}{|}}{C}\!=\!\overset{+}{N}H_2$$

$$R\underset{\underset{Ar}{|}}{\overset{+}{C}}\!=\!\overset{+}{N}H_2 + H_2O \rightarrow Ar\overset{\overset{O}{\|}}{C}R$$

36. H. E. Ungnade and E. W. Crandall, *J. Am. Chem. Soc.* **71**, 2209 (1949).

Scheme 7.4. Friedel–Crafts

A. Intermolecular Reactions

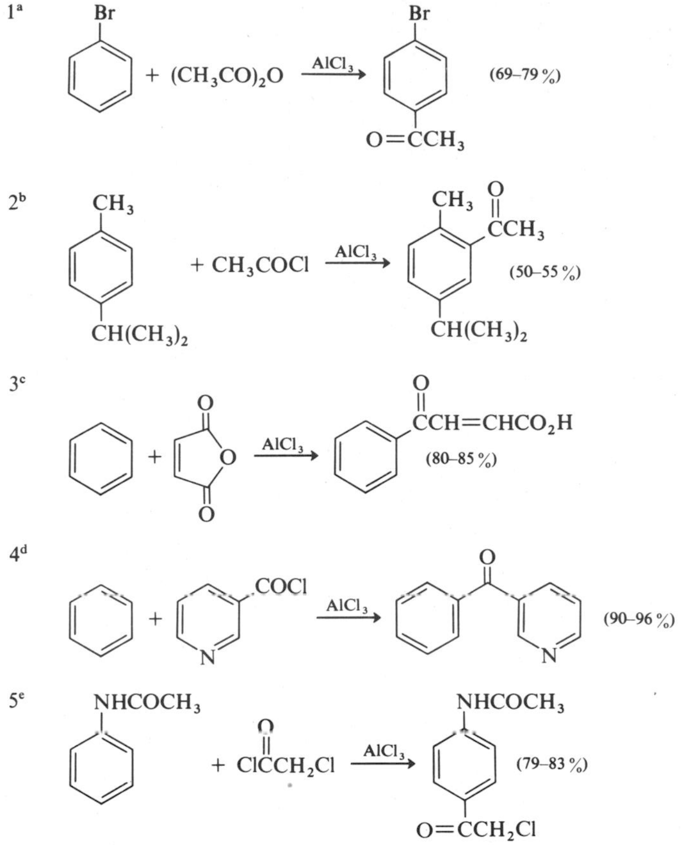

1[a]

(69–79%)

2[b]

(50–55%)

3[c]

(80–85%)

4[d]

(90–96%)

5[e]

(79–83%)

a. R. Adams and C. R. Noller, *Org. Synth* **I**, 109 (1941).
b. C. F. H. Allen, *Org. Synth.* **II**, 3 (1943).
c. O. Grummitt, E. I. Becker, and C. Miesse, *Org. Synth.* **III**, 109 (1955).
d. F. J. Villani and M. S. King, *Org. Synth.* **IV**, 88 (1963).
e. J. L. Leiserson and A. Weissberger, *Org. Synth.* **III**, 183 (1955).

    Many specific examples of these reactions are recorded in reviews in the *Organic Reactions* series.[37]

    Another useful method for introducing formyl and acyl groups onto aromatic nuclei involves the use of *N,N*-dialkylamides and phosphorus oxychloride; this is the *Vilsmeier–Haack reaction.* The active electrophile shown in the mechanistic outline is a product of phosphorylation of the amide at oxygen:

37. N. N. Crounse, *Org. React.* **5**, 290 (1949); W. E. Truce, *Org. React.* **9**, 37 (1957); P. E. Spoerri and A. S. DuBois, *Org. React.* **5**, 387 (1949).

Acylation Reactions

271

SECTION 7.1.
ELECTROPHILIC
AROMATIC
SUBSTITUTION

6[f]

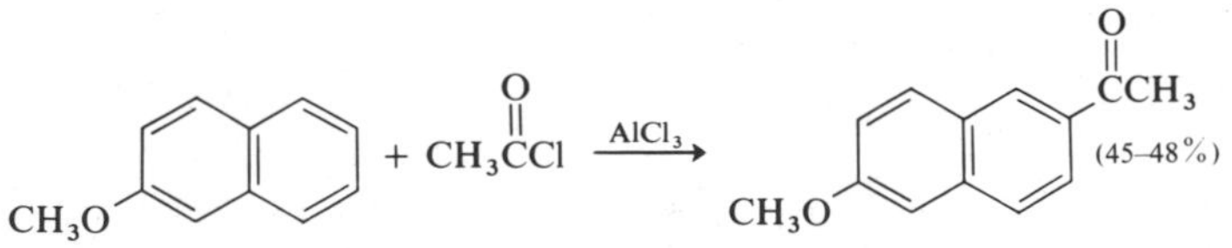

(45–48%)

B. Intramolecular Friedel–Crafts Acylations

7[g]

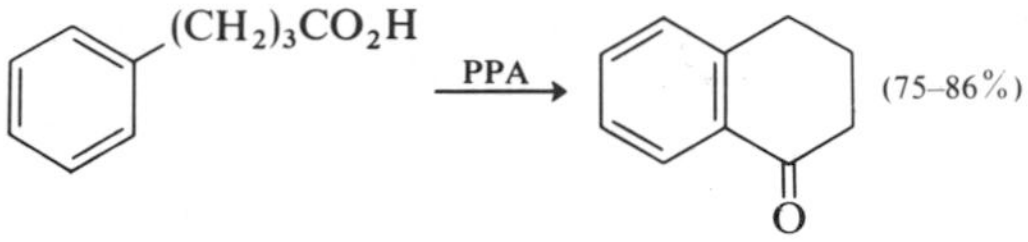

(75–86%)

8[h]

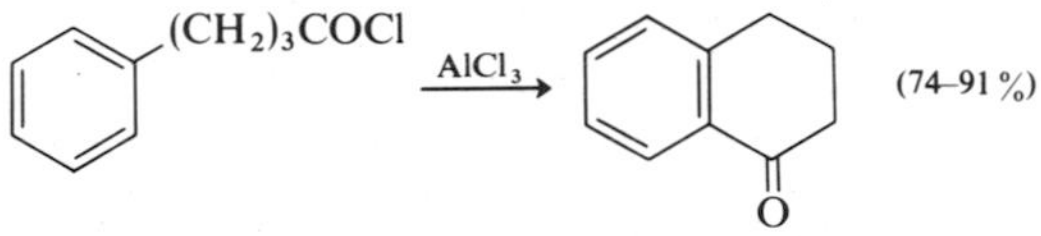

(74–91%)

9[i]

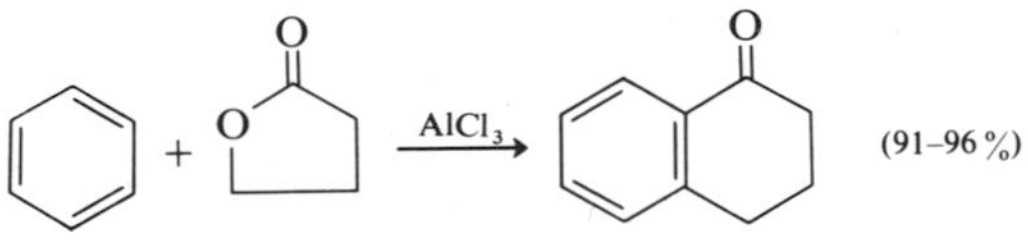

(91–96%)

10[j]

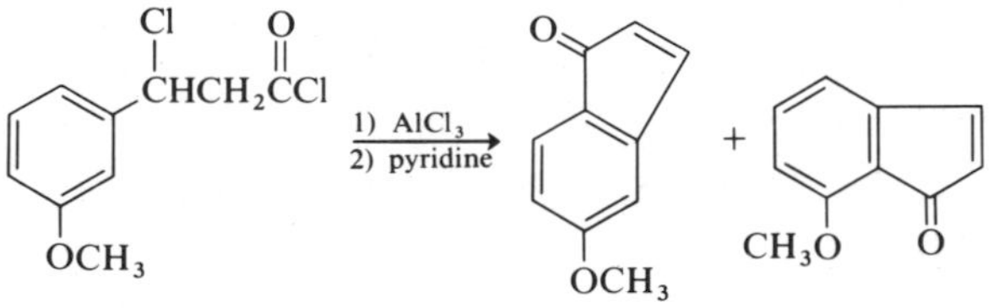

(85% total yield)

f. L. Arsenijevic, V. Arsenijevic, A. Horeau, and J. Jacques, *Org. Synth.* **53**, 5 (1973).
g. H. R. Snyder and F. X. Werber, *Org. Synth.* **III**, 798 (1955).
h. E. L. Martin and L. F. Fieser, *Org. Synth.* **II**, 569 (1943).
i. C. E. Olson and A. F. Bader, *Org. Synth.* **IV**, 898 (1963).
j. M. B. Floyd and G. R. Allen, Jr., *J. Org. Chem.* **35**, 2647 (1970).

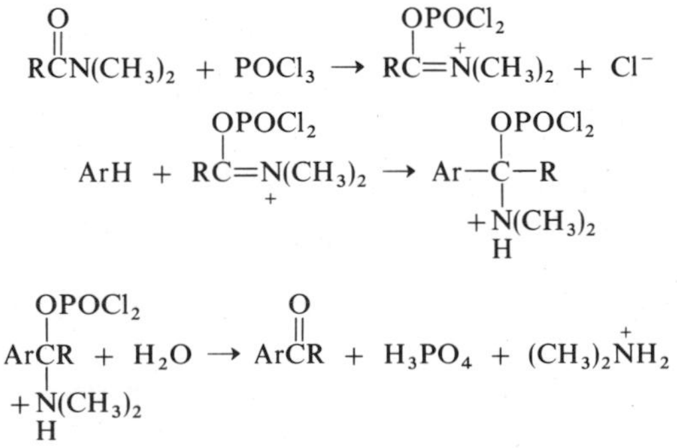

Chloromethylation

1[a]

(74–77%)

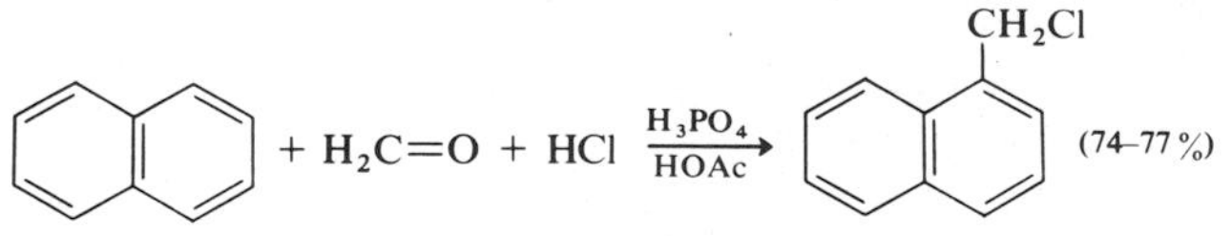

Formylation with Carbon Monoxide

2[b]

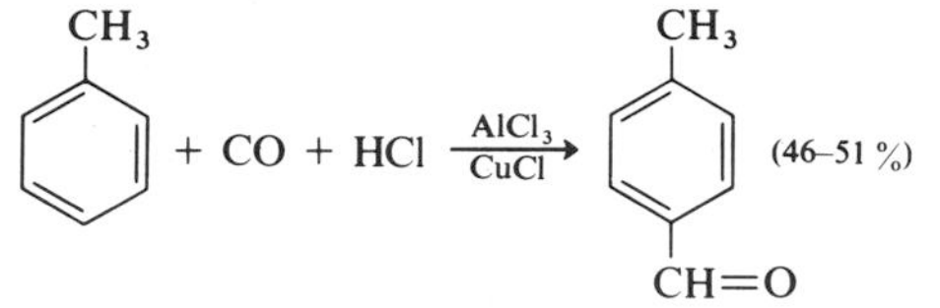

(46–51%)

Acylation with Cyanide and Nitriles

3[c]

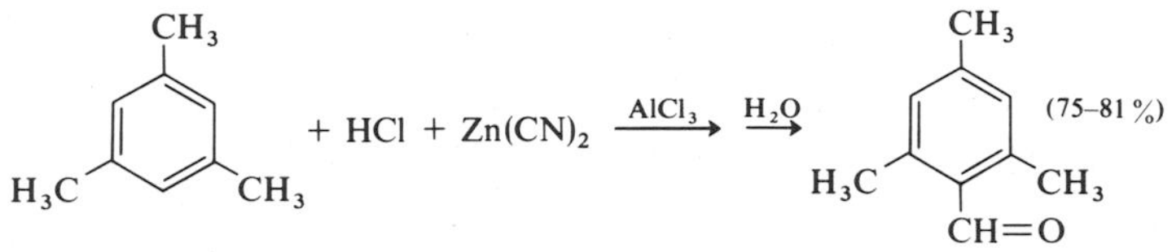

(75–81%)

4[d]

(74–87%)

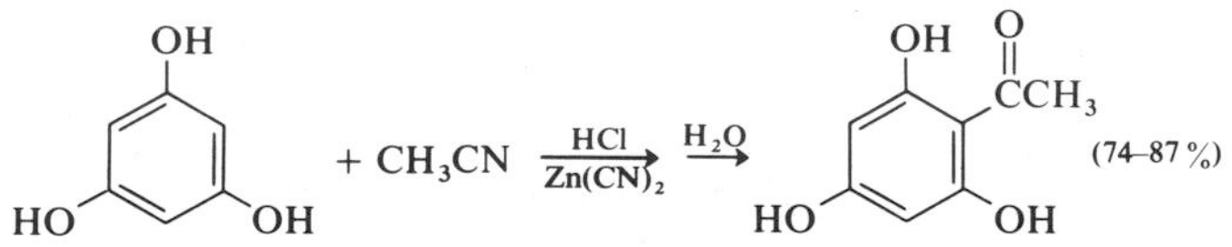

This species is not as reactive an electrophile as an acylium ion, and the aromatic ring must have at least one activating substituent for reaction to proceed effectively.

Examples of these various alternative methods for acylating aromatic rings are given in Scheme 7.5.

7.1.4. Electrophilic Metalation

Aromatic compounds react with mercuric salts to give arylmercury compounds.[38] The reaction shows substituent effects characteristic of electrophilic substitution, and is accelerated by electron-releasing substituents.[39] Mercuration is one of the few electrophilic aromatic substitutions in which proton loss from the σ-complex is rate-determining. The reaction exhibits an isotope effect of about

38. W. Kitching, *Organomet. Chem. Rev.* **3**, 35 (1968).
39. H. C. Brown and C. W. McGary, Jr., *J. Am. Chem. Soc.* **77**, 2300, 2306, 2310 (1955); A. J. Kresge and H. C. Brown, *J. Org. Chem.* **32**, 756 (1967).

Related to the Friedel–Crafts Reaction

273

SECTION 7.1.
ELECTROPHILIC
AROMATIC
SUBSTITUTION

Vilsmeier–Haack Acylation

5^e

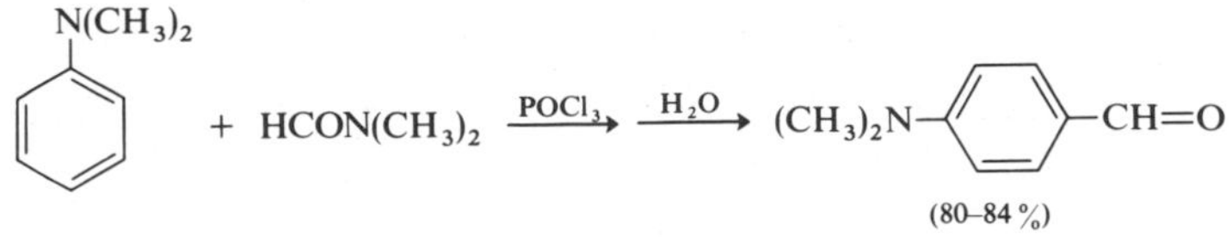

(80–84 %)

6^f

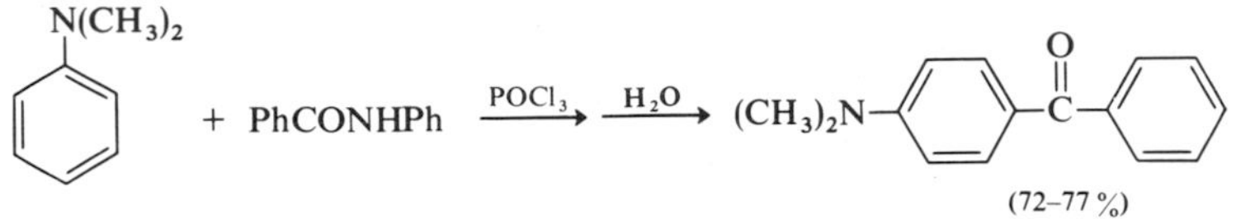

(72–77 %)

7^g

(74–84 %)

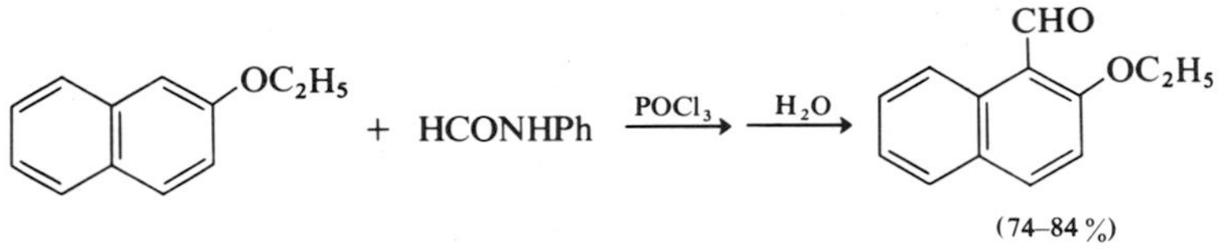

a. O. Grummitt and A. Buck, *Org. Synth.* **III**, 195 (1955).
b. G. H. Coleman and D. Craig, *Org. Synth.* **II**, 583 (1955).
c. R. C. Fuson, E. C. Horning, S. P. Rowland, and M. L. Ward, *Org. Synth.* **III**, 549 (1955).
d. K. C. Gulati, S. R. Seth, and K. Venkataraman, *Org. Synth.* **II**, 522 (1943).
e. E. Campaigne and W. L. Archer, *Org. Synth.* **IV**, 331 (1963).
f. C. D. Hurd and C. N. Webb, *Org. Synth.* **I**, 217 (1941).
g. J. H. Wood and R. W. Bost, *Org. Synth.* **IV**, 98 (1955).

$k_H/k_D = 6.$[40] This indicates that the σ-complex must be reversibly formed. A variety

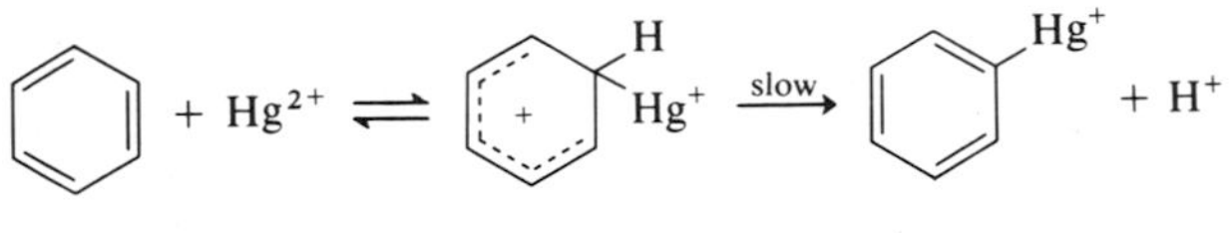

of mercury species are present under typical mercuration conditions. The reactive species are usually considered to be ion pairs of Hg^{2+} or HgX^+.[41] Undissociated species such as $Hg(OAc)_2$ are less reactive. It has been found that trifluoroacetic acid is an excellent solvent for electrophilic mercuration.[42] In this medium, mercuration gives clean second-order kinetics with benzene and its simple alkyl derivatives. The organomercury compounds prepared by electrophilic mercuration have the same range of reactivity as those discussed in Chapter 5. Thus, mercuration can be

40. C. Perrin and F. H. Westheimer, *J. Am. Chem. Soc.* **85**, 2773 (1963); A. J. Kresge and J. F. Brennan, *J. Org. Chem.* **32**, 752 (1967).
41. A. J. Kresge, M. Dubeck and H. C. Brown, *J. Org. Chem.* **32**, 745 (1967).
42. H. C. Brown and R. A. Wirkkala, *J. Am. Chem. Soc.* **88**, 1447, 1453, 1456, (1966).

employed in synthesis by taking advantage of the reactivity of the organometallic product.

In recent years, some interesting reactions that involve metal-exchange reactions of arylmercury compounds with Pd(II) salts have been developed. The resulting organopalladium species serve as a source of reactive intermediates that lead to products. Some of the types of compounds that can be formed via the unstable palladium intermediates are illustrated below:

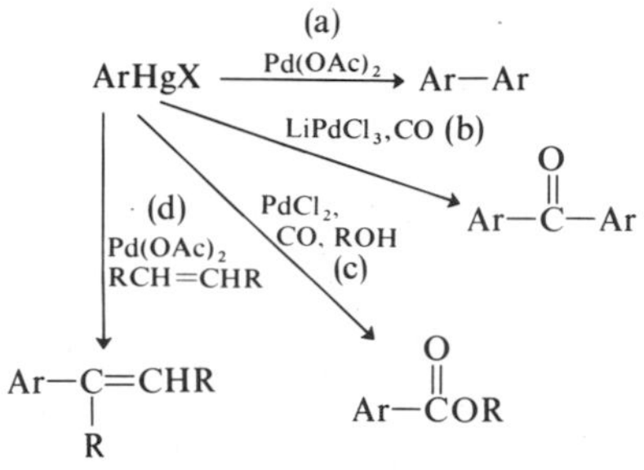

$$
\begin{array}{c}
\text{(a)} \\
ArHgX \xrightarrow{Pd(OAc)_2} Ar{-}Ar
\end{array}
$$

Ref. 43

The reactions with carbon monoxide involve addition of the organopalladium intermediate to carbon monoxide:

$$ ArPdX \rightarrow Ar\overset{O}{\underset{\|}{C}}PdX $$

The acyl palladium intermediate reacts with alcohols to give esters. In other solvents it decomposes to an acid chloride or reacts with remaining arylpalladium halide to give diaryl ketone. The reaction with alkenes involves addition of the organopalladium intermediate to the alkene, followed by decomposition of the alkylpalladium species by a β-elimination which is typical of alkylpalladium species:

$$ ArPdOAc \xrightarrow{RCH=CHR} Ar\overset{R}{\underset{}{CH}}{-}\overset{PdOAc}{\underset{}{CHR}} \rightarrow Ar\overset{}{\underset{R}{C}}{=}CHR + [HPdOAc] $$

Attention has also been called to thallium(III) salts for electrophilic metalation, and a number of procedures for synthesis of aromatic compounds via thallium intermediates have been reported.[44] A route to aryl iodides involves electrophilic substitution by thallium trifluoroacetate, followed by reaction with iodide ion:

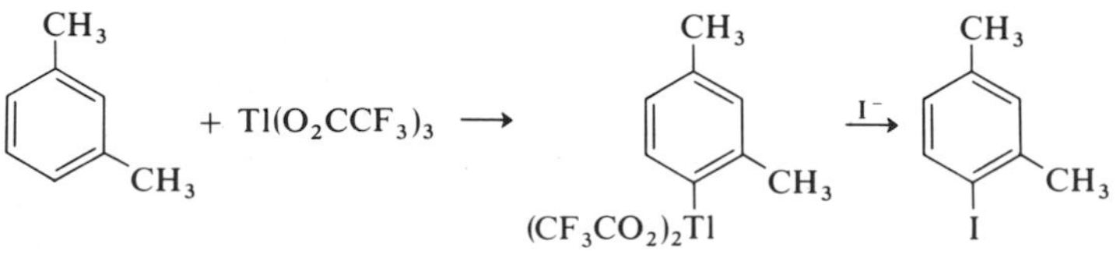

43a. M. O. Unger and R. A. Fouty, *J. Org. Chem.* **34**, 18 (1969).
   b. R. F. Heck, *J. Am. Chem. Soc.* **90**, 5546 (1968).
   c. P. M. Henry, *Tetrahedron Lett.*, 2285 (1968).
   d. R. F. Heck, *J. Am. Chem. Soc.* **91**, 6707 (1969).
44. A. McKillop and E. C. Taylor, *Adv. Organomet. Chem.* **11**, 147 (1973).

The thallium is initially introduced into the position that is most reactive toward electrophiles, but the reaction is reversible, so that heating gives a mixture having a composition governed by the relative thermodynamic stability of the various possible isomers. In some cases, then, thallation followed by isomerization can lead to a product with an orientation different from that obtained by direct electrophilic substitution.[45] The arylthallium compounds can also be converted to phenols via a procedure that involves oxidation and hydrolysis[46]:

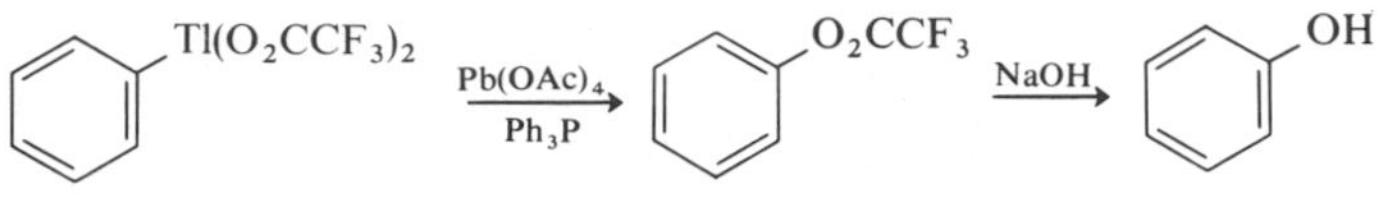

7.2. Nucleophilic Aromatic Substitution

There are a number of important reactions of aromatic compounds that involve departure of a group with its bonding pair of electrons. Unlike nucleophilic substitution at saturated carbon, aromatic substitution rarely, if ever, involves a single-step reaction; instead, discrete intermediates are usually involved. From a synthetic point of view, the aryl diazonium compounds, in which molecular nitrogen serves as the leaving group, are the most important.

7.2.1. Nucleophilic Aromatic Substitution via Diazonium Ions

One of the most versatile classes of intermediates for the synthesis of other aromatic compounds are the aryl diazonium ions formed by diazotization of anilines in acidic solution.[47] Unlike aliphatic diazonium ions, which decompose very rapidly to molecular nitrogen and the aliphatic carbonium ion, aryl diazonium ions have sufficient stability to permit controlled reactions, and in some cases salts of the diazonium ions can be isolated. The usual procedure for nitrosation involves reaction with nitrous acid in aqueous solution:

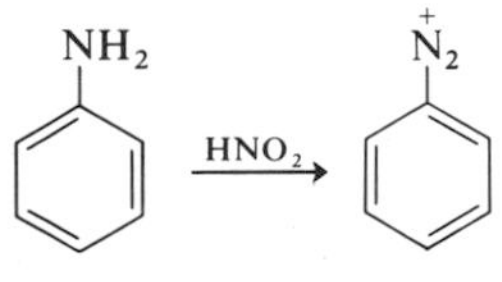

Alkyl nitrites can also serve as the nitrosating agent. The steps in nitrosation are reasonably well understood.[48] Under typical conditions, the active nitrosating

45. A. McKillop, J. D. Hunt, M. J. Zelesko, J. S. Fowler, E. C. Taylor, G. McGillivaray, and F. Kienzle, *J. Am. Chem. Soc.* **93**, 4841 (1971); E. C. Taylor, F. Kienzle, R. L. Robey, A. McKillop, and J. D. Hunt, *J. Am. Chem. Soc.* **93**, 4845 (1971).
46. E. C. Taylor, H. W. Altland, R. H. Danforth, G. McGillivaray, and A. McKillop, *J. Am. Chem. Soc.* **92**, 3520 (1970).
47. H. Zollinger, *Azo and Diazo Chemistry*, Interscience, New York, NY, 1961.
48. J. H. Ridd, *Q. Rev. Chem. Soc.* **15**, 418 (1961); C. D. Ritchie and D. J. Wright, *J. Am. Chem. Soc.* **93**, 2425 (1971).

$$ArNH_2 + HONO \xrightarrow{H^+} Ar\overset{H}{N}-N=O + H_2O$$

$$ArNH-N=O \rightarrow ArN=N-OH \underset{}{\overset{H^+}{\rightleftharpoons}} Ar\overset{+}{N}\equiv N + H_2O$$

species is protonated nitrous acid. In alkaline solution, aryl diazonium salts are converted to diazoate anions[49]:

$$Ar\overset{+}{N}\equiv N + 2\,\overset{-}{O}H \rightarrow ArN=N-\overset{-}{O} + H_2O$$

The great usefulness of the aryl diazonium ions as synthetic intermediates results from the excellence of N_2 as a leaving group. A large driving force for substitution therefore exists, in contrast to the case with other, less reactive, potential leaving groups. There are at least three basic mechanisms by which substitution can occur. One involves first-order decomposition of the diazonium ion, followed by capture of the resulting aryl cation by nucleophiles present in solution. This cation is not very stable, since the vacant orbital is in the plane of the ring and not aligned for

$$C_6H_5-\overset{+}{N}\equiv N \rightarrow C_6H_5{}^+ + N_2$$

$$C_6H_5{}^+ + X^- \rightarrow C_6H_5-X$$

delocalization of the charge into the π-system. Nevertheless, several kinetic studies have indicated that nucleophilic substitution reactions of aryl diazonium ions are first-order and independent of the concentration of the nucleophilic species.[50] Solvent effects, isotope effects, and substituent effects are also in agreement with a rate-determining unimolecular decomposition of the aryl diazonium ion.[51] In other reactions, an adduct of the nucleophile and diazonium ion is a distinct intermediate. Substitution results when nitrogen is eliminated from the

$$C_6H_5-\overset{+}{N}\equiv N + X^- \rightleftharpoons C_6H_5-N=N-X \rightarrow C_6H_5-X + N_2$$

adduct. Finally, substitution can occur via radical reactions involving electron transfer. This mechanism is particularly likely to operate in reactions where copper catalysts are used[52]:

$$C_6H_5-\overset{+}{N}\equiv N + Cu(I)X_2^- \rightarrow C_6H_5-\overset{\cdot\cdot}{N}=\overset{\cdot\cdot}{N}\!: + Cu(II)X_2$$

49. E. S. Lewis and M. P. Hanson, *J. Am. Chem. Soc.* **89**, 6268 (1967).
50. These studies are summarized by H. G. Richey and J. M. Richey, in *Carbonium Ions*, Vol. II, G. A. Olah and P. von R. Schleyer (eds.), Wiley–Interscience, New York, NY, 1970, pp. 922–931.
51. C. G. Swain, J. E. Sheats, and K. G. Harbison, *J. Am. Chem. Soc.* **97**, 783 (1975).
52. T. Cohen, R. J. Lewarchik, and J. Z. Tarino, *J. Am. Chem. Soc.* **96**, 7753 (1974).

$$\text{Ph}-\ddot{N}=\underset{\cdot}{N}: \ + \ Cu(II)X_2 \ \longrightarrow \ \text{Ph}-X \ + \ Cu(I)X \ + \ N_2$$

Examples of the three mechansims are, respectively: (a) hydrolysis of aryl diazonium salts to phenols[53]; (b) reaction of aryl diazonium ions with N_3^- to give the aryl azides[54]; and (c) the Sandmeyer reaction, involving cuprous chloride or bromide for synthesis of aryl halides.[55] Specific synthetically important substitution processes are considered in the succeeding sections.

Replacement of the diazonium group by hydrogen is occasionally a useful synthetic transformation. This transformation is often used when the directing or activating influence of a nitro or amino substituent has been used to control the introduction of a new substituent and must subsequently be removed. Replacement of the diazonium function by hydrogen is best accomplished by reaction with hypophosphorous (H_3PO_2) acid[56] or sodium borohydride,[57] although many older procedures involve heating the diazonium salt in alcohol. The sodium borohydride reduction has been shown to proceed via an aryl diazene, which decomposes with loss of nitrogen[58]:

$$Ar\overset{+}{N}\equiv N \ + \ NaBH_4 \ \longrightarrow \ Ar-N=N-H \ \xrightarrow{-N_2} \ Ar-H$$

A similar reduced species may be involved in the reduction by H_3PO_2. The reduction by alcohols is a radical chain reaction involving hydrogen-atom abstraction from the alcohol by the phenyl radical[59]:

$$Ar\cdot \ + \ CH_3OH \ \longrightarrow \ ArH \ + \ \cdot CH_2OH$$

$$\cdot CH_2OH \ + \ CH_3O^- \ \rightleftharpoons \ \cdot CH_2O^- \ + \ CH_3OH$$

$$\cdot CH_2O^- \ + \ Ar\overset{+}{N}_2 \ \longrightarrow \ Ar\cdot \ + \ N_2 \ + \ CH_2{=}O$$

Hydrolysis of aryl diazonium ions gives phenols, usually with formation of some by-products resulting from introduction of anions, such as halide, present in solution.[60]

Replacement of a diazonium group by halide is a valuable alternative to direct halogenation for the introduction of halogen substituents on aromatic rings. It has the advantage over halogenation that a single isomer is formed. The replacement of a diazonium group by chloride or bromide is effected by use of the corresponding cuprous salt. The mechanism is the electron-transfer process, which was outlined

53. E. S. Lewis, L. D. Hartung, and B. M. McKay, *J. Am. Chem. Soc.* **91**, 419 (1969).
54. C. D. Ritchie and D. J. Wright, *J. Am. Chem. Soc.* **93**, 2429 (1971); C. D. Ritchie and P. O. I. Virtanen, *J. Am. Chem. Soc.* **94**, 4966 (1972).
55. J. K. Kochi, *J. Am. Chem. Soc.* **79**, 2942 (1957); S. C. Dickerman, K. Weiss, and A. K. Ingberman, *J. Am. Chem. Soc.* **80**, 1904 (1958).
56. N. Kornblum, *Org. React.* **2**, 262 (1944).
57. J. B. Hendrickson, *J. Am. Chem. Soc.* **83**, 1251 (1961).
58. C. E. McKenna and T. G. Traylor, *J. Am. Chem . Soc.* **93**, 2313 (1971).
59. J. F. Bunnett and H. Takayama, *J. Org. Chem.* **33**, 1924 (1968).
60. E. S. Lewis, L. D. Hartung, and B. M. McKay, *J. Am. Chem. Soc.* **91**, 419 (1969).

278

CHAPTER 7
AROMATIC
SUBSTITUTION
REACTIONS

Scheme 7.6. Substitution

A. Replacement by Hydrogen

1[a]

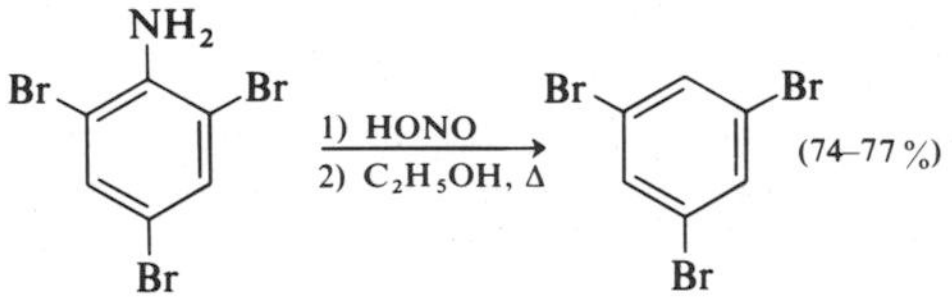

(74–77 %)

2[b]

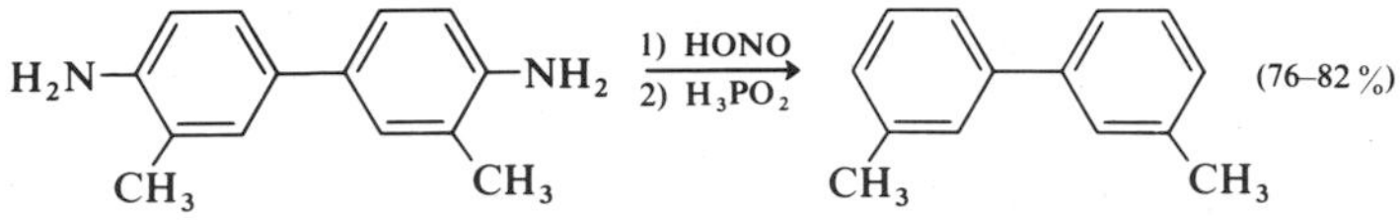

(76–82 %)

3[c]

(70–80 %)

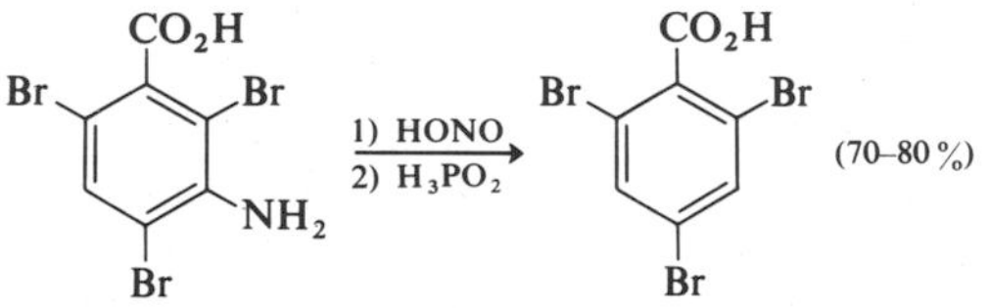

B. Replacement by Hydroxyl

4[d]

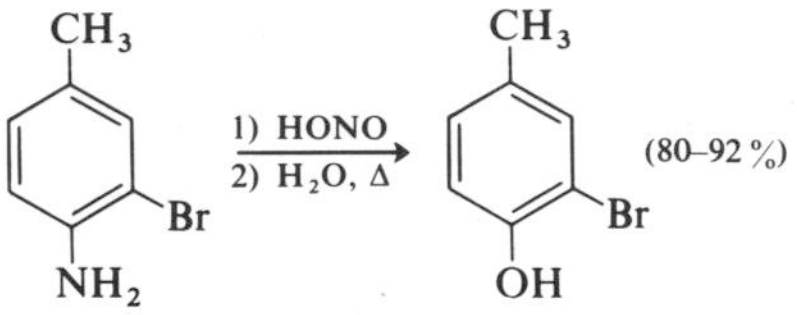

(80–92 %)

5[e]

(60 %)

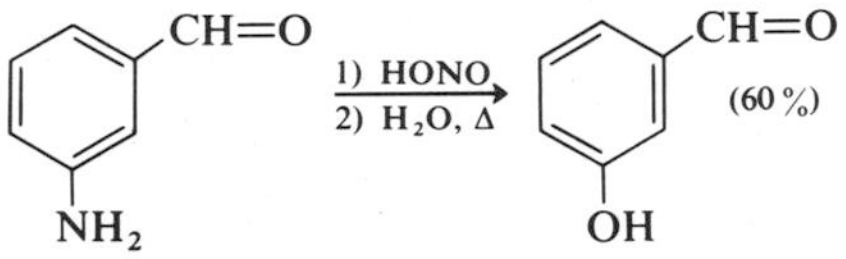

C. Replacement by Halogen

6[f]

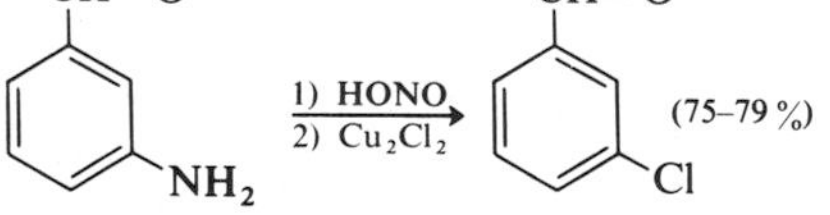

(75–79 %)

7[g]

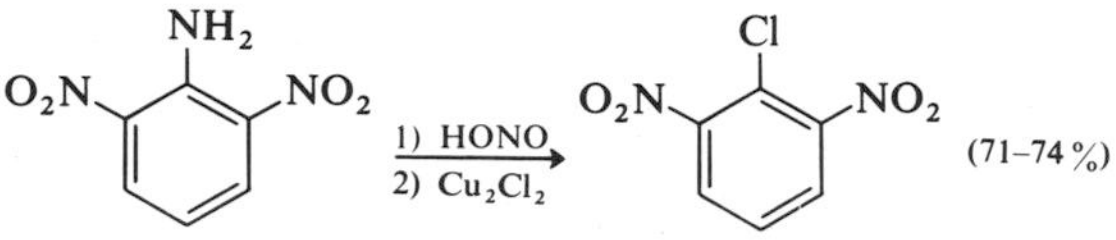

(71–74 %)

8[h]

(89–95 %)

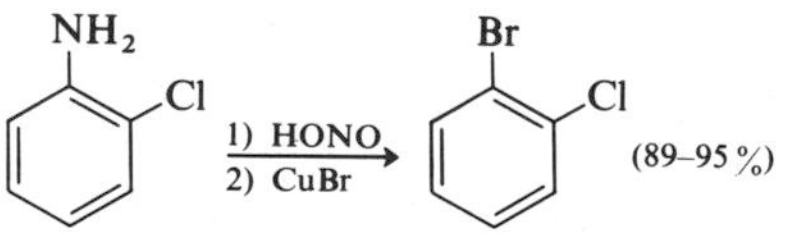

9^i

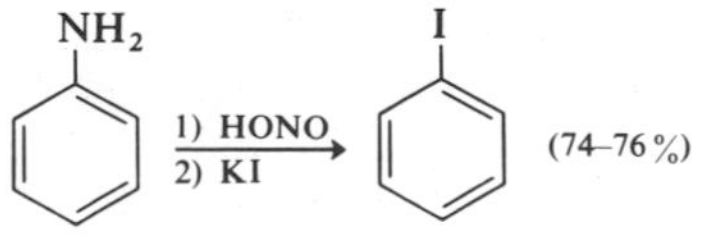

(74–76%)

10^j

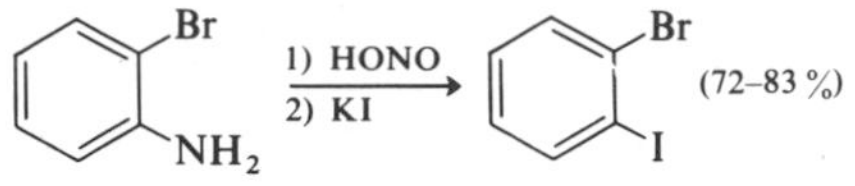

(72–83%)

11^k

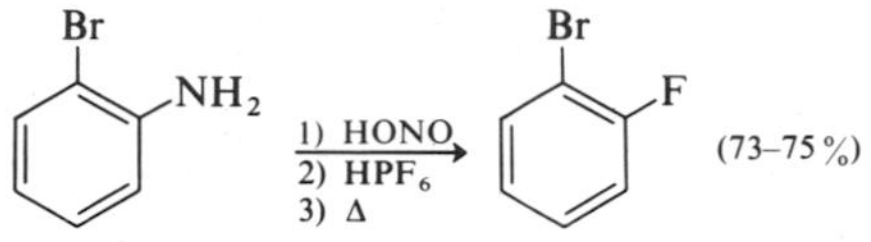

(73–75%)

12^l

(54–56%)

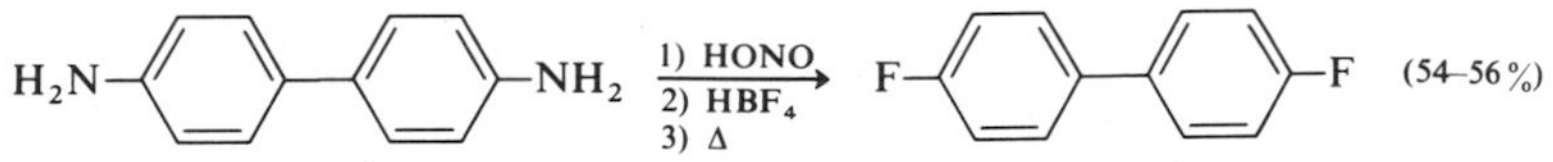

D. Replacement by Other Small Anions

13^m

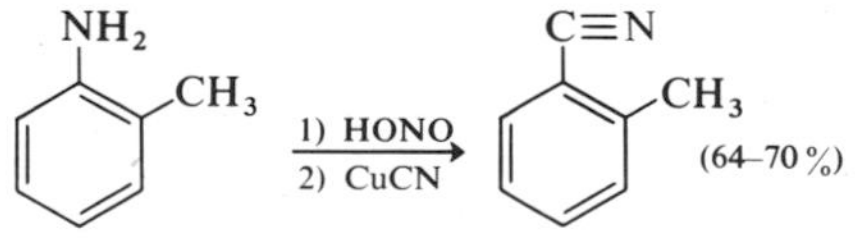

(64–70%)

14^n

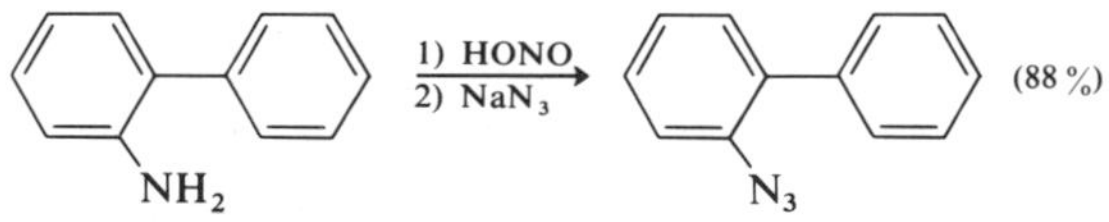

(88%)

15°

(67–82%)

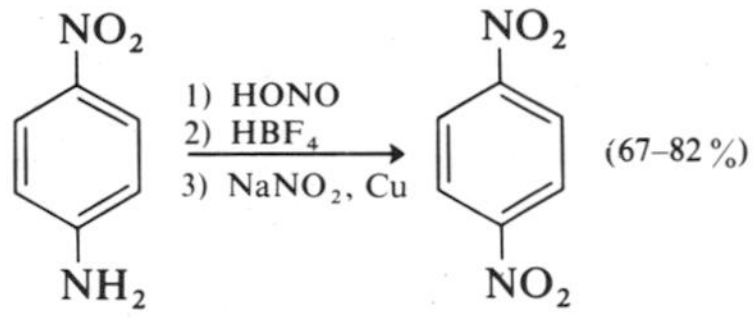

a. G. H. Coleman and W. F. Talbot, *Org. Synth.* **II**, 592 (1943).
b. N. Kornblum, *Org. Synth.* **III**, 295 (1955).
c. M. M. Robison and B. L. Robison, *Org. Synth.* **IV**, 947 (1963).
d. H. E. Ungnade and E. F. Orwoll, *Org. Synth.* **III**, 130 (1955).
e. R. N. Icke, C. E. Redemann, B. B. Wisgarver, and G. A. Alles, *Org. Synth.* **III**, 564 (1953).
f. J. S. Buck and W. S. Ide, *Org. Synth.* **II**, 130 (1943).
g. F. D. Gunstone and S. H. Tucker, *Org. Synth.* **IV**, 160 (1963).
h. J. L. Hartwell, *Org. Synth.* **III**, 185 (1955).
i. H. J. Lucas and E. R. Kennedy, *Org. Synth.* **II**, 351 (1943).
j. H. Heaney and I. T. Millar, *Org. Synth.* **40**, 105 (1960).
k. K. G. Rutherford and W. Redmond, *Org. Synth.* **43**, 12 (1963).
l. G. Schiemann and W. Winkelmüller, *Org. Synth.* **II**, 188 (1943).
m. H. T. Clarke and R. R. Read, *Org. Synth.* **I**, 514 (1941).
n. P. A. S. Smith and B. B. Brown, *J. Am. Chem. Soc.* **73**, 2438 (1957).
o. E. B. Starkey, *Org. Synth.* **II**, 225 (1943).

earlier as one of the general substitution mechanisms of diazonium intermediates.[55,61] Copper catalysis is not necessary for introduction of iodine. Reaction of aryl diazonium salts with solutions of potassium or sodium iodide occurs readily, giving the aryl iodides, usually in good yield. This reaction does not seem to have been closely investigated mechanistically, but it is suspected that both the reducing ability of iodide and the formation of triiodide ion may play a role in the facility of the reaction, compared with that of the other halide ions.

Introduction of fluorine into aromatic rings is carried out using isolated diazonium tetrafluoroborate salts. These compounds are then thermally decomposed, giving the aryl fluoride. This process is known as the *Schiemann reaction.*[62]

$$Ar\overset{+}{N}\equiv N \ \ BF_4^- \ \rightarrow \ ArF \ + \ N_2 \ + \ BF_3$$

The mechanism involves the aryl cation, which abstracts fluoride from the BF_4^- anion, generating the reaction products.[63] The hexafluorophosphate salts behave similarly.[64]

The cyano group can be introduced via reaction of diazonium intermediates with cuprous cyanide. All indications are that an electron-transfer mechanism like that in the reaction with cuprous halides operates.

Another useful reaction employs aryl diazonium ions and alkali metal azides. Aryl azides result in good yield. There have been kinetic studies of the reaction of aryl diazonium ions with azide ion.[54] Analysis of these data indicates that two intermediates are formed, both of which eventually decompose to the azide:

$$Ar\overset{+}{N}\equiv N \ + \ N_3^- \ \rightarrow \ Ar-N\underset{N=N}{\overset{N}{\diagdown}}N \ + \ Ar-N=N-N=\overset{+}{N}=\overset{-}{N}$$

$$\text{slow}\searrow \qquad \qquad \swarrow\text{fast}$$

$$ArN_3 \ + \ N_2$$

This reaction, then, represents an example of the second general mechanism for substitution in diazonium compounds, i.e., substitution via an unstable adduct.

A number of examples of substitution via diazonium intermediates are collected in Scheme 7.6.

7.2.2. Nucleophilic Aromatic Substitution by Addition–Elimination

The addition of a nucleophile to an aromatic ring, followed by elimination of a substituent, results in nucleophilic substitution. The major energetic requirement for

55. See p. 277.
61. S. C. Dickerman, D. J. DeSousa, and N. Jacobson, *J. Org. Chem.* **34**, 710 (1969).
62. A. Roe, *Org. React.* **5**, 193 (1949).
63. C. G. Swain and R. J. Rogers, *J. Am. Chem. Soc.* **97**, 799 (1975).
64. M. S. Newman and R. H. B. Galt, *J. Org. Chem.* **25**, 214 (1960).
54. See p. 277.

this mechanism is the formation of the addition intermediate. The addition step is greatly facilitated by strongly electron-withdrawing substituents, so that nitroaromatics are the best substrates for nucleophilic aromatic substitution. Other electron-withdrawing groups such as cyano, acetyl, and trifluoromethyl also increase reactivity, but to a lesser extent than the nitro group. The intermediate adducts have

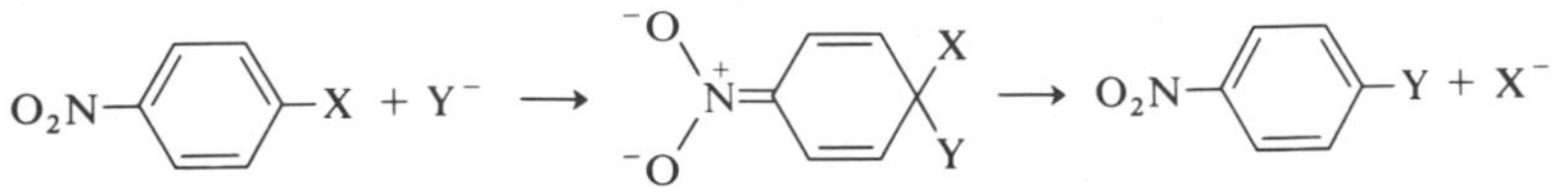

appreciable stability under certain conditions, and are frequently referred to as *Meisenheimer complexes.*[65] Those derived from nitroaromatics are highly colored. Good examples are the compounds obtained by reaction of alkoxide ions with alkyl 2,4,6-trinitrophenyl ethers:

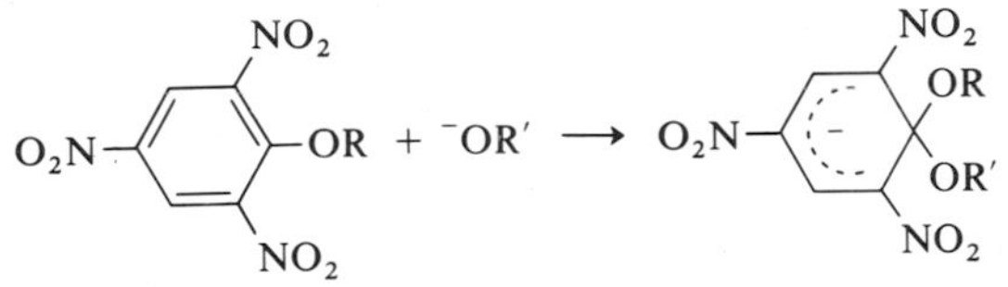

A wide variety of such compounds have been characterized, and recent example include even compounds in which the formal nucleophile is an alkyl group[66]:

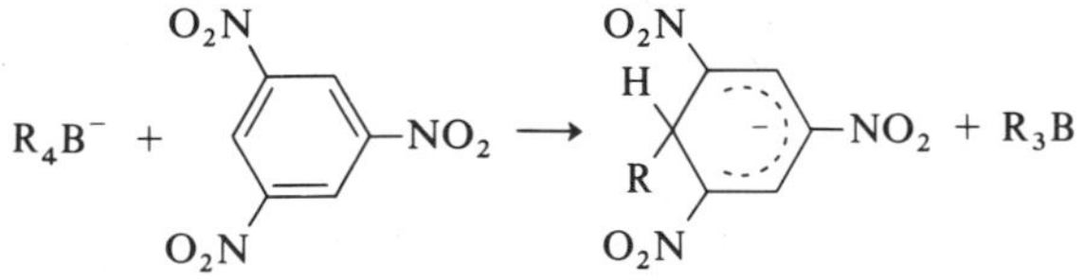

Many other nucleophiles, such as cyanide, amines, thiolates, and enolates, add to nitroaromatics to give similar complexes.

Nucleophilic substitution occurs through similar intermediates when the aromatic ring contains a potential leaving group. The most common case involves displacement of halide, but alkoxy, nitro, and cyano groups can also be displaced by the addition–elimination mechanism. It is noteworthy that leaving-group ability in such reactions does not parallel that found for nucleophilic substitution at saturated carbon. As a particularly striking example, fluoride is often a better leaving group than the other halogens in nucleophilic aromatic substitution. The relative reactivity of the p-halonitrobenzenes toward sodium methoxide at $50°$ is F (312) $\gg$ Cl (1) $>$ Br (0.74) $>$ I (0.36).[67] A principal reason for the order I $>$ Br $>$ Cl $>$ F in S_N2 reactions is the carbon–halogen bond strength, which increases from I to F. Bond strength is not an important factor in nucleophilic aromatic substitution, because

65. M. J. Strauss, *Chem. Rev.* **70**, 667 (1970).
66. R. P. Taylor, *J. Org. Chem.* **35**, 3578 (1970).
67. G. P. Briner, J. Miller, M. Liveris, and P. G. Lutz, *J. Chem. Soc.*, 1265 (1954).

bond-breaking is not ordinarily part of the rate-determining step. The highly electronegative fluorine stabilizes the addition transition state more effectively than the other halogens and therefore has an accelerating effect.

There have been a large number of detailed studies, especially involving kinetic measurements, that have helped to determine the reactivity of various nucleophiles, solvent effects, and the finer details of aromatic nucleophilic substitutions proceeding via the addition–elimination mechanism. We will not attempt to summarize these results here, since reviews are available.[68] Carbanions, alkoxides, and amines are all reactive in nucleophilic aromatic substitution and provide most of the cases in which this reaction has been used preparatively. Some examples are given in Scheme 7.7.

7.2.3. Nucleophilic Aromatic Substitution by Elimination–Addition

The elimination–addition mechanism involves a highly unstable intermediate, which is referred to as *dehydrobenzene* or *benzyne*[69]:

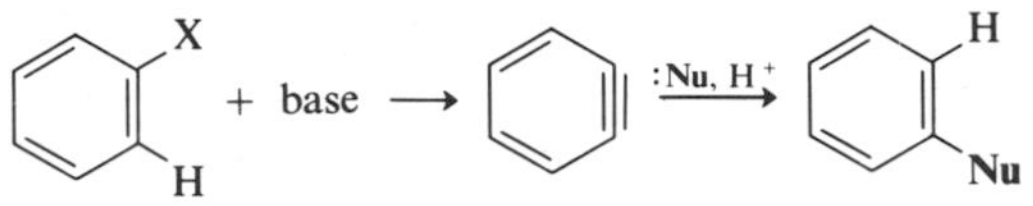

A characteristic feature of this mechanism is the substitution pattern in the product. The entering nucleophile need not always enter the ring at the carbon to which the leaving group was bound:

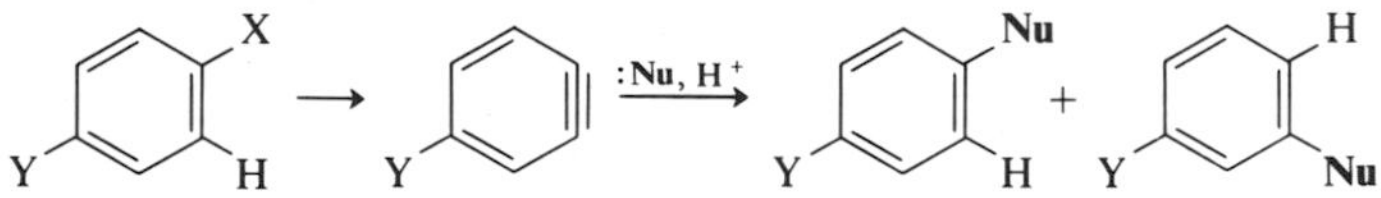

Benzyne has been observed spectroscopically in an inert solid matrix at very low temperatures.[70] For these studies, the molecule was generated photolytically:

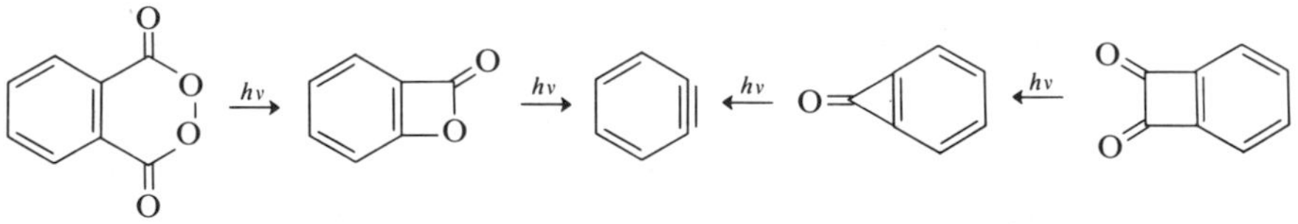

There have been several structural representations of benzyne. The one most generally used represents benzyne as similar to benzene, but with a weak π-bond in the plane of the ring formed using two sp^2-orbitals.[71] Molecular-orbital calculations

68. J. Miller, *Aromatic Nucleophilic Substitution*, Elsevier, Amsterdam, London, New York, 1968; F. Pietra, *Q. Rev. Chem. Soc.* **23**, 504 (1969); C. F. Bernasconi, in *Aromatic Compounds*, H. Zollinger (ed.), University Park Press, Baltimore, MD, 1973, Chapter 2.
69. R. W. Hoffmann, *Dehydrobenzene and Cycloalkynes*, Academic Press, New York, NY, 1967.
70. O. L. Chapman, K. Mattes, C. L. McIntosh, J. Pacansky, G. V. Calder, and G. Orr, *J. Am. Chem. Soc.* **95**, 6134 (1973).
71. H. E. Simmons, *J. Am. Chem. Soc.* **83**, 1657 (1961).

Scheme 7.7. Nucleophilic Aromatic Substitution

1[a]

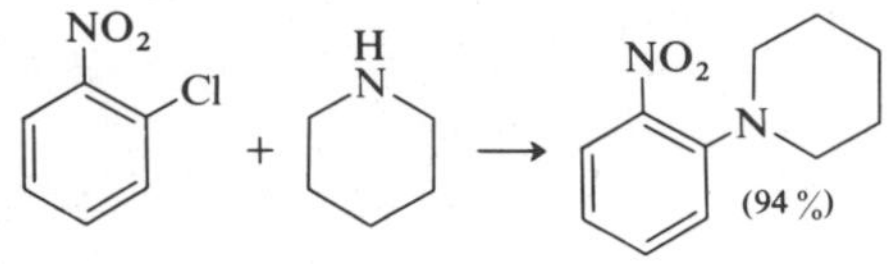

(94 %)

2[b]

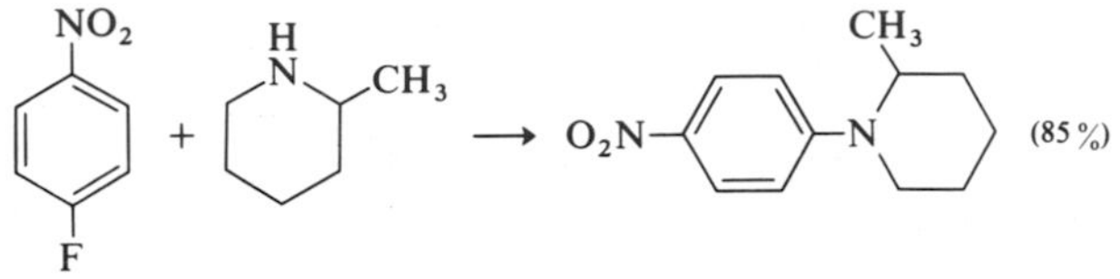

(85 %)

3[c]

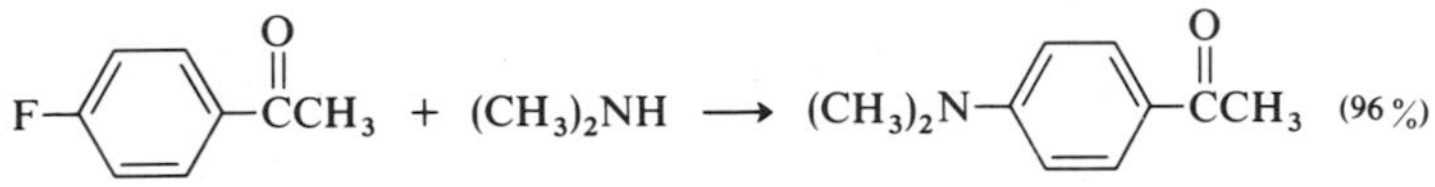

$$F-\text{C}_6H_4-\overset{O}{\overset{\|}{C}}CH_3 + (CH_3)_2NH \longrightarrow (CH_3)_2N-\text{C}_6H_4-\overset{O}{\overset{\|}{C}}CH_3 \quad (96\%)$$

4[d]

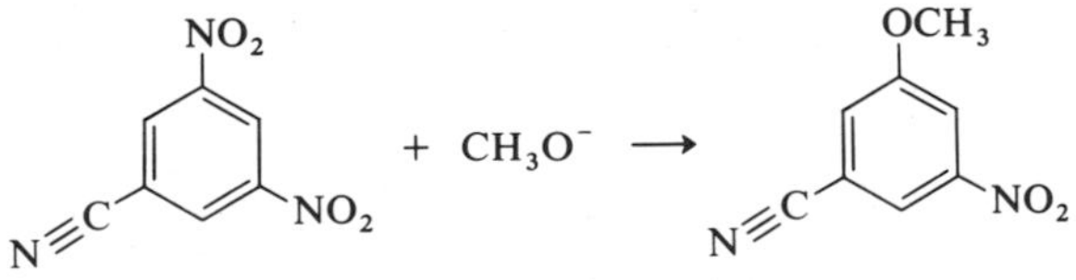

$+ CH_3O^- \longrightarrow$

5[e]

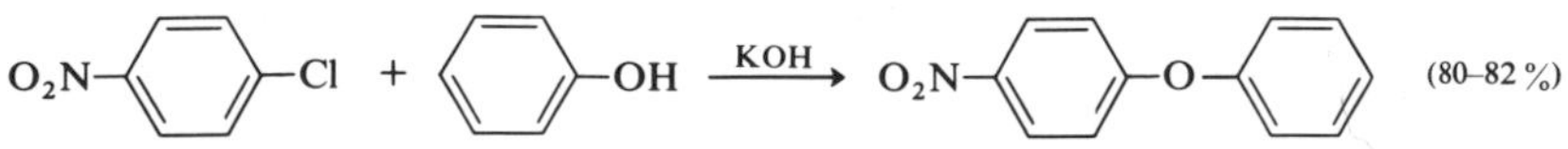

$$O_2N-\text{C}_6H_4-Cl + \text{C}_6H_5-OH \xrightarrow{KOH} O_2N-\text{C}_6H_4-O-\text{C}_6H_5 \quad (80\text{--}82\%)$$

6[f]

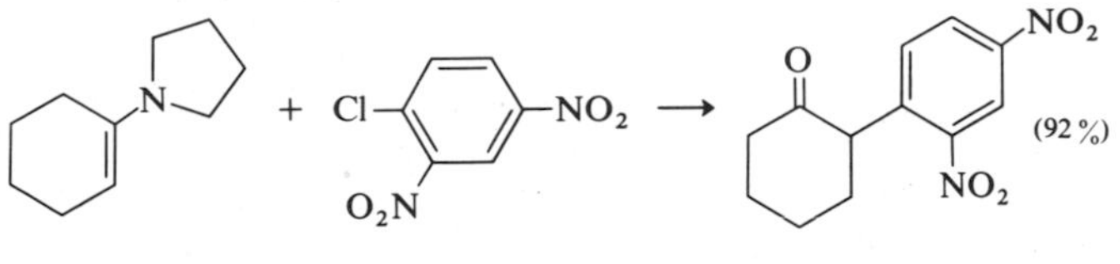

(92 %)

7[g]

$$N\equiv CCHCO_2C_2H_5 + Cl-\text{C}_6H_3(NO_2)(O_2N) \longrightarrow H_5C_2O_2CCH-\text{C}_6H_3(NO_2)(O_2N)$$

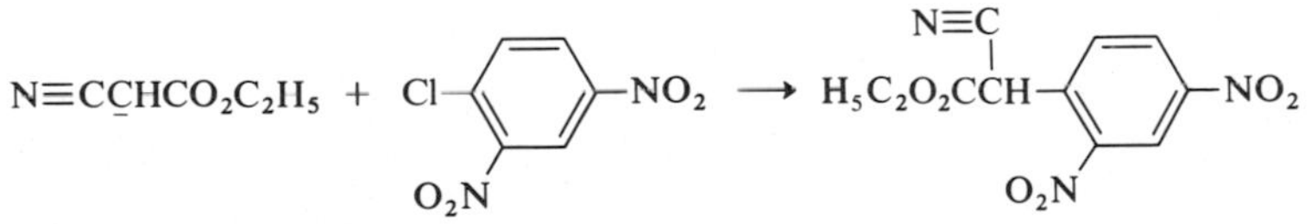

a. S. D. Ross and M. Finkelstein, *J. Am. Chem. Soc.* **85**, 2603 (1963).
b. F. Pietra and F. Del Cima, *J. Org. Chem.* **33**, 1411 (1968).
c. H. Bader, A. R. Hansen, and F. J. McCarty, *J. Org. Chem* **31**, 2319 (1966).
d. E. J. Fendler, J. H. Fendler, N. L. Arthur, and C. E. Griffin, *J. Org. Chem.* **37**, 812 (1972).
e. R. O. Brewster and T. Groening, *Org. Synth.* **II**, 445 (1943).
f. M. E. Kuehne, *J. Am. Chem. Soc.* **84**, 837 (1962).
g. H. R. Snyder, E. P. Merica, C. G. Force, and E. G. White, *J. Am. Chem. Soc.* **80**, 4622 (1958).

indicate that there is additional bonding between the "dehydro" carbons, though the strength of the bond is much less than a normal triple bond.[72]

An early case in which the existence of benzyne as a reaction intermediate was established was in the reaction of chlorobenzene with potassium amide. ^{14}C-Label in the starting material was found to be distributed in the aniline as expected for a benzyne intermediate.[73]

The elimination–addition mechanism is facilitated by electronic effects that favor removal of a hydrogen from the ring as a proton. Relative reactivity also depends on the halide. The order $Br > I > Cl \gg F$ has been established in the reaction of aryl halides with KNH_2 in liquid ammonia.[74] This order has been interpreted as representing a balance of two effects. The inductive order favoring proton removal would be $F > Cl > Br > I$, but this is largely overwhelmed by the order of leaving-group ability $I > Br > Cl > F$, which reflects bond strengths. With organometallic bases in aprotic solvents, the acidity of the hydrogen is the dominant factor, and the reactivity order is $F > Cl > Br > I$.[75]

Addition of nucleophiles such as ammonia or alcohols or their conjugate bases to benzynes takes place very rapidly. These nucleophilic additions are believed to involve capture of the nucleophile by benzyne, followed by protonation to give the substituted benzene.[76] Some evidence for the two-step mechanism can be drawn

from the effect of substituent groups on the direction of nucleophilic addition. Electron-attracting groups tend to favor addition of the nucleophile at the more distant end of the "triple bond," since this permits maximum stabilization of the developing negative charge. Selectivity is usually not high, however, and formation of both possible products from monosubstituted benzynes is common.[77]

72. R. Hoffman, A. Imamura, and W. J. Hehre, *J. Am. Chem. Soc.* **90**, 1499 (1968); D. L. Wilhite and J. L. Whitten, *J. Am. Chem. Soc.* **93**, 2858 (1971).

73. J. D. Roberts, D. A. Semenow, H. E. Simmons, Jr., and L. A. Carlsmith, *J. Am. Chem. Soc.* **78**, 601 (1956).

74. F. W. Bergstrom, R. E. Wright, C. Chandler, and W. A. Gilkey, *J. Org. Chem.* **1**, 170 (1936).

75. R. Huisgen and J. Sauer, *Angew. Chem.* **72**, 91 (1960).

76. J. F. Bunnett, D. A. R. Happer, M. Patsch, C. Pyun, and H. Takayama, *J. Am. Chem. Soc.* **88**, 5250 (1966); J. F. Bunnett and J. K. Kim, *J. Am. Chem. Soc.* **95**, 2254 (1973).

77. E. R. Biehl, E. Nieh, and K. C. Hsu, *J. Org. Chem.* **34**, 3595 (1969).

There are several methods for generation of benzyne in addition to base-catalyzed elimination of hydrogen halide from a halobenzene, and some of these are more generally applicable for preparative work. Probably the most useful method is diazotization of *o*-aminobenzoic acids.[78] Concerted loss of nitrogen and carbon dioxide follows diazotization and generates benzyne. Benzyne can be formed in this manner in the presence of a variety of compounds with which it reacts rapidly. Some specific examples will be given shortly.

$$\text{(o-aminobenzoic acid)} \xrightarrow{\text{HONO}} \text{(diazonium carboxylate)} \longrightarrow \text{(benzyne)} + CO_2 + N_2$$

Oxidation of 1-aminobenzotriazole also serves as a source of benzyne under mild conditions. An oxidized intermediate decomposes with loss of two molecules of nitrogen[79]:

$$\text{(1-aminobenzotriazole)} \longrightarrow \text{(intermediate)} \longrightarrow \text{(benzyne)} + 2\,N_2$$

Another heterocyclic molecule that can serve as a benzyne precursor is benzothiadiazole-1,1-dioxide, which decomposes with elimination of nitrogen and sulfur dioxide[80]:

$$\text{(benzothiadiazole-1,1-dioxide)} \longrightarrow \text{(benzyne)} + SO_2 + N_2$$

Benzyne can also be generated from *o*-dihaloaromatics. Reaction with lithium–amalgam (or magnesium) results in the formation of an organometallic compound that decomposes with elimination of lithium halide. *o*-Fluorobromobenzene is the usual starting material in this procedure.[81]

$$\text{(o-fluorobromobenzene)} \xrightarrow{\text{Li–Hg}} \left[\text{(o-fluorophenyllithium)} \right] \longrightarrow \text{(benzyne)}$$

The diazotization of substituted anilines by alkyl nitrites in acetic anhydride also

78. M. Stiles, R. G. Miller, and U. Burckhardt, *J. Am. Chem. Soc.* **85**, 1792 (1963); L. Friedman and F. M. Logullo, *J. Org. Chem.* **34**, 3089 (1969).
79. C. D. Campbell and C. W. Rees, *Proc. Chem. Soc. London*, 296 (1964); *J. Chem. Soc. C*, 752.
80. G. Wittig and R. W. Hoffmann, *Org. Synth.* **47**, 4 (1967); G. Wittig and R. W. Hoffmann, *Chem. Ber.* **95**, 2718, 2729 (1962).
81. G. Wittig and L. Pohmer, *Chem. Ber.* **89**, 1334 (1956); G. Wittig, *Org. Synth.* **IV**, 964 (1963).

leads to benzyne.[82] Several steps are involved in this process. The key intermediate is the corresponding N-nitrosoacetanilide, and it has been shown that these compounds, prepared separately, can also serve as benzyne precursors. As described in Part A, Section 12.3, the nitrosoacetanilides can serve as precursors of aryl diazonium–acetate ion pairs:

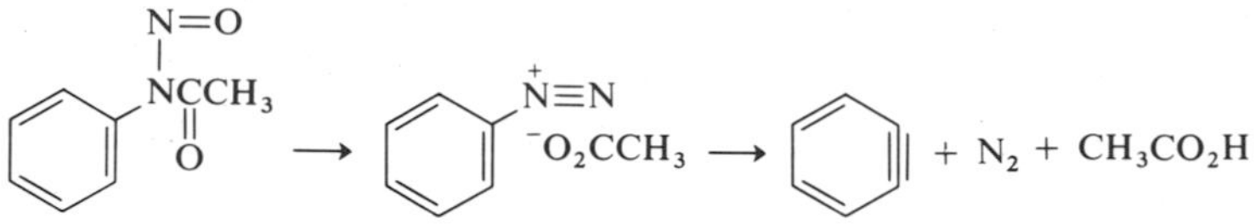

When the reaction conditions are such that acetate acts as a base, removing the proton at an *ortho* position, nitrogen is eliminated and benzyne formed.[83] The o-t-butylanilines are especially good substrates for benzyne formation via the N-nitrosoacetanilide, apparently because the t-butyl group sterically accelerates elimination of nitrogen.[84]

When benzyne is generated in the presence of unsaturated molecules, additions at the highly strained "triple bond" are observed. Benzyne is capable of dimerization, so that in the absence of either nucleophiles or a reactive unsaturated compound, biphenylene is formed.[85] Among the best reagents for reaction with benzyne are

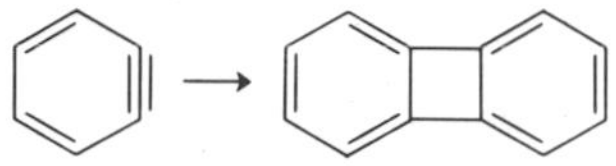

furans and cyclopentadienones, which give $4+2$ cycloaddition products. Anthracene also undergoes cycloaddition to give triptycene.

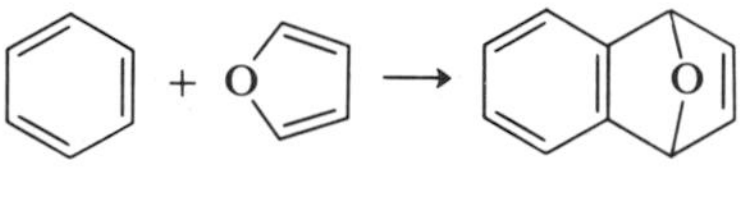

Ref. 86

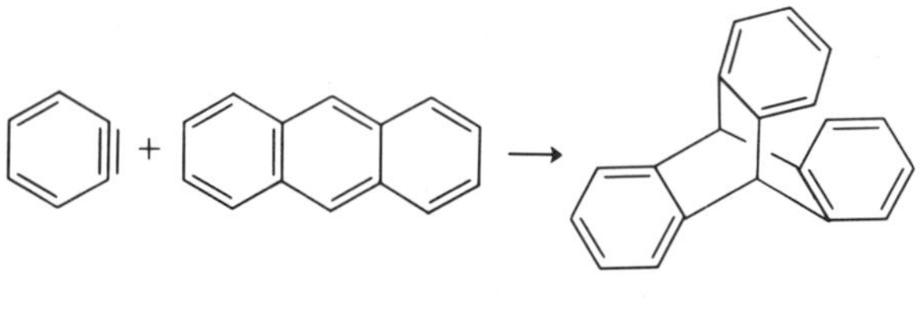

Ref. 87

The stereochemistry of both $2+2$ and $2+4$ cycloadditions has been investigated. For dichloroethylenes, the $2+2$ addition is not stereospecific; for dienes, however, the $4+2$ addition is a stereospecific *cis* addition. Benzyne thus

82. A. Baigrie, J. I. G. Cadogan, J. R. Mitchell, A. K. Robertson, and J. T. Sharp, *J. Chem. Soc. Perkin Trans. I*, 2563 (1972).

83. D. L. Brydon, J. I. G. Cadogan, J. Cook, M. J. P. Harger, and J. T. Sharp, *J. Chem. Soc. B*, 1996 (1971); C. Rüchardt and C. C. Tan, *Angew. Chem. Int. Ed. Engl.* **9**, 522 (1970).

84. R. M. Franck and K. Yanagi, *J. Am. Chem. Soc.* **90**, 5814 (1968).

85 F. M. Logullo, A. H. Seitz, and L. Friedman, *Org. Synth.* **48**, 12 (1968).

86. G. Wittig and L. Pohmer, *Angew. Chem.* **67**, 348 (1955).

87. L. Friedman and F. M. Logullo, *J. Org. Chem.* **34**, 3089 (1969).

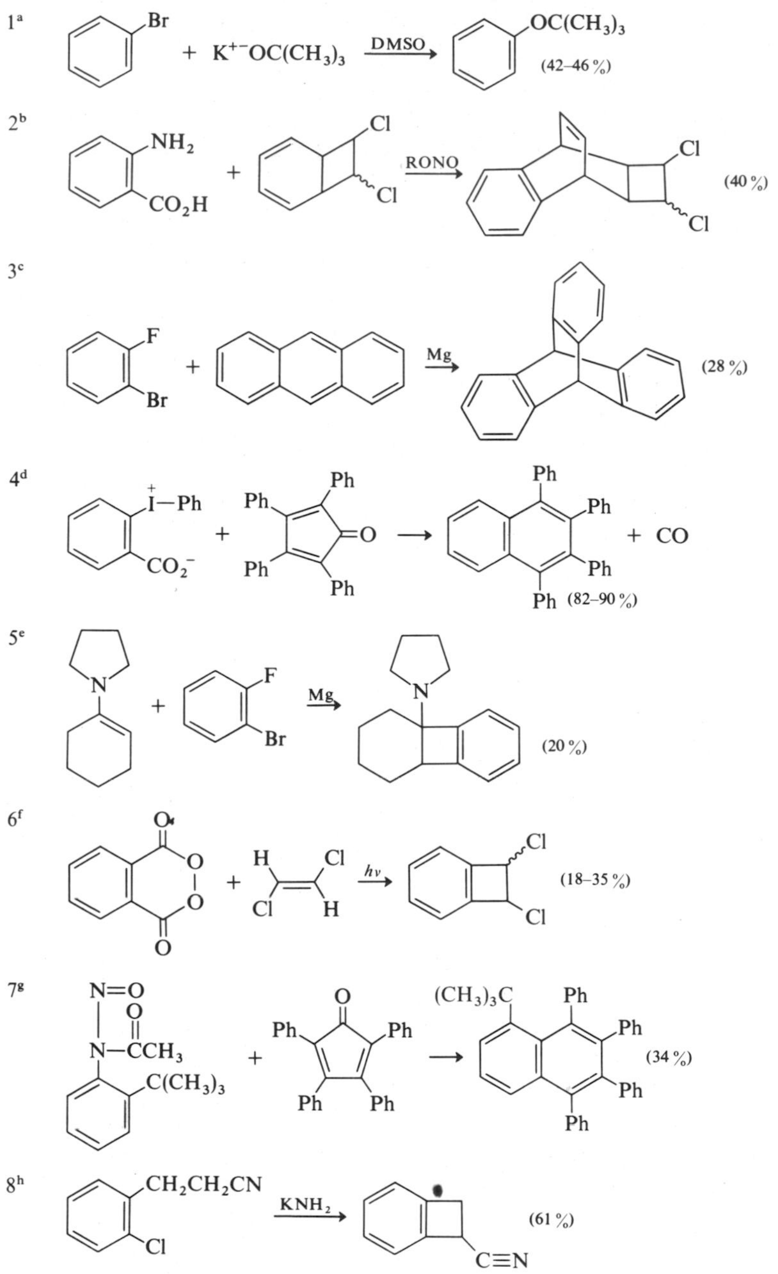

Scheme 7.8. Some Syntheses via Benzyne Intermediates

1[a]

+ K$^+$$^-$OC(CH$_3$)$_3$ $\xrightarrow{\text{DMSO}}$ (42–46%)

2[b] $\xrightarrow{\text{RONO}}$ (40%)

3[c] $\xrightarrow{\text{Mg}}$ (28%)

4[d] + CO (82–90%)

5[e] $\xrightarrow{\text{Mg}}$ (20%)

6[f] $\xrightarrow{h\nu}$ (18–35%)

7[g] (34%)

8[h] $\xrightarrow{\text{KNH}_2}$ (61%)

a. M. R. V. Sahyun and D. J. Cram, *Org. Synth.* **45**, 89 (1965).
b. L. A. Paquette, M. J. Kukla, and J. C. Stowell, *J. Am. Chem. Soc.* **94**, 4920, (1972).
c. G. Wittig, *Org. Synth.* **IV**, 964 (1963).
d. L. F. Fieser and M. J. Haddadin, *Org. Synth.* **46**, 107 (1966).
e. M. E. Kuehne, *J. Am. Chem. Soc.* **84**, 837 (1962).
f. M. Jones, Jr., and M. R. DeCamp, *J. Org. Chem.* **36**, 1536 (1971).
g. D. L. Brydon, J. I. G. Cadogan, J. Cook, M. J. P. Harger, and J. T. Sharp, *J. Chem. Soc. B,* 1996 (1971).
h. J. F. Bunnett and J. A. Skorcz, *J. Org. Chem.* **27**, 3836 (1962).

shows behavior parallel to ground-state ethylenes in undergoing concerted $2+4$ cycloaddition, but nonconcerted $2+2$ additions.[88]

Scheme 7.8 illustrates some of the types of compounds that can be prepared via benzyne intermediates.

7.2.4. Copper-Catalyzed Nucleophilic Aromatic Substitution

One group of nucleophilic aromatic substitution processes does not fit mechanistically into the previous categories. This group is a series of copper-catalyzed displacements of aromatic halogen compounds.

The most important member of the group is an effective synthesis of aryl nitriles that involves heating the aryl halide with cuprous cyanide. Some examples of the reaction are given in Scheme 7.9.

There are several closely related processes involving other nucleophiles that, at least to date, have found less synthetic application. For example, heating aryl halides with cuprous benzoate produces aryl benzoates:

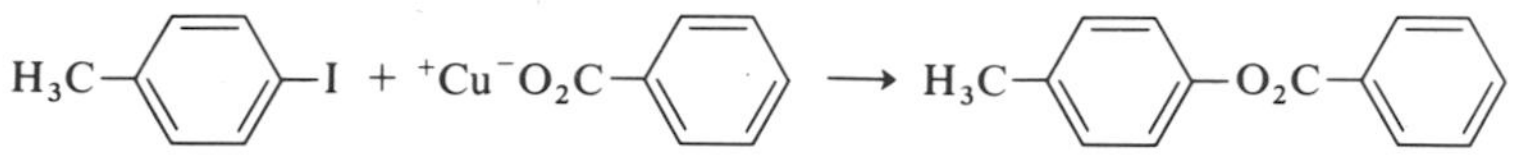

Ref. 89

Another example is the rapid displacement of iodine by ammonia in the presence of cuprous trifluoromethylsulfonate:

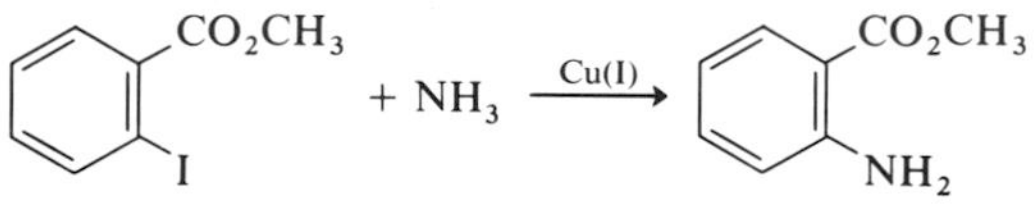

Ref. 90

The scope of these two types of reactions is not yet well documented. All, however, are believed to proceed through an organocopper species formed from the aryl halide.[91]

$$Ar-X + Cu(I) \rightarrow Ar-Cu-X$$

$$Ar-Cu-X + Nu: \rightarrow Ar-Nu + Cu(I) + X^-$$

7.3. Substitutions Involving Aryl Free Radicals

Perhaps the most useful application of aromatic substitution reactions that proceed by way of free radicals is for the synthesis of biphenyls.[92] An aryl radical is

88. M. Jones, Jr., and R. H. Levin, *J. Am. Chem. Soc.* **91**, 6411 (1969).
89. T. Cohen and A. H. Lewin, *J. Am. Chem. Soc.* **88**, 4521 (1966).
90. T. Cohen and J. G. Tirpak, *Tetrahedron Lett.*, 143 (1975).
91. T. Cohen, J. Wood, and A. G. Dietz, Jr., *Tetrahedron Lett.*, 3555 (1974).
92. W. E. Bachmann and R. A. Hoffman, *Org. React.* **2**, 224 (1944).

Scheme 7.9. Preparation of Aryl Cyanides from Haloaromatics

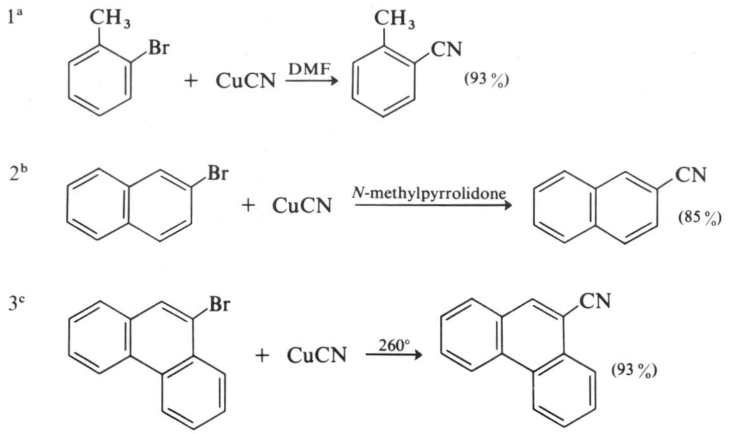

289

SECTION 7.3
SUBSTITUTIONS
INVOLVING
ARYL FREE
RADICALS

1[a]   CH$_3$ Br   + CuCN $\xrightarrow{\text{DMF}}$ CH$_3$ CN   (93%)

2[b]   Br   + CuCN $\xrightarrow{\text{N-methylpyrrolidone}}$ CN   (85%)

3[c]   Br   + CuCN $\xrightarrow{260°}$ CN   (93%)

a. L. Friedman and H. Shechter, *J. Org. Chem.* **26**, 2522 (1961).
b. M. S. Newman and H. Boden, *J. Org. Chem.* **26**, 2525 (1961).
c. J. E. Callen, C. A. Dornfeld, and G. H. Coleman, *Org. Synth.* **III**, 212 (1955).

generated in the presence of an excess of a second aromatic compound, which

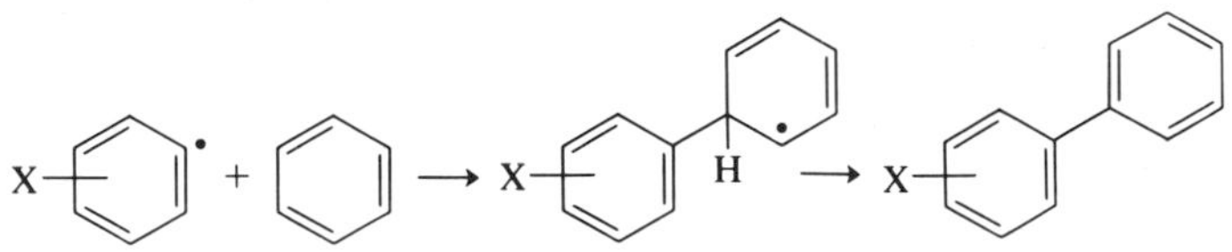

X—⟨ ⟩• + ⟨ ⟩ → X—⟨ ⟩—⟨ ⟩H • → X—⟨ ⟩—⟨ ⟩

undergoes substitution. Any of several sources of aryl radicals can be used. Decomposition of a diazonium ion is probably the most common, but thermal decomposition of *N*-nitrosoacetanilides or aroyl peroxides are alternative possibilities. Some examples are given in section A of Scheme 7.10.

One point that is important in limiting the synthetic utility of aromatic substitutions involving radical intermediates is that substituent-directing effects are not large in radical substitution reactions. Some partial rate factors for substitution have been measured; for nonpolar radicals such as phenyl, these fall within the rather narrow range 0.5–6.5.[93] This means that substituent groups do not strongly stabilize or destabilize the transition states for radical substitution. The practical result is that homolytic aromatic substitution usually gives a mixture of all possible substitution products in comparable amounts. Some of the largest directing effects have been noted for the nitro and cyano groups. In contrast to their strong *meta*-directing effect in electrophilic substitution, both favor substitution at the *ortho* and *para* positions; this difference is due to the ability of these substituents to delocalize the unpaired

93. G. H. Williams, *Homolytic Aromatic Substitution*, Pergamon Press, London, 1960; D. H. Hey, *Adv. Free-Radical Chem.* **2**, 47 (1966).

Scheme 7.10. Aromatic Substitution Involving Radical Intermediates

A. Arylation of Aromatic Compounds

1[a]

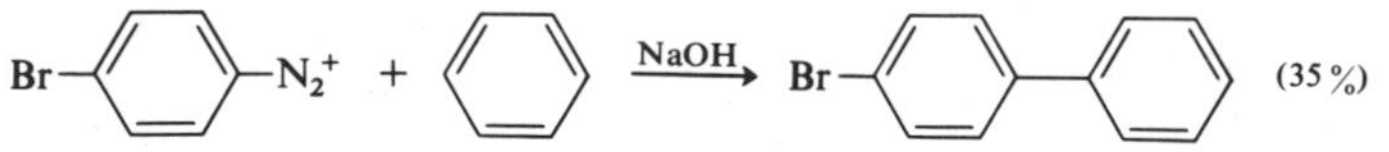

$$Br-C_6H_4-N_2^+ \ + \ C_6H_6 \ \xrightarrow{NaOH} \ Br-C_6H_4-C_6H_5 \quad (35\%)$$

2[b]

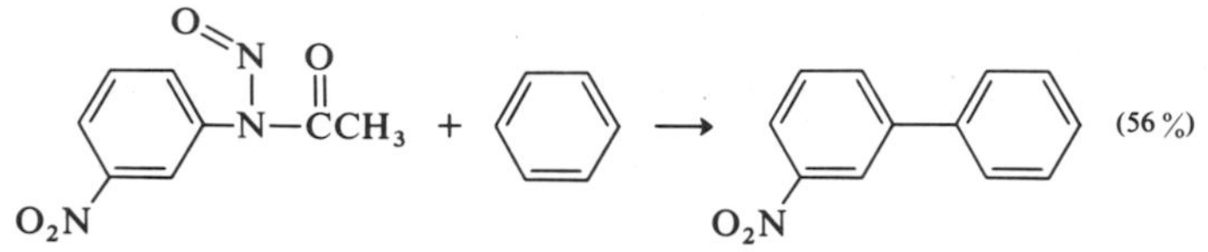

(56%)

3[c]

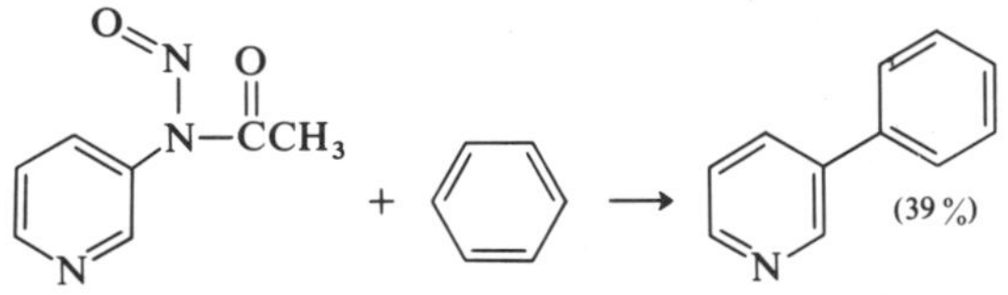

(39%)

4[d]

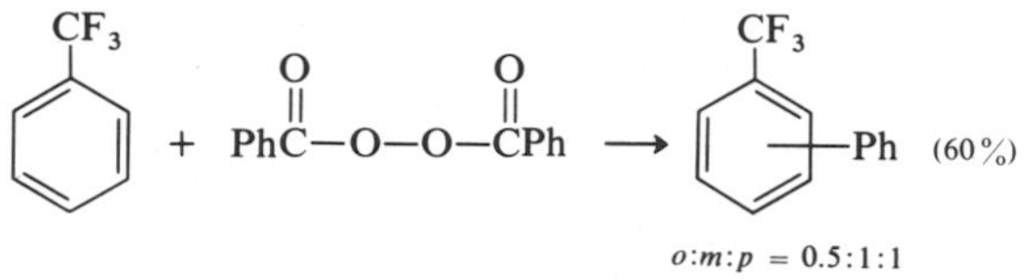

$$CF_3-C_6H_5 \ + \ PhC(O)-O-O-C(O)Ph \ \longrightarrow \ CF_3-C_6H_4-Ph \quad (60\%)$$

$$o:m:p = 0.5:1:1$$

5[e]

$$Cl-C_6H_4-NH_2 \ + \ C_6H_6 \ \xrightarrow{C_5H_{11}ONO} \ Cl-C_6H_4-C_6H_5 \quad (45\%)$$

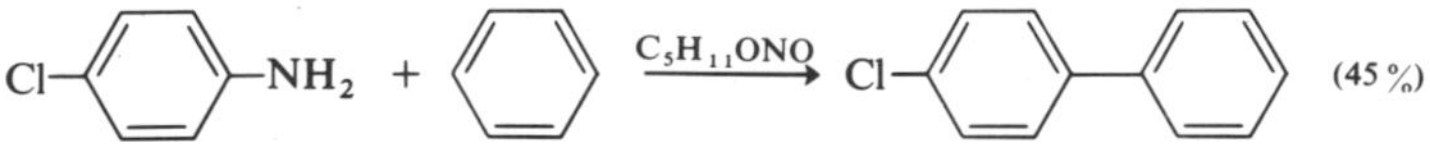

B. Arylation of Alkenes

6[f]

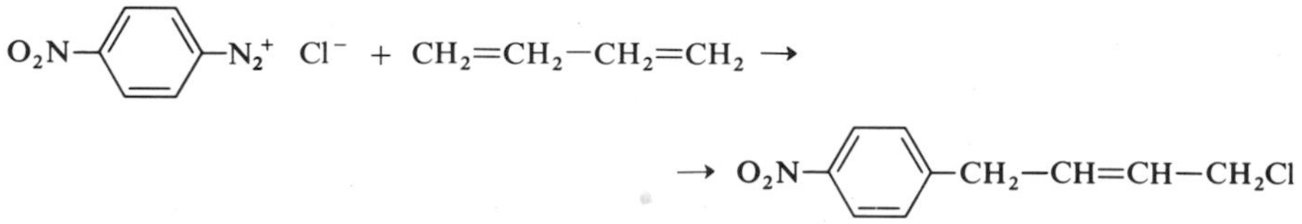

$$O_2N-C_6H_4-N_2^+ \ Cl^- \ + \ CH_2=CH-CH_2=CH_2 \ \longrightarrow$$

$$\longrightarrow \ O_2N-C_6H_4-CH_2-CH=CH-CH_2Cl$$

7[g]

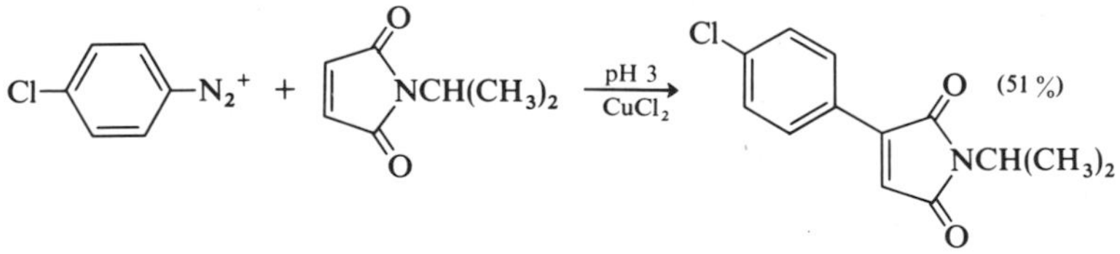

$$Cl-C_6H_4-N_2^+ \ + \ NCH(CH_3)_2 \ \xrightarrow[CuCl_2]{pH\ 3} \quad (51\%)$$

8[h]

$$O_2N-C_6H_4-N_2^+ \ + \ CH_2=CHCN \ \xrightarrow{CuCl_2} \ O_2N-C_6H_4-CH_2CHCN \quad (48\%)$$

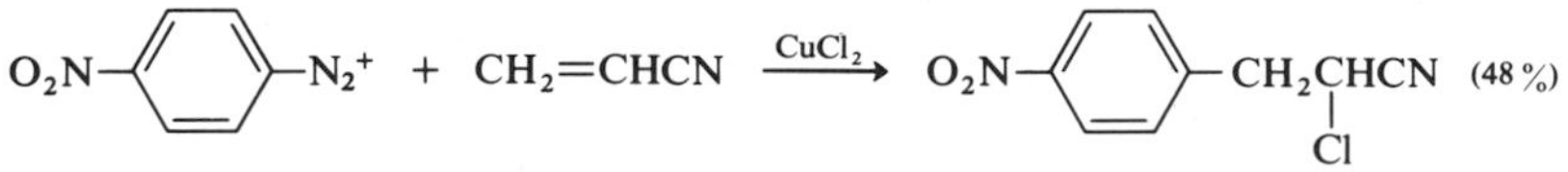

a. M. Gomberg and W. E. Bachmann, *Org. Synth.* **I**, 113 (1941).
b. W. E. Bachmann and R. A. Hoffman, *Org. React.* **2**, 224 (1944).
c. H. Rapoport, M. Look, and G. J. Kelly, *J. Am. Chem. Soc.* **74**, 6293 (1952).
d. C. S. Rondestvedt, Jr., and H. S. Blanchard, *J. Org. Chem.* **21**, 229 (1956).
e. J. I. G. Cadogan, *J. Chem. Soc.*, 4257 (1962).
f. G. A. Ropp and E. C. Coyner, *Org. Synth.* **IV**, 727 (1963).
g. C. S. Rondestvedt, Jr., and O. Vogl, *J. Am. Chem. Soc.* **77**, 2313 (1955).
h. C. F. Koelsch, *J. Am. Chem. Soc.* **65**, 57 (1943).

291

SECTION 7.3.
SUBSTITUTIONS
INVOLVING
ARYL FREE
RADICALS

electron. Because of the weak directing effects, a monosubstituted aromatic compound is likely to give comparable amounts of o-, m-, and p-substitution products when attacked by aryl radicals. Entry 4 in Scheme 7.10 illustrates this point.

Aryl radicals can also be used to introduce alkyl substituents onto aromatic rings. The aryl radical is generated by copper-catalyzed decomposition of an aryl diazonium ion and attacks an alkene. The radical that results is oxidized by Cu(II),

giving a carbonium ion, which gives rise to either an alkene or a halide. The reaction is known as the *Meerwein arylation reaction.*[94] Scheme 7.10 gives several examples.

Homolytic aromatic substitution can also occur via electron-transfer reactions, even in the absence of transition-metal catalysts. For example, irradiation of a liquid ammonia solution of acetone enolate with any of the halobenzenes results in the formation of phenylacetone.[95] The proposed mechanism involves electron transfer,

decomposition of the resulting aromatic radical anion, and combination of the phenyl radical with enolate. A chain reaction results, since the phenylacetone radical anion can transfer an electron to bromobenzene. A similar reaction occurs when

94. C. S. Rondestvedt, Jr., *Org. React.* **11**, 189 (1960).
95. R. A. Rossi and J. F. Bunnett, *J. Org. Chem.* **38**, 1407 (1973).

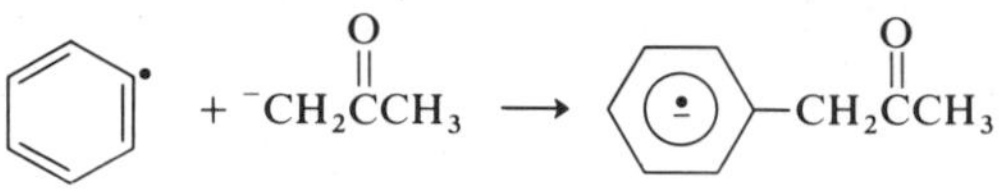

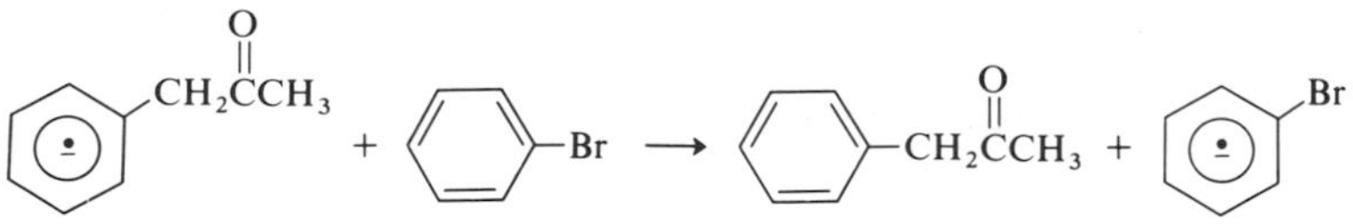

α-cyanocarbanions are employed, but the yields of alkylation product are greatly decreased by formation of by-products resulting from expulsion of cyanide ion from the adduct radical anion.[96]

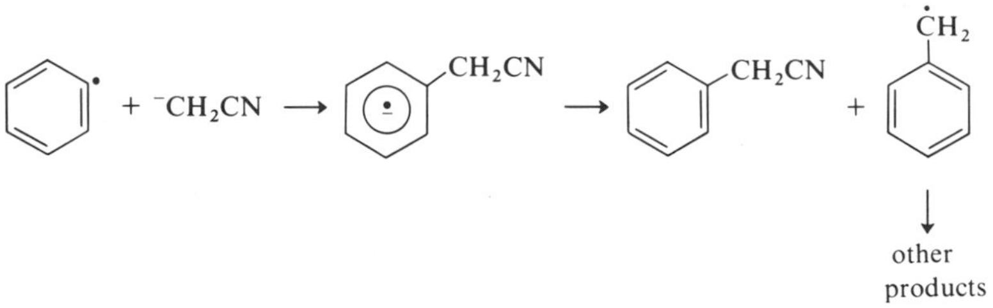

7.4. Reactivity of Polycyclic Aromatics

The types of reactions that can be carried out on the polycyclic aromatic hydrocarbons such as naphthalene, anthracene, phenanthrene, and larger analogs are the same as those discussed in detail for benzene derivatives. An acquaintance with the substitution chemistry of the polycyclic systems therefore requires only some general information concerning the reactivity of these ring systems and their position selectivity. Naphthalene is more reactive than benzene. There is a kinetic

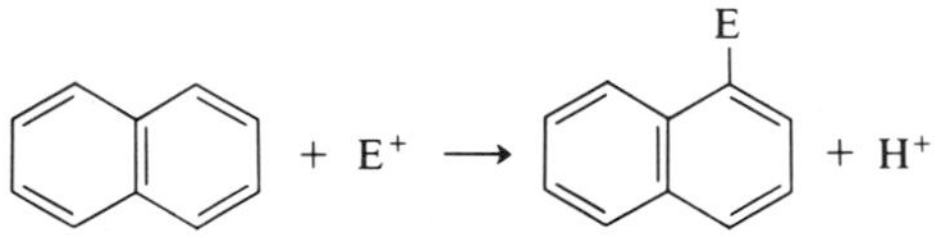

preference for substitution at the 1-position. Two factors can result in dominant substitution at C-2. If the electrophile is very bulky, the hydrogen on the adjacent ring may cause a steric preference for attack at C-2. Under conditions of reversible substitution, where thermodynamic stability is the controlling factor, 2-substitution is preferred. This factor is illustrated by the outcome of sulfonation, where low-temperature reaction gives the 1-sulfonic acid, but high-temperature conditions give the 2-isomer.

96. J. F. Bunnett and B. F. Gloor, *J. Org. Chem.* **38**, 4156 (1973).

Phenanthrene and anthracene are both much more reactive than benzene, and there is a preference for substitution to occur in the center ring. That this behavior would be expected is evident even from simple resonance considerations. The σ-complexes that result from substitution in the center ring have two intact benzene rings. The total resonance stabilization of this intermediate is larger than that of the naphthalene system that results if substitution occurs in one of the terminal rings.

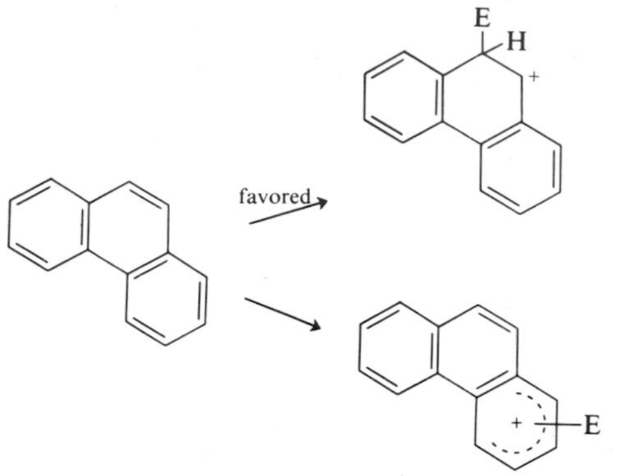
favored
E
H
+
E
(four isomers possible)

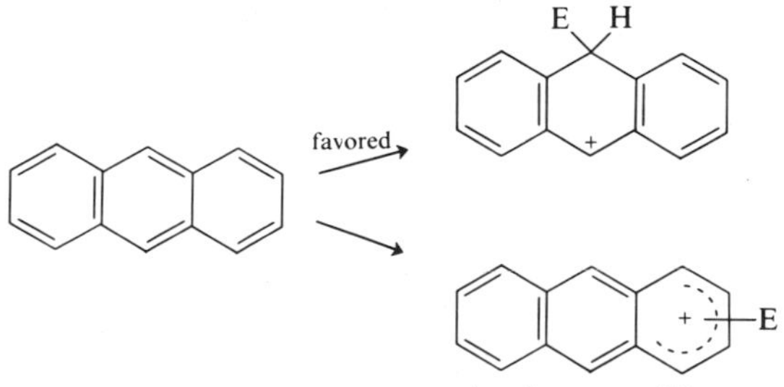
favored
E
H
+
E
(two isomers possible)

Both phenanthrene and anthracene have a tendency to undergo addition reactions under the conditions involved in certain electrophilic substitutions. Halogenation and nitration may proceed in part via addition intermediates:

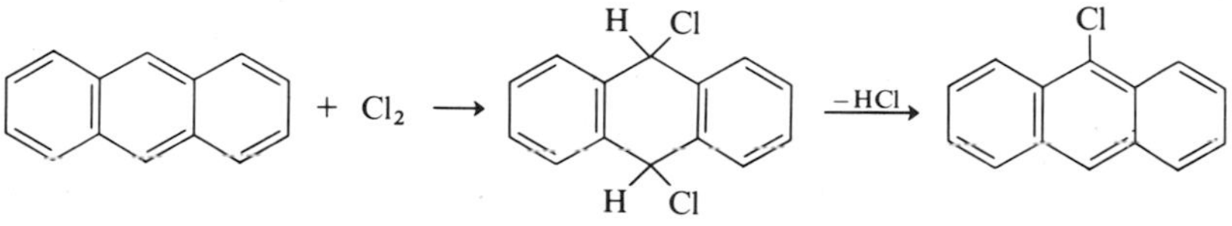
+ Cl2
H
Cl
H
Cl
−HCl
Cl

Ref. 97

The ability of the anthracene ring system to undergo concerted cycloadditions is also well established[98] (see Section 6.1). This tendency toward addition results because little, if any, resonance stabilization is lost on interruption of aromatic conjugation at

97. P. B. D. de la Mare and J. H. Ridd, *Aromatic Substitution,* Academic Press, New York, NY. 1959, p. 174.
98. W. E. Bachmann and L. B. Scott, *J. Am. Chem. Soc.* **70**, 1458 (1948).

the center ring, since two benzene rings possess about as much stabilization as the

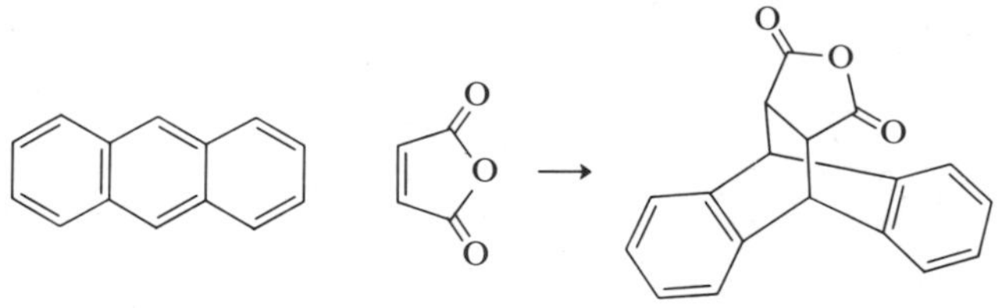

anthracene system. The molecular-orbital calculations of Dewar,[99] for example, assign the resonance stabilization of two benzene rings as 1.74 eV, whereas one anthracene ring is 1.60 eV, suggesting that addition would be accompanied by a slight *gain* in resonance energy.

The reactivity of the numerous larger polycyclic aromatic hydrocarbons has been summarized in a reference work by E. Clar.[100] A dominant feature is the increasing ease of addition reactions in the center portion of the molecule as the size of the molecule increases.

General References

G. A. Olah (ed.), *Friedel–Crafts and Related Reactions*, Vols. I–IV, Interscience, New York, NY, 1962–1964.
G. A. Olah, *Friedel–Crafts Chemistry*, Wiley–Interscience, New York, NY, 1973.
R. O. C. Norman and R. Taylor, *Electrophilic Substitution in Benzenoid Compounds*, Elsevier, Amsterdam, 1965.
L. M. Stock, *Aromatic Substitution Reactions*, Prentice-Hall, Englewood Cliffs, NJ, 1968.

Aromatic Nitration

J. G. Hoggett, R. B. Moodie, J. R. Penton, and K. Schofield. *Nitration and Aromatic Reactivity*, Cambridge University Press, Cambridge, 1971.

Nucleophilic Aromatic Substitution

J. Miller, *Aromatic Nucleophilic Substitution*, Elsevier, Amsterdam, 1968.

Benzyne Intermediates

G. Wittig, *Angew. Chem. Int. Ed. Engl.* **4**, 731 (1965); H. Heaney, *Chem. Rev.* **62**, 81 (1962).

Polycyclic Aromatic Rings

E. Clar, *Polycyclic Hydrocarbons*, Academic Press, New York, NY, 1964.

99. M. J. S. Dewar and C. de Llano, *J. Am. Chem. Soc.* **91**, 789 (1969).
100. E. Clar, *Polycyclic Hydrocarbons*, Vols. 1 and 2, Academic Press, New York, NY, 1964.

(References for these problems will be found on page 502.)

1. Give reaction conditions that would accomplish each of the following transformations. Multistep schemes are not necessary. Be sure to choose conditions that
would afford the desired isomer as the principal product.

(a) 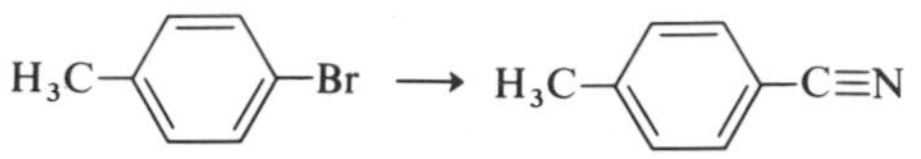 H_3C—⟨ ⟩—Br → H_3C—⟨ ⟩—C≡N

(b)   ![structure: methyl benzoate to 2-iodo methyl benzoate]

(c) $(CH_3)_3C$—⟨ ⟩—$C(CH_3)_3$ → $(CH_3)_3C$—⟨ ⟩—$\overset{O}{\overset{\|}{C}}CH_3$

(d)   ![structure d]

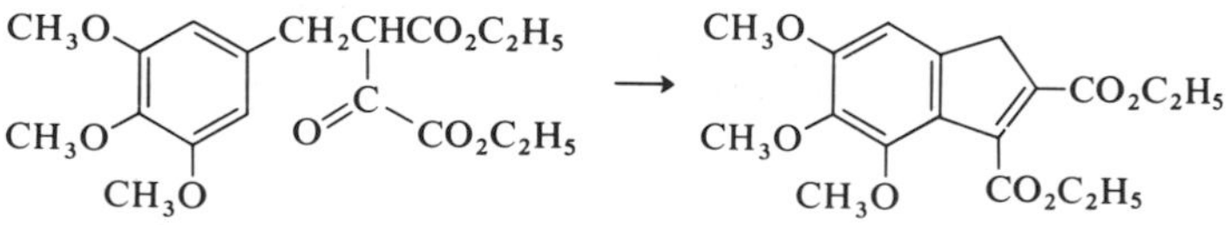

(e) 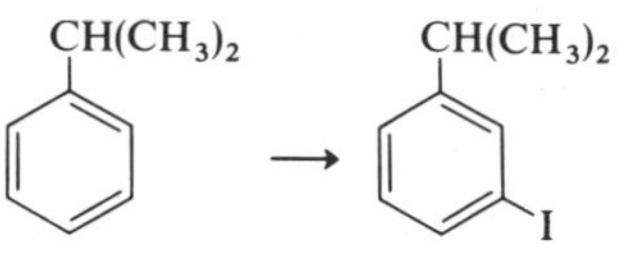   ![structure e: cumene to iodo cumene]

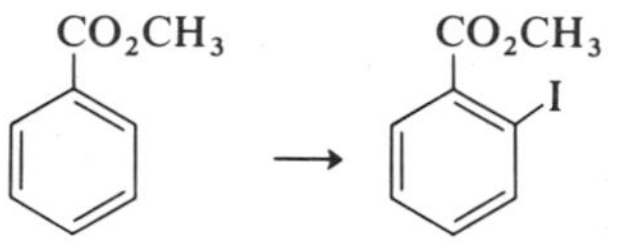

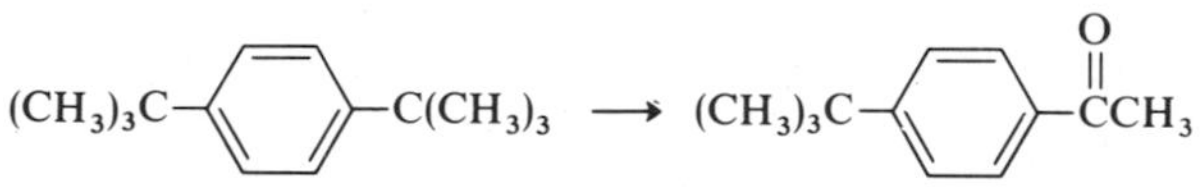

(f) 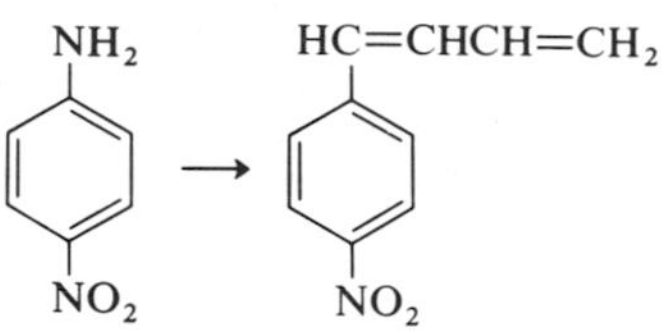   ![structure f: aniline NO2 to allylbenzene NO2]

2. Both *cis*- and *trans*-2-phenylcyclopropanecarbonyl chloride give **A** when
treated with $AlCl_3$ in CH_2Cl_2 suspension. When optically active acid chloride is
used, the product is racemic. Furthermore, when reactions are quenched prior to
completion, the recovered acid chloride is invariably a completely racemic
15:85 *cis–trans* mixture. Account for this behavior.

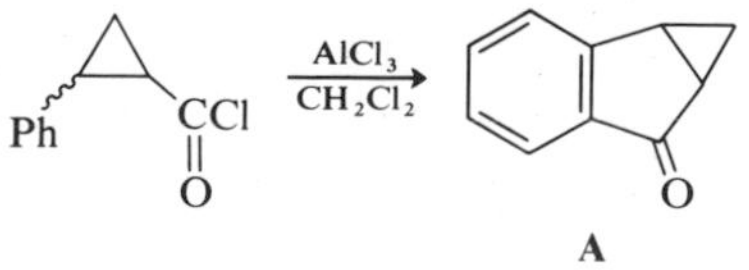

![structure: 2-phenylcyclopropanecarbonyl chloride, AlCl3/CH2Cl2, to A]

A

3. Two examples of less common nucleophilic aromatic substitution reactions are outlined below. Propose a mechanism for each reaction.

(a)

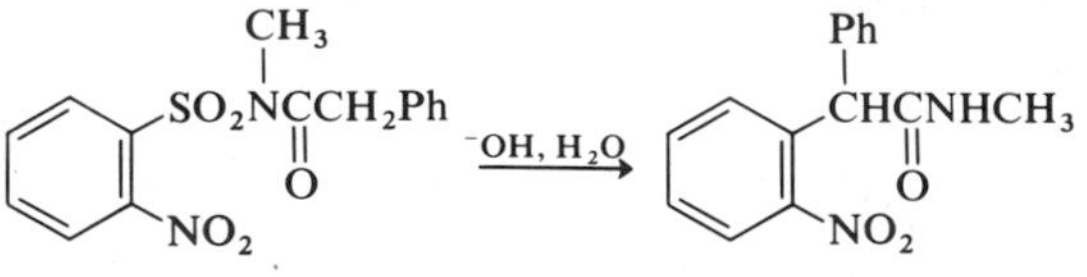

(b)

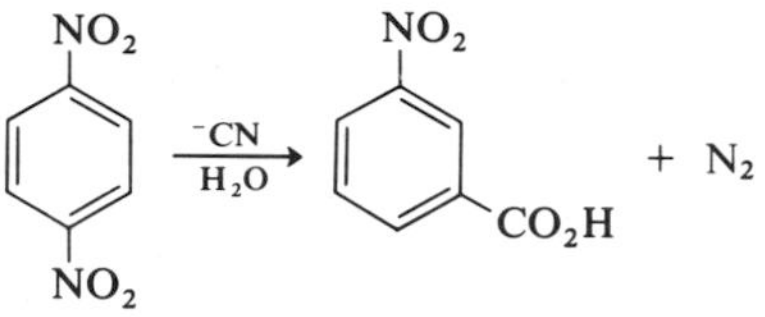

(Other facts: When $H_2^{18}O$ is used as solvent, the CO_2H contains ^{18}O at one half the level present in the $H_2^{18}O$.)

4. When a solution of 2,3-dibromo-2,3-dimethylbutane in benzene is treated with aluminum chloride and carbon monoxide is bubbled through the reaction mixture, a 73% yield of 2,2,3,3-tetramethylindanone is obtained. The same product (54% yield) is obtained if 1,2-dibromo-2,3-dimethylbutane is employed. Write a mechanism for formation of the product from each starting material. Comment on the relative rates of the various processes that are occurring in these reactions.

5. Reaction of 2,4,6-trinitroanisole with methoxide ion initially gives **A** ($k = 950M^{-1}sec^{-1}$) at 25°, but **A** rearranges to **B**. The final K in favor of **B** is very large. Suggest an explanation for the initial formation of **A** and the greater stability of **B**. Relate these observations to the idea of kinetic control versus thermodynamic control.

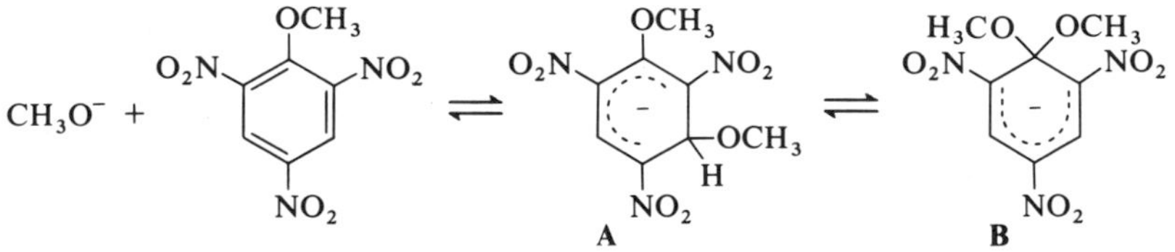

6. A solution of 1,3,5-trinitrobenzene in acetone containing triethylamine rapidly becomes deep red in color. Addition of ether precipitates a red-purple crystalline solid exhibiting intense uv absorption at 464 and 555 nm and a strong band in the ir at 1700 cm^{-1}. The nmr spectrum in acetone-d_6 exhibited the following signals: 8.35 (s, 2H), 6.8 (broad s, 1H), 5.08 (t, $J = 5.5$ Hz, 1H), 3.49 (q, $J = 7$ Hz, 6H), 2.6 (d, $J = 5.5$ Hz, 2H), 2.17 (s, 3H) and 1.42 ppm (t, $J = 7$ Hz, 9H). Suggest a structure for this product.

7. The generation of benzyne in the presence of phenyl isocyanate generates the products shown, although not in high yield. Account for the formation of these products.

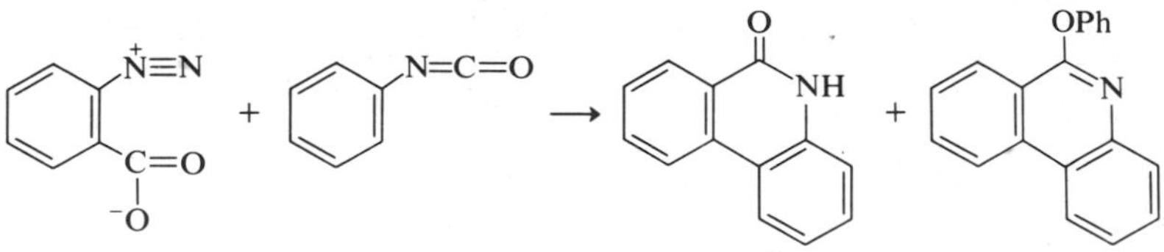

8. Addition of a solution of bromine and potassium bromide to a solution of the carboxylate salt **A** results in the precipitation of a neutral compound having the formula $C_{11}H_{13}BrO_3$. Various spectroscopic data show that the compound is non aromatic. Suggest a structure and discuss the significance of the formation of this product.

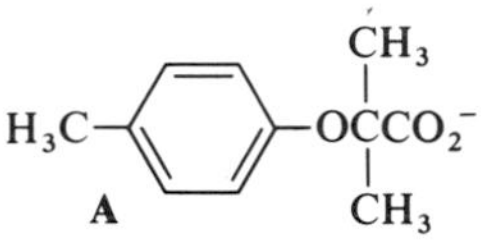

9. Benzaldehyde, benzyl methyl ether, benzoic acid, methyl benzoate, and phenylacetic acid all undergo thallation initially in the *ortho* position. Rationalize this observation.

10. Treatment of 2,6-di-*tert*-butylphenol with bromine in acetic acid produces a crystalline product, **A**, mp 69–71°, isomeric with 4-bromo-2,6-di-*tert*-butylphenol **B**. The product **A** is characterized by strong bands in the ir at 1665 and 1640 cm^{-1} and in the uv at 252 nm ($\varepsilon = 13,400$). On standing in aqueous acetic acid, **A** is converted to **B**. Propose a structure for **A**.

11. Account for the formation of the products of the following reaction:

(total yield 55%, isomer ratio 55:17:23:5)

12. Suggest reaction sequences that would permit synthesis of the following aromatic compounds from the starting material indicated:

(a)

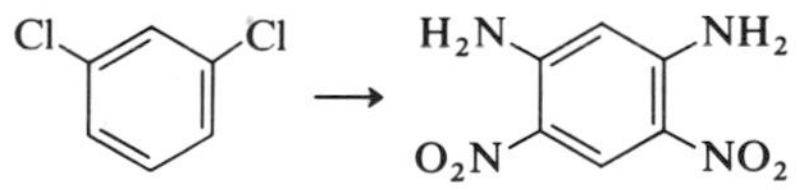

(b)

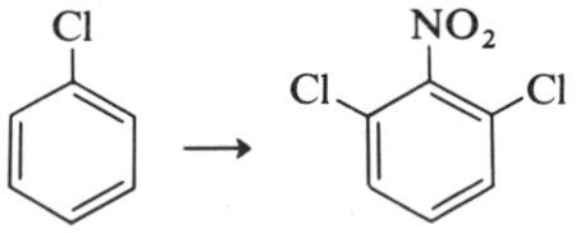

(c)

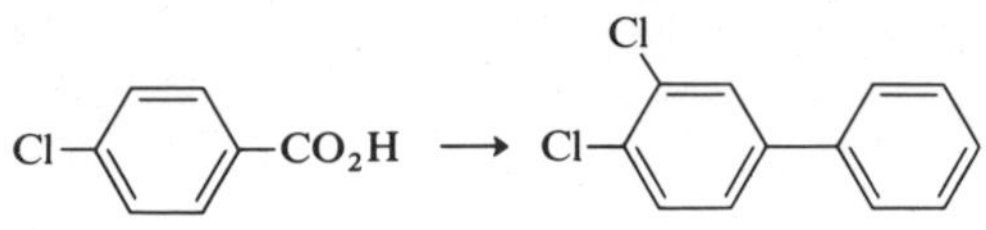

Cl—⟨ ⟩—CO$_2$H → Cl—⟨ ⟩—⟨ ⟩ (with Cl substituent)

(d)

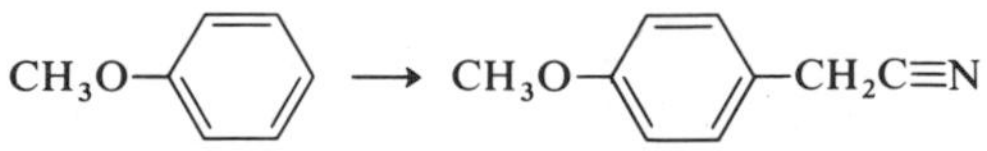

CH$_3$O—⟨ ⟩ → CH$_3$O—⟨ ⟩—CH$_2$C≡N

(e)

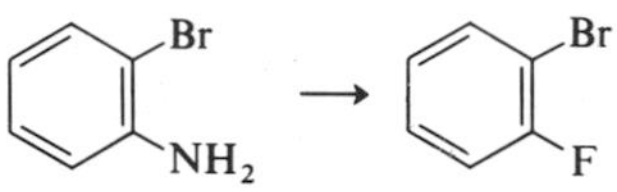

[benzene with Br and NH$_2$] → [benzene with Br and F]

(f)

[benzene with CO$_2$H] → [benzene with CO$_2$H, two Br ortho, one Br para]

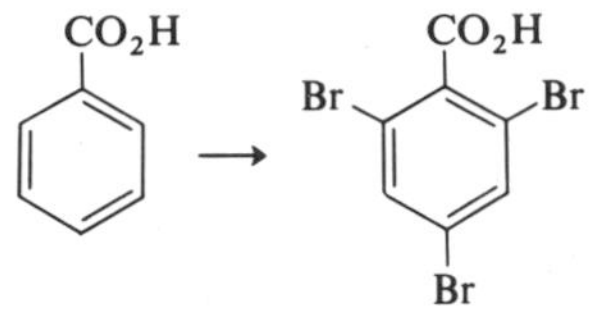

(g)

[benzene with two Br] → [benzene with two S(CH$_2$)$_3$CH$_3$]

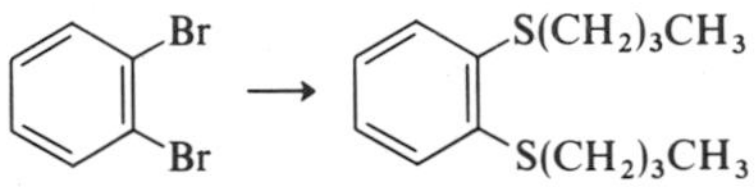

13. Write mechanisms that would account for the following reactions:

(a)

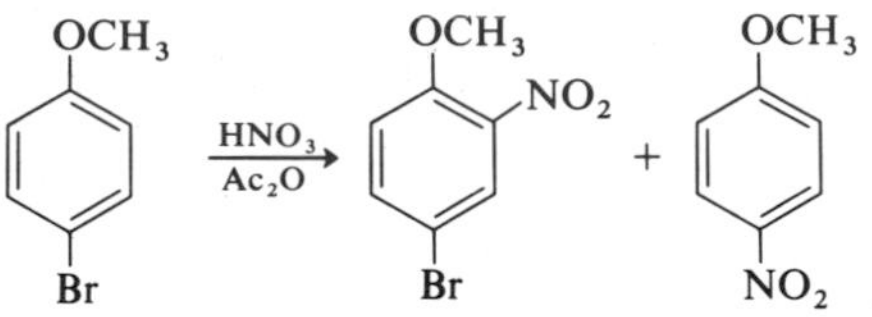

[p-bromoanisole] $\xrightarrow{\text{HNO}_3/\text{Ac}_2\text{O}}$ [OCH$_3$, NO$_2$, Br product] + [OCH$_3$, NO$_2$ para product]

(b)

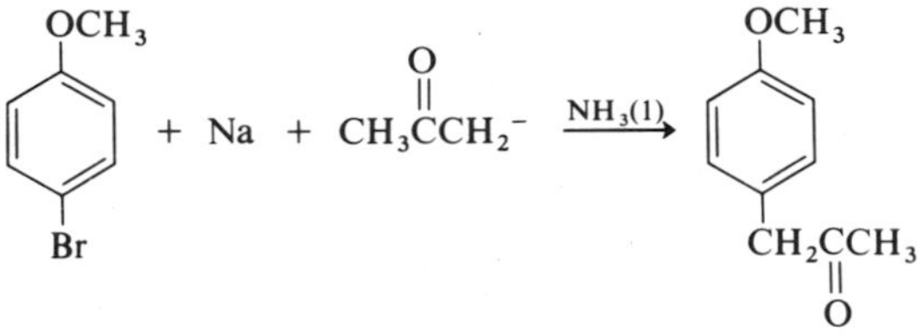

[OCH$_3$, Br benzene] + Na + CH$_3$$\overset{\text{O}}{\overset{\|}{\text{C}}}CH_2^-$ $\xrightarrow{\text{NH}_3(l)}$ [OCH$_3$ benzene with CH$_2$CCH$_3$ (=O)]

(c)

[benzene with CO$_2^-$ and N$_2^+$] + H$_3$C—CH=CH—CH=CH—CH$_3$ →

→ [dihydronaphthalene with CH$_3$ groups] (~74%) + H$_2$C=C(Ph)—CH=C(H)—C(H)=C—CH$_3$ (~6%)

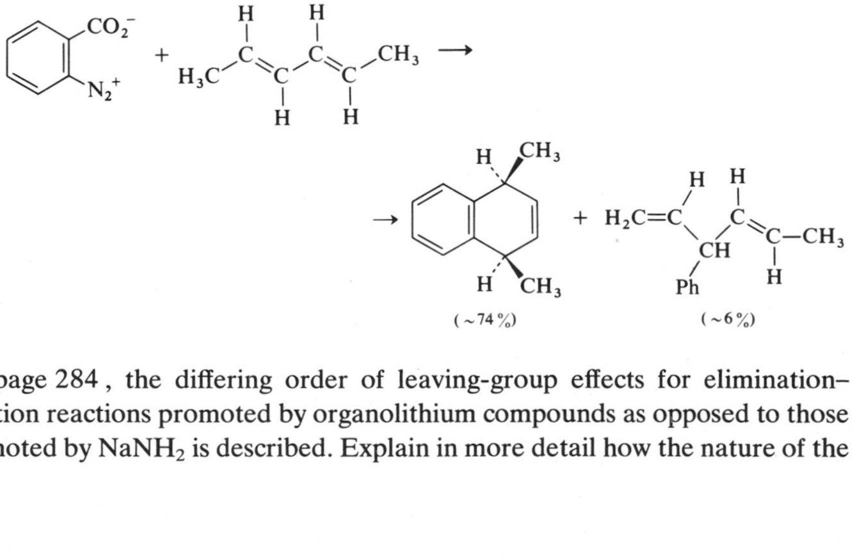

14. On page 284, the differing order of leaving-group effects for elimination–addition reactions promoted by organolithium compounds as opposed to those promoted by NaNH$_2$ is described. Explain in more detail how the nature of the

base that is used might affect the order of reactivity as a function of the leaving group. Use a potential energy versus reaction coordinate diagram to illustrate your explanation.

15. Devise synthetic routes to the following compounds, using readily available starting materials:

(a)

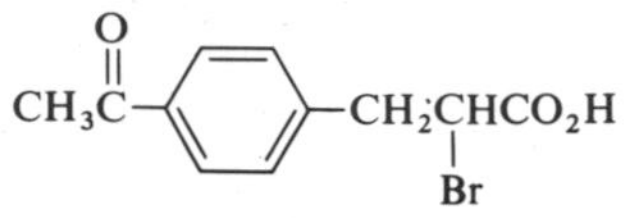

$CH_3\overset{O}{\overset{\|}{C}}$—⬡—$CH_2\overset{|}{\underset{Br}{C}HCO_2H}$

(b)

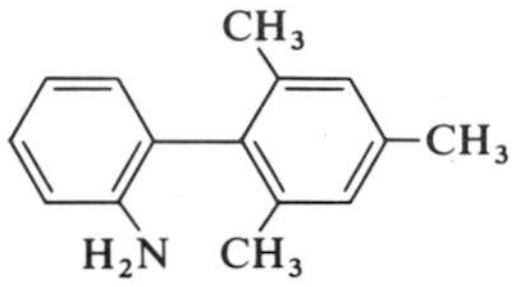

(c)

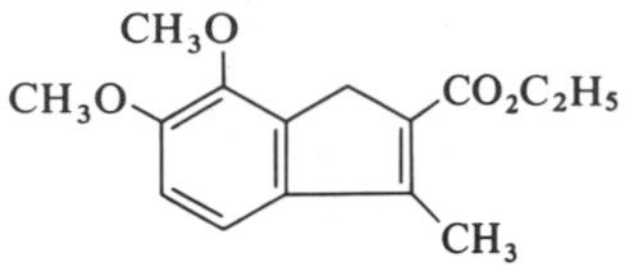

(d)

Br—⬡—$\overset{|}{\underset{CCl_3}{C}HC_6H_5}$

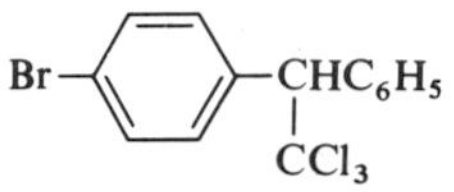

16. Aromatic substitution reactions are key steps in multistep synthetic sequences that effect the following transformations. Suggest reaction sequences that might accomplish the desired syntheses.

(a)

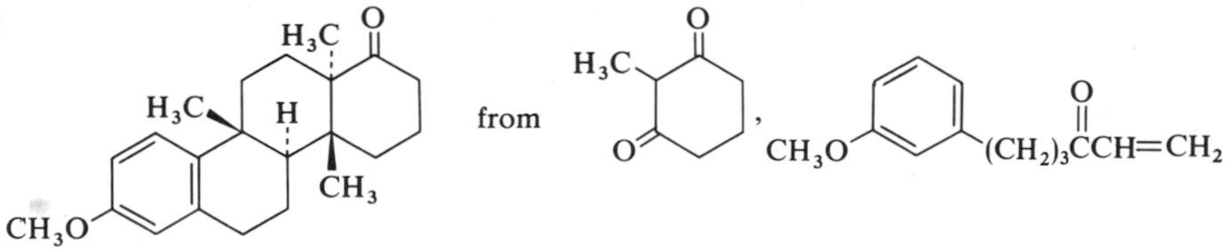

from , CH_3O—⬡—$(CH_2)_3\overset{O}{\overset{\|}{C}}CH=CH_2$

(b)

from , $BrCH_2CO_2CH_3$, $H_3CO_2CCH_2CH_2CO_2CH_3$

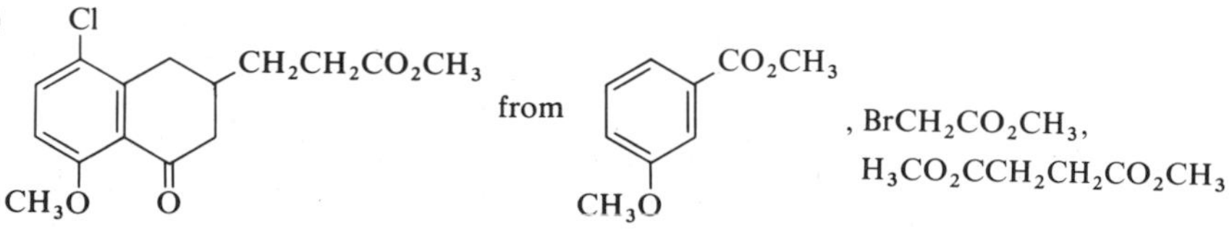

Reactions Involving Carbenes, Nitrenes, and Other Electron-Deficient Intermediates

The reactions to be described in this chapter have in common the formal involvement of even-electron intermediates having unfilled orbitals of low energy. The most familiar of these intermediates are carbonium ions. Sections 8.4 and 8.5 contain examples of fragmentation and rearrangement reactions of carbonium ions that are of synthetic value; the discussion of these reactions supplements the mechanistically oriented discussion presented in Part A, Chapter 5. Also important are the neutral divalent carbon and monovalent nitrogen species, *carbenes* and *nitrenes*, respectively.

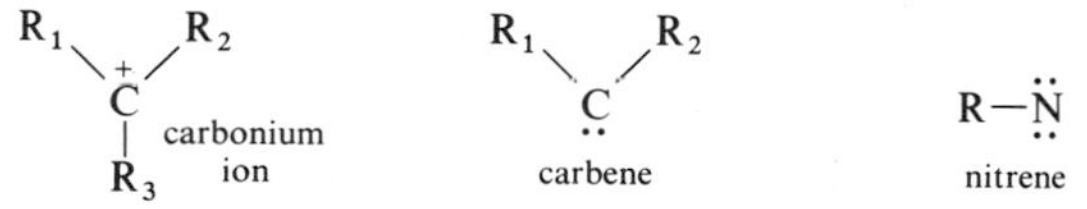

The first two sections of this chapter discuss the structure, generation, and properties of these highly reactive species; Section 8.3 presents synthetically useful reactions in which such intermediates have been suggested. Current organic chemical thought recognizes that many of the reactions ascribed to carbenes and nitrenes may also be available to various intermediates of higher coordination, which are grouped under the nonspecific terms *carbenoids* and *nitrenoids*. It is often difficult to determine exactly the nature of the actual intermediate in a particular process, and a great many of the reactions to be encountered are subject to this ambiguity.

302

CHAPTER 8
REACTIONS
INVOLVING
ELECTRON-
DEFICIENT
INTERMEDIATES

Moreover, some of the reactions to be discussed have been clearly shown not to involve free carbenes or nitrenes, but are included here because the reactivity pattern is closely analogous.

8.1. Carbenes

8.1.1. Structure

A carbene may exist in its ground electronic state as a singlet or a triplet, depending on whether the two nonbonded electrons are, respectively, in the same molecular orbital with paired spins, or in two orbitals of equal energy with parallel spins.

singlet   triplet

These two electronic configurations should be reflected in differing geometries and chemical reactivity, as well as in the important physical property that the singlet is diamagnetic while the triplet is paramagnetic. The triplet may therefore be studied by electron spin resonance spectroscopy.

A qualitative picture of the bonding in the singlet assumes sp^2-hybridization at carbon, with the two electrons in an sp^2-hybridized orbital and an unoccupied p-orbital. The R_1CR_2 angle would be expected to be contracted slightly from the normal 120° angle (provided R_1 and R_2 are small), because the interorbital repulsions will be greatest for the more diffuse lone-pair orbital. The triplet carbene would bond the ligands to carbon sp-hybridized orbitals in a linear array, with the unpaired electrons in two mutually orthogonal p-orbitals.

Molecular-orbital calculations at the *ab initio* level lead to the prediction of HCH angles for methylene (CH_2) of *ca.* 135° for the triplet and *ca.* 105° for the singlet, with the triplet lying some 10 kcal/mol lower in energy than the singlet.[1]

Experimental determinations of the geometry of CH_2 tend to confirm the theoretical predictions. The HCH angle of the triplet state as determined by analysis of its epr spectrum is 125–140°. The HCH angle of the singlet state is determined by electronic spectroscopy as 102°. All the available physical and chemical evidence is consistent with the triplet as the ground state.

Substituents have the effect of perturbing the relative energies of the singlet and triplet states.[2] In general, alkyl groups resemble hydrogen as a substituent, and dialkylcarbenes are ground-state triplets. Lone-pair donors, however, are believed

1. J. F. Harrison, *Acc. Chem. Res.* **7**, 378 (1974).
2. R. Gleiter and R. Hoffmann, *J. Am. Chem. Soc.* **90**, 5457 (1968).

to stabilize the singlet state more than the triplet state by π-donation into the empty *p*-orbital.

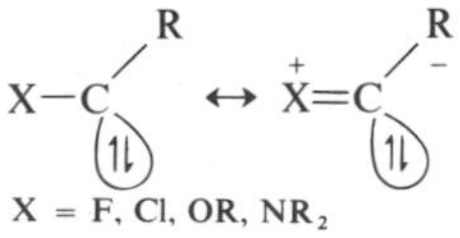

$$X-C\overset{R}{\underset{(\uparrow\downarrow)}{\diagup}} \leftrightarrow \overset{+}{X}=C\overset{R}{\underset{(\uparrow\downarrow)}{\diagup}}\overline{}$$

$$X = F, Cl, OR, NR_2$$

The dihalocarbenes CF_2, CCl_2, and CBr_2 all have singlet ground states with XCX angles of 105° for difluorocarbene (microwave)[3] and $100 \pm 10°$ for dichlorocarbene and dibromocarbene (infrared).[4]

The reactivity of the singlet and triplet states should be different, as proposed by Skell.[5] The triplet state is a diradical, and would be expected to exhibit a selectivity in its reactions toward olefins similar to that of other species having unpaired electrons. The singlet state, with its unfilled *p*-orbital, should be electrophilic and exhibit reactivity toward olefins similar to that of other electrophilic species. In general, carbenes add to alkenes to give cyclopropane derivatives. Dibromocarbene is typical in this respect. This reaction could involve the singlet dibromocarbene in a concerted addition, or the triplet state in a stepwise process:

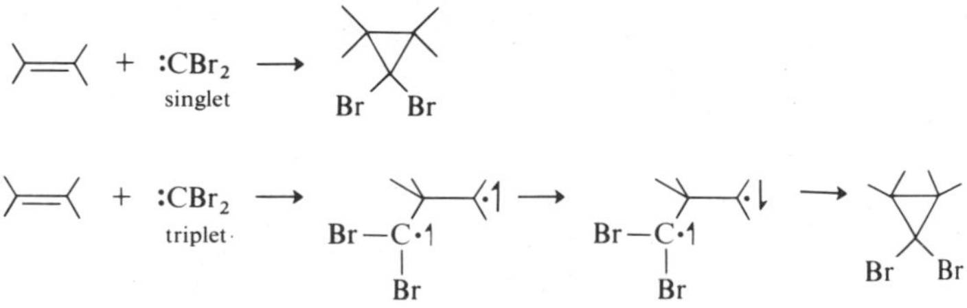

The relative reactivities toward selected olefins (Table 8.1) can be seen to be more in accord with the electrophilic models (bromination, epoxidation) than with the radical model (addition of $\cdot CCl_3$).

Similarly, dichlorocarbene adds to substituted 2-arylpropenes with a ρ of -0.62, which is indicative of an electrophilic species.[6] The magnitude of ρ is similar to that observed for π-complex formation of substituted styrenes by silver ion (-0.77), but significantly less than that for hydration (-4) or bromination (-4.3) of substituted styrenes. The relative reactivity of many carbenes and carbenoids with alkenes has been examined and the data accumulated and critically discussed in a comprehensive review.[7]

Stabilization of the singlet state by electron release from adjacent heteroatoms may be so pronounced as to cause the electrophilic properties of a carbene to vanish.

3. F. X. Powell and D. R. Lide, *J. Chem. Phys.* **45**, 1067 (1966).
4. L. Andrews, *J. Chem. Phys.* **48**, 979 (1968); L. Andrews and T. G. Carver, *J. Chem. Phys.* **49**, 896 (1968).
5. P. S. Skell and A. Y. Garner, *J. Am. Chem. Soc.* **78**, 5430 (1956).
6. D. Seyferth, J. Y-P. Mui, and R. Damrauer, *J. Am. Chem. Soc.* **90**, 6182 (1968).
7. R. A. Moss, in *Carbenes*, M. Jones, Jr., and R. A. Moss (eds.), John Wiley and Sons, New York, NY, 1973, pp. 153–304.

304

CHAPTER 8
REACTIONS
INVOLVING
ELECTRON-
DEFICIENT
INTERMEDIATES

Table 8.1. Relative Rates of Addition to Alkenes[a]

Alkene	·CCl$_3$	CBr$_2$	Br$_2$	Epoxidation
Isobutylene	1.00	1.00	1.00	1.00
Styrene	>19	0.4	0.6	0.1
2-Methylbutene	0.17	3.2	1.9	13.5

a. P. S. Skell and A. Y. Garner, *J. Am. Chem. Soc.* **78**, 5430 (1956).

Dimethoxycarbene, for example, exhibits no electrophilicity toward alkenes.[8] This has been attributed to efficient electron release from oxygen.

$$CH_3\ddot{\underset{..}{O}}-\ddot{C}-\ddot{\underset{..}{O}}CH_3 \leftrightarrow CH_3\overset{+}{\underset{..}{O}}=\bar{C}-\ddot{\underset{..}{O}}CH_3 \leftrightarrow CH_3\ddot{\underset{..}{O}}-\bar{C}=\overset{+}{\underset{..}{O}}CH_3$$

π-Delocalization involving incorporation of a divalent carbon into a conjugated cyclic system has been studied in the case of the interesting species cyclopropenylidene (**A**) and cycloheptatrienylidene (**B**). In these molecules, if the empty orbital is part of the π-system, the electron deficiency would be delocalized over an

aromatic π-system. These carbenes could be described as the conjugate bases of the aromatic cyclopropenium and tropylium ions, respectively. The carbene cyclopropenylidene (**A**) has not yet been generated, although its diphenyl derivative has been.[9] Cycloheptatrienylidene has been generated, however, and its reactivity seems to confirm that the delocalized carbene is not strongly electrophilic. It reacts best with alkenes bearing electron-attracting groups; in reactions with styrenes, the ρ value is about +1.0, indicating that the carbene is acting as a nucleophilic species.[10]

8.1.2. Generation of Carbenes

There are numerous ways of generating carbene intermediates. The most general are summarized in Scheme 8.1, and will be discussed individually in succeeding paragraphs. This discussion will serve to establish some of the limits on the generality of the various procedures.

8. D. M. Lemal, E. P. Gosselink, and S. D. McGregor, *J. Am. Chem. Soc.* **88**, 582 (1966).
9. W. M. Jones, M. E. Stowe, E. E. Wells, Jr., and E. W. Lester, *J. Am. Chem. Soc.* **90**, 1849 (1968).
10. L. W. Christensen, E. E. Waali, and W. M. Jones, *J. Am. Chem. Soc.* **94**, 2118 (1972).

Scheme 8.1. General Methods for Generation of Carbenes

	Precursor	Condition	Products
1[a]	$R_2C=\overset{+}{N}=\overset{-}{N}$ diazoalkanes	photolysis, thermolysis, or metal-ion catalysis	$R_2C:\ +\ N_2$
2[b]	$R_2C=N-\overset{-}{N}SO_2Ar$ salts of sulfonylhydrazones	photolysis or thermolysis: diazoalkanes are intermediates	$R_2C:\ +\ N_2\ +\ ArSO_2^-$
3[c]	$\underset{\text{diazirines}}{R\underset{R}{\diagdown}\overset{\diagup}{\underset{\diagdown}{C}}\overset{N}{\underset{N}{\parallel}}}$	photolysis	$R_2C:\ +\ N_2$
4[d]	$\underset{\text{epoxides}}{\overset{R}{\underset{R}{\diagdown}}C\overset{O}{\diagup\diagdown}C\overset{R}{\underset{R}{\diagup}}}$	photolysis	$R_2C:\ +\ R_2C=O$
5[e]	R_2CH-X halides	strong base or organometallic compounds	$R_2C:\ +\ BH\ +\ X^-$
6[f]	$R_2\underset{X}{\overset{|}{C}}HgR'$ α-halomercury compounds	thermolysis	$R_2C:\ +\ R'HgX$

a. W. J. Baron, M. R. DeCamp, M. E. Hendrick, M. Jones, Jr., R. H. Levin, and M. B. Sohn, in *Carbenes*, M. Jones, Jr., and R. A. Moss (eds.), John Wiley and Sons, New York, NY, 1973, pp. 1–151.
b. W. R. Bamford and T. S. Stevens, *J. Chem. Soc.*, 4735 (1952).
c. H. M. Frey, *Adv. Photochem.* **4**, 225 (1966).
d. G. W. Griffin and N. R. Bertoniere, in *Carbenes*, M. Jones, Jr., and R. A. Moss (eds.), John Wiley and Sons, New York, NY, 1973, pp. 318–332.
e. W. Kirmse, *Carbene Chemistry*, Academic Press, New York, NY, 1971, pp. 96–109, 129–149.
f. D. Seyferth, *Acc. Chem. Res.* **5**, 65 (1972).

Decomposition of diazo compounds to carbenes is a quite general reaction. Examples include the simplest diazo compound, diazomethane, as well as diaryl diazomethanes and diazo compounds in which one or both of the substituents is an acyl group. The inability to synthesize the required diazo compound is sometimes a limitation on the method. The low-molecular-weight diazoalkanes are toxic and unstable, and are usually prepared and used *in situ* rather than isolated. The simple aliphatic diazo compounds are synthesized from derivatives of the corresponding amine. All the common precursors of diazomethane, for example, are derivatives of methylamine. The details of the base-catalyzed decompositions vary somewhat from

$$O=N\ \ \overset{H}{\underset{\displaystyle CH_3N-\overset{\displaystyle N}{\underset{\displaystyle \parallel}{C}}NHNO_2}{\underset{\displaystyle}{|}}} \xrightarrow{\ KOH\ } CH_2N_2 \qquad\qquad \text{Ref. 11}$$

11. M. Neeman and W. S. Johnson, *Org. Synth.* **V**, 245 (1973).

306

CHAPTER 8
REACTIONS
INVOLVING
ELECTRON-
DEFICIENT
INTERMEDIATES

$$CH_3N-CNH_2 \xrightarrow[-OR]{-OH} CH_2N_2 \qquad \text{Ref. 12}$$

$$CH_3N-C-\!\!\!\left\langle\right\rangle\!\!\!-C-NCH_3 \xrightarrow{NaOH} CH_2N_2 \qquad \text{Ref. 13}$$

$$CH_3N-SO_2Ph \xrightarrow{KOH} CH_2N_2 \qquad \text{Ref. 14}$$

case to case, but involve two essential steps.[15] The acyl or sulfonyl substituent migrates from nitrogen to the oxygen of the *N*-nitroso group. This migration is followed by deprotonation of the hydrocarbon substituent and elimination of the oxygen and attached moiety:

$$RCH_2N-Y \;\rightarrow\; RCH-N=N-O-Y \;\rightarrow\; RCH=\overset{+}{N}=\overset{..}{\underset{..}{N}}{}^-$$

Another route to diazo compounds is by oxidation of the corresponding hydrazone. This route is most frequently employed when at least one of the substituents is an aromatic group.

$$Ph_2C=NNH_2 \xrightarrow{HgO} Ph_2C=\overset{+}{N}=\overset{..}{\underset{..}{N}}{}^- \qquad \text{Ref. 16}$$

When an α-diazoketone is needed, the usual synthesis starts with an acid halide. Reaction with a diazoalkane gives the diazoketone as a result of nucleophilic attack, with displacement of the chloride ion:

$$R\overset{O}{\overset{\|}{C}}Cl + RCH=\overset{+}{N}=\overset{..}{\underset{..}{N}}{}^- \rightarrow R-\overset{O}{\overset{\|}{C}}-CR' \atop \underset{:N:^-}{\overset{\|}{N^+}}$$

Such acyl diazo compounds are appreciably more stable than simple diazoalkanes.

The driving force for decomposition of diazo compounds to carbenes is the formation of the very stable nitrogen molecule. Activation energies for the process with diazoalkanes in the gas phase are in the neighborhood of 30 kcal/mol. The requisite energy can also be supplied by photochemical excitation. It is possible to control the photochemical process to give predominantly singlet or triplet carbene. Direct photolysis leads primarily to the singlet intermediate, the reason being that dissociation of the excited diazoalkane is more rapid than intersystem crossing to the

12. F. Arndt, *Org. Synth.* **II**, 165 (1943).
13. Th. J. de Boer and H. J. Backer, *Org. Synth.* **IV**, 250 (1963).
14. J. A. Moore and D. E. Reed, *Org. Synth.* **V**, 351 (1973).
15. W. M. Jones, D. L. Muck, and T. K. Tandy, Jr., *J. Am. Chem. Soc.* **88**, 68 (1966); W. M. Jones and D. L. Muck, *J. Am. Chem. Soc.* **88**, 3798 (1966); R. A. Moss, *J. Org. Chem.* **31**, 1082 (1966).
16. L. I. Smith and K. L. Howard, *Org. Synth.* **III**, 351 (1955).

triplet state. It cannot be assumed, however, that the triplet intermediate is always excluded in direct photolysis experiments. In studies of several types of diazo compounds, the conclusion has been reached that the triplet intermediate is responsible for up to 15–20% of the product on direct photolysis.[17] The triplet carbene can be made the principal intermediate by photosensitized decomposition. Aromatic ketones are frequently employed as photosensitizers.

Addition of certain copper salts to solutions of diazo compounds also leads to evolution of nitrogen and formation of products of the same general types as those formed in thermal and photochemical decompositions of diazoalkanes. The weight of the evidence, however, indicates that free carbene intermediates are not involved in such reactions.[18] Instead, complexes of the carbene unit with the metal ion catalyst seem to be the actual reactants. Such a complex would be an example of a carbenoid species. Although the product suggests the involvement of a carbene-like reactivity, other evidence rules out a completely free carbene of the type generated by photochemical expulsion of a molecule of nitrogen.

The second method in Scheme 8.1, thermal or photochemical decomposition of salts of arenesulfonylhydrazones, is actually a variation on the diazoalkane method, since diazo compounds are intermediates. It is an important and useful method, however, since the ultimate starting materials are ketones, and also because the procedure avoids isolation of the potentially dangerous diazoalkanes. The conditions of the decomposition are usually such that the diazo compound reacts soon after formation, and high concentrations do not build up.[19] The nature of the solvent

$$
\underset{\substack{\| \\ O}}{R\overset{O}{C}R'} + NH_2NHSO_2Ar \rightarrow \underset{R'}{\overset{R}{\diagdown}}C{=}NNSO_2Ar \downarrow \text{base}
$$

$$
{}^-O_2SAr + \underset{R'}{\overset{R}{\diagdown}}C{=}\overset{+}{N}{=}\overset{-}{\underset{..}{N}} \underset{\substack{or \\ \Delta}}{\overset{h\nu}{\longleftarrow}} \underset{R'}{\overset{R}{\diagdown}}C{=}N{-}\overset{-}{N}SO_2Ar
$$

mixture plays an important role in the outcome of tosylhydrazone decompositions. In the presence of proton donors, the diazoalkane intermediates can be diverted to a carbonium-ion pathway by protonation.[20] Aprotic solvents such as dimethoxyethane favor decomposition via the carbene pathway:

$$
R_2C{=}\overset{+}{N}{=}\overset{-}{\underset{..}{N}} \xrightarrow{XOH} R_2\overset{H}{\underset{\smile}{C}}{-}\overset{+}{N}{\equiv}N \rightarrow R_2\overset{H}{\overset{+}{C}} + N_2
$$

17. C. S. Elliott and H. M. Frey, *Trans. Faraday Soc.* **64**, 2352 (1968); C. D. Gutsche, G. L. Bachman, and R. S. Coffey, *Tetrahedron* **18**, 617 (1962).
18. W. R. Moser, *J. Am. Chem. Soc.* **91**, 1135, 1141 (1969).
19. G. M. Kaufman, J. A. Smith, G. G. Van der Stouw, and H. Schechter, *J. Am. Chem. Soc.* **87**, 935 (1965).
20. J. H. Bayless, L. Friedman, F. B. Cook, and H. Schechter, *J. Am. Chem. Soc.* **90**, 531 (1968).

308

CHAPTER 8
REACTIONS
INVOLVING
ELECTRON-
DEFICIENT
INTERMEDIATES

The diazirines (entry 3, Scheme 8.1) are cyclic isomers of diazo compounds. The strain of the small ring, along with the potential for formation of molecular nitrogen, makes these compounds highly reactive toward loss of nitrogen on photoexcitation. Little work has been done on their thermal decomposition. They are, in general, somewhat more difficult to synthesize than either diazo compounds or arenesulfonylhydrazones,[21] and this difficulty limits their use.

Carbenes are also generated when aryl epoxides are photolyzed (entry 4, Scheme 8.1). The other product formed is a carbonyl compound. The photodecomposition of epoxides is not a single-step process; highly colored intermediate species have been detected.[22] The structure assigned these intermediates is the carbonyl ylide, a dipolar valence isomer of the epoxide ring.[23]

It is believed that the decomposition of the ylide is also a photoreaction. It can be seen that an unsymmetrical epoxide can conceivably give rise to two carbenes and two carbonyl compounds. The nature of the substituent groups ordinarily favors one possible mode of cleavage over the other. When R = aryl and R' = alkyl, the aliphatic ketone is generated, and the aryl-substituted carbon is released as the carbene fragment. Electron-withdrawing substituents such as cyano or carbomethoxy favor carbene formation, but an electron-releasing methoxy substituent directs carbene formation to the other oxirane carbon. These substituent effects can be understood by considering which of the two possible resonance structures will be the principal contributor to the carbonyl ylide structure. The relative contribution from each

21. For a review of available synthetic methods: E. Schmitz, *Dreiringe mit Zwei Heteroatomen*, Springer-Verlag, Berlin, 1967, pp. 114–121.

22. R. S. Becker, R. O. Bost, J. Kolc, N. R. Bertoniere, R. L. Smith, and G. W. Griffin, *J. Am. Chem. Soc.* **92**, 1302 (1970).

23. T. Do-Minh, A. M. Trozzolo, and G. W. Griffin, *J. Am. Chem. Soc.* **92**, 1402 (1970).

resonance structure is reflected in the bond order of the C–O bonds. The C–O bond with the greatest double-bond character becomes the carbonyl group, while the weaker C–O bond is cleaved.

The cyclopropane ring can also be photochemically cleaved with elimination of carbene fragments. This reaction has been studied more from the point of view of the photochemistry involved than as a useful method for carbene generation.[24]

The α-elimination of hydrogen halide induced by strong base (entry 5, Scheme 8.1) was the first of the methods for generation of carbenes to receive thorough modern study. The efficient generation of carbenes by α-elimination from halides is, however, restricted to substrates without β-hydrogen, since dehydrohalogenation by β-elimination dominates when it can occur. Classic examples of this method are the generation of dichlorocarbene from chloroform, the formation of chlorocarbene from methylene chloride, and the formation of aryl carbenes from benzyl halides:

$$HCCl_3 + {}^-OR \rightleftharpoons :\bar{C}Cl_3 \rightarrow :CCl_2 + Cl^- \qquad \text{Ref. 25}$$

$$H_2CCl_2 + RLi \rightarrow RH + LiCHCl_2 \rightarrow :CHCl + LiCl \qquad \text{Ref. 26}$$

$$ArCH_2X + RLi \rightarrow RH + \overset{Li}{Ar\ddot{C}HX} \rightarrow Ar\ddot{C}H + LiX \qquad \text{Ref. 27}$$

Potassium t-butoxide has frequently been employed as the base in synthetic work with dichlorocarbene, although a number of other procedures are available.[28]

The exact formulation of the reactive intermediate in α-elimination reactions using organolithium compounds as bases has been difficult. Apart from the free carbene, various carbenoids are possible, including the α-haloorganolithium formed on metalation, and carbene–lithium halide complexes of various degrees of association.[29] In the case of the dichlorocarbene–trichloromethyllithium equilibrium, the

$$Cl_3CLi \rightleftharpoons Cl_2C: + LiCl$$

equilibrium lies heavily to the side of trichloromethyllithium at $-100°$.[30] Reaction with alkenes to afford 1,1-dichlorocyclopropanes, however, appears to involve only CCl_2, and not Cl_3CLi, since the pattern of reactivity versus alkene structure is identical to that observed for free CCl_2 generated in the gas phase.[31]

A general approach to the problem of distinguishing true carbenes from

24. G. W. Griffin and N. R. Bertoniere, in *Carbenes*, Vol. 1, M. Jones, Jr., and R. A. Moss (eds.), John Wiley and Sons, New York, NY, 1973, pp. 306–318.
25. J. Hine, *J. Am. Chem. Soc.* **72**, 2438 (1950); J. Hine and A. M. Dowell, Jr., *J. Am. Chem. Soc.* **76**, 2688 (1954).
26. G. Köbrich, H. Trapp, K. Flory, and W. Drischel, *Chem. Ber.* **99**, 689 (1966); G. Köbrich and H. R. Merkle, *Chem. Ber.* **99**, 1782 (1966).
27. G. L. Closs and L. E. Closs, *J. Am. Chem. Soc.* **82**, 5723 (1960).
28. W. von E. Doering and A. K. Hoffmann, *J. Am. Chem. Soc.* **76**, 6162 (1954).
29. G. Köbrich, *Angew. Chem. Int. Ed. Engl.* **6**, 41 (1967).
30. W. T. Miller, Jr., and D. M. Whalen, *J. Am. Chem. Soc.* **86**, 2089 (1964); D. F. Hoeg, D. I. Lusk, and A. L. Crumbliss, *J. Am. Chem. Soc.* **87**, 4147 (1965).
31. P. S. Skell and M. S. Cholod, *J. Am. Chem. Soc.* **91**, 6035, 7131 (1969); *J. Am. Chem. Soc.* **92**, 3522 (1970).

310

CHAPTER 8
REACTIONS
INVOLVING
ELECTRON-
DEFICIENT
INTERMEDIATES

carbenoids is to compare product distribution and stereochemistry in reactions involving several carbene sources, including unequivocal methods such as diazoalkane decomposition. By such comparisons, it has been concluded, for example, that the intermediates generated from α,α-dibromotoluenes and butyllithium are not free carbenes but, instead, carbenoid reagents in which the incipient carbene remains attached to the elements of LiBr.[32] LiBr is eliminated only when reaction with the substrate occurs.

$$\text{PhCHBr}_2 + \text{BuLi} \rightarrow \overset{\overset{\text{Li}}{|}}{\text{PhCHBr}}$$

A method that provides an alternative route to dichlorocarbene is the decarboxylation of trichloroacetic acid.[33] In essence, this simply constitutes an alternative route to the trichloromethyl anion. Treatment of alkyl trichloroacetates with alkoxide ions is still another way of generating the same carbanion:

The applicability of these methods is restricted to polyhalogenated compounds, since the inductive effect of the three halogen atoms is necessary both for easy decarboxylation and for elimination of a carbanion from the tetrahedral intermediate generated in the alkoxide-cleavage procedure.

The principle underlying the use of organomercury compounds for carbene generation (entry 6, Scheme 8.1) is again the α-elimination mechanism. The carbon–mercury bond is much more covalent than the C–Li bond, however, so that the mercury systems are generally stable at room temperature and easily isolated. They then decompose to the carbene when heated in solution with an appropriate alkene.[34] The decomposition appears to be a reversible unimolecular reaction, and

$$\text{PhHg}-\overset{\overset{\text{Cl}}{|}}{\underset{\underset{\text{Cl}}{|}}{\text{C}}}-\text{Br} \rightleftarrows \text{:CCl}_2 + \text{PhHgBr}$$

the rate is not greatly influenced by the alkene. This observation implies that a free carbene is generated from the precursor.[35] Synthesis of a variety of organomercurials has provided compounds that are appropriate sources for substituted carbenes. For

32. G. L. Closs and R. A. Moss, *J. Am. Chem. Soc.* **86**, 4042 (1964).
33. W. E. Parham and E. E. Schweizer, *Org. React.* **13**, 55 (1963).
34. D. Seyferth, J. M. Burlitch, R. J. Minasz, J. P-P. Mui, H. D. Simmons, Jr., A. J. H. Treiber, and S. R. Dowd, *J. Am. Chem. Soc.* **87**, 4259 (1965).
35. D. Seyferth, J. Y-P. Mui, and J. M. Burlitch, *J. Am. Chem. Soc.* **89**, 4953 (1967).

example, carbenes with a carbomethoxy or trifluoromethyl substituent can be generated from appropriate organomercury precursors[36]:

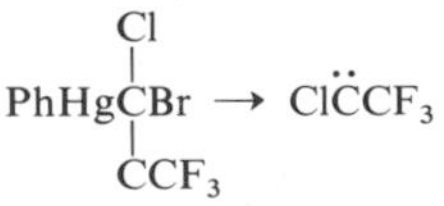

$$\underset{\underset{\overset{|}{CCF_3}}{\overset{\overset{|}{Cl}}{}}}{PhHgCBr} \longrightarrow Cl\ddot{C}CF_3$$

$$PhHgCCl_2CO_2CH_3 \longrightarrow Cl\ddot{C}CO_2CH_3$$

8.1.3. Reactions

8.1.3.1. Addition Reactions with Alkenes and Other Unsaturated Centers

The addition reaction with alkenes is the best-studied reaction of carbene intermediates, both from the point of view of understanding carbene mechanisms and for synthetic applications. The usual course of reaction of a carbene with an alkene results in the formation of a cyclopropane, an observation that is true for both the singlet and the triplet state of most carbenes. The alternative electronic states

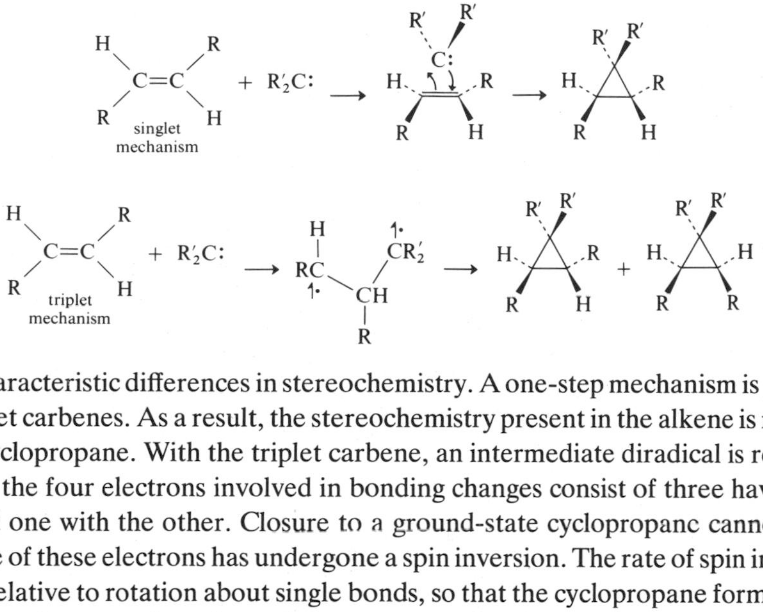

show characteristic differences in stereochemistry. A one-step mechanism is possible for singlet carbenes. As a result, the stereochemistry present in the alkene is retained in the cyclopropane. With the triplet carbene, an intermediate diradical is required, because the four electrons involved in bonding changes consist of three having one spin and one with the other. Closure to a ground-state cyclopropane cannot occur until one of these electrons has undergone a spin inversion. The rate of spin inversion is slow relative to rotation about single bonds, so that the cyclopropane formed from a triplet carbene need not have the same stereochemistry as the starting olefin.[5,37] Usually, the cyclopropanes formed from triplet carbenes are mixtures of the two possible stereoisomers.

Application of the principles of orbital-symmetry control to the concerted

36. D. Seyferth, D. C. Mueller, and R. L. Lambert, Jr., *J. Am. Chem. Soc.* **91**, 1562 (1969).
 5. See p. 303.
37. P. S. Skell and R. C. Woodworth, *J. Am. Chem. Soc.* **78**, 4496 (1956).

312

CHAPTER 8
REACTIONS
INVOLVING
ELECTRON-
DEFICIENT
INTERMEDIATES

addition of a singlet carbene predicts that reaction will occur by a transition state such

as that depicted in **C**.[38] The more symmetrical approach shown in **D** leads to a high-energy transition state; i.e., it is "forbidden" in orbital-symmetry terminology.

The stereospecificity of carbene additions has been widely used in experiments designed to determine the multiplicity of a carbene generated under given conditions. Stereospecific addition is taken as evidence for the involvement of the singlet species. The alternate result, formation of a mixture of the possible isomers, requires additional information for complete interpretation. Lack of stereospecificity implies some triplet carbene, but permits simultaneous involvement of the singlet species.

These generalizations about stereochemistry are not applicable to gas-phase reactions. The reason is that most simple carbene–alkene-addition processes are highly exothermic, since two bonds are formed and none is broken. The initial adduct may be formed with sufficient energy to undergo isomerization, resulting in loss of stereospecificity. In solution, the excess energy is dissipated very rapidly to the medium, and changes in stereochemistry do not usually occur after formation of the product.

The most important synthetic use of carbene intermediates is in the synthesis of cyclopropanes. Most of the general methods of carbene generation introduced in Scheme 8.1 lead to cyclopropane formation if the carbene is generated in the presence of an olefin. A number of examples of the use of carbenes and carbenoid reagents in cyclopropane synthesis are given in Scheme 8.2.

A very effective means for conversion of alkenes to cyclopropanes by transfer of a CH_2 unit involves the system methylene iodide and zinc–copper couple, commonly referred to as the *Simmons–Smith reagent*.[39] The active species is believed to be iodomethylzinc iodide in equilibrium with (bis)iodomethylzinc.[40] The transfer of

$$2\,ICH_2Zn \rightleftharpoons (ICH_2)_2Zn + ZnI_2$$

methylene occurs from the organometallic and is stereospecific. Free CH_2 is not an intermediate. It is also observed that in molecules with polar substituents, especially hydroxyl groups, the CH_2 unit is introduced on the side of the double bond in closer proximity to the hydroxyl group.[41] This observation implies that the attacking

38. R. Hoffmann, *J. Am. Chem. Soc.* **90**, 1475 (1968); R. Hoffmann, D. M. Hayes, and P. S. Skell, *J. Phys. Chem.* **76**, 664 (1972).
39. H. E. Simmons and R. D. Smith, *J. Am. Chem. Soc.* **80**, 5323 (1958); **81**, 4256 (1959); H. E. Simmons, T. L. Cairns, S. A. Vladuchick, and C. M. Hoiness, *Org. React.* **20**, 1 (1973).
40. E. P. Blanchard and H. E. Simmons, *J. Am. Chem. Soc.* **86**, 1337 (1964); H. E. Simmons, E. P. Blanchard, and R. D. Smith, *J. Am. Chem. Soc.* **86**, 1347 (1964).
41. J. H.-H. Chan and B. Rickborn, *J. Am. Chem. Soc.* **90**, 6406 (1968); J. A. Staroscik and B. Rickborn, *J. Org. Chem.* **37**, 738 (1972).

reagent may be complexed at the hydroxyl group before reaction with the carbon–carbon double bond. Entries 8 and 11 in Scheme 8.2 illustrate this stereodirective effect of the hydroxyl group.

Some carbenes are sufficiently reactive that they will add to aromatic rings. Cycloheptatriene is a major product of the photolysis of diazomethane in the presence of benzene[42]:

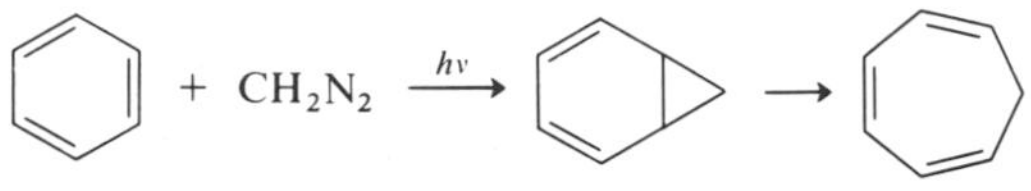

Substituted diazoalkanes such as dicyanodiazomethane and ethyldiazoacetate react in a similar fashion:

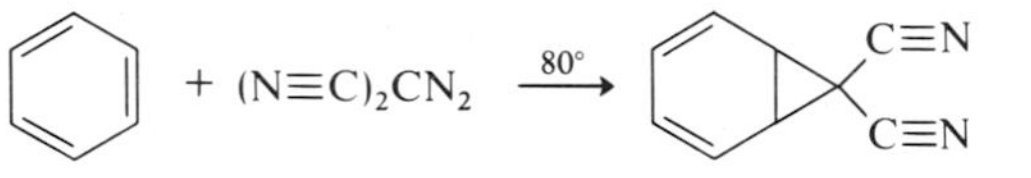

Ref. 43

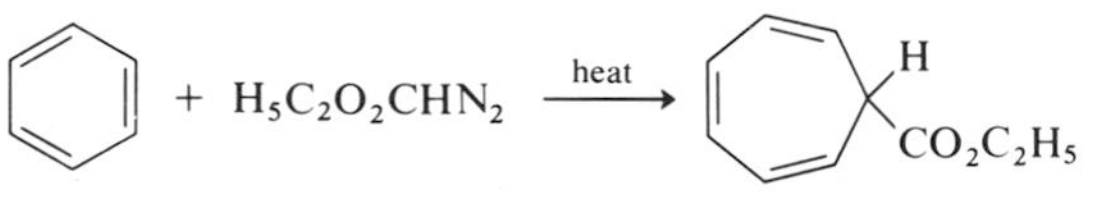

Ref. 44

Carbene additions can also occur intramolecularly, and such reactions are an important route to some highly strained molecules. Entries 12–14 in Scheme 8.2 illustrate this facet of carbene chemistry.

8.1.3.2. Insertion Reactions

A unique feature of the chemistry of carbenes is their ability to undergo insertion reactions. The term is used here in a mechanistic sense to refer to a *one-step* process in which the carbene carbon is quite literally inserted into a single bond:

$$\ddot{C}H_2 + H-CR_3 \longrightarrow H \cdots\cdots\overset{\cdot CH_2\cdots}{}CR_3 \longrightarrow H-CH_2-CR_3$$

It is not always easy to establish that an authentic insertion reaction has occurred. An alternative route to the same product is available; it involves hydrogen-atom abstraction by the triplet carbene, followed by recombination of the fragments:

$$\cdot \dot{C}H_2 + H-CR_3 \rightarrow CH_3\cdot + \cdot CR_3 \rightarrow CH_3-CR_3$$

Intermolecular carbene-insertion reactions are seldom desirable synthetic reactions. The reason is that most carbenes are not selective, and all types of C–H bonds react at comparable rates, leading to product mixtures. The distribution of insertion product from heptane, for example, is almost exactly what would be calculated on a

42. G. A. Russell and D. G. Hendry, *J. Org. Chem.* **28**, 1933 (1963).
43. E. Ciganek, *J. Am. Chem. Soc.* **89**, 1454 (1967).
44. J. E. Baldwin and R. A. Smith, *J. Am. Chem. Soc.* **89**, 1886 (1967).

314

CHAPTER 8
REACTIONS
INVOLVING
ELECTRON-
DEFICIENT
INTERMEDIATES

Scheme 8.2. Synthesis of Cyclopropanes

A. Intermolecular Additions

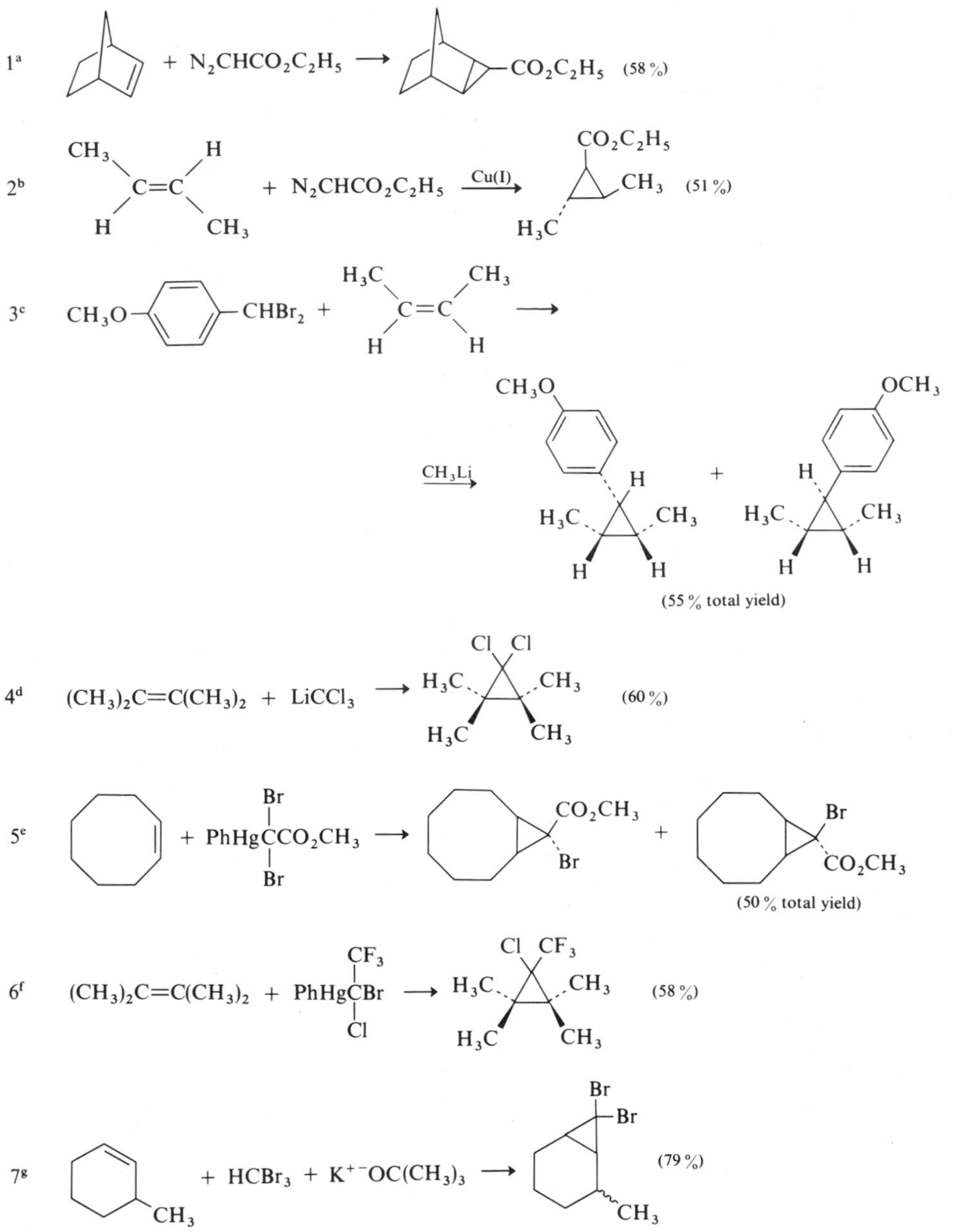

1^a $+ N_2CHCO_2C_2H_5 \longrightarrow$ $\text{---}CO_2C_2H_5$ (58%)

2^b $+ N_2CHCO_2C_2H_5 \xrightarrow{\text{Cu(I)}}$ (51%)

3^c $CH_3O\text{---}$ $\text{---}CHBr_2 +$ $\longrightarrow$

$\xrightarrow{CH_3Li}$

(55% total yield)

4^d $(CH_3)_2C{=}C(CH_3)_2 + LiCCl_3 \longrightarrow$ (60%)

5^e $+ PhHg\overset{Br}{\underset{Br}{C}}CO_2CH_3 \longrightarrow$

(50% total yield)

6^f $(CH_3)_2C{=}C(CH_3)_2 + PhHg\overset{CF_3}{\underset{Cl}{C}}Br \longrightarrow$ (58%)

7^g $+ HCBr_3 + K^{+-}OC(CH_3)_3 \longrightarrow$ (79%)

a. R. R. Sauers and P. E. Sonnett, *Tetrahedron* **20**, 1029 (1964).
b. R. G. Salomon and J. K. Kochi, *J. Am. Chem. Soc.* **95**, 3300 (1973).
c. G. L. Closs and R. A. Moss, *J. Am. Chem. Soc.* **86**, 4042 (1964).
d. D. F. Hoeg, D. I. Lusk, and A. L. Crumbliss, *J. Am. Chem. Soc.* **87**, 4147 (1965).
e. D. Seyferth, D. C. Mueller, and R. L. Lambert, Jr., *J. Am. Chem. Soc.* **91**, 1562 (1969).
f. D. Seyferth and D. C. Mueller, *J. Am. Chem. Soc.* **93**, 3714 (1971).
g. L. A. Paquette, S. E. Wilson, R. P. Henzel, and G. R. Allen, Jr., *J. Am. Chem. Soc.* **94**, 7761 (1972).

8[h]   (66 %)

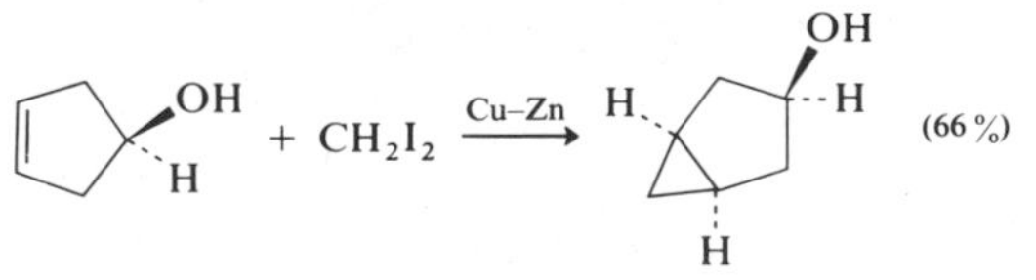

9[i]

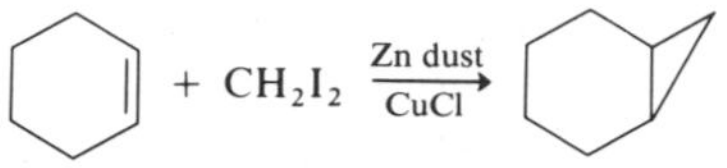

10[j]

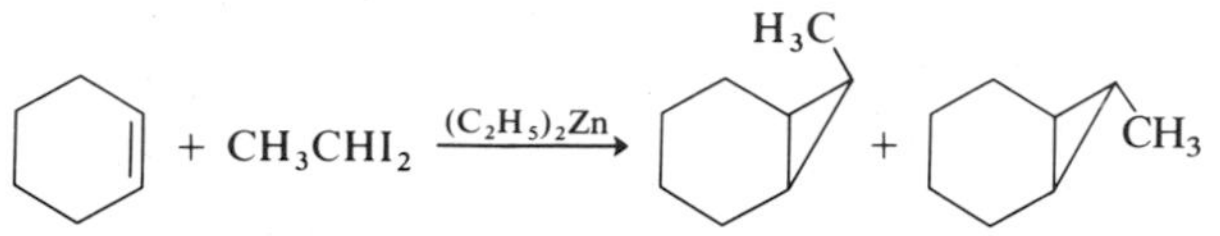

11[k]

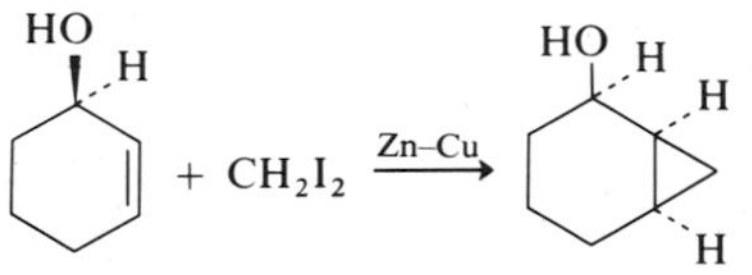

B. Intramolecular Reactions

12[l]   (66 %)

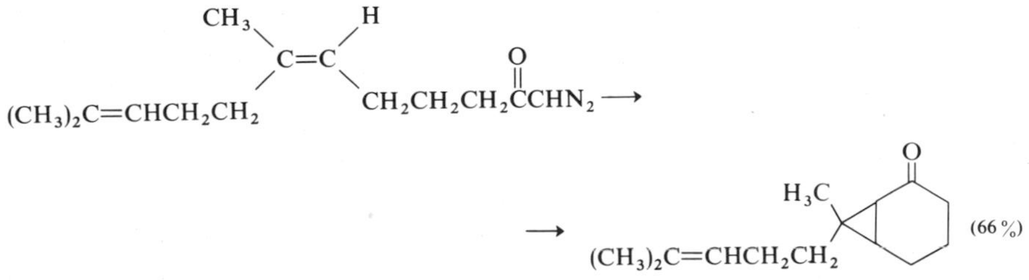

13[m]

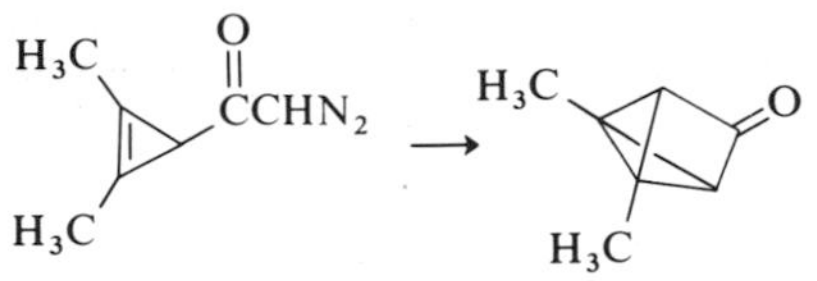

14[n]   (50 %)

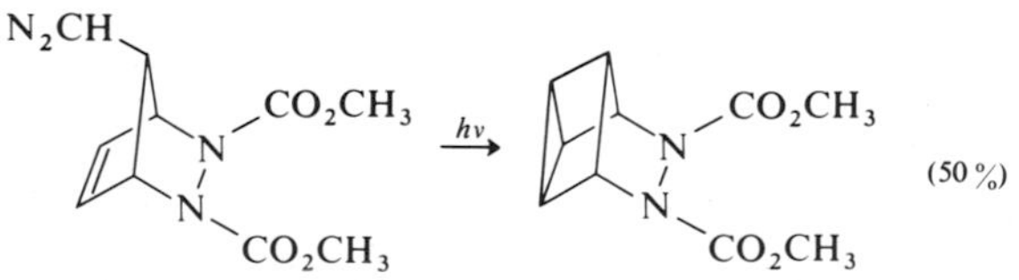

h. S. Winstein and J. Sonnenberg, *J. Am. Chem. Soc.* **83**, 3235 (1961).
i. R. J. Rawson and I. T. Harrison, *J. Org. Chem.* **35**, 2057 (1970).
j. J. Nishimura, N. Kawabata, and J. Furukawa, *Tetrahedron* **25**, 2647 (1969).
k. J. H.-H. Chan and B. Rickborn, *J. Am. Chem. Soc.* **90**, 6406 (1968).
l. U. T. Bhalerao, J. J. Plattner, and H. Rapoport, *J. Am. Chem. Soc.* **92**, 3429 (1970).
m. W. von E. Doering and M. Pomerantz, *Tetrahedron Lett.*, 961 (1964).
n. B. M. Trost and R. M. Cory, *J. Am. Chem. Soc.* **93**, 5572 (1971).

316

CHAPTER 8
REACTIONS
INVOLVING
ELECTRON-
DEFICIENT
INTERMEDIATES

$$CH_3CH_2CH_2CH_2CH_2CH_2CH_3 \xrightarrow[hv]{CH_2N_2} CH_3(CH_2)_6CH_3 + \underset{\underset{CH_3}{|}}{CH_3CH(CH_2)_4CH_3}$$
$$(38\%) \qquad\qquad (25\%)$$

$$+ \underset{\underset{CH_3}{|}}{CH_3CH_2CH(CH_2)_3CH_3}$$
$$(24\%)$$

$$+ \underset{\underset{CH_3}{|}}{(CH_3CH_2CH_2)_2CH}$$
$$(13\%)$$

statistical basis.[45] Some selectivity is observed with functionally substituted carbenes, but it is still not high enough to prevent formation of mixtures. Carbethoxycarbene, for example, inserts at tertiary C–H bonds about 3 times as fast as at primary C–H bonds.[46] In reactions with compounds where both addition and insertion are possible, addition is usually the dominant process. Photolysis of diazomethane in the presence of benzene gives more cycloheptatriene than toluene[42]:

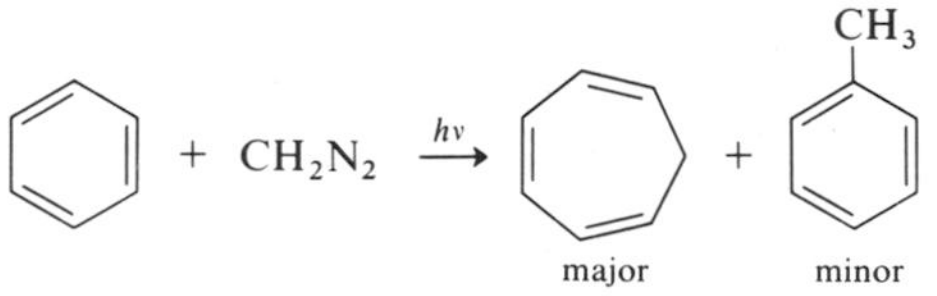

major   minor

Intramolecular insertion reactions are much more selective because the relative proximity of the various C–H bonds to the carbene center plays a major role in determining product distribution. For this reason, the reaction has synthetic utility and is especially valuable for generating derivatives of highly strained ring systems. Cyclopropanes formed by intramolecular insertion reactions are often major products from alkyl carbenes. A competing process for aliphatic carbenes is stabilization

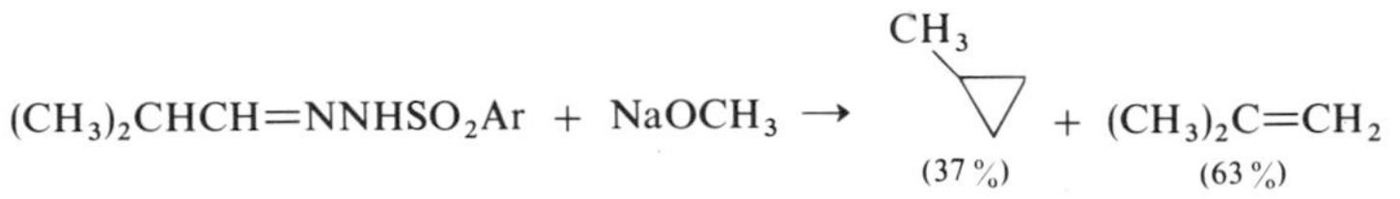

$$(CH_3)_2CHCH=NNHSO_2Ar + NaOCH_3 \rightarrow \underset{(37\%)}{\overset{CH_3}{\triangle}} + \underset{(63\%)}{(CH_3)_2C=CH_2}$$

by hydrogen migration. We will consider this reaction in more detail in the following section on rearrangement reactions. Scheme 8.3 provides some examples of intramolecular insertions that proceed in sufficiently high yield to be synthetically attractive.

8.1.3.3. Rearrangement Reactions

The most common rearrangement reaction of carbenes is the shift of a hydrogen, generating an alkene. This mode of stabilization predominates to the exclusion of most intermolecular reactions of aliphatic and alicyclic carbenes, and

45. D. B. Richardson, M. C. Simmons, and I. Dvoretzky, *J. Am. Chem. Soc.* **83**, 1934 (1961).
46. W. von E. Doering and L. H. Knox, *J. Am. Chem. Soc.* **83**, 1989 (1961).
42. See p. 313.

1[a]

2[b]

3[c]

4[d]

5[e]

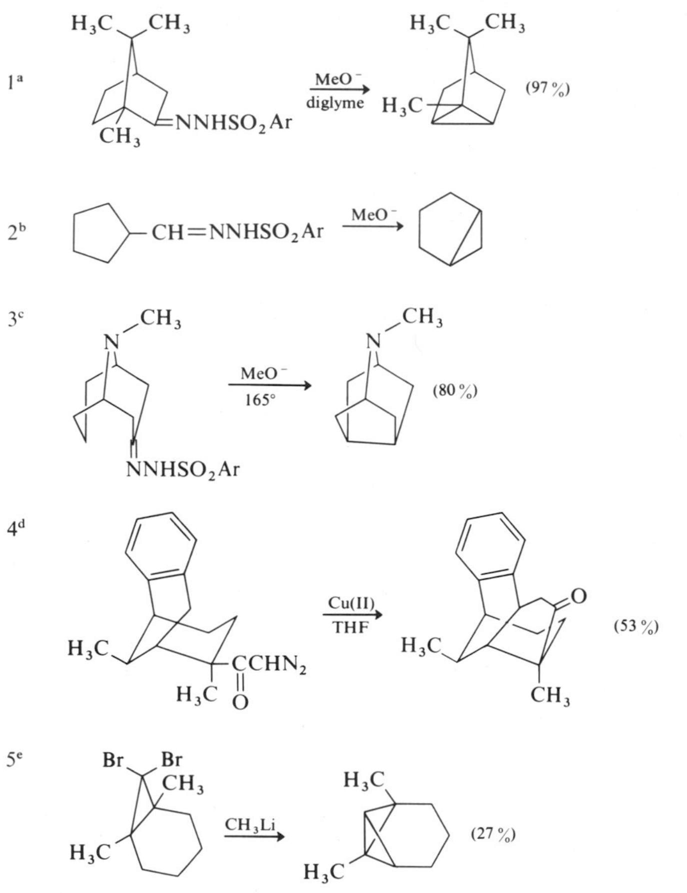

a. R. H. Shapiro, J. H. Duncan, and J. C. Clopton, *J. Am. Chem. Soc.* **89**, 1442 (1967).
b. W. Kirmse and G. Wächtershäuser, *Tetrahedron* **22**, 63 (1966).
c. T. Sasaki, S. Eguchi, and T. Kiriyama, *J. Am. Chem. Soc.* **91**, 212 (1969).
d. U. R. Ghatak and S. Chakrabarty, *J. Am. Chem. Soc.* **94**, 4756 (1972).
e. L. A. Paquette, S. E. Wilson, R. P. Henzel, and G. R. Allen, Jr., *J. Am. Chem. Soc.* **94**, 7761 (1972).

often competes successfully with intramolecular insertion reactions. For example, the carbene generated by decomposition of the tosylhydrazone of 2-methylcyclohexanone gives mainly 1- and 3-methylcyclohexene, and only a trace of

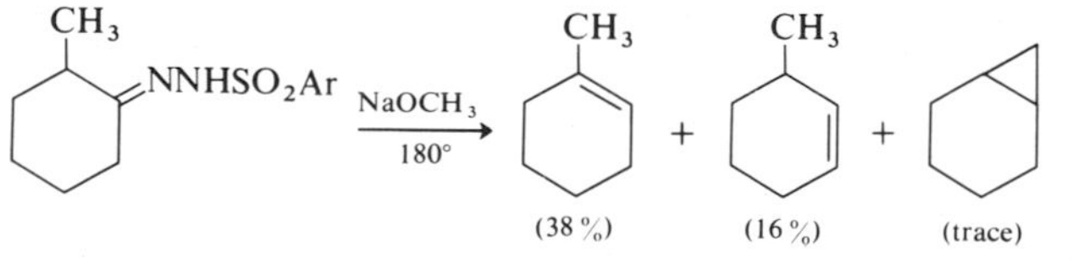

47. J. W. Wilt and W. J. Wagner, *J. Org. Chem.* **29**, 2788 (1964).

318

CHAPTER 8
REACTIONS
INVOLVING
ELECTRON-
DEFICIENT
INTERMEDIATES

the intramolecular insertion product. Carbenes can also be stabilized by migration of alkyl or aryl groups. Compound **1** provides a case in which phenyl migration, methyl migration, and intramolecular insertion are all observed:

$$\underset{\underset{\underset{1}{CH_3}}{|}}{\overset{\overset{CH_3}{|}}{PhCCHN_2}} \xrightarrow{60°} (CH_3)_2C{=}CHPh + Ph\overset{\overset{CH_3}{|}}{C}{=}CHCH_3 + Ph{-}\overset{\overset{CH_3}{|}}{C}\underset{CH_2}{\overset{}{\diagdown}}CH_2 \qquad \text{Ref. 48}$$

$$\text{(50\%)} \qquad\qquad \text{(9\%)} \qquad\qquad \text{(41\%)}$$

A useful synthetic application of the rearrangement of carbenoid species to alkenes is a procedure for converting ketones to olefins. The *p*-toluenesulfonylhydrazone derivative of the ketone is treated with an excess of an alkyllithium reagent:

$$(CH_3)_3C\overset{\overset{NNHSO_2Ar}{||}}{C}CH_3 \xrightarrow{\text{2 BuLi}} (CH_3)_3CCH{=}CH_2 \qquad \text{Ref. 49}$$

$$\text{(100\%)}$$

A mechanism involving a lithiated diazoalkane has been suggested[50]:

$$R\overset{\overset{NNHSO_2Ar}{||}}{C}CH_2R' \xrightarrow{\text{2 R''Li}} R\underset{\underset{Li}{|}}{\overset{\overset{Li^+N\overset{-}{N}SO_2Ar}{||}}{C}}CHR' \rightarrow RC{=}CHR' \rightarrow R\underset{\underset{Li}{|}}{C}{=}CHR' \rightarrow RCH{=}CHR'$$

Scheme 8.4 lists some examples of the use of this reaction in synthesis.

Phenylcarbene undergoes a rearrangement to cycloheptatrienylidene in the gas phase above 250°. Some of the details of this reaction will be encountered by working

problem 10; the references cited for that problem provide more of the story, which is somewhat more complicated than can be covered here.

Certain types of carbenes have special modes of stabilization available to them. Cyclopropylidenes open to allenes:

Ref. 51

48. H. Philip and J. Keating, *Tetrahedron Lett.*, 523 (1961).
49. G. Kaufman, F. Cook, H. Shechter, J. Bayless, and L. Friedman, *J. Am. Chem. Soc.* **89**, 5736 (1967).
50. R. H. Shapiro and M. J. Heath, *J. Am. Chem. Soc.* **89**, 5734 (1967).
51. W. M. Jones, J. W. Wilson, Jr., and F. B. Tutwiler, *J. Am. Chem. Soc.* **85**, 3309 (1963).

Scheme 8.4. Conversion of Ketones to Olefins via Sulfonylhydrazones

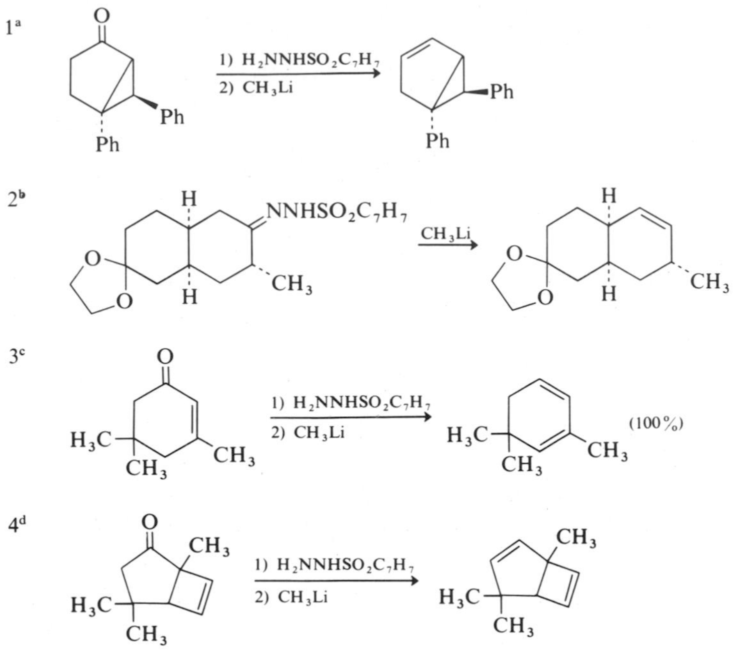

1[a]

2[b]

3[c]   (100%)

4[d]

a. H. E. Zimmerman and G. A. Epling, *J. Am. Chem. Soc.* **94**, 8749 (1972).
b. W. L. Scott and D. A. Evans, *J. Am. Chem. Soc.* **94**, 4779 (1972).
c. W. G. Dauben, M. E. Lorber, N. D. Vietmeyer, R. H. Shapiro, J. H. Duncan, and K. Tomer, *J. Am. Chem. Soc.* **90**, 4762 (1968).
d. J. E. Baldwin and M. S. Kaplan, *J. Am. Chem. Soc.* **94**, 668 (1972).

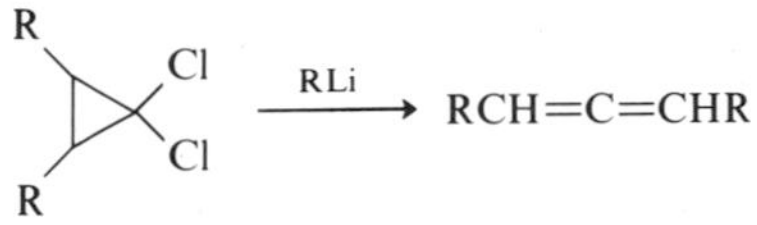

$$RCH=C=CHR \qquad \text{Ref. 52}$$

Carbene centers adjacent to double bonds (vinylcarbenes) cyclize to cyclopropenes[53]:

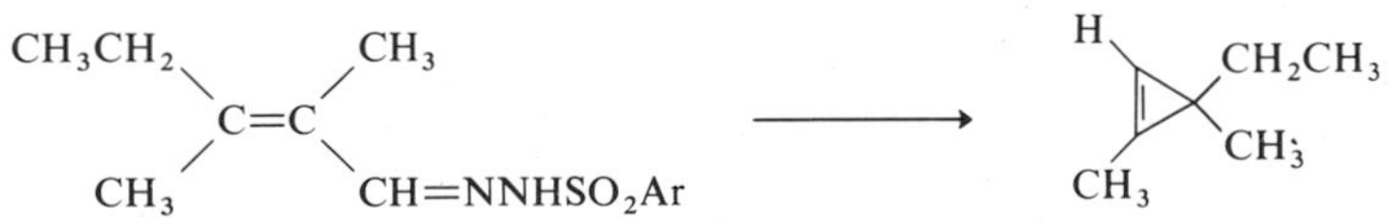

Ref. 54

Both these processes can be viewed as electrocyclic reactions, the former leading to ring-opening, the latter to ring closure:

52. W. R. Moore and H. R. Ward, *J. Org. Chem.* **25**, 2073 (1960).
53. G. L. Closs, L. E. Closs, and W. A. Böll, *J. Am. Chem. Soc.* **85**, 3796 (1963).
54. E. J. York, W. Dittmar, J. R. Stevenson, and R. G. Bergman, *J. Am. Chem. Soc.* **95**, 5680 (1973).

320

CHAPTER 8
REACTIONS
INVOLVING
ELECTRON-
DEFICIENT
INTERMEDIATES

8.2. Nitrenes

The nitrogen analogs of carbenes are called *nitrenes*. As with the carbenes, both singlet and triplet electronic states are possible. The triplet state is usually the ground state, but either species can be involved in reactions. There are many similarities between nitrene and carbene chemistry, but of course there are also significant differences.

By far the most commonly applied means of generating a nitrene is photolysis or thermolysis of an azide. This method is clearly analogous to formation of a carbene from a diazo compound. The types of azides in which the decomposition has been

$$R-\overset{..}{\underset{..}{N}}-\overset{+}{N}{\equiv}N \xrightarrow[\text{or } h\nu]{\Delta} R-\overset{..}{N}: \; + \; N_2$$

studied extensively are those where R = alkyl,[55] aryl,[56] acyl,[57] and sulfonyl.[58] The characteristic reaction of alkyl nitrenes is migration of one of the substituents to nitrogen, giving an imine:

$$R_3C-\overset{..}{\underset{..}{N}}-\overset{+}{N}{\equiv}N \xrightarrow[\text{or } h\nu]{\Delta} \overset{R}{\underset{R}{\diagdown}}C{=}N{-}R$$

R = H or alkyl

Intermolecular insertion and addition reactions are almost unknown for alkyl nitrenes. In fact, it is not clear that the nitrenes exist as discrete species. The conformation of the azide group seems to determine which substituent migrates in the decomposition of alkyl azides. This observation implies that migration begins before the nitrogen molecule has become completely detached from the incipient

55. F. D. Lewis and W. H. Saunders, Jr., in *Nitrenes*, W. Lwowski (ed.), Interscience, New York, NY, 1970, pp. 47–98.
56. P. A. S. Smith, in *Nitrenes*, W. Lwowski (ed.), Interscience, New York, NY, 1970, pp. 99–162.
57. W. Lwowski, in *Nitrenes*, W. Lwowski (ed.), Interscience, New York, NY, 1970, pp. 185–224.
58. D. S. Breslow, in *Nitrenes*, W. Lwowski (ed.), Interscience, New York, NY, 1970, pp. 245–303; R. A. Abramovitch and R. G. Sutherland, *Fortschr. Chem. Forsch.* **16**, 1 (1970).

nitrene center,[59] since after the nitrogen has departed, the three potential migrating groups are all stereochemically equivalent.

At present, the chemistry of aryl nitrenes defies simple systematic description. In general terms, the course of reactions seems to be governed by transformations characteristic of the nitrene and a cyclic azirine isomer. The dominant species seems to depend on the nature of the substituents on the benzene ring. Aryl nitrenes are,

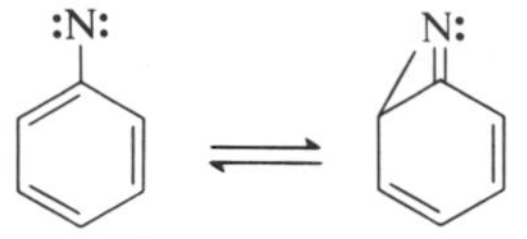

however, involved in several efficient intramolecular reactions, such as the formation of carbazole from o-biphenylnitrene on photolysis or thermolysis.[60]

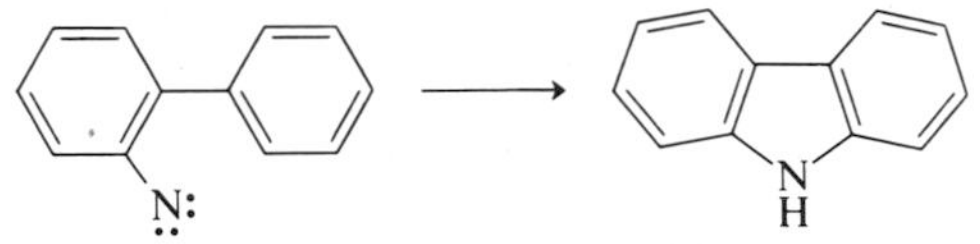

Perhaps the "best behaved" of the nitrenes are the carboalkoxynitrenes.[57]

$$RO-\overset{\overset{\displaystyle O}{\|}}{C}-N_3 \xrightarrow[\text{or } h\nu]{\Delta} RO-\overset{\overset{\displaystyle O}{\|}}{C}-\ddot{N}:$$

These intermediates undergo addition reactions with alkenes and insertion reactions with saturated systems. Addition also occurs on reaction with benzene.[61] Many of the

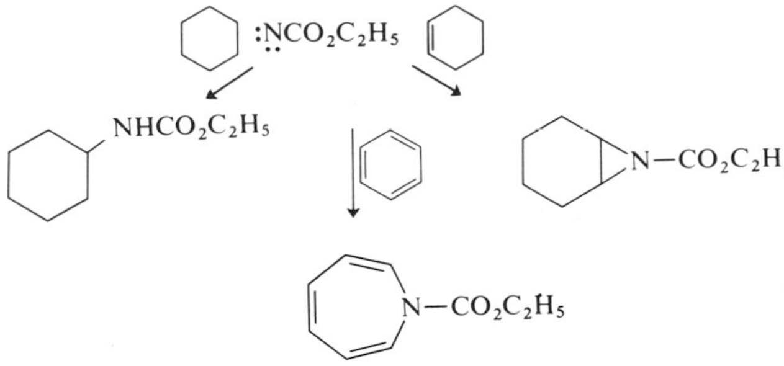

reactions of carbethoxynitrene have been studied in detail. For example, data on the stereospecificity and selectivity of the nitrene in its reactions with various types of hydrocarbons are available, and have been summarized and interpreted.[61] This

59. R. M. Moriarty and R. C. Reardon, *Tetrahedron* **26**, 1379 (1970); R. A. Abramovitch and E. P. Kyba, *J. Am. Chem. Soc.* **93**, 1537 (1971); R. M. Moriarty and P. Serridge, *J. Am. Chem. Soc.* **93**, 1534 (1971).
60. P. A. S. Smith and B. B. Brown, *J. Am. Chem. Soc.* **73**, 2435, 2438 (1951); J. S. Swenton, T. J. Ikeler, and B. H. Williams, *J. Am. Chem. Soc.* **92**, 3103 (1970).
57. See p. 320.
61. W. Lwowski, *Angew. Chem. Int. Ed. Engl.* **6**, 897 (1967).

322

CHAPTER 8
REACTIONS
INVOLVING
ELECTRON-
DEFICIENT
INTERMEDIATES

nitrene is somewhat more selective than simple carbenes, showing selectivities of roughly $1:10:40$ for the primary, secondary, and tertiary positions in 2-methyl-butane in insertion reactions. The relationship between nitrene multiplicity and stereospecificity in addition to olefins is analogous to that described for carbenes. The singlet gives stereospecific addition, while the triplet gives non-stereospecific addition products.

Acyl azides are well-known compounds. Their role in the thermal Curtius rearrangement, a reaction that apparently does not involve a nitrene, will be discussed in subsection 8.3.2. Photochemical decomposition of acyl azides elicits nitrene reactivity. In particular, intramolecular C–H-insertion reactions have been observed, but not usually in high yield.[62]

Sulfonylnitrenes are formed by thermal decomposition of sulfonyl azides. Insertion reactions occur with saturated hydrocarbons.[64] With aromatic rings, addition is believed to occur, but the main products are sulfonanilides, which result from ring-opening of the addition intermediate.

8.3. Rearrangements of Electron-Deficient Intermediates

The reactions grouped in this section are all characterized by skeletal rearrangement. Some are generally considered to be concerted, and thus discrete electron-deficient intermediates (carbonium ion, carbene, nitrene, etc.) are bypassed. Reactions in which migration is to carbon or nitrogen are the most important; these are considered separately in the sections that follow.

8.3.1. Migration to Carbon

Diazoketones give rearranged products when decomposed thermally or photochemically. The reaction is known as the *Wolff rearrangement* and is of synthetic importance, since it constitutes a convenient method for one-carbon homologation

62. O. E. Edwards, in *Nitrenes*, W. Lwowski (ed.), Interscience, New York, NY, 1970, pp. 225–243.
63. I. Brown and O. E. Edwards, *Can. J. Chem.* **45**, 2599 (1967).
64. D. S. Breslow, M. F. Sloan, N. R. Newburg, and W. B. Renfrow, *J. Am. Chem. Soc.* **91**, 2273 (1969).
65. R. A. Abramovitch, G. N. Knaus, and V. Uma, *J. Org. Chem.* **39**, 1101 (1974).

of carboxylic acids known as the *Arndt–Eistert reaction.*[66] Mechanistic studies have

323

SECTION 8.3.
REARRANGE-
MENTS OF
ELECTRON-
DEFICIENT
INTERMEDIATES

$$RCO_2H \rightarrow RCOCl \xrightarrow{CH_2N_2} R\overset{O}{\overset{\|}{C}}CHN_2 \xrightarrow{\Delta} RCH=C=O \xrightarrow{H_2O} RCH_2CO_2H$$

been aimed at determining whether or not migration is concerted with loss of nitrogen. If the mechanism is a two-step process, it would presumably be a carbene reaction. A related issue is whether the carbene, if involved, is in equilibrium with a ring-closed isomer, the oxirene. The conclusion that has emerged from these studies

is that the carbene is generated in photochemical reactions, but there is no general agreement about the thermal process. The existence of the acyl carbene–oxirene equilibrium has been established in the photochemical reactions.[67] This aspect of the mechanism has been demonstrated by isotopic labeling. The distribution of radioactive label between the two carbons is explained by the symmetrical oxirene inter-

mediate. A carbene intermediate that did not reach the symmetrical oxirene

structure would be expected to give product labeled only in the carboxyl group.

In actual practice, the "thermal" reaction is often catalyzed by silver salts. Under these conditions, there is no detectable kinetic carbon-isotope effect when the migrating carbon is ^{14}C.[68] A concerted process would be expected to show an isotope effect. This result favors a two-step mechanism only if it is known that the migration is part of the rate-determining process. Kinetic studies and substituent effects, both in the presence and in the absence of silver catalysts, have also been interpreted in

66. W. E. Bachman and W. S. Struve, *Org. React.* **1**, 38 (1942); L. L. Rodina and I. K. Korobitsyna, *Russ. Chem. Rev.* (English translation) **36**, 260 (1967).
67. S. A. Matlin and P. G. Sammes, *J. Chem. Soc. Perkin Trans. 1*, 2623 (1972); J. Fenwick, G. Frater, K. Ogi, and O. P. Strausz, *J. Am. Chem. Soc.* **95**, 124 (1973).
68. Y. Yukawa and T. Ibata, *Bull. Chem. Soc. Jpn.* **42**, 802 (1969).

324

CHAPTER 8
REACTIONS
INVOLVING
ELECTRON-
DEFICIENT
INTERMEDIATES

terms of a two-step mechanism.[69] The matter has not been the subject of such complete study as to be beyond question, however. Despite the mechanistic ambiguities, the reaction is of general synthetic value. Some examples of the synthetic application of the reaction are given in Scheme 8.5. The last two entries are examples of the use of the rearrangement to effect ring contraction of cyclic diazoketones.

α-Haloketones undergo a skeletal change when treated with base. The mechanism is somewhat different from most of the other rearrangement processes discussed in this section. The overall structural change that is effected is similar, however, to that caused by rearrangement of α-diazoketones; the reaction can therefore be conveniently discussed at this point. The most commonly used bases are alkoxide ions, with the use of which esters are formed. If the ketone is cyclic, a ring contraction

$$RCH_2\overset{O}{\overset{\|}{C}}\underset{X}{C}HR' \xrightarrow{CH_3O^-} CH_3O\overset{O}{\overset{\|}{C}}\underset{CH_2R}{C}HR' + X^-$$

occurs. This base-catalyzed conversion of α-haloketones to carboxylic acid derivatives is known as the *Favorskii reaction*.[70] The reaction has been subjected to extensive mechanistic studies. There is strong evidence that the rearrangement involves the open 1,3-dipolar form of a cyclopropanone and/or the cyclopropanone as a reaction intermediate.[71] There is also a related mechanism that can operate in

$$RCH_2\overset{O}{\overset{\|}{C}}\underset{X}{C}HR' \underset{}{\overset{-OR''}{\rightleftarrows}} RC\overset{O}{\overset{\|}{\underset{-}{C}}}\underset{X}{C}HR' \rightarrow RC\overset{O}{\overset{\|}{\underset{-}{C}}}\underset{+}{C}HR' \leftrightarrow RC\overset{O}{\overset{\|}{\underset{+}{C}}}\underset{-}{C}HR'$$

$$RCHCH_2R' + RCH_2CHR' \longleftarrow \underset{RHC-CHR'}{\overset{\overset{O^-}{\underset{\|}{C}}-OR''}{}} \xleftarrow{\bar{O}R''} \underset{RHC-CHR'}{\overset{\overset{O}{\underset{\|}{C}}}{}}$$
$$\overset{|}{CO_2R''} \qquad \overset{|}{CO_2R''}$$

the absence of an acidic α'-hydrogen; it is known as the "semibenzilic" rearrangement:

$$R\overset{O}{\overset{\|}{C}}\underset{X}{C}HR' \xrightarrow{R''O^-} R\underset{OR''}{\overset{O^-}{\underset{|}{C}}}\underset{X}{C}HR' \rightarrow R''O-\overset{O}{\overset{\|}{C}}-\underset{R}{C}HR'$$

The net structural change is the same for both mechanisms. The energy requirements of the cyclopropanone and semibenzilic mechanisms may be fairly closely balanced, as instances of the operation of the semibenzilic mechanism have been reported even

69. Y. Yukawa, Y. Tsuno, and T. Ibata, *Bull. Chem. Soc. Jpn.* **40**, 2613, 2618 (1967); W. Jugelt and D. Schmidt, *Tetrahedron* **25**, 969 (1969).

70. A. S. Kende, *Org. React.* **11**, 261 (1960); A. A. Akhrem, T. K. Ustynyuk, and Y. A. Titov, *Russ. Chem. Rev.* (English translation) **39**, 732 (1970).

71. F. G. Bordwell, R. G. Scamehorn, and W. R. Springer, *J. Am. Chem. Soc.* **91**, 2087 (1969); F. G. Bordwell and J. G. Strong, *J. Org. Chem.* **38**, 579 (1973).

Scheme 8.5. Wolff Rearrangement of α-Diazoketones

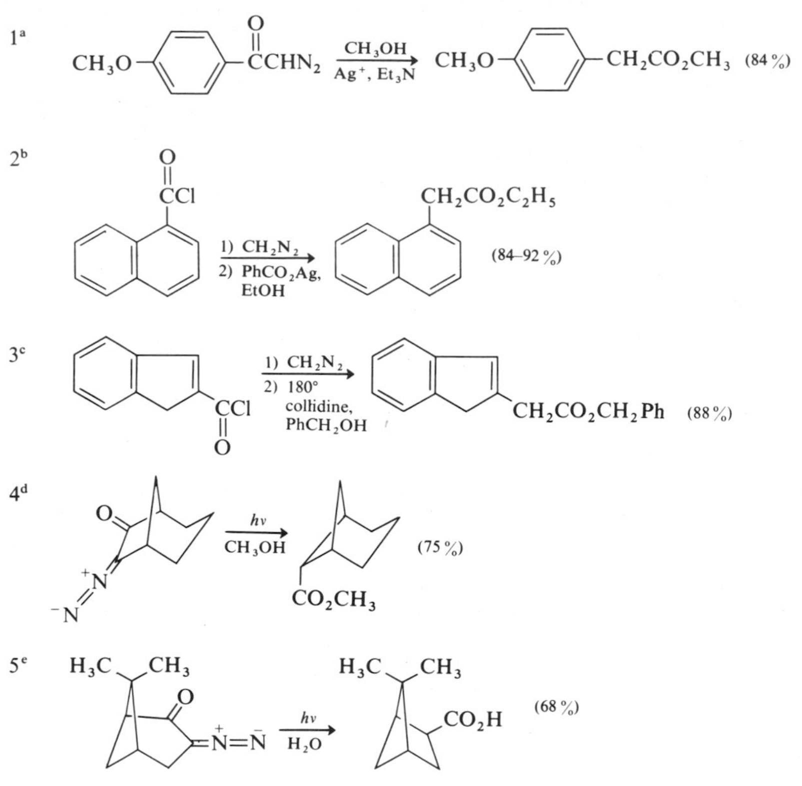

325

SECTION 8.3.
REARRANGE-
MENTS OF
ELECTRON-
DEFICIENT
INTERMEDIATES

1[a]

$$CH_3O-C_6H_4-\overset{O}{\overset{\|}{C}}CHN_2 \xrightarrow[Ag^+, Et_3N]{CH_3OH} CH_3O-C_6H_4-CH_2CO_2CH_3 \quad (84\%)$$

2[b]

$$\xrightarrow[\substack{2)\ PhCO_2Ag,\\ EtOH}]{1)\ CH_2N_2} \quad CH_2CO_2C_2H_5 \quad (84-92\%)$$

3[c]

$$\xrightarrow[\substack{2)\ 180°\\ collidine,\\ PhCH_2OH}]{1)\ CH_2N_2} \quad CH_2CO_2CH_2Ph \quad (88\%)$$

4[d]

$$\xrightarrow[CH_3OH]{hv} \quad CO_2CH_3 \quad (75\%)$$

5[e]

$$\xrightarrow[H_2O]{hv} \quad CO_2H \quad (68\%)$$

a. M. S. Newman and P. F. Beal, III, *J. Am. Chem. Soc.* **72**, 5163 (1956).
b. V. Lee and M. S. Newman, *Org. Synth.* **50**, 77 (1970).
c. E. D. Bergmann and E. Hoffmann, *J. Org. Chem.* **26**, 3555 (1961).
d. K. B. Wiberg and B. A. Hess, Jr., *J. Org. Chem.* **31**, 2250 (1966).
e. J. Meinwald and P. G. Gassman, *J. Am. Chem. Soc.* **82**, 2857 (1960).

for compounds with hydrogen available for enolization.[72] Included in the evidence that the cyclopropanone mechanism usually operates, in preference to the semibenzilic mechanism, is the demonstration, in several instances, that a symmetrical intermediate is involved. The isomeric chloroketones **2** and **3**, for example, lead to the same ester, and it has been shown that the two ketones are not interconverted

$$PhCH(Cl)\overset{O}{\overset{\|}{C}}CH_3 \xrightarrow{\bar{O}CH_3} PhCH_2CH_2CO_2CH_3 \xleftarrow{\bar{O}CH_3} PhCH_2\overset{O}{\overset{\|}{C}}CH_2Cl \qquad \text{Ref. 71}$$

2   **4**   **3**

72. E. W. Warnoff, C. M. Wong, and W. T. Tai, *J. Am. Chem. Soc.* **90**, 514 (1968).
71. See p. 324.

326

CHAPTER 8
REACTIONS
INVOLVING
ELECTRON-
DEFICIENT
INTERMEDIATES

under the conditions of the rearrangement. A common intermediate, such as the cyclopropanone, can explain this observation. The occurrence of a symmetrical intermediate has also been demonstrated by ^{14}C labeling in the case of 2-chlorocyclohexanone.[73]

Because of the operation of the cyclopropanone mechanism, the structure of the ester product cannot be predicted directly from the structure of the reacting haloketone. Instead, the identity of the product is governed by the direction of ring-opening of the cyclopropanone intermediate. The dominant mode of ring-opening would be expected to be that which forms the more stable of the two possible ester enolates. For this reason, a phenyl substituent favors breaking the bond to the substituted carbon, but an alkyl substituent directs the cleavage to a less substituted carbon.[74] That both **2** and **3** (page 325) give the same ester, **4**, is compatible with the directing effect the phenyl group would have on the ring-opening step.

α-Alkoxyketones are common by-products of Favorskii rearrangements catalyzed by alkoxide ions. Generally, these by-products are not formed by a direct S_N2 displacement, since a symmetrical intermediate appears to be involved. The most satisfactory mechanism suggests that the enol form of the chloroketone is the precursor of the alkoxyketones.[75]

There have also been a number of investigations of the stereochemistry of the Favorskii rearrangement. Although several compounds show stereoselectivity, the general result is that these reactions give the stereoisomeric reaction products that

73. P. B. Loftfield, *J. Am. Chem. Soc.* **73**, 4707 (1951).
74. C. Rappe, L. Knutsson, N. J. Turro, and R. B. Gagosian, *J. Am. Chem. Soc.* **92**, 2032 (1970).
75. F. G. Bordwell and M. W. Carlson, *J. Am. Chem. Soc.* **92**, 3377 (1970).

Scheme 8.6. Base-Catalyzed Rearrangements of α-Haloketones

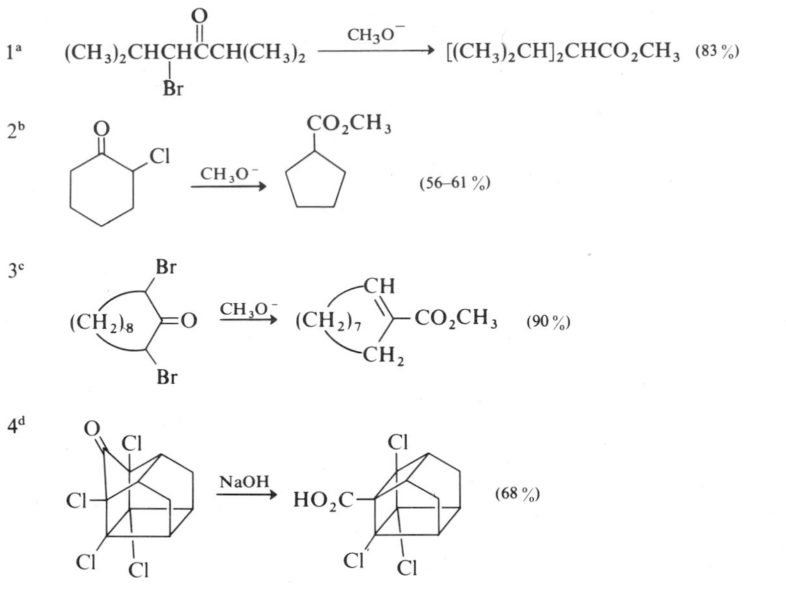

327

SECTION 8.3.
REARRANGE-
MENTS OF
ELECTRON-
DEFICIENT
INTERMEDIATES

1[a] $(CH_3)_2CHCHCCH(CH_3)_2$ $\xrightarrow{CH_3O^-}$ $[(CH_3)_2CH]_2CHCO_2CH_3$ (83%)

2[b] $\xrightarrow{CH_3O^-}$ (56–61%)

3[c] $\xrightarrow{CH_3O^-}$ (90%)

4[d] $\xrightarrow{NaOH}$ (68%)

a. S. Sarel and M. S. Newman, *J. Am. Chem. Soc.* **78**, 416 (1956).
b. D. W. Goheen and W. R. Vaughan, *Org. Synth.* **IV**, 594 (1963).
c. E. W. Garbisch, Jr., and J. Wohllebe, *J. Org. Chem.* **33**, 2157 (1968).
d. R. J. Stedman, L. S. Miller, L. D. Davis, and J. R. E. Hoover, *J. Org. Chem.* **35**, 4169 (1970).

would be expected if a planar enolate and/or dipolar intermediate were generated at some point.

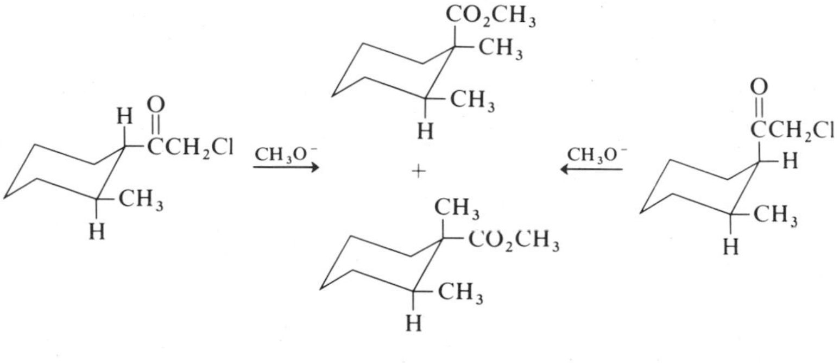

Ref. 76

The Favorskii reaction has been used to effect ring contraction in the course of synthesis of strained ring systems. Entry 4 in Scheme 8.6 illustrates this application of the reaction. With α,α'-dihaloketones, the rearrangement is accompanied by dehydrohalogenation to yield an α,β-unsaturated ester, as illustrated by entry 3 in Scheme 8.6.

76. H. O. House and F. A. Richey, Jr., *J. Org. Chem.* **32**, 2151 (1967).

328

CHAPTER 8
REACTIONS
INVOLVING
ELECTRON-
DEFICIENT
INTERMEDIATES

α-Halosulfones undergo a related rearrangement. The carbanion formed by deprotonation forms an unstable thiirane dioxide[77]:

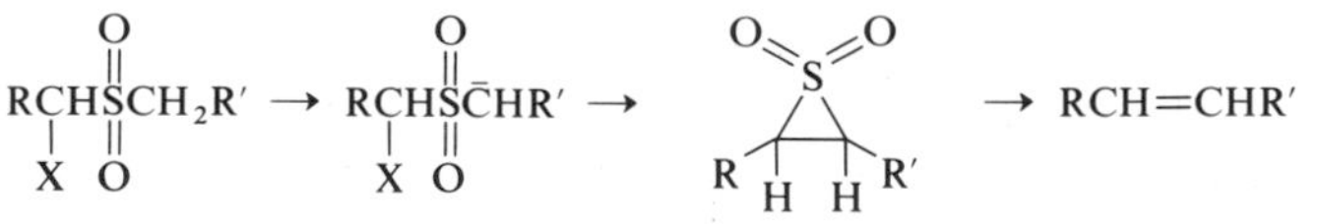

$$\underset{\underset{X}{|}\; \underset{O}{\|}}{RCHSCH_2R'} \rightarrow \underset{\underset{X}{|}\; \underset{O}{\|}}{RCHS\bar{C}HR'} \rightarrow \underset{\underset{R}{}\underset{H}{}\;\underset{H}{}\underset{R'}{}}{\overset{O\diagdown S\diagup O}{\triangle}} \rightarrow RCH{=}CHR'$$

The thiirane dioxides decompose with elimination of sulfur dioxide. The reaction is useful for the synthesis of certain types of olefins.

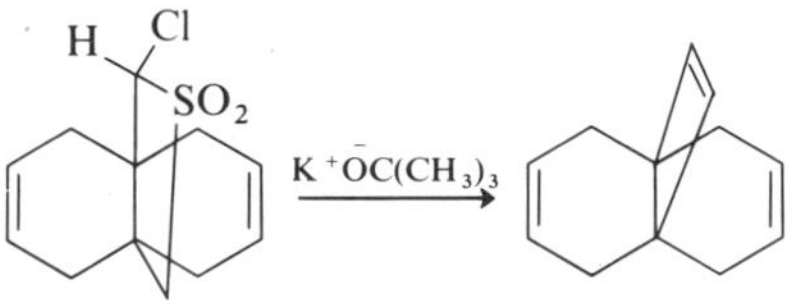

Ref. 78

8.3.2. Migration to Nitrogen

The most important reactions in this group are the *Beckmann rearrangement*, which converts oximes to amides; a family of reactions that convert carboxylic acid derivatives to amines with loss of the carbonyl carbon; and the reaction of ketones with hydrazoic acid to give amides (the *Schmidt reaction*).

The Beckmann rearrangement of oximes is a very general reaction that has been studied over a long period of time. A variety of protonic and Lewis acids can cause the reaction to occur. Early studies established two significant points about the

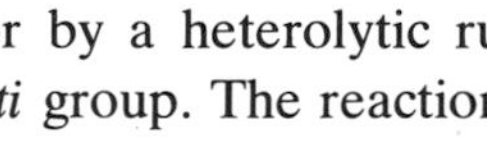

$$\underset{RCR'}{\overset{N-OH}{\|}} \rightarrow \underset{}{\overset{O}{\underset{\|}{RNHCR'}}}$$

stereochemistry of the reaction that provide an insight into the mechanism. First, the group that migrates is the one *anti* to the hydroxyl group on the C=N double bond, so that in the structure shown, R migrates in preference to R'. Second, the stereochemical configuration of the migrating group is retained. These stereochemical features are accounted for by a heterolytic rupture of the N–O bond with concerted migration of the *anti* group. The reaction is completed by addition of a

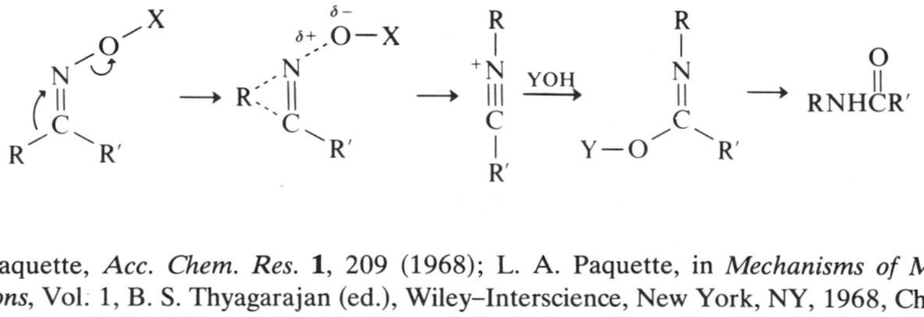

77. L. A. Paquette, *Acc. Chem. Res.* **1**, 209 (1968); L. A. Paquette, in *Mechanisms of Molecular Migrations*, Vol. 1, B. S. Thyagarajan (ed.), Wiley–Interscience, New York, NY, 1968, Chap. 3.
78. L. A. Paquette, J. C. Philips, and R. E. Wingard, Jr., *J. Am. Chem. Soc.* **93**, 4516 (1971).

nucleophile to the resulting nitrilium ion and, eventually, hydrolysis and tautomerism to the stable amide.[79,80]

329

SECTION 8.3.
REARRANGE-
MENTS OF
ELECTRON-
DEFICIENT
INTERMEDIATES

A wide variety of reagents—including sulfuric acid, hydrochloric acid, polyphosphoric acid, phosphorus pentachloride, phosphorus oxychloride, and arenesulfonyl halides—can cause the rearrangement to occur. All function by converting the $-OH$ group to a more reactive leaving group, thereby facilitating rupture of the N–O bond. Under some conditions, inversion of oxime configuration is competitive with the rearrangement step. When this competition occurs, a mixture of amides is formed. The methods that are least likely to promote oxime isomerism, and resultant formation of a mixture of amides, are treatment of the oxime with phosphorus pentachloride or solvolysis of the p-toluenesulfonate ester of the oxime.[81,82] Catalysis of the rearrangement by protic acids is more likely to cause prior equilibration of the oxime isomers.

A variation in the Beckmann rearrangement occurs if one of the groups R or R′ can give rise to a relatively stable carbonium ion. In this circumstance, fragmentation of the molecule occurs, as will be discussed more fully in Section 8.4.

$$\underset{R'}{\overset{R}{\diagdown}}C=N\overset{O-X}{\diagup} \longrightarrow R^+ + R'-C\equiv N + {}^-OX$$

Scheme 8.7 records a number of examples of the Beckmann rearrangement. A review article in *Organic Reactions*[79] contains an extensive survey of reactions reported prior to 1960.

A second useful reaction involving rearrangement to an electron-deficient nitrogen center is the thermal decomposition of acyl azides, known as the *Curtius rearrangement*. The initial product is an isocyanate that may be isolated or react further, depending on the reaction solvent. This rearrangement shares with the

$$\underset{}{\overset{O}{\overset{\|}{RC}}}-\overset{..}{\underset{}{N}}-\overset{+}{N}\equiv N \longrightarrow O=C=N-R + N_2$$

$$H_2O \diagup \qquad \diagdown ROH$$

$$[R-\overset{H}{N}-\overset{\overset{O}{\|}}{C}-OH] \qquad R-\overset{H}{N}-\overset{\overset{O}{\|}}{C}-OR'$$

$$\downarrow$$

$$RNH_2 + CO_2$$

Beckmann rearrangement the feature that the migrating group retains its stereochemical configuration. The migration is usually considered to be concerted

79. I. G. Donaruma and W. Z. Heldt, *Org. React.* **11**, 1 (1960); P. A. S. Smith, *Open Chain Nitrogen Compounds*, Vol. II, W. A. Benjamin, New York, NY, 1966, pp. 47–54.
80. P. A. S. Smith, in *Molecular Rearrangements*, Vol. I, P. de Mayo (ed.), Interscience, New York, NY, 1963, pp. 483–507.
81. R. F. Brown, N. M. van Gulick, and G. H. Schmid, *J. Am. Chem. Soc.* **77**, 1094 (1955).
82. J. C. Craig and A. R. Naik, *J. Am. Chem. Soc.* **84**, 3410 (1962).

330

CHAPTER 8
REACTIONS
INVOLVING
ELECTRON-
DEFICIENT
INTERMEDIATES

Scheme 8.7. Beckmann Rearrangement Reactions

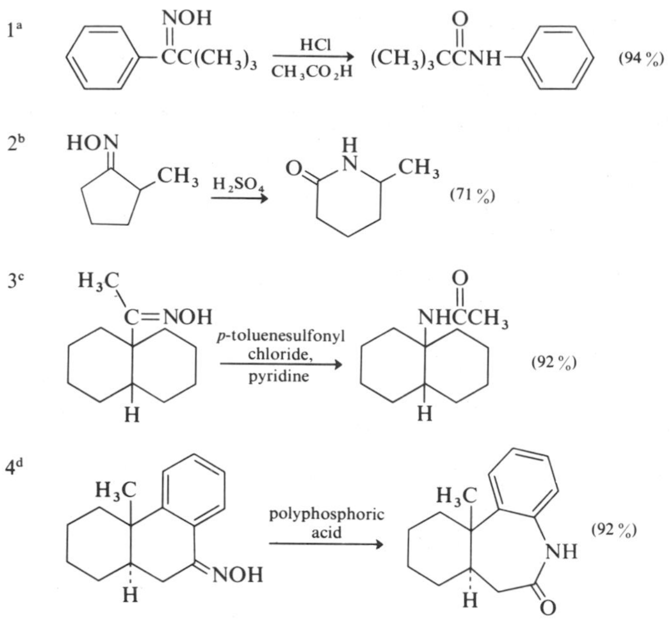

a. R. F. Brown, N. M. van Gulick, and G. H. Schmid, *J. Am. Chem. Soc.* **77**, 1094 (1955).
b. J. G. Hildebrand, Jr., and M. T. Bogert, *J. Am. Chem. Soc.* **58**, 650 (1936).
c. R. K. Hill and O. T. Chortyk, *J. Am. Chem. Soc.* **84**, 1064 (1962).
d. R. A. Barnes and M. T. Beachem, *J. Am. Chem. Soc.* **77**, 5388 (1955).

with the loss of molecular nitrogen, because reactions attributable to a nitrene

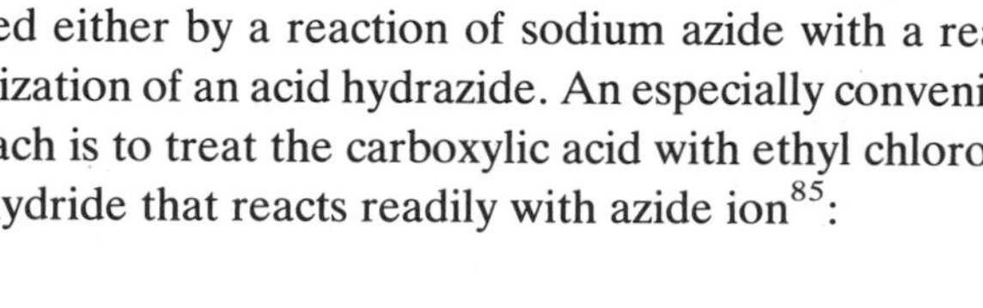

intermediate are not observed.[83] The temperatures required to bring about decomposition of acyl azides are quite low, usually being in the vicinity of 100°.[84] The acyl azides are obtained either by a reaction of sodium azide with a reactive acylating agent or by diazotization of an acid hydrazide. An especially convenient variation of the former approach is to treat the carboxylic acid with ethyl chloroformate, which gives a mixed anhydride that reacts readily with azide ion[85]:

83. S. Linke, G. T. Tisue, and W. Lwowski, *J. Am. Chem. Soc.* **89**, 6308 (1967).
84. P. A. S. Smith, *Org. React.* **3**, 337 (1946); P. A. S. Smith, *Open Chain Nitrogen Compounds*, Vol. 2, W. A. Benjamin, New York, NY, 1966, pp. 219–221.
85. J. Weinstock, *J. Org. Chem.* **26**, 3511 (1961).

331

SECTION 8.3.
REARRANGE-
MENTS OF
ELECTRON-
DEFICIENT
INTERMEDIATES

$$RCO_2H \xrightarrow[R_3N]{ClCOEt} RCOCOEt \xrightarrow{N_3^-} RCN_3$$

$$RCNHNH_2 \xrightarrow[H^+]{NaNO_2} RCN_3$$

Some examples of the use of the Curtius reaction are given in Scheme 8.8.

Another reaction that finds an occasional use for conversion of carboxylic acids to the corresponding amine with loss of carbon dioxide is the hypobromite oxidation of amides. The N-bromoamide is presumably an intermediate.[86] This reaction is known as the *Hofmann rearrangement.* The migration is believed to be concerted,

$$RCNH_2 + {}^-OBr \rightarrow RCNHBr + {}^-OH$$

$$R{-}C{-}N{-}Br \rightarrow O{=}C{=}N{-}R + Br^- \xrightarrow{H_2O} RNH_2 + CO_2$$

with elimination of bromide ion. It can be used to replace carboxyl groups with amino groups, a synthetically useful transformation, particularly for the synthesis of certain aromatic amines.

$$\xrightarrow[KOH]{Br_2}$$

Ref. 87

Carboxylic acids and esters can also be converted to amines with loss of the carbonyl group by reaction with hydrazoic acid, HN_3. This is known as the *Schmidt reaction.*[88] The mechanism of the process is related to that of the Curtius reaction, in that an electron-deficient center is developed by expulsion of nitrogen from an azido group. The intermediate is generated by addition of hydrazoic acid to the carbonyl group. The migrating group retains its configuration.

$$RCO_2H + HN_3 \rightarrow HO{-}C{-}N{-}N{\equiv}N \rightarrow HOCNR + N_2 \xrightarrow{H^+} RNH_3 + CO_2$$

When the Schmidt reaction is applied to α,β-unsaturated carboxylic acids, the product of the rearrangement is an enamine, which hydrolyzes to the corresponding ketone. Entry 3 in Scheme 8.9 is an example of this type of reaction. Reaction with hydrazoic acid converts ketones to amides, and is thus an alternative method for effecting the same transformation as the Beckmann rearrangement. A drawback of

86. E. S. Wallis and J. F. Lane, *Org. React.* **3**, 267 (1946).
87. G. C. Finger, L. D. Starr, A. Roe, and W. J. Link, *J. Org. Chem.* **27**, 3965 (1962).
88. H. Wolff, *Org. React.* **3**, 307 (1946); P. A. S. Smith, in *Molecular Rearrangements*, P. de Mayo (ed.), Vol. 1, Interscience, New York, NY, 1963, pp. 507–527.

332

CHAPTER 8
REACTIONS
INVOLVING
ELECTRON-
DEFICIENT
INTERMEDIATES

Scheme 8.8. Curtius Reactions

1^a

$$CH_3(CH_2)_{10}\overset{O}{\underset{\|}{C}}Cl \xrightarrow[\text{2) benzene, 70°}]{\text{1) NaN}_3} CH_3(CH_2)_{10}N=C=O$$

2^b

$$H_5C_2O_2C(CH_2)_4CO_2C_2H_5 \xrightarrow[\substack{\text{2) benzene} \\ \text{3) H}_2\text{O, HCl}}]{\text{1) N}_2\text{H}_4} Cl^-\overset{+}{H_3}N(CH_2)_4\overset{+}{N}H_3Cl^-$$

3^c

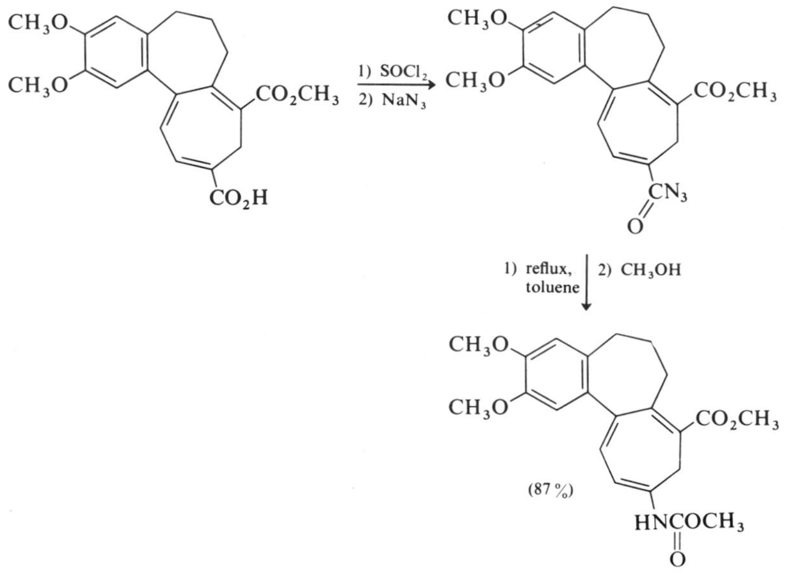

1) reflux, toluene   2) CH$_3$OH

(87 %)

4^d

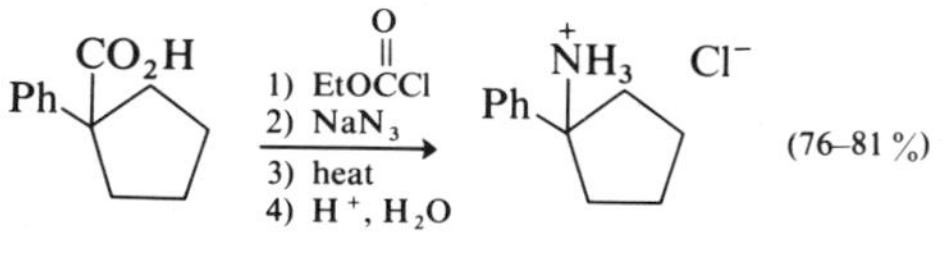

1) EtOCCl
2) NaN$_3$
3) heat
4) H$^+$, H$_2$O

(76–81 %)

5^e

1) SOCl$_2$, pyridine
2) NaN$_3$, xylene

(66 %)

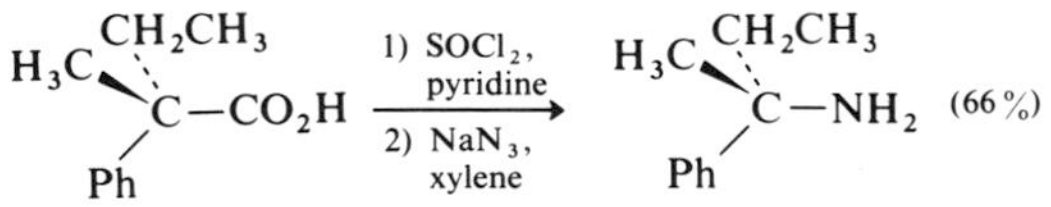

a. C. F. H. Allen and A. Bell, *Org. Synth.* **III**, 846 (1955).
b. P. A. S. Smith, *Org. Synth.* **IV**, 819 (1963).
c J. Schreiber, W. Leimgruber, M. Pesaro, P. Schudel, T. Threlfall, and A. Eschenmoser, *Helv. Chim. Acta* **44**, 540 (1961).
d. C. Kaiser and J. Weinstock, *Org. Synth.* **51**, 48 (1971).
e. D. J. Cram and J. S. Bradshaw, *J. Am. Chem. Soc.* **85**, 1108 (1963).

this method is that the features that determine which of the ketone branches will

$$\underset{RCR'}{\overset{O}{\|}} \xrightarrow{HN_3} \underset{RCNHR'}{\overset{O}{\|}} + \underset{RNHCR'}{\overset{O}{\|}}$$

migrate are not completely understood. For this reason, the composition of the product mixture to be expected from unsymmetrical ketones cannot always be predicted. As in the case of carboxylic acids, the key intermediate is generated by addition of hydrazoic acid to the carbonyl group. The decomposition may, in some cases, be preceded by a dehydration step. Fragmentation to a nitrile and carbonium

$$\underset{RCR'}{\overset{O}{\|}} + HN_3 \rightleftarrows \underset{\underset{\overset{|}{\text{-}N-N\equiv N}}{R-C-R'}}{\overset{OH}{|}} \xrightarrow{H^+} \underset{\underset{\overset{|}{N-N\equiv N}}{R-C-R'}}{\overset{OH}{|}} \rightarrow \underset{RCNHR}{\overset{O}{\|}}$$

ion can occur when one of the ketone substituents is capable of generating a stable carbonium ion[89–91]:

$$\underset{R_3CCR'}{\overset{O}{\|}} \xrightarrow{HN_3} \underset{R_3CCR'}{\overset{N-N\equiv N}{\|}} \rightarrow N\equiv C-R' + R_3C^+$$

Some examples of the Schmidt reaction are given in Scheme 8.9.

8.4. Fragmentation Reactions

The name "fragmentation" has been given to reactions in which the carbon skeleton is cleaved when an electron deficiency develops. A structural feature that permits fragmentation to occur readily is the presence of a carbon beta to the developing electron deficiency that can readily accept carbonium-ion character. This type of reaction occurs particularly readily when the γ-atom is a heteroatom, such as

$$\underset{\gamma \quad \beta \quad \alpha}{Y-C-C-A-X} \rightarrow \overset{+}{Y}=C + C=A + X^-$$

nitrogen or oxygen, having unshared electron pairs available for stabilization of the new carbonium-ion center.[92] The fragmentation can occur as a concerted process or stepwise. The concerted mechanism is characterized by accelerated reaction rates and is restricted to a molecular geometry appropriate for continuous overlap

89. R. K. Hill, R. T. Conley, and O. T. Chortyk, *J. Am. Chem. Soc.* **87**, 5646 (1965).
90. R. M. Palmere, R. T. Conley, and J. L. Rabinowitz, *J. Org. Chem.* **37**, 4095 (1972).
91. R. T. Conley and B. E. Nowak, *J. Org. Chem.* **26**, 692 (1961).
92. C. A. Grob, *Angew. Chem. Int. Ed. Engl.* **8**, 535 (1969).

334

CHAPTER 8
REACTIONS
INVOLVING
ELECTRON-
DEFICIENT
INTERMEDIATES

Scheme 8.9. Schmidt Reactions

A. Carboxylic Acids and Esters

1[a] $\quad$ PhCH$_2$CO$_2$H $\xrightarrow[\text{polyphosphoric acid}]{\text{NaN}_3}$ PhCH$_2$NH$_2$ $\quad$ (67%)

2[b]

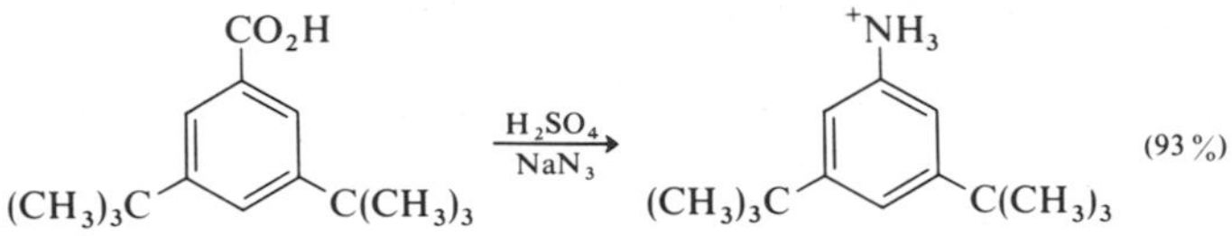

(93%)

3[c]

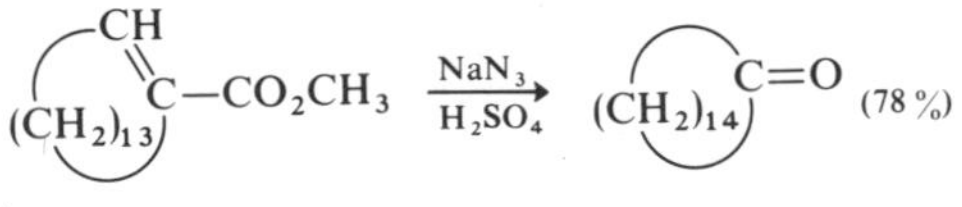

(78%)

B. Ketones

4[d]

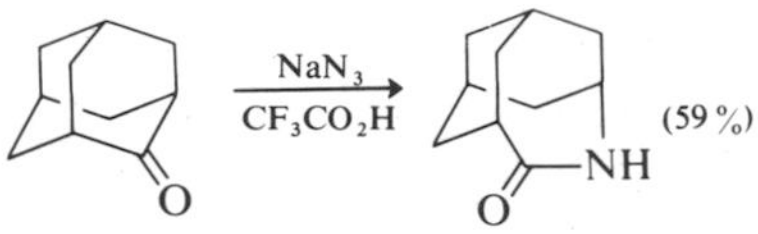

(59%)

5[e]

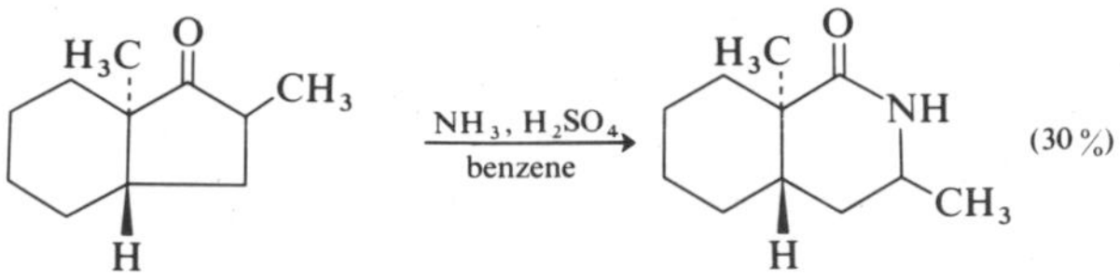

(30%)

6[f]

(44%)

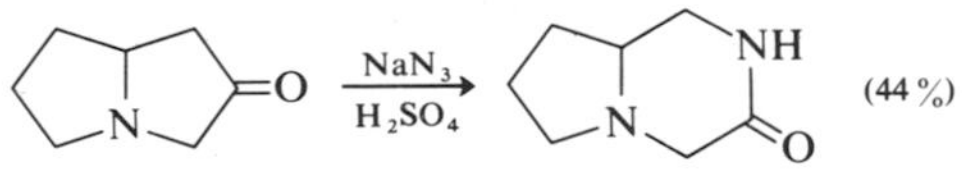

a. R. M. Palmere and R. T. Conley, *J. Org. Chem.* **35**, 2703 (1970).
b. J. W. Elder and R. P. Mariella, *Can. J. Chem.* **41**, 1653 (1963).
c. B. D. Mookherjee, R. W. Trenkle, and R. P. Patel, *J. Org. Chem.* **36**, 3266 (1971).
d. T. Sasaki, S. Eguchi, and T. Toru, *J. Org. Chem.* **35**, 4109 (1970).
e. E. J. Moriconi and M. A. Stemniski, *J. Org. Chem.* **37**, 2035 (1972).
f. L. A. Paquette and M. K. Scott, *J. Org. Chem.* **33**, 2379 (1968).

of the participating orbitals. An example is the solvolysis of 4-chloropiperidine, which is more rapid than solvolysis of chlorocyclohexane and occurs with fragmentation[93]:

The Beckmann rearrangement reaction provides a number of examples of fragmentation. The products that result from fragmentation are a carbonium ion and

93. R. D'Arcy, C. A. Grob, T. Kaffenberger, and V. Krasnobajew, *Helv. Chim. Acta* **49**, 185 (1966).

Scheme 8.10. Fragmentation Reactions

A. Solvolytic Fragmentation Promoted by Heteroatoms

1[a] CH_3SO_2O — [structure with O^- and CH_3] $\xrightarrow{K^+\bar{O}C(CH_3)_3}$ $H_2C=CH$— [bicyclic ketone structure with CH_3]   (64 %)

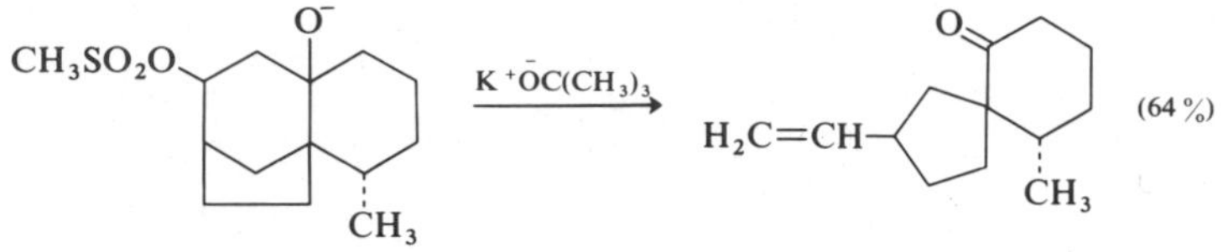

2[b]   [decahydroquinoline structure with OSO_2Ar and N—CH_2Ph] $\xrightarrow[\text{of NaBH}_4]{\substack{\text{solvolysis}\\\text{in the presence}}}$ [ring structure with N—CH_2Ph]   (44–58 %)

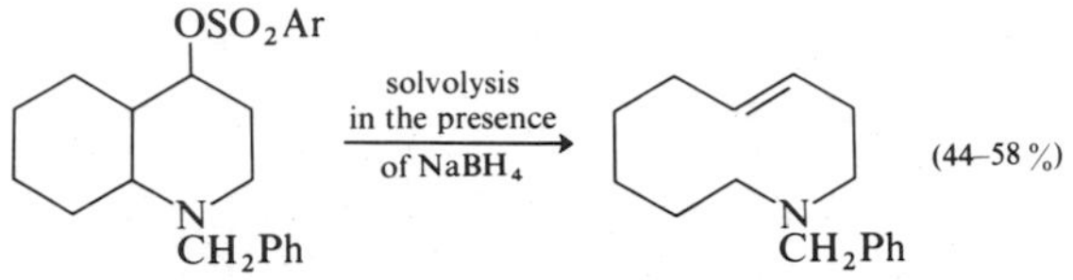

3[c]   [decalin structure with H_3C, OSO_2CH_3, OH] $\xrightarrow{LiAlH_4}$ [H_3C ring structure with OH]   (71 %)

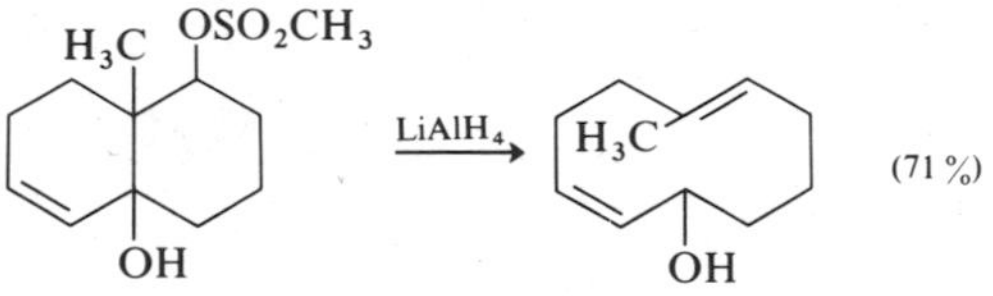

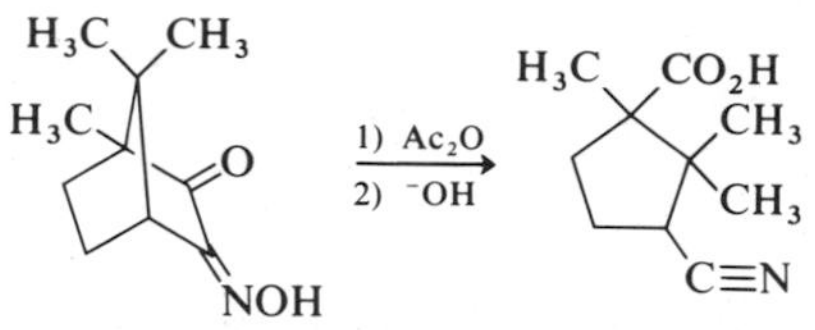

B. Fragmentation in Beckmann Rearrangements

4[d]   [bicyclic structure with $=NOH$, H_3C, CH_3] $\xrightarrow{PCl_5}$ [benzene ring with $CH_2CH_2C\equiv N$, $C=CH_2$, CH_3]   (93 %)

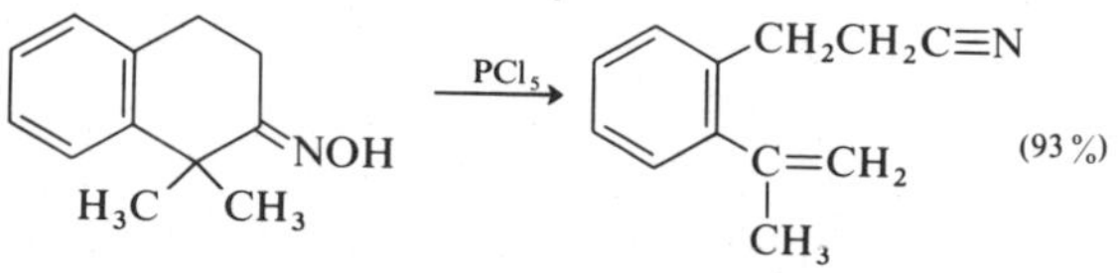

5[e]   [H_3C, CH_3, H_3C bicyclic structure with O, $=NOH$] $\xrightarrow[\text{2) }^-OH]{\text{1) Ac}_2O}$ [cyclopentane with H_3C, CO_2H, CH_3, CH_3, $C\equiv N$]

6[f]   [cyclohexane with $=NOH$, OCH_3] $\xrightarrow{SOCl_2}$ $CH_3OCH(CH_2)_4C\equiv N$ with Cl below CH

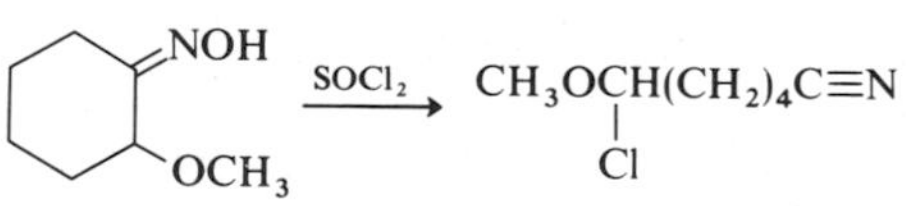

a. J. A. Marshall and S. F. Brady, *J. Org. Chem.* **35**, 4068 (1970).
b. J. A. Marshall and J. H. Babler, *J. Org. Chem.* **34**, 4186 (1969).
c. J. A. Marshall, W. F. Huffman, and J. A. Ruth, *J. Am. Chem. Soc.* **94**, 4691 (1972).
d. R. T. Conley and R. J. Lange, *J. Org. Chem.* **28**, 210 (1963).
e. A. Hassner, W. A. Wentworth, and I. H. Pomerantz, *J. Org. Chem.* **28**, 304 (1963).
f. M. Ohno and I. Terasawa, *J. Am. Chem. Soc.* **88**, 5683 (1966).

336

CHAPTER 8
REACTIONS
INVOLVING
ELECTRON-
DEFICIENT
INTERMEDIATES

a nitrile. Fragmentation is very likely to occur if X is a nitrogen, oxygen, or sulfur

$$\ddot{X} \overset{\frown}{-} C \overset{R}{\underset{\frown}{-C}} = N \overset{\frown}{\underset{\frown}{-}} OY \rightarrow \overset{+}{X} = C + RC \equiv N + {}^-OY$$

atom. It is also likely to occur when one of the oxime substituents is a tertiary alkyl group. Some examples of fragmentation reactions under the conditions of the Beckmann rearrangement are given in Scheme 8.10.

Fragmentation processes can be of synthetic value as a means of obtaining a single large ring from two fused rings. Entry 3 in Scheme 8.10 is an example of this type. Organoboranes have been shown to undergo fragmentation if a good leaving group is present on the δ-carbon.[94] The reactive intermediate is probably the tetrahedral intermediate formed by addition of hydroxide ion at boron.

8.5. Some Synthetically Useful Carbonium-Ion Reactions

Skeletal rearrangements that accompany S_N1 substitution reactions will not be emphasized here, since a discussion of these types of reactions was presented in Part A, Chapter 5. Rather, the focus will be on reactions with special structural features that promote a particularly useful skeletal change.

One such reaction is the acid-catalyzed conversion of diols to ketones. This reaction is sometimes referred to as the *pinacol rearrangement*.[95] The classic example of the reaction is the conversion of 2,3-dimethylbutane-2,3-diol (pinacol) to methyl *t*-butyl ketone (pinacolone):

$$\underset{\underset{HO \quad OH}{|\quad\quad|}}{(CH_3)_2C - C(CH_3)_2} \overset{H^+}{\rightarrow} \overset{O}{\overset{\|}{CH_3CC(CH_3)_3}} \quad (67\text{-}72\%) \qquad \text{Ref. 96}$$

The mechanism involves carbonium ion formation and substituent migration, assisted by electron release from the remaining hydroxyl group:

94. J. A. Marshall and G. L. Bundy, *Chem. Commun.*, 854 (1967); P. S. Wharton, C. E. Sundin, D. W. Johnson, and H. C. Kluender, *J. Org. Chem.* **37**, 34 (1972).
95. C. J. Collins, *Q. Rev. Chem. Soc.* **14**, 357 (1960).
96. G. A. Hill and E. W. Flosdorf, *Org. Synth.* **I**, 451 (1932).

In some cases, part of the reaction also passes through an epoxide intermediate[97]:

337

SECTION 8.5.
SOME
SYNTHETICALLY
USEFUL
CARBONIUM-
ION REACTIONS

$$R_2C-CR_2 \longrightarrow R_2C\underset{O}{\overset{}{-}}CR_2 \rightleftharpoons R_2C\underset{O}{\overset{}{-}}CR_2$$

with HO, $+OH_2$, H substituents as drawn

$$R_2\overset{+}{C}-CR_2 \longrightarrow R_3CCR$$
with OH and O (carbonyl)

The traditional conditions for carrying out the pinacol rearrangement have involved treating the glycol with a strong acid. Under these conditions, the more easily ionized C–O bond generates the carbonium ion, and migration of one of the groups from the adjacent carbinol site ensues. Both stereochemistry and "migratory aptitude" can be factors in determining the extent of migration of two unlike groups.

Another method for carrying out the same net rearrangement involves synthesis of a glycol monosulfonate ester. These compounds rearrange under the influence of

$$R_2C-CR_2 \longrightarrow B-H + RC\overset{R}{-}CR_2 \longrightarrow RCCR_3$$
with B, HO, OSO_2R' and O, OSO_2R and carbonyl O

base. This method can be used to change the nature of rearrangement from that expected in an acid-catalyzed reaction of the corresponding glycol. In the case of a glycol that contains one secondary and one tertiary hydroxyl, for example, the secondary group will be preferentially sulfonylated. In contrast, the tertiary hydroxyl will be more easily ionized under the acid-catalyzed conditions. This method has been of some value in rearranging ring systems, especially in the area of terpene synthesis, as illustrated by entries 4 and 5 in Scheme 8.11.

Aminomethylcarbinols yield ketones when treated with nitrous acid. This reaction has been used synthetically to form ring-expanded cyclic ketones, a procedure known as the *Tiffeneau–Demjanov reaction*.[98] The diazotization procedure generates the same type of β-hydroxycarbonium ion that is formed in the

$$R_2\overset{OH}{C}CH_2NH_2 \xrightarrow{HONO} R_2\overset{OH}{C}CH_2\overset{+}{N}{\equiv}N \rightarrow R\overset{OH}{\underset{R}{C}}{-}\overset{+}{C}H_2 \rightarrow R\overset{O}{C}CH_2R$$

HO   CH₂NH₂ →(HONO)→ O (ring-expanded ketone)   (61%)   Ref. 99

97. S. Wold, *Acta Chem. Scand.* **23**, 1266 (1969).
98. P. A. S. Smith and D. R. Baer, *Org. React.* **11**, 157 (1960).
99. F. F. Blicke, J. Azuara, N. J. Dorrenbos, and E. B. Hotelling, *J. Am. Chem. Soc.* **75**, 5418 (1953).

CHAPTER 8
REACTIONS
INVOLVING
ELECTRON-
DEFICIENT
INTERMEDIATES

Scheme 8.11. Some Examples of Pinacol Rearrangements

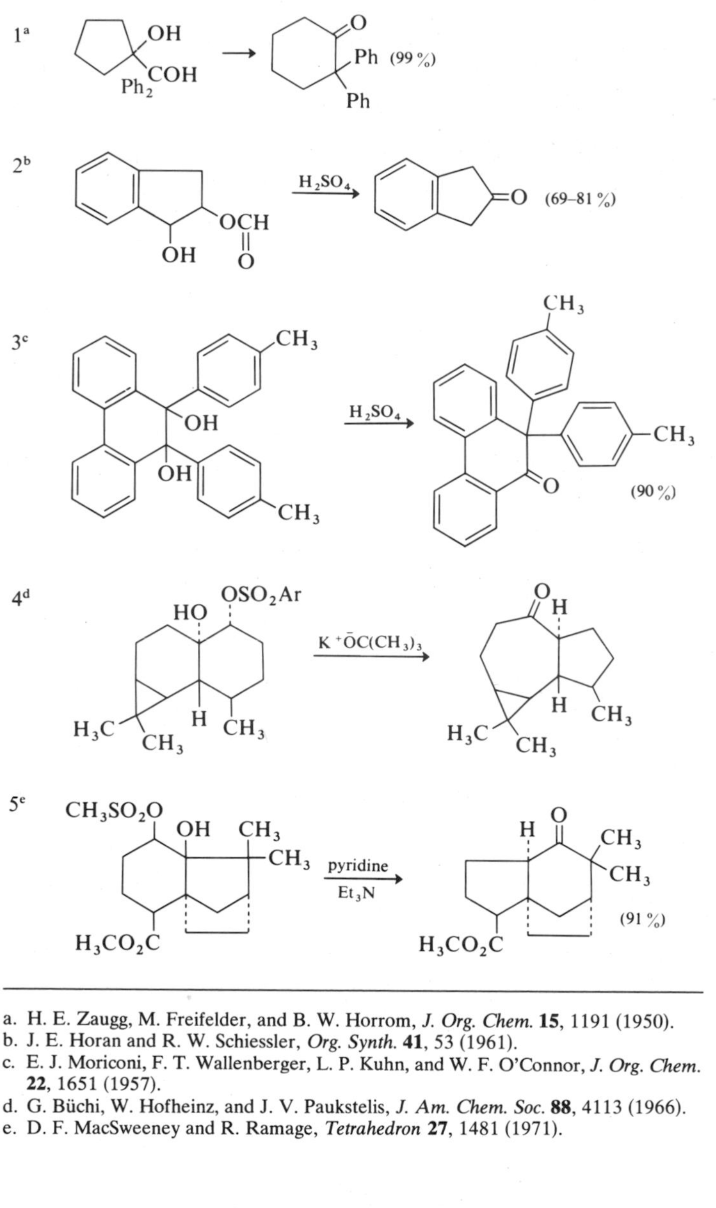

a. H. E. Zaugg, M. Freifelder, and B. W. Horrom, *J. Org. Chem.* **15**, 1191 (1950).
b. J. E. Horan and R. W. Schiessler, *Org. Synth.* **41**, 53 (1961).
c. E. J. Moriconi, F. T. Wallenberger, L. P. Kuhn, and W. F. O'Connor, *J. Org. Chem.*
   22, 1651 (1957).
d. G. Büchi, W. Hofheinz, and J. V. Paukstelis, *J. Am. Chem. Soc.* **88**, 4113 (1966).
e. D. F. MacSweeney and R. Ramage, *Tetrahedron* **27**, 1481 (1971).

pinacol rearrangement. A long-standing method for obtaining the required aminomethylcarbinols involves cyanohydrin formation, followed by reduction:

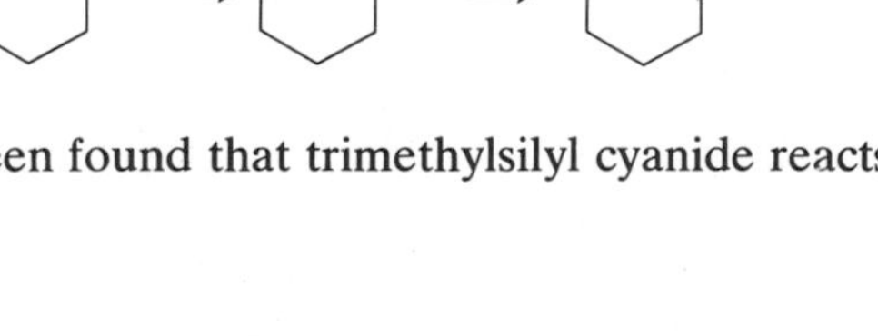

More recently, it has been found that trimethylsilyl cyanide reacts with ketones to give trimethylsilyl ethers of cyanohydrins. These compounds can be directly reduced to the aminomethylcarbinol by lithium aluminum hydride[100]:

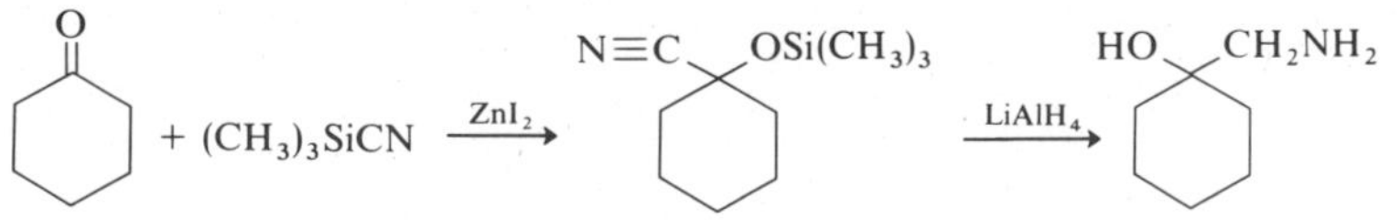

339

SECTION 8.5.
SOME
SYNTHETICALLY
USEFUL
CARBONIUM-
ION REACTIONS

Ref. 100

Another method for the synthesis of aminomethylcycloalkanols involves reduction of the adducts of nitromethane and cyclic ketones[101]:

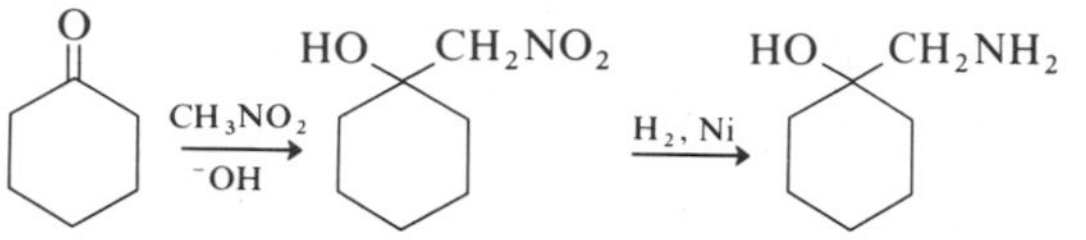

Ref. 101

Diazotization of aminomethylcycloalkanes can also lead to ring expansion, but the absence of participation by a hydroxyl group in the rearrangement makes this a much less specific reaction. Mixtures of alkenes and alcohols, both rearranged and unrearranged, are usually obtained and the reaction is seldom synthetically attractive.[98]

The reactions of ketones with diazoalkanes sometimes lead to a ring-expanded product in synthetically useful yields.[102] The reaction occurs by addition of diazoalkane, followed by elimination of nitrogen and migration:

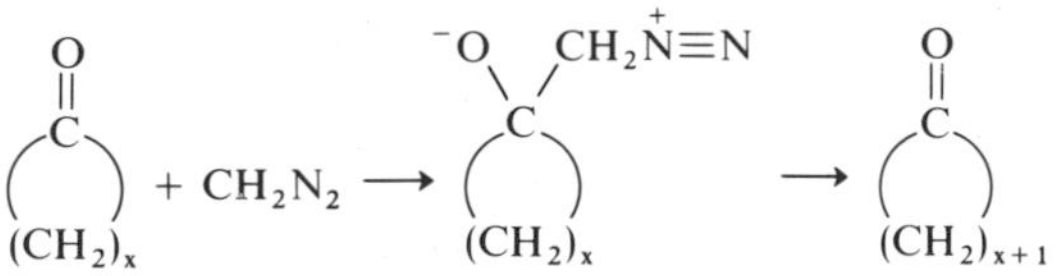

In protic solvents, at least, the reaction proceeds via essentially the same intermediate that is involved in the Tiffenau–Demjanov reaction. Since the product is also a ketone, subsequent addition of diazomethane can lead to higher homologs. The best yields are obtained when the starting ketone is more reactive than the product. For this reason, strained ketones often work well in this reaction. Higher diazoalkanes can also be employed.[103] The reaction has been found to be accelerated by alcoholic solvents. This effect probably involves the hydroxyl group acting as a

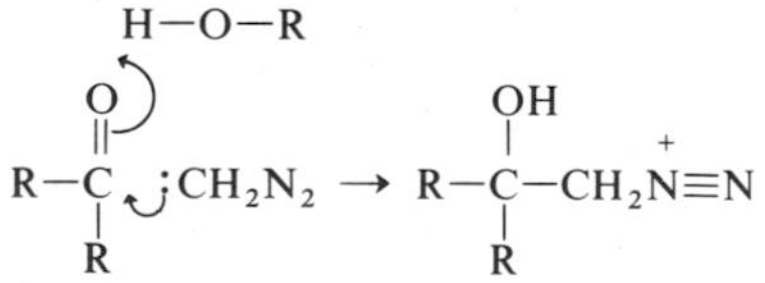

100. D. A. Evans, G. L. Carroll, and L. K. Truesdale, *J. Org. Chem.* **39**, 914 (1974).
101. W. E. Noland, J. F. Kneller, and D. E. Rice, *J. Org. Chem.* **22**, 695 (1957).
 98. See p. 337.
102. C. D. Gutsche, *Org. React.* **8**, 364 (1954).
103. J. A. Marshall and J. J. Partridge, *J. Org. Chem.* **33**, 4090 (1968).

340

CHAPTER 8
REACTIONS
INVOLVING
ELECTRON-
DEFICIENT
INTERMEDIATES

proton donor and thus facilitating the addition step.[104] A side product sometimes encountered in these reactions is the oxirane formed by addition of a CH_2 unit to the carbonyl group:

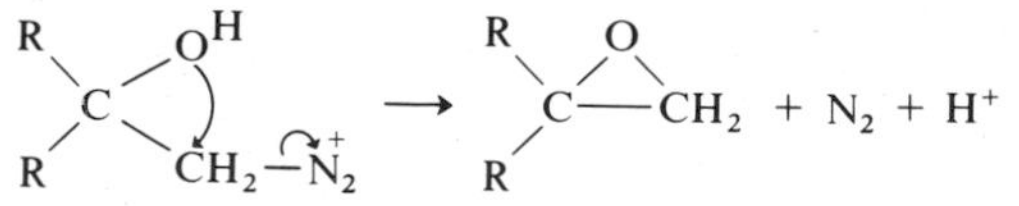

Intramolecular reactions between diazo groups and carbonyl centers can be used to construct bicyclic ring systems:

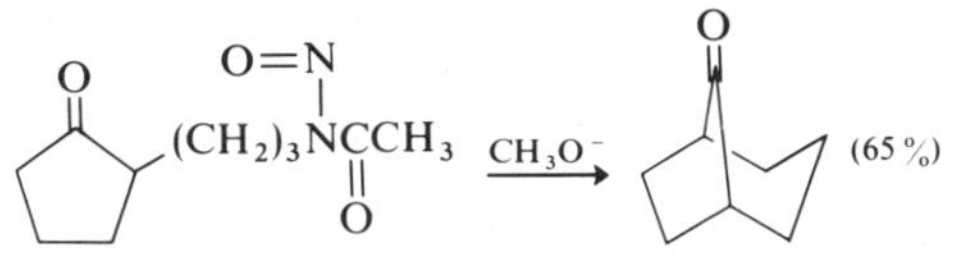

Ref. 105

The treatment of polyolefins with protic or Lewis acids is sometimes an effective method of ring synthesis. The reaction proceeds through an electrophilic attack, and requires that the olefin units that are to participate in the cyclization be properly juxtaposed. For example, compound **5** is converted quantitatively to **6** on treatment with formic acid. The reaction is initiated by protonation and ionization of the allylic alcohol.

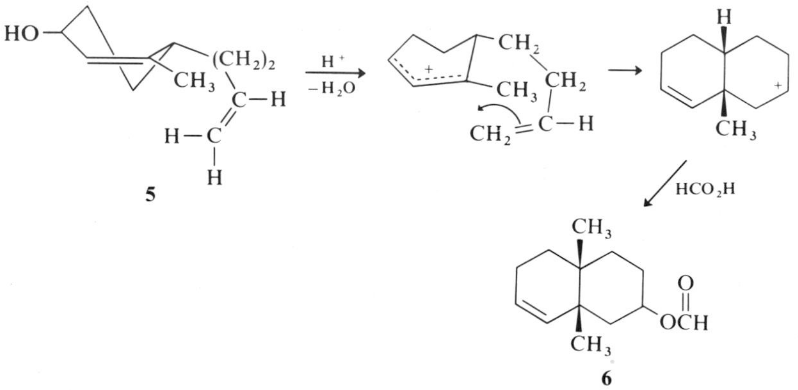

Ref. 106

More extended polyolefins can cyclize to polycyclic systems:

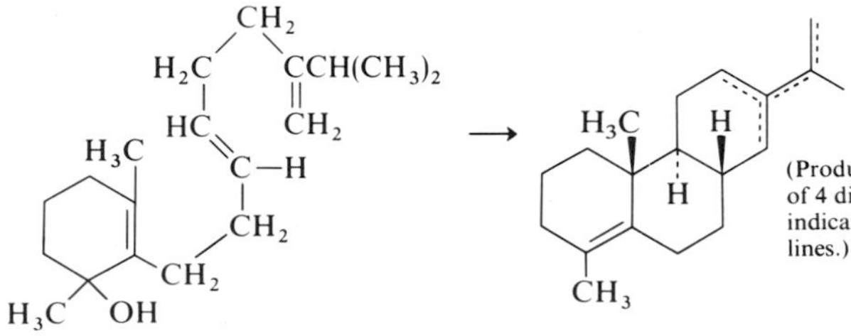

(Product is a mixture of 4 diene isomers indicated by dotted lines.)

Ref. 107

104. J. N. Bradley, G. W. Cowell, and A. Ledwith, *J. Chem. Soc.*, 4334 (1964).
105. C. D. Gutsche and D. M. Bailey, *J. Org. Chem.* **28**, 607 (1963).
106. W. S. Johnson, P. J. Neustaedter, and K. K. Schmiegel, *J. Am. Chem. Soc.* **87**, 5148 (1965).
107. W. S. Johnson, N. P. Jensen, J. Hooz, and E. J. Leopold, *J. Am. Chem. Soc.* **90**, 5872 (1968).

341

SECTION 8.5.
SOME
SYNTHETICALLY
USEFUL
CARBONIUM-
ION REACTIONS

The reactions are usually highly stereoselective, with the stereochemical out-come being predictable on the basis of reactant conformation.[108] The stereochemis-try of cyclization products in the decalin family can be predicted by assuming chair conformations for the developing cyclohexane rings. The stereochemistry at ring junctures is that expected from *anti* attack at the participating double bond(s):

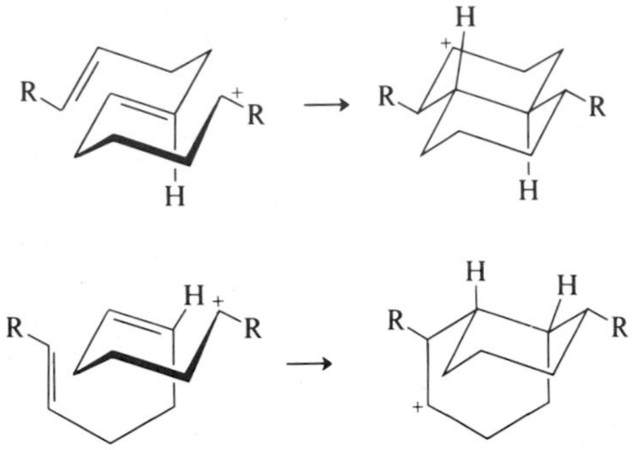

To be of maximum synthetic value, the generation of the cationic site that initiates cyclization should involve mild reaction conditions. The most versatile systems to date have been allylic alcohols, which are readily cyclized in acidic media. Formic acid and Lewis acids such as stannic chloride have proved to be effective reagents for cyclizing polyolefinic allylic alcohols. Acetals generate α-alkoxy car-bonium ions in acidic solutions and can initiate the cyclization of appropriate polyolefinic systems[109]:

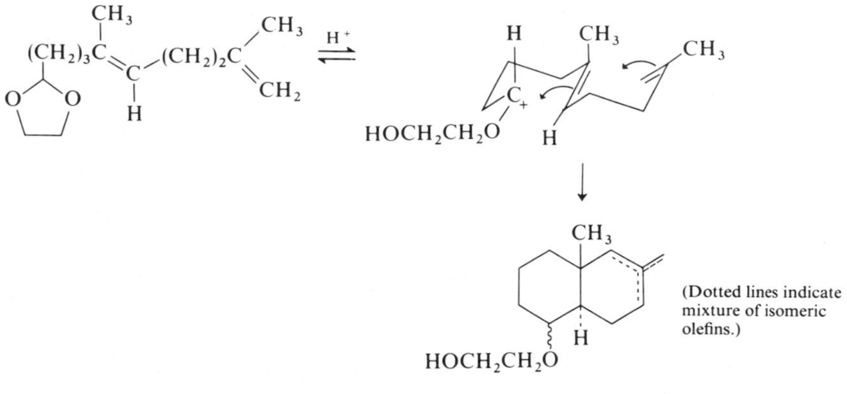

(Dotted lines indicate mixture of isomeric olefins.)

Another significant method of generating the requisite electrophilic site is acid-catalyzed epoxide ring-opening.[110]

Since the immediate product of the cyclization step is a carbonium ion, it is often found that the product consists of a mixture of closely related compounds, all resulting from the same carbonium ion. These usually include products from the

108. W. S. Johnson, *Acc. Chem. Res.* **1**, 1 (1968).
109. A. van der Gen, K. Wiedhaup, J. J. Swoboda, H. C. Dunathan, and W. S. Johnson, *J. Am. Chem. Soc.* **95**, 2656 (1973).
110. E. E. van Tamelen and R. G. Nadeau, *J. Am. Chem. Soc.* **89**, 176 (1967).

342

CHAPTER 8
REACTIONS
INVOLVING
ELECTRON-
DEFICIENT
INTERMEDIATES

capture of the carbonium ion by nucleophilic solvent and a mixture of the various alkenes resulting from loss of a proton.

Polyolefin cyclizations have been of substantial value in the synthesis of polycyclic natural products of the terpene type. To a large extent, these syntheses probably resemble the processes by which polycyclic compounds are assembled in nature from linear polyolefins. The most dramatic example of biological synthesis of a polycyclic skeleton from an acyclic intermediate is the conversion of squalene oxide to the steroid lanosterol and then to other steroids. Scheme 8.12 gives some

Scheme 8.12. Polyolefin Cyclizations

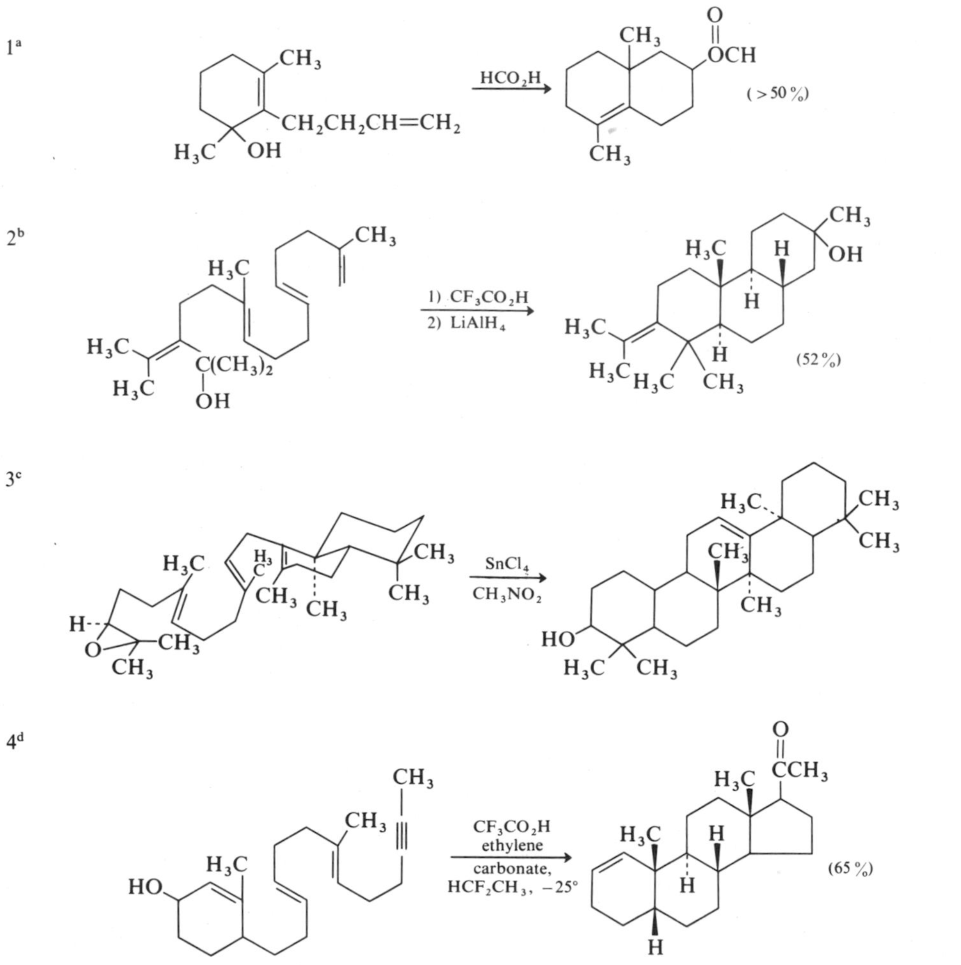

a. J. A. Marshall, N. Cohen, and A. R. Hochstetler, *J. Am. Chem. Soc.* **88**, 3408 (1966).
b. W. S. Johnson and T. K. Schaaf, *Chem. Commun.* 611 (1969).
c. E. E. van Tamelen, R. A. Holton, R. E. Hopla, and W. E. Konz, *J. Am. Chem. Soc.* **94**, 8228 (1972).
d. B. E. McCarry, R. L. Markezich, and W. S. Johnson, *J. Am. Chem. Soc.* **95**, 4416 (1973).

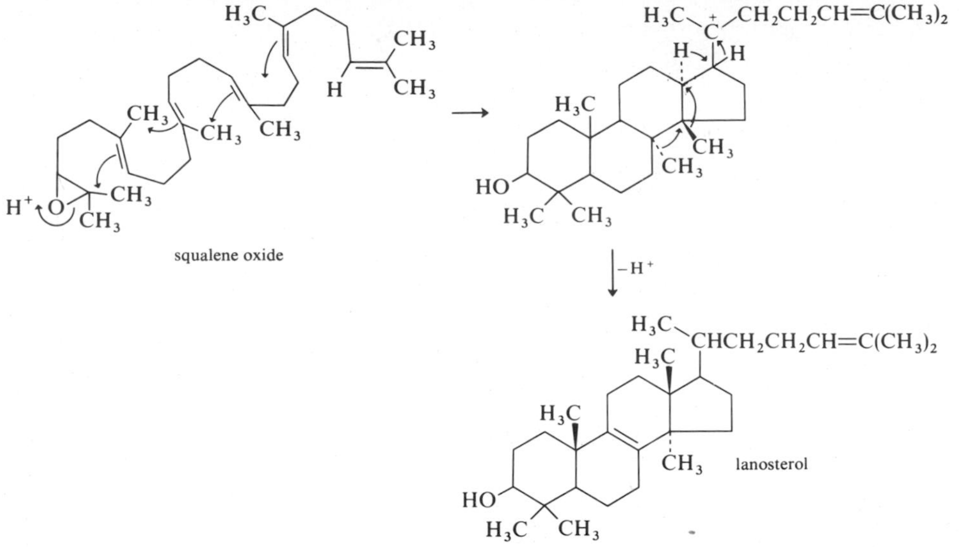

squalene oxide

lanosterol

representative examples of laboratory syntheses involving polyolefin cyclizations.

General References

S. P. McManus (ed.), *Organic Reactive Intermediates,* Academic Press, New York, NY, 1973.

Carbenes

M. Jones, Jr., and R. A. Moss, *Carbenes*, John Wiley and Sons, New York, NY, 1973.
D. Bethell, *Prog. Phys. Org. Chem.* **7**, 153 (1968).
W. Kirmse, *Carbene Chemistry*, Academic Press, New York, NY, 1971.

Nitrenes

W. Lwowski (ed.), *Nitrenes*, Interscience, New York, NY, 1970.
R. A. Abramovitch and E. P. Kyba, in *The Chemistry of the Azido Group*, S. Patai (ed.), Interscience, New York, NY, 1971, pp. 331–395.
G. L'Abbe, *Chem. Rev.* **69**, 345 (1969).

Rearrangements

P. de Mayo (ed.), *Molecular Rearrangements*, Vols. 1 and 2, Interscience, New York, NY, 1963.
B. S. Thyagarajan (ed.), *Mechanisms of Molecular Migrations*, Vols. 1–4, Wiley–Interscience, New York, NY, 1968–1971.
D. Redmore and C. D. Gutsche, *Adv. Alicyclic Chem.* **3**, 1 (1971).

Problems

(References for these problems will be found on page 503.)

1. Give the products expected from the following reactions:

344

CHAPTER 8
REACTIONS
INVOLVING
ELECTRON-
DEFICIENT
INTERMEDIATES

(a)

$$\text{(cyclopentane)}-\overset{\overset{\displaystyle O}{\|}}{C}Cl \xrightarrow[\text{2) Ag}_2\text{O, H}_2\text{O}]{\text{1) CH}_2\text{N}_2}$$

(b)

$$\underset{H \quad C(CH_3)_3}{\overset{N=N}{\diagdown\diagup}} \xrightarrow[\text{nitrobenzene}]{\text{heat}}$$

(c)

$$\underset{H}{\overset{CH_3}{\diagdown}}C=C\underset{CH_3}{\overset{H}{\diagup}} \;+\; N_2CHCO_2C_2H_5 \xrightarrow{\text{CuO}_2\text{CCF}_3}$$

(d)

$$PhCH_2\overset{\overset{\displaystyle O}{\|}}{C}\underset{\underset{\displaystyle Br}{|}}{C}HCH_3 \;+\; CH_3CH_2O^- \longrightarrow$$

(e)

$$\text{(bicyclic)}\;\overset{CH_2NH_2}{\diagup} \xrightarrow[\text{CH}_3\text{CO}_2\text{H}]{\text{NaNO}_2}$$

(f)

$$\text{(adamantane)} \;+\; CH_3O\overset{\overset{\displaystyle O}{\|}}{C}N_3 \xrightarrow{\text{heat}}$$

(g)

$$PhC\equiv CPh \;+\; PhCHCl_2 \xrightarrow{\text{K}^+\bar{\text{O}}\text{C(CH}_3)_3}$$

(h)

$$(CH_3)_3SiO\text{—(arene)—}CH_2CH_2\underset{\overset{\displaystyle |}{H}}{\overset{\overset{\displaystyle H}{|}}{C}}=CCH_2CH_2\text{—(cyclopentene, HO, CH}_3) \xrightarrow[\text{CH}_2\text{Cl}_2]{\text{SnCl}_4}$$

2. Each of the following carbenes has been predicted to have a singlet ground state, either as the result of qualitative structural considerations or on the basis of theoretical calculations. Indicate what structural feature in each case might lead to stabilization of the singlet state.

(a) $\triangleright\!:$

(c) (cycloheptatriene) $:$

(b)

$$CH_3CH_2O\overset{\overset{\displaystyle O}{\|}}{C}\ddot{C}H$$

(d)

$$\underset{N}{\overset{N}{\diagup}}\text{(imidazolidinylidene)}\;\overset{CH_3}{\underset{CH_3}{}}C\!:$$

3. Irradiation of diphenyldiazomethane in 3-methylbutene gives three principal products. Account for the formation of each product.

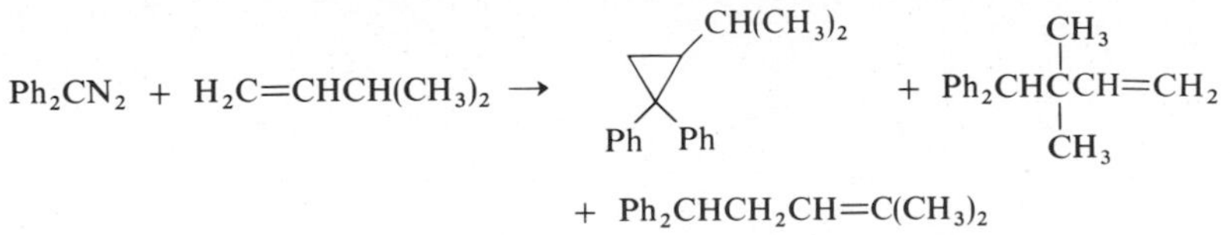

$$Ph_2CN_2 + H_2C{=}CHCH(CH_3)_2 \longrightarrow$$

$$+ Ph_2CHCH_2CH{=}C(CH_3)_2$$

4. The hydroxyl group in *trans*-cycloocten-3-ol determines the stereochemistry of reaction of this compound with the Simmons–Smith reagent. By examining a model, predict the stereochemistry of the resulting product.

5. Treatment of $(CH_3S)_2C{=}NNHSO_2Ar$ with sodium hydride in tetrahydrofuran in the presence of triphenylphosphine gives a yellow suspension. Addition of *p*-nitrobenzaldehyde to the reaction mixture results in the formation of compound **A** in 64% yield. Explain.

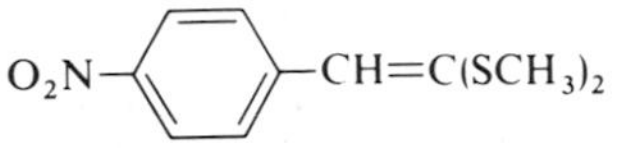

A

6. Suggest a mechanism for each of the following reactions:

(a)

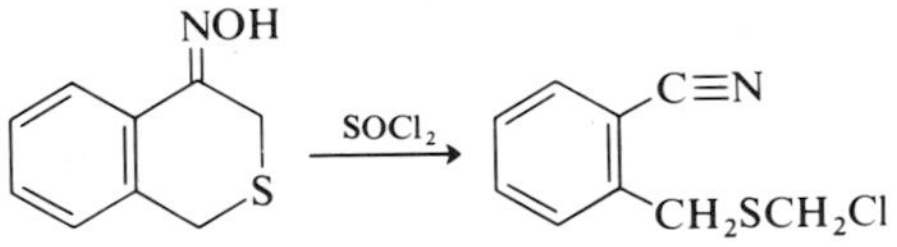

(b)

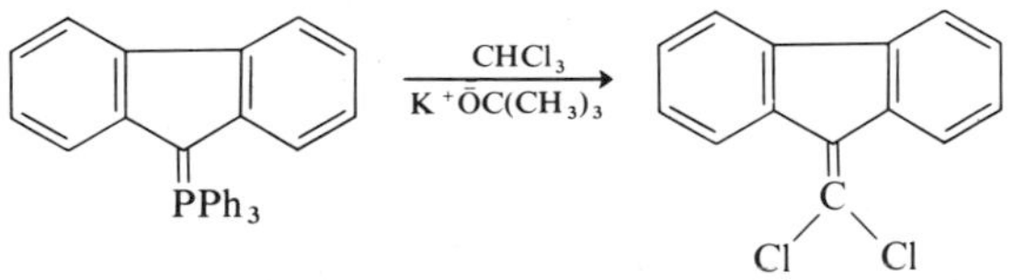

(c)

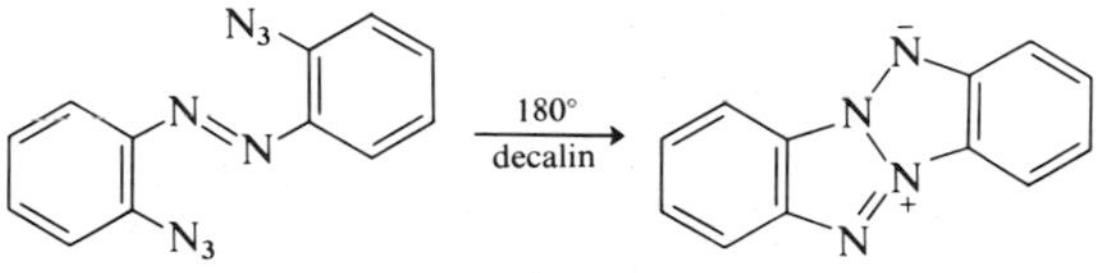

(d)

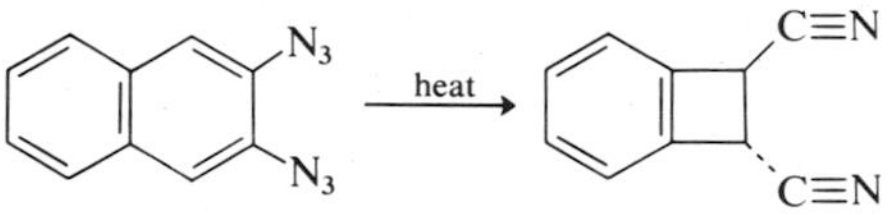

(e)

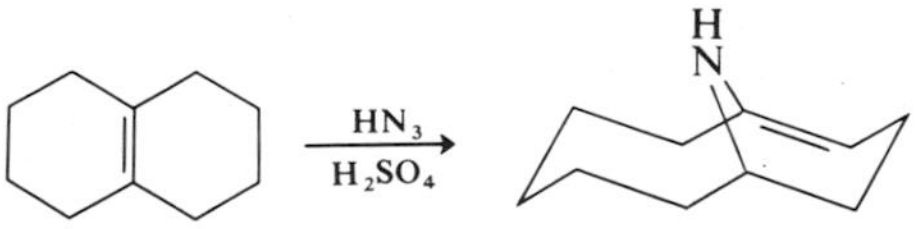

(f)

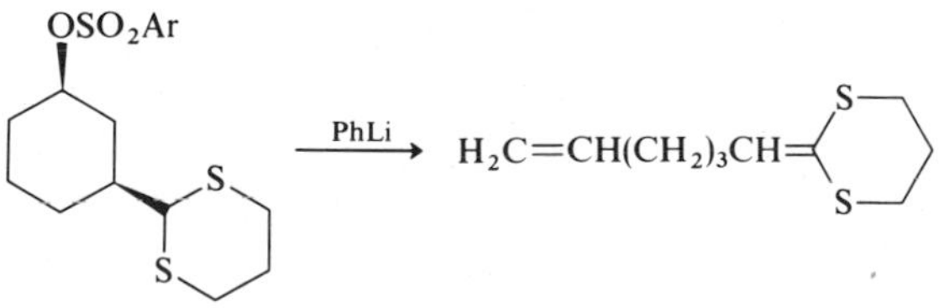

346

CHAPTER 8
REACTIONS
INVOLVING
ELECTRON-
DEFICIENT
INTERMEDIATES

(g)

$$\xrightarrow[\text{H}_2\text{O, 100°}]{\text{KOH}}$$

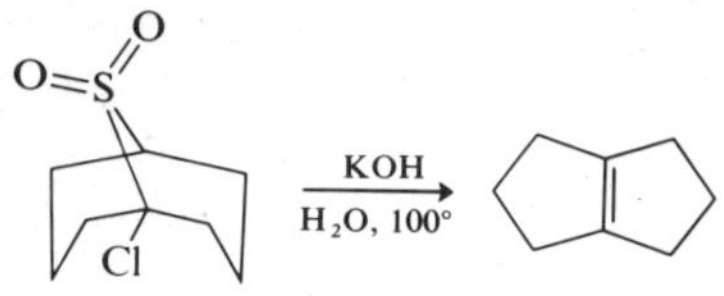

(h) $N_2C(CO_2CH_3)_2 \xrightarrow{280°} H_2C=CHCO_2CH_3$

(i) 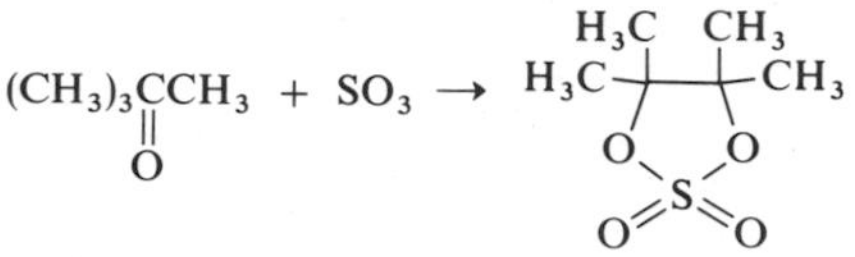 $(CH_3)_3CCH_3 + SO_3 \rightarrow$

(j)

$+ N_3C\equiv N \rightarrow$

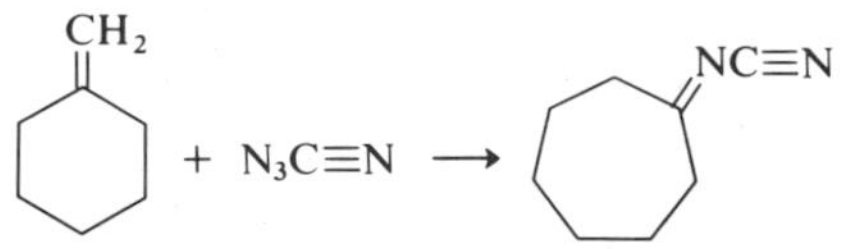

(k)

$$\xrightarrow{\text{KOC(CH}_3)_3}$$

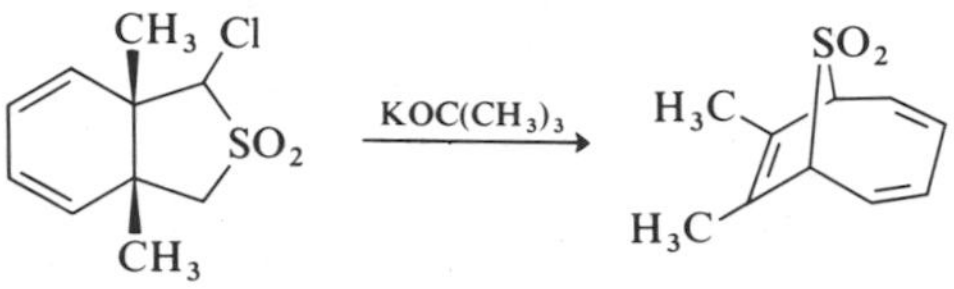

7. The reactions of atomic carbon in its ground state (^3P) with alkenes parallel those of simple carbenes and afford spiropentanes:

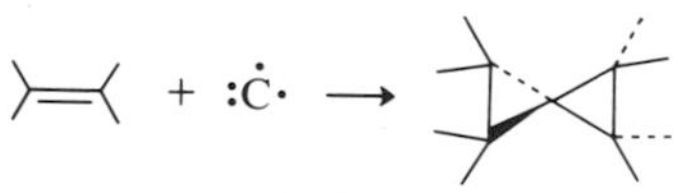

Discuss the observations that reaction of atomic carbon with *trans*-2-butene affords three diastereomeric tetramethylspiropentanes in a ratio of 38:37:25, while reaction with *cis*-2-butene affords two diastereomeric tetramethyl-spiropentanes in equal amounts. One of the products from *trans*-2-butene is the same as one of the products from *cis*-2-butene.

8. Discuss the significance of the relationships between substrate stereochemistry and product composition exhibited by the reactions shown below:

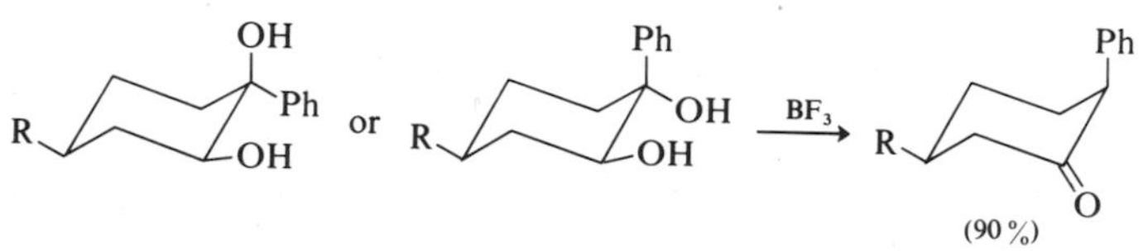

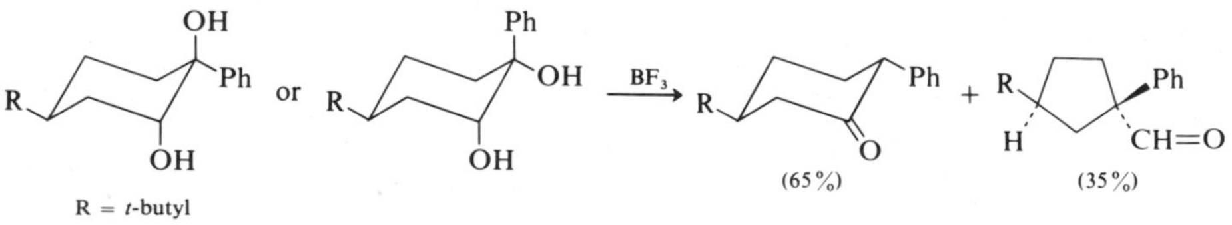

R = *t*-butyl

9. The stereoisomeric chlorodecalones **A** and **B** give different product mixtures on treatment with sodium methoxide in dimethoxyethane. Discuss this result in relationship to the existence of dipolar and cyclopropanone intermediates in Favorskii rearrangements.

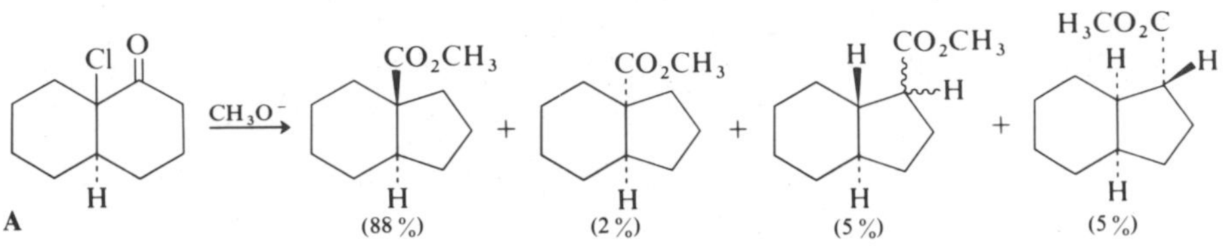

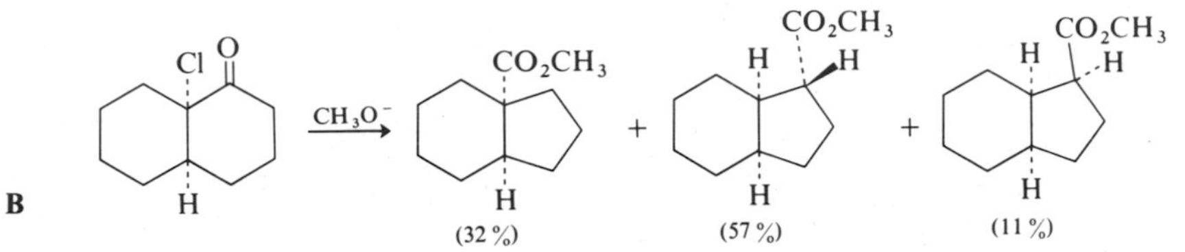

10. Pyrolysis of the sodium salt of benzaldehyde *p*-toluenesulfonylhydrazone gives heptafulvalene **A**:

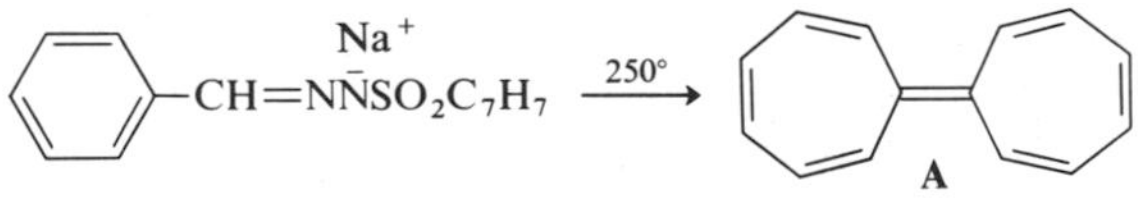

Photolysis of *m*-tolyldiazomethane or *p*-tolyldiazomethane gives product mixtures that include styrene and benzocyclobutene:

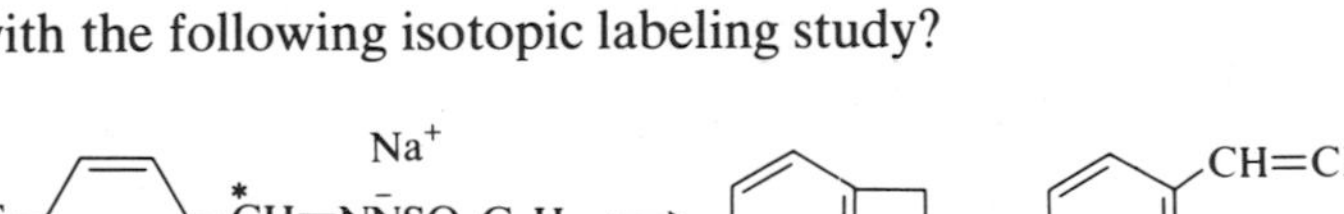

The same products are formed by pyrolysis of the *o*-, *m*-, or *p*-methylbenzaldehyde *p*-toluenesulfonylhydrazones. Suggest mechanisms that could account for these skeletal rearrangements. Are any of your mechanisms consistent with the following isotopic labeling study?

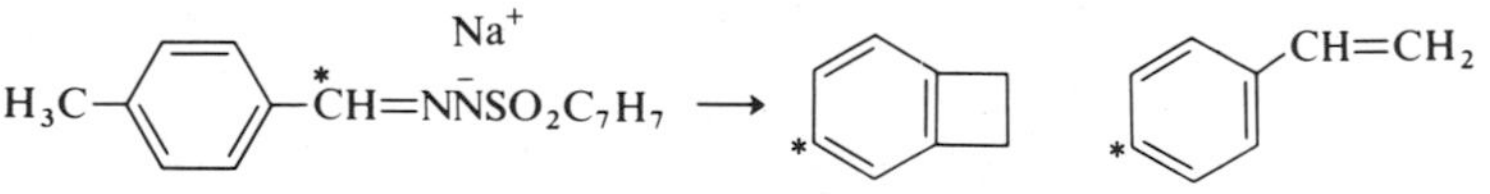

11. Suggest the principal products that would result from photolysis of 3-cyclopentenyldiazomethane.

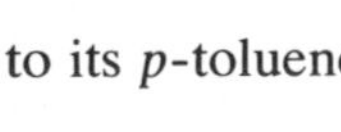

12. The aldehyde **A** was converted to its *p*-toluenesulfonylhydrazone and then to

348

CHAPTER 8
REACTIONS
INVOLVING
ELECTRON-
DEFICIENT
INTERMEDIATES

the corresponding sodium salt, which was then pyrolyzed. Two products were isolated. Suggest possible structures.

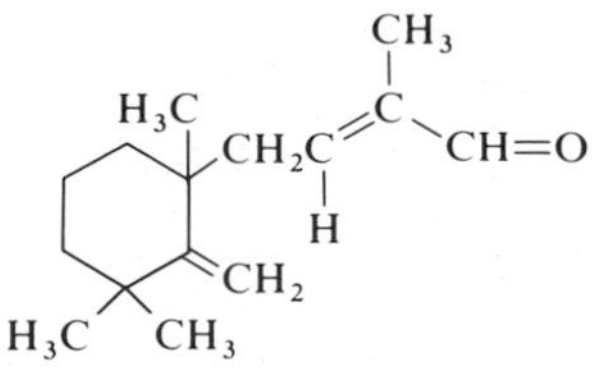

A

13. Suggest a carbene or carbenoid intermediate that could be expected to lead to each of the following strained-ring systems. Suggest a practical means for synthesizing an appropriate precursor for generation of the intermediate.

(a)

(c)

(b)

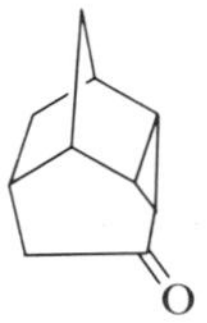

(d)

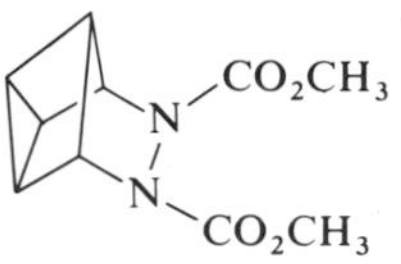

14. Some multistep syntheses are outlined below. Suggest reagents or reaction sequences that could accomplish the lettered steps. In each case, 1 to 3 steps may be required.

(a)

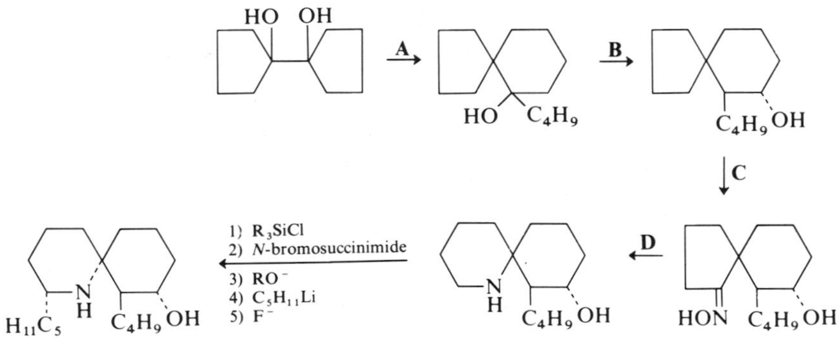

(b)

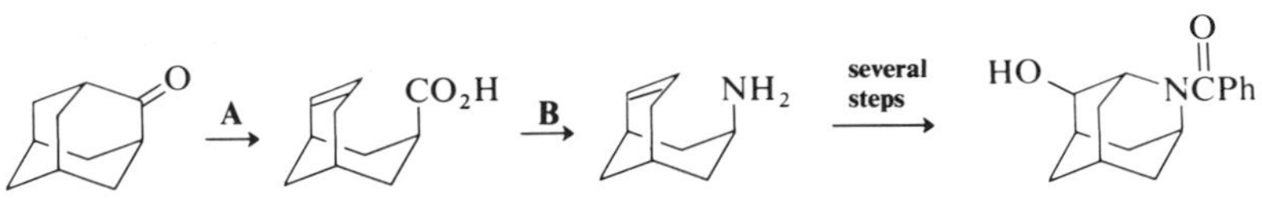

(c)

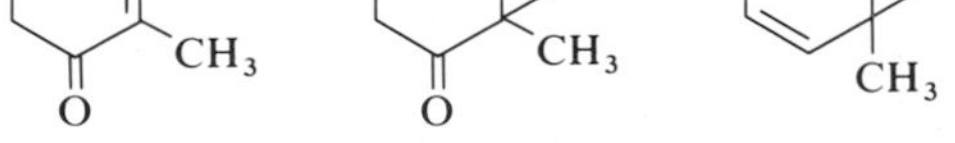

(d)

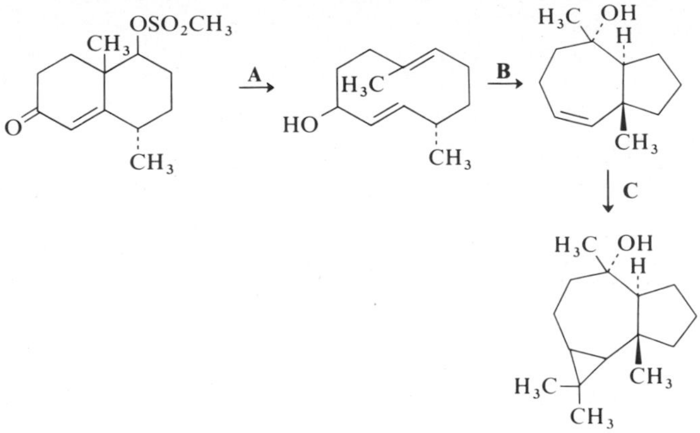

(e)

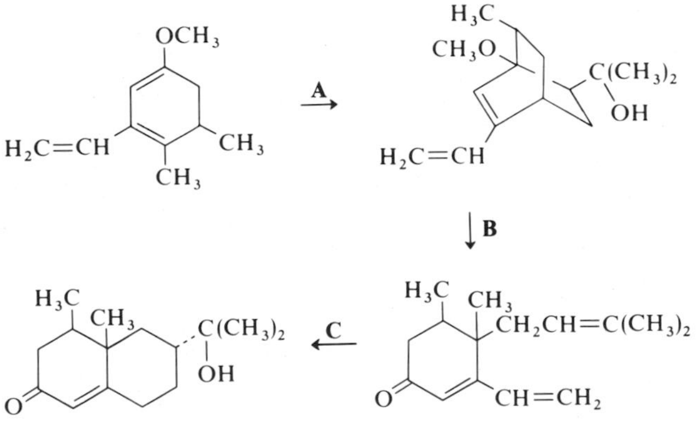

15. A convenient synthesis of neopentyl alcohol consists of stirring diisobutylene (2,4,4-trimethyl-1-pentene) with aqueous hydrogen peroxide in concentrated sulfuric acid. An organic layer separates and is removed and stirred with additional sulfuric acid. The resulting organic layer is distilled, giving neopentyl alcohol in 40% yield. Indicate the course of this synthetic method.

16. Suggest a means of obtaining the desired product from the specified starting material and other easily available compounds. More than one step may be required in some cases.

(a)

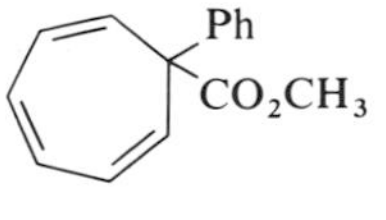

from benzene

(b)

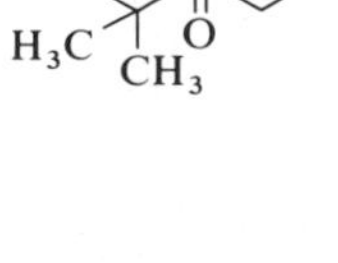

from

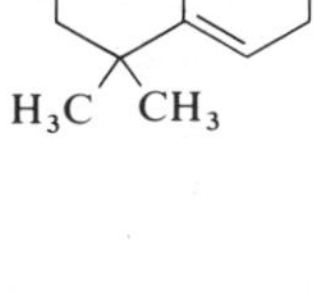

350

CHAPTER 8
REACTIONS
INVOLVING
ELECTRON-
DEFICIENT
INTERMEDIATES

(c) 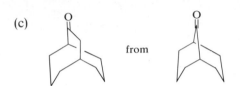 from

(d) [structure with $CO_2C_2H_5$] from [structure with $=O$, $-CH_2CH_2C\equiv N$, $CO_2C_2H_5$]

(e) [cage structure] from [cyclohexadiene structure]

(f) [cyclohexane with CH_3O and $CH_3O^{\cdot\cdot}$] from $H_3CO_2CCH_2CH_2\overset{\displaystyle H_3CO}{\underset{\displaystyle OCH_3}{CHCHCH_2}}CH_2CO_2CH_3$

(*meso* isomer)

(g) [bicyclic structure with $O=$, OCH_3, H] from [cyclopentadiene structure]

(h) [bicyclic structure with CH_3, $O=$, H_3C] from [structure with H_3C, dioxolane, CH_3CH-, OH, OH]

(i) $N\equiv C(CH_2)_7CH=O$ from cyclooctene

(j) [decalin structure with H, H] from [decalin dione structure with H, O, H, O]

17. Give the structure of the expected Favorskii rearrangement product of compound **A**:

[structure of A: bicyclic ketone with Br] $\xrightarrow{CH_3O^-}$

A

Experimentally, it has been found that the above ketone can rearrange by either the cyclopropanone or the semibenzilic mechanism, depending on the reaction conditions. Devise two experiments that would permit you to determine which mechanism was operating under a given set of circumstances.

Oxidations

This chapter will be concerned with transformation of a functional group to a more highly oxidized derivative. The reactions have been chosen for discussion on the basis of their general utility in organic synthesis. As the mechanisms of the reactions are considered, it will become evident that the material in this chapter scans an appreciably wider variety of mechanistic patterns than is true in most of the earlier chapters. Because of this variety in mechanistic patterns, the chapter has been organized along the lines of the functional group transformation that is accomplished. This method of organization facilitates comparison of the methods available for effecting a given synthetic transformation, but has the less desirable result of scattering the reactions of a given oxidant, such as the permanganate ion, into several sections. In general, oxidants have been grouped into three classes: transition-metal derivatives; oxygen, ozone, and peroxides; and other oxidizing agents.

9.1. Oxidation of Alcohols to Aldehydes, Ketones, or Carboxylic Acids

9.1.1. Transition-Metal Oxidants

Probably the most commonly employed of the oxidants based on transition metals are the Cr(VI) oxidants. The dominant form of Cr(VI) in aqueous solution depends upon concentration and pH. In dilute solution, the monomeric acid chromate ion is present; as concentration increases, however, the dichromate ion dominates. The extent of protonation of these ionic species varies with pH. In

$$2\,HO-\underset{\underset{O}{\|}}{\overset{\overset{O}{\|}}{Cr}}-O^{\ominus} \rightleftharpoons {}^-O-\underset{\underset{O}{\|}}{\overset{\overset{O}{\|}}{Cr}}-O-\underset{\underset{O}{\|}}{\overset{\overset{O}{\|}}{Cr}}-O^- + H_2O$$

solutions prepared from CrO_3 and hydroxylic organic solvents, the chromium exists in species that are essentially esters of chromic acid[1]:

$$CH_3CO_2H \ + \ CrO_3 \ \rightarrow \ CH_3CO_2\overset{\overset{O}{\|}}{\underset{\underset{O}{\|}}{Cr}}-OH \ \rightleftharpoons \ CH_3CO_2\overset{\overset{O}{\|}}{\underset{\underset{O}{\|}}{Cr}}O_2CCH_3$$

$$(CH_3)_3COH \ + \ CrO_3 \ \rightarrow \ (CH_3)_3CO\overset{\overset{O}{\|}}{\underset{\underset{O}{\|}}{Cr}}-OH \ \rightleftharpoons \ (CH_3)_3CO\overset{\overset{O}{\|}}{\underset{\underset{O}{\|}}{Cr}}OC(CH_3)_3$$

In pyridine, chromium is present as a complex involving Cr–N bonding:

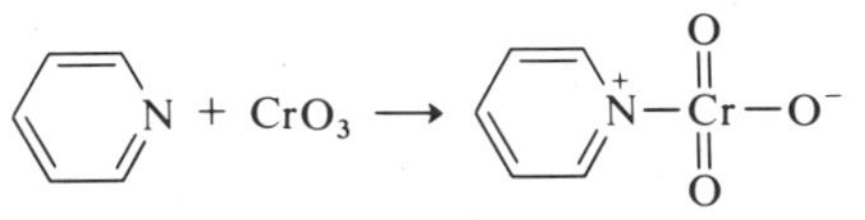

The oxidation state of Cr in each of these compounds is (VI), and they are powerful oxidants. The precise reactivity, however, depends on the solvent and the form of Cr present, so that substantial selectivity can be achieved by choice of the particular reagent and conditions. The transformation most often effected with CrO_3-based oxidants is the conversion of alcohols to the corresponding ketone or aldehyde. The mechanism that is believed to be operative in alcohol oxidations is outlined below. The kinetics of the reaction also indicate a contribution from a variation including an additional proton in the rate-determining transition state. An

$$R_2CHOH \ + \ HCrO_4^- \ + \ H^+ \ \longrightarrow \ R_2CHOCrO_3H \ + \ H_2O$$

$$R_2\underset{\underset{H}{|}}{C}{-}OCrO_3H \ \xrightarrow{\text{slow}} \ R_2C{=}O \ + \ HCrO_3^- \ + \ H^+$$

important piece of evidence pertinent to identification of the rate-determining step is the fact that a large isotope effect is observed when the α-H is replaced by deuterium.[2] The Cr(IV) oxidation state that is produced is not stable, and this species is capable of contributing further to the oxidation. Although other schemes have received consideration and the matter is still a subject of active research, it is believed that a part of the substrate is oxidized via a free-radical intermediate resulting from oxidation by Cr(IV).[3-6]

$$R_2CHOH \ + \ H_2CrO_3 \ + \ 3\,H^+ \ \rightarrow \ R_2\dot{C}OH \ + \ Cr^{+3} \ + \ 3\,H_2O$$

$$R_2\dot{C}OH \ + \ H_2CrO_4 \ \rightarrow \ R_2C{=}O \ + \ HCrO_3 \ + \ H_2O$$

$$R_2CHOH \ + \ HCrO_3 \ + \ 3\,H^+ \ \rightarrow \ R_2C{=}O \ + \ Cr^{+3} \ + \ 3\,H_2O$$

1. K. B. Wiberg, *Oxidation in Organic Chemistry*, Part A, Academic Press, New York, NY, pp. 69–72.
2. F. H. Westheimer and N. Nicolaides, *J. Am. Chem. Soc.* **71**, 25 (1949).
3. M. Rahman and J. Roček, *J. Am. Chem. Soc.* **93**, 5462 (1971).
4. P. M. Nave and W. S. Trahanovsky, *J. Am. Chem. Soc.* **92**, 1120 (1970).
5. K. B. Wiberg and S. K. Mukhergee, *J. Am. Chem. Soc.* **96**, 1884 (1974).
6. M. Doyle, R. J. Swedo, and J. Roček, *J. Am. Chem. Soc.* **95**, 8352 (1973).

With substrates that can generate a relatively stable radical by homolytic fragmentation alpha to the hydroxyl function, this process occurs in competition with the normal oxidation. The process can be formulated as involving an alkoxy radical formed by one-electron oxidation.

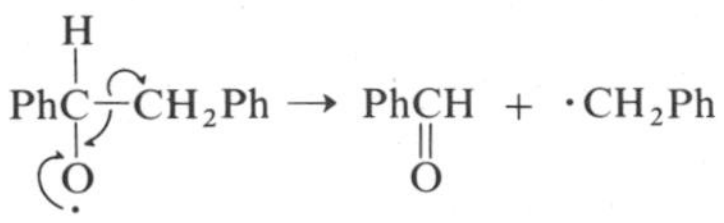

$$PhC\overset{H}{\underset{\underset{\cdot}{O}}{|}}CH_2Ph \rightarrow Ph\underset{O}{\overset{||}{C}}H + \cdot CH_2Ph$$

A variety of reaction conditions have been used for effecting oxidations of alcohols by Cr(VI) on a synthetic scale. The generally preferred method involves addition of an acidic aqueous solution containing chromic acid (known as *Jones' reagent*) to an acetone solution of the compound to be oxidized. Oxidation normally occurs quite rapidly, and overoxidation can be minimized. Often, the reduced chromium salts precipitate, and the acetone solution can be decanted, thus facilitating workup. Entries 2 and 5 in Scheme 9.1 are examples of this technique.

The complex formed by chromium trioxide and pyridine has been found useful in situations where other functional groups in the molecule, especially carbon–carbon double bonds, might be susceptible to oxidation by Cr(VI),[7] or when the molecule is sensitive to acid. The reagent has found application particularly in the steroid field.[8] A procedure for utilizing the CrO₃–pyridine complex originated by Collins[9] has become quite widely accepted in the past several years. The CrO₃–pyridine complex is isolated and dissolved in dichloromethane. With an excess of reagent, oxidation of simple primary and secondary alcohols proceeds to completion in a few minutes, giving aldehydes or ketones, respectively, in high yields. A procedure that avoids isolation of the complex can further simplify the experimental operations.[10] Chromium trioxide is added to pyridine in dichloromethane. Subsequent addition of the alcohol to this solution results in rapid oxidation in high yields. Entries 6–10 in Scheme 9.1 demonstrate the excellent results that have been reported using the CrO₃–pyridine complex in dichloromethane.

Use of the CrO₃–pyridine reagent seems to circumvent a side reaction leading to esters that can be a competing process in the oxidation of primary alcohols. Esters are formed when aldehyde and unreacted alcohol form a hemiacetal.[11] The mechanism

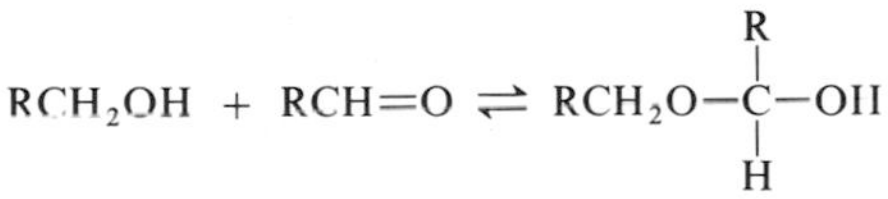

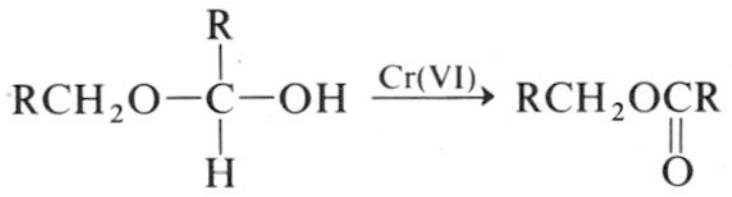

$$RCH_2OH + RCH{=}O \rightleftharpoons RCH_2O{-}\overset{R}{\underset{H}{\overset{|}{\underset{|}{C}}}}{-}OH$$

$$RCH_2O{-}\overset{R}{\underset{H}{\overset{|}{\underset{|}{C}}}}{-}OH \xrightarrow{Cr(VI)} RCH_2O\underset{O}{\overset{||}{C}}R$$

7. G. I. Poos, G. E. Arth, R. E. Beyler, and L. H. Sarett, *J. Am. Chem. Soc.* **75**, 422 (1953).
8. W. S. Johnson, W. A. Vredenburgh, and J. E. Pike, *J. Am. Chem. Soc.* **82**, 3409 (1960); W. S. Allen, S. Bernstein, and R. Littell, *J. Am. Chem. Soc.* **76**, 6116 (1954).
9. J. C. Collins, W. W. Hess, and F. J. Frank, *Tetrahedron Lett.*, 3363 (1968).
10. R. Ratcliffe and R. Rodehorst, *J. Org. Chem.* **35**, 4000 (1970).
11. W. A. Mosher and D. M. Preiss, *J. Am. Chem. Soc.* **75**, 5605 (1953).

A. Chromic Acid Solutions

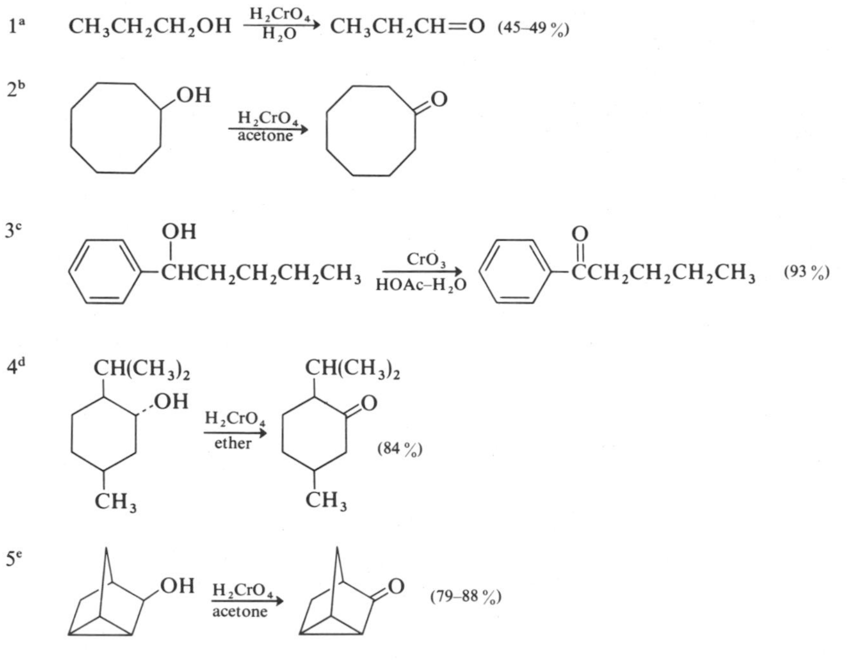

1[a] $CH_3CH_2CH_2OH \xrightarrow[H_2O]{H_2CrO_4} CH_3CH_2CH{=}O$ (45–49 %)

2[b] $\xrightarrow[\text{acetone}]{H_2CrO_4}$

3[c] $\xrightarrow[\text{HOAc–}H_2O]{CrO_3}$ (93 %)

4[d] $\xrightarrow[\text{ether}]{H_2CrO_4}$ (84 %)

5[e] $\xrightarrow[\text{acetone}]{H_2CrO_4}$ (79–88 %)

a. C. D. Hurd and R. N. Meinert, *Org. Synth.* **II**, 541 (1943).
b. E. J. Eisenbraun, *Org. Synth.* **45**, 28 (1965).
c. H. A. Neidig, D. L. Funck, R. Uhrich, R. Baker, and W. Kreiser, *J. Am. Chem. Soc.* **72**, 4617 (1950).
d. H. C. Brown, C. P. Garg, and K.-T. Liu, *J. Org. Chem.* **36**, 387 (1971).
e. J. Meinwald, J. Crandall, and W. E. Hymans, *Org. Synth.* **45**, 77 (1965).

of the oxidation of the hemiacetal is presumably analogous to that operating in the oxidation of simple alcohols.

Potassium permanganate has found little application in the oxidation of alcohols to ketones and aldehydes. The reagent is less selective than Cr(VI), and overoxidation is a problem. On the other hand, manganese (IV) dioxide has found appreciable use. This reagent preferentially attacks allylic and benzylic hydroxyl groups, and therefore possesses a useful degree of selectivity. The precise reactivity of MnO_2[12] depends on its mode of preparation. Manganese dioxide is precipitated by reaction of $Mn(II)SO_4$ with $KMnO_4$ and sodium hydroxide. The activity of the product depends somewhat on the extent of drying, and azeotropic drying by benzene has been recommended.[13] Scheme 9.2 illustrates the various classes of alcohols that are most susceptible to MnO_2 oxidation.

12. D. G. Lee, in *Oxidation*, Vol. 1, R. L. Augustine (ed.), Marcel Dekker, New York, NY, 1969, pp. 66–70.
13. I. M. Goldman, *J. Org. Chem.* **34**, 1979 (1969).

B. Chromium Trioxide–Pyridine

6[f]

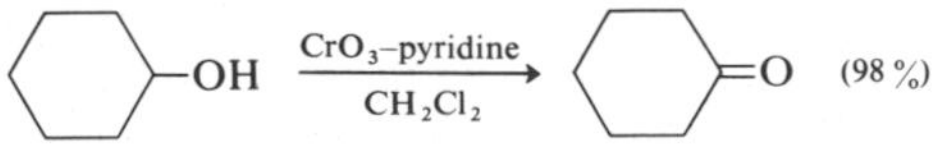

$$\text{cyclohexanol} \xrightarrow[\text{CH}_2\text{Cl}_2]{\text{CrO}_3\text{–pyridine}} \text{cyclohexanone} \quad (98\%)$$

7[g]

$$\underset{\overset{|}{\text{CH}_3}}{\text{CH}_3\text{CH}_2\text{CH}}(\text{CH}_2)_4\text{CH}_2\text{OH} \xrightarrow[\text{CH}_2\text{Cl}_2]{\text{CrO}_3\text{–pyridine}} \underset{\overset{|}{\text{CH}_3}}{\text{CH}_3\text{CH}_2\text{CH}}(\text{CH}_2)_4\text{CH}{=}\text{O} \quad (69\%)$$

8[h]

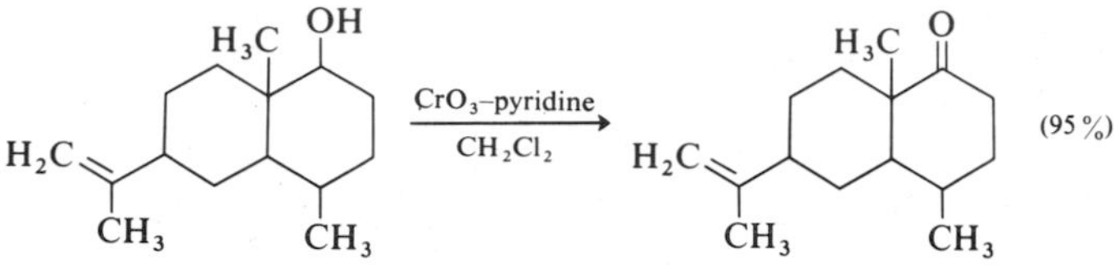

$$\xrightarrow[\text{CH}_2\text{Cl}_2]{\text{CrO}_3\text{–pyridine}} \quad (95\%)$$

9[i]

$$\xrightarrow[\text{CH}_2\text{Cl}_2]{\text{CrO}_3\text{–pyridine}}$$

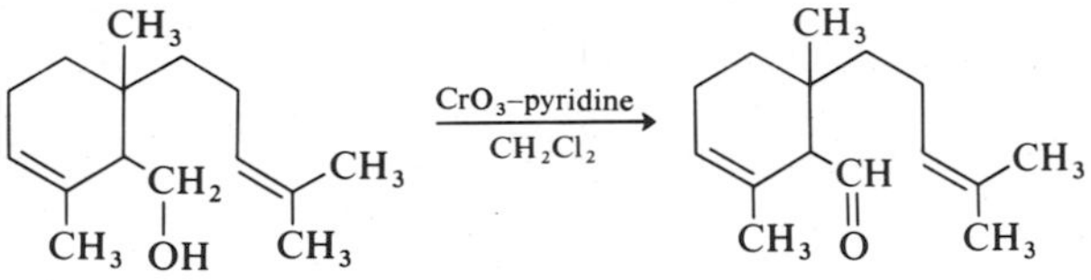

10[j]

$$\text{CH}_3(\text{CH}_2)_5\text{CH}_2\text{OH} \xrightarrow[\text{CH}_2\text{Cl}_2]{\text{CrO}_3\text{–pyridine}} \text{CH}_3(\text{CH}_2)_5\text{CH}{=}\text{O} \quad (70\text{–}84\%)$$

f. J. C. Collins, W. W. Hess, and F. J. Frank, *Tetrahedron Lett.*, 3363 (1968).
g. J. I. DeGraw and J. O. Rodin, *J. Org. Chem.* **36**, 2902 (1971).
h. R. Ratcliffe and R. Rodehorst, *J. Org. Chem.* **35**, 4000 (1970).
i. M. A. Schwartz, J. D. Crowell, and J. H. Musser, *J. Am. Chem. Soc.* **94**, 4361 (1972).
j. J. C. Collins and W. W. Hess, *Org. Synth.* **52**, 5 (1972).

Another reagent that has applicability to oxidation of alcohols to ketones is ruthenium tetroxide. For example, the oxidation of **1** to **2** was successfully achieved with this reagent after a number of other methods failed. This is a potent oxidant, however, and it readily attacks carbon–carbon double bonds.[14]

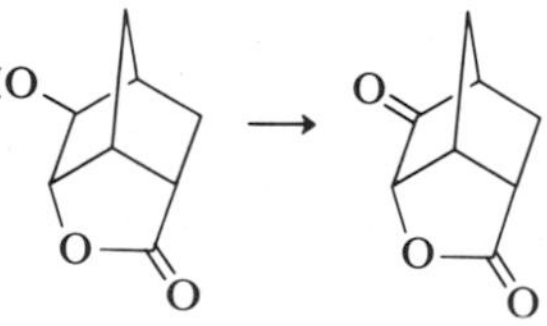

Ref. 15

14. J. L. Courtney and K. F. Swansborough, *Rev. Pure Appl. Chem.* **22**, 47 (1972); D. G. Lee and M. van den Engh, in *Oxidation*, Part B, W. S. Trahanovsky (ed.), Academic Press, New York, NY, 1973, Chap. IV.
15. R. M. Moriarty, H. Gopal, and T. Adams, *Tetrahedron Lett.*, 4003 (1970).

Scheme 9.2. Oxidations of Alcohols with Manganese Dioxide

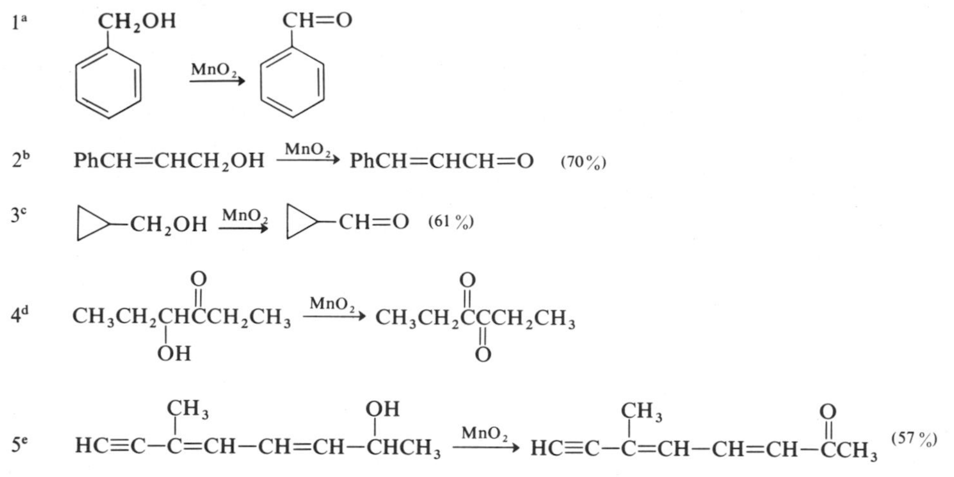

1[a]

2[b] $PhCH=CHCH_2OH \xrightarrow{MnO_2} PhCH=CHCH=O$ (70%)

3[c] $-CH_2OH \xrightarrow{MnO_2}$ $-CH=O$ (61%)

4[d]

5[e]

a. E. F. Pratt and J. F. Van De Castle, *J. Org. Chem.* **26**, 2973 (1961).
b. I. M. Goldman, *J. Org. Chem.* **34**, 1979 (1969).
c. L. Crombie and J. Crossley, *J. Chem. Soc.,* 4983 (1963).
d. E. P. Papadopoulos, A. Jarrar, and C. H. Issidorides, *J. Org. Chem.* **31**, 615 (1966).
e. J. Attenburrow, A. F. B. Cameron, J. H. Chapman, R. M. Evans, B. A. Hems, A. B. A. Jansen, and T. Walker, *J. Chem. Soc.* 1094 (1952).

9.1.2. Oxygen, Ozone, and Peroxides

The alcohol functional group is not rapidly attacked by oxygen or peroxides. Alcohols are attacked by ozone, but no major preparative procedures have been developed.

9.1.3. Other Oxidants

A very useful group of procedures for oxidation of alcohols to ketones have been developed that involve dimethyl sulfoxide (DMSO) and any one of a number of electrophilic molecules, particularly dicyclohexylcarbodiimide, acetic anhydride, and sulfur trioxide.[16] The initial work involved the DMSO–dicyclohexylcarbodiimide system.[17] The utility of the method has been greatest in the oxidation of molecules that are highly sensitive to more powerful oxidants and therefore cannot tolerate alternative methods. The mechanism of the oxidation involves formation of intermediate **A** by nucleophilic attack of DMSO on the carbodiimide, followed by reaction of this species with the alcohol. A major portion

16. Reviews: J. G. Moffat, in *Oxidation*, Vol. 2, R. L. Augustine and D. J. Trecker (eds.), Marcel Dekker, New York, NY, 1971, pp. 1–64; W. W. Epstein and F. W. Sweat, *Chem. Rev.* **67**, 247 (1967).
17. K. E. Pfitzner and J. G. Moffatt, *J. Am. Chem. Soc.* **87**, 5661, 5670 (1965).

of the driving force for the reaction is derived from the conversion of the carbodiimide to a urea, with formation of an amide carbonyl.[18]

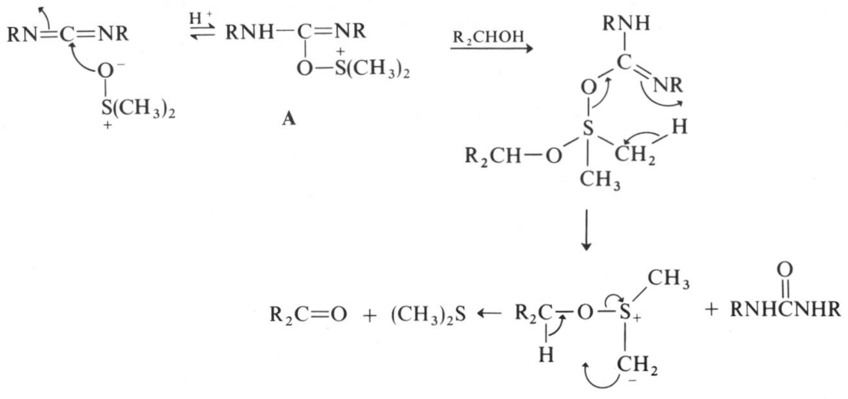

$$RN = C = NR \quad \underset{H^+}{\rightleftharpoons} \quad RNH - C = NR \quad \xrightarrow{R_2CHOH} \quad$$

A

$$R_2C=O + (CH_3)_2S \leftarrow R_2C-O-S_+ + RNHCNHR$$

The role of activating DMSO toward the nucleophilic addition step can be accomplished by other electrophilic species. A method that appears to have certain advantages of convenience is use of the pyridine complex of SO_3.[19] A mechanism similar to that established for the cyclohexylcarbodiimide system can be written:

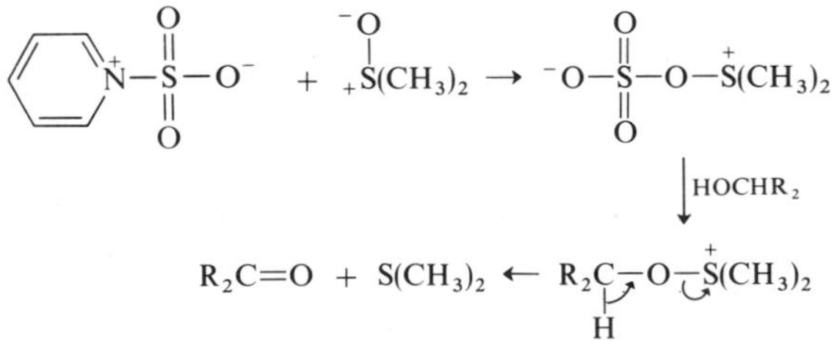

$$\begin{array}{c} N-S-O^- \end{array} + \; _+S(CH_3)_2 \rightarrow \; {}^-O-S-O-\overset{+}{S}(CH_3)_2$$

$$\downarrow HOCHR_2$$

$$R_2C=O + S(CH_3)_2 \leftarrow R_2C-O-\overset{+}{S}(CH_3)_2$$

Other reagents that have been found capable of activating DMSO toward nucleophilic attack include acetic anhydride[20] and phosphorus pentoxide.[21]

Oxidation of alcohols under extremely mild conditions can be effected using a procedure that is mechanistically related to the DMSO method. Dimethyl sulfide is converted to a sulfonium derivative by reaction with N-chlorosuccinimide. This sulfur species reacts readily with alcohols, generating the same kind of alkoxysulfonium salts that are involved in the DMSO procedures. In the presence of mild base, elimination of dimethyl sulfide completes the oxidation.[22]

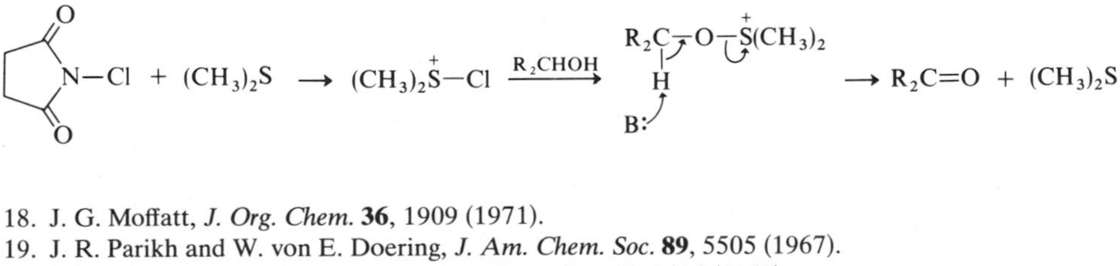

$$N-Cl + (CH_3)_2S \rightarrow (CH_3)_2\overset{+}{S}-Cl \xrightarrow{R_2CHOH} \rightarrow R_2C=O + (CH_3)_2S$$

18. J. G. Moffatt, *J. Org. Chem.* **36**, 1909 (1971).
19. J. R. Parikh and W. von E. Doering, *J. Am. Chem. Soc.* **89**, 5505 (1967).
20. J. D. Albright and L. Goldman, *J. Am. Chem. Soc.* **89**, 2416 (1967).
21. K. Onodera, S. Hirano, and N. Kashimura, *J. Am. Chem. Soc.* **87**, 4651 (1965).
22. E. J. Corey and C. U. Kim, *J. Am. Chem. Soc.* **94**, 7586 (1972).

Scheme 9.3. Oxidations of Alcohols Based on Sulfur Reagents

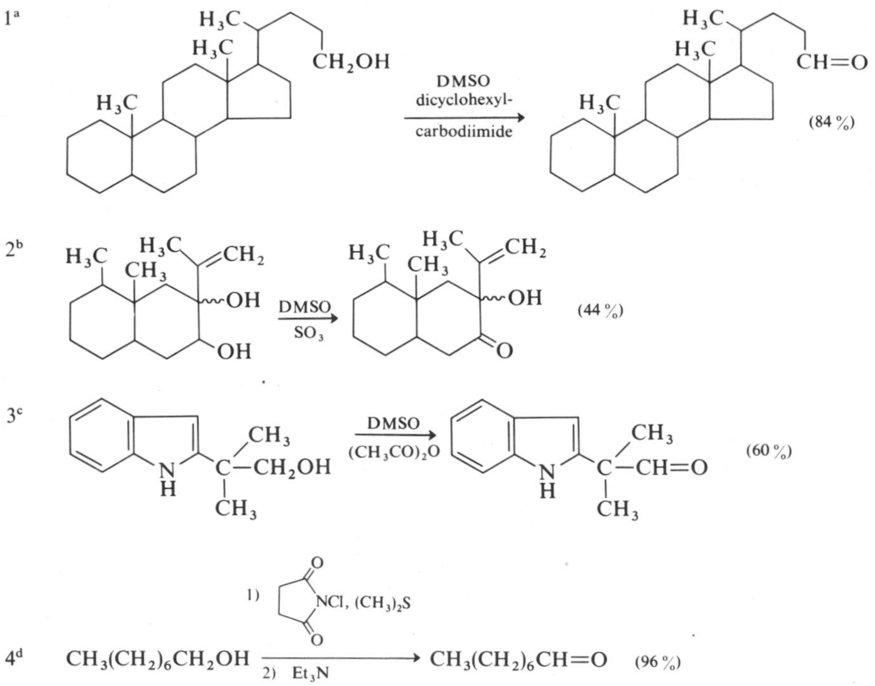

a. J. G. Moffatt, *Org. Synth.* **47**, 25 (1967).
b. J. A. Marshall and G. M. Cohen, *J. Org. Chem.* **36**, 877 (1971).
c. E. Houghton and J. E. Saxton, *J. Chem. Soc. C*, 595 (1969).
d. E. J. Corey and C. U. Kim, *J. Am. Chem. Soc.* **94**, 7586 (1972).

Similarly, reaction of chlorine and DMSO at low temperature gives an adduct that reacts with alcohols, presumably by displacement of chloride from sulfur, to give the ketone and DMSO[23]:

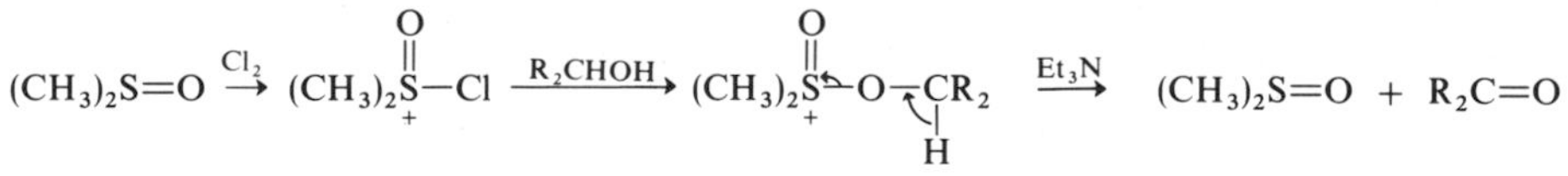

Some examples of oxidations of alcohols with sulfur reagents are presented in Scheme 9.3.

The development of the CrO_3–pyridine complex and the DMSO-based systems has decreased the number of instances in which older oxidation techniques are used. One such method, the *Oppenauer oxidation*,[24] is the reverse of the Meerwein–Ponndorf–Verley reduction (Chapter 3). It involves heating the alcohol to be

23. E. J. Corey and C. U. Kim, *Tetrahedron Lett.*, 919 (1973).
24. C. Djerassi, *Org. React.* **6**, 207 (1951).

oxidized with an aluminum alkoxide in the presence of a carbonyl compound, which acts as the hydrogen acceptor. The reaction is an equilibrium process and proceeds through a cyclic transition state. It can be driven in the desired direction by choosing

$$R_2CHOH + Al[OCH(CH_3)_2]_3 \longrightarrow R_2CHOAl[OCH(CH_3)_2]_2 + (CH_3)_2CHOH$$

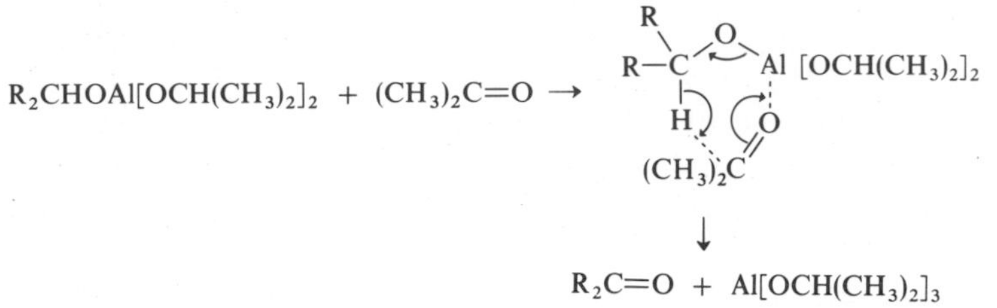

$$R_2C=O + Al[OCH(CH_3)_2]_3$$

a carbonyl compound that is a strong hydrogen acceptor. Quinone and fluorenone have been utilized for this purpose. Alternatively, use of an excess of the hydride acceptor will favor complete oxidation of the reactant. Since the reaction conditions

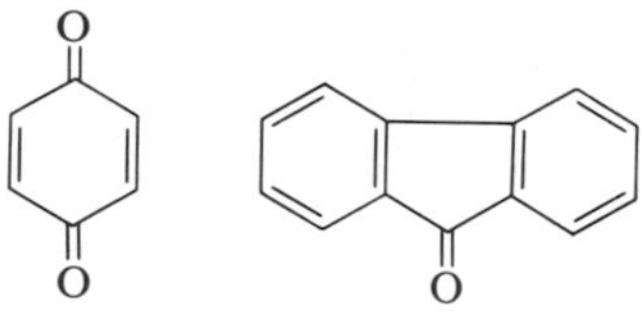

quinone   fluorenone

are nonacidic, this method can be valuable for substances that would not tolerate acidic conditions or the presence of transition-metal ions.

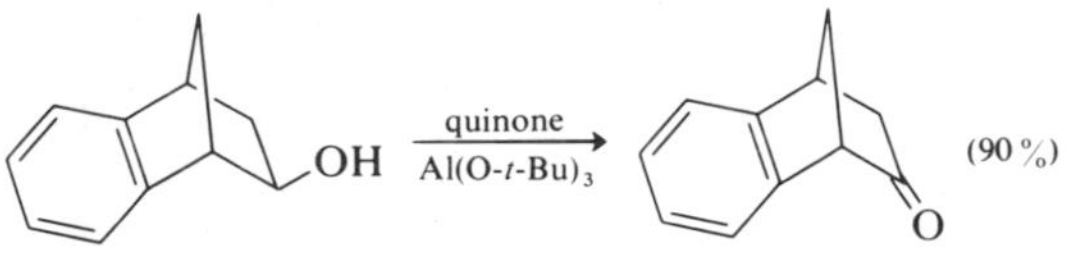

Ref. 25

9.2. Addition of Oxygen at Carbon–Carbon Double Bonds

9.2.1. Transition-Metal Oxidants

Among the high oxidation states of transition metals, permanganate ion and osmium tetroxide are the most effective reagents for addition of oxygen atoms at a double bond. Mild reaction conditions with potassium permanganate can permit relatively high-yield conversion of olefins to glycols. This oxidant is, however, capable of further oxidizing the glycol to a ketol or of cleaving the olefin with formation of carboxylic acids, so that careful control of reaction conditions is essential for efficient oxidation. A cyclic manganese ester is an intermediate in these oxidations. In view of the cyclic nature of this intermediate, it would be expected that

25. P. D. Bartlett and W. P. Giddings, *J. Am. Chem. Soc.* **82**, 1240 (1960).

glycols would be formed with *cis* stereochemistry, and this is indeed the case. An

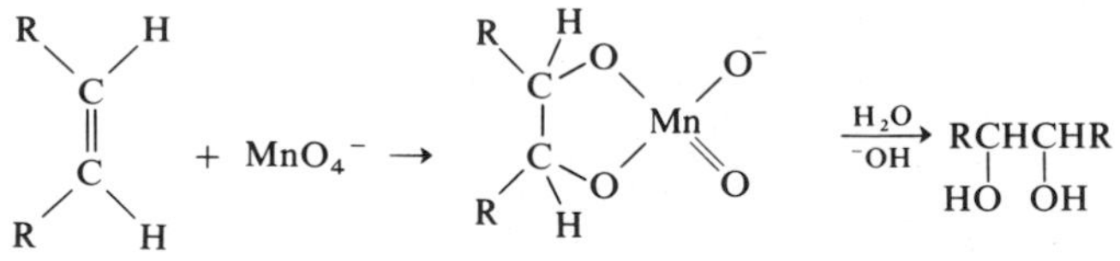

alternative reagent for *syn*-dihydroxylation of olefinic bonds that has the advantage of minimizing the problem of overoxidation, but the disadvantages of being both expensive and toxic, is osmium tetroxide. Here, again, a cyclic intermediate is formed and can be isolated. The glycol is liberated from the cyclic osmate ester by such reagents as hydrogen sulfide or bisulfite ion. Scheme 9.4 gives some examples of *cis*-hydroxylations of olefins.

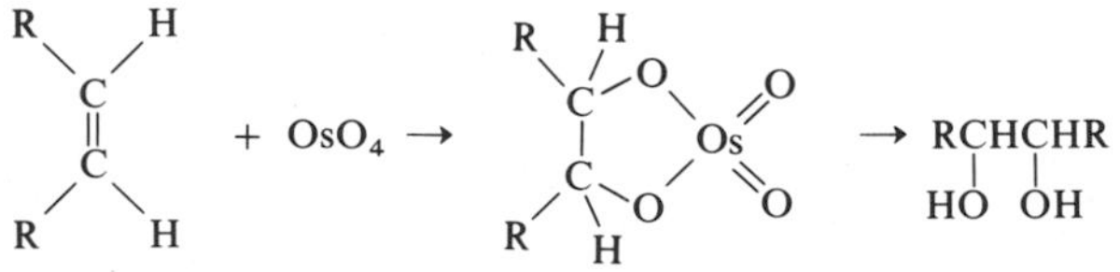

A variety of products derived from addition of oxygen at the double bond have been observed using Cr(VI) reagents. Epoxides and ketols are commonly encountered examples, but these reactions are seldom sufficiently clean to be of use for synthesis.[26]

Chromyl chloride, a volatile liquid with a formal Cr oxidation state of VI, has been found to add oxygen to terminal olefins with formation of aldehydes[27]:

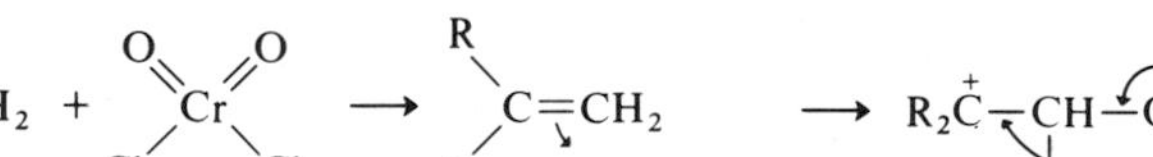

$$(CH_3)_3CCH_2C=CH_2 \xrightarrow[\text{2) Zn, H}_2\text{O}]{\text{1) CrO}_2\text{Cl}_2} (CH_3)_3CCH_2CHCH=O$$
$$\underset{CH_3}{|} \qquad\qquad\qquad \underset{CH_3}{|}$$

Ref. 28

The mechanism that has been proposed involves an electrophilic attack on the carbon–carbon double bond, presumably leading to a species with carbonium-ion character.[29] It is possible that epoxides, which subsequently rearrange to the

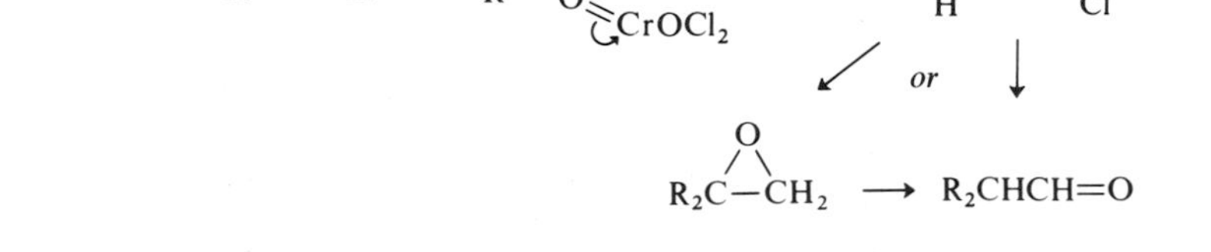

26. K. B. Wiberg, *Oxidation in Organic Chemistry,* Academic Press, New York, NY, pp. 125–142.
27. F. Freeman, P. J. Cameron, and R. H. DuBois, *J. Org. Chem.* **33**, 3970 (1968).
28. F. Freeman, R. H. DuBois, and T. G. McLaughlin, *Org. Synth.* **51**, 4 (1971).
29. F. Freeman, P. D. McCart, and N. J. Yamachika, *J. Am. Chem. Soc.* **92**, 4621 (1970); F. Freeman and K. W. Arledge, *J. Org. Chem.* **37**, 2665 (1972).

Scheme 9.4. Oxidations Involving Addition of Oxygen at Carbon–Carbon
Double Bonds

361

SECTION 9.2.
ADDITION
OF OXYGEN

1[a] + KMnO$_4$ ⟶ (45%)

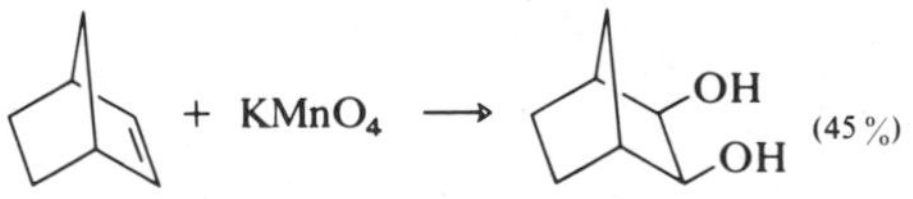

2[b]   CH$_2$=CHCH(OC$_2$H$_5$)$_2$ + KMnO$_4$ → HOCH$_2$CHCH(OC$_2$H$_5$)$_2$ (67%)
                                            |
                                            OH

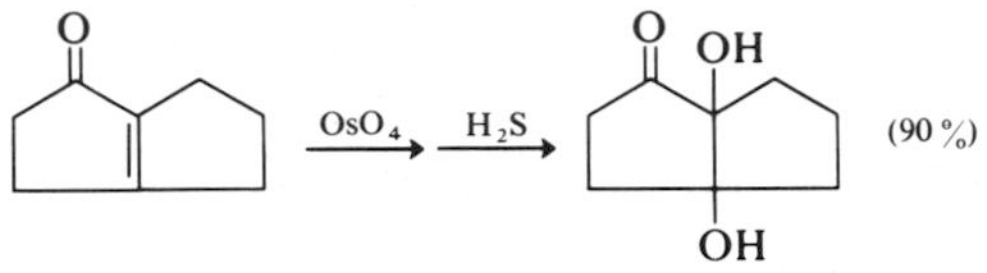

3[c] $\xrightarrow{\text{OsO}_4 \quad \text{H}_2\text{S}}$ (90%)

4[d] $\xrightarrow{\text{OsO}_4 \quad \text{HSO}_3^-}$ (56%)

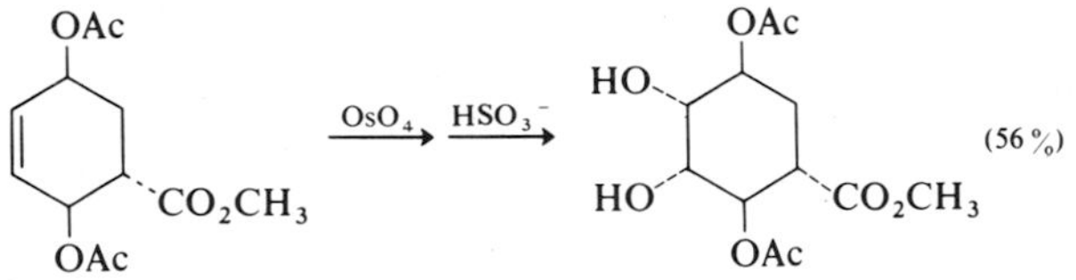

a. K. B. Wiberg and K. A. Saegebarth, *J. Am. Chem. Soc.* **79**, 2822 (1957).
b. E. J. Witzeman, W. L. Evans, H. Haas, and E. F. Schroeder, *Org. Synth.* **II**, 307 (1943).
c. E. Ghera, R. Szpigielman, and E. Wenkert, *J. Chem. Soc. C.*, 1479 (1966).
d. E. E. Smissman, J. T. Suh, M. Oxman, and R. Daniels, *J. Am. Chem. Soc.* **84**, 1040 (1962).

carbonyl compounds, are intermediates. With internal olefins, chromyl chloride gives α-chloroketones[30]:

$$\text{RCH=CHR} \rightarrow \underset{\underset{\text{Cl}}{|}}{\overset{\overset{\text{O}}{\|}}{\text{RCCHR}}}$$

This reaction can be formulated as involving addition of chromyl chloride to the olefin, followed by decomposition of the chromate ester in the same way that decomposition occurs in other Cr(VI) oxidations.

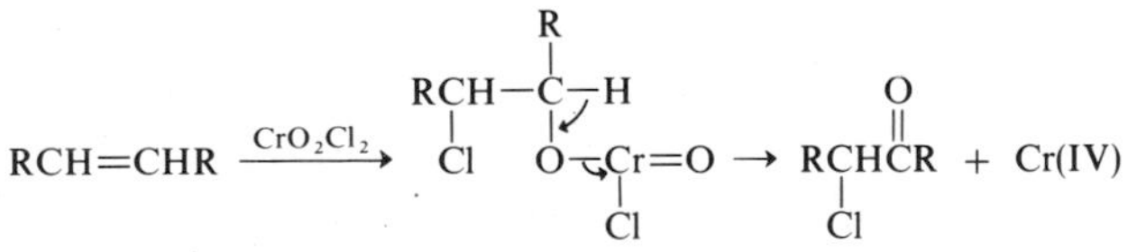

$$\text{RCH=CHR} \xrightarrow{\text{CrO}_2\text{Cl}_2} \underset{\underset{\text{Cl}}{|} \quad \underset{\text{Cl}}{|}}{\overset{\overset{\text{R}}{|}}{\text{RCH}-\overset{|}{\text{C}}-\text{H}} \quad \text{O}-\text{Cr=O}} \rightarrow \underset{\underset{\text{Cl}}{|}}{\overset{\overset{\text{O}}{\|}}{\text{RCHCR}}} + \text{Cr(IV)}$$

30. K. B. Sharpless and A. Y. Teranishi, *J. Org. Chem.* **38**, 185 (1973).

9.2.2. Epoxides from Olefins and Peroxidic Reagents

The most widely used reaction in this category is the conversion of olefins to epoxides.[31] In most cases, one of several peroxycarboxylic acids is used to effect oxidation. *m*-Chloroperoxybenzoic acid, which is commercially available or readily prepared[32] and is a quite stable solid, is the most popular reagent. Peroxyacetic acid, peroxybenzoic acid, peroxytrifluoroacetic acid, and others, however, have also been used frequently.

It has been demonstrated that ionic intermediates are not involved in the epoxidation process. The reaction rates are not directly related to solvent polarity.[33] *Syn* addition with retention of the stereochemistry of the olefin substituents is consistently observed. The reaction is therefore generally believed to be a concerted process. The most satisfactory representation of the transition state for the reaction is illustrated below:

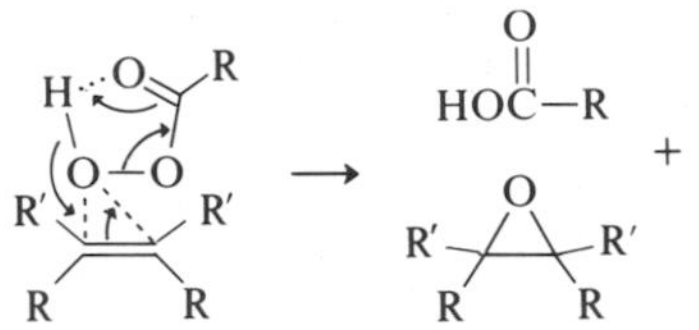

The salient structure–reactivity relationships that have been established include the fact that olefin reactivity is increased by electron-donating alkyl substituents and that peroxyacids with electron-attracting substituents are more reactive than alkyl peroxyacids. This order of reactivity demonstrates that the peroxyacids act as electrophilic species in the oxidation. The transition state cannot be highly polar, however, as the reaction shows fairly low sensitivity to substituent effects. The ρ value for oxidation of *trans*-stilbene by a series of peroxybenzoic acids is about -0.8.[34] Very low reactivity is encountered with olefins conjugated with carbonyl or other strongly electron-attracting substituents, and very strongly oxidizing peroxyacids, such as trifluoroperoxyacetic acid, are required for successful oxidation.[35] Such compounds are also epoxidized by alkaline hydrogen peroxide. A quite different mechanism, initiated by conjugate nucleophilic addition by the hydroperoxide anion, operates in this case.[36]

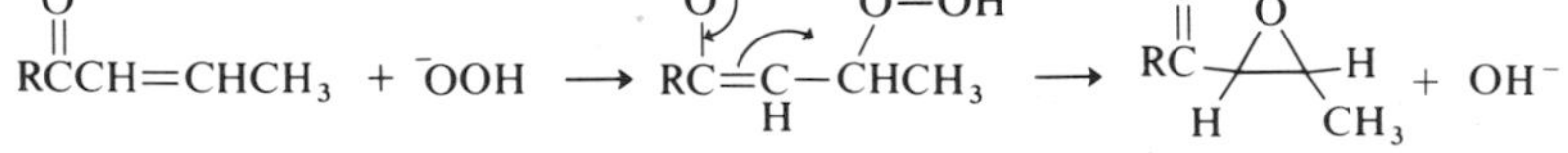

A process that is effective for epoxidation and that completely avoids acidic

31. D. Swern, *Organic Peroxides*, Vol. II, Wiley–Interscience, New York, NY, 1971, pp. 355–533.
32. R. N. McDonald, R. N. Steppel, and J. E. Dorsey, *Org. Synth.* **50**, 15 (1970).
33. N. N. Schwartz and J. N. Blumbergs, *J. Org. Chem.* **29**, 1976 (1964).
34. B. M. Lynch and K. H. Pausacker, *J. Chem. Soc.* 1525 (1955).
35. W. D. Emmons and A. S. Pagano, *J. Am. Chem. Soc.* **77**, 89 (1955).
36. C. A. Bunton and G. J. Minkoff, *J. Chem. Soc.* 665 (1949).

reaction conditions involves reaction of an olefin, a nitrile, and hydrogen peroxide.[37] The nitrile and hydrogen peroxide react, forming a peroxyimidic acid that epoxidizes olefins, presumably by a mechanism similar to that proposed for peroxyacids. This reaction also receives a substantial driving force from the formation of the stable amide carbonyl group.

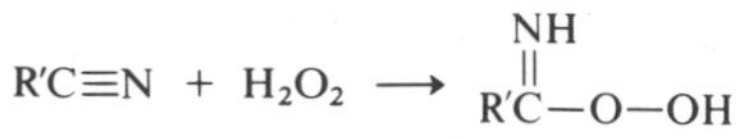

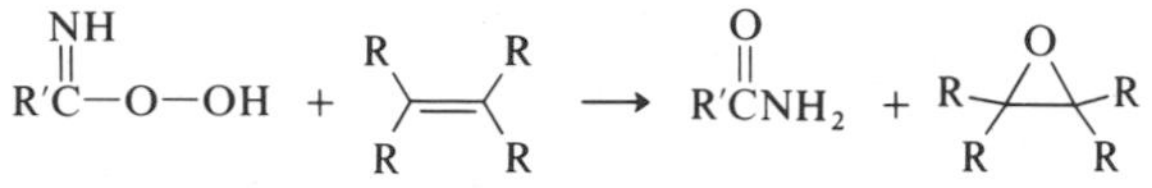

$$R'C\equiv N \ + \ H_2O_2 \ \longrightarrow \ \underset{R'C-O-OH}{\overset{NH}{\overset{\|}{}}}$$

$$\underset{R'C-O-OH}{\overset{NH}{\overset{\|}{}}} + \ \underset{R \quad R}{\overset{R \quad R}{\diagdown=\diagup}} \ \longrightarrow \ \underset{R'CNH_2}{\overset{O}{\overset{\|}{}}} + \ \underset{R \quad R}{R\triangle R}$$

The stereochemistry of epoxidation with peroxycarboxylic acids has been well studied. Attack and addition of oxygen occurs preferentially from the less hindered side of the molecule. Norbornene, for example, gives a 96 : 4 *exo* : *endo* ratio.[38] Other examples where this general trend is followed are summarized in a review by Swern.[31] In molecules where two potential modes of approach are not greatly different, a mixture of products is to be expected. For example, unhindered double bonds exocyclic to six-membered rings yield the epoxides resulting from both the equatorial and axial directions of attack[39]:

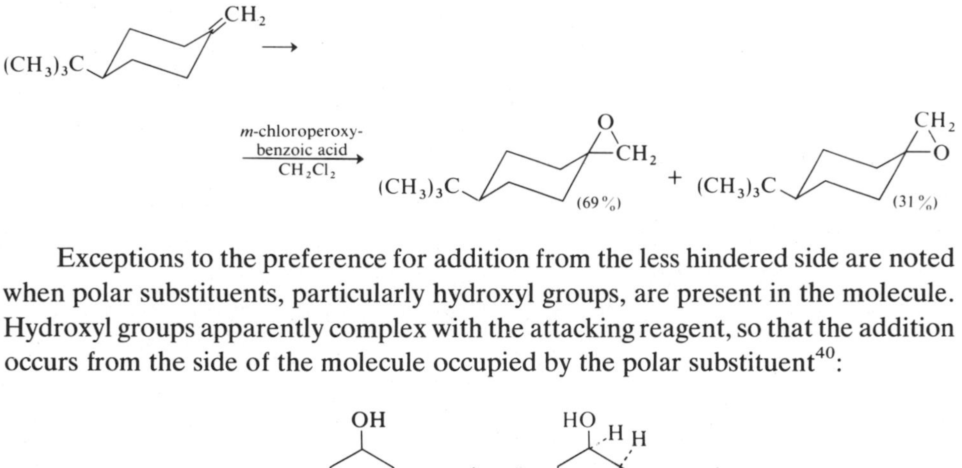

Exceptions to the preference for addition from the less hindered side are noted when polar substituents, particularly hydroxyl groups, are present in the molecule. Hydroxyl groups apparently complex with the attacking reagent, so that the addition occurs from the side of the molecule occupied by the polar substituent[40]:

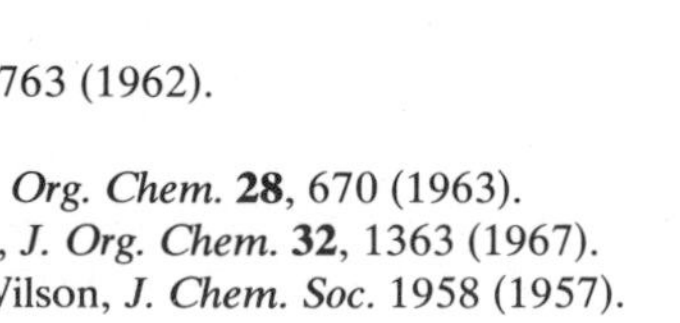

Representative examples of epoxidation reactions are given in Scheme 9.5.

37. G. B. Payne, *Tetrahedron* **18**, 763 (1962).
31. See p.362.
38. H. Kwart and T. Takeshita, *J. Org. Chem.* **28**, 670 (1963).
39. R. G. Carlson and N. S. Behn, *J. Org. Chem.* **32**, 1363 (1967).
40. H. B. Henbest and R. A. L. Wilson, *J. Chem. Soc.* 1958 (1957).

Scheme 9.5. Epoxidation

A. Oxidation of Alkenes with Peroxyacids

1[a]

CH=CH₂
peroxybenzoic acid
H
O
H
H
(69–75 %)

2[b]

peroxybenzoic acid
O

3[c]

CH₃CH=CHCO₂C₂H₅
CF₃CO₃H
H₃C
O
CO₂C₂H₅
H
H
(73 %)

4[d]

CH₃
CH₃
m-chloroperoxybenzoic acid
CH₃
O
CH₃

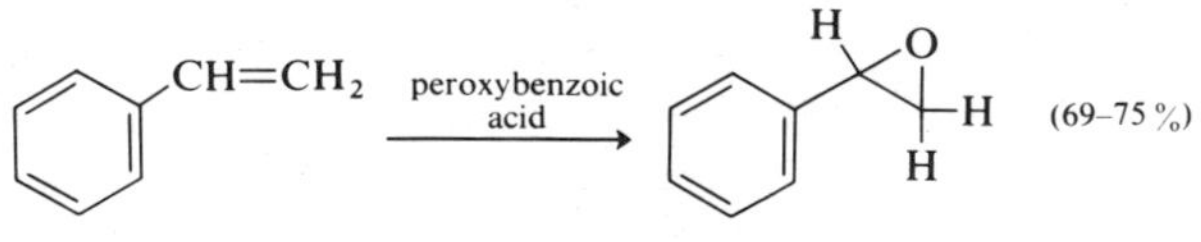

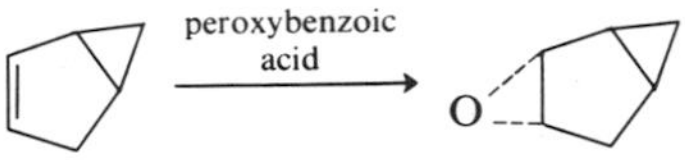

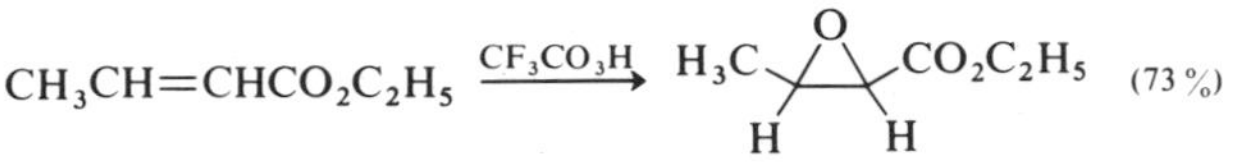

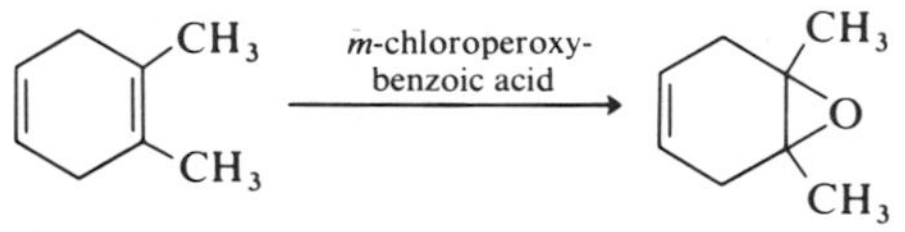

B. Epoxidation of Electrophilic Alkenes

5[e]

O
H₃C
CH₃
CH₃
H₂O₂, ⁻OH
O
H₃C
O
CH₃
CH₃

6[f]

Ph
C≡N
C=C
H
Ph
+ (CH₃)₃CO₂H
Triton B
Ph
O
C≡N
H
Ph
(76 %)

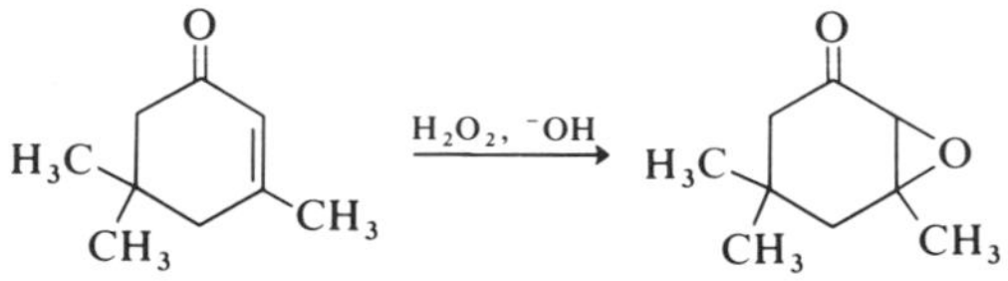

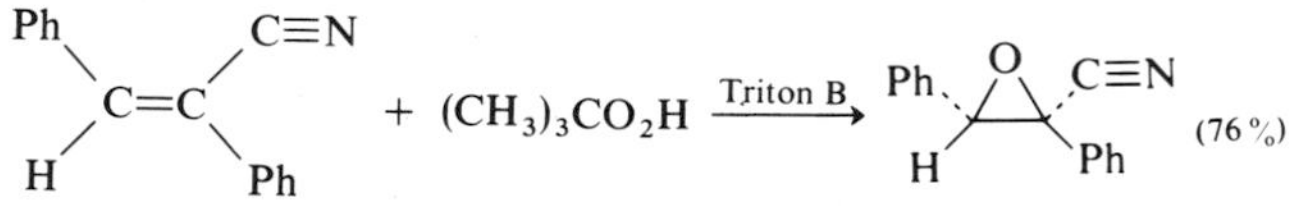

a. H. Hibbert and P. Burt, *Org. Synth.* **I**, 481 (1932).
b. E. J. Corey and R. L. Dawson, *J. Am. Chem. Soc.* **85**, 1782 (1963).
c. W. D. Emmons and A. S. Pagano, *J. Am. Chem. Soc.* **77**, 89 (1955).
d. L. A. Paquette and J. H. Barrett, *Org. Synth.* **49**. 62 (1969).
e. R. L. Wasson and H. O. House, *Org. Synth.* **IV**, 552 (1963).
f. G. B. Payne and P. H. Williams, *J. Org. Chem.* **26**, 651 (1961).

Epoxidation is often preliminary to solvolytic or nucleophilic ring-opening in synthetic sequences. In acidic aqueous media, epoxides are opened to give diols by an *anti* addition process. In cyclic systems, ring-opening occurs to give the diaxial diol.

CH₃
O
CH₃
H⁺
H₂O
OH
CH₃
CH₃
OH

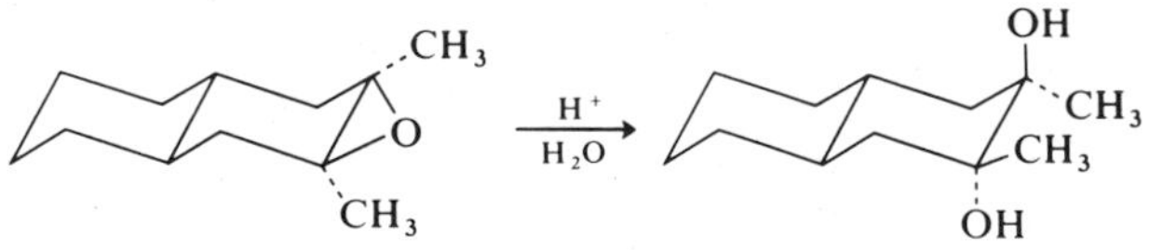

Ref. 41

41. B. Rickborn and D. K. Murphy, *J. Org. Chem.* **34**, 3209 (1969).

7[g]

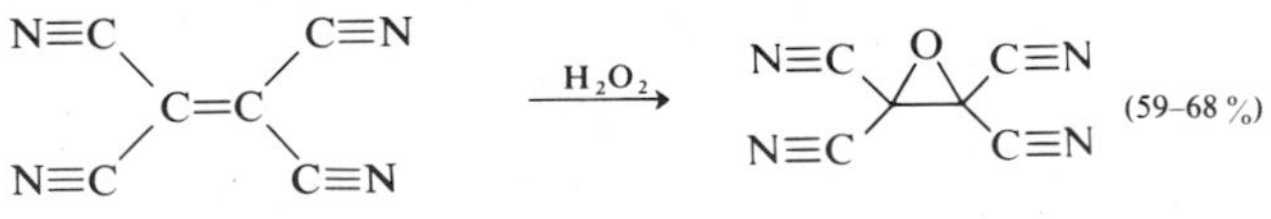

(59–68 %)

8[h]

(67 %)

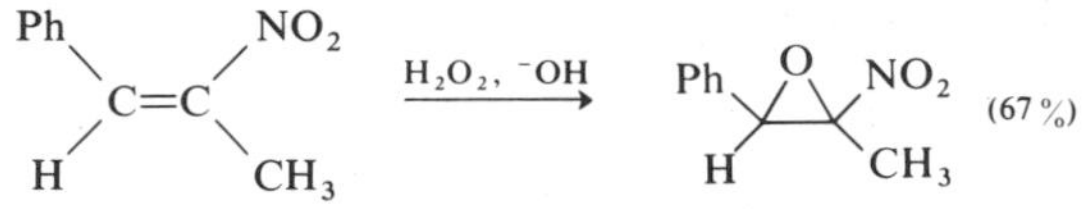

C. Oxidation with Solvolysis of the Intermediate Epoxide

9[i]

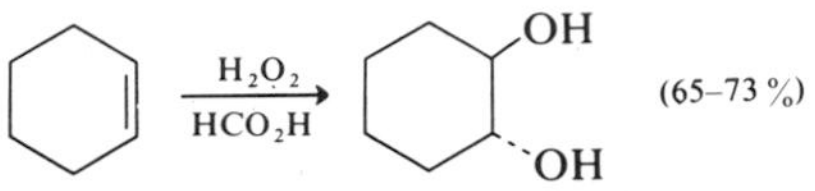

(65–73 %)

10[j]

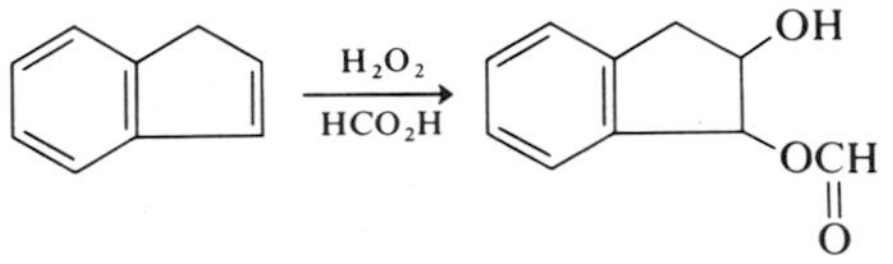

11[k]

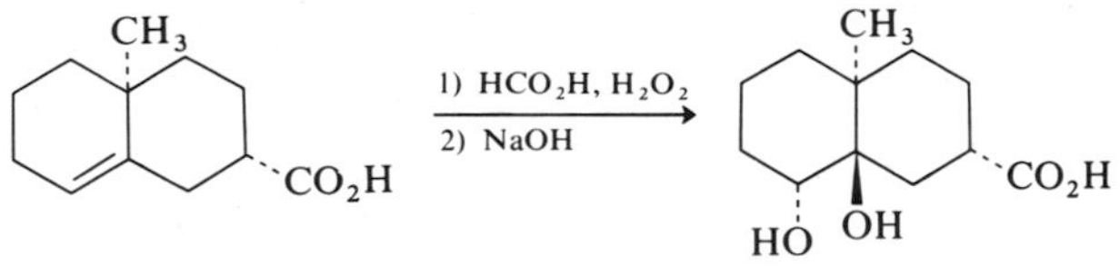

12[l]

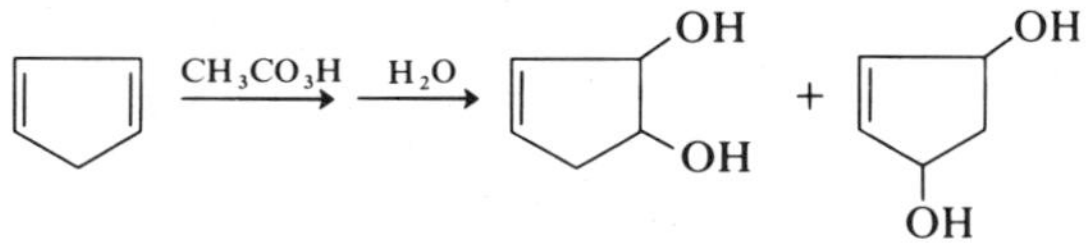

g. W. J. Linn, *Org. Synth.* **49**, 103 (1969).
h. H. Newman and R. B. Angier, *Tetrahedron* **26**, 825 (1970).
i. A. Roebuck and H. Adkins, *Org. Synth.* **III**, 217 (1955).
j. J. E. Horan and R. W. Schiessler, *Org. Synth.* **41**, 53 (1961).
k. T. R. Kelly, *J. Org. Chem.* **37**, 3393 (1972).
l. M. Korach, D. R. Nielsen, and W. H. Rideout, *Org. Synth.* **42**, 50 (1962).

Base-catalyzed epoxide ring-openings, in which the nucleophile provides the driving force for ring-opening, usually involve breaking the bond to the less substituted carbon, since this is the position most open to nucleophilic attack.[42] The situation in acid-catalyzed reactions is more complex. The bonding of a proton to the oxygen weakens the C–O bond, facilitating its rupture by weak nucleophiles. If the C–O bond is largely intact at the transition state, the nucleophile will become attached to the less substituted position for the same steric reasons that were cited in the case of

42. R. E. Parker and N. S. Isaacs, *Chem. Rev.* **59**, 737 (1959).

A. Acid-Catalyzed and Electrophilic Ring-Opening Reactions

1[a]

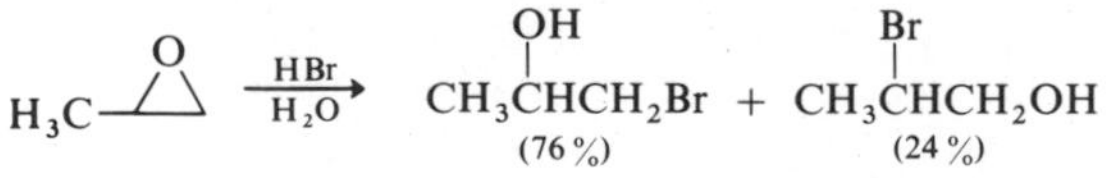

$$H_3C \overset{O}{\triangle} \xrightarrow[H_2O]{HBr} \underset{(76\%)}{CH_3\overset{OH}{\underset{|}{C}}HCH_2Br} + \underset{(24\%)}{CH_3\overset{Br}{\underset{|}{C}}HCH_2OH}$$

2[b]

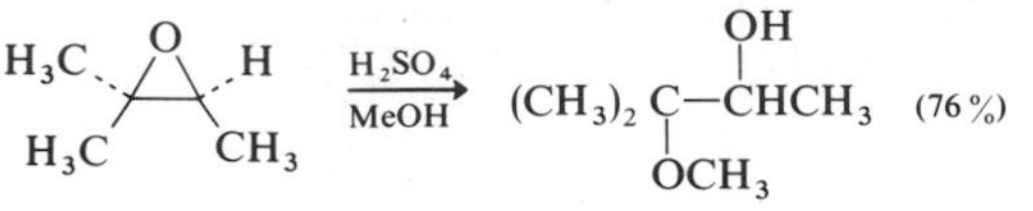

$$\underset{H_3C}{\overset{H_3C}{\diagdown}}\overset{O}{\underset{CH_3}{\diagup}}\overset{H}{\diagdown} \xrightarrow[MeOH]{H_2SO_4} (CH_3)_2\underset{OCH_3}{\overset{OH}{\underset{|}{C}}}-\overset{}{C}HCH_3 \quad (76\%)$$

3[c]

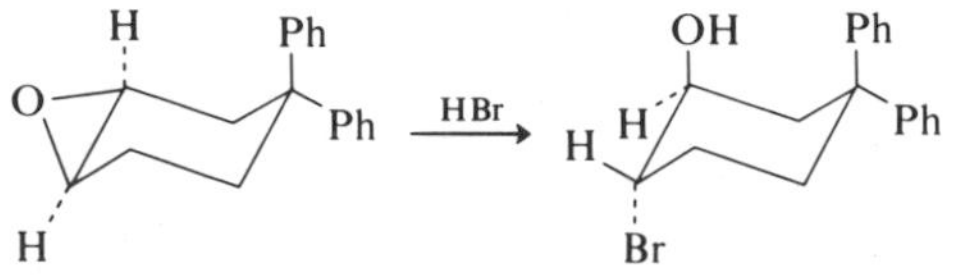

4[d]

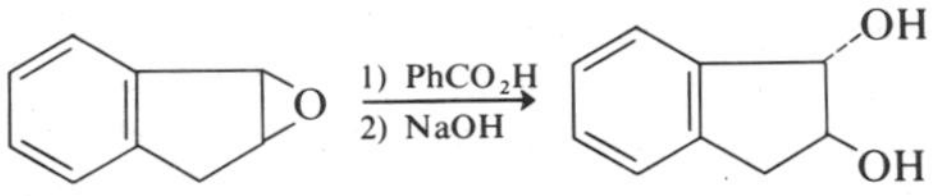

$$\xrightarrow[2)\ NaOH]{1)\ PhCO_2H}$$

5[e]

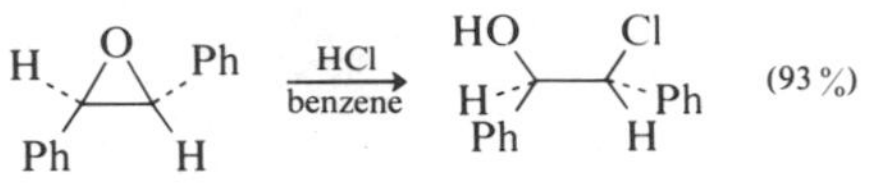

$$\xrightarrow[benzene]{HCl} \quad (93\%)$$

6[f]

$$\xrightarrow[H_2O]{HClO_4} \quad (100\%)$$

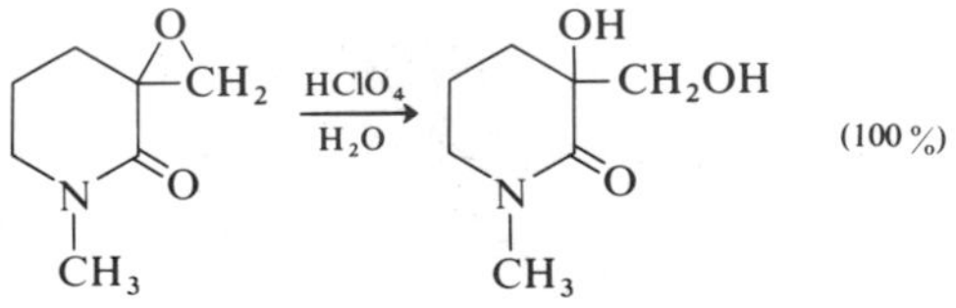

nucleophilic ring-opening. If, on the other hand, C–O rupture is nearly complete when the transition state is reached, the opposite orientation will be observed because of the greater ability of the more substituted carbon to bear the developing

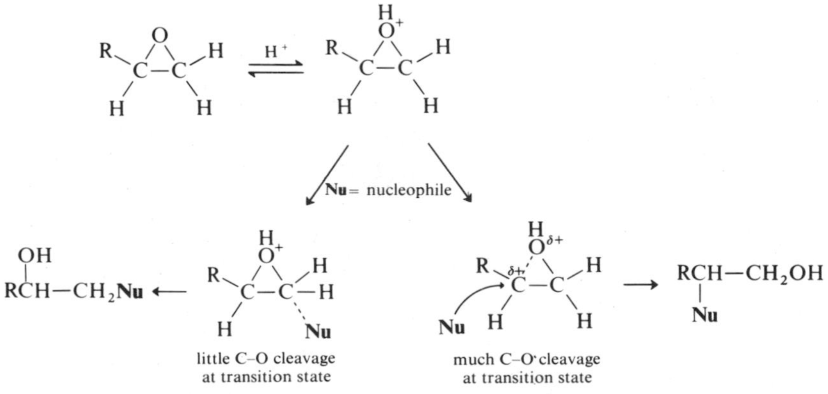

little C–O cleavage
at transition state

much C–O cleavage
at transition state

B. Nucleophilic Ring-Opening Reactions

7[b] H_3C $\overset{O}{\triangle}$ H + CH_3O^- → $(CH_3)_2\overset{OH}{\underset{OCH_3}{C}}CHCH_3$ (53%)
         H_3C CH_3

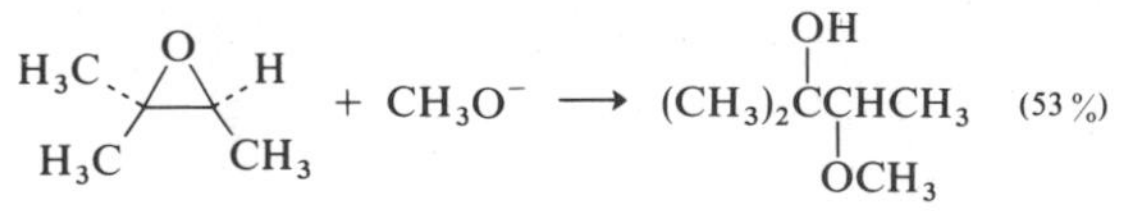

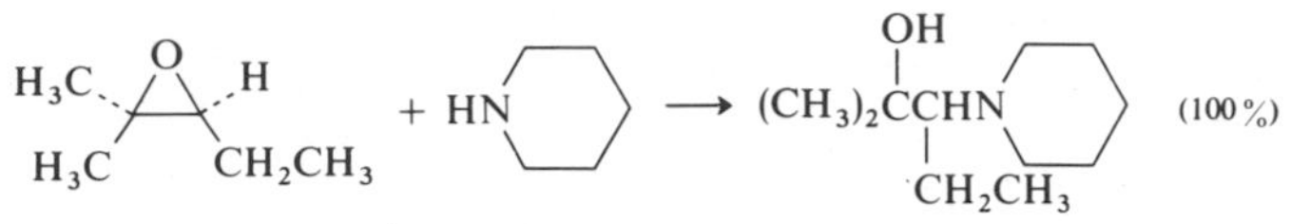

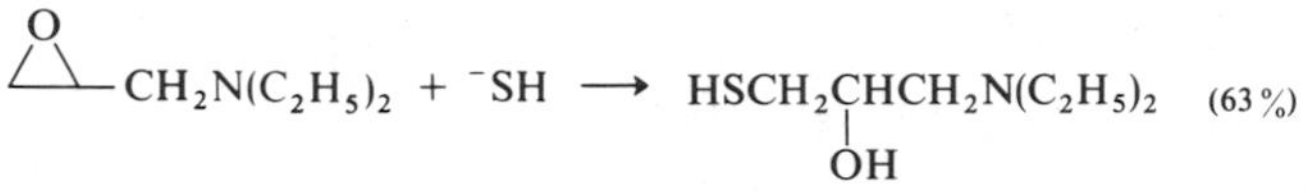

8[g] H_3C $\overset{O}{\triangle}$ H + $HN\langle\bigcirc\rangle$ → $(CH_3)_2\overset{OH}{\underset{CH_2CH_3}{C}}CHN\langle\bigcirc\rangle$ (100%)
         H_3C CH_2CH_3

9[h] $\overset{O}{\triangle}$—$CH_2N(C_2H_5)_2$ + ^-SH → $HSCH_2\underset{OH}{CH}CH_2N(C_2H_5)_2$ (63%)

10[i] H_3C $\overset{O}{\triangle}$ CH_3 + NaN_3 → $CH_3\overset{OH}{\underset{N_3}{CH}}CHCH_3$

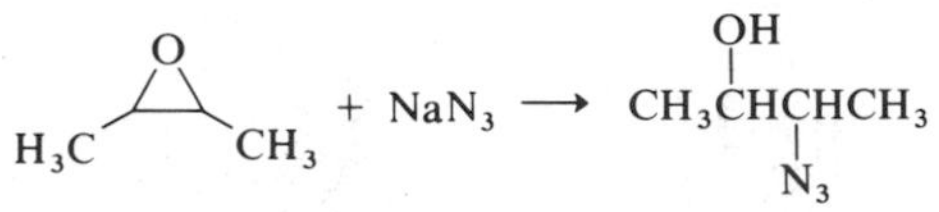

a. C. A. Stewart and C. A. VanderWerf, *J. Am. Chem. Soc.* **76**, 1259 (1954).
b. S. Winstein and L. L. Ingraham, *J. Am. Chem. Soc.* **74**, 1160 (1952).
c. F. G. Bordwell, R. R. Frame, and J. G. Strong, *J. Org. Chem.* **33**, 3385 (1968).
d. A. Gagis, A. Fusco, and J. T. Benedict, *J. Org. Chem.* **37**, 3181 (1972).
e. G. Berti, F. Bottari, P. L. Ferrarini, and B. Macchia, *J. Org. Chem.* **30**, 4091 (1965).
f. M. L. Rueppel and H. Rapoport, *J. Am. Chem. Soc.* **94**, 3877 (1972).
g. T. Colclough, J. I. Cunneen, and C. G. Moore, *Tetrahedron* **15**, 187 (1961).
h. D. M. Burness and H. O. Bayer, *J. Org. Chem.* **28**, 2283 (1963).
i. C. A. VanderWerf, R. Y. Heisler, and W. E. McEwen, *J. Am. Chem. Soc.* **76**, 1231 (1954).

positive charge. When simple aliphatic epoxides such as propylene oxide react with hydrogen halides, the dominant mode of reaction introduces halide at the less substituted primary carbon.[43] Substituents that would further stabilize a carbonium-ion intermediate lead to reversal of the mode of addition.[44] Some examples of both acid-catalyzed and nucleophilic ring-opening are shown in Scheme 9.6.

Epoxides can be reduced to alcohols. Lithium aluminum hydride effects reduction by nucleophilic attack, and hydride is therefore added at the less substituted carbon atom. Lithium triethylborohydride is more reactive than LiAlH$_4$ and is superior for epoxides that are resistant to reduction.[45] A good deal of work has been done on the reduction of epoxides with species generated from reaction of aluminum

43. C. A. Stewart and C. A. VanderWerf, *J. Am. Chem. Soc.* **76**, 1259 (1954).
44. S. Winstein and L. L. Ingraham, *J. Am. Chem. Soc.* **74**, 1160 (1952).
45. S. Krishnamurthy, R. M. Schubert and H. C. Brown, *J. Am. Chem. Soc.* **95**, 8486 (1973).

Scheme 9.7. Alcohols by Reduction of Epoxides

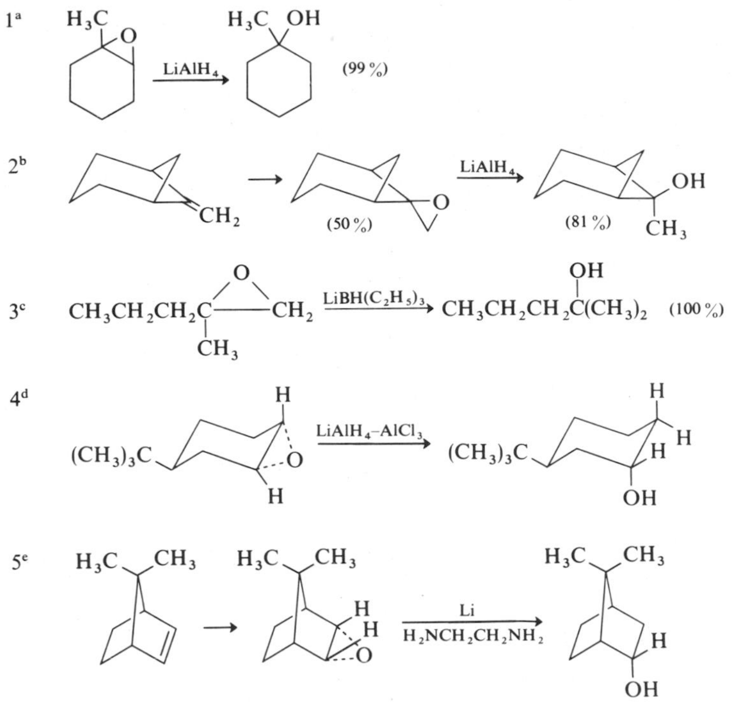

1[a] H_3C epoxide structure $\xrightarrow{\text{LiAlH}_4}$ H_3C OH cyclohexanol (99 %)

2[b]   methylene decalin structure $\rightarrow$ epoxide (50 %) $\xrightarrow{\text{LiAlH}_4}$ OH ... CH_3 (81 %)

3[c] $CH_3CH_2CH_2\overset{O}{\underset{CH_3}{C}}{-}CH_2$ $\xrightarrow{\text{LiBH(C}_2\text{H}_5)_3}$ $CH_3CH_2CH_2\overset{OH}{\underset{|}{C}}(CH_3)_2$ (100 %)

4[d] $(CH_3)_3C$ epoxide with H $\xrightarrow{\text{LiAlH}_4\text{-AlCl}_3}$ $(CH_3)_3C$ alcohol with OH

5[e] H_3C CH_3 alkene structure $\rightarrow$ H_3C CH_3 epoxide with H, O $\xrightarrow[\text{H}_2\text{NCH}_2\text{CH}_2\text{NH}_2]{\text{Li}}$ H_3C CH_3 alcohol with OH

a. D. K. Murphy, R. L. Alumbaugh, and B. Rickborn, *J. Am. Chem. Soc.* **91**, 2649 (1969); H. C. Brown and N. M. Yoon, *J. Am. Chem. Soc.* **88**, 1464 (1966).
b. K. B. Wiberg and W.-F. Chen, *J. Org. Chem.* **37**, 3235 (1972).
c. S. Krishnamurthy, R. M. Schubert, and H. C. Brown, *J. Am. Chem. Soc.* **95**, 8486 (1973).
d. B. Rickborn and J. Quartucci, *J. Org. Chem.* **29**, 3185 (1964).
e. H. C. Brown, S. Ikegami, and J. H. Kawakami, *J. Org. Chem.* **35**, 3243 (1970); see also E. M. Kaiser, C. G. Edmonds, S. D. Grubb, J. W. Smith, and D. Tramp, *J. Org. Chem.* **36**, 330 (1971).

chloride and lithium aluminum hydride.[46] The active reagents in these reactions are alane (AlH_3) and chloroalanes. These species have considerable Lewis acid character, and hydride is often added at the more substituted carbon of the expoxide. In some cases, migration of substituents occurs, presumably via carbonium ions generated by electrophilic opening of the epoxide rings.

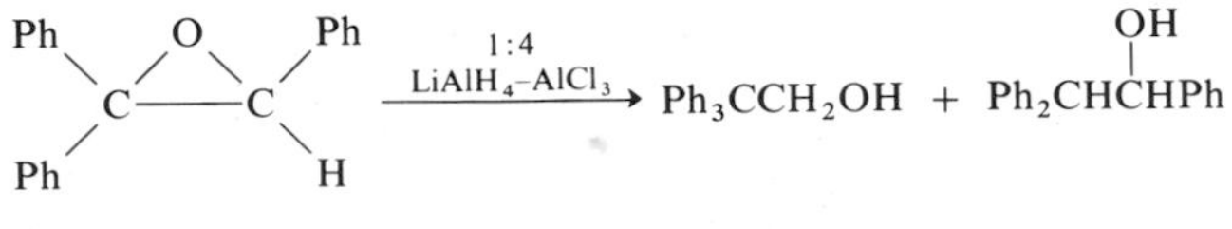

$\underset{Ph}{\overset{Ph}{{>}}}C\overset{O}{-}C\underset{H}{\overset{Ph}{{<}}}$ $\xrightarrow[\text{LiAlH}_4\text{-AlCl}_3]{1:4}$ Ph_3CCH_2OH + $Ph_2CH\overset{OH}{\underset{|}{C}}HPh$

46a. M. N. Rerick and E. L. Eliel, *J. Am. Chem. Soc.* **84**, 2356 (1962).
   b. E. C. Ashby and J. Prather, *J. Am. Chem. Soc.* **88**, 729 (1966).
   c. P. T. Lansbury, D. J. Scharf, and V. A. Pattison, *J. Org. Chem.* **32**, 1748 (1967).
   d. B. Rickborn and W. E. Lamke, II, *J. Org. Chem.* **32**, 537 (1967).
   e. D. K. Murphy, R. L. Alumbaugh, and B. Rickborn, *J. Am. Chem. Soc.* **91**, 2649 (1969).

Diborane in tetrahydrofuran reduces epoxides, but the yields are low, and other products are formed by pathways that result from the electrophilic character of diborane.[47] Diborane reduction occurs in better yield in the presence of BH_4^-, but the electrophilic role played by diborane is still evident because the dominant product is that resulting from addition of hydride at the more substituted carbon.[48] Dissolving metals, particularly lithium in ethylenediamine, give synthetically useful yields of alcohols from epoxides.[49]

Scheme 9.7 presents several examples of preparation of alcohols by reduction of epoxides. Coupled with alkene epoxidation, these reduction methods constitute a route for conversion of alkenes to alcohols.

Base-catalyzed ring-opening of epoxides constitutes a route to allylic alcohols:

$$RCH_2CH \overset{O}{\overbrace{}} CH_2 \rightarrow RCH{=}CHCH_2OH$$

Strongly basic reagents, such as the lithium salt of dialkylamines, are required to promote the reaction, which presumably involves concerted proton-abstraction and ring-opening. The stereochemistry of the ring-opening has been investigated by deuterium labeling. The proton *cis* to the epoxide ring is selectively removed.[50] A

transition state that could account for this stereochemistry is shown as **A**. Such a transition state could be favored by ion-pairing, which would require a close

association of the amide anion and the lithium cation. If the lithium is also coordinated with the epoxide oxygen, a *syn* elimination would result. In other epoxides, a different mechanism, which involves formation of a carbenoid species by α-elimination, becomes competitive.[51,52]

47. D. J. Pasto, C. C. Cumbo, and J. Hickman, *J. Am. Chem. Soc.* **88**, 2201 (1966).
48. H. C. Brown and N. M. Yoon, *J. Am. Chem. Soc.* **90**, 2686 (1968).
49. H. C. Brown, S. Ikegami, and J. H. Kawakami, *J. Org. Chem.* **35**, 3243 (1970).
50. R. P. Thummel and B. Rickborn, *J. Am. Chem. Soc.* **92**, 2064 (1970).
51. A. C. Cope, G. A. Berchtold, P. E. Peterson, and S. H. Sharman, *J. Am. Chem. Soc.* **82**, 6370 (1970).
52. J. K. Crandall, *J. Org. Chem.* **29**, 2830 (1964); J. K. Crandall and L. H. C. Lin, *J. Am. Chem. Soc.* **89**, 4526, 4527 (1967).

Scheme 9.8. Base-Catalyzed Epoxide Ring-Opening

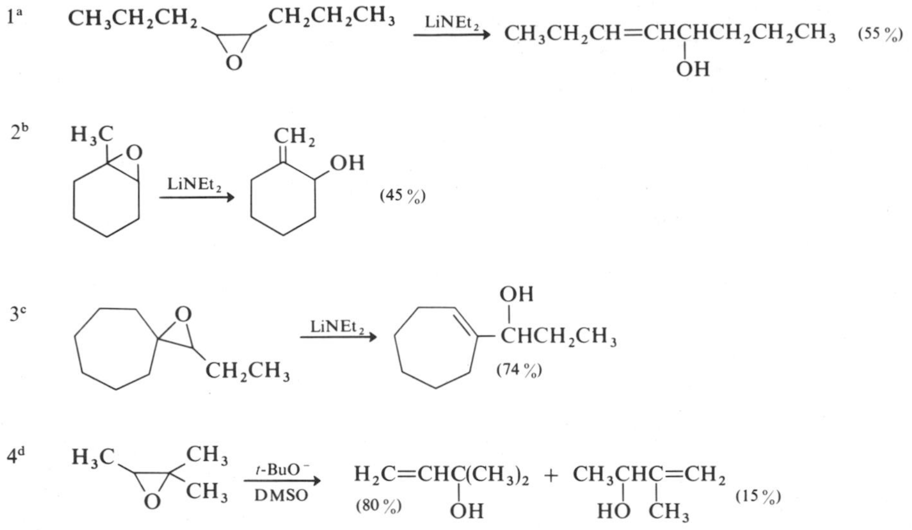

1[a] $CH_3CH_2CH_2$ ⟍ ⟋ $CH_2CH_2CH_3$ $\xrightarrow{\text{LiNEt}_2}$ $CH_3CH_2CH=CHCHCH_2CH_2CH_3$ (55%)
             O |
                                                                 OH

2[b] $\xrightarrow{\text{LiNEt}_2}$ (45%)

3[c] $\xrightarrow{\text{LiNEt}_2}$ (74%)

4[d] $\xrightarrow[\text{DMSO}]{t\text{-BuO}^-}$ $H_2C=CHC(CH_3)_2$ + $CH_3CHC=CH_2$ (15%)
                                                   | | |
                                             (80%) OH   HO CH_3

a. A. C. Cope and J. K. Heeren, *J. Am. Chem. Soc.* **87**, 3125 (1965).
b. J. K. Crandall and L.-H. C. Lin, *J. Org. Chem.* **33**, 2375 (1968).
c. R. P. Thummel and B. Rickborn, *J. Org. Chem.* **36**, 1365 (1971).
d. C. C. Price and D. D. Carmelite, *J. Am. Chem. Soc.* **88**, 4039 (1966).

Scheme 9.8 depicts some base-catalyzed ring-openings of epoxides.

Epoxides can be isomerized to carbonyl compounds by strong Lewis acids.[53] Boron trifluoride has been studied most extensively. Carbonium ions appear to be involved, and the structure and stereochemistry of the product are determined by conformational factors governing the substituent migration that follows carbonium-ion formation. Lithium perchlorate also catalyzes ring-opening and rearrangement to carbonyl compounds via carbonium ions.[54] Olefins can be oxidized directly to carbonyl compounds by using a reaction medium containing both peroxy-trifluoroacetic acid and boron trifluoride.[55] A substituent migration is involved, and high yields of a single product are anticipated only when the olefin is highly symmetrical, as in the case of 2,3-dimethylbutene, or when some structural feature favors migration of a particular olefin substituent.

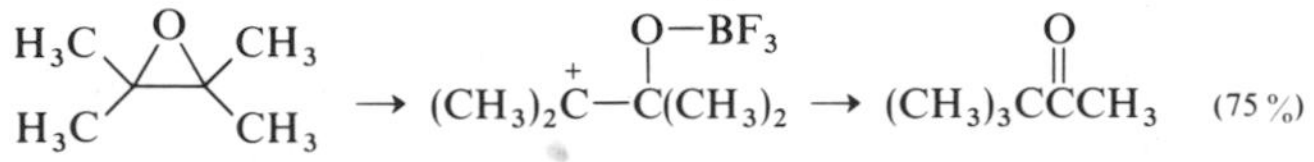

Olefinic groups carrying oxygen and halogen substituents are susceptible to

53. J. M. Coxon, M. P. Hartshorn, and W. J. Rae, *Tetrahedron* **26**, 1091 (1970).
54. B. Rickborn and R. M. Gerkin, *J. Am. Chem. Soc.* **93**, 1693 (1971).
55. H. Hart and L. R. Lerner, *J. Org. Chem.* **32**, 2669 (1967).

epoxidation, and the reactive epoxides that are thereby generated serve as inter-mediates in some useful synthetic transformations. Vinyl chlorides furnish haloepox-ides, which can be rearranged to α-haloketones:

371

SECTION 9.3.
CLEAVAGE
OF CARBON–
CARBON
DOUBLE
BONDS

Ref. 56

Enol acetates form epoxides, which are rearranged to α-acetoxyketones:

Ref. 57

Ref. 58

The stereochemistry of the rearrangement of the acetoxyepoxides involves inversion at the carbon to which the acetoxy group migrates.[59] The reaction probably proceeds through a cyclic transition state.

9.3. Cleavage of Carbon–Carbon Double Bonds

9.3.1. Transition-Metal Oxidants

The most selective methods for cleaving organic molecules at carbon–carbon double bonds are based on procedures in which glycols are intermediates. Oxidation of alkenes to glycols was discussed in Section 9.2. Cleavage of olefins can be carried

56. R. N. McDonald and T. E. Tabor, *J. Am. Chem. Soc.* **89**, 6573 (1967).
57. K. L. Williamson, J. I. Coburn and M. F. Herr, *J. Org. Chem.* **32**, 3934 (1967).
58. R. G. Carlson and J. K. Pierce, *J. Org. Chem.* **36**, 2319 (1971).
59. K. L. Williamson and W. S. Johnson, *J. Org. Chem.* **26**, 4563 (1961).

out in one operation under mild conditions by using solutions containing periodate ion and a catalytic amount of permanganate.[60] The permanganate effects olefin hydroxylation, and the glycol is then cleaved by reaction with periodate. A cyclic intermediate is believed to be involved in the periodate oxidation. Permanganate is continuously regenerated by the oxidizing action of periodate. The reaction has

$$IO_3^- + H_2O + 2 RCH{=}O$$

obvious potential in degradative structure-determination work, but also finds some synthetic application. Osmium tetroxide used in combination with sodium periodate can also effect alkene cleavage cleanly.[61,62] Successful oxidative cleavage of double bonds using ruthenium tetroxide and sodium periodate has also been reported.[63]

Examples of these reactions can be found in Scheme 9.9.

Cr(VI) attacks alkenes to give both products of allylic attack (Section 9.6) and products that involve addition of oxygen at the double bond. The tendency for formation of product mixtures detracts from the synthetic utility of the reaction. Evidence that at least part of the oxidation occurs via the corresponding epoxide has been obtained by demonstrating that in the presence of trapping agents, products derived from epoxides are obtained.[64] For example, it is known that solvolysis of cyclohexene oxide in acetic acid gives cis- and trans-2-acetoxycyclohexanol and a small amount of cyclopentanecarboxaldehyde. These same products are isolated when cyclohexene is oxidized by chromic acid under comparable conditions[65]:

+ other products

Despite these mechanistic complexities, the oxidation can be useful for the synthesis of dicarboxylic acids, since the intermediate oxidation products are susceptible to further oxidation. Obviously, this method cannot be applied if there are sensitive groups elsewhere in the molecule. Entries 4 and 5 in Scheme 9.9 are illustrative.

60. R. U. Lemieux and E. von Rudloff, *Can. J. Chem.* **33**, 1701, 1710 (1955); E. von Rudloff, *Can. J. Chem.* **33**, 1714 (1955).
61. R. Pappo, D. S. Allen, Jr., R. U. Lemieux, and W. S. Johnson, *J. Org. Chem.* **21**, 478 (1956).
62. H. Vorbrueggen and C. Djerassi, *J. Am. Chem. Soc.* **84**, 2990 (1962).
63. W. G. Dauben and L. E. Friedrich, *J. Org. Chem.* **37**, 241 (1972).
64. J. Roček and J. C. Drozd, *J. Am. Chem. Soc.* **92**, 6668 (1970).
65. A. K. Awasthy and J. Roček, *J. Am. Chem. Soc.* **91**, 991 (1969).

Scheme 9.9. Oxidative Cleavage of Alkenes

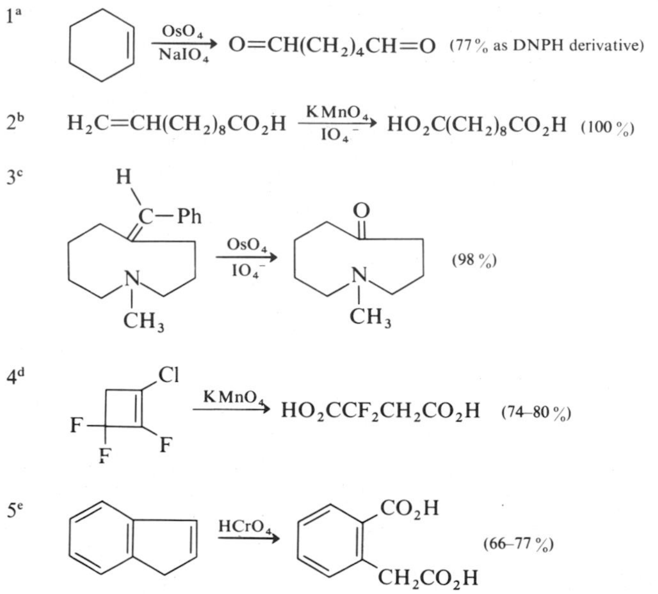

373

SECTION 9.3.
CLEAVAGE
OF CARBON–
CARBON
DOUBLE
BONDS

1[a]

$\xrightarrow[\text{NaIO}_4]{\text{OsO}_4}$ O=CH(CH$_2$)$_4$CH=O (77% as DNPH derivative)

2[b]   H$_2$C=CH(CH$_2$)$_8$CO$_2$H $\xrightarrow[\text{IO}_4^-]{\text{KMnO}_4}$ HO$_2$C(CH$_2$)$_8$CO$_2$H (100%)

3[c]

$\xrightarrow[\text{IO}_4^-]{\text{OsO}_4}$

(98%)

4[d]

$\xrightarrow{\text{KMnO}_4}$ HO$_2$CCF$_2$CH$_2$CO$_2$H (74–80%)

5[e]

$\xrightarrow{\text{HCrO}_4}$

(66–77%)

---

a. R. Pappo, D. S. Allen, Jr., R. U. Lemieux, and W. S. Johnson, *J. Org. Chem.* **21**, 478 (1956).
b. R. U. Lemieux and E. von Rudloff, *Can. J. Chem.* **33**, 1701 (1955).
c. M. G. Reinecke, L. R. Kray, and R. F. Francis, *J. Org. Chem.* **37**, 3489 (1972).
d. M. S. Raasch and J. E. Castle, *Org. Synth.* **42**, 44 (1962).
e. O. Grummitt, R. Egan, and A. Buck, *Org. Synth.* **III**, 449 (1955).

9.3.2. Ozonolysis

The reaction of olefins with ozone constitutes an important method of cleavage of carbon–carbon double bonds.[66] This reaction has been a degradative tool for several decades and finds occasional use in synthesis. Recent years have seen the application of low-temperature and spectroscopic techniques to the study of the rather unstable species that are intermediates in the ozonolysis process. These studies have put early mechanistic ideas on a firmer basis and have elaborated many additional details of the reaction mechanism.

A generalized scheme showing reactions believed to occur during ozonolysis is given in Scheme 9.10. The first step in ozonolysis is believed to be formation of a 1 : 1 cyclic adduct containing a three-membered ring. These intermediates are highly unstable, and direct evidence for their existence is lacking. This species is believed to rearrange to one containing a four-membered ring, as shown in Scheme 9.10.[67] This

66a. P. S. Bailey, *Chem. Rev.* **58**, 925 (1958).
  b. R. W. Murray, *Acc. Chem. Res.* **1**, 313 (1968).
67. P. R. Story, E. A. Whited, and J. A. Alford, *J. Am. Chem. Soc.* **94**, 2143 (1972).

Scheme 9.10. Generalized Ozonolysis Mechanism

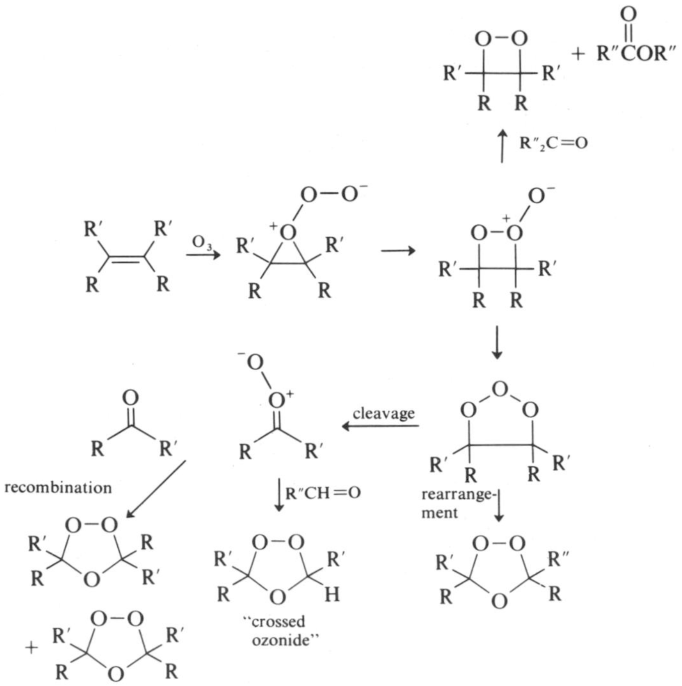

adduct, sometimes referred to as the "molozonide," is also highly unstable and can be detected only by the modified course of the reaction in solvents with which it reacts rapidly (specifically aldehydes and ketones). The next intermediate, the 1,2,3-trioxolane or "initial ozonide," is the earliest intermediate to have been characterized spectrally.[68-70] This adduct decomposes, leading to the final ozonolysis product, the ozonide or 1,2,4-trioxolane, and other peroxidic products.[71,72] There is general acceptance of the idea that a cleavage–recombination pathway is usually the major contributing route to ozonide formation. There have been repeated[73] observations of a dependence of the stereoisomeric ratio of the ozonides on the configuration of the initial olefin. Such a result can be accommodated by the cleavage–recombination mechanism if the zwitterionic intermediates can exist as nonequilibrating *syn* and *anti* isomers. Stereoselectivity in both the formation and

68. P. S. Bailey, J. A. Thompson, and B. A. Shoulders, *J. Am. Chem. Soc.* **88**, 4098 (1966).
69. L. A. Hull, I. C. Hisatsune, and J. Heicklen, *J. Am. Chem. Soc.* **94**, 4856 (1972).
70. L. J. Durham and F. L. Greenwood, *J. Org. Chem.* **33**, 1629 (1968).
71. P. R. Story, J. A. Alford, W. C. Ray, and J. R. Burgess, *J. Am. Chem. Soc.* **93**, 3044 (1971).
72. F. L. Greenwood and L. J. Durham, *J. Org. Chem.* **34**, 3363 (1969).
73. Several references are given by S. Fliszár and J. Carles, *J. Am. Chem. Soc.* **91**, 2637 (1969).

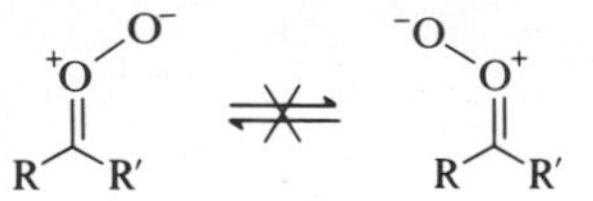

375

SECTION 9.3.
CLEAVAGE
OF CARBON–
CARBON
DOUBLE
BONDS

recombination steps can then account for a dependence of ozonide stereochemistry on olefin geometry. A detailed rationale for the observed stereoselectivity in terms of conformational analysis is available.[74,75]

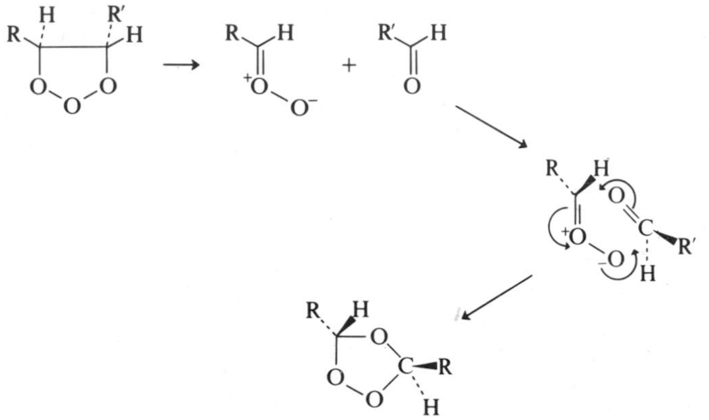

When appreciable concentrations of carbonyl compounds are present before complete formation of ozonide, "crossed ozonides" are formed. This occurs when the added carbonyl compound traps the zwitterion formed in the cleavage step. When *cis*-stilbene is subjected to ozonolysis in the presence of ^{18}O-labeled benzaldehyde, the label is incorporated into the ether rather than the peroxide portion of the ozonide:

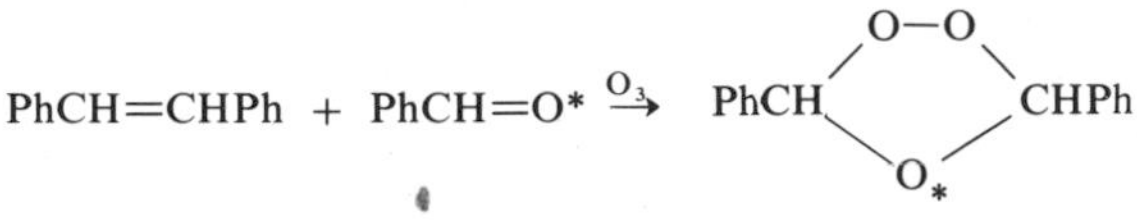

This result is consistent with formation of the crossed ozonide via the cleavage–recombination mechanism[76]:

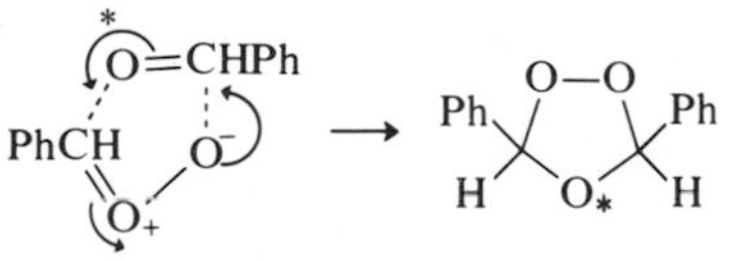

Reactive solvent molecules can modify the course of ozonolysis reactions. We have already mentioned that when ozonolysis is carried out using certain carbonyl compounds as solvents, the reaction is diverted from its normal course. Under these

74. N. L. Bauld, J. A. Thompson, C. E. Hudson, and P. S. Bailey, *J. Am. Chem. Soc.* **90**, 1822 (1968); R. P. Lattimer, R. L. Kuczkowski, and C. W. Gilles, *J. Am. Chem. Soc.* **96**, 348 (1974).
75. J. Renard and S. Fliszár, *J. Am. Chem. Soc.* **92**, 2628 (1970).
76. S. Fliszár and J. Carles, *J. Am. Chem. Soc.* **91**, 2637 (1969).

conditions, dioxetanes are formed.[67,71] The proposed mechanism for this process is shown below:

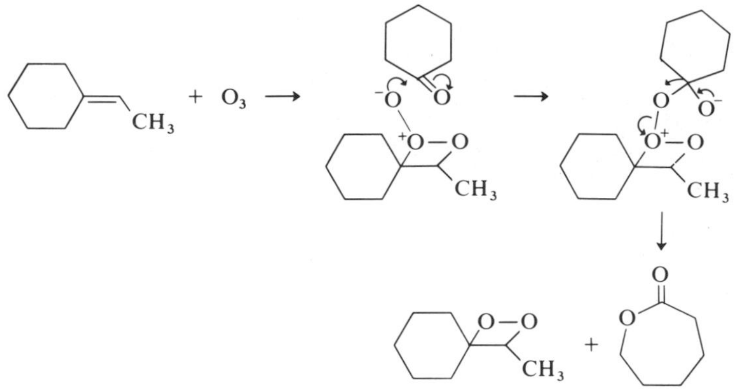

When ozonolysis is performed in alcoholic solvents, the zwitterionic cleavage intermediates are trapped as α-hydroperoxy ethers.[77] Recombination is then prevented, and the carbonyl compound formed in the cleavage step can also be isolated under these conditions.

$$R_2C\overset{+}{=}\overset{}{O}-O^- + CH_3OH \rightarrow R_2\underset{\underset{OCH_3}{|}}{C}OOH$$

$$PhCH=CH_2 \xrightarrow[CH_3OH]{O_3} Ph\underset{\underset{OCH_3}{|}}{C}HOOH + \underset{\underset{OCH_3}{|}}{C}H_2OOH + PhCH=O + CH_2=O$$

$$(31\%) \qquad (23\%) \qquad (26\%) \qquad (27\%)$$

Despite the complexities in the mechanism of ozonolysis, the reaction constitutes a high-yield method for cleavage of carbon–carbon double bonds. The oxidation states of the products that are actually isolated depend on the conditions used in processing the reaction mixture. If the carbonyl products are desired, it is advantageous to carry out ozonolysis in methanol, resulting in the formation of α-methoxyalkyl hydroperoxides. The reaction mixture is then treated with dimethyl sulfide, which reduces the hydroperoxide and permits isolation of the carbonyl compounds in good yield.[78] This procedure prevents oxidation of the carbonyl products, especially aldehydes, by the peroxidic compounds present at the conclusion of ozonolysis.

$$RCH=CH_2 \xrightarrow[CH_3OH]{O_3} R\underset{\underset{OOH}{|}}{C}HOCH_3 \xrightarrow{(CH_3)_2S} RCH=O + (CH_3)_2S=O$$

Other reducing agents that have been employed for the same purpose include

67. See p. 373.
71. See p. 374.
77. W. P. Keaveney, M. G. Berger, and J. J. Pappas, *J. Org. Chem.* **32**, 1537 (1967).
78. J. J. Pappas, W. P. Keaveney, E. Gancher, and M. Berger, *Tetrahedron Lett.*, 4273 (1966).

triphenylphosphine,[79] trimethyl phosphite,[80] tris(dimethylamino)phosphine,[81] sodium sulfite,[82] and zinc.[83] A review[66a] records examples of still other types of reducing agents that have been employed.

377

SECTION 9.4.
SELECTIVE
OXIDATIVE
CLEAVAGES

If the alcohol corresponding to the reduction of the carbonyl cleavage product is desired, the reaction mixture can be reduced with hydride reducing agents.[84] Carboxylic acids are formed in good yields from aldehydes when the ozonolysis reaction mixture is worked up in the presence of excess hydrogen peroxide to insure complete oxidation of aldehyde groups.[83]

Scheme 9.11 illustrates some cases where ozonolysis has been used in synthetic procedures.

9.4. Selective Oxidative Cleavages at Other Functional Groups

9.4.1. Cleavage of Glycols

As indicated in connection with cleavage reactions that originate by oxidative attack on carbon–carbon double bonds, the glycol unit is susceptible to mild oxidative cleavage. The most commonly used reagent for oxidative cleavage of glycols is the periodate anion.[85] Mechanistic work has indicated that the key intermediate is a cyclic complex of the glycol and oxidant. Various investigations of

$$R-\underset{\underset{HO}{|}}{\overset{\overset{H}{|}}{C}}-\underset{\underset{OH}{|}}{\overset{\overset{H}{|}}{C}}-R \rightarrow R-\underset{O}{\overset{H}{\underset{|}{C}}}-\!\!\!-\!\!\!-\underset{O}{\overset{H}{\underset{|}{C}}}-R \rightarrow 2\,RCH + H_2O + IO_3^- $$

the relationship between glycol stereochemistry and rate of oxidation have established that steric features that would retard formation of a cyclic intermediate decrease the oxidation rate. For example, cis-1,2-dihydroxycyclohexane is substantially more reactive than the *trans* isomer.[86] The rate retardation can be attributed to increased strain in the cyclic ester derived from the *trans*-diol. Several glycols in

79. J. J. Pappas, W. P. Keaveney, M. Berger, and R. V. Rush, *J. Org. Chem.* **33**, 787 (1968).
80. W. S. Knowles and Q. E. Thompson, *J. Org. Chem.* **25**, 1031 (1960).
81. A. Furlenmeier, A. Fürst, A. Langemann, G. Waldvogel, P. Hocks, U. Kerb, and R. Wiechert, *Helv. Chim. Acta.* **50**, 2387 (1967).
82. R. H. Callighan and M. H. Wilt, *J. Org. Chem.* **26**, 4912 (1961).
83. A. L. Henne and P. Hill, *J. Am. Chem. Soc.* **65**, 752 (1943).
66a. See p. 373.
84. F. L. Greenwood, *J. Org. Chem.* **20**, 803 (1955).
85. C. A. Bunton, in *Oxidation in Organic Chemistry*, K. B. Wiberg (ed.), Academic Press, New York, NY, pp. 367–388; A. S. Perlin, in *Oxidation*, R. L. Augustine (ed.), Marcel Dekker, New York, NY, 1969, pp. 189–204.
86. C. C. Price and M. Knell, *J. Am. Chem. Soc.* **64**, 552 (1942).

Scheme 9.11. Ozonolysis Reactions

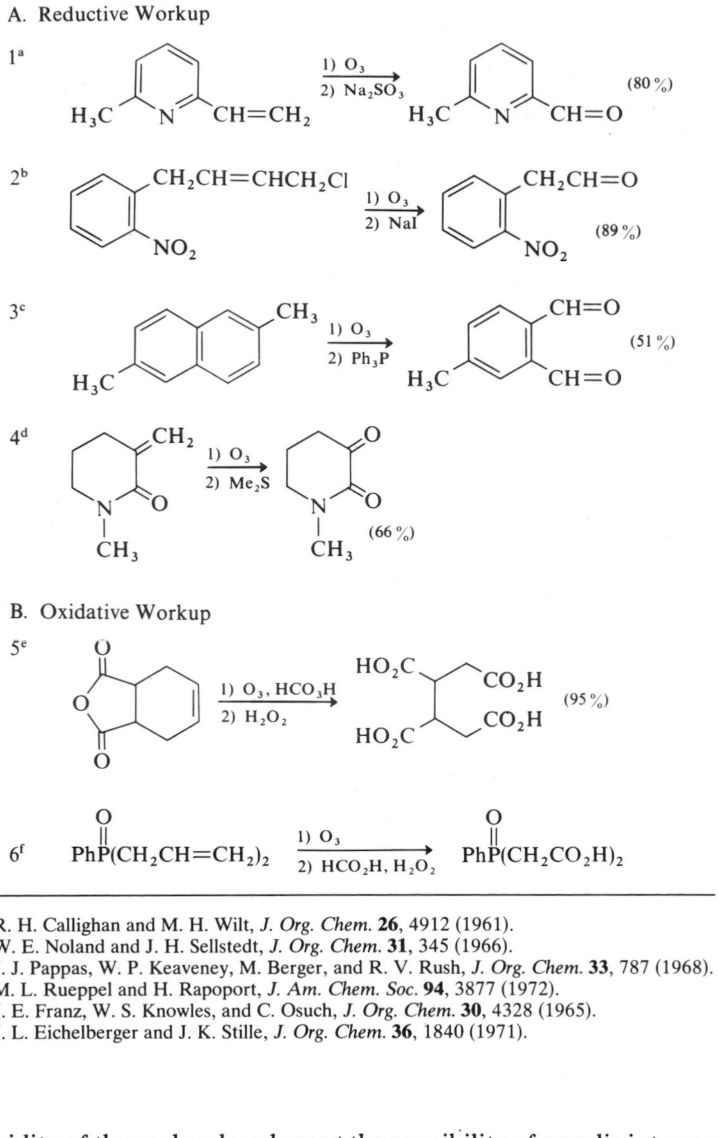
A. Reductive Workup
1a
1) O3
2) Na2SO3
H3C
N
CH=CH2
H3C
N
CH=O
(80%)
2b
CH2CH=CHCH2Cl
NO2
1) O3
2) NaI
CH2CH=O
NO2
(89%)
3c
CH3
H3C
1) O3
2) Ph3P
CH=O
CH=O
H3C
(51%)
4d
CH2
N
O
CH3
1) O3
2) Me2S
O
N
O
CH3
(66%)
B. Oxidative Workup
5e
O
O
O
1) O3, HCO3H
2) H2O2
HO2C
CO2H
HO2C
CO2H
(95%)
6f
O
PhP(CH2CH=CH2)2
1) O3
2) HCO2H, H2O2
O
PhP(CH2CO2H)2

a. R. H. Callighan and M. H. Wilt, *J. Org. Chem.* **26**, 4912 (1961).
b. W. E. Noland and J. H. Sellstedt, *J. Org. Chem.* **31**, 345 (1966).
c. J. J. Pappas, W. P. Keaveney, M. Berger, and R. V. Rush, *J. Org. Chem.* **33**, 787 (1968).
d. M. L. Rueppel and H. Rapoport, *J. Am. Chem. Soc.* **94**, 3877 (1972).
e. J. E. Franz, W. S. Knowles, and C. Osuch, *J. Org. Chem.* **30**, 4328 (1965).
f. J. L. Eichelberger and J. K. Stille, *J. Org. Chem.* **36**, 1840 (1971).

which the rigidity of the molecule rules out the possibility of a cyclic intermediate are essentially inert to periodate. Occasionally, by comparison of reaction rates with model compounds of known stereochemistry, it has been possible to base stereochemical assignments on relative reactivity toward periodate.

Certain other systems containing adjacent functional groups that are capable of formation of a cyclic intermediate are also cleaved by periodate. Diketones are cleaved to carboxylic acids, and it has been proposed that a reactive cyclic intermediate is formed by nucleophilic attack on the diketone.[87] α-Hydroxyketones and α-aminoalcohols are subject to similar oxidative cleavage.

87. C. A. Bunton and V. J. Shiner, Jr., *J. Chem. Soc.* 1593 (1960).

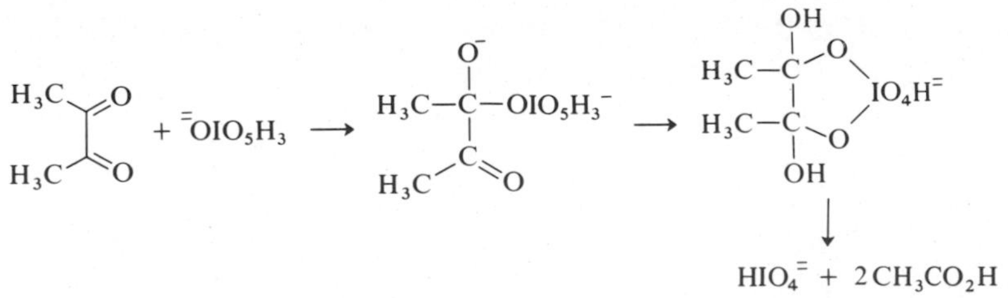

$$H_3C-C(=O)-C(=O)-CH_3 \ + \ ^=OIO_5H_3 \ \longrightarrow \ H_3C-\underset{\underset{C(-CH_3)=O}{|}}{\overset{\overset{O^-}{|}}{C}}-OIO_5H_3^- \ \longrightarrow \ \underset{\underset{H_3C-C-OH}{|}}{\overset{\overset{OH}{|}}{H_3C-C}}\overset{O}{\underset{O}{\diagdown}}IO_4H^=$$

$$\downarrow$$

$$HIO_4^= \ + \ 2\,CH_3CO_2H$$

Lead tetraacetate is an alternative to periodate for glycol cleavage. It is particularly useful for glycols that have low solubility in the aqueous media used for periodate reactions. A cyclic mechanism is indicated by the same kind of stereochemistry–reactivity relationships discussed for periodate.[88] Unlike periodate, however, glycols that cannot form cyclic intermediates are eventually oxidized. For

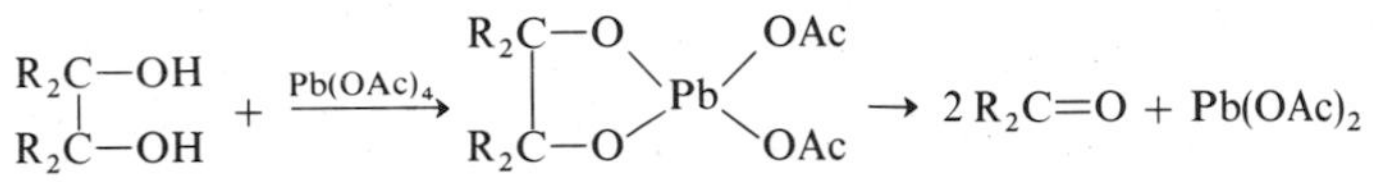

$$\underset{\underset{R_2C-OH}{|}}{R_2C-OH} \ + \ \xrightarrow{Pb(OAc)_4} \ \underset{\underset{R_2C-O}{|}}{R_2C-O}\overset{OAc}{\underset{OAc}{\diagup}}Pb \ \longrightarrow \ 2\,R_2C{=}O \ + \ Pb(OAc)_2$$

example, *trans*-9,10-dihydroxydecalin is oxidized, although the rate for the *cis* isomer is 100 times greater[89]:

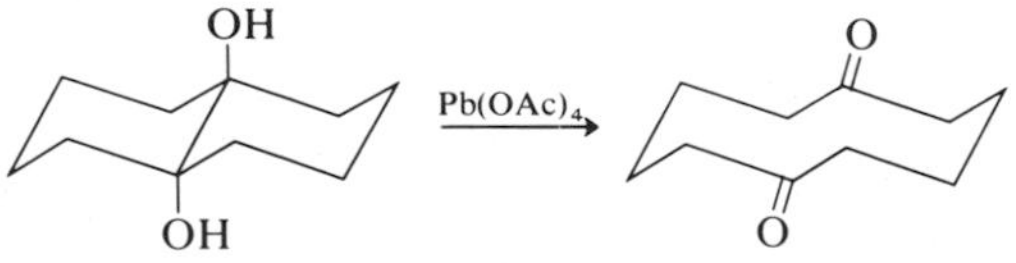

Thus, while a cyclic transition state appears to provide the lowest energy pathway for this oxidative cleavage, other mechanisms are possible.

9.4.2. Oxidative Decarboxylation

Carboxylic acids are oxidized by lead tetraacetate. Decarboxylation occurs, and the product may be an alkene, alkane, acetate ester, or, under modified conditions, an alkyl halide. A free-radical mechanism operates, and the product composition

$$Pb(OAc)_4 \ + \ RCO_2H \ \rightleftharpoons \ RCO_2Pb(OAc)_3 \ + \ CH_3CO_2H$$

$$RCO_2Pb(OAc)_3 \ \rightarrow \ R\cdot \ + \ CO_2 \ + \ Pb(OAc)_3$$

$$R\cdot \ + \ Pb(OAc)_4 \ \rightarrow \ R^+ \ + \ Pb(OAc)_3 \ + \ CH_3CO_2^-$$

and

$$R\cdot \ + \ Pb(OAc)_3 \ \rightarrow \ R^+ \ + \ Pb(OAc)_2 \ + \ CH_3CO_2^-$$

88. C. A. Bunton, in *Oxidation in Organic Chemistry*, K. Wiberg (ed.), Academic Press, New York, NY, 1965, pp. 398–405; W. S. Trahanovsky, J. A. Gilmore, and P. C. Heaton, *J. Org. Chem.* **38**, 760 (1973).
89. R. Crieger, E. Höger, G. Huber, P. Kruck, F. Markischeffel, and H. Schellenberger, *Justus Liebigs Ann. Chem.* **599**, 81 (1956).

depends on the fate of the radical intermediate.[90] The reaction is catalyzed by cupric salts, which function by oxidizing the intermediate radical to the carbonium ion. Cu(II) is much more reactive than Pb(OAc)$_4$ in this step.

Alkanes are formed when the intermediate radical abstracts hydrogen from solvent faster than it is oxidized to the carbonium ion. This reductive process is promoted by good hydrogen-donor solvents. It is also most favorable for primary alkyl radicals because of the higher activation energy associated with formation of primary carbonium ions. The most favorable conditions for alkane formation involve photochemical decomposition of the carboxylic acid in chloroform solution:

Ref. 91

Normally, the dominant products are the alkene and ester. These arise from the carbonium-ion intermediate by, respectively, elimination of a proton and capture of an acetate ion. The presence of copper acetate increases the alkene : ester ratio.[92]

When oxidation is carried out in the presence of halide salts, alkyl halides are formed in good yield. The halide is believed to be introduced at the radical stage by a ligand-transfer reaction.

$$RCO_2Pb(OAc)_3 \rightarrow R\cdot + CO_2 + Pb(OAc)_3$$
$$R\cdot + PbX_n(OAc)_{4-n} \rightarrow RX + PbX_{n-1}(OAc)_{4-n}$$

A second method for conversion of carboxylic acids to bromides with decarboxylation is the *Hunsdiecker reaction*.[93] The most convenient method for carrying out this transformation involves heating the carboxylic acid with mercuric oxide and bromine:

Ref. 94

The overall transformation can also be accomplished by reaction of thallium(I) carboxylate salts with bromine.[95]

1,2-Dicarboxylic acids undergo bis-decarboxylation on reaction with lead

Ref. 96

90. R. A. Sheldon and J. K. Kochi, *Org. React.* **19**, 279 (1972).
91. J. K. Kochi and J. D. Bacha, *J. Org. Chem.* **33**, 2746 (1968).
92. J. D. Bacha and J. K. Kochi, *Tetrahedron* **24**, 2215 (1968).
93. C. V. Wilson, *Org. React.* **9**, 332 (1957).
94. J. S. Meek and D. T. Osuga, *Org. Synth.* **V**, 126 (1973).
95. A. McKillop, D. Bromley, and E. C. Taylor, *J. Org. Chem.* **34**, 1172 (1969).
96. E. Grovenstein, Jr., D. V. Rao, and J. W. Taylor, *J. Am. Chem. Soc.* **83**, 1705 (1961).

381

SECTION 9.5.
OXIDATIONS
OF KETONES
AND
ALDEHYDES

tetraacetate to yield olefins. This reaction has occasionally found use in the preparation of strained olefins. The reaction can be formulated as occurring via a concerted process initiated by a two-electron oxidation:

$$\text{(structure)} \rightarrow \text{(structure)} + 2\,CO_2 + Pb(OAc)_2 + HOAc$$

A concerted mechanism is also possible in the oxidation of α-hydroxycarboxylic acids, and these compounds readily undergo oxidative decarboxylation[97]:

$$R_2C\text{(structure)} \rightarrow R_2C{=}O + CO_2 + Pb(OAc)_2 + CH_3CO_2H$$

α-Dicarbonyl compounds are attacked by alkaline hydrogen peroxide. The most frequent preparative use of this reaction is in the oxidation of α-ketoacids to carboxylic acids, with loss of carbon dioxide[98]:

$$R\overset{O}{C}CO_2H \xrightarrow{\;^-OOH\;} R{-}\overset{\!^-O}{\underset{O}{\overset{|}{C}}}{-}\overset{O}{C}{-}O^- \rightarrow RCO_2{}^- + CO_2 + {}^-OH$$

9.5. Oxidations of Ketones and Aldehydes

9.5.1. Transition-Metal Oxidants

Ketones are cleaved oxidatively by Cr(VI) or Mn(VIII) reagents. The reaction is sometimes of utility in the synthesis of difunctional molecules by ring cleavages. The mechanism for both reagents is believed to involve reaction of an enolic intermediate,[99] although in neither case have all the details been established. A study involving both kinetic data and accurate product-yield determination has permitted a fairly complete description of the Cr(VI) oxidation of benzyl phenyl ketone (desoxybenzoin).[100] The products include both oxidative-cleavage products and

97. R. Criegee and E. Buchner, *Chem. Ber.* **73**, 563 (1940).
98. J. E. Leiffler, *J. Org. Chem.* **16**, 1785 (1951).
99. K. B. Wiberg and R. D. Geer, *J. Am. Chem. Soc.* **87**, 5202 (1965); J. Roček and A. Riehl, *J. Am. Chem. Soc.* **89**, 6691 (1967).
100. K. B. Wiberg, O. Aniline, and A. Gatzke, *J. Org. Chem.* **37**, 3229 (1972).

benzil, which is derived from oxidation alpha to the carbonyl. In addition, the coupling product **3**, which is suggestive of radical intermediates, was formed under some conditions:

$$PhCH_2\overset{O}{\underset{\parallel}{C}}Ph \xrightarrow{Cr(VI)} Ph\overset{O}{\underset{\parallel}{C}}\overset{O}{\underset{\parallel}{C}}Ph + PhCH(=O) + PhCO_2H + PhCH-CHPh$$

with the coupling product showing PhCO and COPh groups labeled **3**.

Both the diketone and cleavage products were shown to be formed from benzoin, which is an intermediate. This, in turn, arises from oxidation of the enol of the

$$PhCH_2\overset{O}{\underset{\parallel}{C}}Ph \rightleftharpoons PhCH=\underset{OH}{\overset{}{C}}Ph \xrightarrow{H_2CrO_4} Ph-CH=CH-Ph$$

$$H_2O^- \qquad OCrO_3H$$

$$\downarrow$$

$$products \longleftarrow PhCH-CPh + Cr(IV)$$
$$\qquad\qquad OH \ O$$

starting ketone. The coupling product is considered to involve an intermediate formed by one-electron oxidation, probably effected by Cr(IV):

$$PhCH_2\overset{O}{\underset{\parallel}{C}}Ph \rightarrow Ph\overset{O}{\underset{\parallel}{C}}HCPh \rightarrow 3$$

Studies in the case of cyclohexanone have indicated the intermediacy of 2-hydroxycyclohexanone, which is further oxidized to cyclohexanedione and cleaved to dicarboxylic acids, principally adipic acid.[101] Because of the efficient oxidation of

$$\text{cyclohexanone} \xrightarrow{Cr(VI)} \text{2-hydroxycyclohexanone} \rightarrow \text{cyclohexanedione} \rightarrow \begin{array}{c} CO_2H \\ CO_2H \end{array}$$

alcohols to ketones, the alcohols can be used as the actual starting materials in oxidative cleavages. These reactions require considerably more vigorous conditions than the alcohol-to-ketone transformation.

Lead tetraacetate reacts with ketones to form α-acetoxyketones.[102] Reported yields are seldom high, however. The reaction has been successfully applied to keto steroids. Boron trifluoride can be used to catalyze these oxidations. It is presumed to function by catalysis of enolization, and it is assumed that the enol is the reactive species.[103]

101. J. Roček and A. Riehl, *J. Org. Chem.* **32**, 3569 (1967).
102. R. Criegee, in *Oxidation in Organic Chemistry*, K. B. Wiberg (ed.), Academic Press, New York, NY, 1965, pp. 305–312.
103. J. D. Cocker, H. B. Henbest, G. H. Phillips, G. P. Slater, and D. A. Thomas, *J. Chem. Soc.*, 6 (1965).

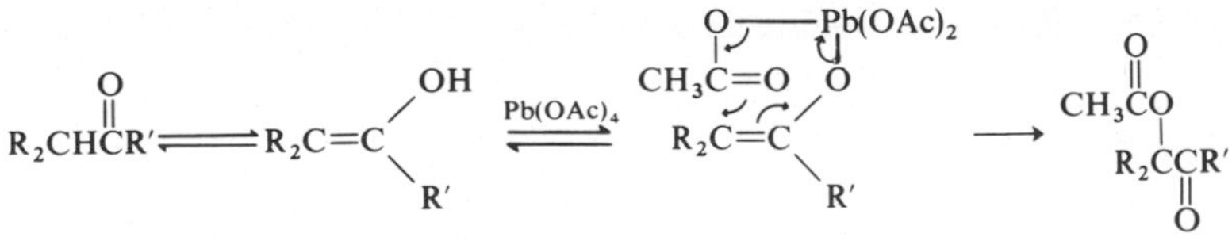

383

SECTION 9.5.
OXIDATIONS
OF KETONES
AND
ALDEHYDES

Molybdenum peroxide oxidizes enolates to α-hydroxyketones; both ketone and ester enolates can be oxidized.[104] This reagent is prepared by dissolving

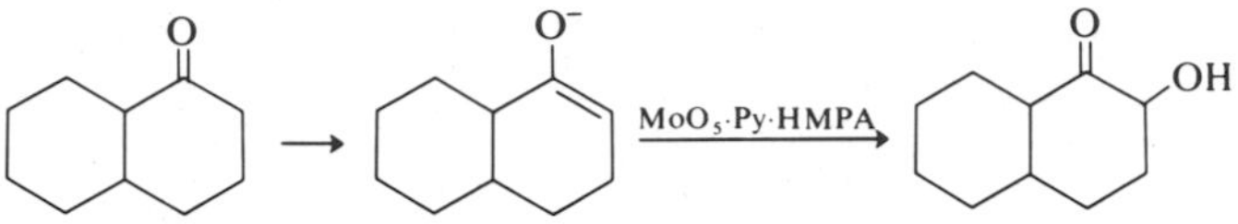

$Mo(VI)O_3$ in hydrogen peroxide, followed by addition of hexamethylphos-phoramide. The resulting precipitate is converted to a pyridine complex and used in that form.

Aldehydes can be oxidized to carboxylic acids by both Mn(VII) and Cr(VI). Fairly detailed mechanistic studies have been completed in the case of Cr(VI). A chromate ester of the aldehyde hydrate is believed to be formed, and this species decomposes in the rate-determining step by a mechanism similar to that which is

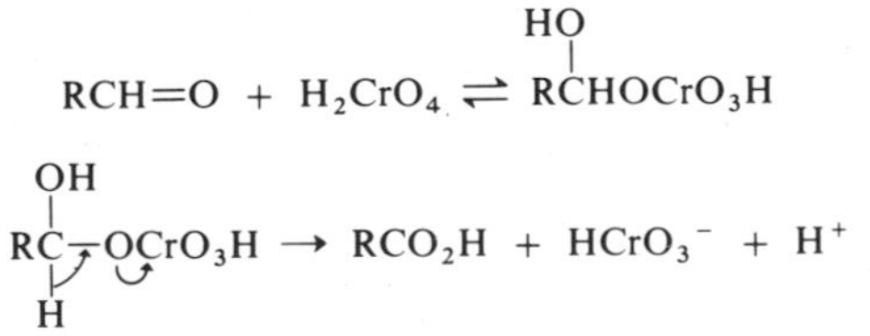

$$RCH{=}O \ + \ H_2CrO_4 \ \rightleftarrows \ \overset{\displaystyle HO}{\underset{}{R}}CHOCrO_3H$$

$$\overset{\displaystyle OH}{\underset{\displaystyle H}{R}}C{-}OCrO_3H \ \rightarrow \ RCO_2H \ + \ HCrO_3^- \ + \ H^+$$

operative in alcohol oxidations.[105] An alternative reagent for carrying out the aldehyde to carboxylic acid oxidation is silver oxide:

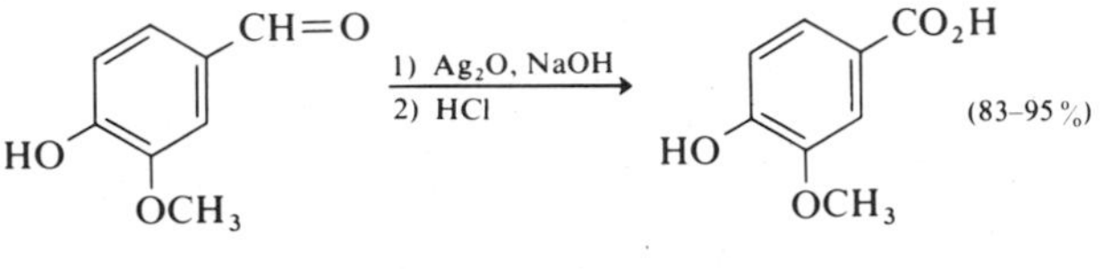

(83–95 %)   Ref. 106

9.5.2. Oxidation of Ketones and Aldehydes by Peroxidic Compounds and Oxygen

In the presence of acid catalysts, peroxy compounds are capable of oxidizing carbonyl compounds in a manner involving formal insertion of an oxygen atom into one of the carbon–carbon bonds at the carbonyl group. This insertion is accomplished by a sequence of steps involving addition to the carbonyl group and migration

104. E. Vedejs, *J. Am. Chem. Soc.* **96**, 5945 (1974).
105. K. B. Wiberg, *Oxidation in Organic Chemistry*, Academic Press, New York, NY, 1965, pp. 172–178.
106. I. A. Pearl, *Org. Synth.* **IV**, 972 (1963).

to oxygen, as outlined in the mechanism below. The concerted O–O heterolysis–

$$\underset{RCR}{\overset{O}{\underset{\|}{}}} + \underset{R'COOH}{\overset{O}{\underset{\|}{}}} \rightleftarrows R-\overset{OH}{\underset{O-O-C=O}{\underset{|}{\overset{|}{C}}}}-R \rightarrow \underset{RCOR}{\overset{O}{\underset{\|}{}}} + R'CO_2H$$

migration is usually the rate-determining step.[107] The reaction is known as the *Baeyer–Villiger oxidation*.[108]

When the reaction involves an unsymmetric ketone, the structure of the product depends on which alkyl group migrates. A number of studies have been directed at ascertaining the basis of migratory aptitude in the Baeyer–Villiger oxidation. From these studies, a general order of likelihood of migration, or "migratory aptitude," has been established: *tert*-alkyl, *sec*-alkyl > benzyl, phenyl > *pri*-alkyl > cyclopropyl > methyl.[109] Thus, methyl ketones are uniformly found to give acetate esters resulting from migration of the larger group.[110] As is generally true of migration to an electron-deficient center, the configuration of the migrating group is retained in Baeyer–Villiger oxidations.

The precise factors that govern migratory aptitude are not completely clear. The electronic nature of the substituents surely contributes; in benzophenones, for example, the relative migratory aptitude decreases in the order $CH_3O > CH_3 > H > Cl > NO_2$ for *para* substituents.[111] It is believed that steric and conformational factors also come into play.[112] The selectivity in the migration also appears to depend on the identity of the peroxyacid, with peroxytrifluoroacetic acid showing somewhat less selectivity than weaker oxidants.

At the present time, peroxytrifluoroacetic acid and *m*-chloroperoxybenzoic acid are most often used to accomplish Baeyer–Villiger oxidation of ketones for synthetic purposes. Some typical examples are shown in Scheme 9.12. A tabulation of work prior to the mid-1950's provides many examples of the use of peroxyacetic acid, peroxybenzoic acid, and hydrogen peroxide with strong acids.[113] Peroxysulfuric acid is also an effective reagent, at least for simple ketones.[114]

Although ketones are essentially inert to molecular oxygen, enolate anions are susceptible to oxidation. The combination of oxygen and a base has found synthetic utility in permitting introduction of an oxygen function at a potential carbanion site.[115] Hydroperoxides are believed to be the initial products of such oxidations, but

107. Y. Ogata and Y. Sawaki, *J. Org. Chem.* **37**, 2953 (1972).
108. C. H. Hassall, *Org. React.* **9**, 73 (1957).
109. H. O. House, *Modern Synthetic Reactions*, Second Edition, W. A. Benjamin, Menlo Park, CA, 1972, p. 325.
110. P. A. S. Smith, in *Molecular Rearrangements*, P. de Mayo (ed.), Interscience, New York, NY, p. 584.
111. W. E. Doering and L. Speers, *J. Am. Chem. Soc.* **72**, 5515 (1950).
112. M. F. Hawthorne, W. D. Emmons, and K. S. McCallum, *J. Am. Chem. Soc.* **80**, 6393 (1958); J. Meinwald and E. Frauenglass, *J. Am. Chem. Soc.* **82**, 5235 (1960).
113. C. H. Hassall, *Org. React.* **9**, 73 (1957).
114. N. C. Deno, W. E. Billups, K. E. Kramer, and R. R. Lastomirsky, *J. Org. Chem.* **35**, 3080 (1970).
115. J. N. Gardner, T. L. Popper, F. E. Carlon, O. Gnoj, and H. L. Herzog, *J. Org. Chem.* **33**, 3695 (1968).

Scheme 9.12. Baeyer–Villiger Oxidations

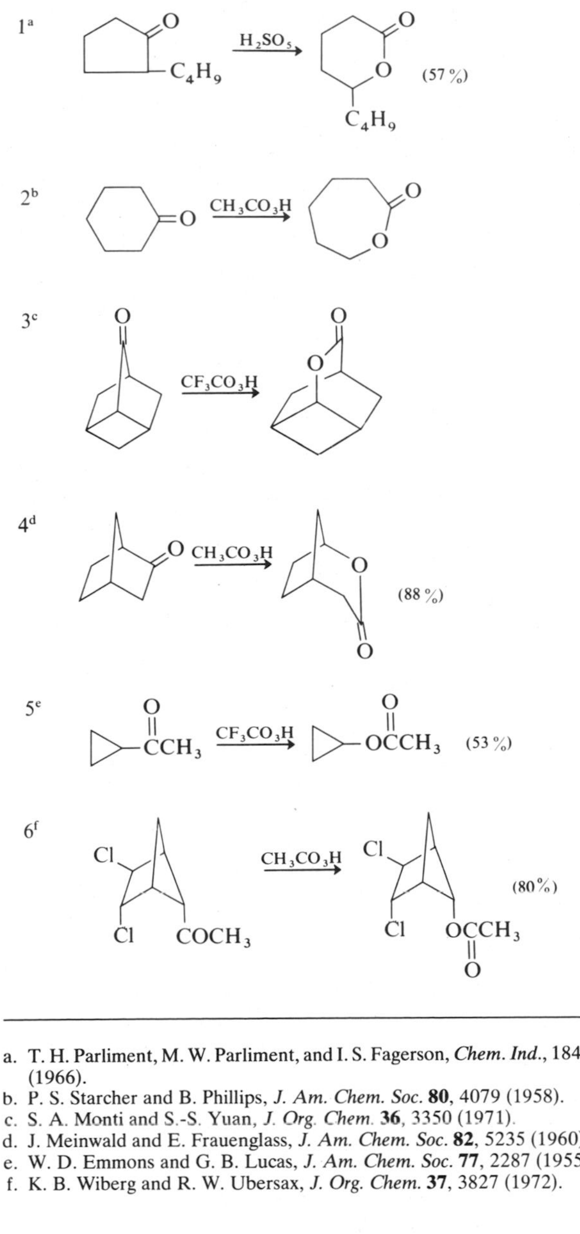

385

SECTION 9.5.
OXIDATIONS
OF KETONES
AND
ALDEHYDES

a. T. H. Parliment, M. W. Parliment, and I. S. Fagerson, *Chem. Ind.*, 1845
   (1966).
b. P. S. Starcher and B. Phillips, *J. Am. Chem. Soc.* **80**, 4079 (1958).
c. S. A. Monti and S.-S. Yuan, *J. Org. Chem.* **36**, 3350 (1971).
d. J. Meinwald and E. Frauenglass, *J. Am. Chem. Soc.* **82**, 5235 (1960).
e. W. D. Emmons and G. B. Lucas, *J. Am. Chem. Soc.* **77**, 2287 (1955).
f. K. B. Wiberg and R. W. Ubersax, *J. Org. Chem.* **37**, 3827 (1972).

when DMSO or some other substance capable of reducing the hydroperoxide is
present, the alcohol is isolated. A procedure that has met with success for steroids
involves oxidation in the presence of a trialkyl phosphite.[116] The intermediate
hydroperoxide is efficiently reduced by the phosphite. The mechanism of oxidation,

116. J. N. Gardner, F. E. Carlon, and O. Gnoj, *J. Org. Chem.* **33**, 3294 (1968).

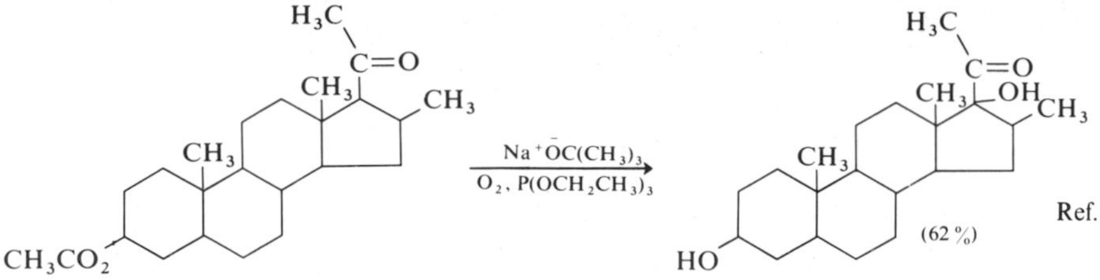

Ref. 116

(62%)

at least for some carbanions, has been shown to be initiated by electron transfer, and this mechanism may be generally applicable.[117] The subsequent steps are probably chain reactions similar to those encountered in other oxidations by molecular oxygen, which were discussed in Part A, Chapter 12. Hydrogen peroxide has been

$$R_3C:^- + O_2 \rightarrow R_3C\cdot + O_2^{\pm}$$

$$R_3C\cdot + O_2 \rightarrow R_3C-O-O\cdot$$

$$R_3C-O-O\cdot + R_3C:^- \rightarrow R_3C\cdot + R_3C-O-O^-$$

used as the oxidant in several cases.[118,119] In such instances, the product alcohol is formed directly.

$$R_3C^- + {}^-OOH \rightarrow R_3CO^- + {}^-OH$$

9.5.3. Oxidation with Other Reagents

Selenium dioxide has been used quite widely to effect oxidation of ketones and aldehydes to α-dicarbonyl compounds. The reaction often gives high yields of products when there is a single type of CH_2 group adjacent to the carbonyl group. In unsymmetrical ketones, oxidation usually occurs at the CH_2 that is most readily enolized.[120] The mechanism of the reaction is thought to involve formation of a

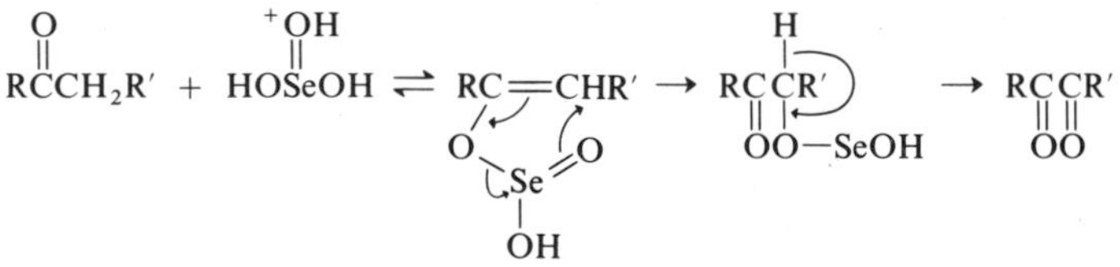

selenium ester of the enol.[121] Other aspects of the mechanism of selenium dioxide oxidations will be discussed further in subsection 9.6.3.

116. See p. 385.
117. G. A. Russell and A. G. Bemis, *J. Am. Chem. Soc.* **88**, 5491 (1966).
118. G. Büchi, K. E. Matsumoto, and H. Nishimura, *J. Am. Chem. Soc.* **93**, 3299 (1971).
119. R. Volkmann, S. Danishefsky, J. Eggler, and D. M. Solomon, *J. Am. Chem. Soc.* **93**, 5576 (1971).
120. E. N. Trachtenberg, in *Oxidation*, R. L. Augustine (ed.), Marcel Dekker, New York, NY, 1969.
121. E. J. Corey and J. P. Schaefer, *J. Am. Chem. Soc.* **82**, 918 (1960).

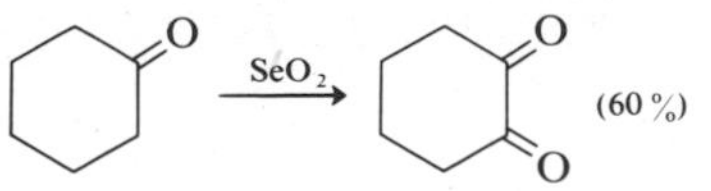

Ref. 122

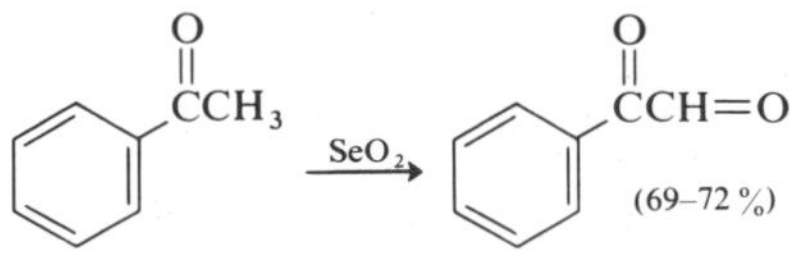

Ref. 123

Methyl ketones are degraded to the next lower carboxylic acid by reaction with hypochlorite or hypobromite ions. The initial step in these reactions involves base-catalyzed halogenation. The haloketones are more reactive than their precursors, and rapid halogenation to the trihalo compound results. Trihalomethyl ketones

$$(CH_3)_3CCCH_3 \xrightarrow[Br_2]{NaOH} (CH_3)_3CCO_2H \quad (71\text{–}74\%)$$

Ref. 124

$$(CH_3)_2C=CHCOCH_3 \xrightarrow{KOCl} \xrightarrow{H^+} (CH_3)_2C=CHCO_2H \quad (49\text{–}53\%)$$

Ref. 125

are susceptible to alkaline cleavage. This lability toward alkaline cleavage is the

$$RCCH_3 \underset{slow}{\rightleftharpoons} RC=CH_2 \xrightarrow{OBr^-} RCCH_2Br \xrightarrow{-OH} RC=CHBr \xrightarrow{fast} RCCBr_3$$

$$RCCBr_3 \rightleftharpoons RC-CBr_3 \rightarrow RCO_2H + {}^-CBr_3 \rightleftharpoons RCO_2^- + HCBr_3$$

result of the inductive stabilization provided by the halogen atoms.

9.6. Allylic Oxidation of Olefins

9.6.1. Transition-Metal Oxidants

Olefins are subject to oxidation both at the double bond and at the allylic position. The CrO_3–pyridine reagent in methylene chloride appears to be the most satisfactory reagent for allylic oxidation among the Cr(VI) reagents.[126] Several pieces of mechanistic information indicate that allylic radicals or cations are intermediates in these oxidations. Thus, ^{14}C in cyclohexene is distributed in the product cyclohexenone in a manner indicating that a symmetrical allylic intermediate is

122. C. C. Hach, C. V. Banks, and H. Diehl, *Org. Synth.* **IV**, 229 (1963).
123. H. A. Riley and A. R. Gray, *Org. Synth.* **II**, 509 (1943).
124. L. T. Sandborn and E. W. Bonsquet, *Org. Synth.* **I**, 512 (1932).
125. L. I. Smith, W. W. Prichard, and L. J. Spillane, *Org. Synth.* **III**, 302 (1955).
126. W. G. Dauben, M. Lorber, and D. S. Fullerton, *J. Org. Chem.* **34**, 3587 (1969).

involved at some stage.[127] In many allylic oxidations, the double bond is found in a

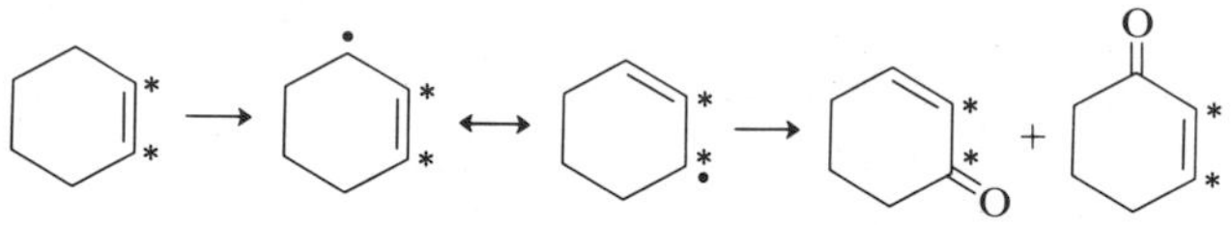

position indicating that an "allylic shift" occurs in the course of the oxidation. When

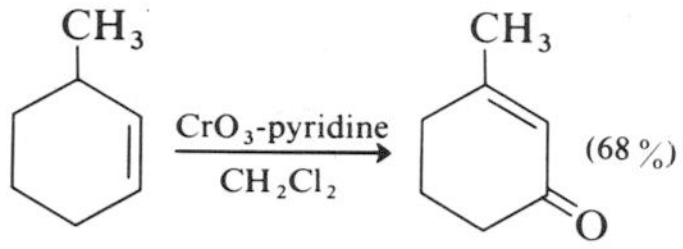

Ref. 126

more than one allylic methylene group is present in an alkene, a mixture of products usually results. Oxidation at allylic methyl groups appears to be much slower than at more substituted positions, and has been observed only rarely.

9.6.2. Oxygen, Ozone, and Peroxides

Olefins are attacked by the singlet excited state of oxygen to form hydroperoxides. This reaction always proceeds with allylic shift of the double bond. A concerted mechanism explains this shift, as well as the observation that the hydrogen that is abstracted is *cis* relative to the newly formed C–O bond.[128] There has been

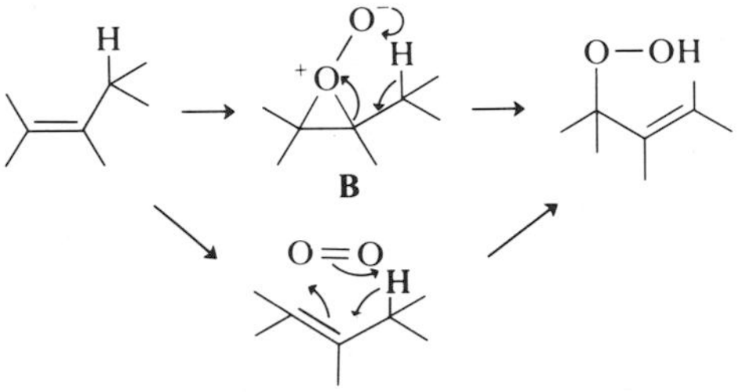

disagreement about whether there might be a very unstable intermediate involved in the reaction. Most attention has focused on the perepoxide intermediate **B**. It will be noted that this intermediate bears some similarity to the first intermediate proposed in the ozonolysis mechanism. The existence of such an intermediate is compatible with the stereochemistry and allylic shift that characterize the oxidation. The perepoxide intermediate would be expected to have a very short lifetime, and it has not been directly detected. Several types of solvent-trapping experiments have been interpreted in terms of the perepoxide intermediate.[129] None of the results finally

127. K. B. Wiberg and S. D. Nielsen, *J. Org. Chem.* **29**, 3353 (1964).
126. See p. 381.
128. A. Nickon and J. F. Bagli, *J. Am. Chem. Soc.* **83**, 1498 (1961).
129. A. P. Schaap and G. R. Faler, *J. Am. Chem. Soc.* **95**, 3381 (1973); N. M. Hasty and D. R. Kearns, *J. Am. Chem. Soc.* **95**, 3380 (1973); W. Fenical, D. R. Kearns, and P. Radlick, *J. Am. Chem. Soc.* **95**, 7888 (1973).

Scheme 9.13. Generation of Singlet Oxygen

1[a] Photosensitizer $+ h\nu \rightarrow$ ^{1}[Photosensitizer]*

^{1}[Photosensitizer]* $\rightarrow$ ^{3}[Photosensitizer]*

^{3}[Photosensitizer]* $+ \ ^3O_2 \rightarrow \ ^1O_2 +$ Photosensitizer

2[b] $H_2O_2 + \ ^-OCl \rightarrow \ ^1O_2 + H_2O + Cl^-$

3[c] $(RO)_3P + O_3 \rightarrow (RO)_3P\underset{O}{\overset{O}{<}}\Big|O \rightarrow (RO)_3P{=}O + \ ^1O_2$

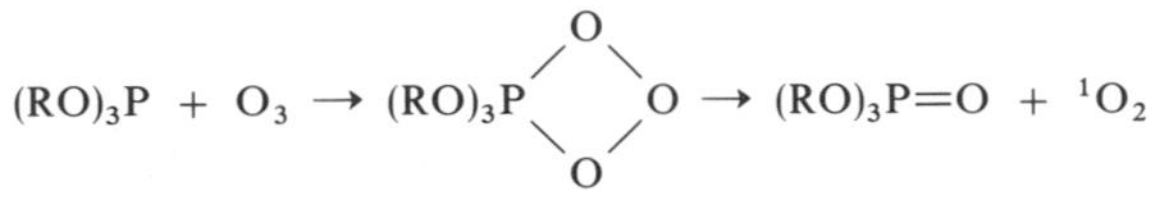

4[d]

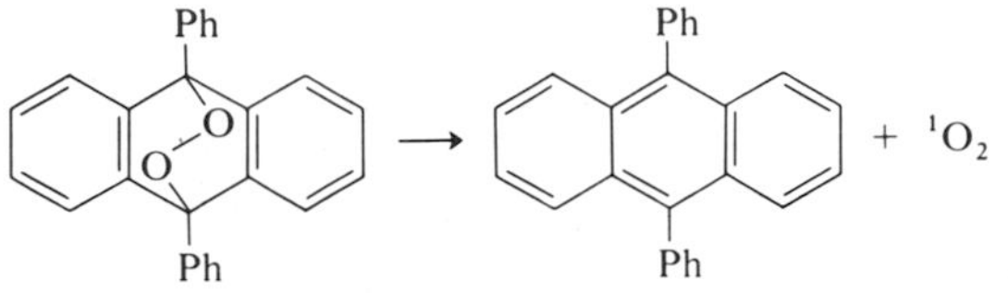

a. C. S. Foote and S. Wexler, *J. Am. Chem. Soc.* **86**, 3880 (1964).
b. C. S. Foote and S. Wexler, *J. Am. Chem. Soc.* **86**, 3879 (1964).
c. R. W. Murray and M. L. Kaplan, *J. Am. Chem. Soc.* **90**, 537 (1964).
d. H. H. Wasserman, J. R. Scheffer, and J. L. Cooper, *J. Am. Chem. Soc.* **94**, 4991 (1972).

exclude a one-step concerted mechanism, however, and it may be the mechanism that generally operates.[130]

There are several means of generating the oxidant.[131] From a preparative viewpoint, the most important are photosensitization of triplet ground-state oxygen to the excited singlet state, oxidation of hydrogen peroxide by hypochlorite, decomposition of 9,10-diphenylanthracene peroxide, and decomposition of the adducts formed at low temperature from phosphite esters and ozone. The close correspondence among product mixtures obtained with these reagents leads to the conclusion that the same chemical species is the active oxidant in each case. These methods of generating singlet oxygen are presented in Scheme 9.13.

The excited oxygen molecule will decay to the ground-state triplet if it does not encounter an olefin suitable for reaction. The rate of this decay process has been shown to depend strongly on the identity of the solvent.[132] Measured lifetimes range from roughly 700 μsec in carbon tetrachloride to 2 μsec in water. It is evident that the solvent can then have a pronounced effect on the efficiency of oxidation; the longer the excited-state lifetime, the more likely is a productive encounter with an alkene substrate.

The reactivity order of olefins is that expected for attack by a relatively electrophilic reagent. Reactivity increases with the number of alkyl substituents on

130. C. S. Foote, T. T. Fujimoto, and Y. C. Chang, *Tetrahedron Lett.*, 45 (1972).
131. D. R. Kearns, *Chem. Rev.* **71**, 395 (1971).
132. P. B. Merkel and D. R. Kearns, *J. Am. Chem. Soc.* **94**, 1029, 7244 (1972).

the olefin.[133,134] Terminal olefins are relatively inert and are usually not converted to product in significant amount. The reaction is also prevented when the olefinic group is severely sterically hindered.[128] Steric effects govern the direction of approach of the oxygen, and the dominant mode of attack is from the less hindered side of the molecule. Alkenes, such as norbornene, that cannot accommodate the concerted mechanism for stereoelectronic reasons are unreactive toward singlet oxygen.[135]

Certain olefins react with singlet oxygen in a different manner, giving the cyclic dioxetane adduct[136–138]:

$$\underset{R}{\overset{R}{>}}\!\!=\!\!\underset{R}{\overset{R}{<}} \;+\; {}^1O_2 \;\longrightarrow\; \begin{array}{c} O-O \\ R\!-\!\!\!\!\!\fbox{}\!\!\!\!\!-R \\ R \quad R \end{array}$$

This reaction is not usually a major factor with olefins bearing only alkyl substituents, but becomes important with vinyl ethers, for example.

Singlet oxygen undergoes $4+2$ cycloaddition reactions with dienes, generating peroxides:

Ref. 139

Ref. 140

Often, the hydroperoxides generated in the oxidation of alkenes are not the ultimately desired product. They can be reduced without extensive prior purification to the corresponding allylic alcohol. The sequence of singlet-oxygen oxidation followed by reduction constitutes a reasonably general method of synthesis of allylic alcohols from olefins *with allylic migration of the double bond*. Scheme 9.14 records some specific examples.

9.6.3. Other Oxidants

Selenium dioxide attacks allylic positions in preference to carbon–carbon double bonds and is a very useful reagent for allylic oxidation. A good deal of

133. K. R. Kopecky and H. J. Reich, *Can. J. Chem.* **43**, 2265 (1965).
134. C. S. Foote and R. W. Denny, *J. Am. Chem. Soc.* **93**, 5162 (1971).
128. See p.388.
135. F. A. Litt and A. Nickon, *Oxidation of Organic Compounds—III*, Advances in Chemistry Series, No. 77, American Chemical Society, Washington, DC, 1968, pp. 118–132.
136. W. Fenical, D. R. Kearns, and P. Radlick, *J. Am. Chem. Soc.* **91**, 3396 (1969).
137. S. Mazur and C. S. Foote, *J. Am. Chem. Soc.* **92**, 3225 (1970).
138. P. D. Bartlett and A. P. Schaap, *J. Am. Chem. Soc.* **92**, 3223 (1970).
139. C. S. Foote, S. Wexler, W. Ando, and R. Higgins, *J. Am. Chem. Soc.* **90**, 975 (1968).
140. C. H. Foster and G. A. Berchtold, *J. Am. Chem. Soc.* **94**, 7939 (1972).

Scheme 9.14. Oxidation of Olefins with Singlet Oxygen

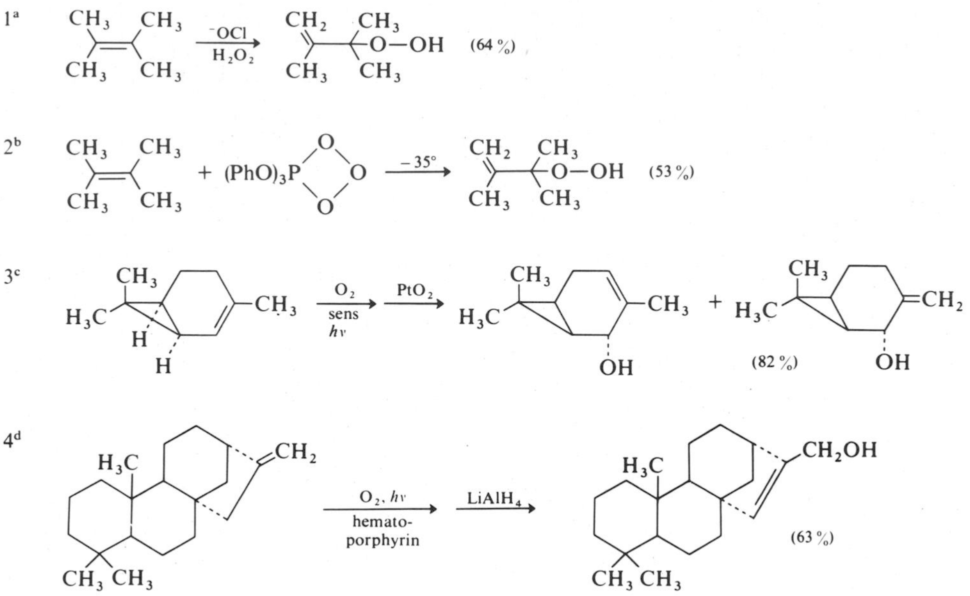

391

SECTION 9.6.
ALLYLIC
OXIDATION
OF OLEFINS

1[a]

(64 %)

2[b]

(53 %)

3[c]

(82 %)

4[d]

(63 %)

a. C. S. Foote, S. Wexler, W. Ando, and R. Higgins, *J. Am. Chem. Soc.* **90**, 975 (1968).
b. R. W. Murray and M. L. Kaplan, *J. Am. Chem. Soc.* **91**, 5358 (1968).
c. K. Gollnick and G. Schade, *Tetrahedron Lett.*, 2335 (1966).
d. R. A. Bell, R. E. Ireland, and L. N. Mander, *J. Org. Chem.* **31**, 2536 (1966).

information about the selectivity of this reagent toward various types of allylic sites is available and has been summarized.[141] The reaction mechanism cannot be considered to have been established in detail, but a general outline of the steps implicated is useful in discussing the types of products normally encountered in selenium dioxide oxidation. The steps below constitute one of the simplest sequences of

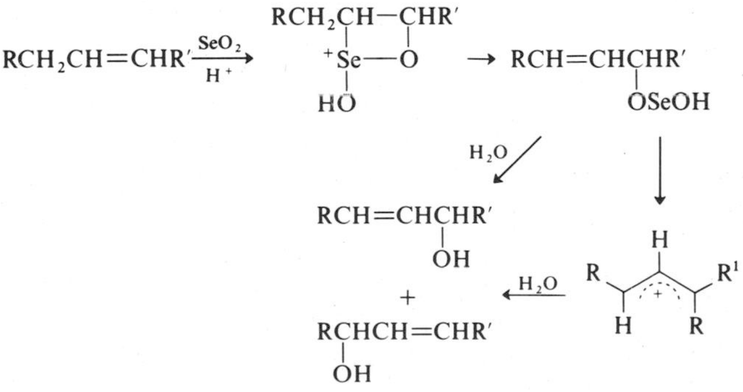

141. E. N. Trachtenberg, in *Oxidation*, Vol. 1, R. L. Augustine (ed.), Marcel Dekker, New York, NY, 1969, pp. 119–187.

reactions that can rationalize the reactivity patterns.[142,143] The allyl selenium esters may also arise via other pathways than that shown.[144,145] The alcohols that are the initial oxidation products are susceptible to further oxidation by SeO_2 to the corresponding carbonyl compound; this further oxidation occurs under the normal reaction conditions, so it is usually the carbonyl compounds that are isolated. Alcohols can be isolated only at the expense of low conversion of the olefin. If the alcohol is the desired product, the oxidation can be run in acetic acid as solvent. This procedure affords acetate esters, which are inert to SeO_2 oxidation.

In terms of selectivity among several allylic positions in a molecule, the order of reactivity is $CH_2 > CH_3 > CH$ in the case of trisubstituted alkenes, but becomes $CH > CH_2 > CH_3$ for disubstituted compounds. The proposed mechanism suggests that when the allylic intermediate is unsymmetrical, a mixture of reaction products will be found. Such a mixture has been observed in many instances. With terminal olefins, products resulting from double-bond migration are often found.

$$CH_2=CH(CH_2)_4\overset{O}{\overset{\|}{O}}CCH_3 \xrightarrow[CH_3CO_2H]{SeO_2} CH_3\overset{O}{\overset{\|}{C}}OCH_2CH=CH(CH_2)_3\overset{O}{\overset{\|}{O}}CCH_3 \quad (27\%)$$

Ref. 146

$$+ \; H_2C=CH\underset{\underset{O}{\overset{\|}{O}}CCH_3}{\overset{|}{C}H}(CH_2)_3O\overset{O}{\overset{\|}{C}}CH_3 \quad (17\%)$$

Another fairly general observation that can be reconciled with the intervention of allylic intermediates is that dienes are usually obtained rather than the alcohol, if the latter would be tertiary. As illustrated by the last entry in Scheme 9.15, the reaction reveals high stereoselectivity when applied to trisubstituted *gem*-dimethyl olefins.[144] In other cases where the stereochemistry of selenium dioxide oxidations has been examined,[143,144] mixtures of the possible stereoisomers are found.

Phenylselenyl bromide is an alternative reagent for allylic oxidation of alkenes.[147] Addition takes place, and reaction of the adduct with acetic acid, followed by oxidation, generates the allylic acetate.

$$CH_3(CH_2)_2CH=CH(CH_2)_2CH_3 \xrightarrow[\substack{3) \; H_2O_2}]{\substack{1) \; PhSeBr \\ 2) \; CH_3CO_2H}} CH_3CH_2CH=CHCH\underset{\underset{O}{\overset{\|}{O}}CCH_3}{\overset{|}{C}}H_2CH_2CH_3 \quad (85\%)$$

Several examples of selenium dioxide oxidations are given in Scheme 9.15.

142. J. P. Schaefer, B. Horvath, and H. P. Klein, *J. Org. Chem.* **33**, 2647 (1968).
143. E. N. Trachtenberg, C. H. Nelson, and J. R. Carver, *J. Org. Chem.* **35**, 1653 (1970).
144. U. T. Bhalerao and H. Rapoport, *J. Am. Chem. Soc.* **93**, 4835 (1971).
145. K. B. Sharpless and R. F. Lauer, *J. Am. Chem. Soc.* **94**, 7154 (1972).
146. J. Colonge and M. Reymermier, *Bull. Soc. Chim. Fr.*, 1531 (1955).
147. K. B. Sharpless and R. F. Lauer, *J. Org. Chem.* **39**, 429 (1974).

Scheme 9.15. Selenium Dioxide Oxidations

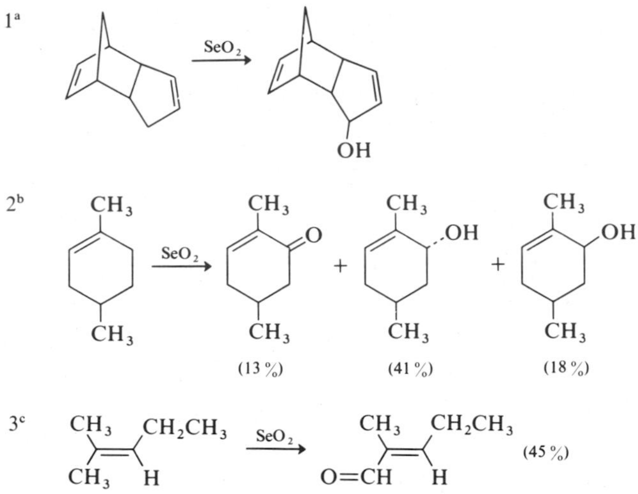

393

SECTION 9.7.
OXIDATIONS
AT UNFUNCTION-
ALIZED CARBON
ATOMS

1[a]

2[b]

(13 %) (41 %) (18 %)

3[c]

(45 %)

a. R. B. Woodward and T. J. Katz, *Tetrahedron* **5**, 70 (1959).
b. E. N. Trachtenberg and J. R. Carver, *J. Org. Chem.* **35**, 1646 (1970).
c. U. T. Bhalerao and H. Rapoport, *J. Am. Chem. Soc.* **93**, 4835 (1971).

9.7. Oxidations at Unfunctionalized Carbon Atoms

Attempts to achieve selective oxidations of hydrocarbons or other compounds when the desired site of attack is remote from an activating functional group are faced with difficulties. With the powerful transition-metal oxidants, the initial oxidation products are almost always more susceptible to oxidation than the starting material. Once a hydrocarbon is attacked, it is likely to be oxidized to a carboxylic acid, with chain cleavage by successive rapid oxidation of alcohol and carbonyl intermediates. There are a few special circumstances that make oxidations of hydrocarbons synthetically useful processes. A large group involve catalytic industrial processes. Much work has been expended on the development of such systems, and several have attained economic importance. Since the mechanisms are often obscured by limited understanding of heterogeneous catalysis, however, we will not devote additional attention to these reactions.

Perhaps the most familiar and useful hydrocarbon oxidation is the oxidation of aromatic side chains. Two factors enter into making this a high-yield procedure despite the use of powerful oxidants: First, the benzylic site is activated to oxidation. Either radical or carbonium intermediates can be especially easily formed here because of the potential resonance stabilization. Second, the aromatic ring is inert to attack by the Mn(VII) and Cr(VI) oxidants that attack the alkyl side chain.

It has been difficult to formulate detailed reaction mechanisms for these oxidations, since several metal oxidation states are undoubtedly involved during the course of the reaction. In the case of permanganate, it is considered likely that the initial attack involves hydrogen-atom abstraction, probably followed by collapse to an ester of Mn(V).[148]

$$Ar-\overset{\overset{\displaystyle R}{|}}{\underset{\underset{\displaystyle R}{|}}{C}}-H \ + \ MnO_4^- \ \rightarrow \ \left[Ar-\overset{\overset{\displaystyle R}{\cdot}}{\underset{\displaystyle R}{C}} \ + \ MnO_4H^- \right]$$

$$\downarrow$$

$$\text{further reactions} \ \leftarrow \ Ar-\overset{\overset{\displaystyle R}{|}}{\underset{\underset{\displaystyle R}{|}}{C}}-O-\overset{\overset{\displaystyle O}{||}}{\underset{\underset{\displaystyle O}{||}}{Mn}}-OH$$

Several examples of oxidation of aromatic side chains are given in Scheme 9.16. Entries 3 and 5 are cases where the oxidation is terminated at stages short of the carboxylic acid oxidation state.

A second class of hydrocarbon substrates in which circumstances permit selective oxidations are the bicyclic hydrocarbons.[149] Here, the bridgehead position is the preferred site of initial attack because of the order of reactivity of C–H bonds, which is $3° > 2° > 1°$. The initial oxidation products, however, are not further oxidized with ease. The geometry of the bicyclic rings (Bredt's rule) prevents dehydration of the alcohol. The tertiary bridgehead hydroxyls, of course, cannot be converted to ketones; oxidation that begins at a bridgehead position therefore stops at the alcohol stage. Since the competing methylene groups tend to be less reactive than the bridgehead positions, selective oxidation is possible. Chromic acid oxidation has been the most useful reagent for functionalizing unstrained bicyclic hydrocarbons. The reaction fails for strained bicyclics such as norbornane because the reactivity of the bridgehead position is lowered by the unfavorable energy of either radical or carbonium-ion intermediates.

Other successful selective oxidations of hydrocarbons by Cr(VI) reagents have been reported—for example, the oxidation of *cis*-decalin to the corresponding alcohol—but careful attention to reaction conditions is required, and even then there are few hydrocarbons that have been reported to give high yields of oxidation products.

Ref. 150

148. R. Stewart, in *Oxidation in Organic Chemistry*, K. B. Wiberg (ed.), Academic Press, New York, NY, 1965, pp. 36–41; J. I. Brauman and A. J. Pandell, *J. Am. Chem. Soc.* **92**, 329 (1970).
149. R. C. Bingham and P. v. R. Schleyer, *J. Org. Chem.* **36**, 1198 (1971).
150. K. B. Wiberg and G. Foster, *J. Am. Chem. Soc.* **83**, 423 (1961).

Scheme 9.16. Side-Chain Oxidation of Aromatic Compounds

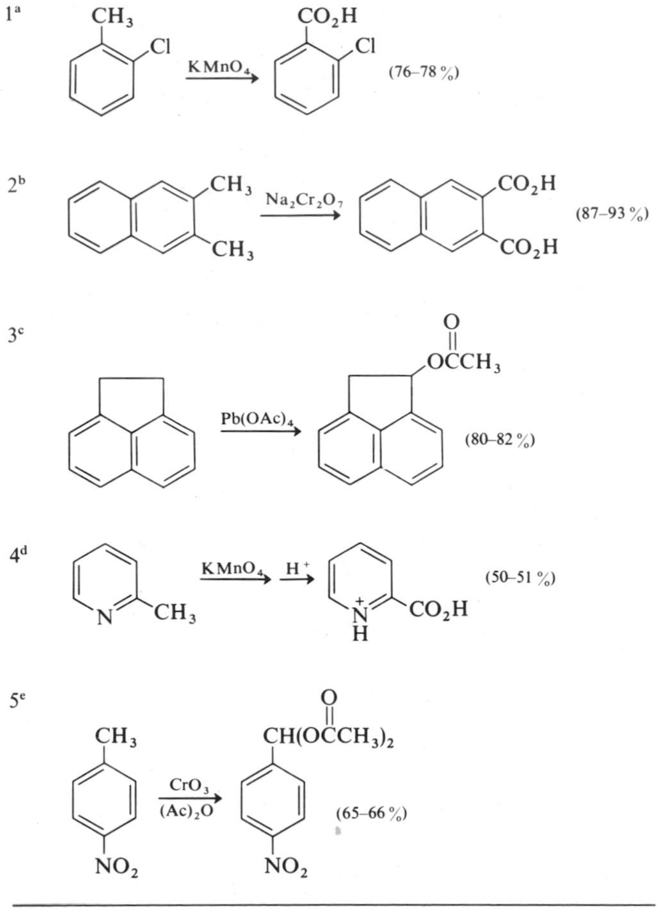

395

SECTION 9.7.
OXIDATIONS
AT UNFUNCTION-
ALIZED CARBON
ATOMS

1[a]

CH₃ → CO₂H with Cl substituent

KMnO₄

(76–78%)

2[b]

Na₂Cr₂O₇

(87–93%)

3[c]

Pb(OAc)₄

(80–82%)

4[d]

KMnO₄, H⁺

(50–51%)

5[e]

CrO₃ / (Ac)₂O

(65–66%)

---

a. H. T. Clarke and E. R. Taylor, *Org. Synth.* **II**, 135 (1943).
b. L. Friedman, *Org. Synth.* **43**, 80 (1963); L. Friedman, D. L. Fishel, and H. Shechter, *J. Org. Chem.* **30**, 1453 (1965).
c. J. Cason, *Org. Synth.* **III**, 3 (1955).
d. A. W. Singer and S. M. McElvain, *Org. Synth.* **III**, 740 (1955).
e. T. Nishimura, *Org. Synth.* **IV**, 713 (1963).

Lead tetraacetate effects oxidations at unfunctionalized C–H bonds, but this process intimately involves a functional group (hydroxyl) elsewhere in the molecule.[151] This reaction is believed to involve an alkoxylead intermediate.[152] The

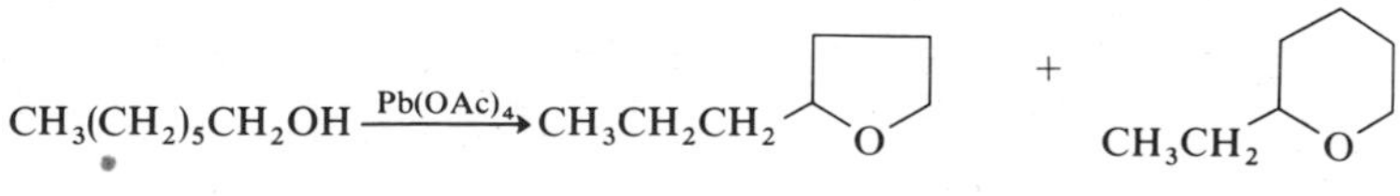

CH₃(CH₂)₅CH₂OH —Pb(OAc)₄→ CH₃CH₂CH₂— (tetrahydrofuran ring) + CH₃CH₂— (tetrahydropyran ring)

151. V. M. Mićović, R. I. Mamuzić, D. Jeremić, and M. L. Mihailović, *Tetrahedron* **20**, 2279 (1964).
152. K. Heusler and J. Kalvoda, *Angew. Chem.* **76**, 518 (1964).

crucial step for functionalization of the saturated chain involves an intramolecular hydrogen-atom abstraction by an alkoxy radical:

$$RCH_2(CH_2)_3OH \xrightarrow{Pb(OAc)_4} RCH_2(CH_2)_3OPb(OAc)_3 \rightarrow RCH_2(CH_2)_3O\cdot + Pb(OAc)_3$$

$$RCH_2(CH_2)_3O\cdot \rightarrow R\overset{\cdot}{C}H(CH_2)_3OH \xrightarrow{Pb(OAc)_4} R\overset{+}{C}H(CH_2)_3OH + Pb(OAc)_3 + {}^-OAc$$

$$R\overset{+}{C}H(CH_2)_3OH \rightarrow \quad + H^+$$

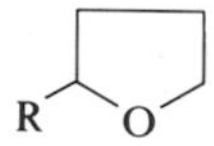

The preferred transition state for this type of abstraction is six-centered, so the major products are tetrahydrofuran derivatives. Smaller amounts of six-membered ring ethers are also often encountered.

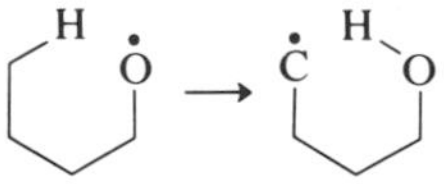

Section 12.5. in Part A describes some additional reactions that are of synthetic value and involve intramolecular hydrogen abstraction at unactivated groups. Of particular value is the nitrite photolysis method developed by Barton, which has been important in the functionalization of methyl groups in steroids.[153]

General References

K. B. Wiberg (ed.), *Oxidation in Organic Chemistry*, Part A, Academic Press, New York, NY, 1965.
W. S. Trahanovsky (ed.), *Oxidation in Organic Chemistry*, Part B, Academic Press, New York, NY, 1973.
R. L. Augustine (ed.), *Oxidation*, Vol. 1, Marcel Dekker, New York, NY, 1969.
R. L. Augustine and D. J. Trecker (eds.), *Oxidation*, Vol. 2, Marcel Dekker, New York, NY, 1971.
L. J. Chinn, *Selection of Oxidants in Synthesis*, Marcel Dekker, New York, NY, 1971.

Problems

(*References for these problems will be found on page 504.*)

1. Give the products expected for each reaction:

(a)

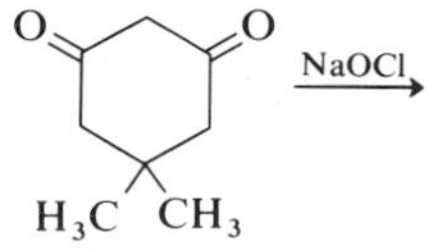

(b)

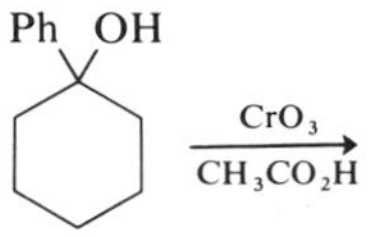

153. D. H. R. Barton, J. Beaton, L. E. Geller, and M. M. Pechet, *J. Am. Chem. Soc.* **83**, 4076 (1961).

(c)

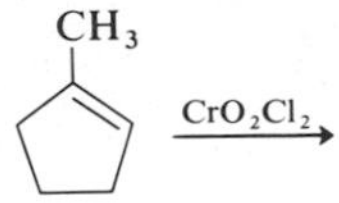

(d)

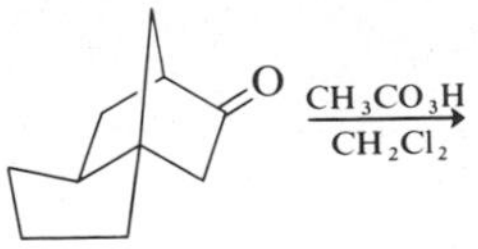

(e)

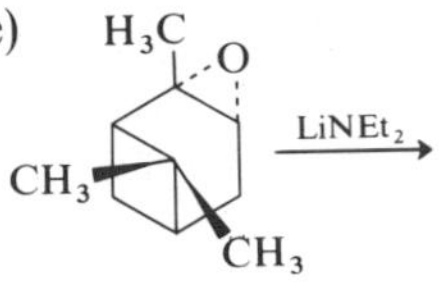

(f)

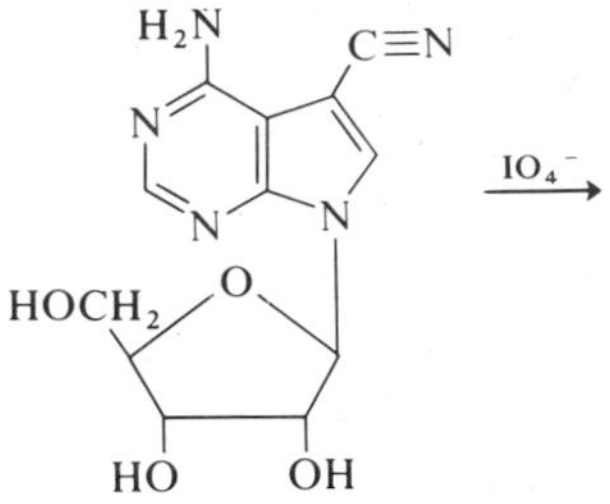

2. When styrenes of the general structure **A** are subjected to photooxidation, followed by reduction, the product is a mixture of alcohols **B** and **C**. The **B** : **C** ratio is almost invariant at 2.6 : 1 for nine substituted styrenes, with X ranging from *p*-methoxy to *p*-cyano in electron-donating capacity. What conclusions about the transition state in photooxidation do you draw from this result?

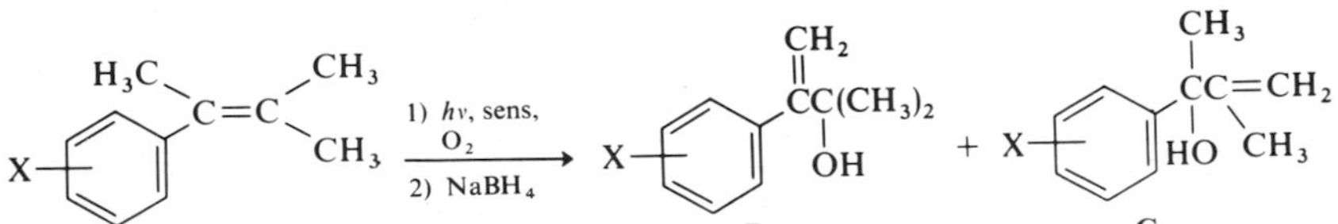

3. 4-Methylcyclohexene shows little stereoselectivity in epoxidation, giving a 54 : 46 ratio favoring the *trans*-epoxide. *cis*-4,5-Dimethylcyclohexene, on the other hand, gives predominantly (87 : 13) *trans,trans*-4,5-dimethylcyclohexene oxide. Explain.

4. List oxidizing agents that could be successfully employed for each of the following oxidations:

(a)

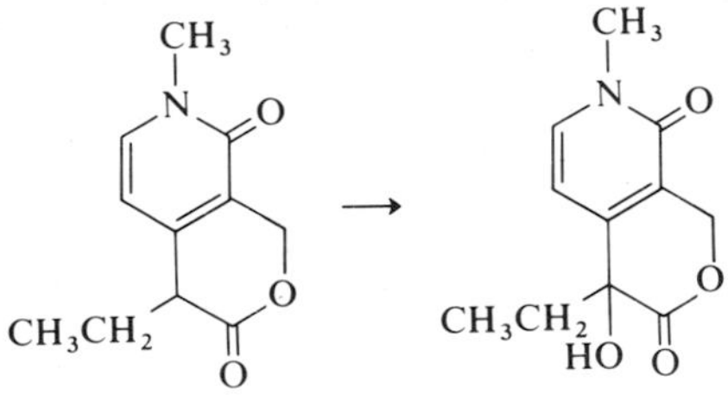

(b)

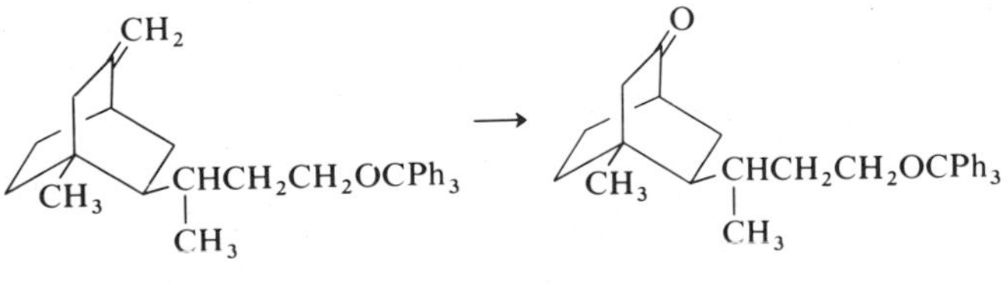

(c)

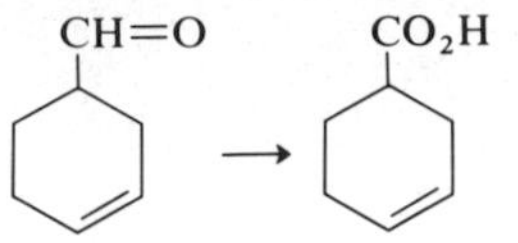

(d)

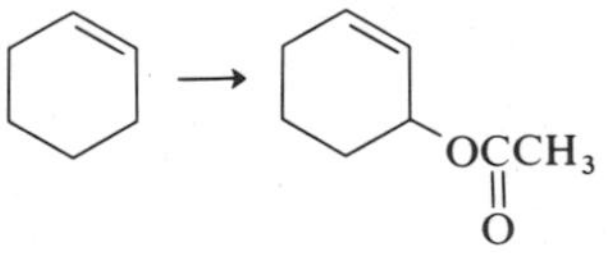

(e)

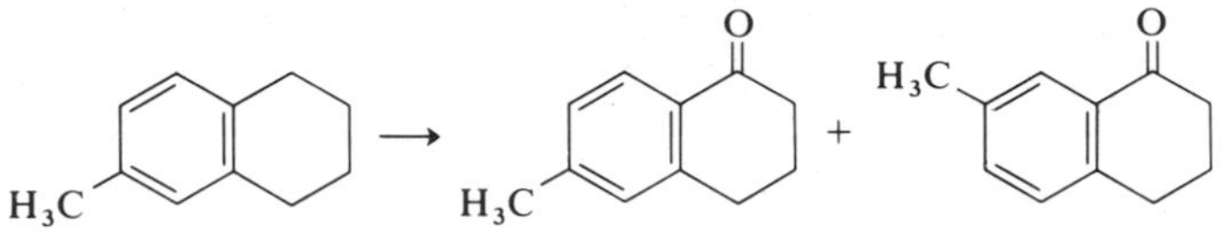

5. Suggest reasonable mechanisms for the following oxidations, which are effected by Tl(III) salts:

(a)

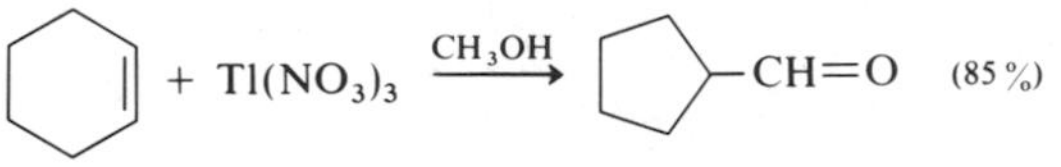

(b)

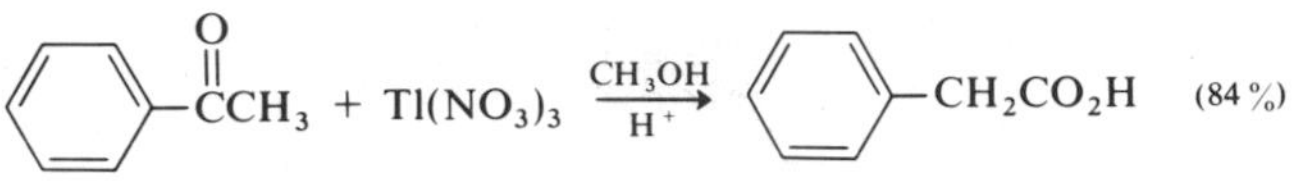

(c)

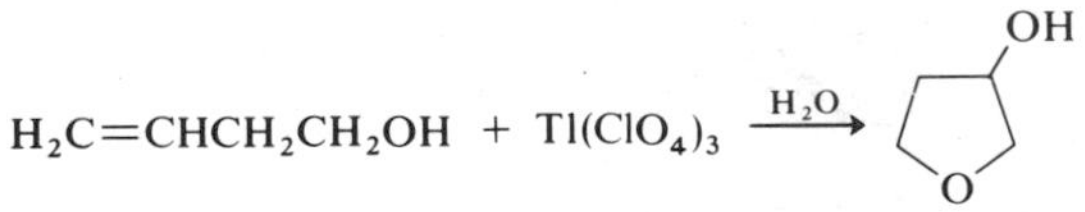

$$H_2C=CHCH_2CH_2OH + Tl(ClO_4)_3 \xrightarrow{H_2O}$$

6. Predict the product of the following oxidation reactions. Be sure to specify the stereochemistry when stereoisomers are possible.

(a)

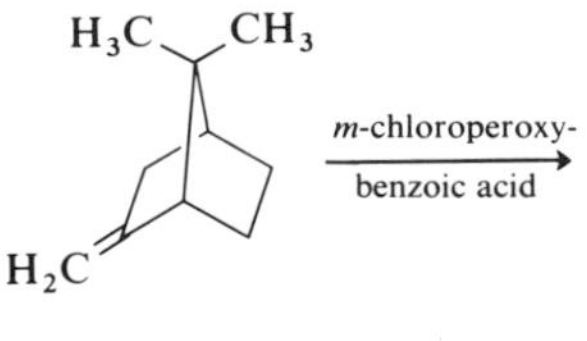

(b)

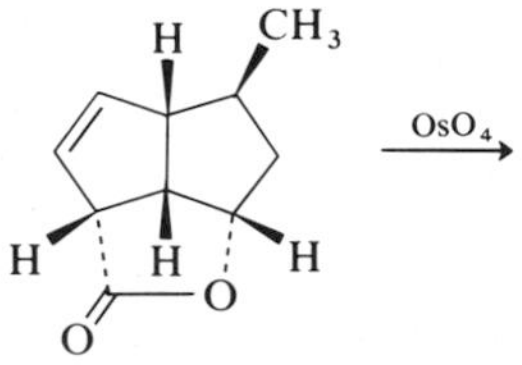

(c)

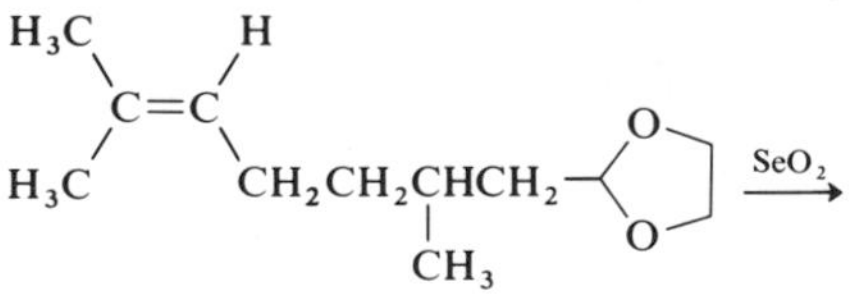

(d)

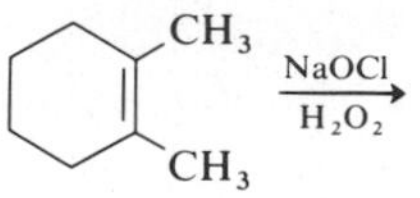

CH_3

CH_3

$\xrightarrow[\text{H}_2\text{O}_2]{\text{NaOCl}}$

(e)

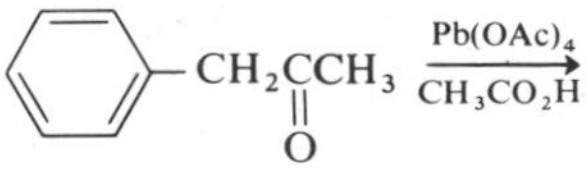

$-CH_2CCH_3$ $\xrightarrow[\text{CH}_3\text{CO}_2\text{H}]{\text{Pb(OAc)}_4}$

O

(f) 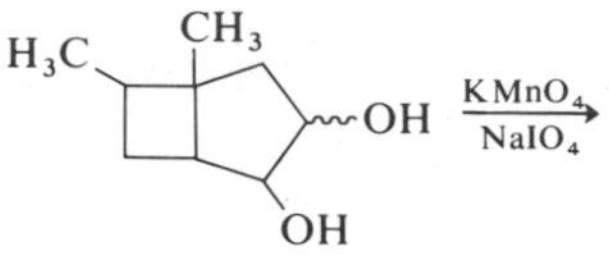 H_3C CH_3

$\sim\text{OH}$ $\xrightarrow[\text{NaIO}_4]{\text{KMnO}_4}$

OH

7. Each of the following molecules would require careful control of oxidation conditions to prevent changes at other points in the molecule. Suggest the best reagent for achieving the desired selectivity in each case. Explain the basis of your selection.

(a)

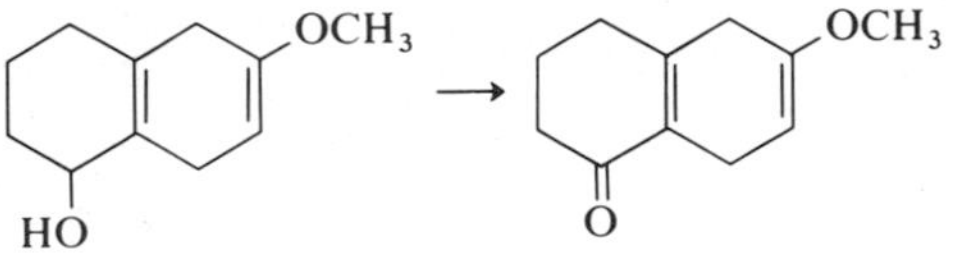

OCH_3 $\rightarrow$ OCH_3

HO O

(b)

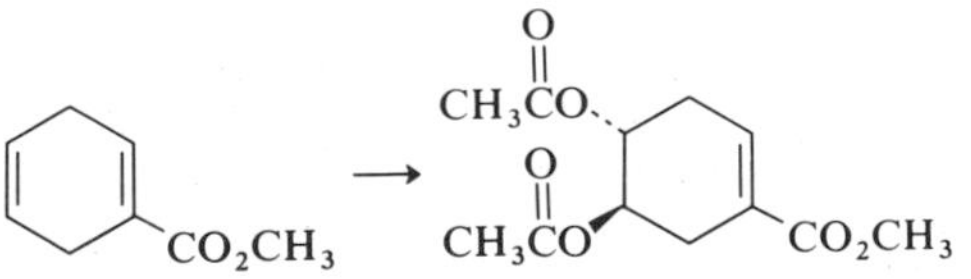

CO_2CH_3 $\rightarrow$

CH_3CO O

CH_3CO CO_2CH_3

(c)

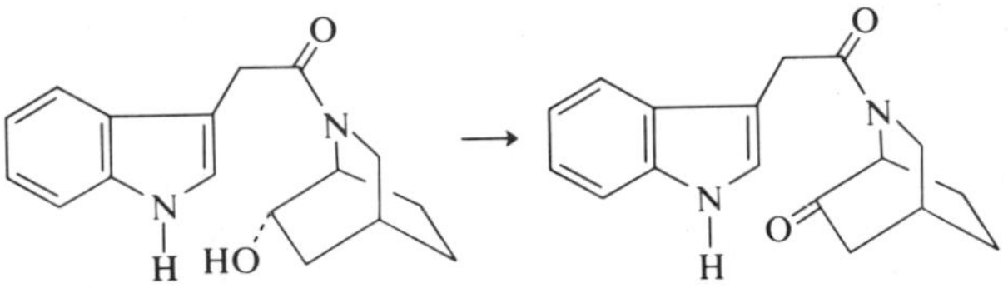

$\rightarrow$

N HO H

N O H

(d)

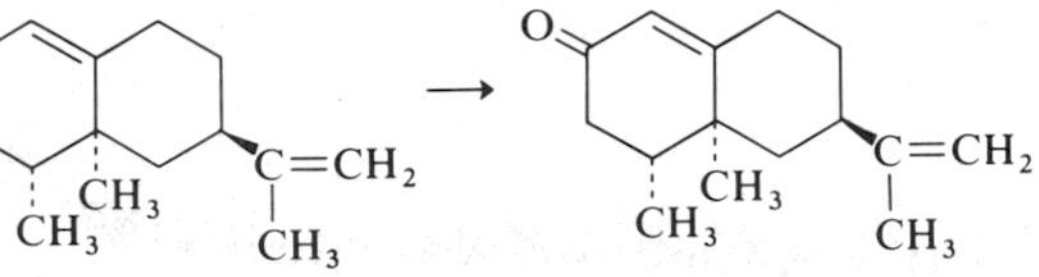

$\rightarrow$

CH_3 $C=CH_2$

CH_3 CH_3

CH_3 $C=CH_2$

CH_3 CH_3

(e)

H O

H $CH=O$ $\rightarrow$ H O

H CO_2H

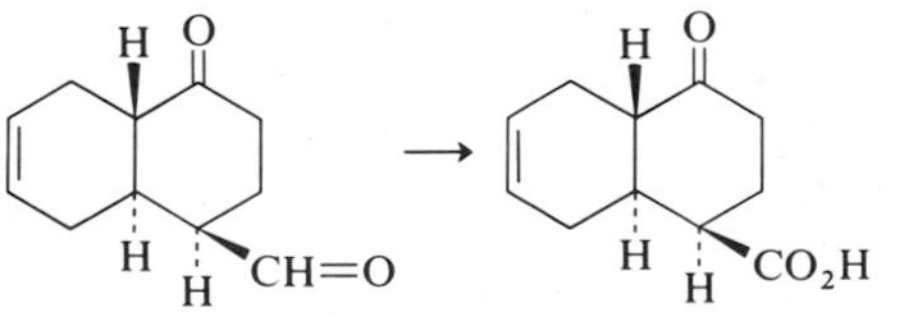

8. In chromic acid oxidation of isomeric cyclohexanols it is usually found that axial hydroxyl groups react more rapidly than equatorial groups. For example, *trans*-4-*t*-butylcyclohexanol is less reactive (by a factor of 3.2) than the *cis* isomer. An even larger difference is noted with *cis*- and *trans*-3,3,5-trimethylcyclohexanol.

The *trans* alcohol is more than 35 times more reactive than the *cis*. Are these data compatible with the mechanism given on page 352? What additional detail do these data provide about the reaction mechanism? Explain.

9. Formulate a mechanism to account for each of the following reactions:

(a)

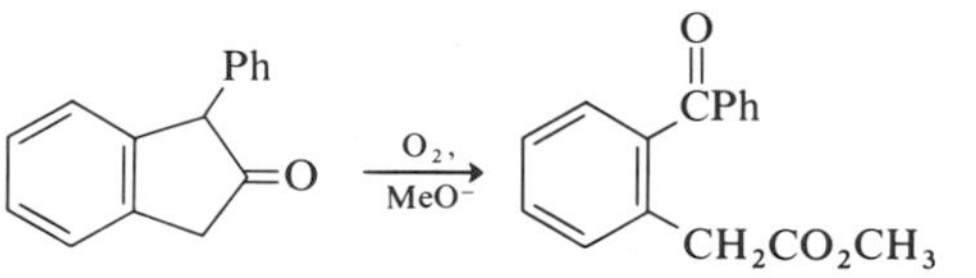

(b)

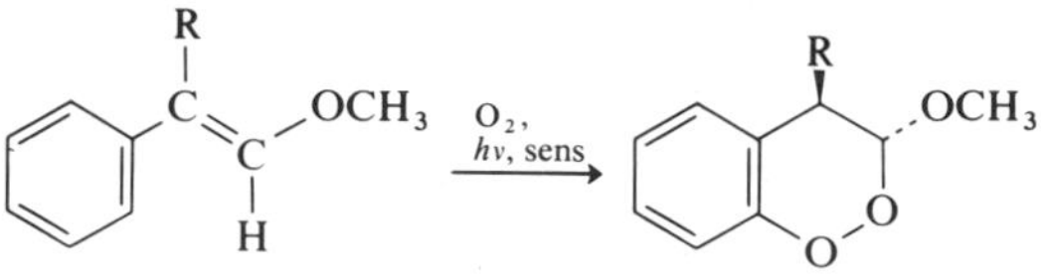

(c)

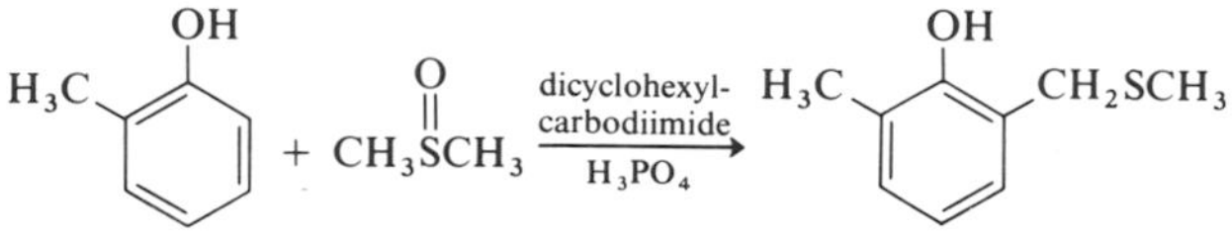

(None of the *para* isomer is formed.)

(d)

(e)

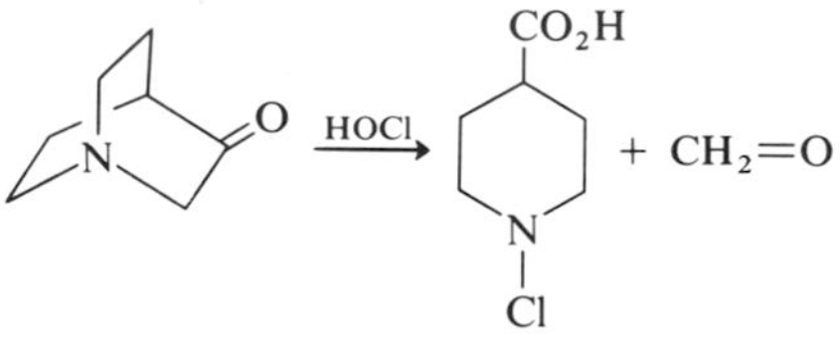

10. A direct method for conversion of aldehydes to esters has been reported. It is applicable to α,β-unsaturated aldehydes and aromatic aldehydes. The process involves stirring the aldehyde for several hours with sodium cyanide and manganese dioxide in a solution of methanol containing some acetic acid. The product is the methyl ester of the α,β-unsaturated or aromatic carboxylic acid. Indicate the basis for the success of this procedure.

11. Besides Pb(OAc)$_4$ and periodate, another reagent that brings about cleavage of glycols is ceric(IV) ammonium nitrate (CAN). A number of parallel studies with Pb(OAc)$_4$ and CAN have been carried out in acetic acid. Examine the structure–reactivity data for the two oxidants given below. What conclusions about the similarities and differences of the mechanisms of oxidation do you find on examining the data? Can you suggest a mechanism for the CAN oxidation?

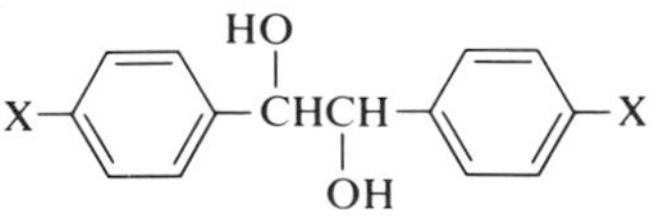

X	Relative rate	
	Pb(OAc)$_4$	CAN
p-CH$_3$	2.6	1.0
m-CH$_3$	1.8	1.8
H (meso)	1.0	1.0
H (dl)	8.3	0.9
p-Cl	0.40	1.8
p-Br	0.45	1.1
m-Cl	0.24	1.4
p-NO$_2$	0.062	1.2
H (monomethyl ether of diol)	0.17	0.05

12. Suggest methods for effecting the following synthetic transformations. Several steps may be required, but in each case at least one oxidation is necessary.

(a)

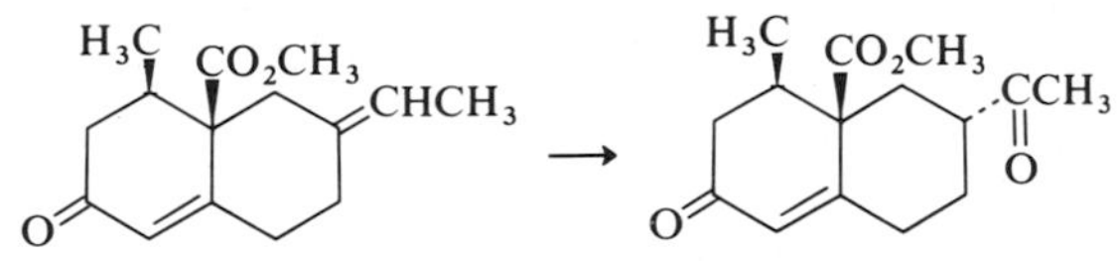

(b)

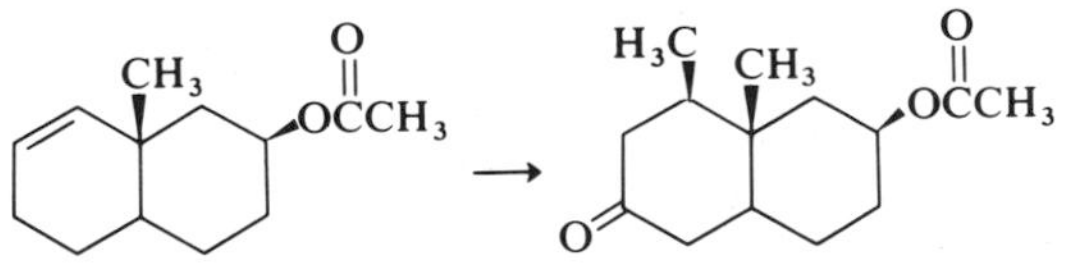

(c)

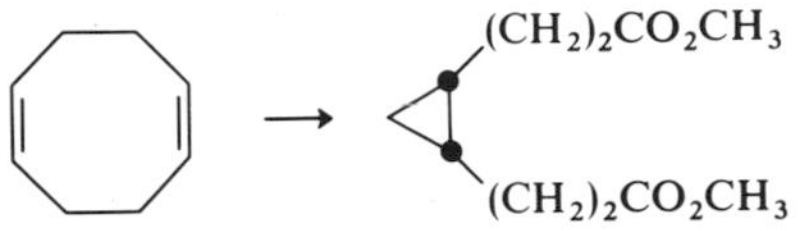

(d)

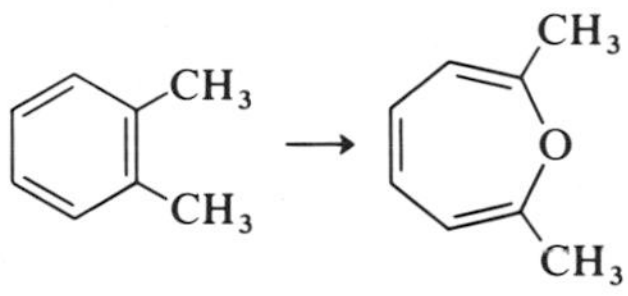

(e)

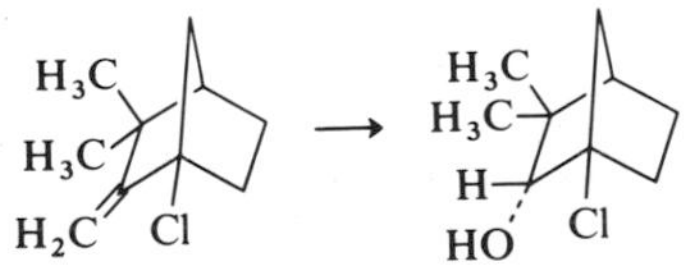

13. Several multistage syntheses in which oxidations are required are outlined below. Give reagents or reaction sequences that could effect each of the lettered transformations. Up to four steps may be required.

(a)

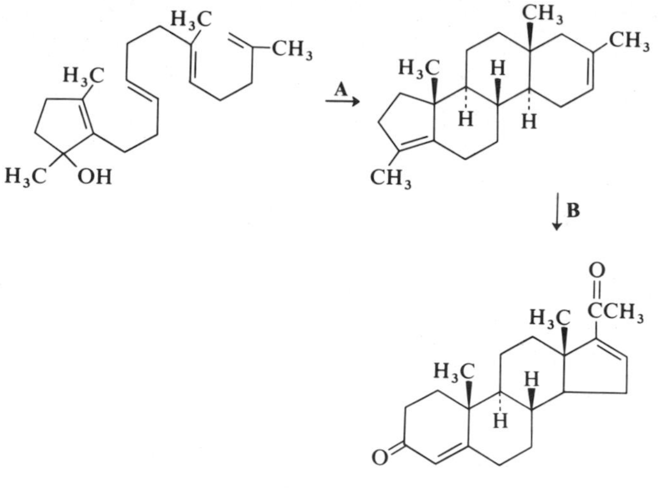

(b)

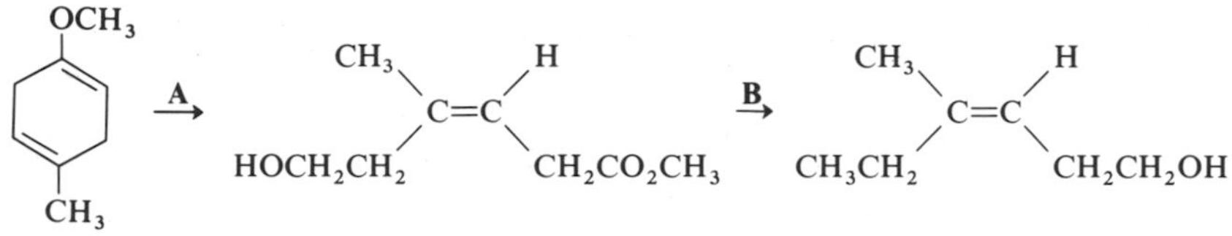

(c)

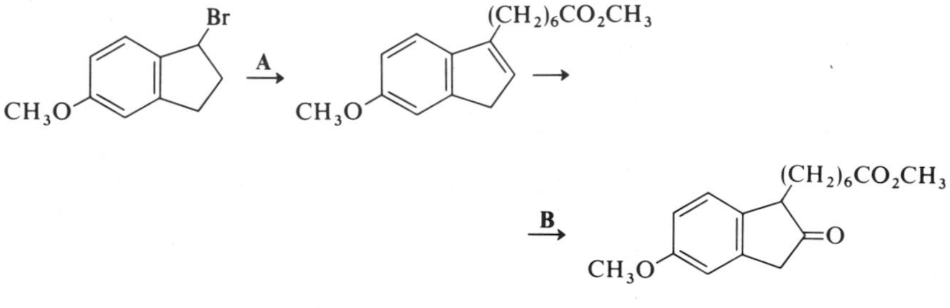

(d)

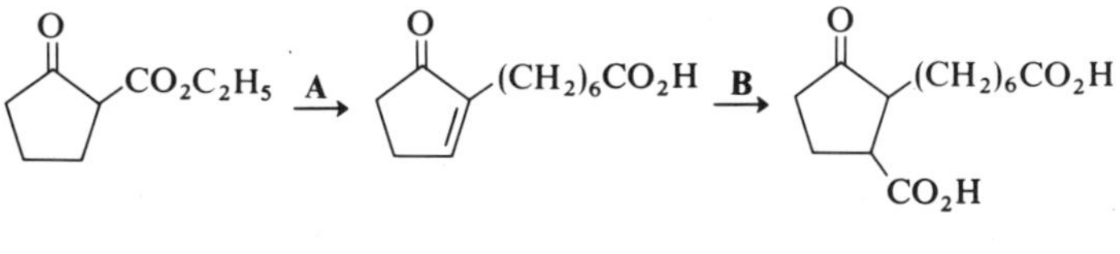

(e)

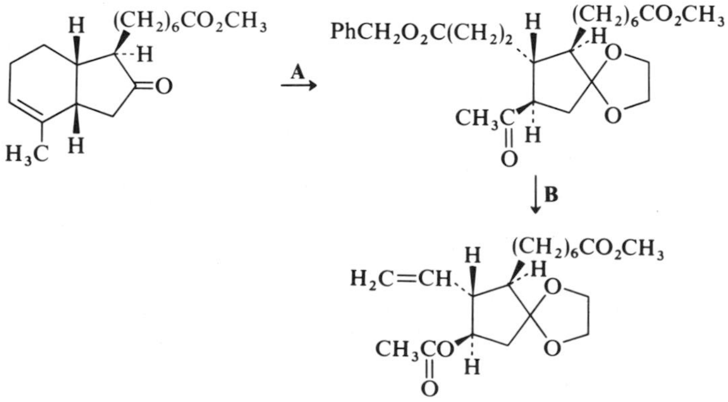

(f)

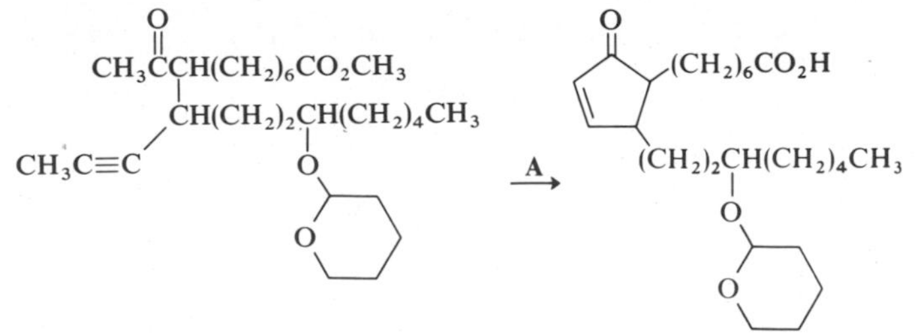

(g)

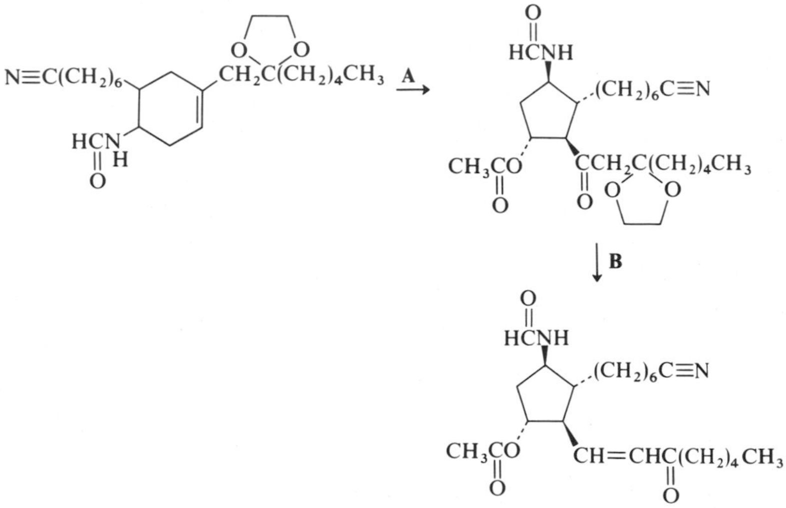

14. The singlet-oxygen oxidation of the optically pure alkene **A** has been studied. The major product is the alcohol **B** (after reduction of the hydroperoxide intermediate). The product of R configuration contains hydrogen and no deuterium at C-4. The product of S configuration contains only deuterium at C-4. The two products are formed in equal amounts. Are these results in better accord with the "concerted" or with the "perepoxide" mechanism for singlet-oxygen oxidation?

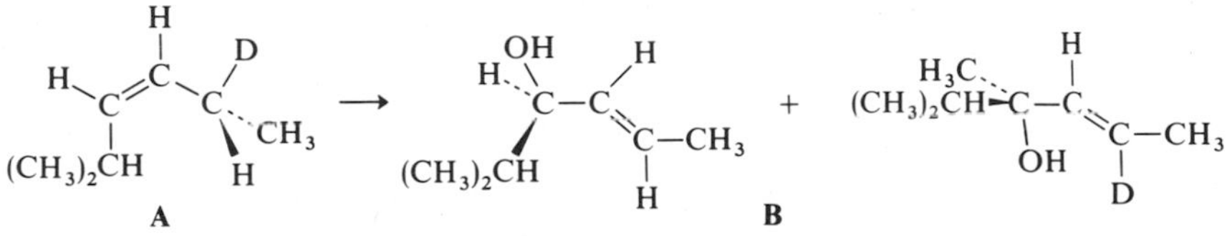

15. Low-temperature photooxidation of 2,5-dimethylhexa-2,4-diene gives an unstable adduct that decomposes at slightly above room temperature to acetone and 4-methylpent-2-ene. Suggest a structure for this product. Is this type of product formed from dienes generally? If not, why might the present compound be unusual?

16. A method for introducing carbon–carbon double bonds α,β to a carbonyl group has been developed. The ketone, aldehyde, or ester is first converted to an

α-phenylseleno derivative, usually by reaction with PhSeCl. The seleno derivative is then treated with hydrogen peroxide or periodate:

$$RCH_2CH_2\overset{\overset{O}{\|}}{C}R' \rightarrow PhCH_2\underset{\underset{Se}{|}}{\overset{\overset{O}{\|}}{C}}HCR' \overset{[O]}{\rightarrow} PhCH{=}CH\overset{\overset{O}{\|}}{C}R'$$

Ph

Suggest in general outline a mechanism that might explain this reaction.

17. A method for synthesis of ozonides that involves no ozone has been reported. It
consists of photosensitized oxidation of solutions of diazo compounds and
aldehydes. Suggest a mechanism.

$$Ph_2CN_2 + PhCH{=}O \xrightarrow[hv]{O_2,\ sens} Ph_2C\underset{O-O}{\overset{O}{\diagup\diagdown}}CHPh$$

18. Overoxidation of carbonyl products during ozonolysis can be prevented by
addition of tetracyanoethylene to the reaction mixture. The stoichiometry of the
reaction is then:

$$R_2C{=}CR_2 + (N{\equiv}C)_2C{=}C(C{\equiv}N)_2 + O_3 \rightarrow 2\,R_2C{=}O + \underset{NC}{\overset{NC}{\diagdown}}C\underset{}{\overset{\overset{O}{\diagup\diagdown}}{-}}C\underset{CN}{\overset{}{\diagup}}$$

Propose a reasonable mechanism that would account for the effect of tetracyanoethylene. Does your mechanism suggest that tetracyanoethylene would
be a particularly effective alkene for this purpose? Explain.

19. A study of the mechanism of singlet-oxygen oxidation has focused on adamantylideneadamantane, **A**. The reaction gives the products shown when carried
out in methyl t-butyl ketone as solvent. What relevance do these results have to
the question of a perepoxide intermediate in singlet-oxygen oxidations?

20. Formation of methylthiomethyl ethers is often observed in oxidation of alcohols
by dimethyl sulfoxide–acetic anhydride. Suggest a mechanism that would
account for the formation of these by-products.

$$R_2CHOH \xrightarrow[Ac_2O]{DMSO} R_2C{=}O + R_2CHOCH_2SCH_3$$

21. The following quantitative data concerning the ozonolysis of substituted
styrenes have been obtained in CCl$_4$ solution: $\rho = -0.91$; $\Delta E_a \cong 11.5$ kcal/
mol; $\Delta S^{\ddagger} \approx 18 \pm 2$ eu. The reaction is first-order in both styrene and ozone

when the reaction is followed by ozone disappearance. Do these data provide an indication about which step in the ozonolysis mechanism outlined in Scheme 9.10 is rate-determining?

22. It has been noted that when unsymmetrical olefins are ozonized in methanol, there is often a large preference for one cleavage mode over the other. For example:

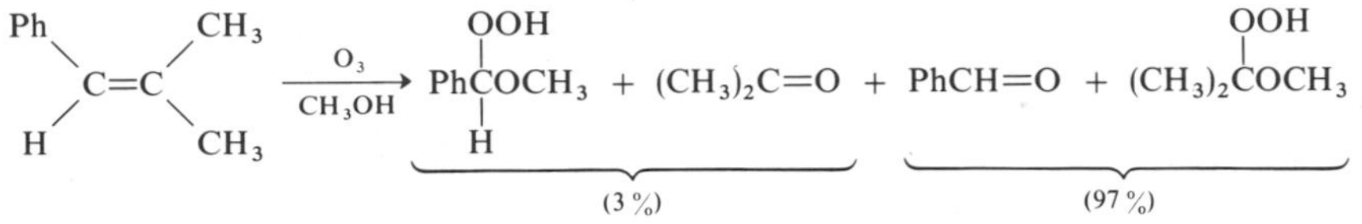

How would you explain this example of regioselective cleavage?

23. The two stereoisomers **A** and **B** have been found to differ greatly in reactivity toward singlet oxygen. **A** is oxidized to **C**, but **B** is unreactive under similar conditions. Explain.

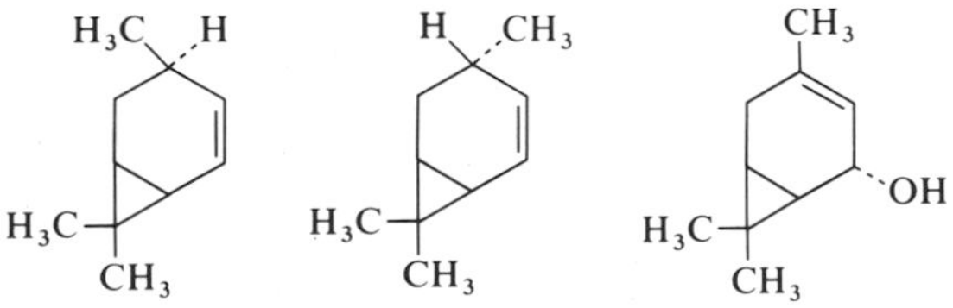

24. Treatment of epoxy ester **A** with hydrochloric acid in ether gives the products shown. Write a mechanism that would account for the formation of the major products and suggest feasible experiments that could test your mechanism.

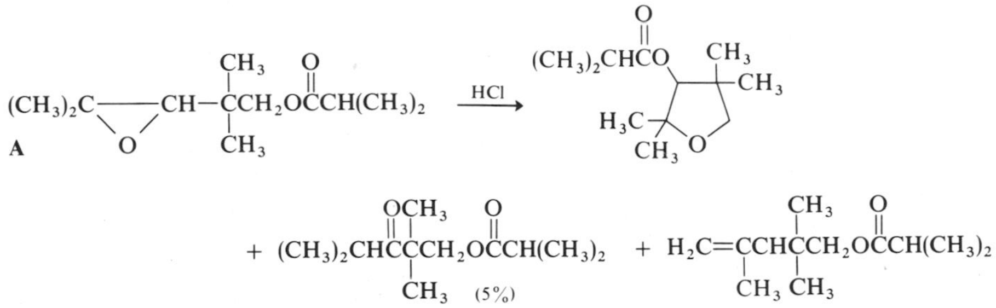

25. Predict the products from opening of the two stereoisomeric epoxides derived from limonene by reaction with (a) acetic acid, (b) dimethylamine, and (c) lithium aluminum hydride.

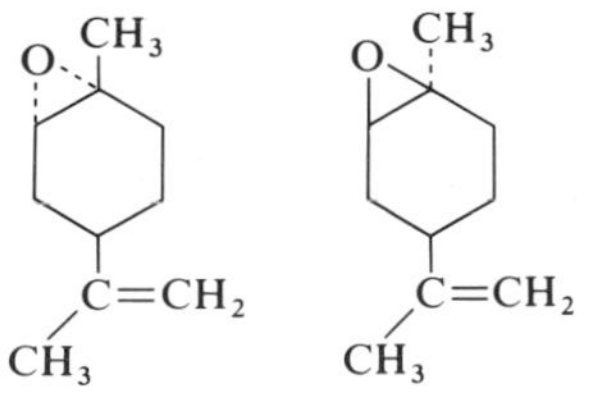

26. The oxidation of t-butylphenylcarbinol by Cr(VI) reagents results in the formation of benzaldehyde and t-butanol, as well as the major product t-butyl phenyl ketone. Very low concentrations of Ce(III) and Ce(IV) greatly suppress the formation of these products and decrease the reaction rate somewhat. Propose a mechanism that would account for the effect of the added Ce ions on the product composition.

Multistep Syntheses

Many reactions that are important tools for organic synthesis have been introduced in the preceding chapters. Although many of the problems in these earlier chapters have dealt with synthesis, little has been said explicitly about the subject of multistep synthesis. In this chapter, the focus of attention will be on multistep syntheses, using as essential background the knowledge of chemical reactions gained in the preceding chapters. The kinds of synthetic problems that are a challenge to organic synthetic techniques at present generally involve *multifunctional molecules*. For this reason, a multistep synthesis requires planning for the compatibility of proposed reactions with the various functional groups in the molecule.

In Sections 10.1 and 10.2, "Protective Groups" and "Synthetic Equivalent Groups," we will consider ways of temporarily modifying functional groups that would interfere with reactions at other points in the molecule. In Section 10.3, "Asymmetric Syntheses," we will illustrate how stereochemistry at one point in a molecule can influence the stereochemical outcome of reactions at other points. In Section 10.4, "Synthetic Strategy," we will illustrate the planning and execution of multistep syntheses, using examples from the recent literature.

10.1. Protective Groups

In the best of circumstances, each step in a synthetic sequence will involve only those sites in a molecule that are important to the desired transformation. It often happens, however, that a reagent required for a particular transformation will react not only with the target functional group, but also with another site in the molecule. It then becomes necessary to temporarily modify the troublesome functional group in a way that makes it inert to the reagent, yet capable of being regenerated in a subsequent step.

Three considerations are important in choosing an appropriate protective group: (1) the nature of the group requiring protection; (2) the reaction conditions under which the protective group must exert its masking effect and, accordingly, to which it must be stable; and (3) the conditions that can be tolerated during removal of the protecting group. No universal protecting groups exist even for a single functionality. The state of the art has been developed to a high level, however, and the many mutually complementary protective groups provide a great deal of flexibility in allowing the design of syntheses of complex molecules.

The sections that follow describe some of the protective groups that are often used for some of the most common functional groups.

10.1.1. Hydroxyl-Protecting Groups

A common requirement in synthesis is that a hydroxyl group be masked as some derivative lacking an active hydrogen. An example of this requirement is in reactions involving a Grignard or other organometallic reagent. The active hydrogen of a hydroxyl group will destroy an equivalent of a strongly basic organometallic reagent, and possibly further adversely affect the reaction.

Since an ether linkage is usually inert to organometallic reagents, conversion of the alcohol to an ether is an attractive means of protecting the hydroxyl function. The choice of appropriate ether group is largely dictated by the conditions that can be tolerated in subsequent cleavage of the protecting group. An important method that is applicable when mildly acidic hydrolysis is an appropriate method for deprotection is to form the tetrahydropyranyl ether[1]:

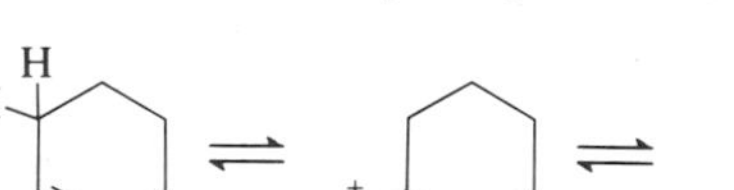

The protective group is introduced by an acid-catalyzed addition to the vinyl ether moiety in dihydropyran. It can be removed by dilute aqueous acid. The chemistry involved in the introduction and removal of the tetrahydropyranyl group is the reversible acid-catalyzed formation and hydrolysis of an acetal (see Part A, Section

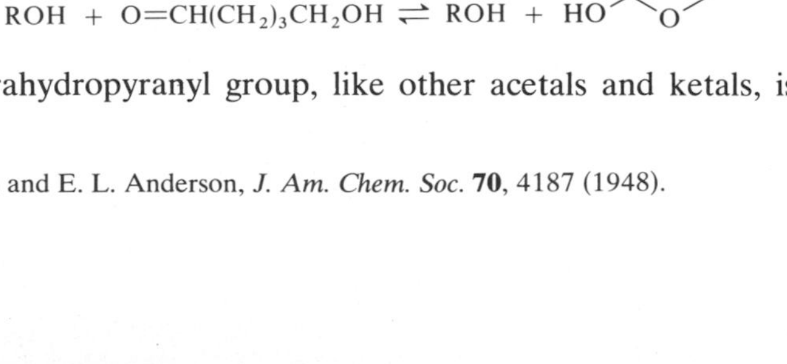

$$ROH + O=CH(CH_2)_3CH_2OH \rightleftharpoons ROH + HO\text{—}\overset{\frown}{O}$$

8.1). The tetrahydropyranyl group, like other acetals and ketals, is inert to nu-

---

1. W. E. Parham and E. L. Anderson, *J. Am. Chem. Soc.* **70**, 4187 (1948).

cleophilic reagents and is unchanged under such conditions as hydride reduction, organometallic reactions, or base-catalyzed reactions in aqueous solution.

A major disadvantage of the tetrahydropyranyl ether as a protecting group is that an asymmetric center is produced at C-2 of the tetrahydropyran ring on reaction with the alcohol. This asymmetry presents no difficulties if the alcohol is achiral, since a racemic mixture results. If the alcohol has an asymmetric center anywhere in the molecule, however, condensation with dihydropyran can afford a mixture of diastereomeric tetrahydropyranyl ethers, which may complicate purification and characterization. One way of surmounting this problem is to use methyl 2-propenyl ether, rather than dihydropyran. No asymmetric center is introduced, and the acetal offers the further advantage of being hydrolyzed under milder conditions than those required for tetrahydropyranyl ethers.[2] Ethyl vinyl ether is also useful as a hydroxyl-

$$ROH + CH_2=\underset{\underset{CH_3}{|}}{C}-OCH_3 \xrightarrow{H^+} ROC(CH_3)_2OCH_3$$

protecting group although it, like dihydropyran, also gives rise to diastereomers when a chiral alcohol is used.

The simple alkyl groups are not very useful for protecting alcohol functions as ethers. Although they can be introduced readily enough by alkylation, subsequent cleavage requires drastic conditions, such as concentrated hydrobromic acid, which can cause further reactions of the alcohol group. The t-butyl group is an exception, and has found some use as a protecting group. Because of the stability of the t-butyl cation, t-butyl ethers are cleaved under moderately acidic conditions.

The triphenylmethyl (trityl) group is removed under even milder conditions, and is an important hydroxyl-protecting group. This group is introduced by reaction of the alcohol with triphenylmethyl chloride. The introduction of the group, of course, involves an S_N1 substitution. Hot aqueous acetic acid suffices to remove the trityl group. The ease of removal can be increased by addition of electron-releasing substituents. The p-methoxy derivatives have been employed for this purpose.[3] The triarylmethyl groups are sufficiently bulky that they can usually be introduced only at primary alcohol centers.

The benzyl group can serve as an alcohol-protecting group when acidic conditions for ether cleavage cannot be tolerated. The benzyl C–O bond is cleaved by catalytic hydrogenolysis,[4] or with sodium in liquid ammonia.[5] Allyl ethers are an alternative to benzyl ethers as a protecting group. Allyl ethers may be isomerized quantitatively by potassium t-butoxide in dimethyl sulfoxide to propenyl ethers, which are quite labile to dilute acid.[6]

$$ROCH_2CH=CH_2 \xrightarrow[\text{DMSO}]{\text{KO-}t\text{-Bu}} ROCH=CHCH_3 \xrightarrow{H_3O^+} ROH + CH_3CH_2CH=O$$

2. A. F. Kluge, K. G. Untch, and J. H. Fried, *J. Am. Chem. Soc.* **94**, 7827 (1972).
3. M. Smith, D. H. Rammler, I. H. Goldberg, and H. G. Khorana, *J. Am. Chem. Soc.* **84**, 430 (1962).
4. W. H. Hartung and R. Simonoff, *Org. React.* **7**, 263 (1953).
5. E. J. Reist, V. J. Bartuska, and L. Goodman, *J. Org. Chem.* **29**, 3725 (1964).
6. R. Gigg and C. D. Warren, *J. Chem. Soc. C*, 1903 (1968).

Silyl ethers have an important role as hydroxyl-protecting groups.[7] Alcohols can be easily converted to trimethylsilyl ethers by reaction with trimethylsilyl chloride in the presence of an amine or by heating with hexamethyldisilazane. Although these are useful compounds when the objective is preparation of a less polar derivative of

$$ROH \ + \ (CH_3)_3SiCl \xrightarrow{R_3N} ROSi(CH_3)_3$$

an alcohol, the trimethylsilyl group is so readily removed by hydrolytic or nucleophilic conditions that its use as a protecting group is somewhat restricted. The increased steric bulk present in the dimethyl-t-butylsilyl group improves the stability of the silyl group enormously and permits its use as a protecting group in such reactions as hydride reduction (diisobutylaluminum hydride) and Cr(VI) oxidation (chromic acid in acetone).[8] This group is attached by using imidazole as a catalyst for the reaction of an alcohol with dimethyl-*tert*-butylsilyl chloride in dimethylformamide. Cleavage of the protecting group is slow under hydrolytic conditions, but fluoride ion (tetra-n-butylammonium fluoride in tetrahydrofuran) effects its removal.

Diols represent a special case in terms of applicable protecting groups. 1,2-Diols and 1,3-diols form cyclic acetals and ketals with aldehydes and ketones unless cyclization is precluded by the geometry of the molecule. The isopropylidene derivatives (acetonides) formed by reaction with acetone are probably the most common example:

$$\underset{\underset{\displaystyle HO \quad OH}{|\quad\ |}}{RCHCHR} \ + \ \underset{\underset{\displaystyle O}{||}}{CH_3CCH_3} \xrightarrow{H^+} RCH-HCR$$

Being a ketal, this protective group shares with tetrahydropyranyl derivatives the property of being resistant to basic and nucleophilic reagents, but is readily removed by aqueous acid. The isopropylidene group can be introduced by acid-catalyzed condensation with acetone, or by acid-catalyzed exchange with 2,2-

$$\underset{\underset{\displaystyle OH}{|}}{RCHCH_2OH} \ + \ \underset{\underset{\displaystyle OCH_3}{|}}{CH_3CCH_3} \xrightarrow{H^+} R-CH-CH_2$$

dimethoxypropane.[9] Formaldehyde, acetaldehyde, and benzaldehyde have all been used as the carbonyl component in formation of cyclic acetals. They function in the

7. J. F. Klebe, in *Advances in Organic Chemistry, Methods and Results*, Vol. 8, E. C. Taylor (ed.), Wiley–Interscience, New York, NY, 1972, pp. 97–178; A. E. Pierce, *Silylation of Organic Compounds*, Pierce Chemical Co., Rockford, Ill, 1968.
8. E. J. Corey and A. Venkateswarlu, *J. Am. Chem. Soc.* **94**, 6190 (1972).
9. M. Tanabe and B. Bigley, *J. Am. Chem. Soc.* **83**, 756 (1961).

same manner as acetone, but usually offer no advantage relative to acetone. A disadvantage is present with acetaldehyde and benzaldehyde when the glycol contains a chiral center. The acetals formed from these aldehydes introduce a new chiral center, and thus can lead to a mixture of diastereomers.

Protection of an alcohol function by esterification sometimes offers advantages over acetal protecting groups such as the tetrahydropyranyl ethers. Generally, acetals are stable in base and labile in acid, while esters are more stable in acid than acetals and are readily hydrolyzed in base. Esters are notably useful in reactions such as oxidations, but are not satisfactory in organometallic reactions. Acetates and benzoates are the most common ester protecting groups; they can be conveniently prepared by reaction of unhindered alcohols with acetic anhydride or benzoyl chloride, respectively, in the presence of pyridine or other tertiary amine. The use of reactive amides such as N-acylimidazoles (imidazolides) allows the reaction to be carried out in the absence of added bases[10]:

$$\text{ROH} + \text{R'C(=O)}-N\diagdown\diagup N \longrightarrow \text{ROCR'} + \text{HN}\diagdown\diagup N$$

Imidazolides are less reactive than the corresponding acid chlorides and exhibit a high degree of selectivity in reactions with molecules possessing several hydroxyl groups:

Ref. 11

(78 %)

Hindered hydroxyl groups require special acylation procedures. The simplest expedient is to increase the reactivity of the hydroxyl group by converting it to an alkoxide ion with strong base (e.g., n-BuLi, EtMgBr). When this conversion is either not feasible or ineffective, more reactive acylating agents are used. Highly reactive acylating agents are generated *in situ* when carboxylic acids are mixed with trifluoroacetic anhydride. The mixed anhydrides exhibit increased reactivity because of the high reactivity of the trifluoroacetate as a leaving group.[12] Dicyclohexylcarbodiimide is another reagent that serves to activate carboxyl groups by forming the iminoanhydride **A** (see Part A, Section 8.6).

10. H. A. Staab, *Angew. Chem.* **74**, 407 (1962).
11. F. A. Carey and K. O. Hodgson, *Carbohydr. Res.* **12**, 463 (1970).
12. R. C. Parish and L. M. Stock, *J. Org. Chem.* **30**, 927 (1965); J. M. Tedder, *Chem. Rev.* **55**, 787 (1955).

1. Tetrahydropyranyl ether[a]

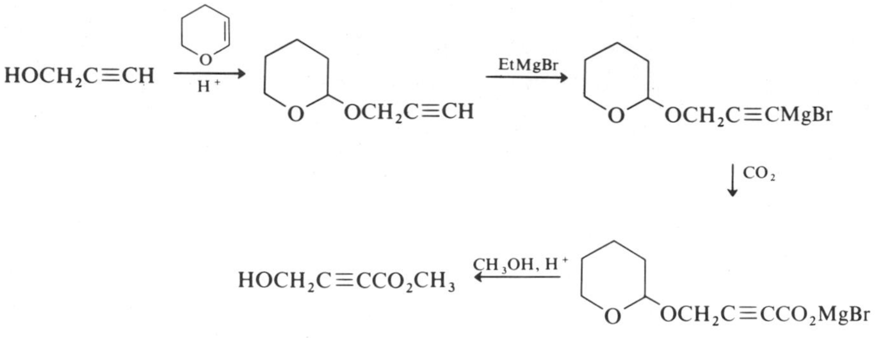

2. Methoxymethyl ether[b]

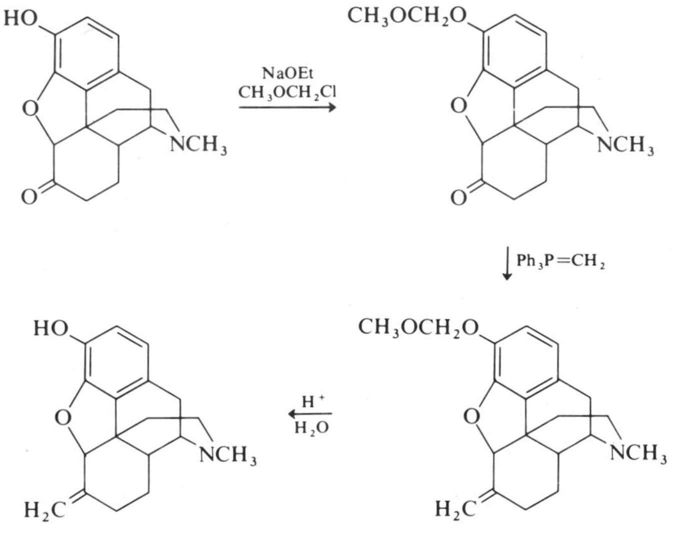

3. Triarylmethyl ether[c]

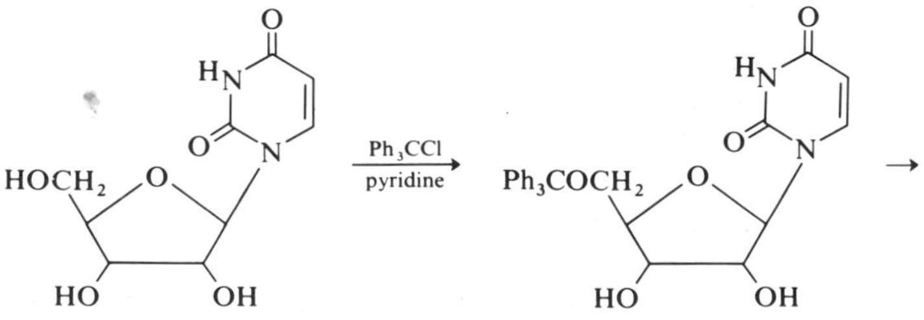

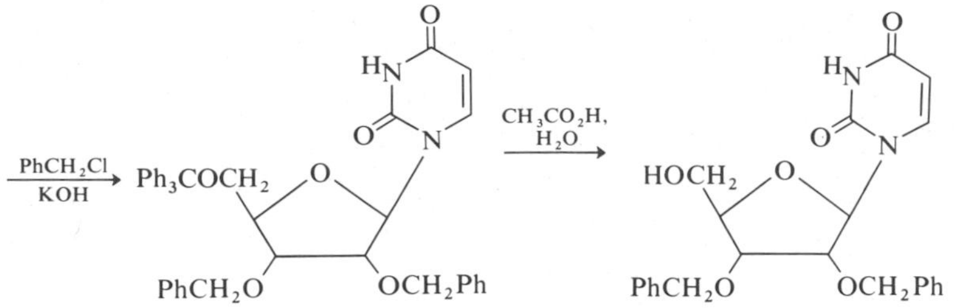

4. Benzyl ether[d]

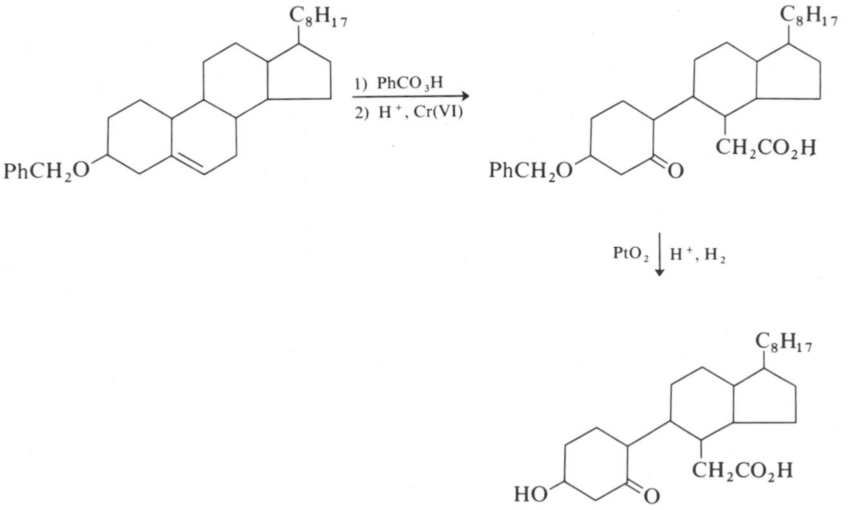

5. Glycol protection by isopropylidene derivative[e]

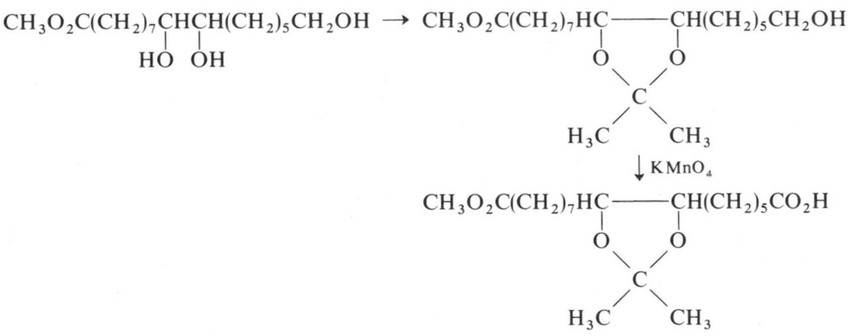

$$CH_3O_2C(CH_2)_7CHCH(CH_2)_5CH_2OH \rightarrow$$
$$\underset{HO\ \ OH}{}$$

a. H. B. Henbest, E. R. H. Jones, and I. M. S. Walls, *J. Chem. Soc.*, 3646 (1950).
b. M. A. Abdel-Rahman, H. W. Elliott, R. Binks, W. Küng, and H. Rapoport, *J. Med. Chem.* **9**, 1 (1965).
c. A. M. Michelson and A. Todd, *J. Chem. Soc.*, 3459 (1956).
d. L. Knof, *Justus Liebigs Ann. Chem.*, **656**, 183 (1962).
e S. D. Sabnis, H. H. Mathur, and S. C. Bhattacharyya, *J. Chem. Soc.*, 2477 (1963).

Cyclic carbonate esters are easily prepared from *vic*-diols and can be used in a complementary fashion to the cyclic acetals previously described. These esters are commonly prepared from *N,N'*-carbonyldiimidazole or by transesterification using diethyl carbonate. These reagents are preferable to phosgene both in efficiency and convenience.

Scheme 10.1 depicts some short synthetic sequences that illustrate the use of several of the important hydroxyl-protecting groups.

10.1.2. Amino-Protecting Groups

Unprotected amino groups are sites of both nucleophilicity and a weakly acidic hydrogen. If a given reaction cannot proceed in the presence of either of these types of reactivity, a protected derivative must be employed. The masking of nucleophilicity can be accomplished by acylation. The importance of temporarily masking amino-group nucleophilicity in synthesis of polypeptides will be discussed in Chapter 11.

The most useful protecting group for this purpose is the carbobenzyloxy group. The utility of this group lies in the ease with which it can be removed. Because of the lability of the benzyl C–O bonds towards hydrogenolysis, the amine can be regenerated from a carbobenzyloxy derivative by hydrogenation, which is accompanied by spontaneous decarboxylation:

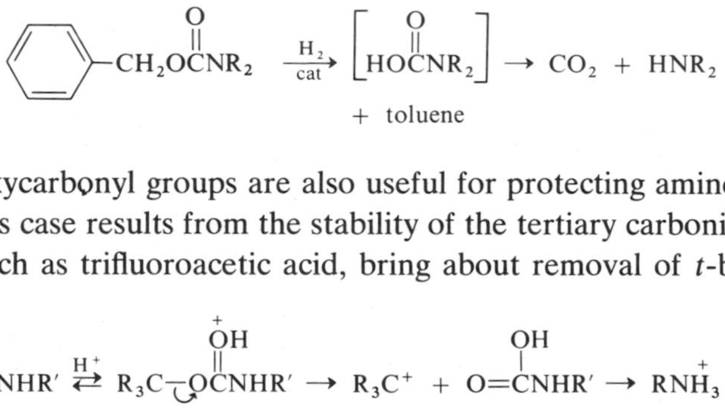

$$\text{C}_6\text{H}_5\text{-CH}_2\text{OCNR}_2 \xrightarrow[\text{cat}]{\text{H}_2} \left[\text{HOCNR}_2 \right] \rightarrow \text{CO}_2 + \text{HNR}_2$$
$$+ \text{ toluene}$$

Tertiary alkoxycarbonyl groups are also useful for protecting amines. The ease of removal in this case results from the stability of the tertiary carbonium ion. Acidic conditions, such as trifluoroacetic acid, bring about removal of *t*-butoxycarbonyl

$$\text{R}_3\text{COCNHR}' \overset{\text{H}^+}{\rightleftarrows} \text{R}_3\text{C-OCNHR}' \rightarrow \text{R}_3\text{C}^+ + \text{O=CNHR}' \rightarrow \text{RNH}_3^+ + \text{CO}_2$$

groups. Simple amides are satisfactory protecting groups only if the molecule can resist the vigorous acidic or alkaline conditions required for hydrolytic removal. Phthaloyl groups have been used to protect primary amine centers. The group can be removed hydrolytically or by treatment with hydrazine. The latter method results in the formation of the cyclic hydrazide of phthalic acid.

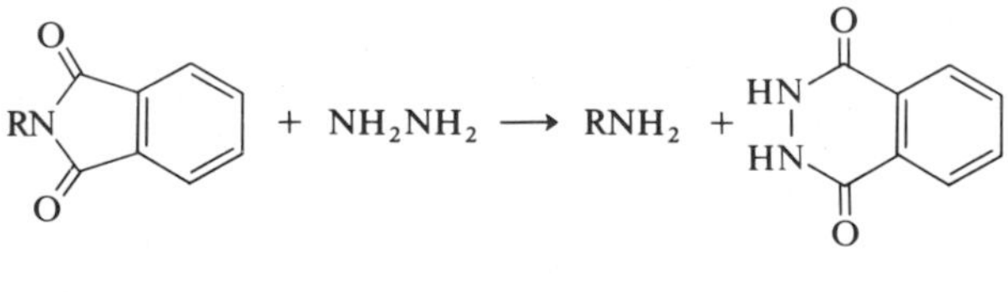

$$\text{RN(CO)}_2\text{C}_6\text{H}_4 + \text{NH}_2\text{NH}_2 \longrightarrow \text{RNH}_2 + \text{(CO)}_2\text{C}_6\text{H}_4(\text{NHNH})$$

Phthaloyl and succinoyl derivatives of ammonia are relatively acidic, and the corresponding anions can be easily formed and alkylated. This is the basis of the

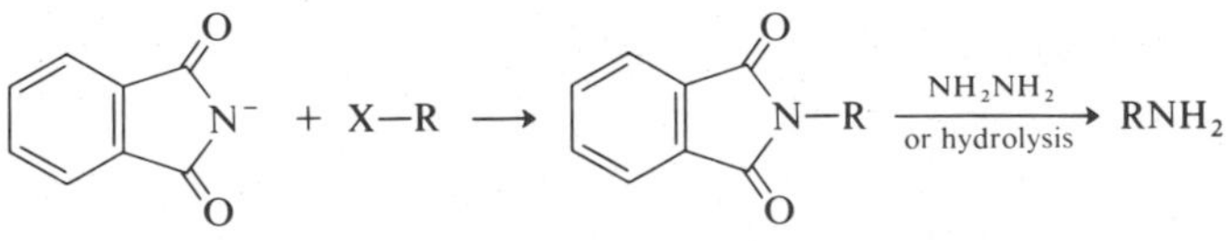

Gabriel synthesis of primary amines.[13] More recently, phthalimide has been used in a procedure that converts alcohols directly to amines[14]:

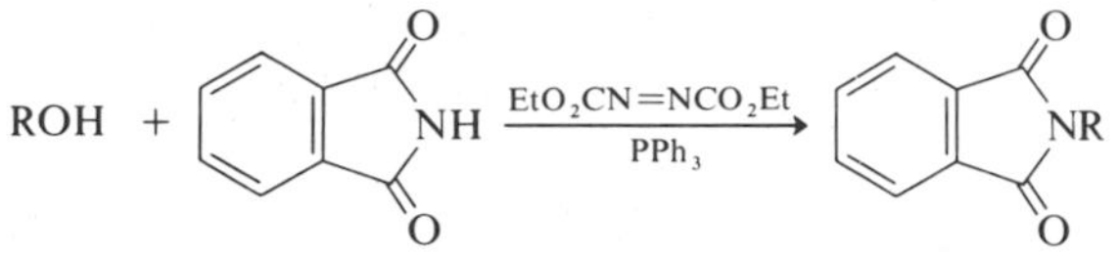

The diethyl azodicarboxylate converts the alcohol to an alkoxyphosphonium salt, which serves as the active alkylating agent:

$$PPh_3 + H_5C_2O_2CN{=}NCO_2C_2H_5 \longrightarrow Ph_3\overset{+}{P}{-}\overset{-}{N}(CO_2C_2H_5)CO_2C_2H_5$$

$$Ph_3\overset{+}{P}{-}\overset{-}{N}(CO_2C_2H_5)CO_2C_2H_5 + ROH \longrightarrow Ph_3\overset{+}{P}{-}OR + H_5C_2O_2CNHNHCO_2C_2H_5$$

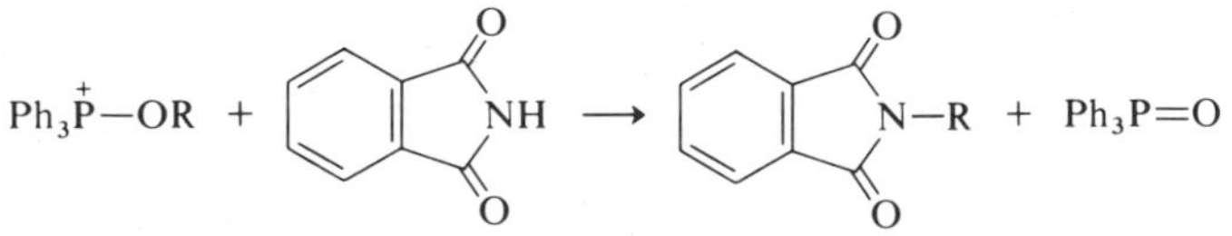

Under some circumstances, it is possible to effect removal of amide groups by selective hydride reduction. Trichloroacetamides are readily cleaved by sodium borohydride in alcohols.[15] Another reductive method that can be applied to benzamides, and probably to other simple amides, involves treatment with diisobutylaluminum hydride. At low temperatures, this reduction stops at the carbinolamine stage. Hydrolysis then yields the amine.[16]

$$R_2N\overset{O}{\overset{\|}{C}}Ph \xrightarrow{R_2AlH} R_2N\underset{H}{\overset{OAlR_2}{C}}Ph \xrightarrow[H_2O]{H^+} R_2NH + PhCHO$$

The trifluoroacetyl group can serve as a useful protecting group under some circumstances. Because of the electron-withdrawing effect of the trifluoromethyl group, the trifluoroacetamides are much more subject to alkaline hydrolysis than most amides, and can be removed under relatively mild conditions.[17]

13. M. S. Gibson and R. W. Bradshaw, *Angew. Chem. Int. Ed. Engl.* **7**, 919 (1968).
14. O. Mitsunobu, M. Wada, and T. Sano, *J. Am. Chem. Soc.* **94**, 679 (1972).
15. F. Weygand and E. Frauendorfer, *Chem. Ber.* **103**, 2437 (1970).
16. J. Gutzwiller and M. Uskoković, *J. Am. Chem. Soc.* **92**, 204 (1970).
17. A. Taurog, S. Abraham, and I. L. Chaikoff, *J. Am. Chem. Soc.* **75**, 3473 (1953).

10.1.3. Carbonyl-Protecting Groups

There is a very general method for protecting aldehyde and ketone carbonyl groups against addition of nucleophiles or reduction when the reaction conditions are nonacidic. This method involves conversion of the carbonyl group to the ketal or acetal. Ethylene glycol, which gives a cyclic dioxolane derivative, is the most frequently employed reagent for this purpose. The derivative is usually prepared by

$$
RCR' + HOCH_2CH_2OH \xrightarrow{H^+} \underset{R'}{\overset{R}{>}}C\underset{O-CH_2}{\overset{O-CH_2}{<}} + H_2O
$$

heating in the presence of a trace of acid with provision for azeotropic removal of water. The dioxolane ring is inert to powerful nucleophiles, including organometallic reagents and the hydride reducing agents. It is readily removed in acidic solution containing water by the general hydrolysis mechanism for acetals and ketals (Part A, Section 8.1).

Acyclic acetals and ketals can also be prepared. A convenient route involves acid-catalyzed exchange with an orthoester[18]:

$$
RCR' + HC(OCH_3)_3 \xrightarrow{H^+} \underset{OCH_3}{\overset{OCH_3}{RCR'}} + HCO_2CH_3
$$

Ketals and acetals can also be prepared from the carbonyl compound and alcohol in the presence of an acidic catalyst. Provision of a means for removal of the water formed ensures that the reaction will go to completion.

If a carbonyl group must be regenerated under conditions other than acid-catalyzed hydrolysis, β-haloalcohols can be used. Reductive cleavage by zinc metal is then possible. Another variation is the use of mercaptoethanol in place of

$$
\underset{BrCH_2}{\overset{R}{\bigvee}} \xrightarrow{Zn} R_2C=O + HOCH_2CH=CH_2 \qquad \text{Ref. 19}
$$

$$
\underset{R'}{\overset{}{RC(OCH_2CCl_3)_2}} \xrightarrow[THF]{Zn} RCR' + CH_2=CCl_2 \qquad \text{Ref. 20}
$$

ethylene glycol. The 1,3-oxathiolane derivatives are readily formed from ketones and mercaptoethanol in the presence of BF_3,[21] or by heating in benzene with p-toluenesulfonic acid catalyst with azeotropic removal of water.[22] The 1,3-

18. C. A. MacKenzie and J. H. Stocker, *J. Org. Chem.* **20**, 1695 (1955).
19. E. J. Corey and R. A. Ruden, *J. Org. Chem.* **38**, 834 (1973).
20. J. L. Isidor and R. M. Carlson, *J. Org. Chem.* **38**, 554 (1973).
21. G. E. Wilson, Jr., M. G. Huang, and W. W. Scholman, Jr., *J. Org. Chem.* **33**, 2133 (1968).
22. C. Djerassi and M. Gorman, *J. Am. Chem. Soc.* **75**, 3704 (1953).

oxathiolanes have an advantage over the 1,3-dioxolanes for applications where nonacidic conditions are required for subsequent removal of the protecting group. The 1,3-oxathiolane group can be removed by treatment with Raney nickel in alcoholic solution, even under slightly alkaline conditions.[23] Removal can also be accomplished by treating with compound **1** (chloramine-T), which probably oxidizes the sulfur to a sulfonium salt. This moiety is a more reactive leaving group, and removal of the protecting group then occurs by a hydrolytic process.[24]

$$X = Cl \text{ or } NSO_2Ar$$

The various derivatives of carbonyl compounds that involve formation of carbon–nitrogen double bonds at the carbonyl center (see Section 8.1) are not very generally useful as protecting groups. This includes oximes, semicarbazones, and hydrazones. The compounds are usually formed easily enough, but mild conditions for removal are not available. An exception involves the removal of the oxime group by conversion to the acetate ester, followed by chromous ion reduction[25]:

$$R_2C=NOH \xrightarrow{Ac_2O} R_2C=N-O-\overset{\overset{\displaystyle O}{\|}}{C}CH_3 \xrightarrow{Cr(II)} R_2C=O$$

10.1.4. Carboxylic Acid Protecting Groups

If only the O–H, as opposed to the carbonyl, group of a carboxyl function needs to be masked, the masking can be readily accomplished by the standard esterification techniques. The more difficult problem of protecting the carbonyl group can be accomplished by conversion to a 2-oxazoline derivative. The most convenient 2-oxazolines are the 4,4- and 5,5-dimethyl derivatives, which can be prepared from the acid and 2-amino-3-methyl-1-propanol and from 2,2-dimethylaziridine, respectively.[26] The carboxyl group is successfully masked by the heterocyclic ring in the presence of Grignard reagents or hydride reducing agents. The carboxyl group can be regenerated by acidic hydrolysis. Alternatively, treatment with acid in an alcohol returns the carboxylic ester.

23. C. Djerassi, E. Batres, J. Romo, and G. Rosenbraz, *J. Am. Chem. Soc.* **74**, 3634 (1952).
24. D. W. Emerson and H. Wynberg, *Tetrahedron Lett.*, 3445 (1971).
25. E. J. Corey and J. E. Richman, *J. Am. Chem. Soc.* **92**, 5276 (1970).
26. A. I. Meyers and D. L. Temple, Jr., *J. Am. Chem. Soc.* **92**, 6644 (1970); D. Haidukewych and A. I. Meyers, *Tetrahedron Lett.*, 3031 (1972).

$$RCO_2H + HOCH_2C(CH_3)_2 \longrightarrow R-C$$

$$RCO_2H + HN \longrightarrow RC-N \xrightarrow{H^+} \longrightarrow R$$

A wide variety of alcohols have been used to convert carboxylic acids to esters. The resulting esters successfully mask the acidic properties of the carboxyl group, but do not usually prevent nucleophilic attack at the carbonyl center, especially by such species as Grignard reagents or organolithium compounds. Methyl and ethyl esters are very commonly employed in synthesis. Alkaline hydrolysis is the most common method for regenerating the carboxyl group. If hydrolytic conditions are not suitable, benzyl esters can be prepared. In this case, it is possible to liberate the acid group by hydrogenolysis. t-Butyl esters are readily cleaved by mildly acidic conditions because of the facile cleavage of the alkyl oxygen bond. 2,2,2-Trichloroethyl esters find some specialized use. This group can be removed with zinc dust in aqueous acetic acid[27]:

$$RCOCH_2CCl_3 \xrightarrow{Zn} RCO_2H + CH_2{=}CCl_2$$

Esters can be prepared by acid-catalyzed esterification or by reaction of the acid chloride with the alcohol. In small-scale syntheses, it is often more convenient to prepare the ester by reaction of the carboxylic acid with the appropriate diazo compound. Diazomethane is routinely used for making methyl esters, but more highly substituted esters can be prepared if the corresponding diazo compound is available. Benzhydryl esters, for example, are readily prepared from carboxylic acids by reaction with diphenyldiazomethane[28]:

$$Ph_2CN_2 + RCO_2H \longrightarrow RCO_2CHPh_2$$

10.2. Synthetic Equivalent Groups

The protecting groups discussed in the previous section are essentially passive in nature during a synthetic sequence. They are introduced and removed at appropriate stages, but do not serve to directly influence the reactivity at the points in the molecule where transformations are being carried out. It is often advantageous to

27. R. B. Woodward, K. Heusler, J. Gosteli, P. Naegeli, W. Oppolzer, R. Ramage, S. Ranganathan, and H. Vorbrüggen, *J. Am. Chem. Soc.* **88**, 852 (1966).
28. A. A. Aboderin, G. R. Delpierre, and J. S. Fruton, *J. Am. Chem. Soc.* **87**, 5469 (1965).

combine the need for masking of a functional group with a molecular transformation that does confer a desirable chemical reactivity to the group in question. The term *synthetic equivalent group* has been used to describe such a masked but nevertheless reactive group.

As an example, suppose the transformation **2** to **3** is necessary:

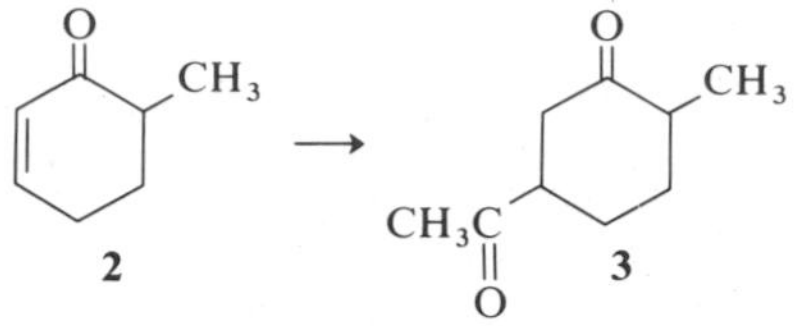

None of the procedures we have discussed will do this in one step. The point at which the acetyl group needs to be attached is potentially reactive toward nucleophiles, but there is no nucleophilic CH_3CO species available. The transformation requires a nucleophilic synthetic equivalent that could undergo conjugate addition to the electrophilic carbon–carbon double bond and then be converted to CH_3CO after the carbon skeleton has been formed. A number of nucleophilic synthetic equivalents for CH_3CO have been developed, of which the one most suitable for the problem under discussion is an ether of the corresponding cyanohydrin. Strong bases, in particular lithium diisopropylamide, convert the alkoxynitrile to a carbanion[29]:

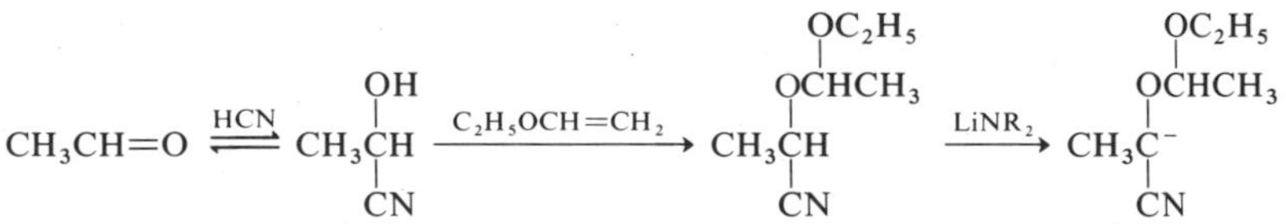

This nucleophilic species can successfully add to cyclohexenone. Subsequent acidic hydrolysis gives the desired product.

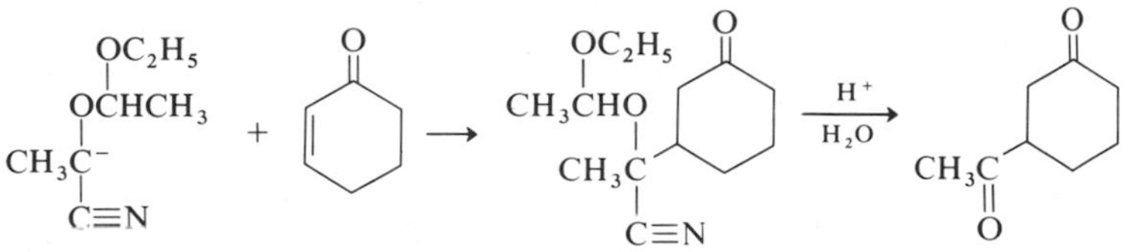

A number of other systems can be utilized as the synthetic equivalent of a nucleophilic carbonyl group, particularly in S_N2 alkylation reactions. Dithiane derivatives of aldehydes can be converted to nucleophiles by lithiation. The resulting anion is easily alkylated. Hydrolysis leads to a ketone.[30] Lithiated vinyl ethers and

29. G. Stork and L. Maldonado, *J. Am. Chem. Soc.* **93**, 5286 (1971); **96**, 5272 (1974).
30. D. Seebach, *Synthesis*, 17 (1969); D. Seebach and E. J. Corey, *J. Org. Chem.*, **40**, 231 (1975).

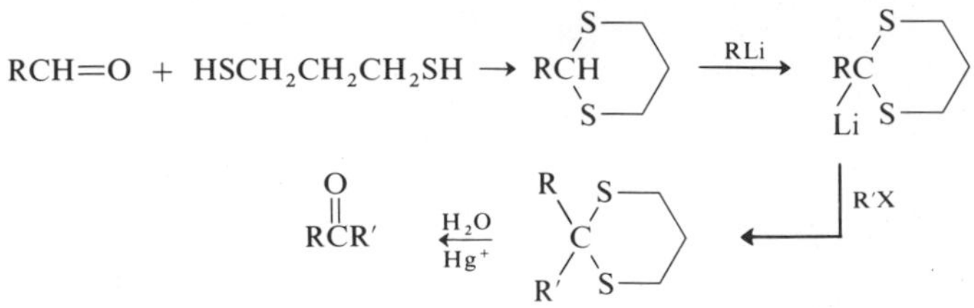

$$RCH{=}O \ + \ HSCH_2CH_2CH_2SH \ \rightarrow \ RCH\begin{array}{c}S{-}\\ \diagdown\\ S{-}\end{array} \xrightarrow{RLi} \ RC\begin{array}{c}S{-}\\ \diagup\diagdown\\ Li \ \ S{-}\end{array}$$

$$\underset{RCR'}{\overset{O}{\parallel}} \ \underset{Hg^+}{\overset{H_2O}{\leftarrow}} \ \underset{R'}{\overset{R}{}}\underset{S{-}}{\overset{S{-}}{C}} \ \xleftarrow{R'X}$$

vinyl thioethers are also synthetic equivalents of a nucleophilic carbonyl group. Methyl vinyl ether, for example, is lithiated by t-butyllithium in THF at $-65°$. The vinyllithium compound undergoes addition to a variety of carbonyl compounds, giving vinyl ethers, which can be easily hydrolyzed to ketones.[31]

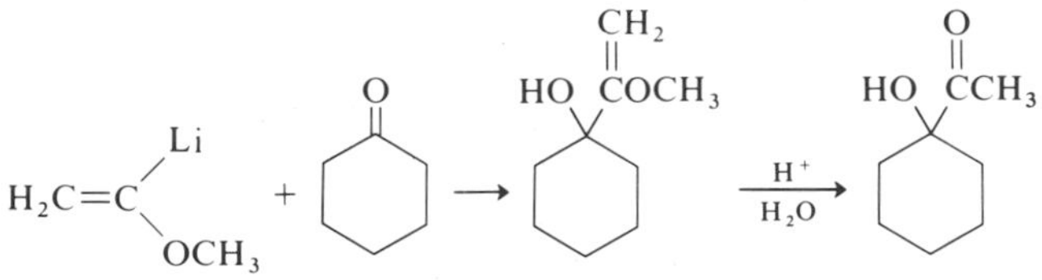

$$H_2C{=}C\begin{array}{c}Li\\ \diagdown\\ OCH_3\end{array} \ + \ \text{(cyclohexanone)} \ \rightarrow \ \text{(cyclohexane ring with } HO, \ \overset{CH_2}{\underset{COCH_3}{\parallel}}) \ \xrightarrow[H_2O]{H^+} \ \text{(cyclohexane ring with } HO, \ \overset{O}{\underset{CCH_3}{\parallel}})$$

α-Lithioaldimines can be prepared by addition of organolithium compounds to isonitriles:

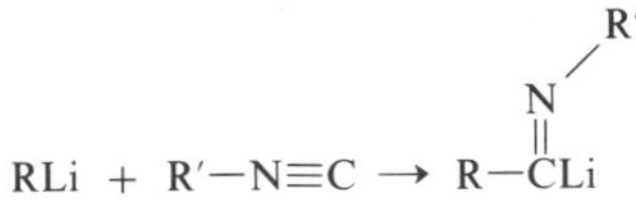

$$RLi \ + \ R'{-}N{\equiv}C \ \rightarrow \ R{-}\overset{\overset{\displaystyle N\diagdown R'}{\parallel}}{C}Li$$

These, too, are nucleophilic species that, after alkylation, can be converted to ketones. For example, α-hydroxyketones are formed when the lithioaldimine adds to a carbonyl group:

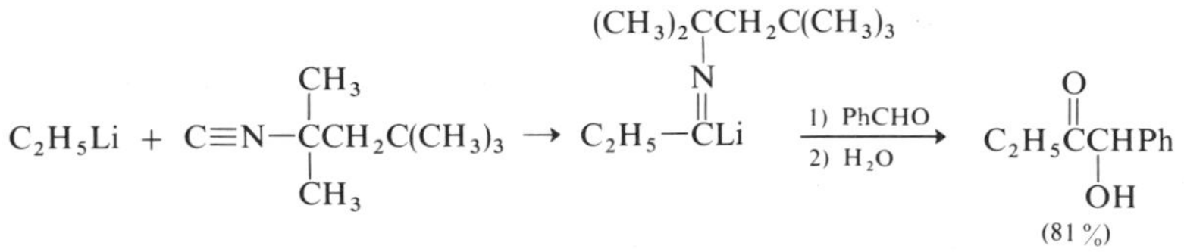

$$C_2H_5Li \ + \ C{\equiv}N{-}\underset{CH_3}{\overset{CH_3}{\underset{\vert}{\overset{\vert}{C}}}}CH_2C(CH_3)_3 \ \rightarrow \ C_2H_5{-}\overset{\overset{\displaystyle N(CH_3)_2CCH_2C(CH_3)_3}{\parallel}}{C}Li \ \xrightarrow[2) \ H_2O]{1) \ PhCHO} \ C_2H_5\underset{OH}{\overset{O}{\underset{\vert}{\overset{\parallel}{C}}}}CHPh \qquad \text{Ref. 32}$$

(81%)

A summary of some nucleophilic synthetic equivalents of carbonyl groups is given in Scheme 10.2.

It is possible to conceive of a wide variety of synthetic transformations based on the underlying principle associated with the use of synthetic equivalents. Any molecular fragment that has a useful range of synthetic reactivity as well as the ability to be efficiently converted to another functionality after reaction can serve as a synthetic equivalent group. Some examples have already been encountered in the earlier chapters. Diels–Alder adducts of α-chloroacrylonitrile can be converted to

31. J. E. Baldwin, G. A. Höfle, and O. W. Lever, Jr., *J. Am. Chem. Soc.* **96**, 7125 (1974).
32. G. E. Niznik, W. H. Morrison, III, and H. M. Walborsky, *J. Org. Chem.* **39**, 600 (1974).

1[a]

$$
\underset{RCH}{\overset{O}{\|}} \rightarrow \underset{\underset{CN}{|}}{\overset{OH}{|}}{RCH} \rightarrow \underset{\underset{CN}{|}}{\overset{OR'}{|}}{RCH} \rightarrow \underset{\underset{CN}{|}}{\overset{OR'}{|}}{RC^-} \xrightarrow{R''X} \underset{\underset{CN}{|}}{\overset{OR'}{|}}{RCR''} \xrightarrow{H^+} \underset{RCR''}{\overset{O}{\|}}
$$

2[b]

3[c]

4[d]

$$
H_2C=CHOCH_3 \xrightarrow{R'Li} \underset{H_2C=\underset{}{\overset{Li}{|}}COCH_3}{} \xrightarrow{R''X} \underset{H_2C=\overset{R''}{\overset{|}{C}}OCH_3}{} \xrightarrow[H_2O]{H^+} \underset{R''CCH_3}{\overset{O}{\|}}
$$

5[e]

$$
CH_3SCH=CHR \xrightarrow{RLi} CH_3\overset{Li}{\underset{}{S}}C=CHR \xrightarrow{R'X} CH_3\underset{\overset{|}{R'}}{SC}=CHR \xrightarrow{Hg^{2+}} RCH_2\overset{O}{\overset{\|}{C}}R
$$

6[f]

$$
R'N\equiv C + RLi \rightarrow R'N=C\underset{Li}{\overset{R}{<}} \xrightarrow{R''X} R'N=C\underset{R''}{\overset{R}{<}} \xrightarrow[H_2O]{H^+} \underset{RCR''}{\overset{O}{\|}}
$$

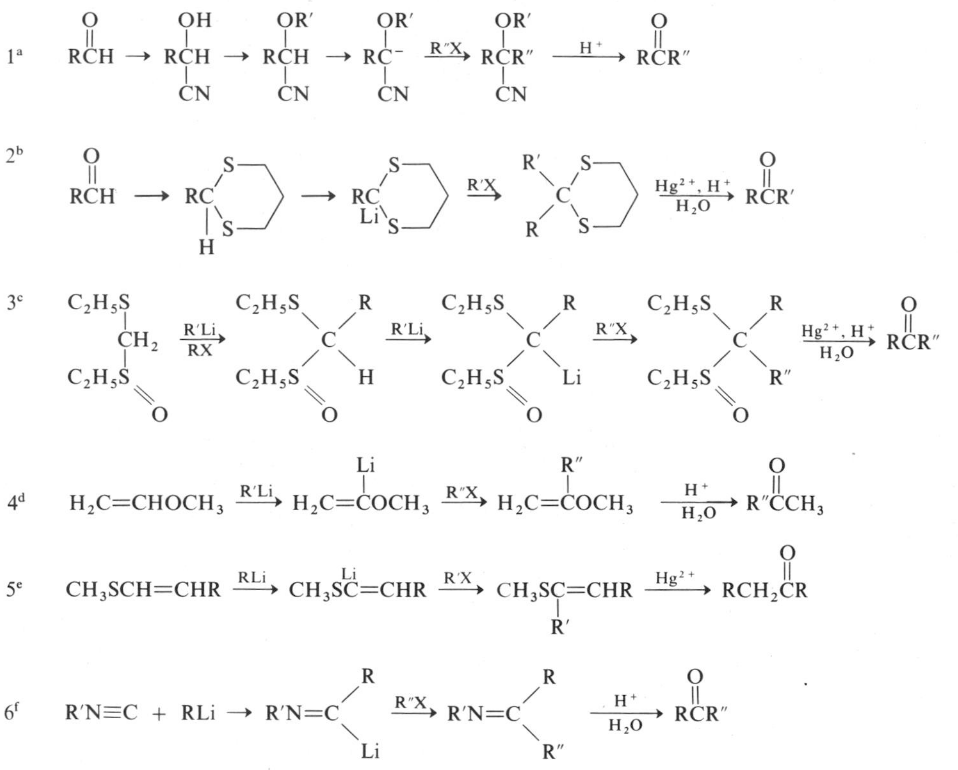

a. G. Stork and L. Maldonado, *J. Am. Chem. Soc.* **93**, 5286 (1971); **96**, 5272 (1974).
b. D. Seebach, *Synthesis*, 17 (1969).
c. J. E. Richman, J. L. Herrmann, and R. H. Schlessinger, *Tetrahedron Lett.*, 3267 (1973); J. L. Herrmann, J. E. Richman, and R. H. Schlessinger, *Tetrahedron Lett.*, 3271 (1973).
d. J. E. Baldwin, G. A. Höfle, and O. W. Lever, Jr., *J. Am. Chem. Soc.* **96**, 7125 (1974).
e. K. Oshimo, K. Shimoji, H. Takahashi, H. Yamamoto, and H. Nozaki, *J. Am. Chem. Soc.* **95**, 2694 (1973).
f. G. E. Niznik, W. H. Morrison, III, and H. M. Walborsky, *J. Org. Chem.* **39**, 600 (1974).

cyclic ketones by hydrolysis. Used under these circumstances, this dienophile is the synthetic equivalent of ketene. It is a useful equivalent, since ketene itself is not a satisfactory dienophile.

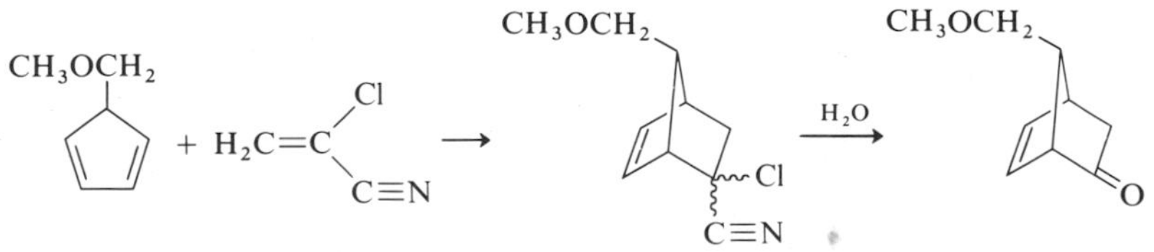

Ref. 33

33. E. J. Corey, N. M. Weinshenker, T. K. Schaaf, and W. Huber, *J. Am. Chem. Soc.* **91**, 5675 (1969).

Scheme 10.3. Examples of Masked Functionalities in Synthesis

No.	Synthon	Reagent	Reaction sequence

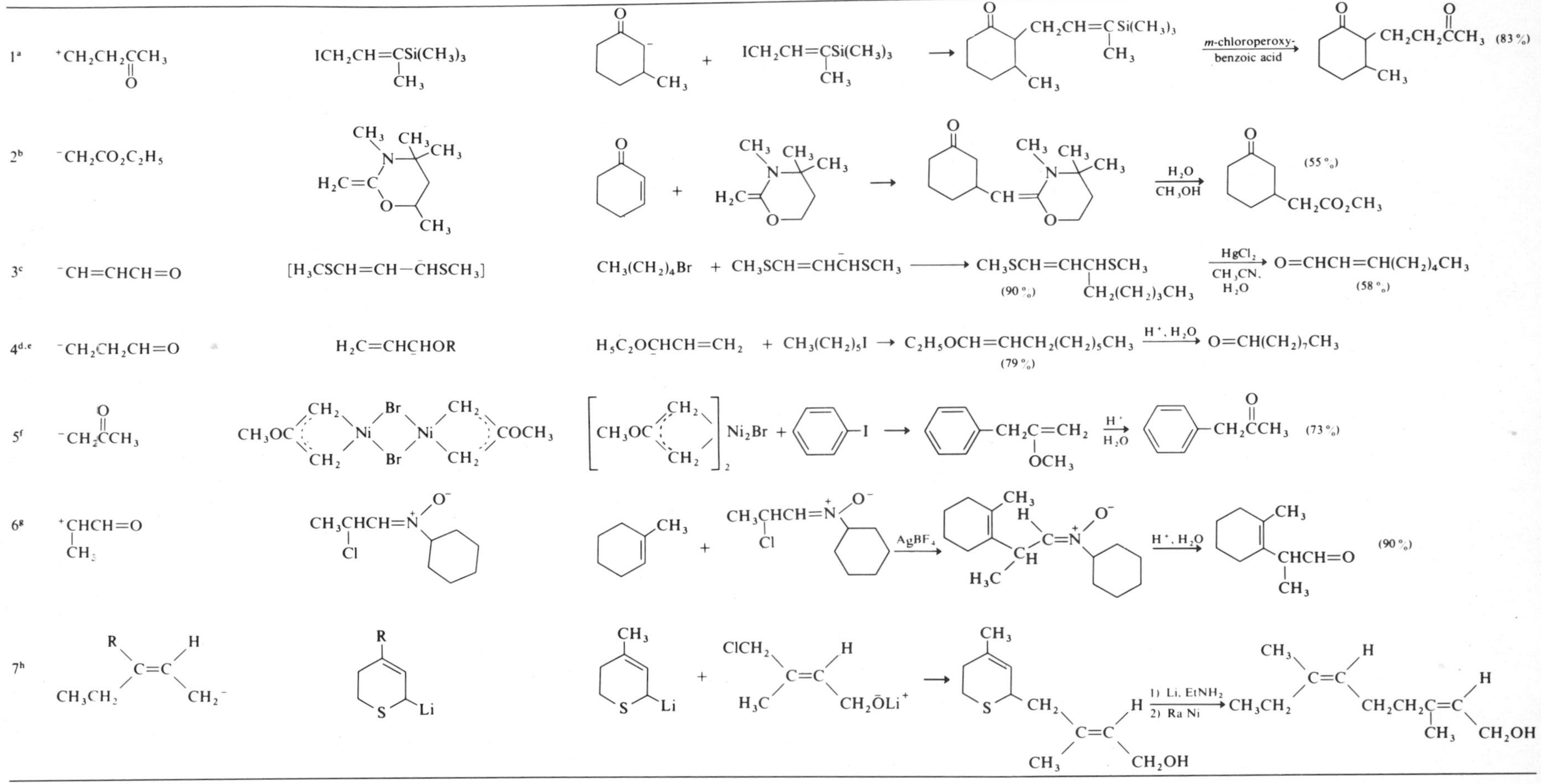

a. G. Stork and M. E. Jung, *J. Am. Chem. Soc.* **96**, 3682 (1974).
b. A. I. Meyers and N. Nazarenko, *J. Org. Chem.* **38**, 175 (1973).
c. E. J. Corey, B. W. Erickson, and R. Noyori, *J. Am. Chem. Soc.* **93**, 1724 (1971).
d. D. A. Evans, G. C. Andrews, and B. Buckwalter, *J. Am. Chem. Soc.* **96**, 5560 (1974).
e. W. C. Still and T. L. Macdonald, *J. Am. Chem. Soc.* **96**, 5561 (1974).
f. L. S. Hegedus and R. K. Stiverson, *J. Am. Chem. Soc.* **96**, 3250 (1974).
g. S. Shatzmiller, P. Gygax, D. Hall, and A. Eschenmoser, *Helv. Chim. Acta* **56**, 2961 (1973).
h. P. L. Stotter and R. E. Hornish, *J. Am. Chem. Soc.* **95**, 4444 (1973).

In Chapter 2, the application of alkoxymethyl and aryloxymethylphosphonium salts in the Wittig reactions was discussed. Since the vinyl ethers formed in such reactions are easily hydrolyzed to aldehydes, the alkoxy- and aryloxymethylene-phosphoranes are synthetic equivalents of nucleophilic formyl groups[34]:

$$ROCH_2\overset{+}{P}Ph_3 \xrightarrow{\text{base}} ROCH{=}PPh_3$$

$$ROCH{=}PPh_3 + R'CH{=}O \longrightarrow ROCH{=}CHR'$$

$$ROCH{=}CHR' \xrightarrow[H_2O]{H^+} O{=}CHCH_2R'$$

A similar procedure using α-methoxylalkylphosphonium salts provides a route to ketones.[35] Masked reactive functionalities have achieved fundamental importance in synthetic methodology.[36] Thus, when one surveys the methods of aldehyde synthesis, for example, reactions that could lead to RCH=CHOR' should be included along with methods that lead directly to aldehydes, because the terminal vinyl ether grouping is easily convertible to an aldehyde. Scheme 10.3 lists some other small molecular fragments that have useful reactivity and constitute synthetic equivalents of various structural units.

10.3. Asymmetric Syntheses

In designing a multistep synthesis, one must consider aspects of stereochemistry as well as functionality. In the chapters dealing with individual reactions, many examples were given in which the aspects of stereochemistry were a direct consequence of the reaction mechanism. For example, hydroboration–oxidation involves a *syn* addition followed by oxidation with retention of configuration. The generalization, widely but not universally correct, that reagents attack molecules from the sterically less hindered side was also illustrated on numerous occasions.

We now wish to consider how these elements of stereochemistry come into play in synthesis. It is important to know how reaction stereochemistry is controlled by structural features of the reactant molecules. This topic can be broadly covered by the term *asymmetric synthesis*,[37] which has been defined as "a reaction in which an achiral unit in an ensemble of substrate molecules is converted by a reactant into a chiral unit in such a manner that the stereoisomeric products are produced in unequal amounts." Thus, we will be dealing with methods for controlling the configuration of newly formed chiral centers. As will be seen, these methods often depend on the fact that reagents attack molecules along the less hindered path.

34. S. G. Levine, *J. Am. Chem. Soc.* **80**, 6150 (1958); G. Wittig, W. Böll, and K.-H. Krück, *Chem. Ber.* **95**, 2514 (1962).
35. C. A. Henrick, F. Schaub, and J. B. Siddall, *J. Am. Chem. Soc.* **94**, 5374 (1972).
36. D. Seebach and D. Enders, *Angew. Chem. Int. Ed. Engl.* **14**, 15 (1975).
37. J. D. Morrison and H. S. Mosher, *Asymmetric Organic Reactions*, Prentice-Hall, Englewood Cliffs, NJ, 1971.

Perhaps the simplest case of asymmetric synthesis to visualize is a reaction between an achiral molecule and a chiral reagent in such a way that a new chiral center is created at the reaction site. One such reaction is the Meerwein–Ponndorf–Verley reduction of ketones with aluminum salts of optically active alcohols. The

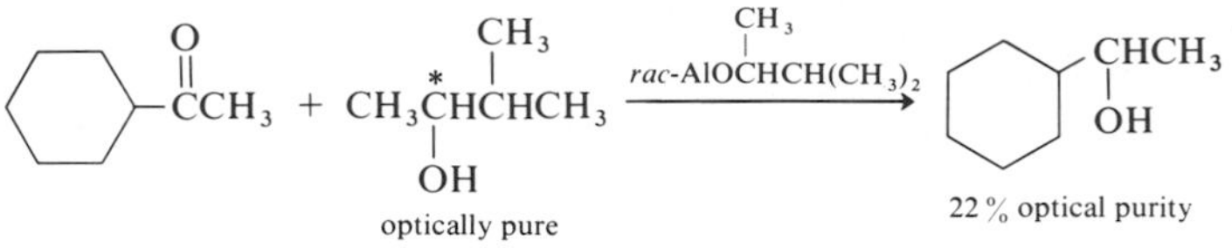

Ref. 38

achiral ketone methyl cyclohexyl ketone is reduced to the corresponding alcohol in such a way that a predominance of one of the enantiomers is formed.

Another case that falls within the scope of the definition is the generation of an excess of one configuration at a new chiral center in the reaction between a chiral molecule and an achiral reagent. The control of the stereochemistry of addition of Grignard reagents to carbonyl centers adjacent to a chiral carbon is exemplary. In

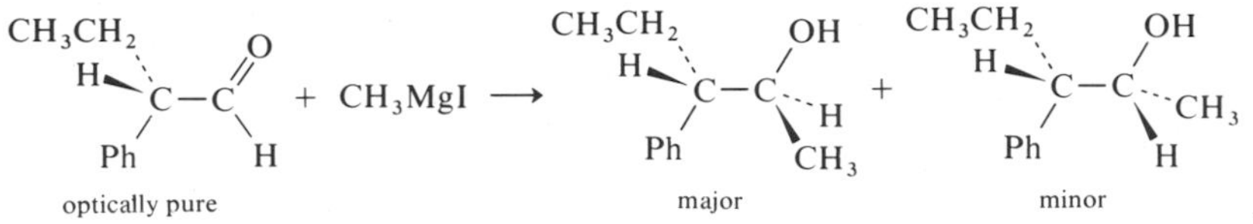

Ref. 39

this case, the configurationally homogeneous ketone gives an excess of one of the two possible diastereomers when treated with an achiral reagent.

Asymmetric synthesis involves diastereomeric transition states. In the case just cited, we can focus on a specific conformation of the reactant and consider the possible directions of approach by the reagent. Let us consider **B** and **C**, two of the

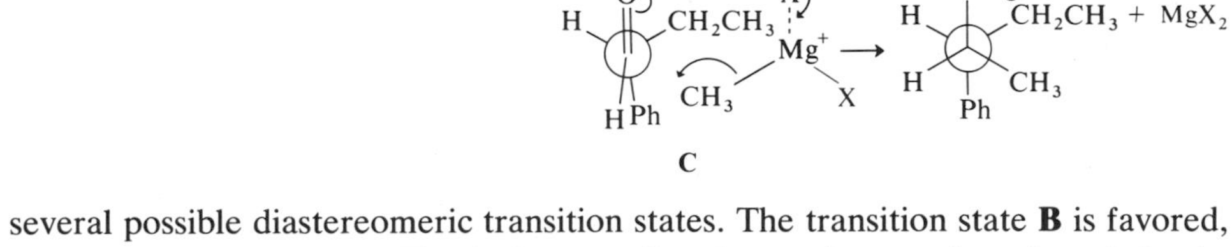

several possible diastereomeric transition states. The transition state **B** is favored, because of steric factors. The hydrogen offers less resistance than the ethyl substituent to the incoming Grignard reagent. As a result, a predominance of the new

38. W. E. Doering and R. W. Young, *J. Am. Chem. Soc.* **72**, 631 (1950).
39. D. J. Cram and F. A. Abd Elhafez, *J. Am. Chem. Soc.* **74**, 5828 (1952).

chiral center is of the S configuration. There are other conformations that could be considered for the transition states; the rationalization for considering the one shown is that it minimizes steric interactions with the carbonyl oxygen. Since organometallic reagents and many of the hydride-transfer reagents coordinate at this oxygen early in the course of the reaction, it is assumed that the oxygen and coordinated groups will prefer to be staggered between the two smallest groups.

Regardless of the details of the mechanism of nucleophilic addition to carbonyl groups, a useful empirical relationship has been established and is known as *Cram's rule of steric control of asymmetric induction*. The major diastereomer will correspond to transfer of the incoming group from the less hindered side of the carbonyl group in the conformation in which the carbonyl group is staggered between the two smallest substituents on the adjacent chiral center.[40]

Many types of chemical reactions have been examined to determine their ability to undergo asymmetric transformations. The extent of selectivity in an asymmetric reaction is expressed in terms of the *percent enantiomeric excess*, thus:

$$
\begin{array}{ccc}
\text{percent} & \text{percent} & \text{percent} \\
\text{enantiomeric} = & \text{major} & - & \text{minor} \\
\text{excess} & \text{enantiomer} & \text{enantiomer}
\end{array}
$$

It is instructive to consider examples of some asymmetric transformations at this point.

It is a fairly simple matter to carry out hydroboration–oxidation with optically active boranes. The readily available optically active terpene $(+)$-2-pinene is converted by diborane to a dialkylborane.[41] This reagent can then add to alkenes, and, on oxidation, optically active alcohols are obtained.[42] Table 10.1 gives some data on the percent enantiomeric excess observed in such reactions.

High selectivity is noted for relatively reactive internal alkenes (entries 3, 5, and 6). Much lower selectivity is found for the more hindered *trans* isomers and in unstrained cyclic alkenes (entries 4, 7, and 9). It is believed that this selectivity results from the more hindered alkenes reacting with a partially dealkylated, and therefore less bulky, borane. Terminal alkenes also show modest selectivity, probably because there is less steric bulk at the reaction site, and the diastereomeric transition states are therefore not very different energetically. Another reason for the diminished selectivity in terminal alkenes is that the chiral center is formed one atom further

40. D. J. Cram and K. R. Kopecky, *J. Am. Chem. Soc.* **81**, 2748 (1959).
41. G. Zeifel and H. C. Brown, *J. Am. Chem. Soc.* **86**, 393 (1964).
42. H. C. Brown, N. R. Ayyangar, and G. Zweifel, *J. Am. Chem. Soc.* **86**, 397 (1964).

**Table 10.1. Stereoselectivity in Some Hydroboration–
Oxidations of Alkenes with Di-3-pinanylborane[a]**

No.	Alkene	Enantiomeric excess, %
1	2-Methyl-1-butene	21
2	2,3-Dimethyl-1-butene	30
3	*cis*-2-Butene	87
4	*trans*-2-Butene	13
5	*cis*-2-Pentene	82
6	*cis*-3-Hexene	91
7	1-Methylcyclohexene	18
8	Norbornene	70
9	Bicyclo[2.2.2]oct-2-ene	17

a. H. C. Brown, N. R. Ayyangar, and G. Zweifel, *J. Am. Chem. Soc.* **86**, 397 (1964); **86**, 1071 (1964); G. Zweifel, N. R. Ayyangar, T. Munekata, and H. C. Brown, *J. Am. Chem. Soc.* **86**, 1076 (1964).

away from the optically active borane. As a general rule, the enantiomeric selectivity

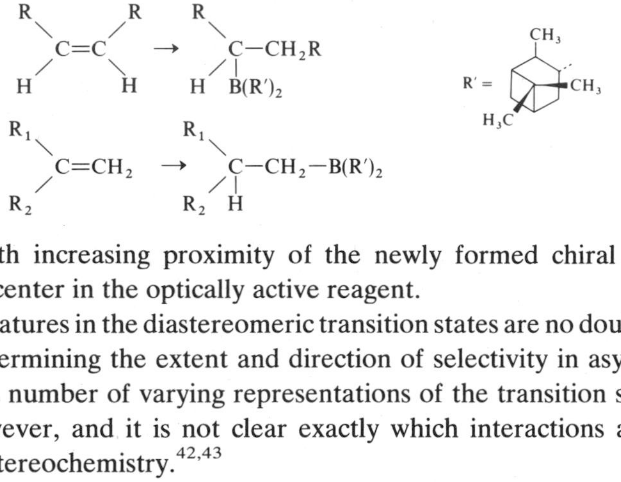

increases with increasing proximity of the newly formed chiral center and the asymmetric center in the optically active reagent.

Steric features in the diastereomeric transition states are no doubt the governing factor in determining the extent and direction of selectivity in asymmetric hydroborations. A number of varying representations of the transition states have been offered, however, and it is not clear exactly which interactions are dominant in controlling stereochemistry.[42,43]

Moderate enantiomeric enrichment has been obtained in Diels–Alder reactions of optically active esters of acrylic acids:

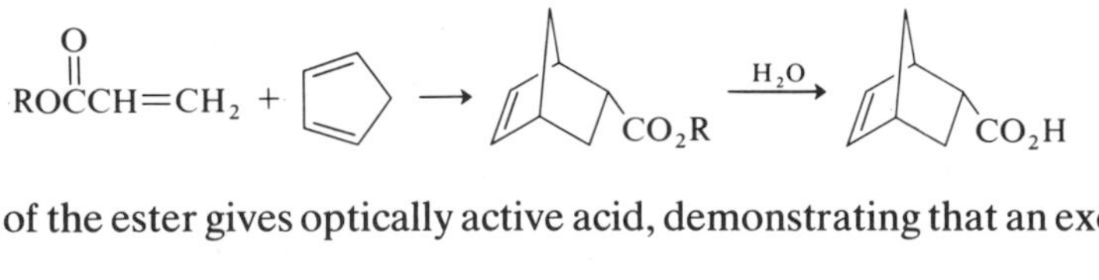

Hydrolysis of the ester gives optically active acid, demonstrating that an excess of one

42. See p. 425.
43. K. R. Varma and E. Caspi, *Tetrahedron* **24**, 6365 (1968); A. Streitwieser, Jr., L. Verbit, and R. Bittman, *J. Org. Chem.* **32**, 1530 (1967); D. R. Brown, S. F. A. Kettle, J. McKenna, and J. M. McKenna, *Chem. Commun.*, 667 (1967); D. J. Sandman, K. Mislow, W. P. Giddings, J. Dirlam, and G. C. Hanson, *J. Am. Chem. Soc.* **90**, 4877 (1968).

Table 10.2. Stereoselectivity in Some Diels–Alder Reactions of Chiral Acrylate Esters[a]

R	BF_3-O(Et)$_2$ $-70°$	Uncatalyzed 25–35°
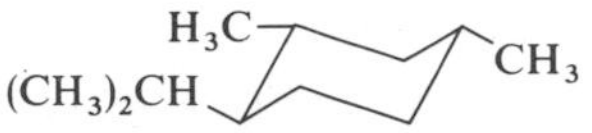 (CH$_3$)$_2$CH, H$_3$C, CH$_3$	82–85	7
CH$_3$(CH$_2$)$_5$CHCH$_3$	27	3
(CH$_3$)$_3$CCHCH$_3$	88	5

a. J. Sauer and J. Kredel, *Tetrahedron Lett.*, 6359 (1966).

of the enantiomers has been formed. The extent of enantiomeric enrichment varies with the chiral group R and with reaction conditions. Specifically, Lewis acid-catalyzed additions at low temperature are much more highly selective than uncatalyzed reactions carried out near room temperature. Some of the data available are presented in Table 10.2.

Asymmetric synthesis of dialkylacetic acids can be done using a chiral oxazoline intermediate **4** derived from a readily available optically active amino alcohol. The

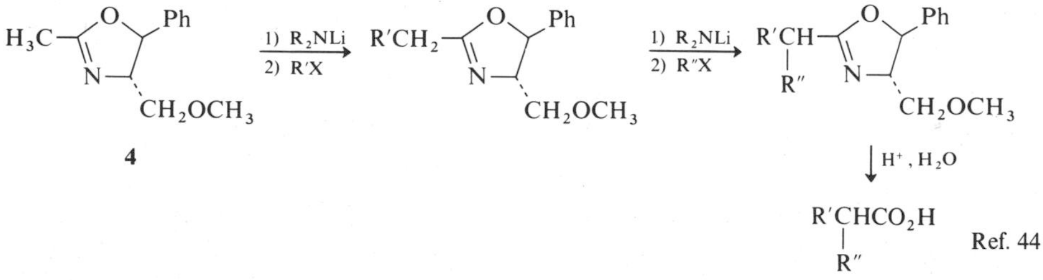

oxazoline is converted to a carbanion and alkylated with one of the two desired substituent groups. A second sequence of carbanion formation and alkylation, followed by hydrolysis of the oxazoline, provides the carboxylic acid in 50–75% optical purity. An interesting facet of this procedure is that the acid can be obtained in predominantly the *R* or *S* configuration, depending on the sequence in which the groups R' and R'' are introduced. The chiral center, of course, is formed in the second alkylation step. The transition state proposed is shown:

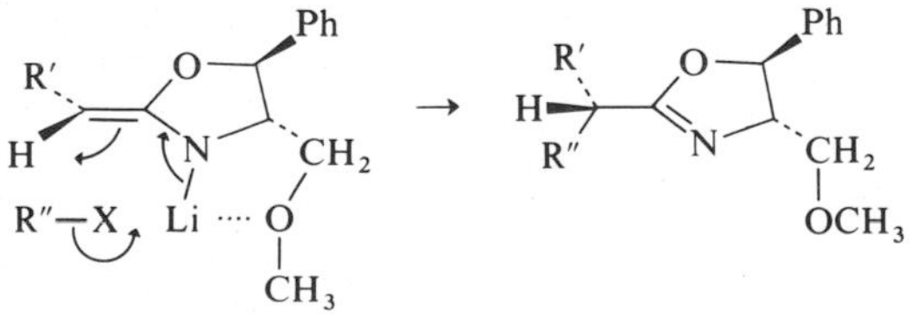

44. A. I. Meyers and G. Knaus, *J. Am. Chem. Soc.* **96**, 6508 (1974).

Introduction of asymmetry in the chiral sense observed requires preferential bottom-side attack and also that the configuration of the exocyclic double bond be as shown. The chelation by the methoxy group can account for preferential bottom-side attack; the steric bulk of the phenyl group may also contribute. The configuration of the double bond is presumably dictated by steric factors.

These transformations serve to illustrate the principles involved in asymmetric synthesis. The requirements for efficient synthetic utilization are: (a) an easily available optically active reagent that can carry out the desired transformation, and (b) reaction conditions that lead to a high percentage of enantiomeric preference. In general, it is also desirable to be able to recover the optically active reagent. The Diels–Alder example is a case where this can be accomplished. Hydrolysis or lithium aluminum hydride reduction gives the product and also returns the original alcohol, which can be reused. Similarly, in the synthesis of dialkylacetic acids, the optically active amino alcohol can be recovered by hydrolysis.

The prediction and rationalization of the direction of stereoselective reactions normally rests on steric factors. Situations in which large differences in steric interactions can be anticipated are likely to give high selectivity. An example that can be cited is the very bulky hydride reducing agent developed by Corey and co-workers to meet the need for a highly selective asymmetric reduction in the course of prostaglandin synthesis.[45] A key step in the synthesis is the reduction of the ketone group in compound **5**:

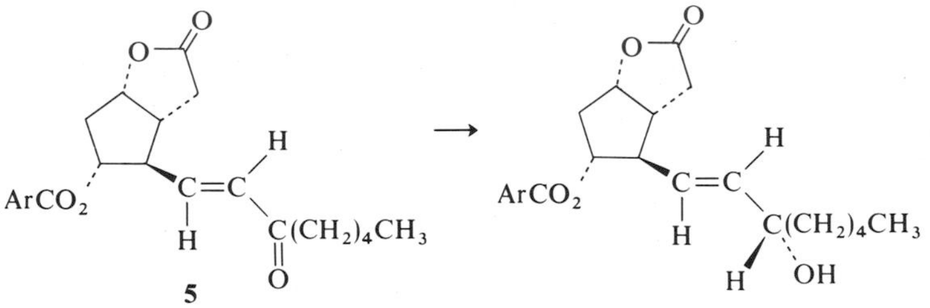

The desired configuration at the chiral center created on reduction is the one shown. Reduction with zinc borohydride is essentially nonstereoselective, giving a 1:1 mixture of the desired alcohol and its diastereomer. A complex hydride prepared from di-3-pinanylborane and methyllithium (**6**) increases the ratio in favor of the

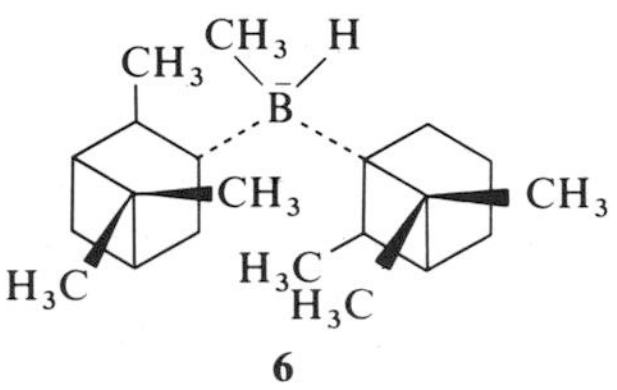

desired diastereomer to ~2:1. An even bulkier complex hydride was prepared by reaction of *t*-hexylborane with limonene:

45. E. J. Corey, S. M. Albonico, U. Koelliker, T. K. Schaaf, and R. K. Varma, *J. Am. Chem. Soc.* **93**, 1491 (1971).

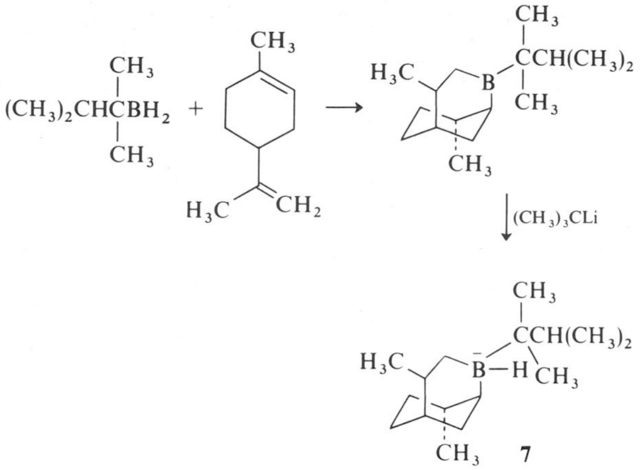

This trialkylborane gave a borohydride ion, **7**, on reaction with *t*-butyllithium. This borohydride increases the ratio in favor of the desired alcohol to 4.5 : 1.

10.4. Synthetic Strategy

The material covered to this point has been primarily a description of the tools the synthetic chemist has at his disposal. The most fundamental component of this armamentarium is the extensive catalog of reactions now available to chemists. The information on reaction conditions, stereochemistry, and efficiency that applies to a given reaction is basic to judging the applicability of the reaction to a given synthetic objective. General mechanistic insight is also an important tool. New synthetic procedures must often be developed to meet particular synthetic challenges, and the usual basis for proposing a new synthetic method is that the suggested reaction appears mechanistically sound. In this chapter, some special tactical tools of synthesis, such as protecting groups and synthetic equivalents, have been considered. All these tools of synthesis, however, require an adequate plan and strategy if they are to be applied successfully to a synthetic problem.

The key to planning the efficient synthesis of an organic compound lies in critical evaluation of alternative reaction sequences that could reasonably be expected to lead to the desired structure from available starting materials. In general, both the number of alternative sequences and the complexity of any single synthetic plan increase with the size of the molecule and with increasing numbers of functional groups and chiral centers. The problem of analyzing synthetic possibilities with a view to choosing the most efficient of several routes is one of recognizing the possible pathways between the acceptable starting materials and the required goal. A suitable pathway of synthetic steps must be laid out.

The restrictions applied to this pathway will depend on the reasons for the synthesis. Rigid control of stereochemistry is necessary if, for example, a biologically active natural product with several chiral centers is the goal of the synthesis. A commercial synthesis of a material used in substantial quantities may impose availability and cost of the starting material as the critical limiting factors in an acceptable synthesis. An industrial synthetic process would bear heavy restrictions as to acceptable by-products, a problem that is not usually important in laboratory-scale syntheses.

The development of a satisfactory plan for the synthetic task at hand is the initial intellectual challenge to a synthetic chemist. This task puts a premium on creativity and imagination. There is no single correct answer that can be arrived at by an established routine.

The initial step in the development of a synthetic plan should involve a *retrosynthetic* analysis. The structure of the molecule should be dissected step by step along reasonable pathways to successively simpler compounds until molecules acceptable as starting material are reached. Several factors must enter into this process, and all are closely interrelated. There is the *molecular framework* that can be built up through a series of *key intermediates*. The initial stage in a retrosynthetic analysis is to recognize key fragments of the molecule that might be combined. At this stage of the analysis, the potential advantages of a *convergent synthesis* should be considered. If, for example, a molecule consists of two major fragments, **G** and **H**, and a side chain, **I**, it is more efficient to synthesize **G** and **H** separately and then combine them, rather than make **G** first and build **H** upon it, step by step. The overall

Convergent Synthesis

$$\text{A} + \text{B} \rightarrow \text{C} \xrightarrow{\text{D}} \text{G}$$
$$\text{E} + \text{F} \rightarrow \text{H}$$
$$\searrow \text{G–H} \xrightarrow{\text{I}} \text{G–H–I}$$

Linear Synthesis

$$\text{A} + \text{B} \rightarrow \text{C} \xrightarrow{\text{D}} \text{G} \xrightarrow{\text{E}} \text{G–E} \xrightarrow{\text{F}} \text{H} \rightarrow \text{G–H} \xrightarrow{\text{I}} \text{G–H–I}$$

yield in a synthetic sequence is the product of the yields of the individual steps. Since yields are almost always less than 100%, total yield tends to decrease with an increasing number of steps in a sequence. A linear sequence maximizes the number of steps to which the original starting materials must be subjected. A convergent synthesis, in contrast, allows one to build up separate fragments and then combine them; the number of steps to which each set of starting materials is subjected is thus decreased.

Once candidates for key intermediates are recognized, the issues of *stereochemistry* must be faced. These take several forms: *cis–trans* isomerism at carbon–carbon double bonds, stereochemistry at ring junctions, and relative configuration at chiral centers. Only those pathways that promise stereospecific or stereoselective formation of the desired compound are likely to be acceptable. For example, a molecule with one possibility for *cis–trans* isomerism and two chiral centers allows for $\frac{1}{2}(2^3) = 4$ diastereomers. Because diastereomers usually have somewhat similar physical properties, the likelihood of obtaining a single pure stereoisomer becomes very low if the stereochemistry is not controlled adequately.

Finally, there is *functionality*. We have discussed in the sections on protecting groups and synthetic equivalents the idea of protected or masked functional groups. It is frequently necessary to interconvert related functional groups. A carbon atom that must be substituted by a hydroxyl group in the final product may be carried through a synthetic sequence as a carbonyl carbon, and then converted to the desired alcohol functionality rather late in the synthetic scheme. Similarly, a cyano group may eventually be converted to an ester group, and so on. The achievement of correct functionality is often somewhat less difficult than the establishment of the overall molecular skeleton and stereochemistry because of the large number of procedures for interconverting many of the common functional groups.

These ideas can be illustrated by considering some examples of successful multistep syntheses. In these examples, we will have the benefit of hindsight, but let us attempt to recognize in particular the stages at which molecular skeleton, stereochemistry, and functionality were faced. The examples chosen are natural products, but the same ideas apply to any synthetic target.

Juvabione. Let us look first at a molecule that has been synthesized in a number of alternative ways. Juvabione is a terpene-derived ketoester that has been isolated from certain plant species. It exhibits juvenile hormone activity; that is, it can modify the process of metamorphosis in certain insects. At least seven syntheses of this

$$(CH_3)_2CHCH_2\overset{\overset{\displaystyle O}{\|}}{C}CH_2 - \overset{\overset{\displaystyle H_3C}{|}}{\underset{H}{C}}\diagup$$

juvabione

material have been reported. They are outlined in Schemes 10.4 through 10.10. Comparison of some of these schemes can indicate alternative approaches to this compound. The syntheses in Schemes 10.6, 10.7, and 10.10 are stereoselective. Those in Schemes 10.4, 10.5, and 10.8 are not. The stereochemistry of the synthesis in Scheme 10.9 is uncertain from the abstract of the patent in which it is reported. The two syntheses in Schemes 10.4 and 10.5 are parallel for much of their course, and we will refer primarily to Scheme 10.4.

In all of these approaches to juvabione, a precursor of the cyclohexene ring is the key intermediate. In Schemes 10.4 and 10.5, an aromatic ring is the precursor of the cyclohexene system, and the reduction of the aromatic system is carried out fairly late in the synthesis. Schemes 10.6 and 10.9 use an optically active natural product, limonene, as the starting material. This provides a useful simplification of stereochemistry, particularly if the product is required in optically active form. The syntheses in Schemes 10.4, 10.6, and 10.10 proceed to build up the side chain, step-by-step, from the key cyclic intermediate. In the syntheses in Schemes 10.7, 10.8, and 10.9, the side chain is almost in final form when it is attached to the cyclic intermediate.

The issue of stereochemistry in these syntheses concerns the relative configuration of the two adjacent chiral carbons. In Schemes 10.4 and 10.5, the

Scheme 10.4. Juvabione Synthesis: K. Mori and M. Matsui[a]

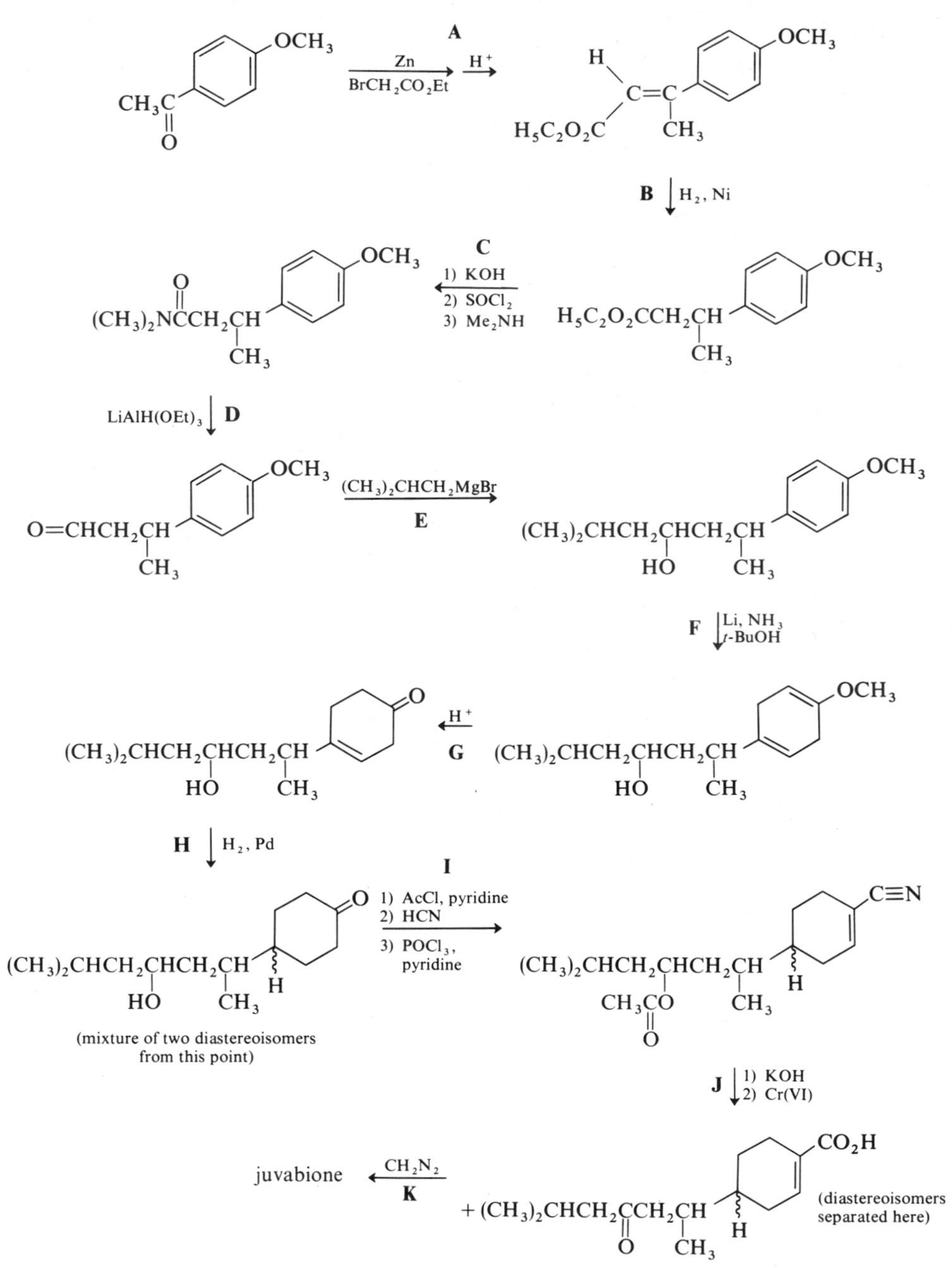

a. K. Mori and M. Matsui, *Tetrahedron* **24**, 3127 (1968).

Scheme 10.5. Juvabione Synthesis: K. S. Ayyar and G. S. K. Rao[a]

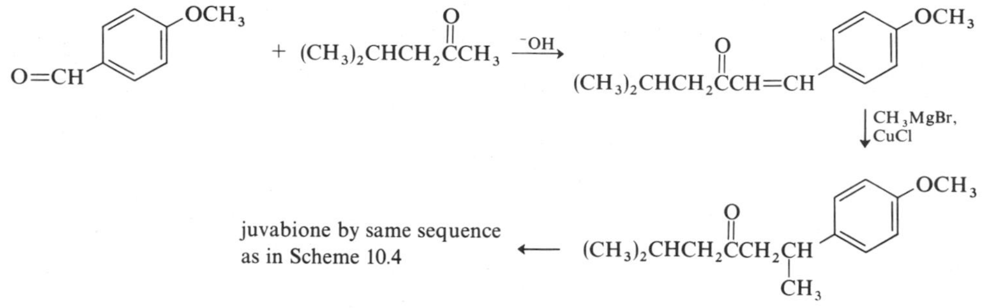

a. K. S. Ayyar and G. S. K. Rao, *Can. J. Chem.* **46**, 1467 (1968).

stereochemistry of the product is established at the reduction step, step H. Both possible diastereomers are formed, and they have to be separated after completion of step J. Similarly, no attempt to control stereochemistry was made in Scheme 10.8, and the product is a mixture of the two diastereomers.

Scheme 10.6. Juvabione Synthesis: B. A. Pawson, H.-C. Cheung, S. Gurbaxani, and G. Saucy[a]

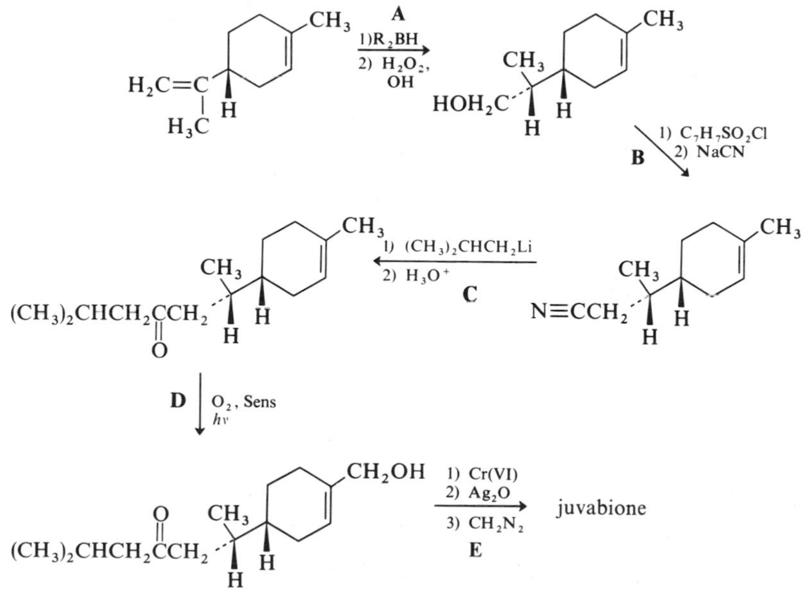

a. B. A. Pawson, H.-C. Cheung, S. Gurbaxani, and G. Saucy, *J. Am. Chem. Soc.* **92**, 336 (1970).

Scheme 10.7. Juvabione Synthesis: A. J. Birch, P. L. Macdonald, and V. H. Powell[a]

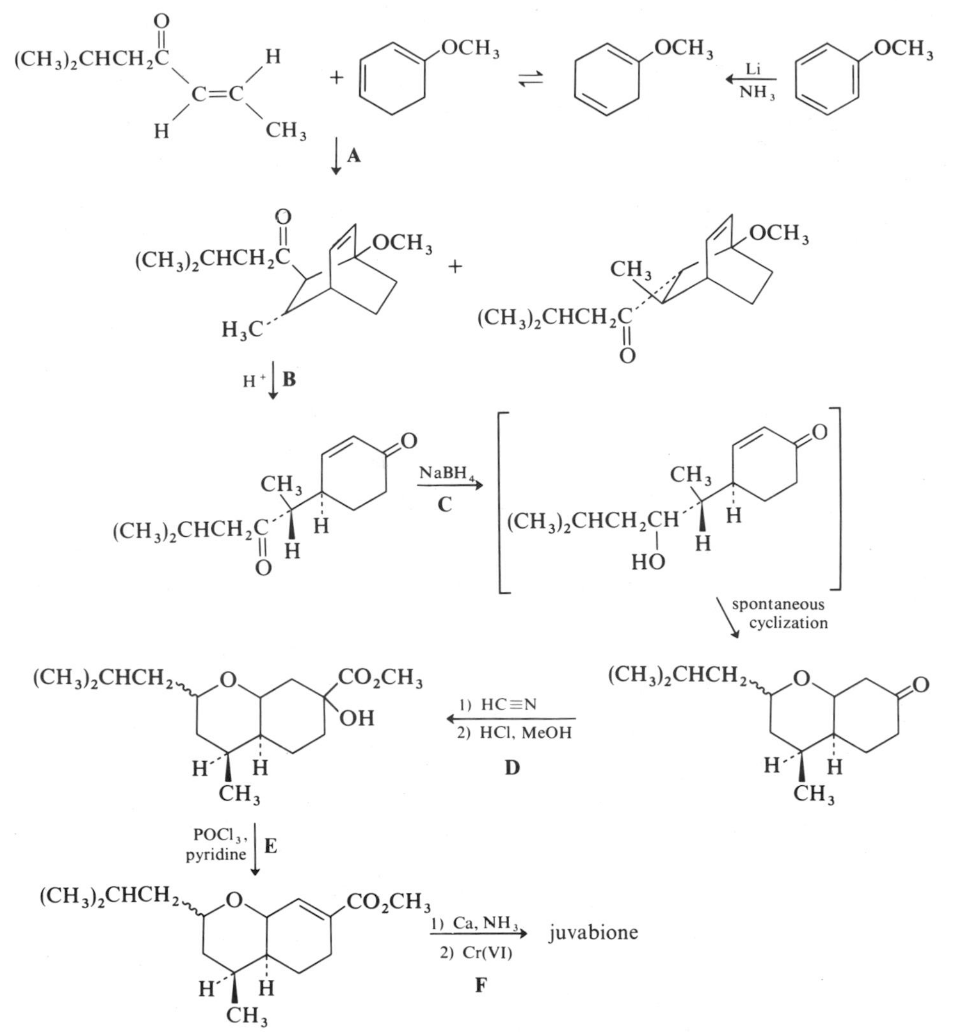

a. A. J. Birch, P. L. Macdonald, and V. H. Powell, *J. Chem. Soc. C.*, 1469 (1970).

The syntheses in Schemes 10.6, 10.7, and 10.10 are stereoselective. In Scheme 10.6, the starting material is an optically active terpene, limonene. The chiral center leads to modest stereoselectivity in the hydroboration (step A), there is a 3:2 preference for the desired stereoisomer; and the intermediate is obtained stereochemically pure after crystallization of a solid derivative. These chiral centers are not involved in any of the further transformations, so the stereochemistry is established early in the synthesis, and all further reactions are free of complications due to separation of stereoisomers. In Scheme 10.7, the stereochemistry is controlled by virtue of the bicyclic nature of the intermediate Diels–Alder adduct. The isomeric

C

juvabione $\longleftarrow$ 1) H_2O, NH_3
2) H^+, H_2O
3) CH_2N_2

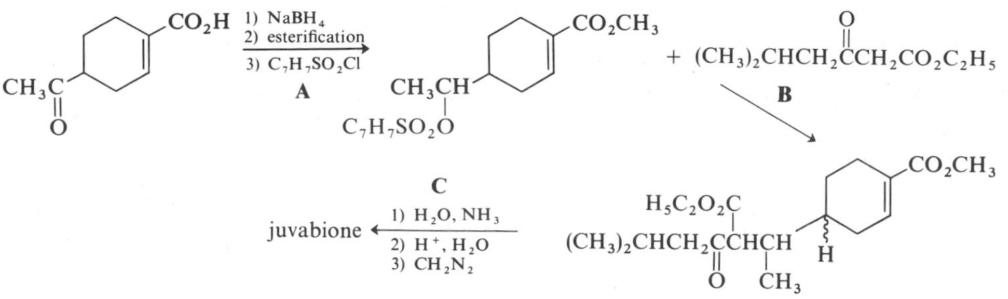

a. A. A. Drabkina and Y. S. Tsizin, *J. Gen. Chem. USSR* (Engl. translation) **43**, 422, 691 (1973).

adduct that is used has the required relative configuration of the two chiral centers, and no further changes in configuration occur at these centers. Step C leads to a mixture of diastereoisomers that is carried through steps D and E, but this chiral center is not present in the final product and so is not a problem with respect to the overall stereochemistry of the synthesis.

In the synthesis described in Scheme 10.10, the stereochemistry is established at an early stage. Stereoselectivity in the protonation of the enamine moiety during the hydrolysis in step B is presumably the result of preferential protonation from the less hindered side of the molecule. None of the other transformations affects these chiral centers, and the overall synthesis is stereoselective. It is interesting to note that step E in this synthesis employs a protected cyanohydrin as a nucleophilic acyl synthetic equivalent.

Scheme 10.9. Juvabione Synthesis: R. J. Crawford[a]

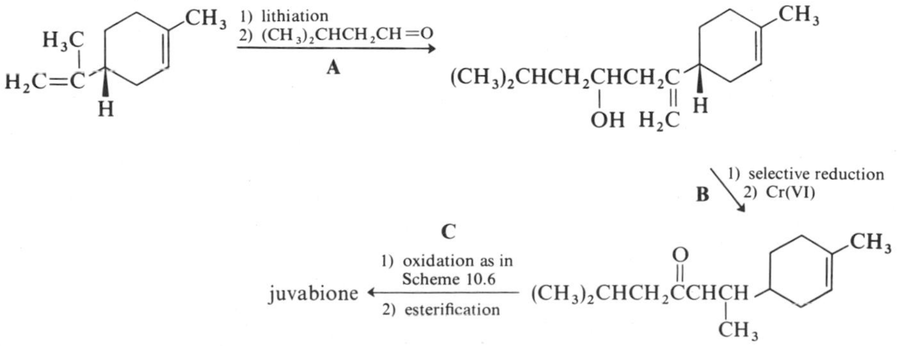

a. R. J. Crawford, US Patent 3,676,506; *Chem. Abstr.* **77**, 113889e (1972).

Scheme 10.10 Juvabione Synthesis: J. Ficini, J. D'Angelo, and J. Noiré[a]

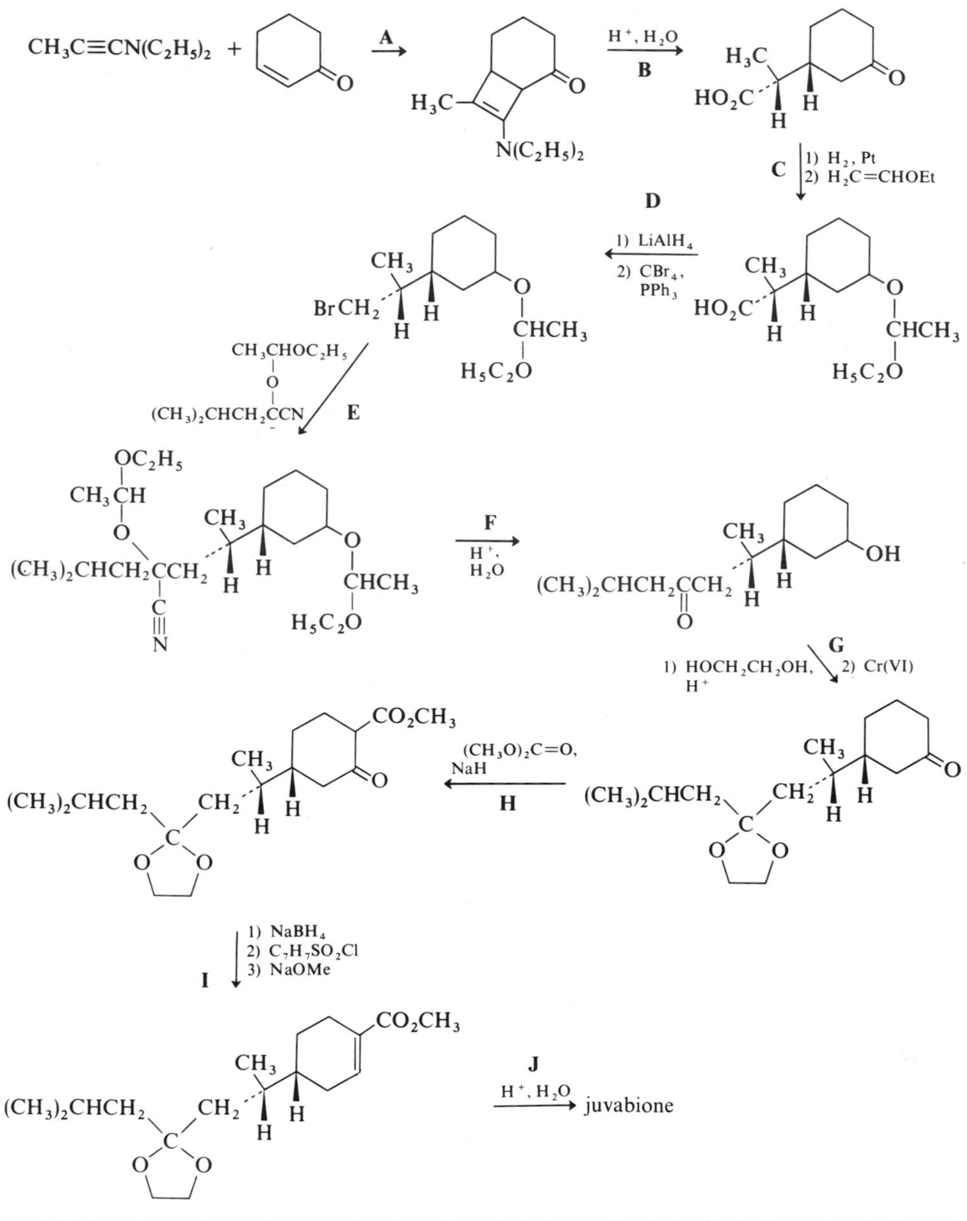

a. J. Ficini, J. D'Angelo, and J. Noiré, *J. Am. Chem. Soc.* **96**, 1213 (1974).

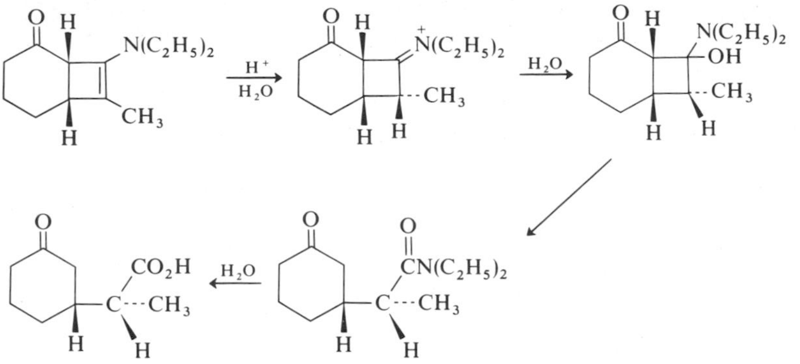

This set of syntheses indicates that there are a variety of ways of elaborating the aliphatic chain required in juvabione. The methods of controlling stereochemistry in the three syntheses where this was done are all different as well. The method of asymmetric synthesis is used in Scheme 10.6. Scheme 10.7 relies on establishing the stereochemistry of a bicyclic intermediate and using the correct isomer for subsequent steps. In Scheme 10.10, a stereoselective protonation establishes the configuration at the chiral centers.

Many syntheses of polyfunctional natural products have been recorded in the literature, and a few are considered in the succeeding paragraphs. The examples have been chosen not because the particular molecules have overriding importance, but with the aim that these examples can serve to indicate how carbon framework, functionality, and stereochemical control are intertwined in the synthesis of moderately complex molecules.

Fumagillol. Fumagillol is an alcohol that occurs in an antibiotic ester fumagillin. Although the molecule is comparable in size to juvabione, the problems of

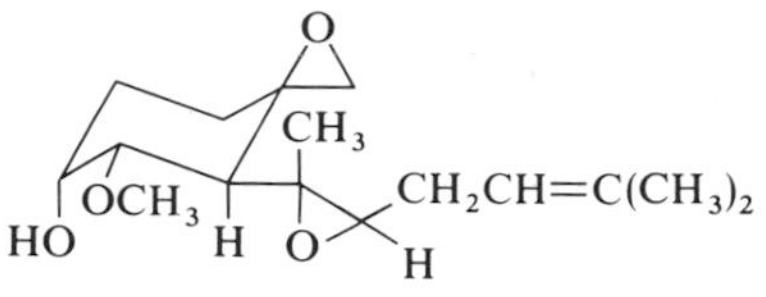

fumagillol

stereochemistry are more severe. There are six chiral centers. The two epoxide functions add to the synthetic challenge. They are relatively reactive groups and in general would therefore need to be introduced fairly late in the synthesis. Alkene oxidation, reaction of a carbonyl compound with a sulfur ylide, and intramolecular nucleophilic cyclization are the most general procedures that might be used to introduce the epoxide functionality at appropriate points. Since a ring is involved, one possible approach to synthesis of fumagillol might be construction of a ring, followed by addition of the side chain. The ring might be formed by intramolecular alkylation or condensation, but the involvement of a six-membered ring also suggests that the Diels–Alder addition should be considered. Can a dienophile be conceived in which X and Y could potentially be converted to the exocyclic epoxide? If so, the Diels–Alder reaction would be attractive in that it would provide a double bond at the position where the hydroxyl groups and methoxy group are needed. These groups might be introduced by a sequence of regioselective and stereospecific reactions at the double bond. If these challenges could be met, the problem would have been reduced to achieving a synthesis of the diene in which R would have the potential of being converted to the required side chain.

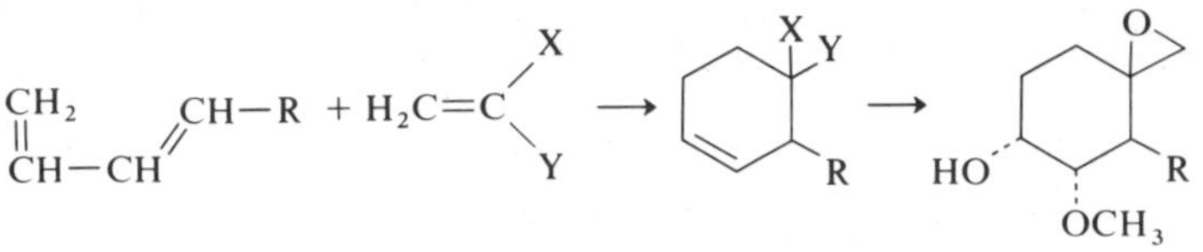

A solution of the synthetic problem along these lines was achieved by Corey and Snider.[46] It is outlined in Scheme 10.11. α-Bromoacrolein functions as a synthetic equivalent of the unstable, and probably unreactive, dienophile allene oxide,

$$CH_2=C-CH_2.$$ The epoxide ring is eventually formed in step H by intramolecular alkylation after the aldehyde group has been reduced to a primary alcohol. Step I is the point at which the stereochemistry of the hydroxyl and methoxy function is established. The osmium tetroxide oxidation insures *syn* addition. The stereochemistry relative to the alkyl side chain is governed by steric factors. The reagent approaches the double bond from the side opposite the alkyl substituent. The selective methylation of the diol in step J is probably the result of the greater accessibility of an equatorial hydroxyl group. It is the equatorial hydroxyl group that is selectively methylated. The precise selectivity of this step is not entirely clear, since

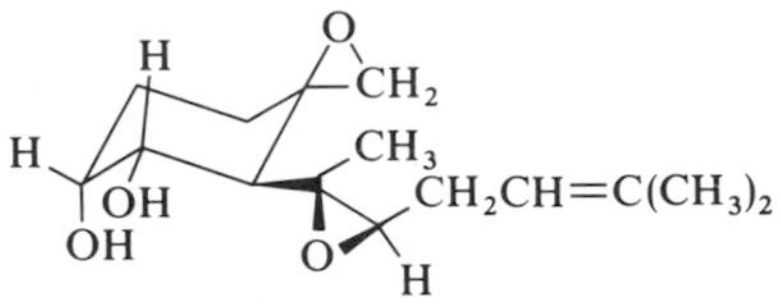

the moderate yield (65%) of the desired product permits the possibility that some of the isomer is formed.

Step G involves a question of both chemical and stereochemical selectivity. The trisubstituted double bond must be oxidized in preference to the endocyclic double bond. This preferred substitution does occur, largely because of the increased reactivity of the more highly substituted alkene linkage in peroxycarboxylic acid oxidations. The double bond oxidized presents two nonidentical (diastereotopic) faces to the oxidant. The observed (and desired) stereochemistry results because of

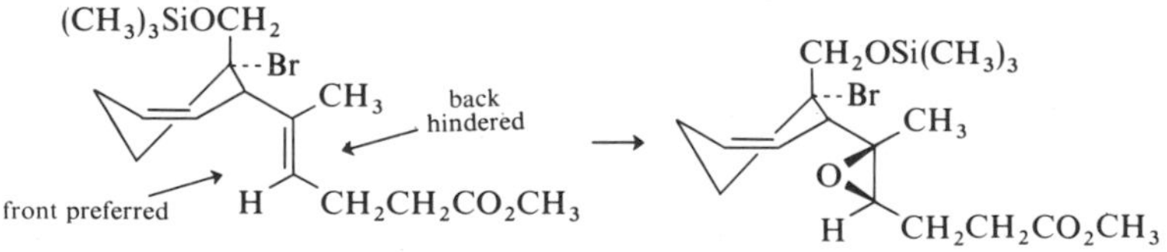

46. E. J. Corey and B. B. Snider, *J. Am. Chem. Soc.* **94**, 2549 (1972).

Scheme 10.11. Synthesis of Fumagillol[a]

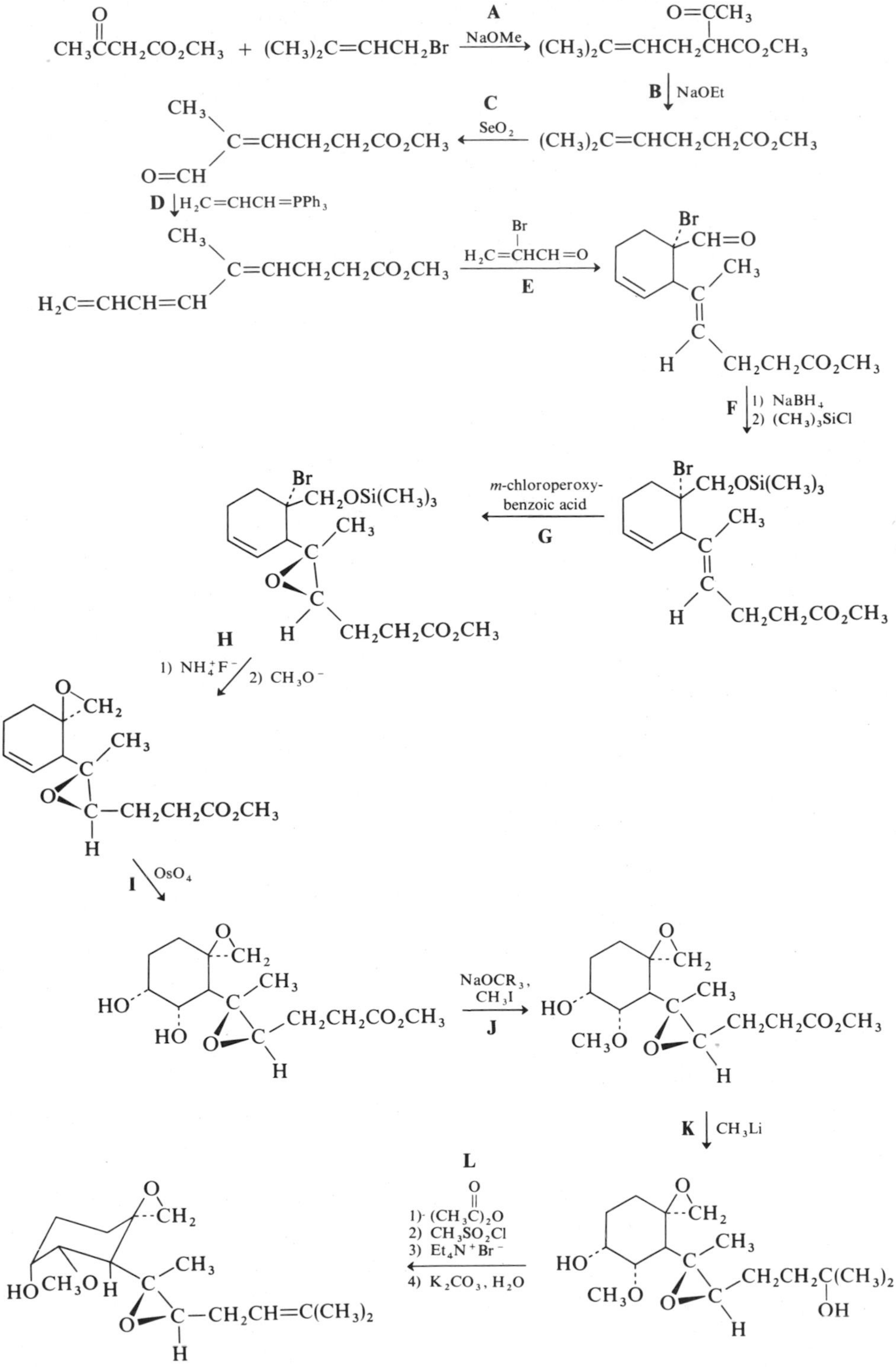

a. E. J. Corey and B. B. Snider, *J. Am. Chem. Soc.* **94**, 2549 (1972).

the steric shielding created on the back side of the double bond by the bromo and trimethylsilyl groups, as pictured.

Caryophyllene. Caryophyllene, **8**, is a sesquiterpene that presents the challenge of the synthesis of fused small (four-membered) and medium (nine-membered) rings. The functionality is not complex, since only alkene groups are present. There are two chiral centers and one alkene unit that must be constructed with *trans* stereochemistry. The successful synthesis is outlined in Scheme 10.12.[47] Several of

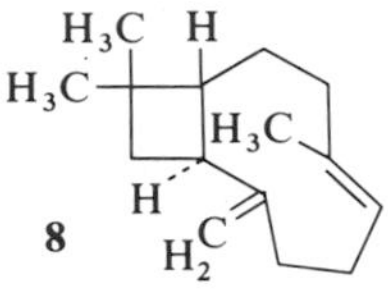

caryophyllene

the reactions are functional group transformations that require no special comment, as the reactions have been discussed at other points in the text. Steps A (Section 6.2), B, C (Sections 1.4 and 2.4), D (Section 5.1), G, H (Section 3.1), I, and L (Section 2.5) are relatively straightforward transformations. A pertinent feature of step D is the introduction of a potential carboxylate group masked as the dimethoxy acetal. After removal of the triple bond, the acetal function is converted to a carboxylic acid group by hydrolysis and oxidation. Lactonization occurs because of the proximity of the hydroxyl group. Step F is a Dieckmann condensation (Section 2.4) with the α-carbon of the lactone ring serving as the nucleophile. A key step is J. Here, a fragmentation reaction (Section 8.5) is used to construct the medium-sized ring. This procedure also permits establishment of the correct stereochemistry at the ring junction, since the adjacent carbonyl group permits epimerization to the more stable *trans* fusion. The fragmentation reaction is stereospecific. Of the two diols separated in step H, the one with a *cis* relationship between the secondary hydroxyl group and the angular methyl group will fragment to the required *trans* alkene.

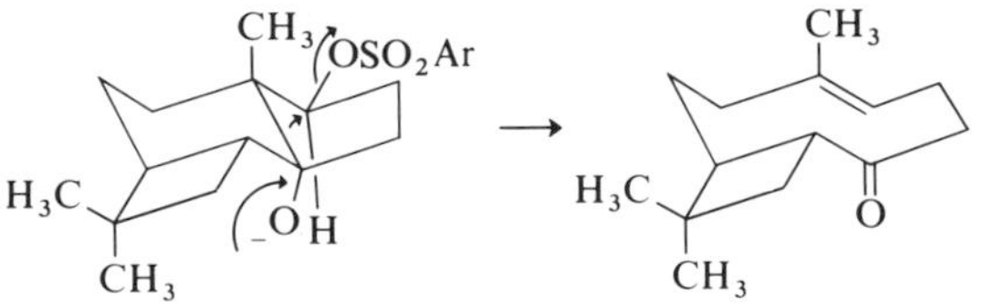

Sirenin. Sirenin is a hormone involved in the sexual reproduction process of the fungus *Allomyces*. The alkene linkage needs to be constructed in such a way as to

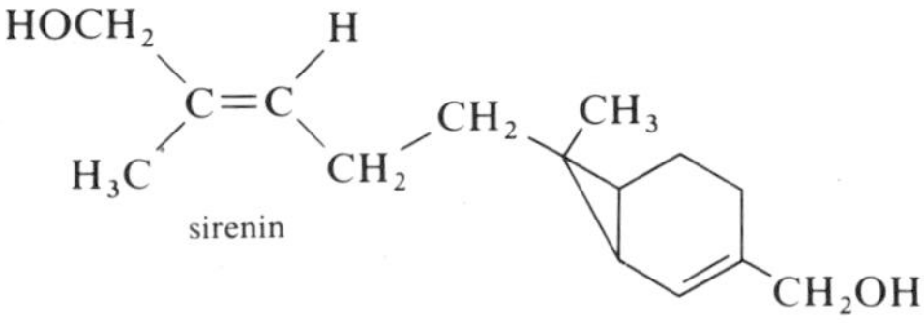

sirenin

47. E. J. Corey, R. B. Mitra, and H. Uda, *J. Am. Chem. Soc.* **86**, 485 (1964).

Scheme 10.12. Synthesis of Caryophyllene[a]

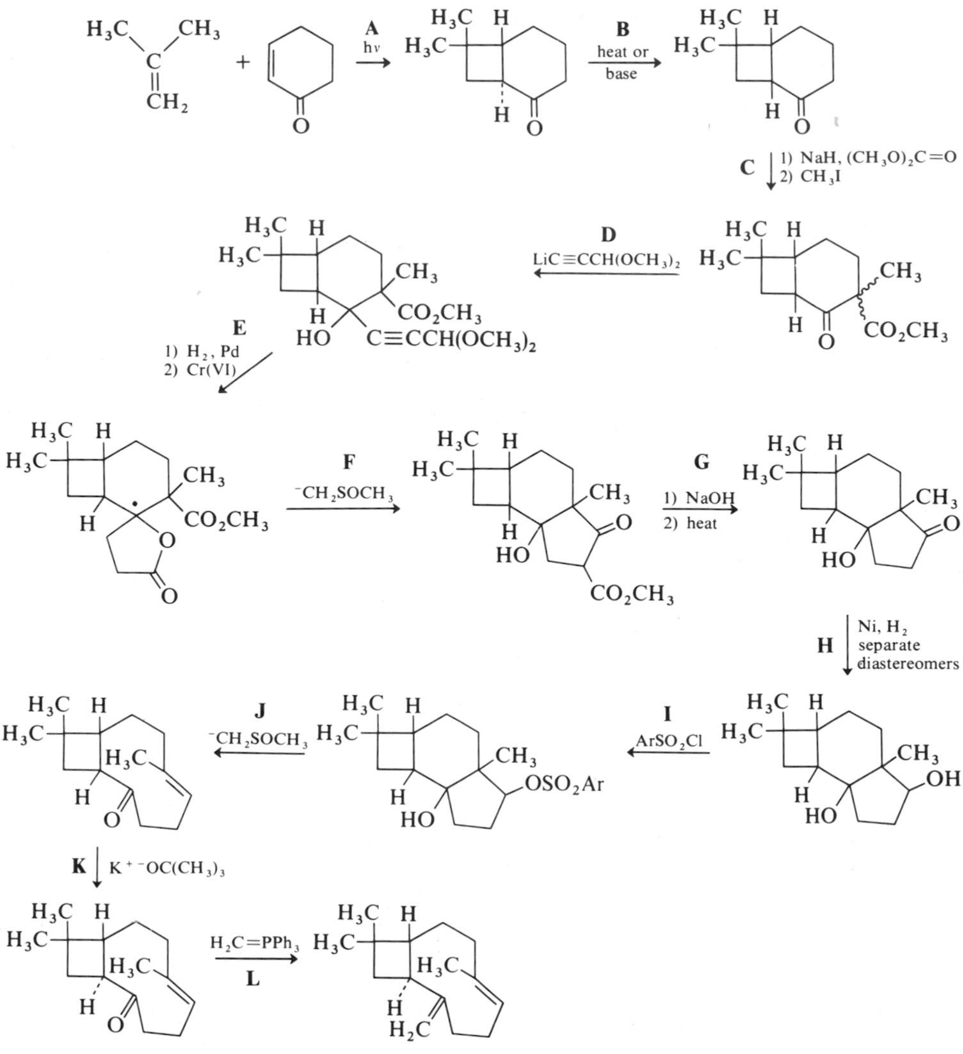

a. E. J. Corey, R. B. Mitra, and H. Uda, *J. Am. Chem. Soc.* **86**, 485 (1964).

place the hydroxymethyl group in the *trans* orientation relative to the alkyl substituent. The synthesis outlined in Scheme 10.13 was carried out by P. A. Grieco.[48] Others have been reported. The starting material is an available natural product, geraniol. An interesting point attaches to each of the steps in the synthesis. Step A is a special procedure involving an S_N2 displacement of arenesulfonate by chloride ion. These particular conditions are necessary to prevent isomeric chlorides from being

48. P. A. Grieco, *J. Am. Chem. Soc.* **91**, 5660 (1969).

Scheme 10.13. Synthesis of Sirenin[a]

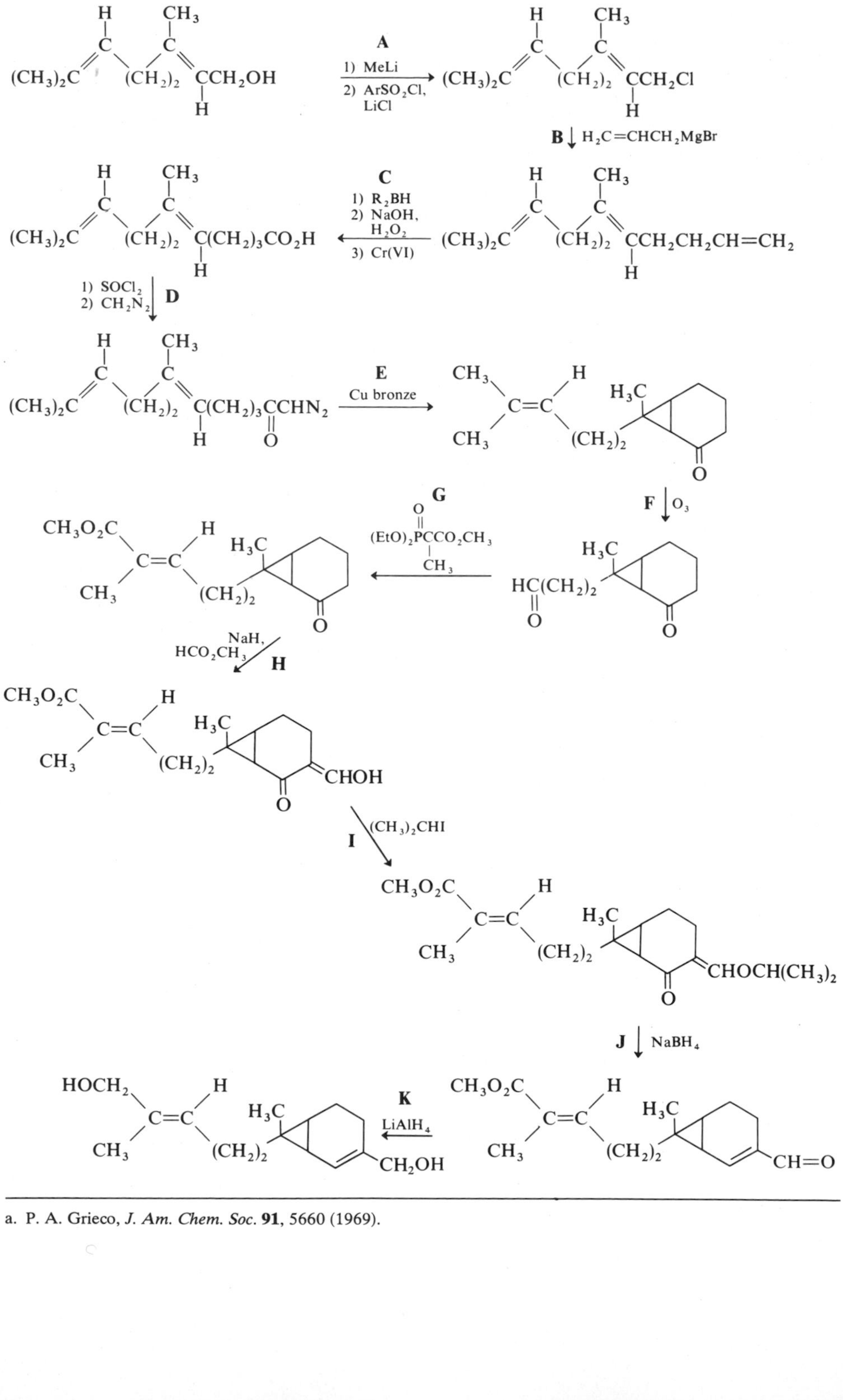

$$\text{(CH}_3\text{)}_2\text{C}=\overset{\text{H}}{\text{C}}\ \ \overset{\text{CH}_3}{\text{C}}=\text{(CH}_2\text{)}_2\ \overset{\text{H}}{\text{C}}\text{CH}_2\text{OH}$$

A
1) MeLi
2) ArSO$_2$Cl, LiCl

$$\text{(CH}_3\text{)}_2\text{C}=\overset{\text{H}}{\text{C}}\ \ \overset{\text{CH}_3}{\text{C}}=\text{(CH}_2\text{)}_2\ \overset{\text{H}}{\text{C}}\text{CH}_2\text{Cl}$$

B ↓ H$_2$C=CHCH$_2$MgBr

C
1) R$_2$BH
2) NaOH, H$_2$O$_2$
3) Cr(VI)

1) SOCl$_2$
2) CH$_2$N$_2$ ↓ **D**

E
Cu bronze

F ↓ O$_3$

G
(EtO)$_2$PCCO$_2$CH$_3$
CH$_3$

NaH,
HCO$_2$CH$_3$ **H**

I (CH$_3$)$_2$CHI

J ↓ NaBH$_4$

K
LiAlH$_4$

a. P. A. Grieco, *J. Am. Chem. Soc.* **91**, 5660 (1969).

formed as a result of the system being allylic. Step B is an organometallic coupling (Section 5.1) that is favorable because of the allylic nature of the reactants. In step C, the triene is selectively hydroborated at the terminal position by using a bulky disubstituted (diisoamylborane) borane, followed by a successful oxidation of a primary alcohol to the carboxylic acid in the presence of carbon–carbon double bonds. In step E, the bicyclic ring is constructed. A routine α-diazoketone synthesis (Section 8.2) is followed by copper-catalyzed decomposition. Ring formation occurs by intramolecular carbenoid addition (Section 8.2). In step F, the synthesis seems to take a step backward, but this is to permit stereospecific introduction of an oxygenated functional group. This introduction is accomplished in the next step using a phosphonate modification of the Wittig reaction, which is known to proceed with the required stereochemistry (Section 2.5). Steps H through J are a sequence of reactions that can accomplish the transformation:

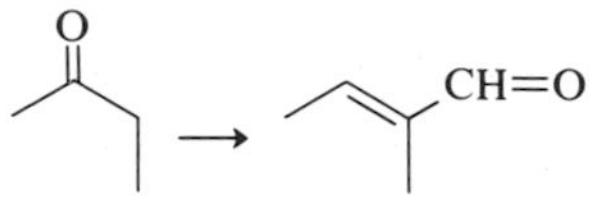

The mechanism is outlined below:

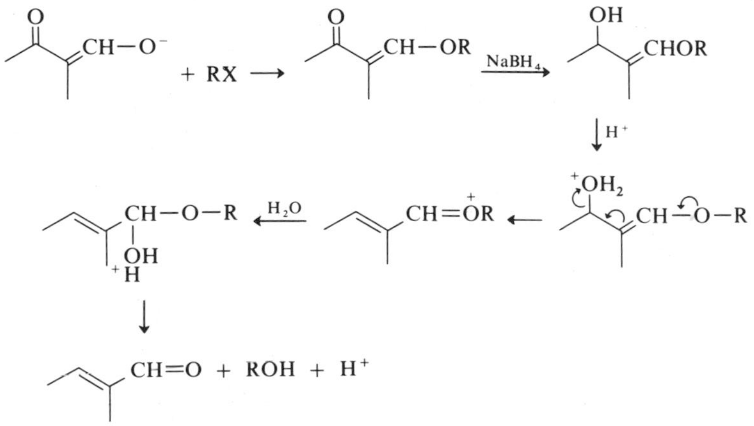

The final step, step H, is a LiAlH$_4$ reduction of both the aldehyde and ester groups to the required primary alcohol.

  Prostaglandin Intermediates. Schemes 10.14 through 10.17 outline some synthetic routes to several derivatives of **9** and **10**. These stereoisomeric compounds

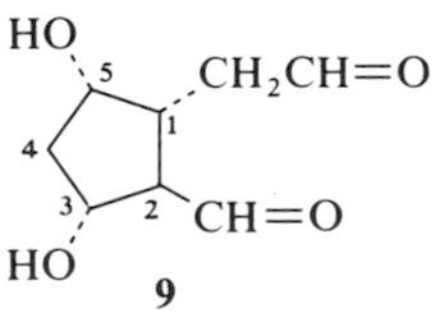

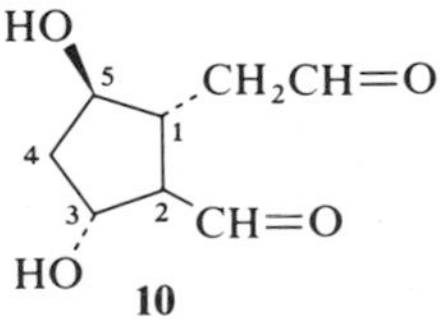

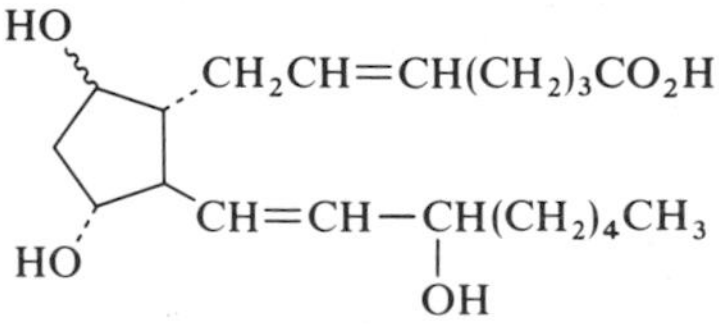

prostaglandin

are intermediates in the synthesis of prostaglandins, a class of natural materials present in tiny amounts in various organs of mammals and other organisms. The

Scheme 10.14. A Synthesis of a Prostaglandin Intermediate[a]

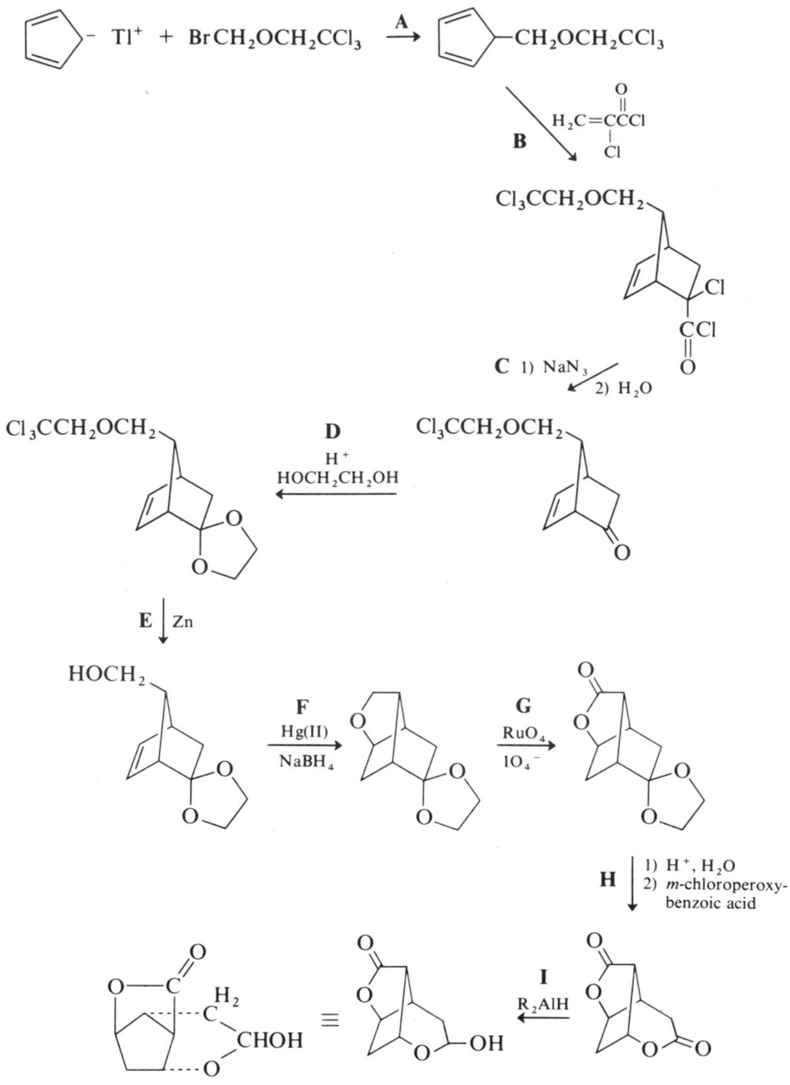

a. J. S. Bindra, A. Grodski, T. K. Schaaf, and E. J. Corey, *J. Am. Chem. Soc.* **95**, 7522 (1973).

synthesis of the compounds is of much current interest because they exhibit a broad range of potent physiological activity, but are not readily available from natural sources. The aldehyde groups in these intermediates are successively generated or unmasked, and the side chains are then introduced, usually using functionalized Wittig reagents. The design of each of these routes is such that the product is obtained in a stereochemically pure form. This purity is of great importance, since with each intermediate having four chiral centers, eight diastereomeric racemic

Scheme 10.15. A Synthesis of a Prostaglandin Intermediate[a]

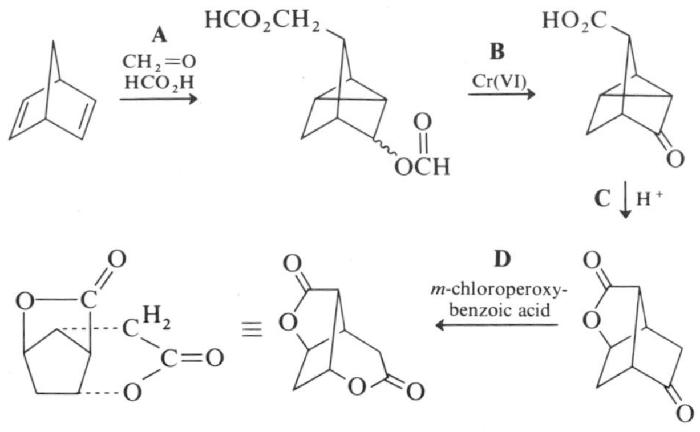

a. J. S. Bindra, A. Grodski, T. K. Schaaf, and E. J. Corey, *J. Am. Chem. Soc.* **95**, 7522 (1973).

compounds could be formed in the absence of any stereochemical control. Each of the syntheses controls the stereochemistry of the product by taking advantage of bicyclic intermediates that establish the relative stereochemistry of the various groups.

In Scheme 10.14, the relative stereochemistry at positions 1 and 3 on the cyclopentane ring is established with the formation of the bicyclo[2.2.1]heptene ring in step B. The geometry of the ring requires that the two new carbon–carbon bonds created in step B be *cis* with respect to one another. The next stereochemically important manipulation of substituents 1 and 3 occurs in step H, where a Baeyer–Villiger oxidation is carried out. Since this reaction proceeds with retention of configuration of the migrating group (Section 9.5), the *cis* relationship of substituents 1 and 3 is preserved. The substituents at C-2 and C-5 must be *cis* to one another, but *trans* to those at C-1 and C-3. The key reaction for establishing this relationship is also step B. The attack of the dienophile is from the less hindered side of the substituted cyclopentadiene. The *cis* relationship of the C-2 substituent to the oxygen at C-5 is established by forming the C–O bond intramolecularly in step F.

Scheme 10.15, which is appreciably shorter than 10.14, uses the same general methods to establish the stereochemical relationships. In step C, the cyclopropane ring is opened by protonation. The incipient carbonium ion is captured intramolecularly, and the *cis* relationship between the C-2 and C-5 substituent is thereby established. The geometry of the bicyclic ring system and the retention of stereochemistry in the Baeyer–Villiger reaction are used to ensure the *cis* stereochemistry of the substituents at C-1 and C-3, as in Scheme 10.14. Step A is an electrophilic substitution that is initiated by protonated formaldehyde. The resulting cation is captured by formic acid, and the primary alcohol is also formylated under the reaction conditions.

$$CH_2=O + H^+ \rightleftharpoons CH_2=\overset{+}{O}H$$

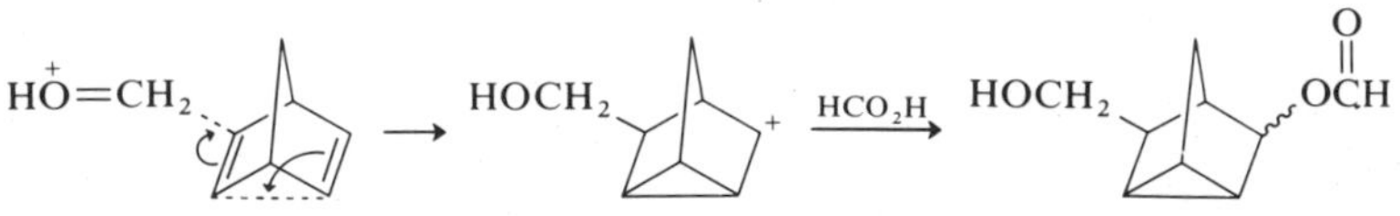

Schemes 10.16 and 10.17 lead to intermediates in which the C-1, C-3, and C-5 substituents are all *cis* to one another. In Scheme 10.16 the *cis* relationship between the C-1 and C-3 groups is again established by the Baeyer–Villiger oxidation, which is carried out as the first step in the synthesis. In step C, the *cis* relationship of the oxygen function at C-5 is established. The iodolactonization is a stereospecific *anti* addition, and the formation of *cis*-fused five-membered rings is preferred to the *trans* ring fusion. The other steps represent reactions that have been discussed previously.

The final route (Scheme 10.17) is radically different from those described in Schemes 10.14 through 10.16. The *cis* stereochemistry of the C-1 carbon and C-5 oxygen is established at step E. This reaction involves solvolysis of the primary alkyl

Scheme 10.16. A Synthesis of a Prostaglandin Intermediate[a]

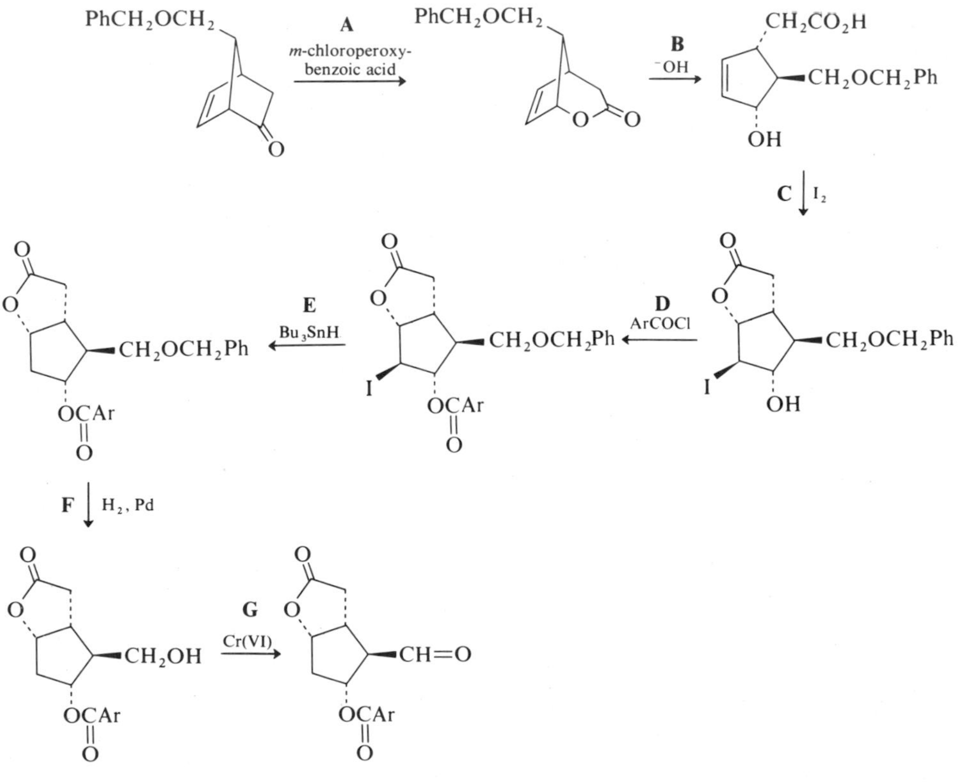

a. E. J. Corey, S. M. Albonico, U. Koelliker, T. K. Schaaf, and R. K. Varma, *J. Am. Chem. Soc.* **93**, 1491 (1971).

Scheme 10.17. A Synthesis of a Prostaglandin Intermediate[a]

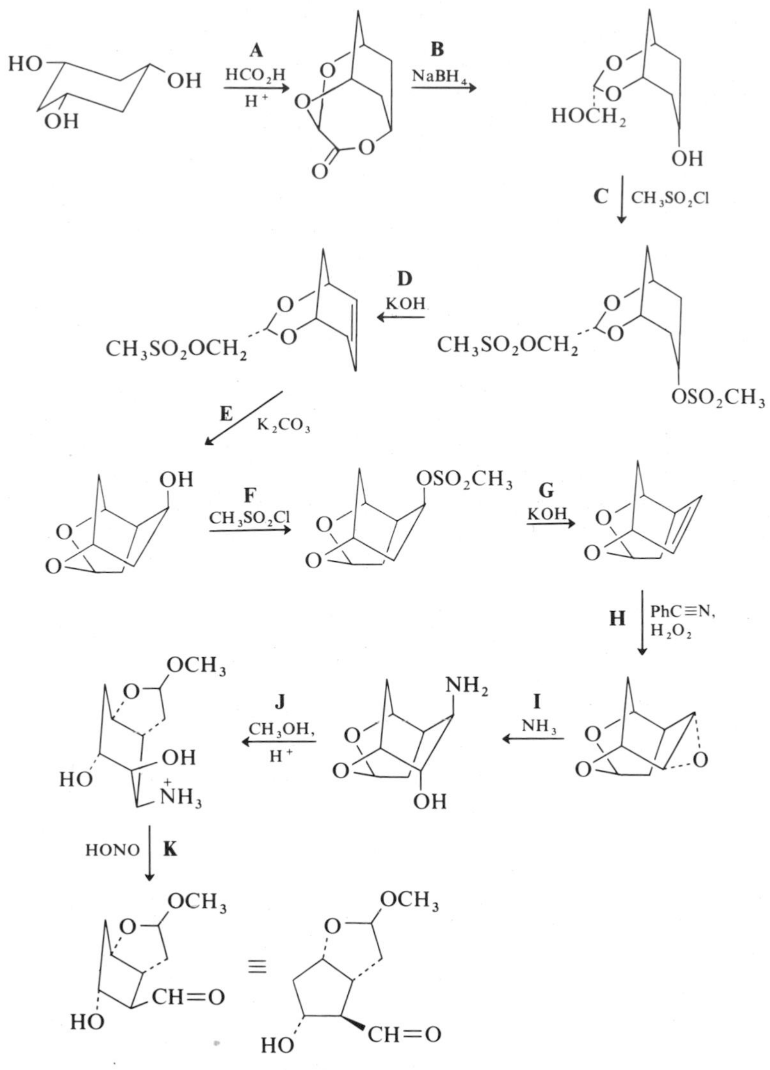

a. R. B. Woodward, J. Gosteli, I. Ernest, R. J. Friary, G. Nestler, H. Raman, R. Sitrin, C. Suter, and J. K. Whitesell, *J. Am. Chem. Soc.* **95**, 6853 (1973).

sulfonate group with participation by a carbon–carbon double bond. The *endo* stereochemistry of the hydroxymethyl group generated in step B ensures the correct stereochemistry. The required *endo* stereochemistry was established by the use of the all *cis*-1,3,5-trihydroxycyclohexane as starting material for the cyclization carried out in step A.

An interesting feature of Scheme 10.17 is that the cyclopentane ring is formed in the final step by a Tiffeneau–Demjanov-type ring contraction (Section 8.6). The stereochemistry of the cyclohexane precursor has been manipulated so that a concerted ring contraction is possible by ensuring that the group that is to migrate is *anti* to the departing nitrogen substituent.

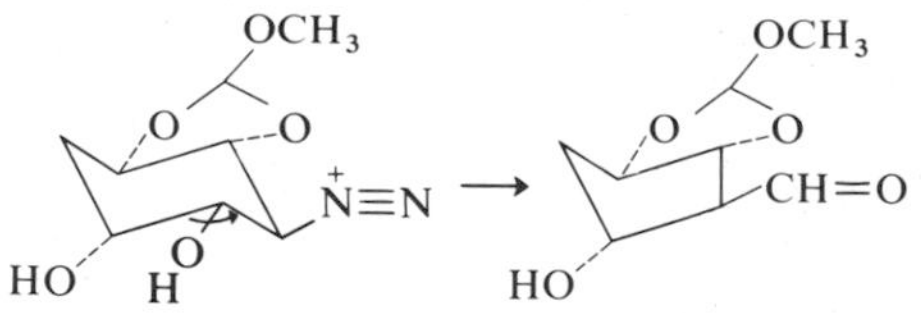

Another interesting feature of this scheme is the presence of the hydroxyl groups from the very beginning of the synthesis. They are masked as a cyclic acetal in steps A through I. One is unmasked in the penultimate step to permit formation of the conformation required for the final rearrangement.

The syntheses we have just considered exemplify the types of syntheses that are now within the capabilities of organic chemists. Indeed, substantially more complex molecules have been synthesized. The complete synthesis of Vitamin B_{12}, for example, was announced in 1972 after an effort that spanned more than 10 years.[49] In terms of complexity of the target molecule and the logistical demands of the synthetic effort, this achievement ranks as the most imposing synthesis that has been accomplished to date.

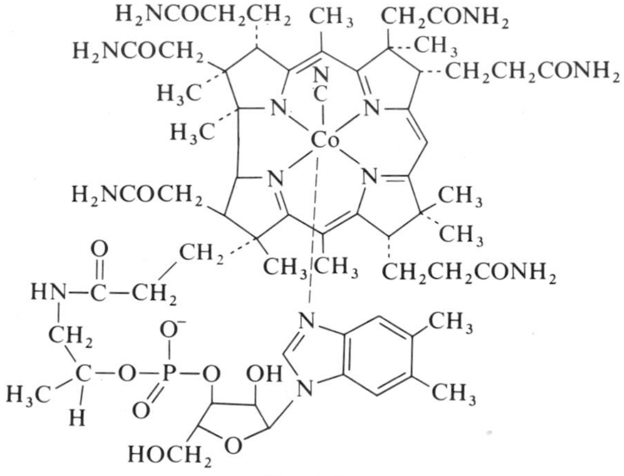

Vitamin B_{12}

The ability to arrive at an acceptable synthetic plan by a rational retrosynthetic analysis depends on a number of factors, among which the chemist's experience, imagination, and insight are foremost. The range of reactions that are within the chemist's grasp are also important. This involves not only a catalog of reactions in one's memory, but also the ability to use the available chemical literature to locate a reaction suitable for a given transformation. The most systematic and wide-ranging source of this type is the series *Synthetic Methods of Organic Chemistry*, published under the editorship of W. Theilheimer. This source, along with a number of other

49. R. B. Woodward, *Peter A. Leermakers Symposium*, Wesleyan University, Middletown, CN, 1972.

compilations of synthetic methods, is listed in the general references. These books are excellent sources of information, especially for functional-group conversions. The problem of arriving at a satisfactory plan for a specific synthetic target is a problem that must be approached for each individual molecule or closely related group of molecules. The problems at the end of the chapter provide some opportunity to test one's ability in this regard.

The possibility that a computer can be programmed to accomplish the task of retrosynthetic analysis is currently being actively explored. This is an attractive alternative, since a systematic catalog of reactions stored in a computer can easily be more complete than that stored in the memory of a chemist. Combining such a catalog of reactions with an efficient program for retrosynthetic analysis is by no means a simple task, however. A report of the general concepts involved in this effort has been published by E. J. Corey and his associates.[50] Improvements in the method are being described as they are accomplished.[51] Basically, the program develops a "tree" of potential intermediates by retrosynthetic analysis of the synthetic target. By working backward through each series of possible precursors, an array of possible synthetic pathways is developed. The term *antithetic transform* has been introduced to describe each step in the process of structural analysis that generates a potential starting material for a given reaction product.

The present program makes use of intensive interaction between the chemist and the computer. Communication between the chemist and the computer is accomplished via graphic structural formulas drawn by the chemist; the computer can also provide the structures of compounds it generates as potential intermediates.[52] No attempt will be made here to describe any of the details of the analytical programs. The system has been used by the Harvard group in synthetic analysis, and as the range of problems that it can handle expands, it will become increasingly powerful. It seems unlikely at this point, however, that computer-assisted analysis will lessen the need for human ability to design synthetic pathways. In 1971, Corey described the task of developing a general procedure for solving complex synthetic problems as "unlikely to be complete in the foreseeable future."[50] It is clear, however, that the systematic nature of this approach will help to define the routines and strategies of retrosynthetic analysis and perhaps help chemists systemize their thinking about synthesis.

General References

J. F. W. McOmie, *Protective Groups in Organic Chemistry*, Plenum Press, New York, NY, 1973.
R. E. Ireland, *Organic Synthesis*, Prentice-Hall, Englewood Cliffs, NJ, 1969.
W. Theilheimer, *Synthetic Methods of Organic Chemistry*, Vols. 1–30, S. Karger, Basel, 1946–1976.

50. E. J. Corey, *Q. Rev. Chem. Soc.* **25**, 455 (1971).
51. E. J. Corey, R. D. Cramer, III, and W. J. Howe, *J. Am. Chem. Soc.* **94**, 440 (1972); E. J. Corey, W. J. Howe, and D. A. Pensak, *J. Am. Chem. Soc.* **96**, 7724 (1974).
52. E. J. Corey, W. T. Wipke, R. D. Cramer, III, and W. J. Howe, *J. Am. Chem. Soc.* **94**, 421, 431 (1972).

S. R. Sandler and W. Karo, *Organic Functional Group Preparations*, Vols. I–III, Academic Press, New York, NY, 1968–1972.

C. A. Buehler and D. E. Pearson, *Survey of Organic Syntheses*, Wiley–Interscience, New York, NY, 1970.

I. T. Harrison and S. Harrison, *Compendium of Organic Synthetic Methods*, Vols. 1 and 2, Wiley, New York, NY, 1971, 1974.

R. B. Wagner and H. D. Zook, *Synthetic Organic Chemistry*, John Wiley and Sons, New York, NY, 1953.

J. ApSimon, *The Total Synthesis of Natural Products*, Vols. 1 and 2, Wiley–Interscience, New York, NY, 1973.

I. Fleming, *Selected Organic Syntheses*, Wiley, New York, NY, 1973.

Problems

(References for these problems will be found on page 505.)

1. In each of the synthetic transformations shown, the reagents are appropriate, but the reactions will not be practical as they are written. What modification would be necessary to permit each transformation to be carried out along the lines indicated?

(a)

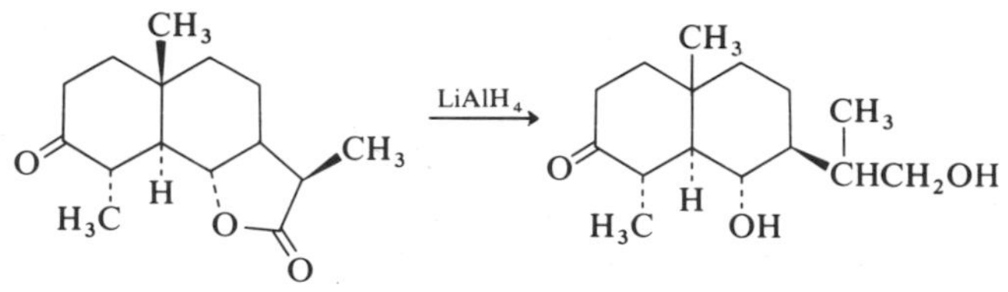

(b)

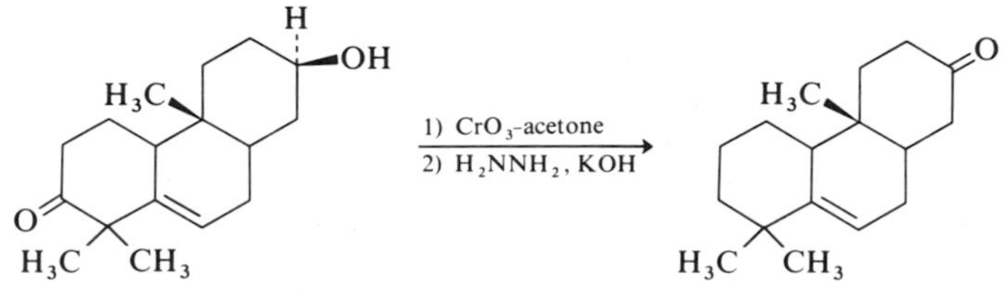

(c)

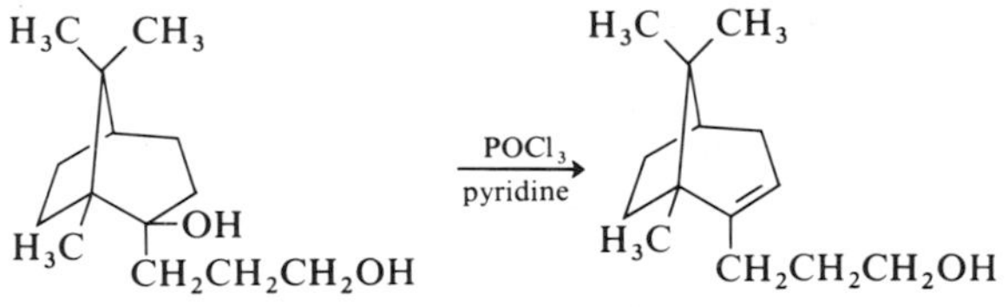

(d)

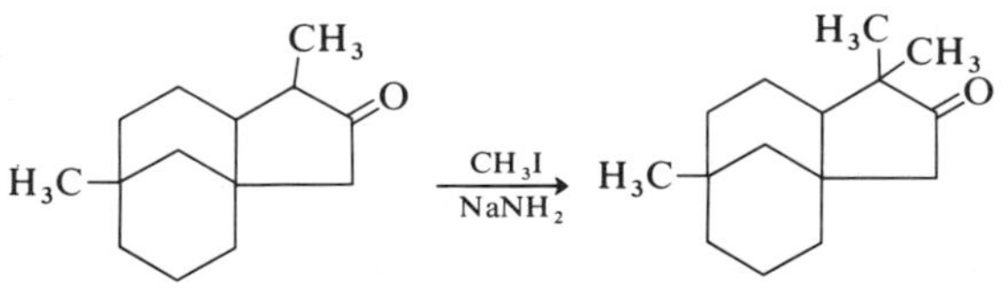

(e)

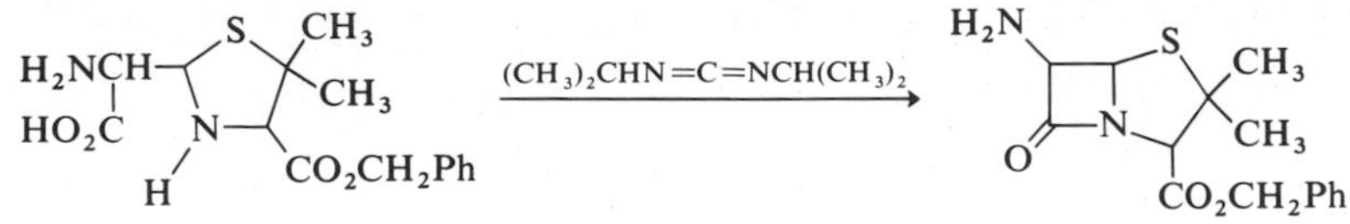
H₂NCH
HO₂C
S
CH₃
CH₃
N
H
CO₂CH₂Ph
(CH₃)₂CHN=C=NCH(CH₃)₂
H₂N
O
CO₂CH₂Ph

(f)

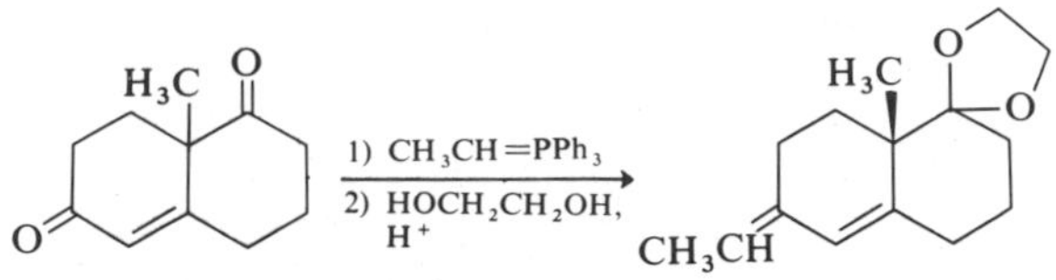
H₃C
O
1) CH₃CH=PPh₃
2) HOCH₂CH₂OH, H⁺
CH₃CH

2. Suggest methods for effecting the following synthetic transformations:

(a)

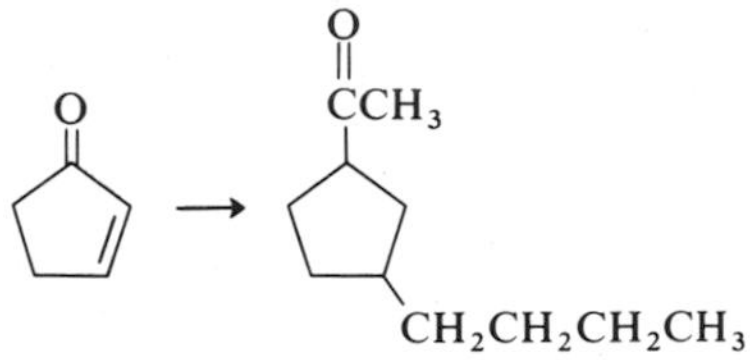
O
CCH₃
CH₂CH₂CH₂CH₃

(b)

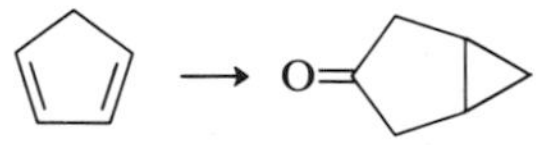
O

(c)

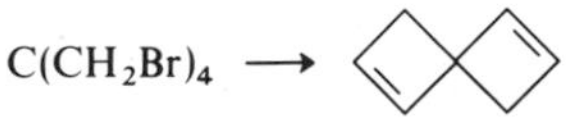
C(CH₂Br)₄

(d)

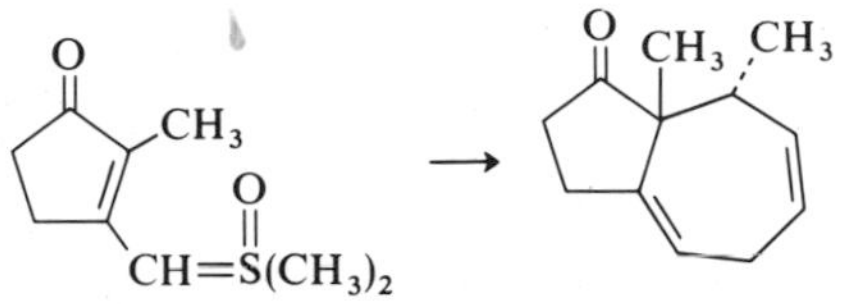
O
CH₃
CH=S(CH₃)₂
CH₃
CH₃

3. Show how, by use of synthetic equivalent reagents such as those in Schemes 10.2 and 10.3 and other required reagents, each of the following transformations could be accomplished. More than one step may be necessary.

(a)

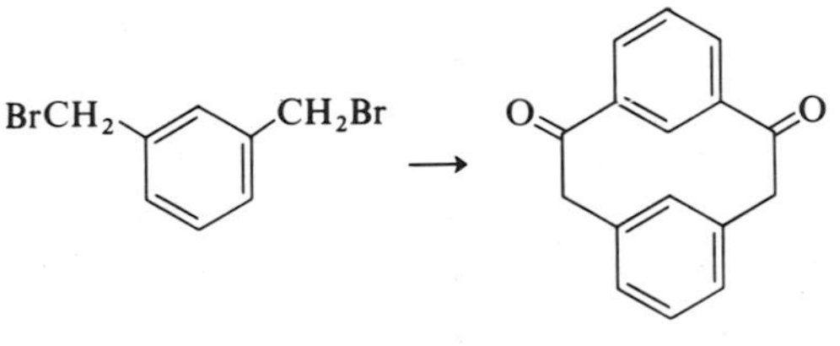
BrCH₂
CH₂Br
O
O

(b)

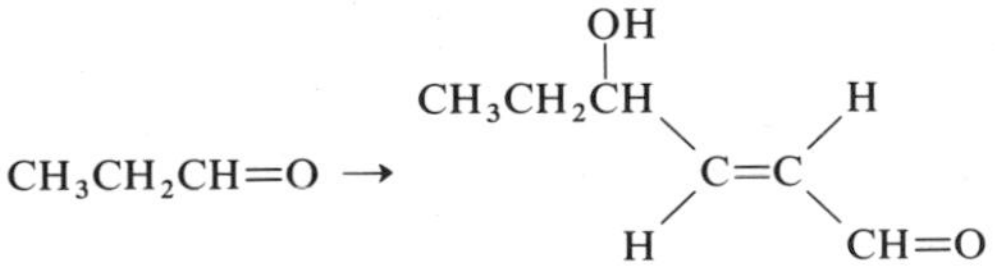
CH₃CH₂CH=O
OH
CH₃CH₂CH
H
C=C
H
CH=O

(c)

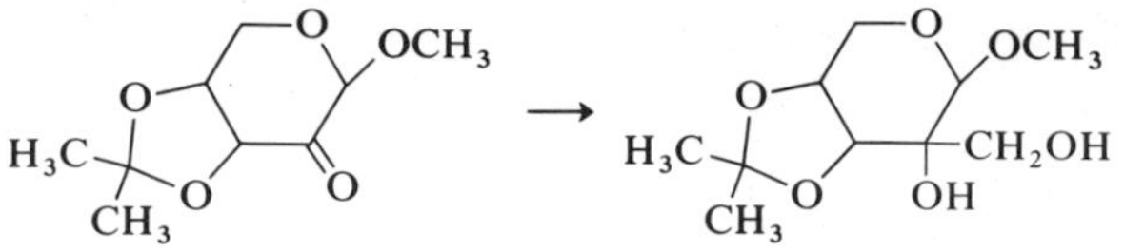

(d)

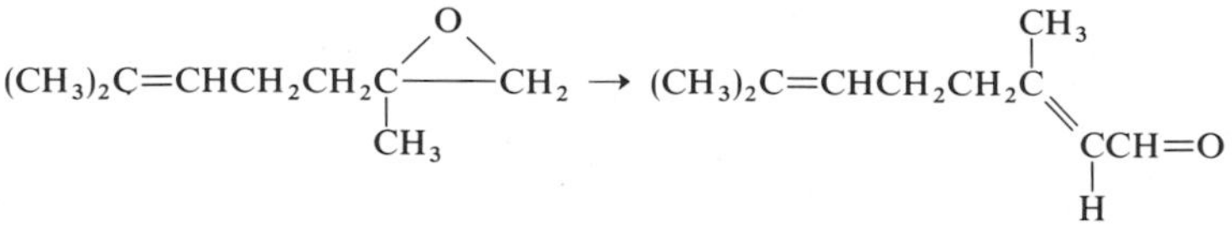

$(CH_3)_2C{=}CHCH_2CH_2C{-}{-}CH_2 \rightarrow (CH_3)_2C{=}CHCH_2CH_2C\overset{CH_3}{\underset{\overset{|}{\underset{H}{CCH{=}O}}}{}}$

(e)

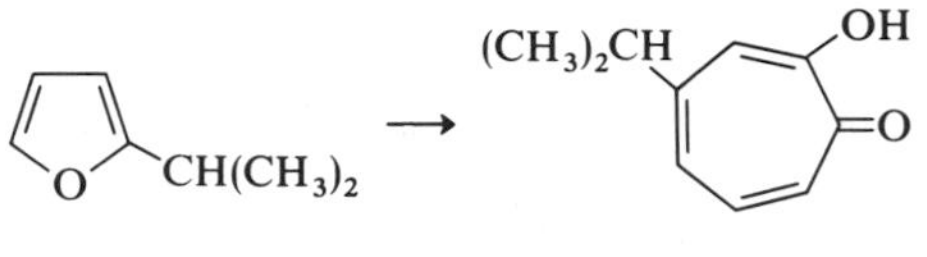

4. Outlines of synthetic schemes for several natural products are given below. In each step marked with a letter, indicate a reagent or series of reactions that would accomplish the desired transformation.

(a) isotelekin

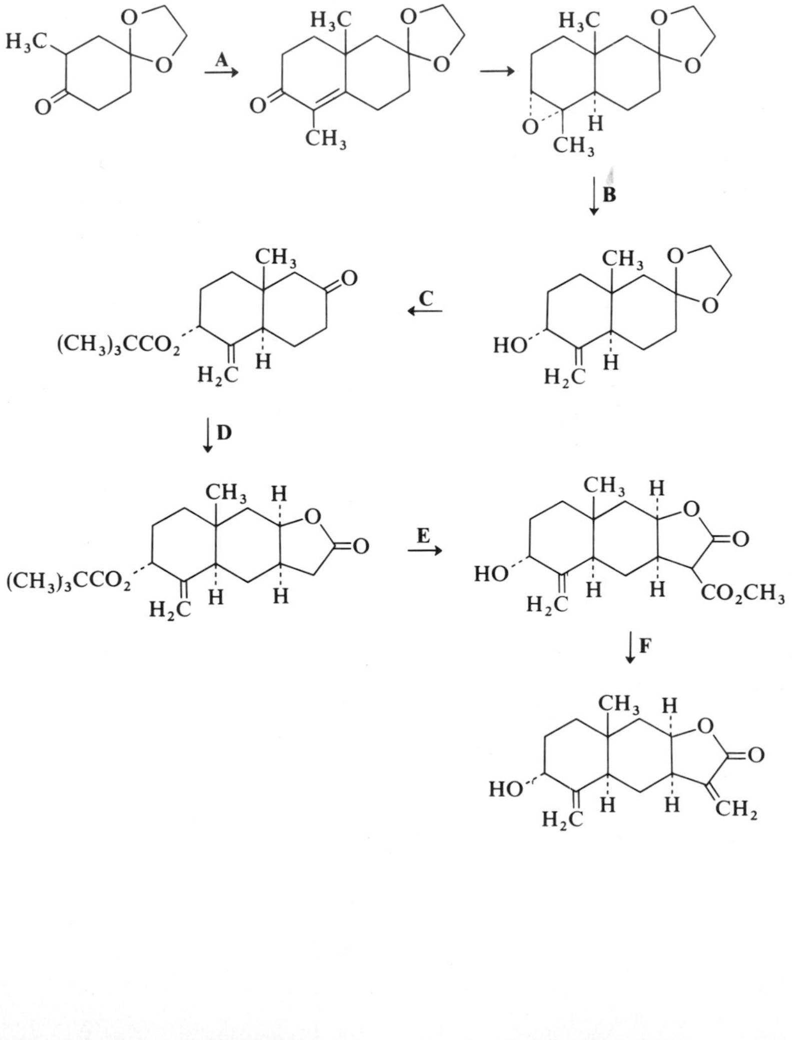

(c) aromandrene

(d) α-bourbonene

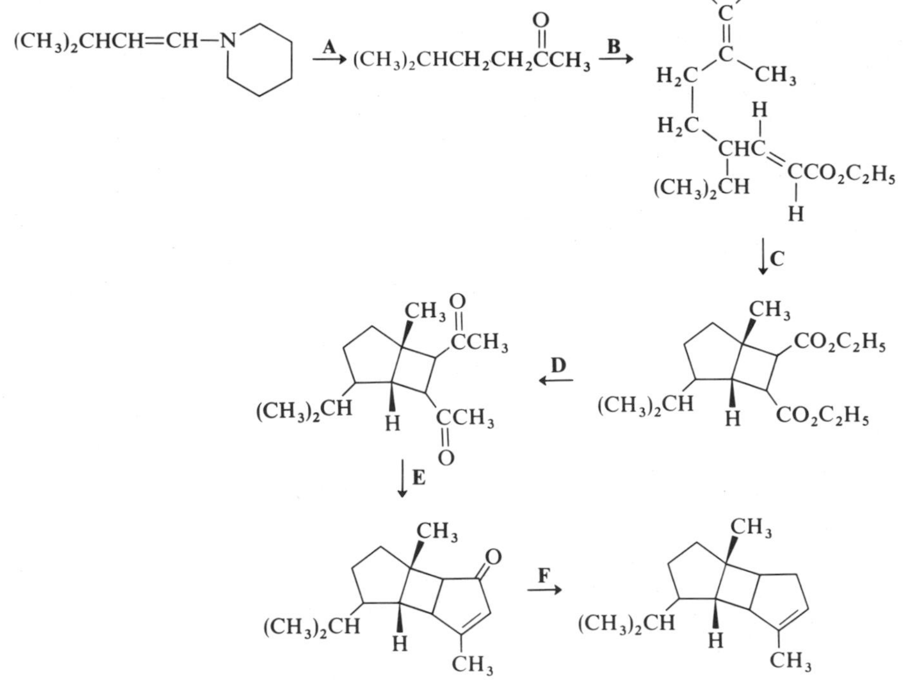

5. A novel and successful route to acetylenic ketones was devised based on the premise that fragmentation of systems such as **A** would occur under mild conditions:

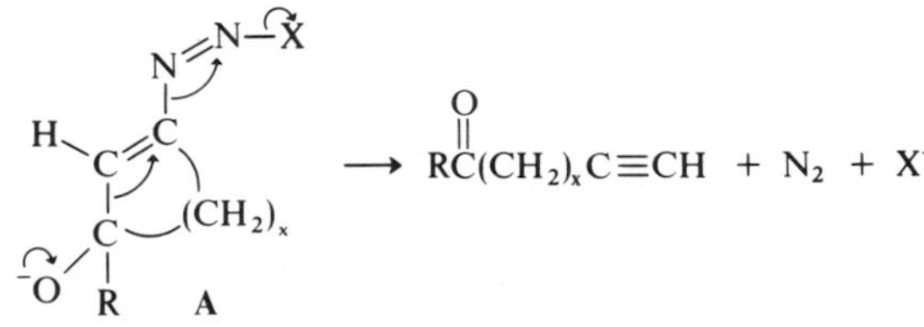

$$\longrightarrow R\overset{O}{\overset{\|}{C}}(CH_2)_xC\equiv CH + N_2 + X^-$$

Suggest a practical means of achieving the required intermediate **A** from an α,β-unsaturated ketone, using readily available reagents.

6. Devise synthetic approaches to the polycyclic ring compounds shown below:

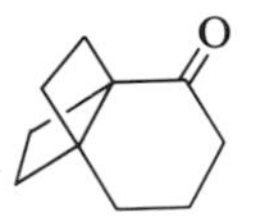

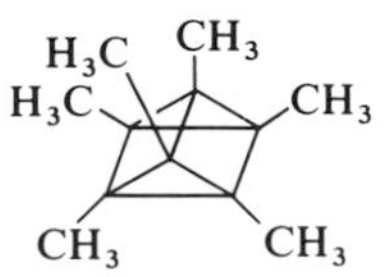

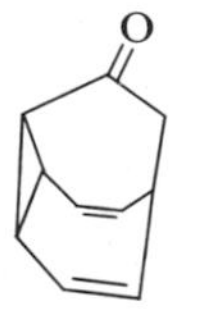

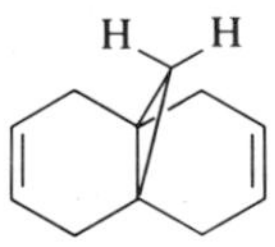

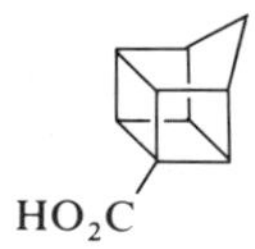

7. Indicate a reagent or short reaction sequence that would accomplish each of the following synthetic transformations:

(a)

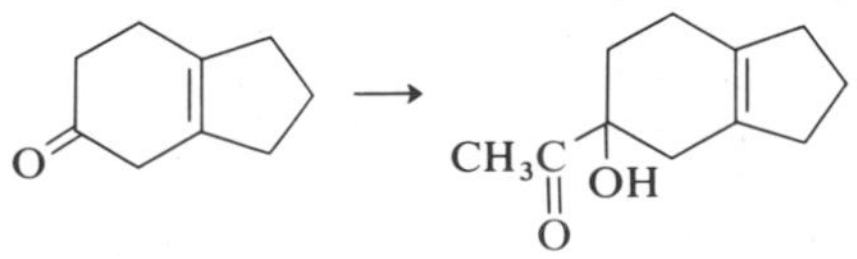

(b)

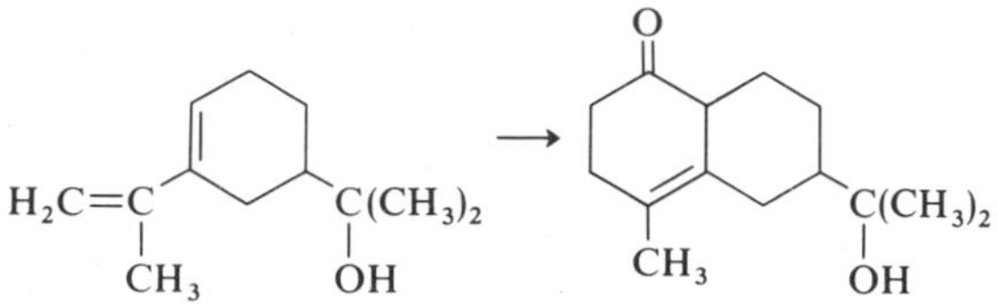

(c)

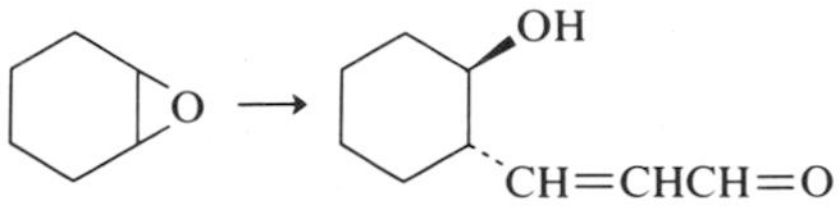

(d)

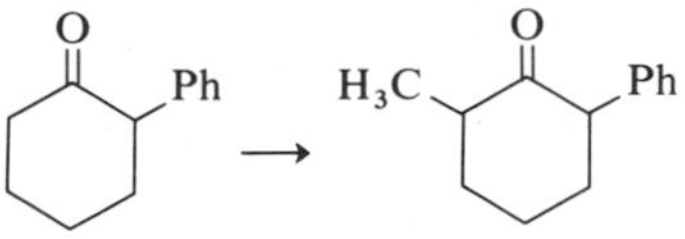

(e)

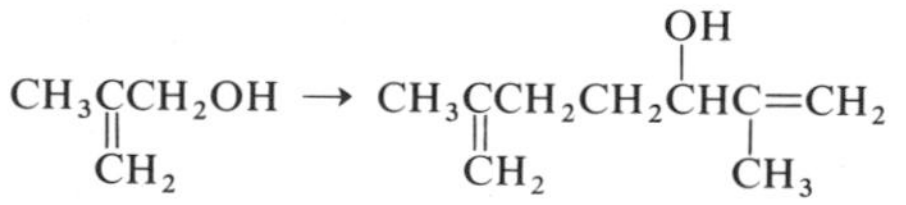

$$CH_3CCH_2OH \rightarrow CH_3CCH_2CH_2CHC=CH_2$$

with $\overset{\|}{CH_2}$ and $\overset{\|}{CH_2}$, $\overset{OH}{\underset{CH_3}{|}}$

(f)

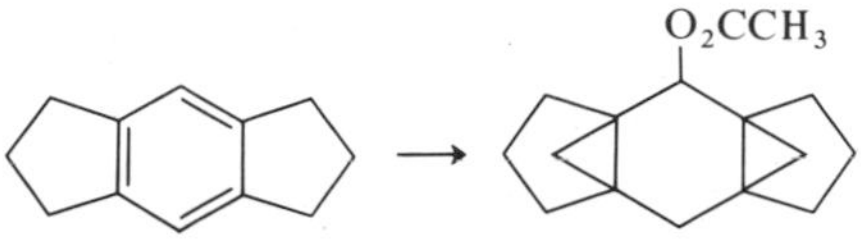

8. Indicate available optically active starting materials that would be useful starting points for the synthesis of each of the following compounds:

(a) firefly luciferin

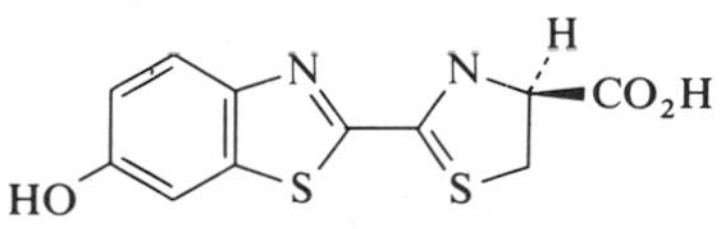

(b) helminthosporal

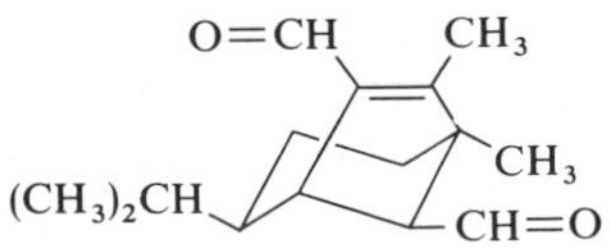

(c) α-cyperone

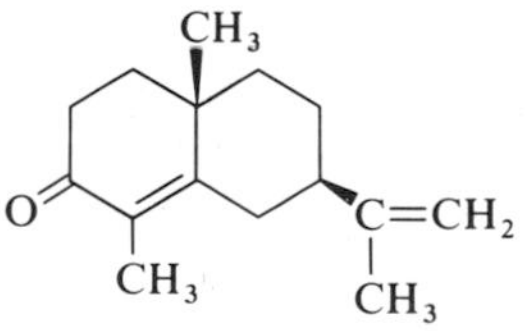

(d) streptose

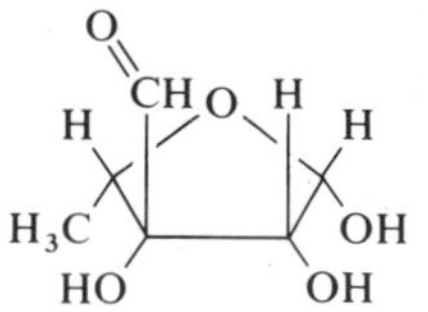

9. Suggest methods for the synthesis of the compounds below in optically active form from the reagents given, plus any of the following available optically pure materials:

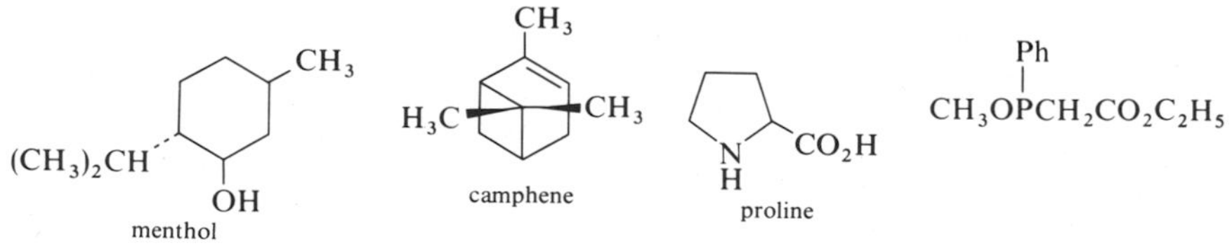

menthol   camphene   proline $CH_3OPCH_2CO_2C_2H_5$

(a)

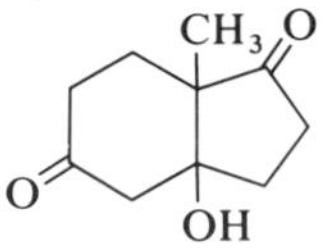

from 2-methylcyclopentane-1,3-dione, methyl vinyl ketone

(b) 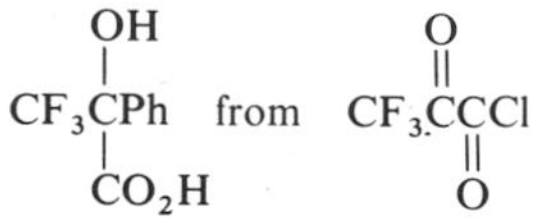   OH
         |
    CF_3CPh from CF_3CCCl
         | ‖
       CO_2H O

(c)

from   (racemic)

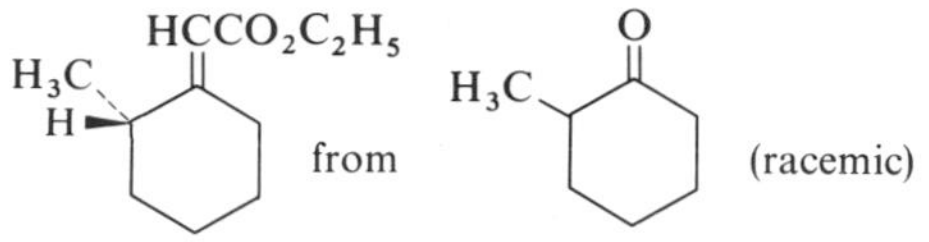

10. Indicate points at which protection might be required in order to accomplish the following synthetic transformations. Suggest possible protecting groups and suitable synthetic schemes for effecting each transformation.

(a)

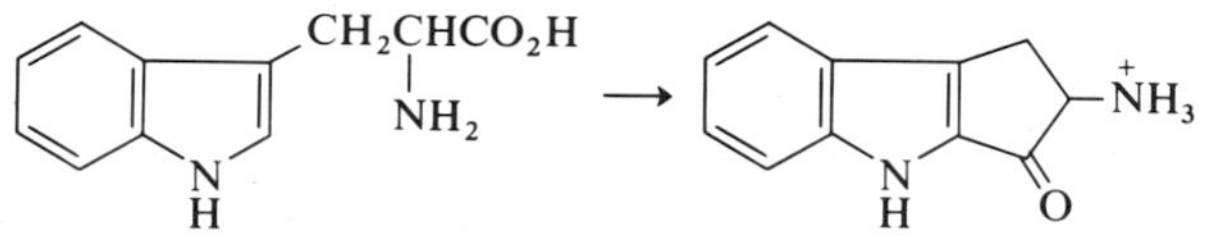

(b) 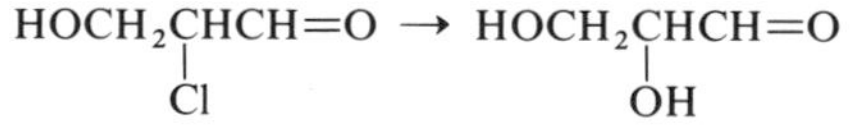 $HOCH_2CHCH=O \rightarrow HOCH_2CHCH=O$
              | |
              Cl   OH

(c)

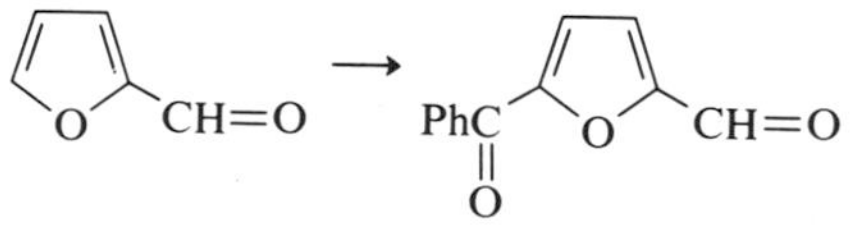

(d)

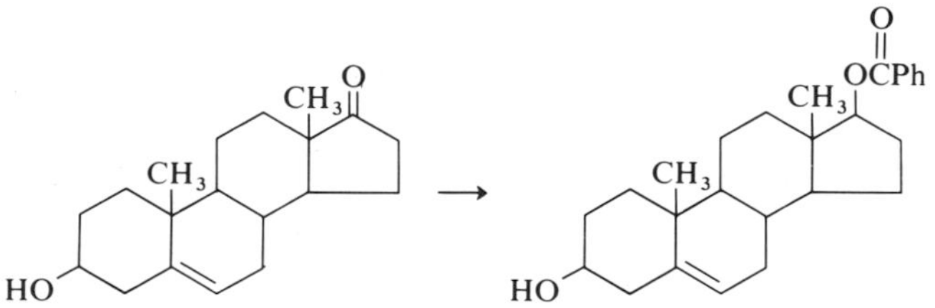

(e)

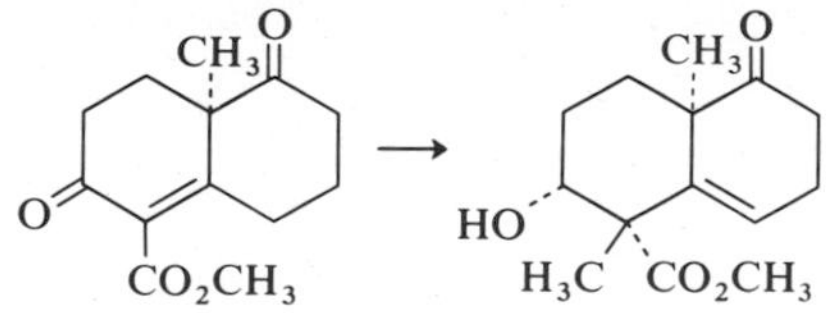

(f)

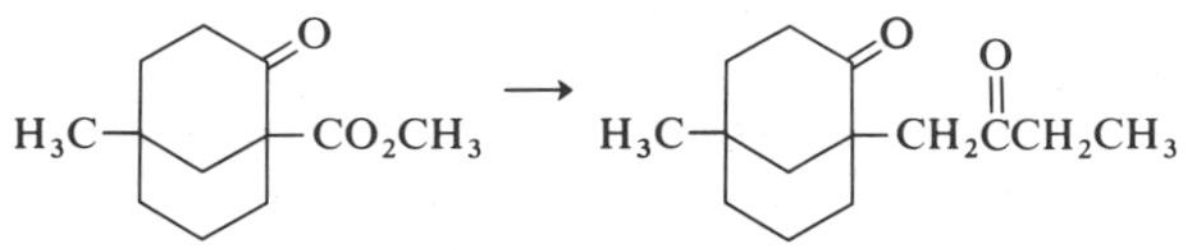

(g)

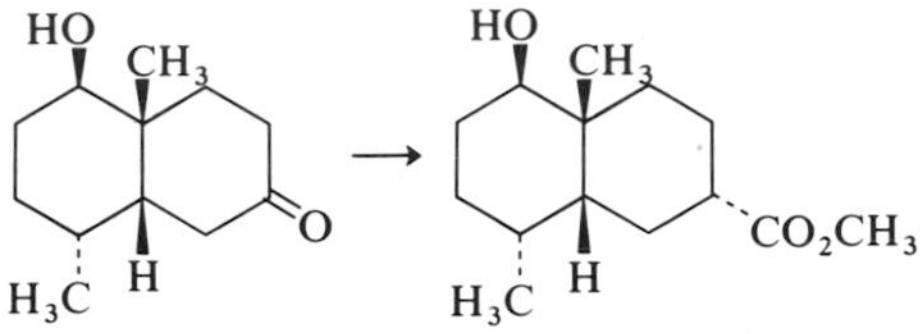

11. Compound **A** was investigated as a potential formylating reagent for organometallics according to the equation:

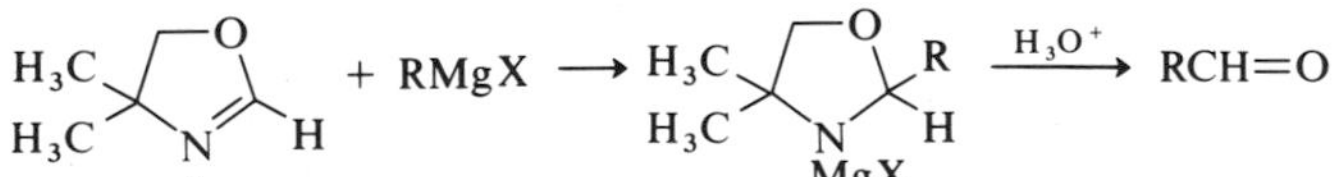

Addition of organometallic compounds to **A** was found to be inefficient. Suggest a modification that might improve the feasibility of this procedure by making the addition step more favorable.

12. Perform a retrosynthetic analysis for each of the following natural products. Develop at least three separate schemes. Discuss the relative merits of the three proposals and describe a fully elaborated synthetic plan based on the most promising approach.

(a)

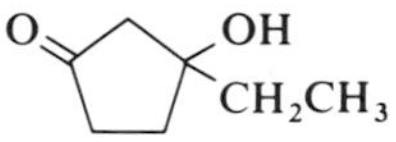

(b)

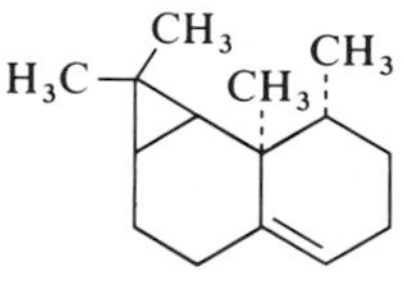

(c)

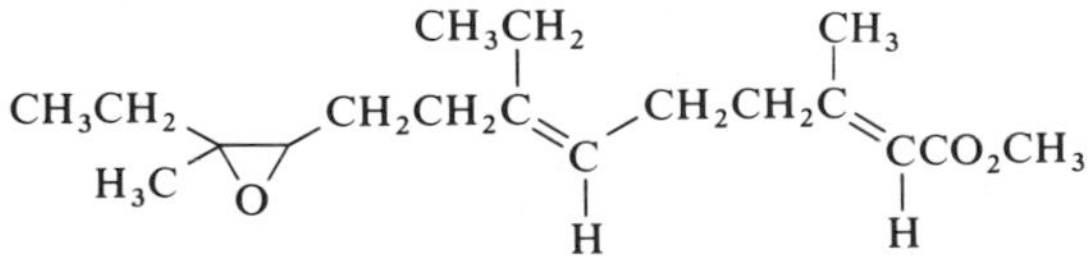

(d)

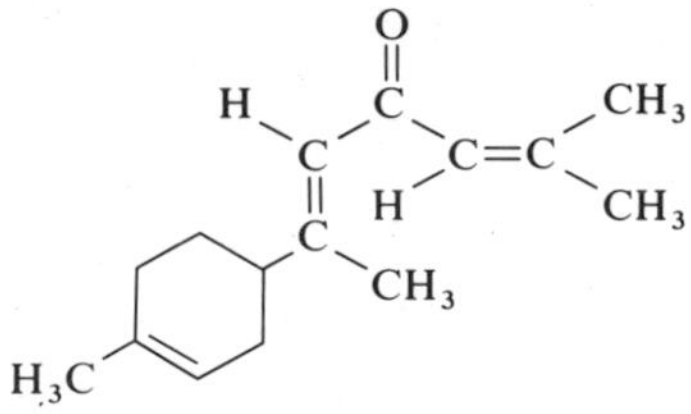

(e)

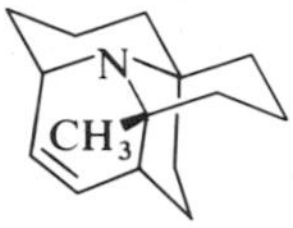

(f)

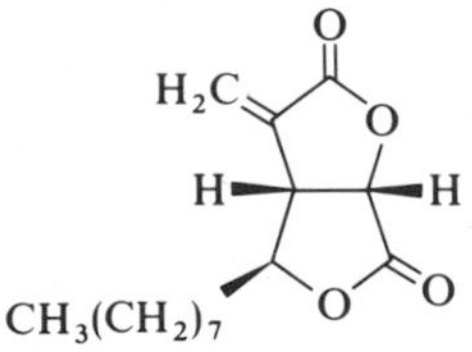

(g)

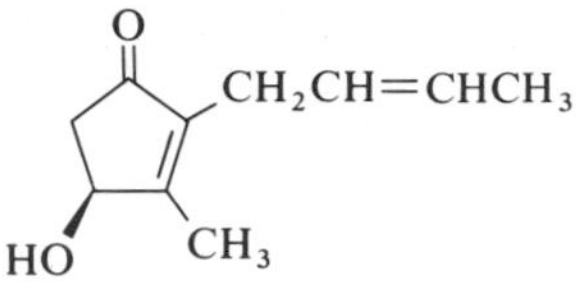

(h)

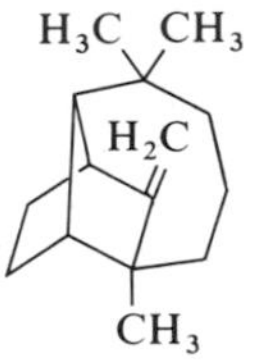

13. Suggest methods that would be expected to effect assymetric syntheses of the following compounds:

(a)

(b)

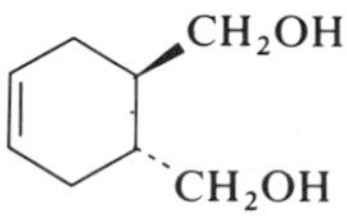

(c)

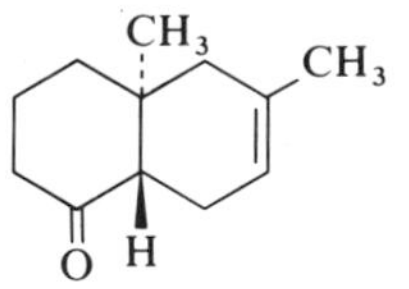

from

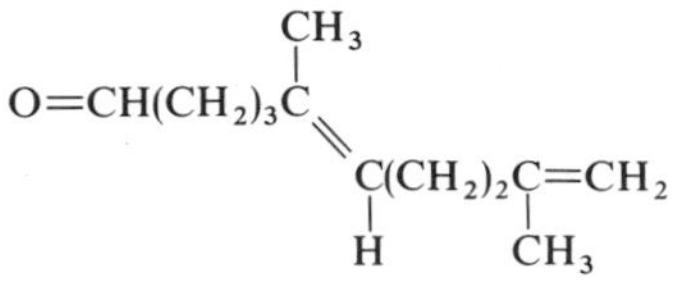

(d)

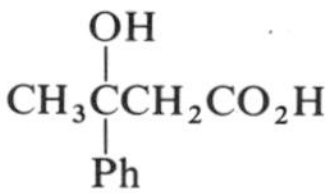

Synthesis of Macromolecules

Much of the effort of organic chemists since about 1930, especially in industrial research laboratories, has been directed toward the synthesis of polymeric materials, and many such substances have come to play a prominent role in industry and commerce. In this chapter, some of the reactions that are utilized to create useful polymers will be illustrated. It will be seen from the discussion that the basic mechanisms of polymerization reactions are the same as those encountered in the reactions of small organic molecules, and that the special features of polymerizations are the result of the high molecular weights of the molecules involved.

Three broad types of polymerization processes are generally recognized: chain-addition polymerization, step-growth polymerization, and coordination polymerization.

Chain-Addition Polymerization. The characteristic pattern of chain-addition polymerization is the one-by-one attachment of individual monomer units to the growing polymer chain. Addition of each successive monomer unit regenerates a

$$R-(M)_n-M_a \;+\; M \;\rightarrow\; R-(M)_{n+1}-M_a \xrightarrow{M} R-(M)_{n+2}-M_a$$

reactive site, M_a, which propagates growth of the chain by reacting with another monomer. The detailed mechanism of the reaction between the growing polymer chain and the monomer molecule may involve a radical, cationic, or anionic intermediate; specific examples of each type will be given later.

Step-Growth Polymerization. In step-growth polymerization, the monomer units are not added one by one. Instead, there is a gradual buildup of chain length and

$$M-M \;+\; M-M-M-M \;\rightarrow\; M-M-M-M-M-M$$

$$M-M-M-M-M-M \;+\; M \;\rightarrow\; M-(M)_5-M$$

$$M-(M)_5-M \;+\; M-(M)_{20}-M \;\rightarrow\; M-(M)_{27}-M$$

$$M-(M)_{27}-M \;+\; M-(M)_{10}-M \;\rightarrow\; M-(M)_{39}-M$$

etc.

459

molecular weight from telomers and small polymers. In many such step-growth polymerizations, a small molecule such as water or an alcohol is formed as a by-product. For this reason, this class of polymerization reactions is sometimes referred to as *condensation polymerization*. By-product formation is not an absolute requirement, however, and the term step-growth polymerization is preferred.

Coordination Polymerization. A third general polymerization type is coordination polymerization. Like addition polymerization, it occurs by addition of monomer units, one by one. Coordination polymerization is a form of addition polymerization, but differs in that the addition of the monomer involves a third molecular species besides the monomer and the growing polymer chain. In coordination polymerization, the addition step takes place with the monomer and polymer coordinated to the third species, which functions to promote the formation of the new bond. Usually, this third species is a metal complex.

11.1. Polymerization

11.1.1. Chain-Addition Polymerization

The most important chain-addition polymerizations are processes in which the reacting end of the polymer is a free radical. This type of polymerization is commonly called *radical-chain polymerization*. The basic mechanistic concepts are those introduced in Part A, Chapter 12, where free-radical chemistry was considered.

Polymerization is normally dependent on an initiator, which begins the chain reactions in the polymerization mixture. Typical initiators are molecules that have rather low thermal stability and generate radicals on decomposition. Other initiator systems depend on photolytic decomposition or on the generation of radical intermediates as the result of redox reactions. The rate constants for the individual propagation steps that follow initiation are usually very large, but polymerizations are normally carried out in such a way that the concentration of the reacting chains is very low ($<10^{-8} M$). As a result, the overall rate of polymerization is moderate.

$$\textit{Initiation:} \quad \text{In} \xrightarrow{k_i} 2\text{In·}$$

$$\textit{Propagation:} \quad \text{In·} + \text{M} \rightarrow \text{In—M·}$$

$$\text{In—M·} + \text{M} \rightarrow \text{In—M—M·}$$

$$\textit{Termination:} \quad 2\text{In—(M)}_n\text{—M·} \xrightarrow{k_t} \text{products}$$

Kinetic analysis of this type of system is possible. The basic equations for the kinetics of radical-chain polymerization are derived by assuming a low steady-state concentration of reacting radicals. Under these conditions, the rate of termination equals that of initiation:

$$2k_i[\text{In}] = 2(k_t)[\text{I(M)}_n\text{M·}]^2$$

$$[IM_nM\cdot] = \left(\frac{k_i}{k_t}\right)^{1/2}[In]^{1/2}$$

$$\text{Rate of Polymerization} = k_p[I(M)_nM\cdot][M] = k_p[M]\left(\frac{k_i}{k_t}\right)[In]^{1/2}$$

$$= k_{obs}[M][In]^{1/2}$$

The term k_t may contain several individual rate constants corresponding to each of several possible termination processes. This kinetic equation is closely analogous to the expressions in Part A, Chapter 12, for the kinetics of free-radical addition reactions.

The molecular weight of a polymer prepared by radical-chain polymerization depends on the relative rate of propagation versus the rates of steps that stop the growth of the polymer chain. One of the processes that stops chain growth is termination, which can occur either by combination of two radical centers or by disproportionation, to give one polymer with a saturated end group and another with a vinyl end group:

$$2\,I(M)_n-CH_2\overset{\cdot}{C}HX \rightarrow I(M)_n-CH_2CH_2X + I(M)_n-CH=CHX$$

Another process that stops chain growth is chain transfer. Chain transfer does not annihilate a reactive radical center, but does stop the growth of the polymer chain and initiate a new center for chain growth. The relative frequency of chain transfer

$$I(M)_n-CH_2\overset{\cdot}{C}HX + R-H \rightarrow I(M)_nCH_2CH_2X + R\cdot$$

and termination will govern the molecular weight of the polymer. A common mechanism by which chain transfer occurs is abstraction of a hydrogen atom from another molecule in the solution. This molecule may be the solvent or some other additive. The addition of compounds that function as chain-transfer agents is a major method of controlling the final distribution of molecular weight in a polymer. The presence of a chain-transfer agent means that growth of a given polymer chain will be stopped, but without a termination that would destroy an active radical. The higher the concentration or reactivity of the chain-transfer agent, the more often chain growth will be interrupted and the lower the average molecular weight of the polymer.

The types of compounds that can be polymerized readily by the radical-chain mechanism are the same types that easily undergo free-radical addition reactions. Alkenes with aryl, ester, nitrile, or halide substituent groups that can stabilize the intermediate radical are most susceptible to radical polymerization. Terminal alkenes are generally more reactive toward radical-chain polymerization than more highly substituted isomers. The dominant mode of addition in radical-chain polymerization is head-to-tail. The reason for this orientation is that each successive addition of monomer takes place in such a way that the most stable possible radical intermediate is formed. For example, the addition to styrene occurs to give the phenyl-substituted radical; to acrylonitrile, to give the cyano-substituted radical:

$$\overset{Ph}{\underset{|}{In(CH_2CH)_n}}CH_2\overset{\cdot}{C}HPh \;+\; CH_2{=}CHPh \;\rightarrow\; \overset{Ph}{\underset{|}{In(CH_2CH)_{n+1}}}CH_2\overset{\cdot}{C}HPh$$

$$\overset{CN}{\underset{|}{In(CH_2CH)_n}}CH_2\overset{\cdot}{C}HCN \;+\; CH_2{=}CHCN \;\rightarrow\; \overset{CN}{\underset{|}{In(CH_2CH)_{n+1}}}CH_2\overset{\cdot}{C}HCN$$

The substituent groups therefore occur regularly on alternating carbons down the chain, rather than randomly.

The technology of polymerization has been extensively studied, and there are several important methods by which polymerization can be carried out. Radical-chain polymerization can be carried out on essentially pure monomer (plus initiator and chain-transfer agents for regulation). This is called *bulk polymerization.* Styrene and methyl methacrylate are examples of monomers that are commercially converted to the corresponding polymers by bulk-polymerization processes. Another technique is *suspension polymerization,* which, as the name implies, involves polymerization beginning with a suspension of the monomer. Vinyl chloride and vinyl acetate are examples of monomers that are polymerized in this manner. The polymerization of dienes to synthetic rubbers is an example of *emulsion polymerization,* a process that differs significantly from suspension polymerization. The initiator in emulsion polymerization is water-soluble, but the monomer is present partially in micelles that are formed in the presence of soaps or similar surfactants. The characteristic feature of emulsion polymerization is that propagation takes place in the micelle near the surface. Because of the very large number and small size of the micelles, they are inoculated only infrequently by an initiating radical, and termination resulting from collision of two radicals is very unlikely. As polymerization proceeds within the micelle, more monomer molecules are dissolved by the micelle. The net result is a very long-chain polymer. Polymerization can also be carried out under conditions where the monomer is initially in solution. If the polymer is soluble in the solvent, it is recovered by evaporation. If not, the polymer is isolated after precipitation from solution.

Many commercially important polymers are copolymers resulting from *copolymerization,* which is the polymerization of mixtures of two (or more) monomers. Copolymerization will be discussed somewhat later, after the other important mechanisms for chain-addition polymerization have been considered.

Polymerization of alkenes can also be initiated by a variety of protonic or Lewis acids. The reactive site on the growing polymer chain is then a carbonium ion. Chains

$$A^+ \;+\; CH_2{=}CHR \;\rightarrow\; A{-}CH_2{-}\overset{+}{C}HR$$

$$A{-}CH_2{-}\overset{+}{C}HR \;+\; CH_2{=}CHR \;\rightarrow\; A\underset{\underset{R}{|}}{CH_2CH}{-}CH_2{-}\overset{+}{C}HR$$

$$etc.\ldots \qquad\qquad \rightarrow\; ACH_2\underset{\underset{R}{|}}{CH}(CH_2\underset{\underset{R}{|}}{CH})_nCH_2\overset{+}{C}HR$$

are terminated by loss of a proton, by hydride-abstraction from solvent or some other molecule present in solution, or by combination with an anion. The kinetic equations

Termination:

$$\mathrm{A(CH_2\underset{\displaystyle R}{\underset{|}{C}}H)_nCH_2\overset{+}{C}HR \longrightarrow A(CH_2\underset{\displaystyle R}{\underset{|}{C}}H)_nCH{=}CHR + H^+}$$

or

$$\mathrm{A(CH_2\underset{\displaystyle R}{\underset{|}{C}}H)_nCH_2\overset{+}{C}HR + HS \longrightarrow A(CH_2\underset{\displaystyle R}{\underset{|}{C}}H)_nCH_2CH_2R + S^+}$$

or

$$\mathrm{A(CH_2\underset{\displaystyle R}{\underset{|}{C}}H)_nCH_2\overset{+}{C}HR + Y^- \longrightarrow A(CH_2\underset{\displaystyle R}{\underset{|}{C}}H)_nCH_2\underset{\displaystyle R}{\underset{|}{C}}HY}$$

for cationic polymerization are derived from the assumption of a low steady-state concentration of actively growing chains and, therefore, equality of chain-initiation and -termination rates:

$$k_i[A^+][M] = k_t[AM^+]$$

$$\frac{k_i}{k_t}[A^+][M] = [AM^+]$$

$$\text{Propagation Rate} = k_p[AM^+][M] = \frac{k_i}{k_t}[A^+][M]^2$$

where $[A^+]$ is the concentration of catalytically active acid.

The alkenes that are most easily polymerized under cationic conditions are those that can lead to relatively stable carbonium-ion intermediates. Isobutylene, vinyl ethers, and alkenes with aromatic substituents are among the types of monomers that yield commercially important polymers by cationic polymerization.

$$\mathrm{CH_2{=}C(CH_3)_2 \xrightarrow{A^+} ACH_2\overset{\displaystyle CH_3}{\overset{|}{\underset{\displaystyle CH_3}{\underset{|}{C^+}}}} \longrightarrow A(CH_2\overset{\displaystyle CH_3}{\overset{|}{\underset{\displaystyle CH_3}{\underset{|}{C}}}})_nCH_2CH(CH_3)_2}$$

$$\mathrm{CH_2{=}CHOR \xrightarrow{A^+} ACH_2\overset{+}{C}HOR \longrightarrow A(CH_2\underset{\displaystyle OR}{\underset{|}{C}}H)_nCH_2CH_2OR}$$

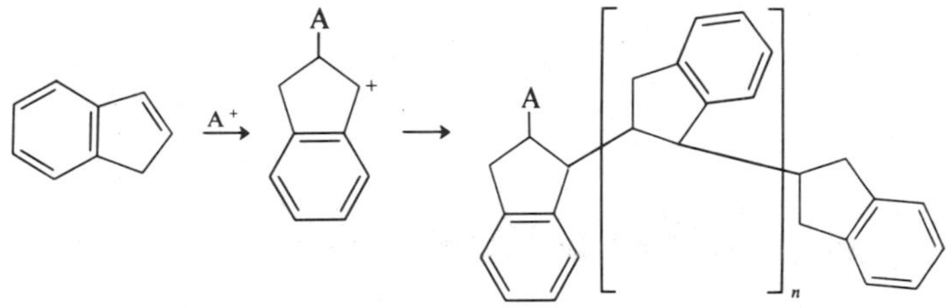

Because the growing polymer chain and monomer will react to give the more stable of the two possible carbonium ions, cationic polymerization, like radical-chain polymerization, leads to regular head-to-tail incorporation of monomer units.

In actual practice, there are, of course, many complexities superimposed on this basic mechanism of cationic polymerization. Many of the catalyst systems consist of two components. In some systems, for example, BF_3 is an effective catalyst only in the presence of some water. It may be that the active catalyst is the adduct, which serves as a proton donor. Alkyl halides in combination with Lewis acids can also

$$BF_3 + H_2O \rightleftharpoons F_3\overset{-\;+}{B}OH_2 \rightleftharpoons F_3\overset{-}{B}OH + H^+$$

serve as initiating catalysts. This system presumably functions by generation of a carbonium ion from the alkyl halide. This mode of generation of carbonium ions was considered in Part A, Chapter 5.

$$R-X + MX_n \rightarrow R^+ + (MX_{n+1})^-$$

Because cationic polymerization involves carbonium-ion intermediates, there are several possible ways in which variation in the polymer structure can occur. Chain branching can occur if a hydride ion is abstracted from an internal position on the polymer chain:

$$A(CH_2CH)_nCH_2CH(CH_2CH)_nH + R'^+ \rightarrow A(CH_2CH)_nCH_2\overset{+}{C}(CH_2CH)_nH$$
$$\underset{R}{|} \quad\quad \underset{R}{|} \quad \underset{R}{|} \quad\quad\quad\quad \underset{R}{|} \quad\quad \underset{R}{|} \quad \underset{R}{|}$$

$$A(CH_2CH)_nCH_2-\overset{+}{\underset{R}{C}}-(CH_2CH)_nH + CH_2{=}CH \rightarrow A(CH_2CH)_nCH_2-\overset{CH_2\overset{+}{C}HR}{\underset{R}{\overset{|}{C}}}-(CH_2CH)_nH$$

With certain monomers, rearrangement is competitive with propagation. When this occurs, the polymer backbone consists of a mixture of rearranged and unrearranged monomer units. A specific example is *t*-butylethylene:

$$AlCl_3 + CH_2{=}CH\overset{CH_3}{\underset{CH_3}{\overset{|}{C}}}-CH_3 \xrightarrow{-130°} CH_3CH-\overset{CH_3}{\underset{CH_3}{\overset{|}{C}}}-\Big(CH_2-CH-\overset{CH_3}{\underset{CH_3}{\overset{|}{C}}}\Big)_n-CH_2-CH-\overset{CH_3}{\underset{CH_3}{\overset{|}{C}}}+$$

Most of the recurring units in the polymer are $-CH_2-\underset{CH_3}{\overset{\overset{\textstyle CH_3}{|}}{\underset{|}{CH}}}-\overset{}{C}-$ rather than $-CH_2-\underset{C(CH_3)_3}{\overset{|}{CH}}-$ units.[1]

Cationic polymerization rates and behavior are much more sensitive to the identity of the initiator than are radical processes. The reason is probably that the cationic active center interacts with counterions and with solvent. The carbonium-ion site may range from being essentially free to being strongly paired with the counterion, depending on the solvent and the identity of the counterion.[2] Such factors come strongly into play in controlling copolymerization. Under conditions where the cationic site is essentially free and highly reactive, it may not strongly discriminate between two available monomers. On the other hand, when reactivity is diminished by strong ion-pairing, selectivity can be high.[3] For example, in the

1. J. P. Kennedy, J. J. Elliott, and B. E. Hudson, Jr., *Makromol. Chem.* **79**, 109 (1964).
2. N. Kanoh, K. Ikeda, A. Gotoh, T. Higashimura, and S. Okamura, *Makromol. Chem.* **86**, 200 (1965).
3. A. V. Tobolsky and R. J. Boudreau, *J. Polym. Sci.* **51**, S53 (1961); C. G. Overberger and V. G. Kamath, *J. Am. Chem. Soc.* **85**, 446 (1963).

copolymerization of isobutylene and p-chlorostyrene catalyzed by $AlBr_3$, selectivity increases as the solvent polarity increases. This increase is attributed to more effective solvation, and diminished reactivity, in the more polar solvent.

Chain polymerization of alkenes can also be catalyzed by anionic reagents. Anionic polymerization occurs most readily when the alkene carries a carbanion stabilizing substituent. Even ethylene can be polymerized anionically, however, if a

$$A^- + CH_2=CHY \rightarrow A-CH_2\overset{-}{C}HY \overset{CH_2=CHY}{\longrightarrow} A-CH_2\underset{\underset{Y}{|}}{C}HCH_2\overset{-}{C}HY$$

sufficiently strong base, such as an alkyllithium, is employed. As in cationic polymerization, the nature of the solvent and counterion are important, since these factors determine just how reactive the anions will be in propagating polymerization.

One important class of anionic-polymerization initiators are the organometallic compounds, especially those of lithium and sodium (see Chapter 5 to review the general properties of the metalloalkyls). Grignard reagents can induce polymerization of acrylonitrile and acrylate esters, but not of hydrocarbons. Initiation by organosodium compounds often involves *in situ* generation of the active species from sodium metal. Sodium metal catalyzes the polymerization of dienes, and the mechanism has been shown to involve a dianion formed by dimerization of the diene radical anion. An alternative initiation system for anionic polymerization involves

$$Na + CH_2=CH-CH=CH_2 \rightarrow [CH_2=CH-CH=CH_2]^{\overline{\cdot}} + Na^+$$

$$2\,Na^+ + 2\,[CH_2=CH-CH=CH_2]^{\overline{\cdot}} \rightarrow \overset{+}{Na}\overset{-}{CH_2}CH=CHCH_2CH_2CH=CHCH_2^-\,Na^+$$

$$or\ [CH_2=CH-CH=CH_2]^{\overline{\cdot}} + Na \rightarrow \overset{+}{Na}\overset{-}{CH_2}-CH=CH-CH_2^-\,Na^+$$

prior formation of a radical anion such as that from naphthalene (Part A, Section 12.1.7). The radical anion then reacts with the monomer, generating the monomer

radical anion, which dimerizes. It is this dianion that begins the propagation phase of polymerization. In highly purified monomer, there is no effective termination

mechanism, since the carbanionic reactive sites do not tend to dimerize or disproportionate, and proton exchange regenerates a carbanionic site.[4] Two consequences of this situation are important:

1. All the initiation steps occur very rapidly because of the speed with which

4. M. Szwarc, M. Levy, and R. Milkovich, *J. Am. Chem. Soc.* **78**, 2656 (1956); C. E. Frank and W. E. Foster, *J. Org. Chem.* **26**, 303 (1961).

electron transfer occurs. The chains then grow at constant rates until monomer is exhausted. For this reason, the molecular weights of the chains formed in anionic polymerizations are all very similar; i.e., the polymer has a very narrow distribution of molecular weight.

2. The reactive chain ends are long-lived. After monomer is exhausted, polymerization begins anew on addition of more monomer because the active carbanionic centers are still present. The phrase "living polymer" has been used to convey this special aspect of these anionic polymerizations.[5] They are quenched, however, by water, alcohols, or any other proton donor that is sufficiently acidic to neutralize the carbanion sites.

Epoxides are an important class of monomer that can be polymerized by the anionic-chain-addition mechanism. The polymerization is catalyzed by strong bases, such as alkoxide ions, and results in polyethers. Each addition step generates a

$$RO^- + CH_2{-}CH_2 \rightarrow ROCH_2CH_2O^- \xrightarrow{nCH_2{-}CH_2} RO(CH_2CH_2O)_nCH_2CH_2OH$$

new alkoxide center, which can react with another monomer molecule by nucleophilic ring-opening.

A special type of addition polymerization that must be considered is *coordination polymerization*. These polymerizations are reactions that take place with both the monomer and polymer coordinated to a catalytic center of some sort. Usually, a heterogeneous catalyst is involved; the best known of these is the Ziegler–Natta catalyst, which is prepared by reaction of a trialkylaluminum with titanium tetrachloride in an inert hydrocarbon solvent. There is a large family of related systems. The mechanism of action of these catalysts is believed to involve two basic functions. The initiating group, and subsequently the growing polymer chain, is considered to be bound to titanium centers on the catalyst surface. The titanium can also accept the monomer into its coordination sphere, presumably as a π-bonded ligand. The two coordinated species then react, resulting in an extended alkyl chain and leaving a site available for π-coordination of another monomer molecule. The exact structure of the active site and the question of just how intimately involved is the other metal present, aluminum, are points of uncertainty.

$$\underset{Ti}{\overset{R'}{|}} + CH_2{=}CHR \rightarrow \underset{Ti-\|}{\overset{R'}{|}}\overset{CH_2}{\underset{CHR}{}} \rightarrow \underset{Ti}{\overset{R'CH_2CHR}{|}}$$

One of the special features of the Ziegler–Natta catalyst is the stereochemistry associated with polymerization. Radical- and cationic-chain polymerization of

5. M. Szwarc, *Carbanions, Living Polymers and Electron Transfer Processes*, Interscience, New York, NY, 1968, Chap. 2.

monosubstituted olefins lead to products having random stereochemical configuration. This is called an *atactic* polymer:

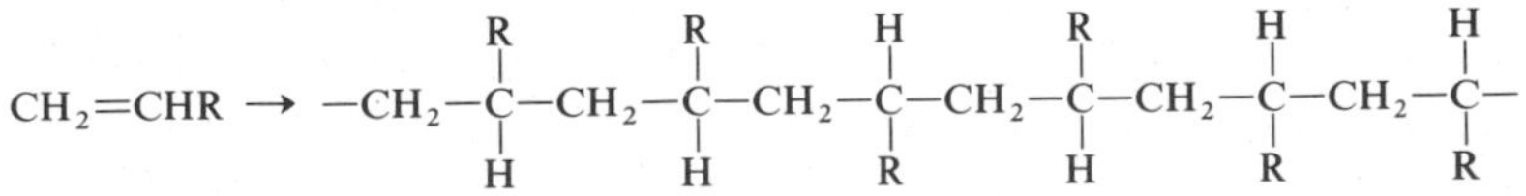

$$CH_2{=}CHR \rightarrow \quad -CH_2-\overset{\displaystyle R}{\underset{\displaystyle H}{C}}-CH_2-\overset{\displaystyle R}{\underset{\displaystyle H}{C}}-CH_2-\overset{\displaystyle H}{\underset{\displaystyle R}{C}}-CH_2-\overset{\displaystyle R}{\underset{\displaystyle H}{C}}-CH_2-\overset{\displaystyle H}{\underset{\displaystyle R}{C}}-CH_2-\overset{\displaystyle H}{\underset{\displaystyle R}{C}}-$$

The regular stereoisomers are possible, and the Ziegler–Natta catalysts promote formation of stereoregular polymers. Two types of regular stereochemistry are shown below. *Isotactic* polymers have the same stereochemistry at each asymmetric carbon, whereas in *syndiotactic* polymers the configuration alternates regularly down the chain:

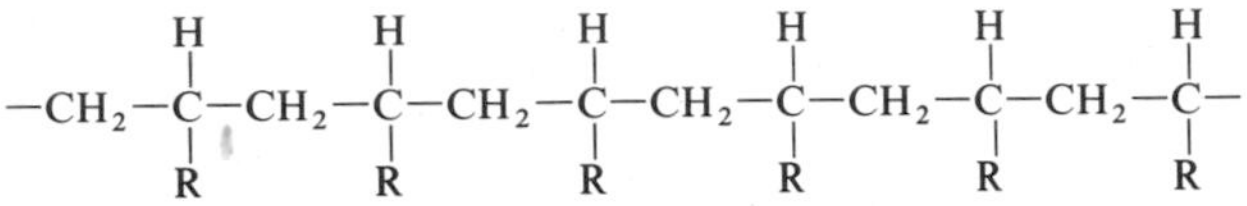

$$-CH_2-\overset{\displaystyle H}{\underset{\displaystyle R}{C}}-CH_2-\overset{\displaystyle H}{\underset{\displaystyle R}{C}}-CH_2-\overset{\displaystyle H}{\underset{\displaystyle R}{C}}-CH_2-\overset{\displaystyle H}{\underset{\displaystyle R}{C}}-CH_2-\overset{\displaystyle H}{\underset{\displaystyle R}{C}}-CH_2-\overset{\displaystyle H}{\underset{\displaystyle R}{C}}-$$

isotactic polymer

$$-CH_2-\overset{\displaystyle R}{\underset{\displaystyle H}{C}}-CH_2-\overset{\displaystyle H}{\underset{\displaystyle R}{C}}-CH_2-\overset{\displaystyle R}{\underset{\displaystyle H}{C}}-CH_2-\overset{\displaystyle H}{\underset{\displaystyle R}{C}}-CH_2-\overset{\displaystyle R}{\underset{\displaystyle H}{C}}-CH_2-\overset{\displaystyle H}{\underset{\displaystyle R}{C}}-CH_2-\overset{\displaystyle R}{\underset{\displaystyle H}{C}}-$$

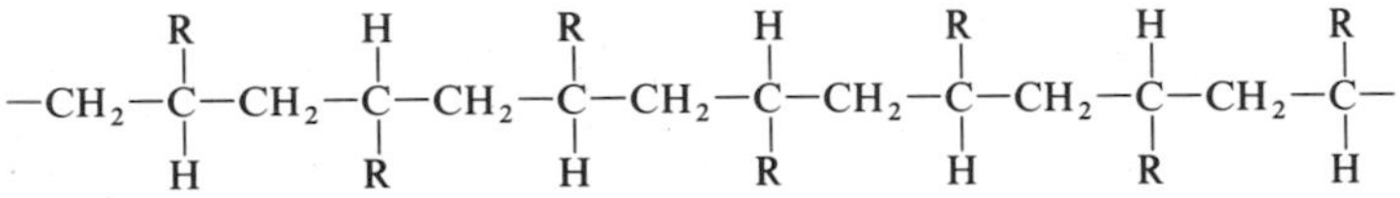

syndiotactic polymer

Usually, isotactic polymers are formed using Ziegler–Natta catalysts, but propylene, for example, can also be caused to polymerize to give the syndiotactic form.[6] Polymers prepared using coordination catalysts are also characterized by minimal amounts of branching. As a result, they have higher crystallinity and density than polymers of the same composition prepared by radical or cationic catalysis. The high degree of stereoregularity achieved by coordination polymerization is presumably a result of steric interactions at the reactive site. A preferred stereochemistry for binding and subsequent reaction of the monomer units can clearly lead to isotactic polymer.

Scheme 11.1 is a summary of major polymers and includes some polymers that are prepared by chain-addition methods.

11.1.2. Step-Growth Polymerization

The second general pattern of polymerization is "step-growth" or condensation polymerization. While terminal alkenes are the most common monomers in chain-growth polymerization, bifunctional molecules are the characteristic monomers for step-growth polymerization. The polyester-, polyamide-, and polyurethane-forming reactions shown below are examples of step-growth polymerizations:

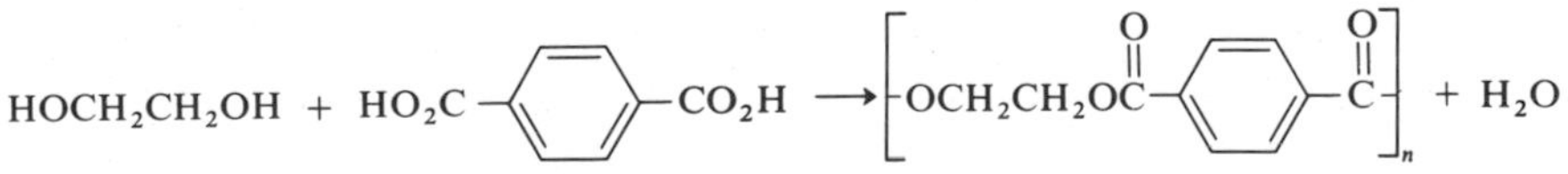

$$HOCH_2CH_2OH \;+\; HO_2C-\!\!\left\langle\!\!\bigcirc\!\!\right\rangle\!\!-CO_2H \;\longrightarrow\; \left[OCH_2CH_2O\overset{O}{\overset{\|}{C}}-\!\!\left\langle\!\!\bigcirc\!\!\right\rangle\!\!-\overset{O}{\overset{\|}{C}}\right]_n \;+\; H_2O$$

6. J. Boor, Jr., *Macromol. Rev.* **2**, 115 (1967).

$$\overset{+}{H_3}N(CH_2)_6\overset{+}{N}H_3 \; + \; \bar{O}_2C(CH_2)_4CO_2^- \;\rightarrow\; \left[\overset{O}{\overset{\|}{C}}(CH_2)_4\overset{O}{\overset{\|}{C}}NH(CH_2)_6NH-\right]_n + H_2O$$

$$HO(CH_2)_nOH \; + \; O{=}C{=}N{-}\bigcirc{-}\bigcirc{-}N{=}C{=}O \;\longrightarrow$$

$$\longrightarrow \left[-O(CH_2)_n O\overset{O}{\overset{\|}{C}}NH{-}\bigcirc{-}\bigcirc{-}NH\overset{O}{\overset{\|}{C}}-\right]_n$$

The reactions involved in these types of polymerizations are not qualitatively different from those of the respective functional groups encountered earlier in the book. Condensation reactions at carbonyl groups are particularly prevalent, but the third example above is a nucleophilic addition to an electron-deficient cumulene system.

The feature that distinguishes step-growth polymerization from chain polymerization is the pattern of polymer molecular-weight change versus time. In the chain polymerizations, only a relatively few chains are growing at a given instant. At any point in time, the reaction mixture consists primarily of polymer and monomer, with only a few chains of intermediate length in the state of growth. The amount of polymer increases and that of monomer decreases with time, but the chain length of the fraction that is polymer is relatively constant. In step-growth polymerization, on the other hand, the monomer disappears quite rapidly in favor of dimer, trimer, tetramer, etc., and as reaction proceeds the average molecular weight increases constantly as units of ever-increasing molecular weight are coupled. The term step-growth, then, comes from this characteristic addition of telomers or oligomers, rather than monomer units, to the polymer.

Reactions that are suitable for step-growth polymerization must proceed to a high degree of conversion with no side reactions in order to achieve acceptable molecular weights. If one of the bifunctional reagents is exhausted prematurely, growth of the polymer chain is terminated at that point for lack of additional reactive functional groups. For example, if A and B are present in equimolar amount, and B is converted to polymer with 98% efficiency and with 2% loss to unreactive side product, the polymerization will stop with a molecular weight of $(A)_{50} (B)_{49}$, on the average. The desired molecular-weight range is usually much above this. Thus, only reactions of very high efficiency are suitable for step-growth polymerization.

A wide variety of difunctional molecules have been converted into polymers by the step-growth process. It is also possible to prepare polymers from a single difunctional molecule if the two functional groups are reactive toward one another. The preeminent example of such a case is the formation of Nylon 6 from caprolactam by ring-opening and subsequent linear polymerization of the resulting amino acid:

$$\overset{H_2O}{\underset{-OH}{\rightleftharpoons}} \; H_2N(CH_2)_5CO_2^- \;\rightarrow\; \left[-NH(CH_2)_5\overset{O}{\overset{\|}{C}}-\right]_n$$

Some of the polymers prepared by step-growth processes are included in Scheme 11.1.

The search for improvements in polymer characteristics has led to a variety of techniques for modification of polymer properties. One with a clearly chemical basis is the process of copolymerization. This can be carried out by polymerization of a mixture of monomers, both of which are susceptible to the operating polymerization mechanism. The extent of incorporation of each monomer under these conditions is a function of their concentration and relative reactivities toward the reactive site on the polymer chain. Certain pairs of monomers copolymerize with alternating structure:

$$n\text{A} + n\text{B} \rightarrow \text{A}{-}\text{B}{-}\text{A}{-}\text{B}{-}\text{A}{-}\text{B}$$

This alternation occurs when the reactive site derived from A, A*, is more reactive toward monomer B than toward A and vice versa. The alternate case, with A* more reactive toward A than toward B, leads to a polymer rich in A:

$$n\text{A} + n\text{B} \rightarrow \text{A}{-}\text{A}{-}\text{A}{-}\text{A}{-}\text{B}{-}\text{A}{-}\text{A}{-}\text{A}{-}\text{A}{-}\text{B}{-}\text{A}{-}\text{A}{-}\text{A} + \text{B}$$

The tendency for a monomer to react with the second monomer in a copolymerization rather than with itself is commonly expressed as the r value, where:

$$r = \frac{\text{rate of homo reaction}}{\text{rate of cross reaction}}$$

Alternating introduction of monomers occurs when the r values for both monomers are near zero. Where r values are near 1, random copolymerization occurs, but r values $\gg 1$ will lead to homopolymerization.

Anionic polymerization that leads to "living polymer" offers a different approach to copolymers. After a monomer has been polymerized, but before it has been exposed to any quenching agent, it is capable of propagating the polymerization of a second monomer. This propagation results in a "block copolymer," since the two

$$\text{X}^- + \text{CH}_2{=}\text{CHR} \rightarrow \text{X}(\text{CH}_2\text{CH})_n\overset{-}{\text{CH}_2\text{CHR}}$$
$$\underset{\text{R}}{|}$$

$$\text{X}(\text{CH}_2\text{CH})_n\,\text{CH}_2\overset{-}{\text{CHR}} + m\,\text{CH}_2{=}\text{CHR}' \rightarrow \text{X}(\text{CH}_2\text{CH})_{n+1}(\text{CH}_2\overset{-}{\text{CH}})_m$$
$$\underset{\text{R}}{|} \qquad\qquad\qquad\qquad\qquad \underset{\text{R}}{|} \quad \underset{\text{R}'}{|}$$

monomer types are found in separate blocks, rather than being randomly dispersed. Block copolymers can also be made by mixing a polymer and monomer having mutually reactive end groups.

$$\text{HO}{-}(\text{CH}_2\text{CH}_2\text{O})_n{-}\text{CH}_2\text{CH}_2\text{OH} + \text{O}{=}\text{C}{=}\text{N}{-}\bigcirc{-}\bigcirc{-}\text{N}{=}\text{C}{=}\text{O} \rightarrow$$

$$\rightarrow \left[{-}(\text{OCCH}_2\text{CH}_2\text{O})_{n+1}{-}\overset{\displaystyle\text{O}}{\overset{\displaystyle\|}{\text{C}}}\text{NH}{-}\bigcirc{-}\bigcirc{-}\underset{\text{H}}{\overset{}{\text{N}}}{-}\overset{\displaystyle\text{O}}{\overset{\displaystyle\|}{\text{C}}}{-}\text{O}{-}(\text{CH}_2\text{CH}_2\text{O})_{n+1}{-}\right]$$

The properties of polymers are also strongly influenced by the extent of

Scheme 11.1. Some

Polymer	Monomer(s)	Principal method
	Chain-Growth Polymerization	
Polyethylene (high density)	$CH_2{=}CH_2$	coordination polymerization
Polyethylene (low density)	$CH_2{=}CH_2$	radical-chain addition polymerization
Polypropylene (isotactic)	$CH_2{=}CHCH_3$	coordination polymerization
Polystyrene	$CH_2{=}CHPh$	peroxide-catalyzed radical-chain polymerization
Polyisoprene (*cis*)	$CH_2{=}CH{-}\overset{\displaystyle CH_3}{\underset{\displaystyle\vert}{C}}{=}CH_2$	coordination or butyllithium-catalyzed anionic polymerization
Polyisobutylene	$CH_2{=}C\overset{\textstyle CH_3}{\underset{\textstyle CH_3}{\big\langle}}$	low-temperature cationic polymerization, BF_3 and $AlCl_3$ catalysis
Polyvinyl chloride	$CH_2{=}CHCl$	redox catalysis, aqueous suspension
Polychloroprene	$CH_2{=}CH{-}\overset{\displaystyle Cl}{\underset{\displaystyle\vert}{C}}{=}CH_2$	radical-chain emulsion polymerization, redox catalysis
Polytetrafluoroethylene (Teflon)	$CF_2{=}CF_2$	radical-chain polymerization in aqueous suspension
Poly(methyl methacrylate)	$CH_2{=}C\overset{\textstyle CO_2CH_3}{\underset{\textstyle CH_3}{\big\langle}}$	radical-chain addition polymerization
Polyacrylonitrile	$CH_2{=}CHCN$	radical-chain addition polymerization in aqueous suspension
Poly(vinyl acetate)	$CH_2{=}CHO\overset{\displaystyle O}{\overset{\displaystyle\|}{C}}CH_3$	radical-chain addition polymerization

Major Polymers

Polymer	Monomer(s)	Principal method
Chain-Growth Polymerization (*continued*)		
Poly(ethylene oxide)	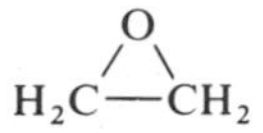 $H_2C\overset{O}{\overline{\diagup\diagdown}}CH_2$	base-catalyzed anionic chain addition
Step-Growth Polymerization		
Polyesters	 $CH_3O_2C\!-\!\langle\text{benzene}\rangle\!-\!CO_2CH_3$ $+HOCH_2CH_2OH$	step-growth polymerization by ester interchange
	$(PhO)_2C{=}O+$ 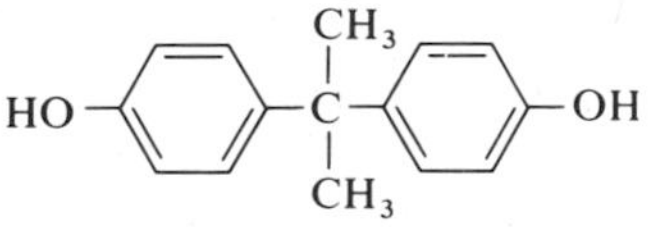 $HO\!-\!\langle\text{benzene}\rangle\!-\!\underset{CH_3}{\overset{CH_3}{C}}\!-\!\langle\text{benzene}\rangle\!-\!OH$	step-growth polymerization by ester interchange
Nylon 66	$^-O_2C(CH_2)_4CO_2^-$ $+H_3\overset{+}{N}(CH_2)_6\overset{+}{N}H_3$	step-growth polymerization by thermal amidation
Nylon 6	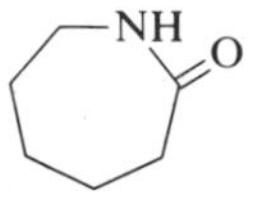 (caprolactam ring with $-NH$ and $=O$)	step-growth polymerization by ring-opening
Polyurethanes	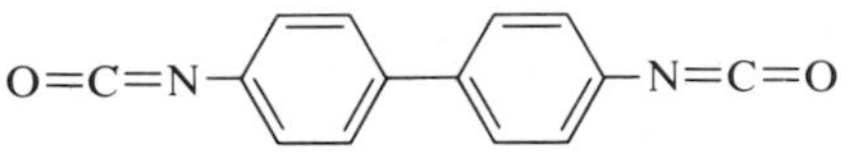 $O{=}C{=}N\!-\!\langle\text{benzene}\rangle\!-\!\langle\text{benzene}\rangle\!-\!N{=}C{=}O$ $+\ HOCH_2CH_2OH$	step-growth polymerization by nucleophilic addition
Epoxy resin	 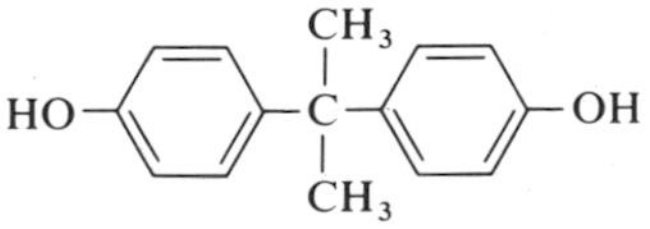 $\overset{O}{\overline{\diagup\diagdown}}\!-\!CH_2Cl\ +$ $HO\!-\!\langle\text{benzene}\rangle\!-\!\underset{CH_3}{\overset{CH_3}{C}}\!-\!\langle\text{benzene}\rangle\!-\!OH$	step-growth polymerization by nucleophilic substitution

cross-linking. The mechanisms that have been discussed so far would result in macromolecules that would involve a single chain. Mechanisms that might operate to cause branching of the chain have been described, but no processes that would make connections between the chains—i.e., bring about cross-linking—have been considered. The presence of such cross-links has very important effects on the physical properties of polymers. In particular, extensive networks of cross-linking tend to confer strength, rigidity, and low solubility on polymers.

Cross-linking often involves a chemical step distinct from the polymerization. Both natural and synthetic rubbers, for example, are cross-linked by heating with sulfur. The reactions that take place depend on the presence of unsaturated groups in the polymer chain. Their presence permits the occurrence of reactions with sulfur that incorporate chemical bonds between adjacent chains.

Often, small amounts of a monomer susceptible to cross-linking are incorporated into a polymer to be used subsequently for purposes of cross-linking. The copolymerization of terminal alkenes with maleic anhydride is an example. At sites on the chain where a maleic anhydride monomer has been introduced, reactions involving the anhydride or derived groups can occur.

In condensation polymerizations involving formation of polyesters or polyurethanes, cross-linking can be achieved by use of some of a trihydroxylic alcohol, such as glycerol, in addition to the dihydroxylic alcohol.

This section has provided a brief glance at some of the basic concepts underlying the synthesis of polymers. The mechanistic principles that govern polymerizations are the same as those that serve to explain small-molecule reactions. The additional challenges that arise in polymer production have to do with controlling the extent of polymerization and modifying the properties of polymers to increase their functional utility. The ever-changing needs for materials with specific properties have to be met by modification of molecular structure or introduction of additives that provide the desired properties. Of the many such properties that are currently sought after, flame resistance and biodegradability can be cited as only two examples.

11.2. Peptide and Protein Synthesis

The synthesis of peptides and proteins involves formation of amide bonds between amino and carboxyl groups of α-amino acids. The synthesis of a polypeptide

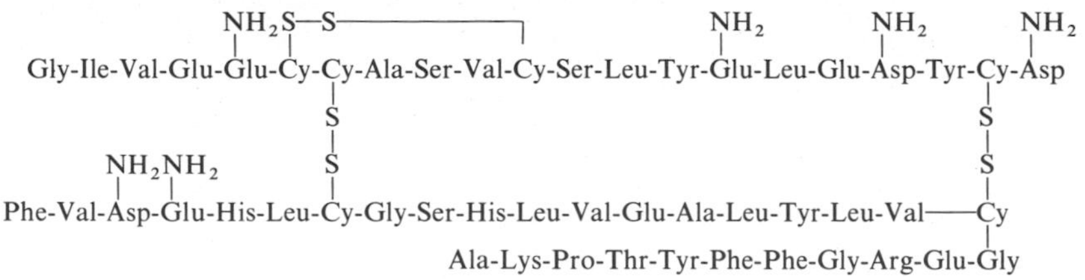

$$\begin{array}{l} \text{NH}_2\text{S——S}\overline{} \qquad\qquad \text{NH}_2 \qquad \text{NH}_2 \qquad \text{NH}_2 \\ \text{Gly-Ile-Val-Glu-Glu-Cy-Cy-Ala-Ser-Val-Cy-Ser-Leu-Tyr-Glu-Leu-Glu-Asp-Tyr-Cy-Asp} \end{array}$$

Gly-Ile-Val-Glu-Glu-Cy-Cy-Ala-Ser-Val-Cy-Ser-Leu-Tyr-Glu-Leu-Glu-Asp-Tyr-Cy-Asp

NH₂NH₂   S   S

Phe-Val-Asp-Glu-His-Leu-Cy-Gly-Ser-His-Leu-Val-Glu-Ala-Leu-Tyr-Leu-Val——Cy

Ala-Lys-Pro-Thr-Tyr-Phe-Phe-Gly-Arg-Glu-Gly

Figure 11.1. Amino acid sequence of bovine insulin.

derived from a single amino acid is therefore somewhat similar in concept to the synthesis of a polyamide such as Nylon 6:

$$\underset{R}{\text{H}_2\text{NCHCO}_2\text{H}} \rightarrow \text{H}_2\text{N}\underset{R}{\text{CH}}\underset{O}{\text{C}}\text{NH}\underset{R}{\text{CH}}\underset{O}{\text{C}}\text{NH}\underset{R}{\text{CH}}\underset{O}{\text{C}}\cdots\text{NH}\underset{R}{\text{CH}}\text{CO}_2\text{H}$$

Polypeptides with single monomer units, however, are not of primary interest for study of the biological aspects of peptides and proteins. What is of interest in the biological realm is a specific sequence of monomer units chosen from the twenty-some amino acids found in natural proteins. Peptide or protein synthesis, therefore, is usually a stepwise process; each specific monomer unit must be added in proper sequence.

The strategy that has been developed to make this possible is built around the use of activating groups and protecting groups. Thus, a typical procedure would start with an amino acid protected at the carboxyl group. The second amino acid would be protected at the amino group. Reaction is brought about by a reagent that converts the free carboxyl group to an active acylating agent. After the first amide bond has been formed, the amino-protecting group of the second unit is removed and the third residue added. The sequence is repeated until the desired chain has been constructed.

There is also a crucial problem of stereochemistry in the synthesis of polypeptides. Natural proteins consist of L-amino acids, and any racemization at the chiral centers has a profound effect on the structure and biological activity. Altered stereochemistry drastically changes the three-dimensional shape of a polypeptide chain, and this three-dimensional structure is normally essential for the biological function of the polypeptide.

The techniques for synthesis of peptides have been developed to a stage of refinement such that in the mid-1960's, several groups of workers were able to report total syntheses of insulin.[7] The sequence of 51 amino acids present in insulin is given in Figure 11.1. Still larger polypeptides have subsequently been synthesized.

The sensitivity of certain of the amino acids and, more generally, the problem of

7. P. G. Katsoyannis, A. Tometsko and C. Zalut, *J. Am. Chem. Soc.* **88**, 166 (1966); Y.-T. Kung *et al.*, *Ko Hsueh Tung Pao*, 941 (1966); *Chem. Abstr.* **66**, 18850z (1967); H. Zahn, J. Meienhofer, and E. Schnabel, *Acta Chim. Acad. Sci. Hung.* **44**, 109 (1965); *Chem. Abstr.* **63**, 13405 (1965).

racemization demand that all the steps in polypeptide synthesis take place under mild conditions. A key requirement of successful polypeptide synthesis, therefore, is the availability of protecting groups that can be removed under mild conditions (see Chapter 10 for a discussion of protecting groups). Reference to Scheme 11.2 shows that both amino- and carboxyl-protecting groups are required. The most widely used of the amino-protecting groups is the benzyloxycarbonyl or carbobenzoxy group[8]:

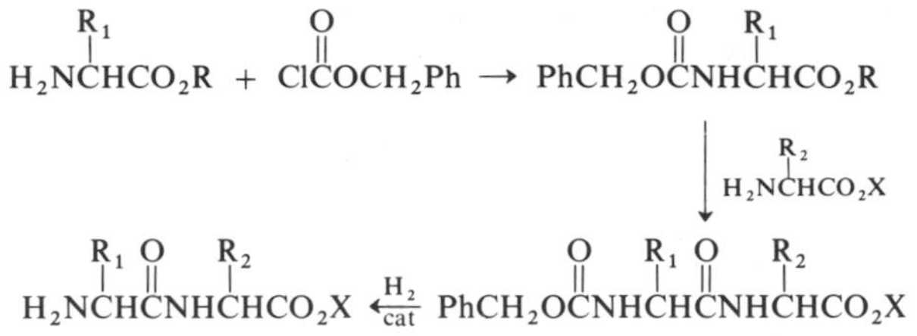

$$\underset{\overset{|}{R_1}}{\text{H}_2\text{NCHCO}_2\text{R}} + \text{Cl}\overset{\text{O}}{\overset{\|}{\text{C}}}\text{OCH}_2\text{Ph} \rightarrow \text{PhCH}_2\text{O}\overset{\text{O}}{\overset{\|}{\text{C}}}\text{NHCHCO}_2\text{R}$$

$$\downarrow \underset{\overset{|}{R_2}}{\text{H}_2\text{NCHCO}_2\text{X}}$$

$$\text{H}_2\text{NCHCNHCHCO}_2\text{X} \xleftarrow[\text{cat}]{\text{H}_2} \text{PhCH}_2\text{OCNHCHCNHCHCO}_2\text{X}$$

It is introduced by N-acylation, using benzyl chloroformate. We have already discussed removal of benzyl groups by hydrogenolysis (Section 10.2). In the case of benzyl carbamates, the hydrogenolysis is followed by decarboxylation of the liberated carbamic acid:

$$\text{PhCH}_2\text{OCNHCHCO}_2\text{X} \rightarrow \text{PhCH}_3 + \text{HO}_2\text{CNHCHCO}_2\text{X} \xrightarrow{-\text{CO}_2} \text{H}_2\text{NCHCO}_2\text{X}$$

Other methods for effecting the removal of the carbobenzoxy group that have found use in polypeptide synthesis include reduction by sodium in liquid ammonia[9] and treatment with concentrated HBr.[10]

The t-butoxycarbonyl group also ranks as an important amino-protecting group. The introduction by N-acylation is usually accomplished using t-butoxycarbonyl azide. Removal is based on the relative stability of the t-butyl cation. Acidic conditions (anhydrous trifluoroacetic acid is often used) result in cleavage of the C–O bond, and decarboxylation follows.[11]

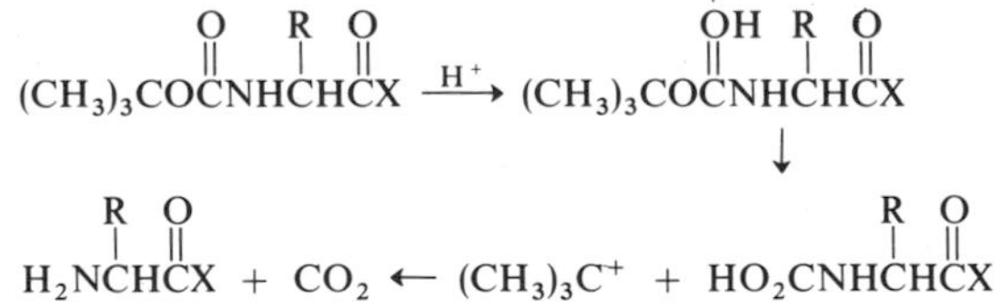

$$(\text{CH}_3)_3\text{COCNHCHCX} \xrightarrow{\text{H}^+} (\text{CH}_3)_3\text{COCNHCHCX}$$

$$\text{H}_2\text{NCHCX} + \text{CO}_2 \leftarrow (\text{CH}_3)_3\text{C}^+ + \text{HO}_2\text{CNHCHCX}$$

The trifluoroacetyl group is also useful as an amino-protecting group. Because of the inductive effect of the fluorine substituents, trifluoroacetyl derivatives are

8. M. Bergmann and L. Zervas, *Chem. Ber.* **65**, 1192 (1932).
9. R. H. Sifferd and V. du Vigneaud, *J. Biol. Chem.* **108**, 753 (1935).
10. D. Ben-Ishai and A. Berger, *J. Org. Chem.* **17**, 1564 (1952); G. W. Anderson, J. Blodinger, and A. D. Welcher, *J. Am. Chem. Soc.* **74**, 5309 (1952).
11. L. A. Carpino, *J. Am. Chem. Soc.* **79**, 98 (1957).

X, Y = protecting groups; Z = reactive leaving group

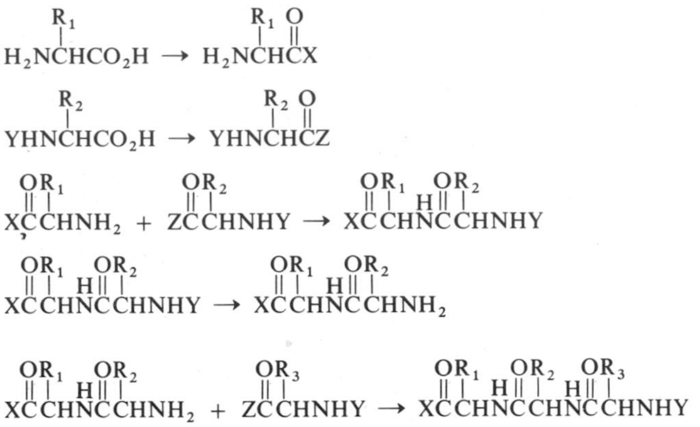

$$
\begin{array}{l}
\underset{\text{H}_2\text{NCHCO}_2\text{H}}{\overset{R_1}{|}} \rightarrow \underset{\text{H}_2\text{NCHCX}}{\overset{R_1\ \ O}{|\ \ \ ||}} \qquad\qquad \text{terminal CO}_2\text{H protection}
\end{array}
$$

$$
\underset{\text{YHNCHCO}_2\text{H}}{\overset{R_2}{|}} \rightarrow \underset{\text{YHNCHCZ}}{\overset{R_2\ \ O}{|\ \ \ ||}} \qquad\qquad \text{carboxyl activation}
$$

$$
\underset{\text{XCCHNH}_2}{\overset{OR_1}{||\,|}} + \underset{\text{ZCCHNHY}}{\overset{OR_2}{||\,|}} \rightarrow \underset{\text{XCCHNCCHNHY}}{\overset{OR_1\ \ \ OR_2}{||\,|\ \ H||\,|}} \qquad \text{coupling}
$$

$$
\underset{\text{XCCHNCCHNHY}}{\overset{OR_1\ \ OR_2}{||\,|\ H||\,|}} \rightarrow \underset{\text{XCCHNCCHNH}_2}{\overset{OR_1\ \ OR_2}{||\,|\ H||\,|}} \qquad \text{amino deprotection}
$$

$$
\underset{\text{XCCHNCCHNH}_2}{\overset{OR_1\ \ OR_2}{||\,|\ H||\,|}} + \underset{\text{ZCCHNHY}}{\overset{OR_3}{||\,|}} \rightarrow \underset{\text{XCCHNCCHNCCHNHY}}{\overset{OR_1\ \ OR_2\ \ OR_3}{||\,|\ H||\,|\ H||\,|}} \qquad \text{coupling}
$$

much more readily hydrolyzed in alkaline solution than are normal amides.[12] They can also be cleaved reductively, using sodium borohydride.[13] Because of these unique conditions for removal, trifluoroacetyl can be used as a protecting group at amino sites that must remain protected during removal of the *t*-butoxycarbonyl or benzyloxycarbonyl groups.

Carboxyl groups can be protected as esters. Methyl and ethyl esters can be removed by hydrolysis with dilute alkali; benzyl esters can be removed by hydrogenolysis. Probably the most widely used of the carboxyl-protecting groups, however, is the *t*-butyl group. Removal is easy because of the stability of the *t*-butyl cation; it can be effected by HBr in acetic acid or with trifluoroacetic acid. The esters are prepared by acid-catalyzed addition of the carboxyl group to isobutylene. This reaction, too, depends on the stability of the *t*-butyl cation. Reference to Scheme 11.2 shows that it is usually necessary to protect the C-terminal carboxyl group with a group that will remain in place during the subsequent removal of amino-protecting groups. The *t*-butyl group can serve this function if the amino-protecting groups are removed by hydrogenolysis.

The synthesis of polypeptide fragments is carried out by successively adding units in which the amino group is protected and the carboxyl group is activated. The peptide chain is deprotected at the amino group before each subsequent activated amino acid is added. The carboxyl activation must be accomplished under mild conditions to avoid racemization or other side reactions. Some of the means for activating carboxyl groups for acylation are considered in the following paragraphs.

12. E. E. Schallenberg and M. Calvin, *J. Am. Chem. Soc.* **77**, 2779 (1955).
13. F. Weygand and T. E. Frauendorfer, *Chem. Ber.* **103**, 2437 (1970).

Although protected amino acid chlorides were used in several of the first peptide syntheses,[14] their use is not common in the more recent work. These compounds have limited stability and are also difficult to purify; these problems have led to their being supplanted by other activated carbonyl systems. Mixed anhydrides are related in concept, but are much more useful in practical terms. They are prepared by reaction of a protected amino acid with an acylating system, usually ethyl chloroformate[15]:

$$\underset{\text{XCNHCHCO}_2\text{H}}{\overset{\text{O} \quad \text{R}}{\parallel \quad |}} + \underset{\text{ClCOC}_2\text{H}_5}{\overset{\text{O}}{\parallel}} \xrightarrow{\text{R}_3\text{N}} \underset{\text{XCNHCHCOCOC}_2\text{H}_5}{\overset{\text{O} \quad \text{R} \quad \text{O} \quad \text{O}}{\parallel \quad | \quad \parallel \quad \parallel}}$$

A related approach is the use of cyclic *N*-carboxy anhydrides of amino acids:

$$\underset{\text{peptide}}{\text{H}_2\text{NCHC—peptide}} \rightarrow \underset{\text{peptide}}{\text{H}_2\text{NCHCNHCHC—peptide}}$$

This approach has the novel feature that the amino group is partially freed from its internal protecting group on reaction with the peptide chain. The subsequent decarboxylation to give the nucleophilic amino group ready for the next coupling occurs under very mild conditions. Much care[16] is required to see that decarboxylation does not occur before desired. If decarboxylation occurs while unreacted anhydride is still present, two or more identical units may be introduced, leading to inhomogeneity in the peptide. It is often very difficult to separate the peptide mixtures that result from such undesired side reactions because of great similarities in the properties of the desired peptide and the impurity.

One of the most common techniques for activating the carboxyl group involves formation of the *p*-nitrophenyl ester[17]:

$$\underset{\text{YHNCHCO}_2\text{H}}{\overset{\text{R}}{|}} \rightarrow \underset{\text{YHNCHCO}_2}{\overset{\text{R}}{|}}\text{—}\langle\bigcirc\rangle\text{—NO}_2$$

These derivatives are stable to extended storage, and yet the *p*-nitrophenolate anion is a sufficiently good leaving group that reaction with free amino groups occurs readily. 2,4,5-Trichlorophenyl esters are occasionally employed in the same fashion.[18]

Another somewhat different approach involves the coupling of the protected

14. E. Fischer, *Chem. Ber.* **39**, 453 (1906); E. Abderhalden and A. Fodor, *Chem. Ber.* **49**, 561 (1916).
15. N. F. Albertson, *Org. React.* **12**, 157 (1962).
16. R. Hirschmann, R. G. Strachan, H. Schwam, E. F. Schoenewaldt, H. Joshua, B. Barkemeyer, D. F. Veber, W. J. Paleveda, Jr., T. A. Jacob, T. E. Beesley, and R. G. Denkewalter, *J. Org. Chem.* **32**, 3415 (1967).
17. M. Bodanszky and V. du Vigneaud, *J. Am. Chem. Soc.* **81**, 5688 (1959).
18. J. Pless and R. A. Boissonnas, *Helv. Chim. Acta* **46**, 1609 (1963).

amino acid with N-hydroxysuccinimide.[19] Like the p-nitrophenyl esters, the acylated N-hydroxysuccinimides can be isolated and purified, but rapidly react with free amino groups. The N-hydroxysuccinimide that is liberated is easily removed because

$$\underset{\substack{\parallel \\ }}{\text{XCNHCHCO}} \ \ \ + \ \ \text{H}_2\text{NCHCY} \rightarrow \text{XCNHCHCNHCHCY} + \text{HO}-\text{N}$$

of its solubility in dilute base. The relative stability of the anion of N-hydroxysuccinimide is also important in the ability of its esters to function as acylating agents toward amino groups.

In each of the preceding methods, the activated carboxyl compound is prepared before the carboxyl unit is brought into contact with the amino component. Other methods are based on activating the carboxyl group in the presence of the amino group. The reagent responsible for carboxyl activation, and thus for peptide bond formation, is referred to as the "coupling reagent." The most commonly employed

$$\text{XCNCHCO}_2\text{H} \ + \ \text{H}_2\text{NCHCY} \xrightarrow[\text{reagent}]{\text{coupling}} \text{XCNCHCNHCHCY}$$

coupling reagent in peptide synthesis is dicyclohexylcarbodiimide. The mechanism by which it effects carboxyl activation was discussed in Part A, Section 8.6. Since its

$$\text{XCHNCHCO}_2\text{H} + \text{C}_6\text{H}_{11}-\text{N}=\text{C}=\text{N}-\text{C}_6\text{H}_{11} \rightarrow$$

$$\text{H}_2\text{NCHCY}$$

$$\text{H}_{11}\text{C}_6\text{NHCNHC}_6\text{H}_{11} + \text{XCNHCHCNHCHCY}$$

introduction in 1955,[20] dicyclohexylcarbodiimide has been of great importance in peptide-bond formation. Relatively few efforts in polypeptide synthesis circumvent the use of this reagent.

A number of other reagents have been examined as coupling reagents. One that is encountered fairly frequently in polypeptide synthesis is N-ethyl-5-phenylisoxazolium sulfonate.[21] Via a somewhat complicated mechanism that we will

19. G. W. Anderson, J. E. Zimmerman, and F. M. Callahan, *J. Am. Chem. Soc.* **86**, 1839 (1964).
20. J. C. Sheehan and G. P. Hess, *J. Am. Chem. Soc.* **77**, 1067 (1955); J. C. Sheehan, M. Goodman, and G. P. Hess, *J. Am. Chem. Soc.* **78**, 1367 (1956).
21. R. B. Woodward and R. A. Olofson, *J. Am. Chem. Soc.* **83**, 1007 (1961); R. B. Woodward, R. A. Olofson, and H. Mayer, *Tetrahedron Suppl.* **8**, 321 (1966).

not consider here, this reagent converts protected amino acids to an activated

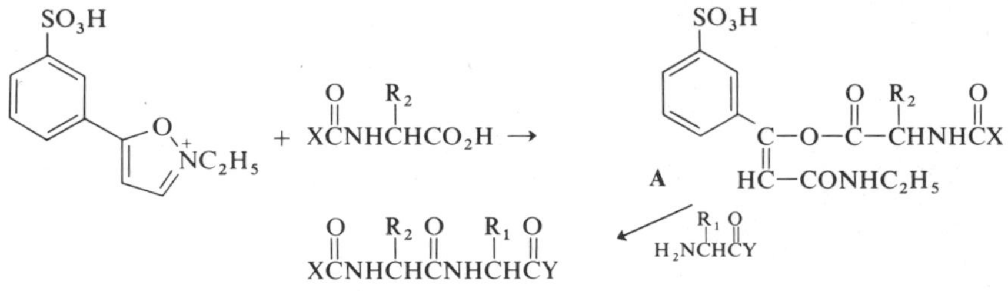

derivative, **A**, which reacts with the amino unit. The reactivity of **A** toward nucleophilic attack is the result of its being an ester of an enol. Such esters are considerably more reactive toward nucleophilic attack than alkyl esters.

The synthesis of polypeptides of significant size is usually completed by the coupling of smaller fragments. This coupling involves the deprotection of the carboxyl terminus of one fragment, followed by activation and reaction with the amino terminus of a second fragment. An important technique for this fragment-coupling involves diazotization of a C-terminal hydrazide. When this method is to be employed, the hydrazide group can also serve as the carboxyl-protecting group during synthesis of the fragments, and is then in place for use in fragment coupling. The acyl hydrazides can be prepared by reaction of a carboxyl ester group with hydrazine. These hydrazides can be carried through multistep procedures protected at the amino group by carbobenzyloxy, *t*-butoxycarbonyl, or other removable acyl

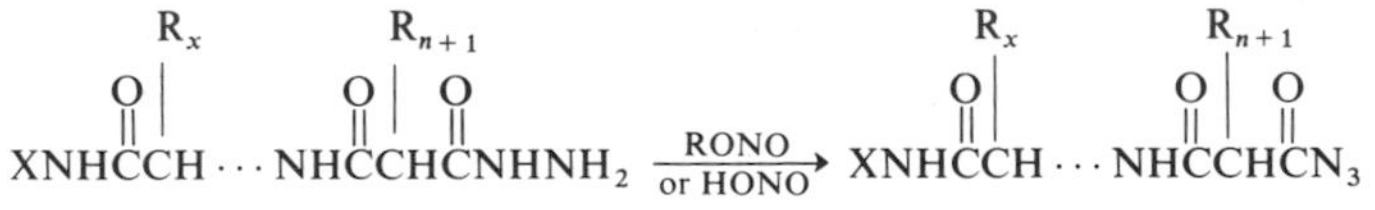

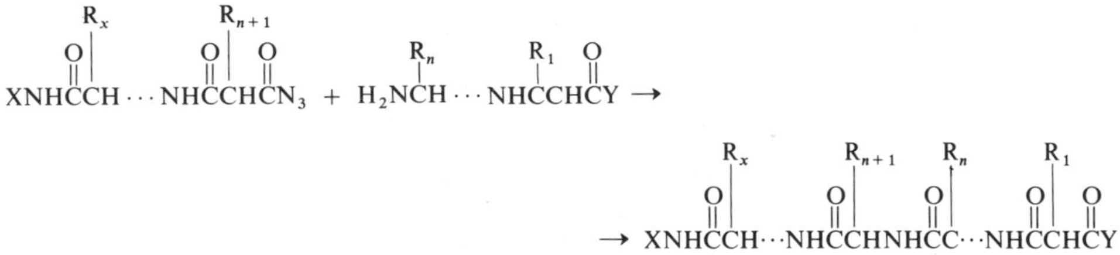

substituents. The acyl azide prepared by diazotization reacts as an activated carbonyl system, resulting in coupling.

Acyl azides can also be used much as mixed·anhydrides are in small peptide synthesis. In this case, the azido group can be introduced via reaction of an amino-protected acid chloride with sodium azide, as well as by hydrazinolysis. There is also a reagent, diphenylphosphoryl azide, that converts carboxyl groups directly to the azide[22]:

$$RCO_2H \ + \ (PhO)_2\overset{O}{\overset{\|}{P}}N_3 \ \rightarrow \ R\overset{O}{\overset{\|}{C}}N_3 \ + \ (PhO)_2\overset{O}{\overset{\|}{P}}OH$$

22. T. Shioiri, K. Ninomiya, and S. Yamada, *J. Am. Chem. Soc.* **94**, 6203 (1972).

A significant feature of the azide coupling method is that it has a minimal tendency to lead to racemization. Although minor amounts of racemization have been detected in certain systems, other coupling methods usually lead to more extensive racemization under comparable conditions.[23]

Most syntheses of a polypeptide containing more than a few amino acid units also involve amino acids such as serine, lysine, and cysteine in which hydroxyl, amino, and sulfhydryl groups, respectively, must be protected during peptide synthesis. Although we will not go into details of these particular operations, the basic principles and requirements are the same as for amino and carboxyl protection. The group must be easily introduced and stable to coupling conditions, but nevertheless removed without damage to the polypeptide at a late stage in the synthesis.

Scheme 11.3 provides reference to recent papers in which syntheses of polypeptides of varying complexity have been reported. Reading any of these papers will provide insight into the practical application of protecting groups in peptide synthesis.

In 1962, a new approach to peptide synthesis was introduced by R. B. Merrifield.[24] The key idea was that the first amino acid residue would be attached to an insoluble polymer. In practice, the attachment was accomplished by introducing chloromethyl groups on polystyrene and alkylating the carboxyl end of the first residue:

$$ClCH_2\text{—}\langle\bigcirc\rangle\text{—}\underset{\underset{|}{CH}}{\overset{\overset{|}{CH_2}}{}} + YHN\overset{\overset{R_1}{|}}{CH}CO_2H \rightarrow YHN\overset{\overset{R_1}{|}}{CH}CO_2CH_2\text{—}\langle\bigcirc\rangle\text{—}\underset{\underset{|}{CH}}{\overset{\overset{|}{CH_2}}{}}$$

Each of the subsequent residues could then be introduced by repeating the sequence: 1) amino deprotection, 2) dicyclohexylcarbodiimide coupling with the next amino-protected residue. The reaction system remains heterogeneous through all the operations, with the growing peptide attached to the insoluble polymer. It was envisioned that this would greatly reduce losses in the individual isolation and purification steps involved in conventional peptide synthesis. It also proved possible to mechanize and automate the addition of the reagents so that the peptide synthesis could proceed rapidly and quickly, with greatly reduced effort in the purification of individual intermediates. When the desired sequence of amino acids has been completed, the peptide is freed from the polymer by reductive cleavage, and then purified by chromatographic techniques. This method is referred to as *solid-phase peptide synthesis*, since the peptide is built up on an insoluble support.[25]

Since the synthesis proceeds without purification of intermediates, the technique demands *very high yields* in each step. Any time a residue is missed, the synthesis proceeds to give a peptide with an erroneous sequence. Simple statistical considerations show that in the synthesis of a 25-residue peptide with a 99% success rate for

23. Y. S. Klausner and M. Bodanszky, *Synthesis*, 549 (1974).
24. R. B. Merrifield, *J. Am. Chem. Soc.* **85**, 2149 (1963).
25. J. M. Stewart and J. D. Young, *Solid Phase Peptide Synthesis*, W. H. Freeman, San Francisco, CA, 1969.

Scheme 11.3. Summary of Protecting Groups and Activation

Polypeptide	Number of amino acid residues in final polypeptide	Protecting group(s)	Activation techniques	Ref.
Positions 81–104 of ribonuclease T_1	24	benzyloxycarbonyl, t-butoxycarbonyl	hydrazide diazotization, nitrophenyl esters, N-hydroxysuccinimide	a
Positions 48–80 of ribonuclease T_1	33	benzyloxycarbonyl, t-butoxycarbonyl	hydrazide diazotization, nitrophenyl esters, N-hydroxysuccinimide	b
Positions 1–47 of ribonuclease T_1	47	benzyloxycarbonyl, t-butoxycarbonyl	hydrazide diazotization, nitrophenyl esters, N-hydroxysuccinimide	c
Positions 1–26 of insulin B-chain	26	benzyloxycarbonyl, t-butoxycarbonyl	nitrophenyl esters, isoxazolium sulfate, N-hydroxysuccinimide	d
B-chain of human insulin	30	benzyloxycarbonyl, t-butoxycarbonyl	nitrophenyl esters, hydrazide diazotization, dicyclohexylcarbodiimide, isoxazolium sulfate	e
Adrenocorticotropin* (ACTH)	39	t-butoxycarbonyl	dicyclohexylcarbodiimide	f
Ribonuclease A*	124†	t-butoxycarbonyl	dicyclohexylcarbodiimide	g

* Synthesis accomplished by solid-phase method.
† Complete homogeneity was not claimed for these materials, only that the synthetic material possessed some of the biological activity of the native protein.

a. K. Kawasaki, R. Camble, G. Dupuis, H. Romovacek, H. T. Storey, C. Yanaihara, and K. Hofmann, *J. Am. Chem. Soc.* **95**, 6815 (1973).
b. R. Camble, G. Dupuis, K. Kawasaki, H. Romovacek, N. Yanaihara, and K. Hofmann, *J. Am. Chem. Soc.* **94**, 2091 (1972).
c. H. T. Storey, J. Beacham, S. F. Cernosek, F. M. Finn, C. Yanihara, and K. Hofmann, *J. Am. Chem. Soc.* **94**, 6170 (1972).
d. P. G. Katsoyannis, J. Ginos, A. Cosmatos, and G. Schwartz, *J. Am. Chem. Soc.* **95**, 6427 (1973).

introduction of each unit, only 75% of the final chains would have the desired sequence. The rest of the product will be closely related peptides lacking one or occasionally two residues. At 100 residues, each chain would be expected to have one defect, on the average, and the yield of "perfect" peptide would be very low. Current research is therefore aimed at identifying steps that do not meet these very

Techniques Employed in Some Polypeptide Syntheses

481

SECTION 11.2.
PEPTIDE AND
PROTEIN
SYNTHESIS

Polypeptide	Number of amino acid residues in final polypeptide	Protecting group(s)	Activation techniques	Ref.
Ribonuclease AS′	104†	t-butoxycarbonyl	N-carboxy anhydrides, N-thiocarboxy anhydrides, N-hydroxysuccinimide	h
Bradykinin*	9	t-butoxycarbonyl	dicyclohexylcarbodiimide	i
Oxytocin	9 (cyclic)		acid chloride, dicyclohexylcarbodiimide, mixed anhydride, nitrophenyl esters	j
ACTH-(1–23)	23	benzyloxycarbonyl	mixed anhydride, carbonyl azide, dicyclohexylcarbodiimide	k

Abbreviations for protecting groups and peptide and activation techniques‡

Protecting groups		Activation techniques	
Z (or Cbz)	Carbobenyloxy	MA	Mixed anhydride
Boc	t-Butoxycarbonyl	DCC	Dicyclohexylcarbodiimide
		NP	Nitrophenyl ester
		HOSO	N-hydroxysuccinimide
		EPIO	N-ethyl-5-phenylisoxazolium sulfonate

‡ J. Biol. Chem. **247**, 977 (1972); M. Bodanszky and M. A. Ondetti, *Peptide Synthesis*, Wiley–Interscience, New York, NY, 1966, pp. 255–256.

e. P. G. Katsoyannis, M. Tilak, J. Ginos, and K. Suzuki, *J. Am. Chem. Soc.* **93**, 5862 (1971); P. G. Katsoyannis, J. Ginos, and M. Tilak, *J. Am. Chem. Soc.* **93**, 5866 (1971); P. G. Katsoyannis, J. Ginos, C. Zalut, M. Tilak, S. Johnson, and A. C. Trakatellis, *J. Am. Chem. Soc.* **93**, 5877 (1971).
f. D. Yamashiro and C. H. Li, *J. Am. Chem. Soc.* **95**, 1310 (1973).
g. B. Gutte and R. B. Merrifield, *J. Am. Chem. Soc.* **91**, 501 (1969).
h. R. Hirschmann, R. F. Nutt, D. F. Veber, R. A. Vitali, S. L. Varga, T. A. Jacob, F. W. Holly, and R. C. Denkewalter, *J. Am. Chem. Soc.* **91**, 507 (1970).
i R. B. Merrifield, *Biochemistry* **3**, 1385 (1964); see also M. Fridkin, A. Patchornik, and E. Katchalski, *J. Am. Chem. Soc.* **90**, 2953 (1968).
j. V. du Vigneaud, C. Ressler, J. M. Swan, C. W. Roberts, and P. G. Katsoyannis, *J. Am. Chem. Soc.* **76**, 3115 (1954); M. Bodanszky and V. du Vigneaud, *J. Am. Chem. Soc.* **81**, 2504, 5688 (1959).
k. K. Hofmann, H. Yajima, T. Liu, and N. Yanaihara, *J. Am. Chem. Soc.* **84**, 4475 (1962).

exacting yield requirements and improving conditions for such steps so that the exceedingly stringent success requirement in the coupling stage can be reliably achieved.

The solid-phase method has been used successfully to build up intermediate-size fragments. These fragments can then be individually purified prior to coupling by normal techniques. Because of the purification of the individual fragments, this approach is less likely to yield a mixture of closely related, but nonidentical, peptides.

Some representative polypeptide syntheses by the solid-phase method are included in Scheme 11.3. This scheme also serves to introduce the abbreviations for protecting groups and activation techniques that are in extensive use in this research area.

11.3. Nucleosides, Nucleotides, and Polynucleotides

The synthesis of the component units and, eventually, the polymeric structures that occur in DNA and RNA is also an important and distinct research area in organic chemistry. The smaller molecules are of interest both because of their biological activity and because they are component parts of the macromolecules.

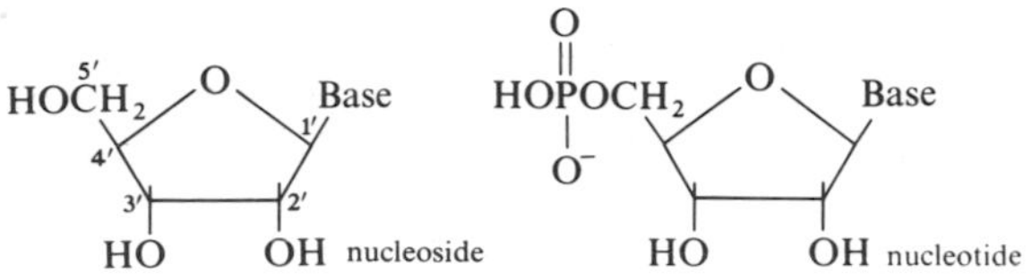

The nucleosides that occur in RNA are derivatives of ribose bonded to one of the four heterocyclic bases, adenine, cytosine, guanine, or uracil. When phosphorylated, they are referred to as nucleotides. The polymers are formed by phosphate linkages between the 3′-hydroxy group and the hydroxymethyl group of the adjacent nucleoside. In DNA, the sugar lacks the 2-hydroxy group (deoxyribose), and

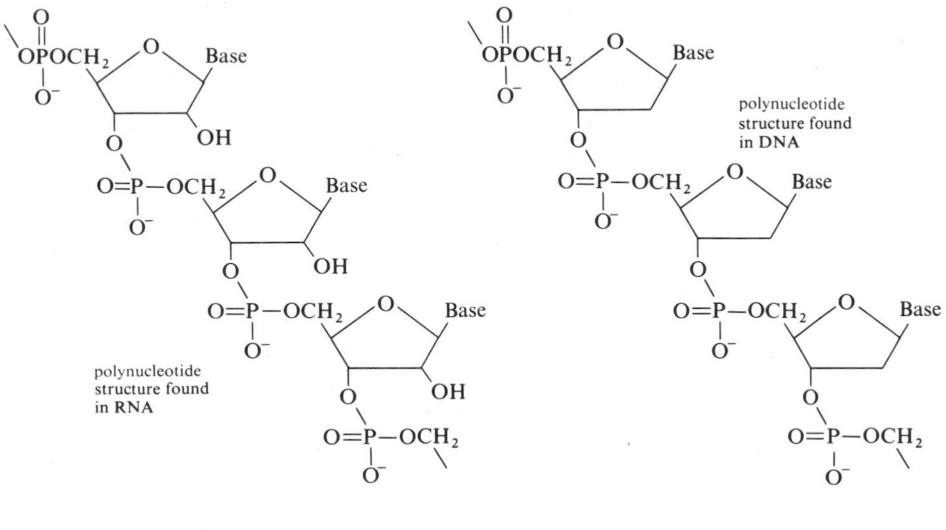

thymine is utilized as a base in place of uracil. Many other nucleosides exist in which other sugars are involved, but these are not encountered in DNA or RNA. Likewise, there is some variation in the structures of the nitrogen heterocycles. The five just named are, however, by far the most prevalent in DNA and RNA.

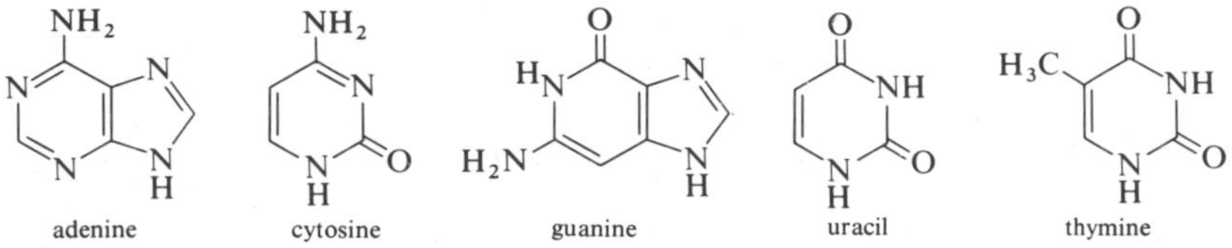

483

SECTION 11.3.
NUCLEOSIDES,
NUCLEOTIDES,
AND POLY-
NUCLEOTIDES

adenine   cytosine   guanine   uracil   thymine

The basic concepts that are applied to the synthesis of polynucleotides are similar to those of polypeptide synthesis. The chain is built up by successive addition of nucleoside units or small oligonucleotides. In practice, the problems involved in such operations have proved to be quite difficult, and the techniques for nucleotide synthesis and polymerization are not so highly developed as in peptide chemistry. Nevertheless, it is an important area in macromolecular synthesis, and some of the reactions that are the basis of present procedures in this area are therefore discussed in the succeeding paragraphs.

A fundamental reaction in nucleoside synthesis is the reaction of halide derivatives of a furanose sugar with the base acting as a nucleophile. The sugar hydroxyl groups are ordinarily protected during these operations as benzyl ethers or as acetyl or benzoyl esters. The ether groups can be removed by hydrogenolysis, while ester groups are removed by reaction with aqueous base or with aqueous ammonia. Most of the examples cited here have been chosen from the ribose and deoxyribose series, but the reactions are, in general, equally applicable to other sugars.

Since most of the bases of interest have more than one potential nucleophilic site, selectivity among these sites is important. This selectivity can be accomplished by protection of one or more of the competing sites or by taking advantage of intrinsic selectivity in the bases. For example, the desired selectivity can be achieved in adenine by using the N-mercury derivative of benzoylated adenine:

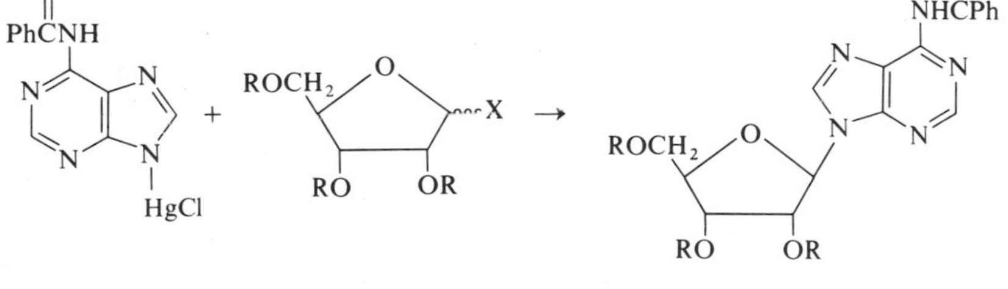

Direct reaction of adenine with the halo sugar results in attachment of the ribose at N-3 of adenine as well as at N-4[26]:

26. N. J. Leonard and R. A. Laursen, *J. Am. Chem. Soc.* **85**, 2026 (1963).

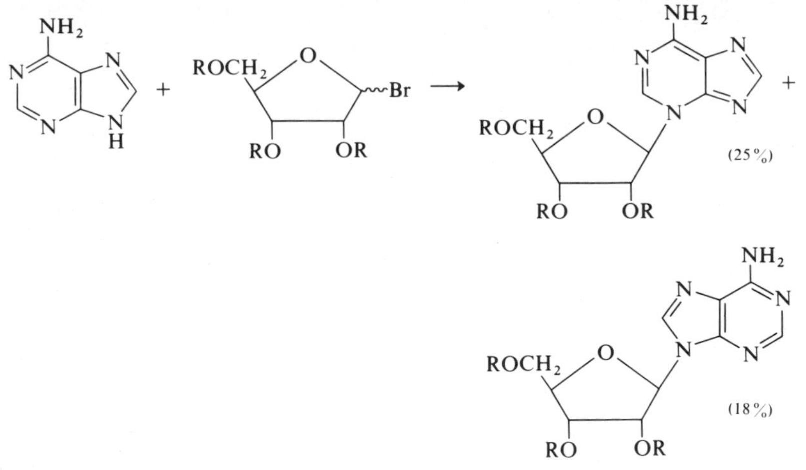

An alternative method of forming the bond linking the base and sugar is the so-called "fusion method." Tetracetylribose and the base are heated together in the absence of solvent. Acetate is, of course, a poorer leaving group than the halide, but at the temperatures used, 130–150°, reaction occurs[27]:

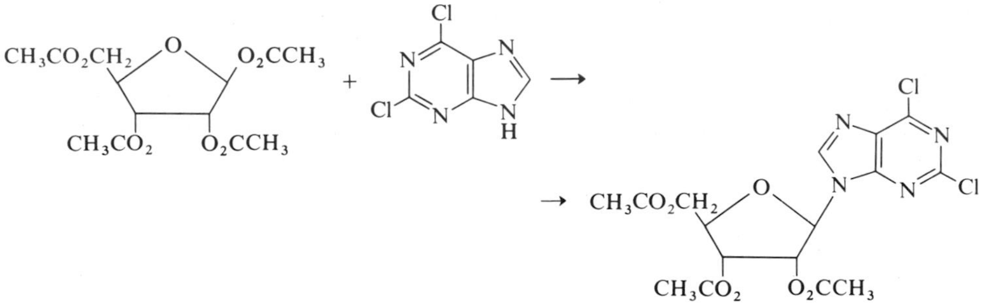

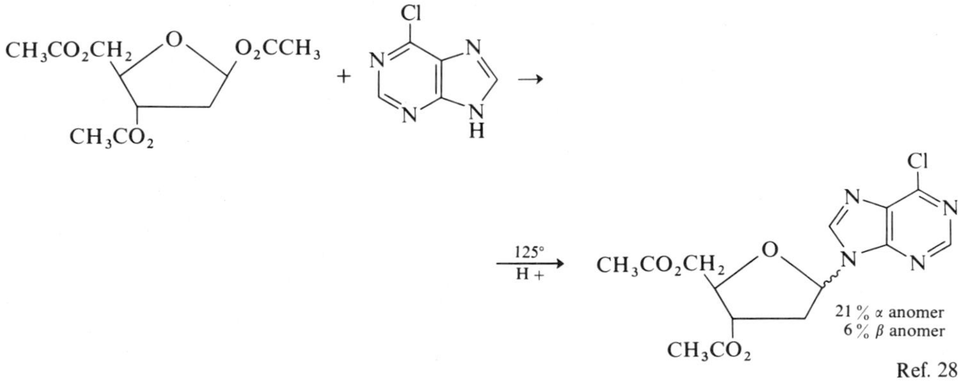

Ref. 28

27. J. A. Montgomery and K. Hewson, *J. Heterocycl. Chem.* **1**, 213 (1964).
28. M. J. Robins and R. K. Robins, *J. Am. Chem. Soc.* **87**, 4934 (1965).

485

SECTION 11.3.
NUCLEOSIDES,
NUCLEOTIDES,
AND POLY-
NUCLEOTIDES

The pyrimidine bases cytosine, uracil, and thymine cannot be directly alkylated by halo sugars. The nucleophilicity of the pyridone-type ring nitrogens is much reduced relative to the pyridine-type nitrogens present in the purine bases. The dialkoxypyrimidines are reactive, and dealkylation occurs at the alkylated nitrogen. This is known as the *Hilbert–Johnson procedure*.[29] A recent modification employs

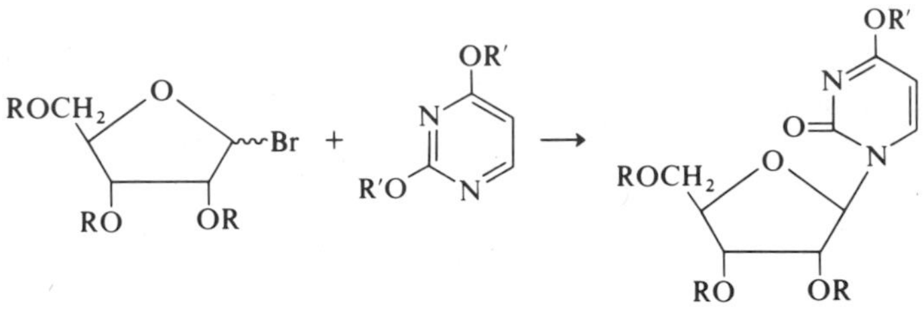

trimethylsilyl ethers of the pyrimidine bases. With this more sensitive group, both oxygens are freed from the protecting group during the alkylation and workup

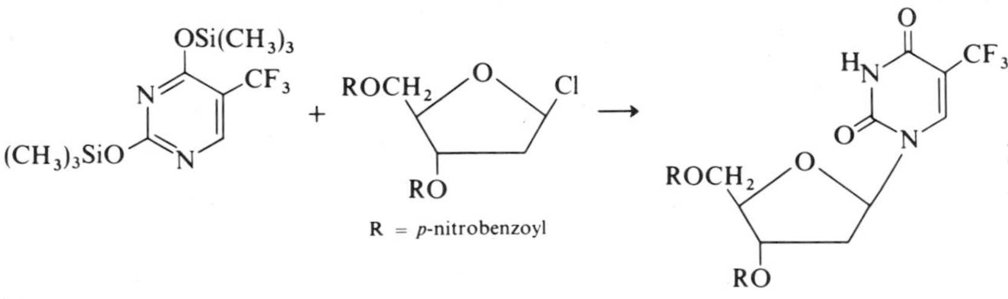

Ref. 30

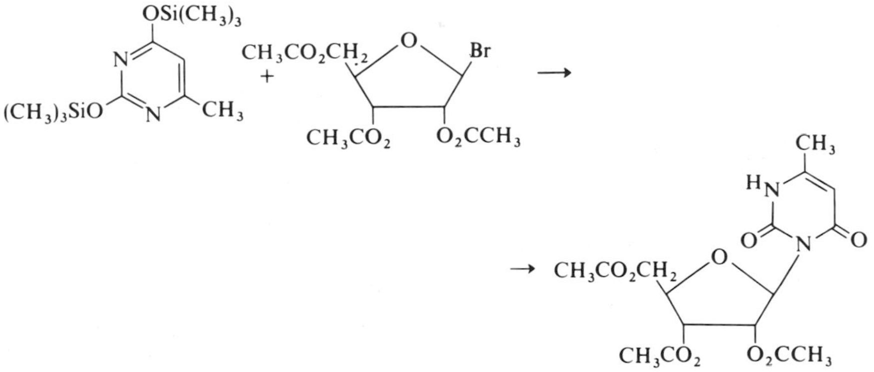

Ref. 31

procedure. Mercury derivatives can also be prepared from pyrimidine bases, and these give nucleosides on reaction with halo sugars.[32]

The stereochemistry of introduction of the base on the furanose ring in nucleoside synthesis can be governed by intramolecular participation of the sugar

29. G. E. Hilbert and T. B. Johnson, *J. Am. Chem. Soc.* **52**, 4489 (1930).
30. K. J. Ryan, E. M. Acton, and L. Goodman, *J. Org. Chem.* **31**, 1181 (1966).
31. M. W. Winkley and R. K. Robins, *J. Org. Chem.* **33**, 2822 (1968).
32. J. J. Fox and I. Wempen, *Adv. Carbohydr. Chem.* **14**, 283 (1959).

substituents. When the furanose ring carries a 2-α-acyloxy substituent, the base usually is introduced with β-stereochemistry because of intramolecular participation[33]:

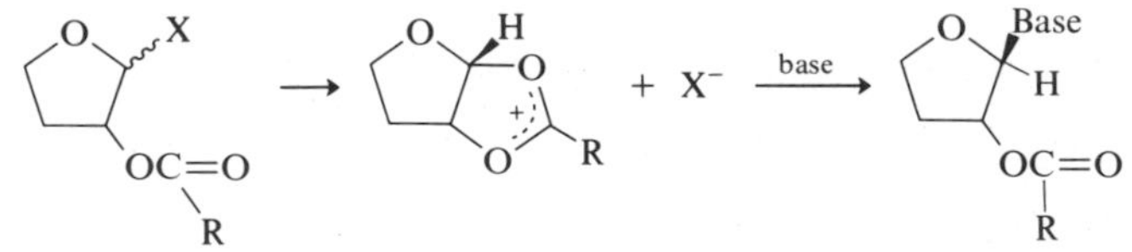

In the absence of this structural feature, a mixture of stereoisomers may be formed.

These general procedures make available a wide variety of nucleosides, as illustrated in Scheme 11.4.

The building blocks for the synthesis of polynucleotides related to RNA are all readily available and have names derived from the base constituent. A successful

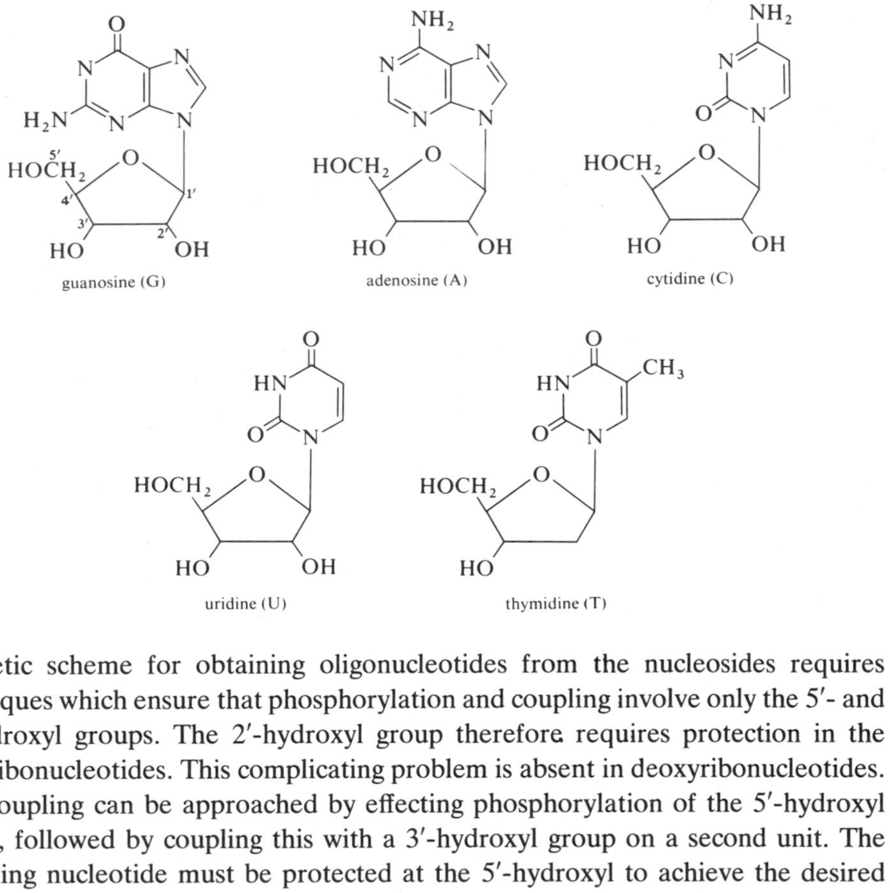

guanosine (G)

adenosine (A)

cytidine (C)

uridine (U)

thymidine (T)

synthetic scheme for obtaining oligonucleotides from the nucleosides requires techniques which ensure that phosphorylation and coupling involve only the 5'- and 3'-hydroxyl groups. The 2'-hydroxyl group therefore requires protection in the oligoribonucleotides. This complicating problem is absent in deoxyribonucleotides. The coupling can be approached by effecting phosphorylation of the 5'-hydroxyl group, followed by coupling this with a 3'-hydroxyl group on a second unit. The incoming nucleotide must be protected at the 5'-hydroxyl to achieve the desired reaction. Alternatively, the 3'-position can be the one initially phosphorylated. A desirable type of protecting group is a masked phosphoryl unit, so that deprotection and phosphorylation can be accomplished in a minimum number of steps.

33. T. L. V. Ulbricht, *Purines, Pyrimidines and Nucleotides*, Macmillan, New York, NY, 1964, pp. 48–50.

Scheme 11.4. Syntheses of Some Nucleosides

487

SECTION 11.3.
NUCLEOSIDES,
NUCLEOTIDES,
AND POLY-
NUCLEOTIDES

1[a]

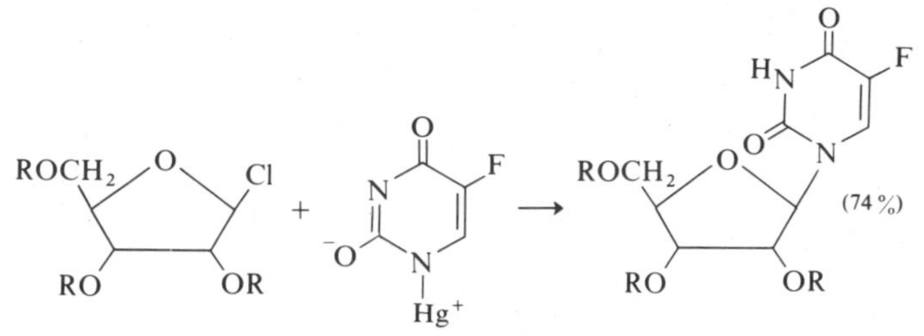

R = benzoyl

(74%)

2[b]

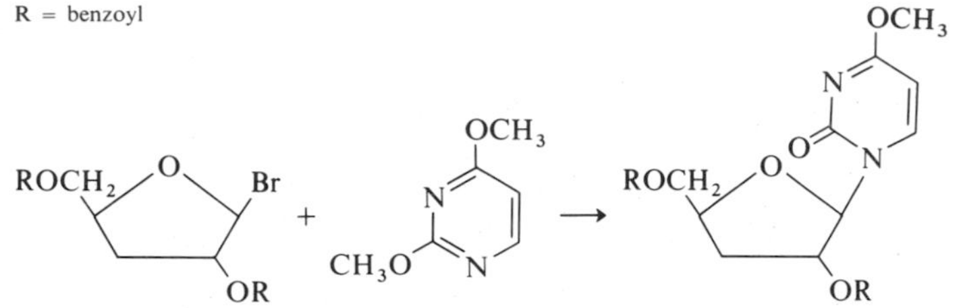

R = p-nitrobenzoyl

3[c]

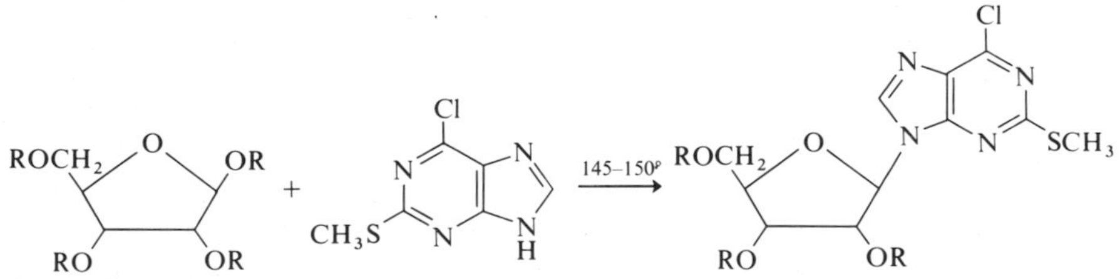

R = acetyl

4[d]

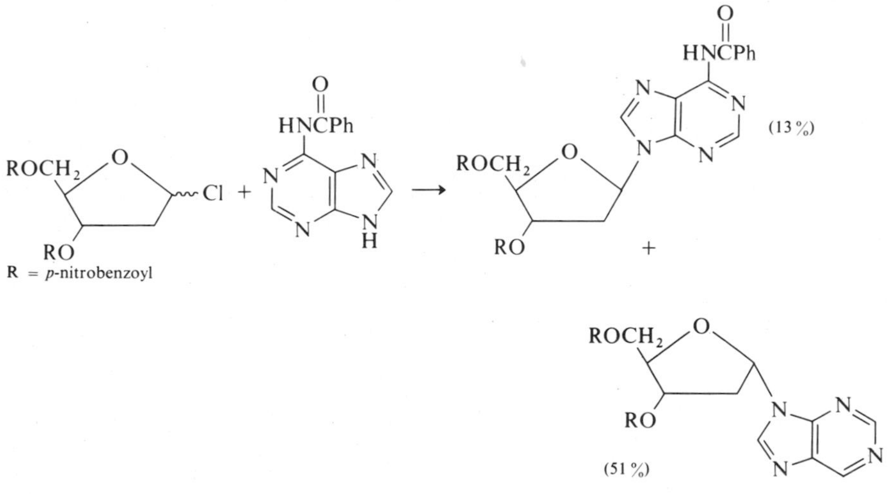

R = p-nitrobenzoyl

(13%)

(51%)

a. N. C. Yung, J. H. Burchenal, R. Fecher, R. Duschinsky, and J. J. Fox, *J. Am. Chem. Soc.* **83**, 4060 (1961).
b. E. Walton, F. W. Holly, G. E. Boxer, and R. F. Nutt, *J. Org. Chem.* **31**, 1163 (1966).
c. T. Sato, in *Synthetic Procedures in Nucleic Acid Chemistry*, W. W. Zorbach and R. S. Tipson (eds.), Interscience, New York, NY, 1968, pp. 264–268.
d. R. K. Ness, in *Synthetic Procedures in Nucleic Acid Chemistry*, W. W. Zorbach and R. S. Tipson (eds.), Interscience, New York, NY, 1968, pp. 183–192.

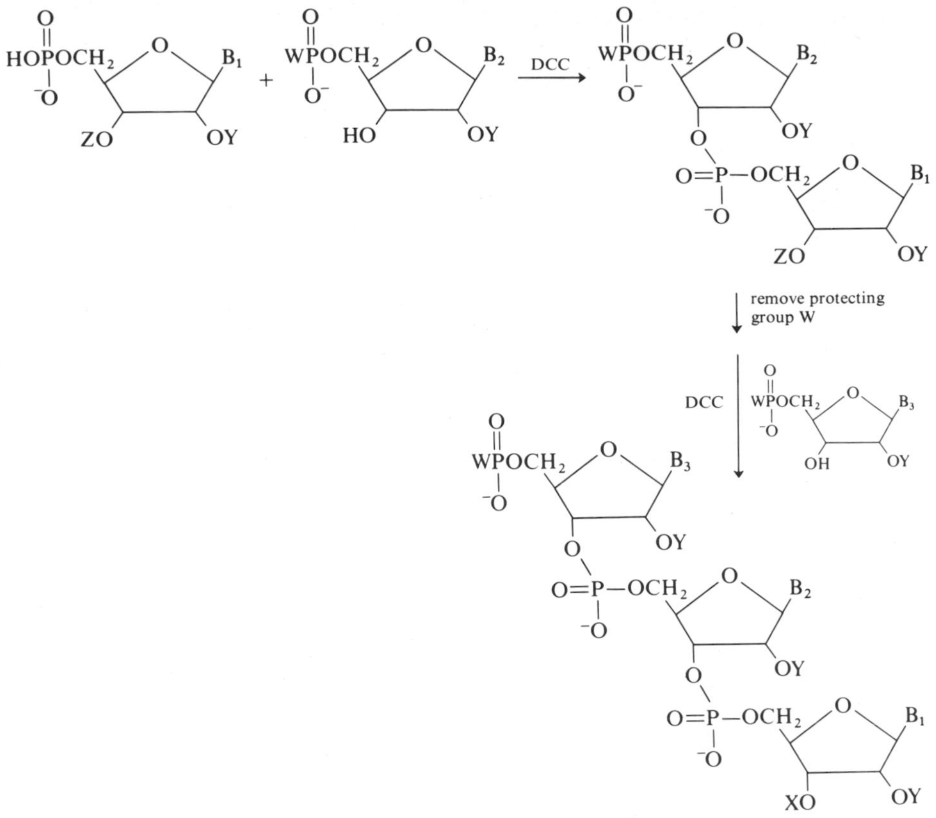

The phosphorylation of a hydroxyl group in the nucleoside is accomplished by coupling of a phosphoric acid derivative with the hydroxyl group. The phosphorus compound used may be an active phosphorylating reagent; alternatively, an alkyl phosphate monoanion and a coupling agent can be used. One common procedure for phosphorylation involves reaction with cyanoethyl phosphate.[34] The cyanoethyl group is then removed with dilute base. The mechanistic basis for its facile removal is a base-catalyzed β-elimination reaction.

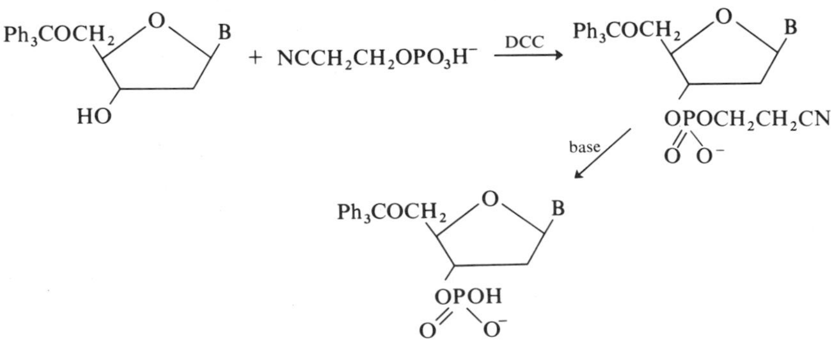

34. G. M. Tener, *J. Am. Chem. Soc.* **83**, 159 (1961); G. Weimann and H. G. Khorana, *J. Am. Chem. Soc.* **84**, 419 (1962).

A technique that illustrates conversion of a masked phosphate to an active phosphorylating agent begins with the phosphate group masked as an *N*-arylphosphoramidate. The phosphate group is liberated by nitrosation, and a coupling agent such as DCC can then be used to bring about phosphorylation of the next nucleotide

489

SECTION 11.3.
NUCLEOSIDES,
NUCLEOTIDES,
AND POLY-
NUCLEOTIDES

unit.[35] It is advantageous for some applications to use large aromatic groups to promote solubility in organic solvents, and *p*-(triphenylmethyl)phenylphosphoramidates have been used for this purpose.[36]

The hydroxyl-protecting groups utilized in oligonucleotide synthesis are all among those introduced in Chapter 10 and need no detailed comment here as to their introduction or removal. Benzoate and acetate esters are frequently used. Trityl, *p*-methoxytrityl, and tetrahydropyranyl groups are also frequently employed.

Dicyclohexylcarbodiimide is often used as a coupling reagent. The mechanism of activation is believed to involve formation of phosphoric anhydrides.

Sulfonyl chlorides are also used as coupling agents. Those most commonly encountered are methanesulfonyl chloride (MS), mesitylsulfonyl chloride, and triisopropylbenzenesulfonyl chloride (TPS). These reagents function by formation of mixed sulfonic–phosphoric anhydrides, which then react with the free nucleophilic hydroxyl group to give phosphorylated product.

The synthesis of a polynucleotide involves a stepwise addition of individual

35. E. Ohtsuka, K. Murao, M. Ubasawa, and M. Ikehara, *J. Am. Chem. Soc.* **92**, 3441 (1970).
36. K. L. Agarwal, A. Yamazoki, and H. G. Khorana, *J. Am. Chem. Soc.* **93**, 2754 (1971).

Scheme 11.5. Syntheses of Some Oligonucleotides*

Sequence synthesized	Protecting groups employed	Coupling agent	Ref.
dTpTpTpApG	monomethoxytrityl, naphthyl isocyanate	TPS	a
All possible dinucleotides	p-triphenylmethylphenylamino, acetate	TPS	b
pTpTpApApTpCpCpApTpApTpGpC	trityl, acetate	MS	c
Various trinucleotides	phenylthioethyl, acetate	MS	d
pApCpC (ribonucleotide)	p-methoxyphenylamino, monomethoxytrityl	DCC	e
CpGpUpCpCpApCpCpA (ribonucleotide)	p-methoxyphenylamino, monomethoxytrityl, benzoate	TPS, DCC	f
ApTpGpCpApCpTpCpTpTpApG	cyanoethyl, monomethoxytrityl	mesitylsulfonyl chloride	g

* For a definition of abbreviations: *J. Biol. Chem.* **241**, 527 (1966).

a. K. L. Agarwal and H. G. Khorana, *J. Am. Chem. Soc.* **94**, 3578 (1972).
b. K. L. Agarwal, A. Yamazaki, and H. G. Khorana, *J. Am. Chem. Soc.* **93**, 2754 (1971).
c. A. F. Cook, E. P. Heimer, M. J. Holman, D. T. Maichuk, and A. L. Nussbaum, *J. Am. Chem. Soc.* **94**, 1334 (1972).
d. S. A. Narang, O. S. Bhanot, J. Goodchild, R. H. Wightman, and S. K. Dheer, *J. Am. Chem. Soc.* **94**, 6183 (1972).
e. E. Ohtsuka, K. Murao, M. Ubasawa, and M. Ikehara, *J. Am. Chem. Soc.* **92**, 3441 (1970).
f. E. Ohtsuka, M. Ubasawa, S. Morioka, and M. Ikehara, *J. Am. Chem. Soc.* **95**, 4725 (1973).
g. A. Kumar and H. G. Khorana, *J. Am. Chem. Soc.* **91**, 2743 (1969).

nucleotide units. Repetitive sequences involving phosphorylation, deprotecting, and coupling are used to introduce each nucleotide unit.

The chemical synthesis of homopolymers of nucleotides can be accomplished more readily than when a specific sequence is required. A ribonucleotide that is unprotected at both a C-5' phosphate and 3'-hydroxyl group will polymerize on exposure to a coupling reagent such as DCC to give a high-molecular-weight material consisting of a single nucleotide.[37]

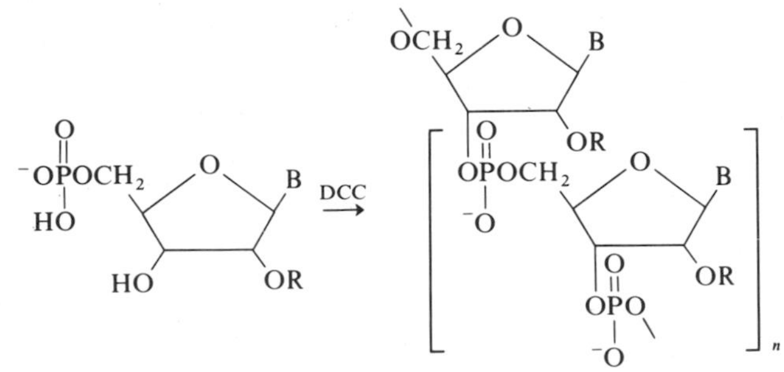

37. Y. Lapidot and H. G. Khorana, *J. Am. Chem. Soc.* **85**, 3857 (1963); C. Coutsogeorgopoulos and H. G. Khorana, *J. Am. Chem. Soc.* **86**, 2926 (1964); D. B. Straus and J. R. Fresco, *J. Am. Chem. Soc.* **95**, 5025 (1973).

Efforts to apply the solid-phase methods developed in protein synthesis to the problem of oligonucleotide synthesis have begun.[38] The requirement for high coupling yields at each stage is harder to meet in nucleotide coupling than in peptide coupling at the present time, however, so the length of chain that can satisfactorily be synthesized by solid-phase techniques is quite limited.

Scheme 11.5 summarizes some recent papers that can serve to illustrate the techniques used in oligonucleotide synthesis. Some of the abbreviations widely used in this field are also included in this scheme.

General References

Synthetic Polymers

F. W. Billmeyer, Jr., *Textbook of Polymer Science*, Wiley–Interscience, New York, NY, 1971.
R. B. Seymour, *Introduction to Polymer Chemistry*, McGraw-Hill, New York, NY, 1971.
G. Odian, *Principles of Polymerization*, McGraw-Hill, New York, NY, 1970.
S. R. Sandler and W. Karo, *Polymer Syntheses*, Vol. I, Academic Press, New York, NY, 1974.

Polypeptide Synthesis

M. Bodanszky and M. A. Ordetti, *Peptide Synthesis*, Interscience, New York, NY, 1966.
J. M. Stewart and J. D. Young, *Solid Phase Peptide Synthesis*, W. H. Freeman, San Francisco, CA, 1969.

Nucleosides, Nucleotides, and Polynucleotides

W. W. Zorbach and R. S. Tipson, *Synthetic Procedures in Nucleic Acid Chemistry*, Interscience, New York, NY, 1968.
T. L. V. Ulbricht, *Purines, Pyrimidines and Nucleotides*, Macmillan, New York, NY, 1964.

Problems

(*References for these problems will be found on page 507.*)

1. List five methods for activating an amino acid carboxyl group so that it can acylate the amino group of an amino acid.

2. From the following list of catalyst systems, choose those that would be expected to catalyze polymerization of each of the vinyl compounds listed below.

Catalysts: $AlCl_3-H_2O$, HF, $(CH_3)_2\underset{NC}{C}=\underset{CN}{C}(CH_3)_2$, PhLi, $(CH_3)_3CO^-K^+$, $Fe(II)-H_2O_2$

(a) $PhCH=CH_2$ (b) $CH_2=CH\underset{CH_3}{C}=CH_2$

38. R. L. Letsinger and V. Mahadevan, *J. Am. Chem. Soc.* **88**, 5319 (1966); E. Ohtsuka, S. Morioka, and M. Ikehara, *J. Am. Chem. Soc.* **94**, 3229 (1972); K. F. Yip and K. C. Tsou, *J. Am. Chem. Soc.* **93**, 3272 (1971); L. R. Melby and D. R. Strobach, *J. Org. Chem.* **34**, 421 (1969).

(c) $CH_2=\underset{\underset{CH_3}{|}}{C}CO_2CH_3$

(d) $CH_2=CH_2$

(e) $CH_2=CHCN$

(f) $ClCH=CHCl$

3. Indicate reaction conditions that could be used to remove the protecting group in each case:

(a) $CH_3O\overset{O}{\overset{||}{C}}CH_2NH\overset{O}{\overset{||}{C}}\underset{\underset{CH_2Ph}{|}}{C}HNH\overset{O}{\overset{||}{C}}OCH_2Ph$

(b)

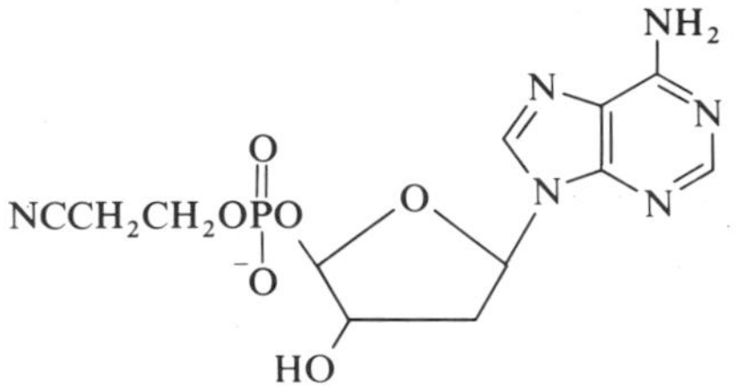

(c)

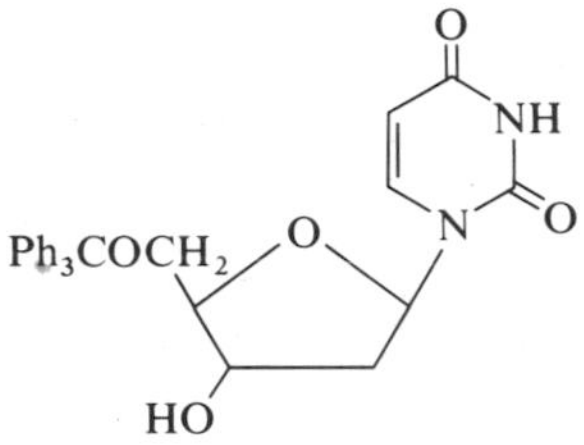

(d) $(CH_3)_3CO\overset{O}{\overset{||}{C}}NHCH_2\overset{O}{\overset{||}{C}}NH\underset{\underset{CH_2Ph}{|}}{C}H\overset{O}{\overset{||}{C}}NH\underset{\underset{CH_3}{|}}{C}H\overset{O}{\overset{||}{C}}NHCH_2CO_2H$

(e)

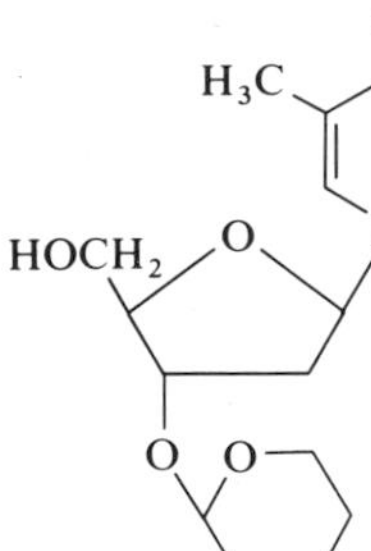

4. In the synthesis of peptides containing lysine, the ε-amino group must be protected. What kind of impurities would arise if this were not done? What special requirements would there be in solid-phase synthesis for a protecting group appropriate for protecting the ε-amino group of lysine? A recent successful approach to this problem used the t-butoxycarbonyl group for protecting α-amino groups during the synthesis, and the 2,4-dichlorobenzyloxycarbonyl group for the protection of the lysine ε-amino group. What properties of the latter group might make it particularly suitable for this use?

5. Suppose you wished to incorporate fire-retardant characteristics into a polyester, and decided that copolymerization using a high-chlorine-content monomer might be one approach. Could you expect random introduction of the chlorinated monomer into the polyester chain? Explain.

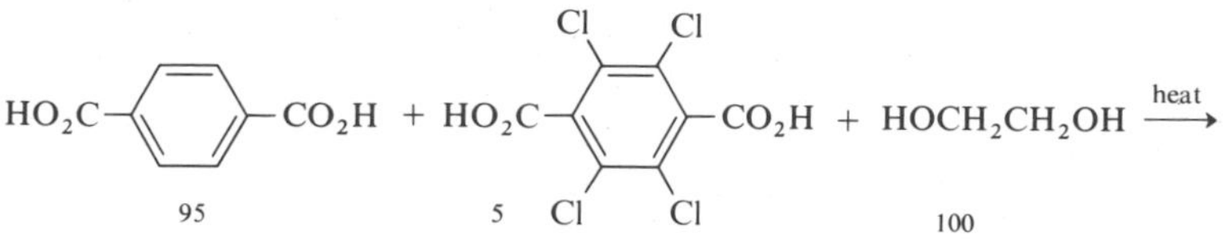

Ratio: 95   5   100

6. Polyurethane foams are made by a process in which polymerization and cross-linking is accompanied by gas evolution, which results in the formation of gas bubbles throughout the polymer. The starting materials for polyurethane foam formation are low-molecular-weight polyethers with hydroxyl end groups and an aromatic diisocyanate. Rapid polymerization and gas evolution occurs when the catalyst, typically a tertiary amine, is added:

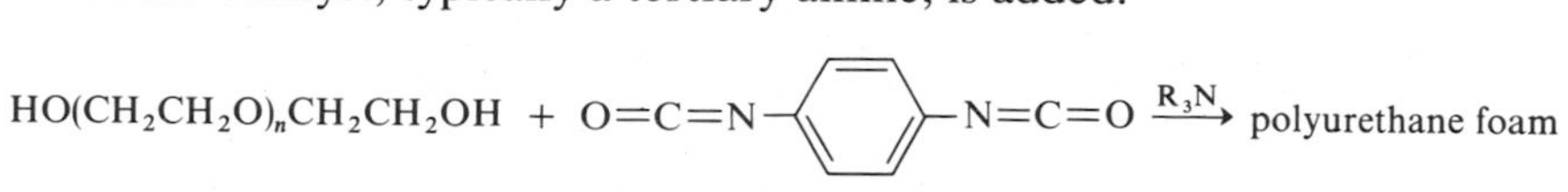

Indicate the final structure of the polymer and the possibilities for cross-links in the polymer. What is the source and identity of the gas that causes "foaming"?

7. Radical-chain polymerization can be accompanied by "backbiting," an internal hydrogen-atom transfer. What would be the effect of the occurrence of this process on the structure and average molecular weight of the polymer?

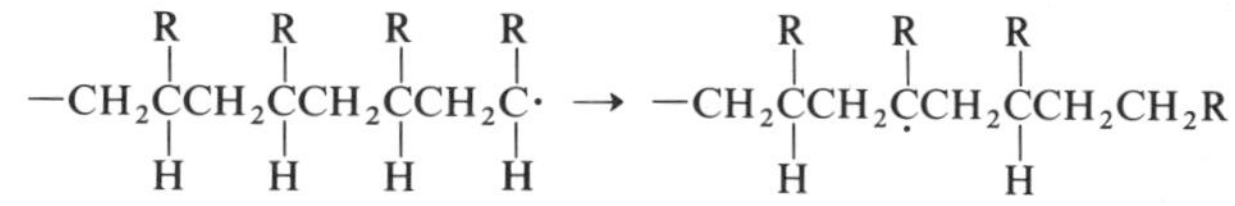

8. The following derivatives of better-known protecting groups have been used where there is a requirement for especially mild conditions for removal of the protecting group. Indicate the standard protecting group to which each derivative is related. Identify the features of the substituted protecting group that would make it especially labile. Indicate the types of reaction conditions that might permit its removal.

(a)   (b) 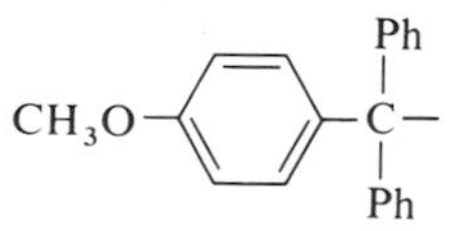   (c)

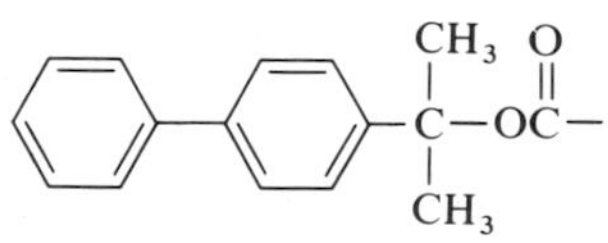

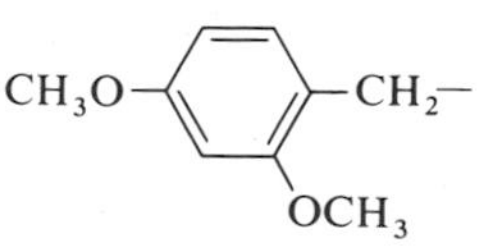

9. The chain transfer constant, C_s, of a polymerization solvent is defined as the reactivity of the solvent with the growing polymer chain relative to the reactivity of the polymer chain with the monomer. For the polymerization of styrene in some aromatic solvents, the values given below have been reported. Can you provide a rationalization for the position of each solvent in the series?

benzene	0.018
t-butylbenzene	0.04
toluene	0.125
ethylbenzene	0.67
isopropylbenzene	0.82

10. Suggest at least two ways in which hydrazine groups could be attached to polystyrene polymer in such a way that a polypeptide chain could be built up from the hydrazine moiety and then cleaved to give the peptide hydrazide. What advantage might there be in synthesizing the peptide in the form of its hydrazide?

11. In the course of a synthesis employing the solid-phase method, it was found that in the presence of carboxylic acids, the dipeptide valylproline was cleaved from the polystyrene residue. The cleavage product was shown to be **A**. Write a mechanism that would account for the cleavage of the dipeptide from the resin.

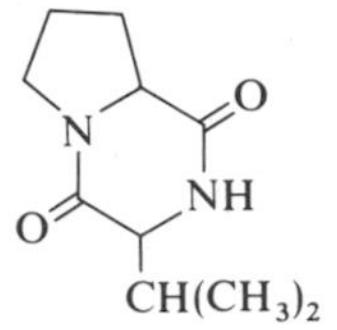

A

12. An early recipe for manufacture of synthetic rubber calls for the following materials to be mixed (parts by weight) at 50°. What is the role of each component in the reaction mixture?

butadiene	75
styrene	25
water	180
soap	5
dodecyl mercaptan	0.5
potassium persulfate	0.3

13. The 2-phenylthioethyl group has been used as an O-protecting group for phosphate linkages in oligonucleotide synthesis. It is stable to the various coupling procedures employed, but is easily removed by treatment first with periodate solution and then 2N NaOH. What is the mechanistic basis for the facile removal of the group?

14. Suggest a mechanism to account for release of phosphoric acid groups from phosphoranilides on treatment with amyl nitrite:

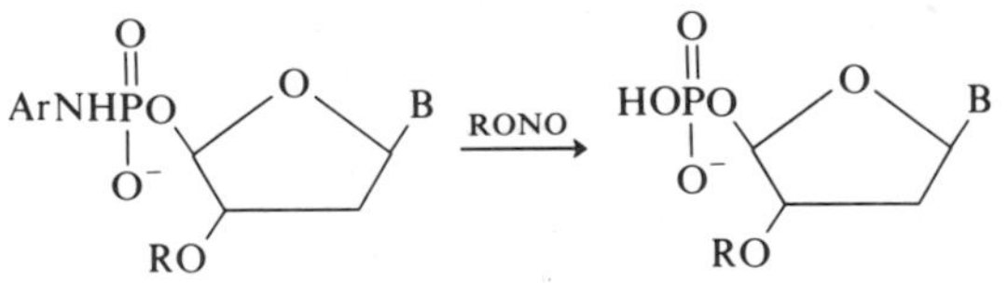

15. A method for solid-phase peptide synthesis has been introduced that attaches a histidine residue to the resin in such a way that the polypeptide chain can be extended from both the amino and carboxyl groups of histidine, neither of which is directly bound to the resin. The sequence of reactions is outlined below. Write equations showing the reactions that occur at each stage.

   (1) Boc-glycine + chloromethylpolystyrene → I
   (2) I + 1,5-difluoro-2,4-dinitrobenzene $\xrightarrow{Et_3N}$ II
   (3) II + Boc-histidine → III
   (4) III + *p*-nitrophenyl trifluoroacetate → IV
   (5) IV + prolinamide → V
   (6) V + trifluoroacetic acid → VI
   (7) VI + pentachlorophenyl pyrrolidin-2-one-5-carboxylate → VII
   (8) VII + *N*-methylmorpholine → VIII
   (9) VIII + HSCH₂CH₂OH → IX

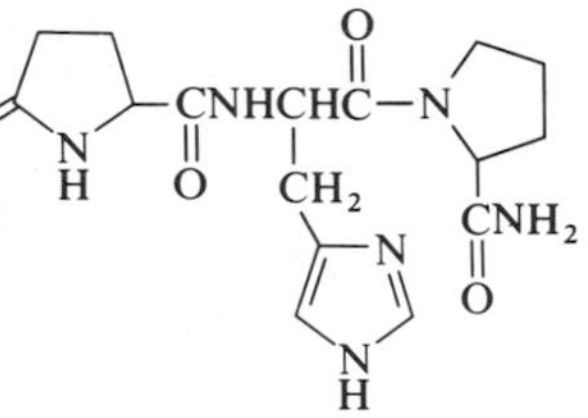

16. The following acetals give polymers when treated with protic or Lewis acids. Suggest structures for the resulting polymers and the mechanism by which polymerization could occur.

   (a)

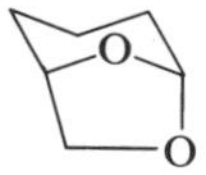

   (b)

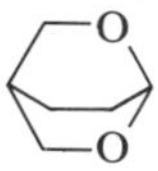

17. Give the structures of polymers that could be formed by reaction of the following monomers. Write the mechanism by which you expect polymerization to occur.

   (a)

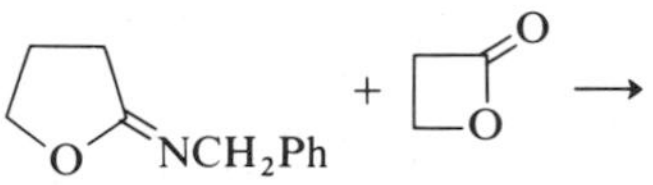

   (b)

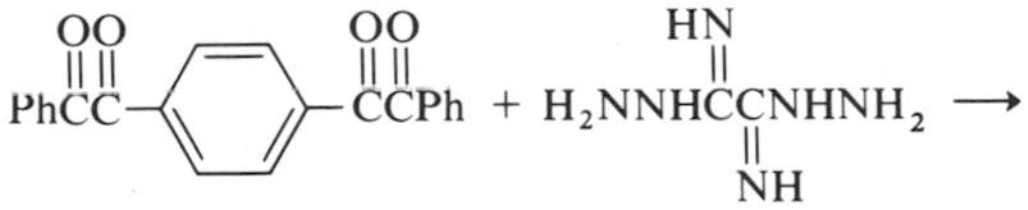

   (c)

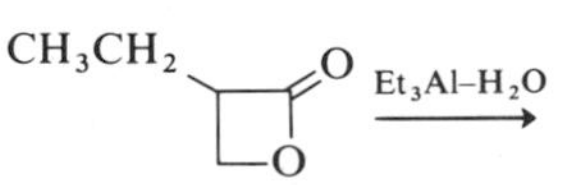

   (d)

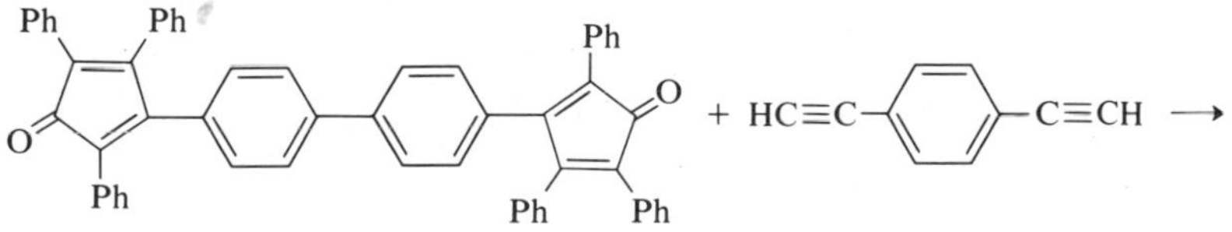

18. Attempted use of DMSO instead of pyridine as the solvent for the dicyclohexylcarbodiimide-promoted polymerization of thymidine 5′-phosphate, **A**, resulted in cleavage to produce thymine in quantitative yield. Suggest an explanation.

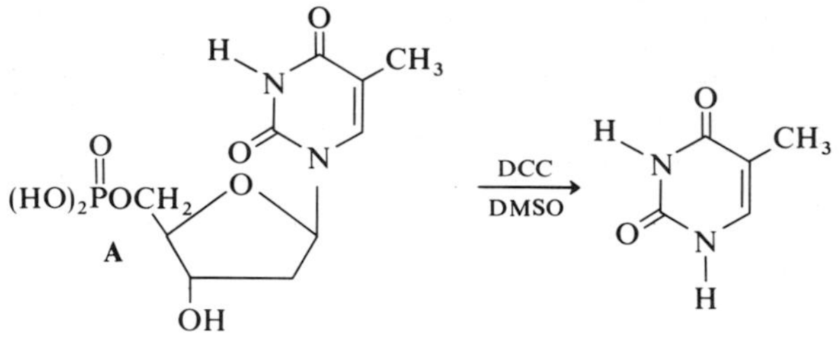

References for Problems

Chapter 1

2. E. D. Bergmann, D. Ginsburg, and R. Pappo, *Org. React.* **10**, 179 (1959).
3a. E. J. Corey and D. S. Watt, *J. Am. Chem. Soc.* **95**, 2302 (1973).
  b. A. Wissner and J. Meinwald, *J. Org. Chem.* **38**, 1697 (1973).
  c. W. J. Gensler and P. H. Solomon, *J. Org. Chem.* **38**, 1726 (1973).
  d. H. W. Whitlock, Jr., *J. Am. Chem. Soc.* **84**, 3412 (1962).
4. J. A. Markisz and J. D. Gettler, *Can. J. Chem.* **47**, 1965 (1969).
5. C. R. Hauser, T. M. Harris, and T. G. Ledford, *J. Am. Chem. Soc.* **81**, 4099 (1959).
6. J. Fried, *Heterocyclic Compounds,* R. C. Elderfield (ed.), Vol. 1, Wiley, New York, NY, 1950, p. 358.
7. R. Chapurlat, J. Huet, and J. Dreux, *Bull. Soc. Chim. Fr.*, 2446, 2450 (1967).
8a. K. Wiesner, K. K. Chan, and C. Demerson, *Tetrahedron Lett.*, 2893 (1965).
  b. C. H. Heathcock, R. A. Badger, and J. W. Patterson, Jr., *J. Am. Chem. Soc.* **89**, 4133 (1967).
  c. C. R. Hauser and W. R. Dunnavant, *Org. Synth.* **40**, 38, (1960).
  d. G. Opitz, H. Milderberger, and H. Suhr, *Justus Liebigs Ann. Chem.* **649**, 47 (1961).
  e. K. Shimo, S. Wakamatsu, and T. Inoüe, *J. Org. Chem.* **26**, 4868 (1961).
  f. T. A. Spencer, K. K. Schmiegel, and K. L. Williamson, *J. Am. Chem. Soc.* **85**, 3785 (1963).
9a. H. Feuer, A. Hirschfeld, and E. D. Bergmann, *Tetrahedron* **24**, 1187 (1968).
  b. A. Baradel, R. Longeray, J. Dreux, and J. Doris, *Bull. Soc. Chim. Fr.*, 255 (1970).
  c. H. H. Baer and K. S. Ong, *Can. J. Chem.* **46**, 2511 (1968).
  d. A. Wettstein, K. Heusler, H. Ueberwasser, and P. Wieland, *Helv. Chim. Acta* **40**, 323 (1957).
10. H. O. House and M. J. Umen, *J. Org. Chem.* **38**, 1000 (1973).
11. S. Masamune, *J. Am. Chem. Soc.* **86**, 288 (1964).
12. E. J. Corey, M. Ohno, R. B. Mitra, and P. A. Vatakencherry, *J. Am. Chem. Soc.* **86**, 478 (1964).
13a. E. Wenkert and D. P. Strike, *J. Org. Chem.* **27**, 1883 (1962).
  b. S. J. Etheredge, *J. Org. Chem.* **31**, 1990 (1966).
  c. R. Deghenghi and R. Gaudry, *Tetrahedron Lett.*, 489 (1962).
  d. P. A. Grieco and C. C. Pogonowski, *J. Am. Chem. Soc.* **95**, 3071 (1973).
  e. E. M. Kaiser, W. G. Kenyon, and C. R. Hauser, *Org. Synth.* **V**, 559 (1973).
14. W. G. Kofron and L. G. Wideman, *J. Org. Chem.* **37**, 555 (1972).
15. M. S. Newman, V. DeVries, and R. Darlak, *J. Org. Chem.* **31**, 2171 (1966).
16. H. D. Zook, W. L. Kelly, and I. Y. Posey, *J. Org. Chem.* **33**, 3477 (1968).
17. M. E. Kuehne, *J. Org. Chem.* **35**, 171 (1970).

Chapter 2

2. A. J. Speziale and D. E. Bissing, *J. Am. Chem. Soc.* **85**, 3878 (1963).
3. R. B. Woodward, F. Sondheimer, D. Taub, K. Heusler, and W. M. McLamore, *J. Am. Chem. Soc.* **74**, 4223 (1952).
4a. W. A. Mosher and R. W. Soeder, *J. Org. Chem.* **36**, 1561 (1971).
 b. T. Nozoe, K. Takase, T. Nakazawa, and S. Fukuda, *Tetrahedron* **27**, 3357 (1971).
 c. J. E. McMurry and T. E. Glass, *Tetrahedron Lett.*, 2575 (1971).
 d. D. J. Cram, A. Langemann, and F. Hauck, *J. Am. Chem. Soc.* **81**, 5750 (1959).
5. T. T. Howarth, G. P. Murphy, and T. M. Harris, *J. Am. Chem. Soc.* **91**, 517 (1969).
6a. M. W. Rathke and D. F. Sullivan, *J. Am. Chem. Soc.* **95**, 3050 (1973).
 b. E. J. Corey, H. Yamamoto, D. K. Herron, and K. Achiwa, *J. Am. Chem. Soc.* **92**, 6635 (1970).
 c. E. J. Corey and D. E. Cane, *J. Org. Chem.* **36**, 3070 (1971).
 d. E. W. Yankee and D. J. Cram, *J. Am. Chem. Soc.* **92**, 6328 (1970).
 e. W. G. Dauben, C. D. Poulter, and C. Suter, *J. Am. Chem. Soc.* **92**, 7408 (1970).
 f. P. A. Grieco and K. Hiroi, *J. Chem. Soc. Chem. Commun.*, 1317 (1972).
 g. T. Mukaiyama, M. Higo, and H. Takei, *Bull. Chem. Soc. Jpn.* **43**, 2566 (1970).
 h. I. Vlattas, I. T. Harrison, L. Tökés, J. H. Fried, and A. D. Cross, *J. Org. Chem.* **33**, 4176 (1968).
 i. A. T. Nielsen and W. R. Carpenter, *Org. Synth.* **V**, 288 (1973).
 j. M. L. Miles, T. M. Harris, and C. R. Hauser, *Org. Synth.* **V**, 718 (1973).
7. G. Stork, S. D. Darling, I. T. Harrison, and P. S. Wharton, *J. Am. Chem. Soc.* **84**, 2018 (1962).
8. Text references 45–47 (pp. 59 and 62).
9a. W. G. Dauben and J. Ipaktschi, *J. Am. Chem. Soc.* **95**, 5088 (1973).
 b. T. J. Curphey and H. L. Kim, *Tetrahedron Lett.*, 1441 (1968).
10. E. Vedejs, K. A. J. Snoble, and P. L. Fuchs, *J. Org. Chem.* **38**, 1178 (1973).
11. K. P. Singh and L. Mandell, *Chem. Ber.* **96**, 2485 (1963).
12a. W. S. Wadsworth, Jr., and W. D. Emmons, *J. Am. Chem. Soc.* **83**, 1733 (1960).
 b. D. Seyferth, S. O. Grim, and T. O. Read, *J. Am. Chem. Soc.* **83**, 1617 (1961).
 c. M. Engelhardt, H. Plieninger, and P. Schreiber, *Chem. Ber.* **97**, 1713 (1964).
 d. K. W. Ratts and R. D. Partos, *J. Am. Chem. Soc.* **91**, 6112 (1969).
13. E. E. Schweizer and G. J. O'Neil, *J. Org. Chem.* **30**, 2082 (1965); E. E. Schweizer, *J. Am. Chem. Soc.* **86**, 2744 (1964).
14. J. Adams, L. Hoffman, Jr., and B. M. Trost, *J. Org. Chem.* **35**, 1600 (1970).
15a. G. Wittig and H.-D. Frommeld, *Chem. Ber.* **97**, 3548 (1964).
 b. R. J. Sundberg, P. A. Bukowick, and F. O. Holcombe, *J. Org. Chem.* **32**, 2938 (1967).
 c. D. R. Howton, *J. Org. Chem.* **10**, 277 (1945).
 d. R. N. McDonald and T. W. Campbell, *J. Org. Chem.* **24**, 1969 (1959).
 e. Y. Chan and W. W. Epstein, *Org. Synth.* **53**, 48 (1973).
16. M. P. Cooke, Jr., and R. Goswami, *J. Am. Chem. Soc.* **95**, 7891 (1973).
17. P. M. McCurry, Jr., and R. K. Singh, *J. Org. Chem.* **39**, 2316 (1974).
18. K. D. Sears, R. L. Casebier, H. L. Hergert, G. H. Stout, and L. E. McCandlish, *J. Org. Chem.* **39**, 3244 (1974).
19a. R. M. Coates and J. E. Shaw, *J. Am. Chem. Soc.* **92**, 5657 (1970).
 b. K. Mitsuhashi and S. Shiotoni, *Chem. Pharm. Bull.* **18**, 75 (1970).
20. G. Büchi and H. Wüest, *Helv. Chim. Acta* **54**, 1767 (1971).

Chapter 3

2. G. V. Smith and R. L. Burwell, Jr., *J. Am. Chem. Soc.* **84**, 925 (1962).
3a. H. W. Thompson, *J. Org. Chem.* **36**, 2577 (1971).
 b. E. Piers, W. de Waal, and R. W. Britton, *J. Am. Chem. Soc.* **93**, 5113 (1971).
4. G. V. Smith and R. L. Burwell, Jr., *J. Am. Chem. Soc.* **84**, 925 (1962).
5. D. J. Pasto and J. A. Gontarz, *J. Am. Chem. Soc.* **93**, 6902 (1971).
6. J. Fried and E. F. Sabo, *J. Am. Chem. Soc.* **79**, 1130 (1957).
7a. H. C. Brown, M. M. Rogić, H. Nambu, and M. W. Rathke, *J. Am. Chem. Soc.* **91**, 2147 (1969); H. C. Brown, H. Nambu, and M. M. Rogić, *J. Am. Chem. Soc.* **91**, 6852 (1969).
 b. H. C. Brown and R. A. Coleman, *J. Am. Chem. Soc.* **91**, 4606 (1969).

c. H. C. Brown and G. W. Kabalka, *J. Am. Chem. Soc.* **92**, 714 (1970).

d. G. Zweifel, R. P. Fisher, J. T. Snow, and C. C. Whitney, *J. Am. Chem. Soc.* **93**, 6309 (1971).

e. H. C. Brown and M. W. Rathke, *J. Am. Chem. Soc.* **89**, 2738 (1967).

f. H. C. Brown and M. M. Rogić, *J. Am. Chem. Soc.* **91**, 2146 (1969).

8. A. Pelter, M. G. Hutchings, and K. Smith, *Chem. Commun.*, 1048 (1971); H. C. Brown, B. A. Carlson, and R. H. Prager, *J. Am. Chem. Soc.* **93**, 2070 (1971).

9. D. J. Pasto and C. C. Cumbo, *J. Am. Chem. Soc.* **86**, 4343 (1964).

10. D. J. Pasto and J. A. Gontarz, *J. Am. Chem. Soc.* **93**, 6902 (1971).

11. H. C. Brown, P. J. Geoghegan, Jr., G. J. Lynch, and J. T. Kurek, *J. Org. Chem.* **37**, 1941 (1972).

12a. E. Kloster-Jensen, E. Kováts, A. Eschenmoser, and E. Heilbronner, *Helv. Chim. Acta* **39**, 1051 (1956).

b. P. N. Rao, *J. Org. Chem.* **36**, 2426 (1971).

c. W. I. Fanta and W. F. Erman, *J. Org. Chem.* **33**, 1656 (1968).

d. W. E. Billups, J. H. Cross, and C. V. Smith, *J. Am. Chem. Soc.* **95**, 3438 (1973).

13a. J. Meinwald and J. K. Crandall, *J. Am. Chem. Soc.* **88**, 1292 (1966).

b. H. Stetter and J. Gärtner, *Chem. Ber.* **99**, 925 (1966).

c. W. Barbieri, A. Consonni, and R. Sciaky, *J. Org. Chem.* **33**, 3544 (1968).

d. P. E. Peterson and J. E. Duddey, *J. Am. Chem. Soc.* **85**, 2865 (1963).

e. L. A. Paquette and G. R. Krow, *Tetrahedron Lett.*, 2139 (1968).

14a. K. Izawa, T. Okuyama, T. Sakagami, and T. Fueno, *J. Am. Chem. Soc.* **95**, 6752 (1973).

b. G. Stork and R. Borch, *J. Am. Chem. Soc.* **86**, 935 (1964).

15. R. L. Burwell, Jr., *Acc. Chem. Res.* **2**, 289 (1969).

16. J. Hooz and D. M. Gunn, *J. Am. Chem. Soc.* **91**, 6195 (1969).

17a. G. Zweifel, J. T. Snow, and C. C. Whitney, *J. Am. Chem. Soc.* **90**, 7139 (1968).

b. H. C. Brown, M. M. Rogić, M. W. Rathke, and G. W. Kabalka, *J. Am. Chem. Soc.* **89**, 5709 (1967).

c. A. Hassner, R. P. Hoblitt, C. Heathcock, J. E. Kropp, and M. Lorber, *J. Am. Chem. Soc.* **92**, 1326 (1970).

d. G. Zweifel and C. C. Whitney, *J. Am. Chem. Soc.* **89**, 2753 (1967).

e. G. Zweifel, R. P. Fisher, J. T. Snow, and C. C. Whitney, *J. Am. Chem. Soc.* **93**, 6309 (1971).

f. R. E. Ireland and P. Bey, *Org. Synth.* **53**, 63 (1973).

g. G. Zweifel, R. P. Fisher, J. T. Snow, and C. C. Whitney, *J. Am. Chem. Soc.* **93**, 6309 (1971).

h. G. Zweifel and C. C. Whitney, *J. Am. Chem. Soc.* **89**, 2753 (1967).

i. J. M. Jerkunica and T. G. Traylor, *Org. Synth.* **53**, 94 (1973).

j. G. Zweifel and H. C. Brown, *Org. Synth.* **52**, 59 (1972).

18. D. G. Garratt, A. Modro, K. Oyama, G. H. Schmid, T. T. Tidwell, and K. Yates, *J. Am. Chem. Soc.* **96**, 5295 (1974).

Chapter 4

1b. W. Parker, R. Ramage, and R. A. Raphael, *J. Chem. Soc.*, 1558 (1962).

d. S. Yamamura, M. Toda, and Y. Hirata, *Org. Synth.* **53**, 86 (1973).

2. D. C. Wigfield and D. J. Phelps, *J. Am. Chem. Soc.* **96**, 543 (1974).

3a. E. J. Corey, T. K. Schaaf, W. Huber, U. Koelliker, and N. M. Weinshenker, *J. Am. Chem. Soc.* **92**, 397 (1970).

b. E. J. Corey and R. Noyori, *Tetrahedron Lett.*, 311 (1970).

c. R. F. Borch, *Org. Synth.* **52**, 124 (1972).

d. D. Seyferth and V. A. Mai, *J. Am. Chem. Soc.* **92**, 7412 (1970).

4a. J. E. Siggins, A. A. Larsen, J. H. Ackerman, and C. D. Carabateas, *Org. Synth.* **53**, 52 (1973).

b. W. Sucrow, *Tetrahedron Lett.*, 4725 (1970).

c. P. A. Grieco, *J. Am. Chem. Soc.* **91**, 5660 (1969).

d. R. S. Lenox and J. A. Katzenellenbogen, *J. Am. Chem. Soc.* **95**, 957 (1973).

e. V. Bažant, M. Čapka, M. Černý, C. Chvalovský, K. Kochloefl, M. Kraus, and J. Málek, *Tetrahedron Lett.*, 3303 (1968).

f. L. I. Zakharkin and I. M. Khorlina, *Tetrahedron Lett.*, 619 (1962).

5a. S. Krishnamurthy, R. M. Schubert, and H. C. Brown, *J. Am. Chem. Soc.* **95**, 8486 (1973).

b. C. W. Jefford, D. Kirkpatrick, and F. Delay, *J. Am. Chem. Soc.* **94**, 8905 (1972).

c. J. San Filippo, Jr., and G. M. Anderson, *J. Org. Chem.* **39**, 473 (1974).

   d. G. R. Wenzinger and J. A. Ors, *J. Org. Chem.* **39**, 2060 (1974).

   e. C. H. Heathcock, R. A. Badger, and J. W. Patterson, Jr., *J. Am. Chem. Soc.* **89**, 4133 (1967).

  6. E. C. Ashby, J. P. Sevenair, and F. R. Dobbs, *J. Org. Chem.* **36**, 197 (1971).

 7a. G. Stork and W. N. White, *J. Am. Chem. Soc.* **78**, 4604 (1956).

   b. H. C. Brown and W. C. Dickason, *J. Am. Chem. Soc.* **92**, 709 (1970).

   c. D. Seyferth, H. Yamazaki, and D. L. Alleston, *J. Org. Chem.* **28**, 703 (1963).

   d. G. Stork and S. D. Darling, *J. Am. Chem. Soc.* **82**, 1512 (1960).

  8. R. O. Hutchins, C. A. Milewski, and B. E. Maryanoff, *J. Am. Chem. Soc.* **95**, 3662 (1973).

  9. R. M. Coates and J. E. Shaw, *J. Org. Chem.* **35**, 2597 (1970).

10a. W. C. Agosta and W. L. Schreiber, *J. Am. Chem. Soc.* **93**, 3947 (1971).

   b. D. Taub, R. D. Hoffsommer, C. H. Kuo, H. L. Slates, Z. S. Zelawski, and N. L. Wendler, *Chem. Commun.*, 1258 (1970).

   c. G. Büchi, J. A. Carlson, J. E. Powell, Jr., and L.-F. Tietze, *J. Am. Chem. Soc.* **92**, 2165 (1970).

 11. P. Y. Johnson and M. A. Priest, *J. Am. Chem. Soc.* **96**, 5619 (1974).

12a. N. M. Yoon, C. S. Pak, H. C. Brown, S. Krishnamurthy, and T. P. Stocky, *J. Org. Chem.* **38**, 2786 (1973).

   b. D. J. Dawson and R. E. Ireland, *Tetrahedron Lett.*, 1899 (1968).

   c. E. J. Corey and H. Yamamoto, *J. Am. Chem. Soc.* **92**, 6636 (1970).

   d. H. Kwart and R. A. Conley, *J. Org. Chem.* **38**, 2011 (1973).

   e. P. Kohn, R. H. Samaritano, and L. M. Lerner, *J. Am. Chem. Soc.* **87**, 5475 (1965).

   f. D. J. Marshall and R. Deghenghi, *Can. J. Chem.* **47**, 3127 (1969).

   g. G. R. Pettit and J. R. Dias, *J. Org. Chem.* **36**, 3207 (1971).

   h. N. M. Yoon, C. S. Pak, H. C. Brown, S. Krishnamurthy, and T. P. Stocky, *J. Org. Chem.* **38**, 2786 (1973).

   i. W. F. Johns, *J. Org. Chem.* **29**, 1490 (1964).

   j. R. O. Hutchins, C. A. Milewski, and B. A. Maryanoff, *J. Am. Chem. Soc.* **95**, 3662 (1973).

   k. M. J. Kornet, P. A. Thio, and S. I. Tan, *J. Org. Chem.* **33**, 3637 (1968).

   l. C. T. West, S. J. Donnelly, D. A. Kooistra, and M. P. Doyle, *J. Org. Chem.* **38**, 2675 (1973).

   m. M. R. Johnson and B. Rickborn, *J. Org. Chem.* **35**, 1041 (1970).

13a. S. Iwaki, S. Manmo, T. Saito, M. Yamada, and K. Katagiri, *J. Am. Chem. Soc.* **96**, 7842 (1974).

   b. R. E. Ireland and S. C. Welch, *J. Am. Chem. Soc.* **92**, 7232 (1970).

   c. M. D. Soffer and M. A. Jevnik, *J. Am. Chem. Soc.* **77**, 1003 (1955).

   d. E. J. Corey and D. S. Watt, *J. Am. Chem. Soc.* **95**, 2303 (1973).

   e. E. J. Corey and G. Moinet, *J. Am. Chem. Soc.* **95**, 6831 (1973).

 14. C. T. West, S. J. Donnelly, D. A. Kooistra, and M. P. Doyle, *J. Org. Chem.* **38**, 2675 (1973).

15a. U. T. Bhalerao, J. J. Plattner, and H. Rapoport, *J. Am. Chem. Soc.* **92**, 3429 (1970).

   b. S. Takahashi and L. A. Cohen, *J. Org. Chem.* **35**, 1505 (1970).

16a. F. A. Carey, D. H. Ball, and L. Long, Jr., *Carbohydr. Res.* **3**, 205 (1966).

   b. D. J. Cram and F. A. Abd Elhafez, *J. Am. Chem. Soc.* **74**, 5828 (1952).

   c. H. O. House, H. C. Müller, C. G. Pitt, and P. P. Wickham, *J. Org. Chem.* **28**, 2407 (1963).

   d. J. Klein, E. Dunkelblum, E. Eliel, and Y. Senda, *Tetrahedron Lett.*, 6127 (1968).

   e. G. R. Pettit and J. R. Dias, *J. Org. Chem.* **36**, 3207 (1971).

17a. N. J. Leonard and S. Gelfand, *J. Am. Chem. Soc.* **77**, 3272 (1955).

   b. P. S. Wharton and D. H. Bohlen, *J. Org. Chem.* **26**, 3615 (1961); W. R. Benn and R. M. Dodson, *J. Org. Chem.* **29**, 1142 (1964).

   c. G. Lardelli and O. Jeger, *Helv. Chim. Acta* **32**, 1817 (1949).

   d. R. J. Petersen and P. S. Skell, *Org. Synth.* **V**, 929 (1973).

18a. E. J. Corey, J. A. Katzenellenbogen, N. W. Gilman, S. A. Roman, and B. W. Erickson, *J. Am. Chem. Soc.* **90**, 5618 (1968).

   b. A. W. Burgstahler and L. R. Worden, *J. Am. Chem. Soc.* **83**, 2587 (1961).

Chapter 5

1a. K. Hafner and H. Kaiser, *Org. Synth.* **V**, 1088 (1973).

   b. D. Seyferth, J. Fogel, and J. K. Heeren, *J. Am. Chem. Soc.* **88**, 2207 (1966).

   c. R. A. Benkeser and F. J. Riel, *J. Am. Chem. Soc.* **73**, 3472 (1951).

   d. J. F. Baldwin, G. A. Hofle, and O. W. Lever, Jr., *J. Am. Chem. Soc.* **96**, 7125 (1974).

   e. G. H. Posner, C. E. Whitten, and J. J. Sterling, *J. Am. Chem. Soc.* **95**, 7788 (1973).

f. D. J. Peterson, *J. Am. Chem. Soc.* **93**, 4027 (1971).

g. R. S. Bly and R. L. Veazey, *J. Am. Chem. Soc.* **91**, 4221 (1969).

2a. B. M. Graybill and D. A. Shirley, *J. Org. Chem.* **31**, 1221 (1966).

b. S. Akiyama and J. Hooz, *Tetrahedron Lett.*, 4115 (1973).

c. C. D. Broaddus, *J. Org. Chem.* **35**, 10 (1970).

d. D. A. Shirley and P. A. Roussel, *J. Am. Chem. Soc.* **75**, 375 (1953).

e. K. P. Klein and C. R. Hauser, *J. Org. Chem.* **32**, 1479 (1967).

f. D. J. Peterson and H. R. Hays, *J. Org. Chem.* **30**, 1939 (1965).

3a. M. P. Dreyfuss, *J. Org. Chem.* **28**, 3269 (1963).

b. P. J. Pearce, D. H. Richards, and N. F. Scilly, *Org. Synth.* **52**, 19 (1972).

c. U. Schöllkopf, H. Küppers, H.-J. Traenckner, and W. Pitteroff, *Justus Liebigs Ann. Chem.* **704**, 120 (1967).

d. J. V. Hay and T. M. Harris, *Org. Synth.* **53**, 56 (1973).

e. E. L. Eliel, R. O. Hutchins, and M. Knoeber, *Org. Synth.* **50**, 38 (1970).

f. J. C. H. Hwa and H. Sims, *Org. Synth.* **V**, 608 (1973).

4a. C. R. Johnson, R. W. Herr, and D. M. Wieland, *J. Org. Chem.* **38**, 4263 (1973).

b. E. J. Corey and P. L. Fuchs, *J. Am. Chem. Soc.* **94**, 4014 (1972).

c. W. G. Dauben and A. J. Kielbania, Jr., *J. Am. Chem. Soc.* **93**, 7345 (1971).

d. C. R. Johnson and G. A. Dutra, *J. Am. Chem. Soc.* **95**, 7777 (1973).

e. R. A. J. Smith and T. A. Spencer, *J. Org. Chem.* **35**, 3220 (1970).

f. H. O. House and M. J. Umen, *J. Org. Chem.* **38**, 3893 (1973).

5. P. S. Skell and G. P. Bean, *J. Am. Chem. Soc.* **84**, 4660 (1962).

6. E. C. Ashby, J. Laemmle, and H. M. Neumann, *J. Am. Chem. Soc.* **93**, 4601 (1971).

7. M. Newcomb and W. T. Ford, *J. Am. Chem. Soc.* **95**, 7186 (1973).

8. R. W. Herr and C. R. Johnson, *J. Am. Chem. Soc.* **92**, 4979 (1970).

9. G. Zon and L. A. Paquette, *J. Am. Chem. Soc.* **96**, 215 (1974).

10. P. G. Gassman, G. R. Meyer, and F. J. Williams, *J. Am. Chem. Soc.* **94**, 7741 (1972); P. G. Gassman and F. J. Williams, *J. Am. Chem. Soc.* **94**, 7733 (1972).

11. R. K. Russell, R. E. Wingard, Jr., and L. A. Paquette, *J. Am. Chem. Soc.* **96**, 7483 (1974).

12a. E. O. Fischer, K. Öfele, H. Essler, W. Fröhlich, J. P. Mortensen, and W. Semmlinger, *Chem. Ber.* **91**, 2763 (1958).

b. R. Pettit, *J. Am. Chem. Soc.* **81**, 1266 (1959).

c. M. Dubeck and A. H. Filbey, *J. Am. Chem. Soc.* **83**, 1257 (1961).

d. M. P. Cooke, Jr., *J. Am. Chem. Soc.* **92**, 6080 (1970).

e. C. H. Wei and L. F. Dahl, *J. Am. Chem. Soc.* **88**, 1821 (1966).

13a. M. F. Semmelhack and H. T. Hall, *J. Am. Chem. Soc.* **96**, 7091 (1974).

b. M. L. H. Green and P. L. I. Nagy, *J. Organometal. Chem.* **1**, 58 (1963).

14a. E. J. Corey and R. L. Carney, *J. Am. Chem. Soc.* **93**, 7318 (1971).

b. W. G. Dauben, G. Ahlgren, T. J. Leitereg, W. C. Schwarzel, and M. Yoshioko, *J. Am. Chem. Soc.* **94**, 8593 (1972).

c. S. W. Pelletier and S. Prabhakar, *J. Am. Chem. Soc.* **90**, 5318 (1968).

d. E. J. Corey, D. E. Cane, and L. Libit, *J. Am. Chem. Soc.* **93**, 7016 (1971).

e. R. D. Clark and C. H. Heathcock, *Tetrahedron Lett.*, 1713 (1974); W. G. Dauben, J. W. McFarland, and J. B. Rogan, *J. Org. Chem.* **26**, 297 (1961).

15a. E. J. Corey, C. U. Kim, R. H. K. Chen, and M. Takeda, *J. Am. Chem. Soc.* **94**, 4395 (1972).

b. J. A. Marshall, W. F. Huffman, and J. A. Ruth, *J. Am. Chem. Soc.* **94**, 4691 (1972).

c. L. Watts, J. D. Fitzpatrick, and R. Pettit, *J. Am. Chem. Soc.* **88**, 623 (1966).

d. H. A. Whaley, *J. Am. Chem. Soc.* **93**, 3767 (1971).

e. J. A. Marshall, W. F. Huffman, and J. A. Ruth, *J. Am. Chem. Soc.* **94**, 4691 (1972).

f. M. F. Semmelhack and P. M. Helquist, *Org. Synth.* **52**, 115 (1972).

g. L. S. Hegedus and R. K. Stiverson, *J. Am. Chem. Soc.* **96**, 3250 (1974).

h. G. D. Prestwich and J. N. Labovitz, *J. Am. Chem. Soc.* **96**, 7103 (1974).

Chapter 6

1. J. E. Baldwin and R. K. Pinschmidt, Jr., *J. Am. Chem. Soc.* **92**, 5247 (1970).

2. H. E. Zimmerman, G. L. Grunewald, R. M. Paufler, and M. A. Sherwin, *J. Am. Chem. Soc.* **91**, 2330 (1969).

3a,b. J. A. Berson and M. Jones, Jr., *J. Am. Chem. Soc.* **86**, 5019 (1964).

 c. G. Büchi and J. E. Powell, Jr., *J. Am. Chem. Soc.* **89**, 4559 (1967).

 d. C. Cupas, W. E. Watts, and P. von R. Schleyer, *Tetrahedron Lett.*, 2503 (1964).

 e. T. C. Jain, C. M. Banks, and J. E. McCloskey, *Tetrahedron Lett.*, 841 (1970).

 f. R. K. Hill and N. W. Gilman, *Chem. Commun.*, 619 (1967).

 4. S. J. Rhoads and J. M. Watson, *J. Am. Chem. Soc.* **93**, 5813 (1971); S. J. Rhoads and C. F. Brandenburg, *J. Am. Chem. Soc.* **93**, 5805 (1971).

 5. D. A. Evans, C. A. Bryan, and C. L. Sims, *J. Am. Chem. Soc.* **94**, 2891 (1972).

 6. C. H. Heathcock and R. A. Badger, *J. Org. Chem.* **37**, 234 (1972).

 7a. B. B. Snider, *J. Org. Chem.* **38**, 3961 (1973).

 b. P. K. Freeman, D. M. Balls, and D. J. Brown, *J. Org. Chem.* **33**, 2211 (1968).

 c. S. Danishefsky and T. Kitahara, *J. Am. Chem. Soc.* **96**, 7807 (1974).

 d. W. Dilthy, W. Schommer, W. Höschen, and H. Dierichs, *Chem. Ber.* **68**, 1159 (1935).

 8. B. D. Challand, H. Hikino, G. Kornis, G. Lange, and P. de Mayo, *J. Org. Chem.* **34**, 794 (1969).

 9. H. W. Gschwend, A. O. Lee, and H.-P. Meier, *J. Org. Chem.* **38**, 2169 (1973).

 10. H.-S. Ryang, K. Shima, and H. Sakurai, *J. Org. Chem.* **38**, 2860 (1973).

 11. W. T. Brady, F. H. Parry, III, and J. D. Stockton, *J. Org. Chem.* **36**, 1486 (1971); W. T. Brady and R. Roe, Jr., *J. Am. Chem. Soc.* **93**, 1662 (1971).

 12. J. A. Berson and S. S. Olin, *J. Am. Chem. Soc.* **91**, 777 (1969).

 13. B. J. Arnold, S. M. Mellows, P. G. Sammes, and T. W. Wallace, *J. Chem. Soc. Perkin Trans. I*, 401 (1974); B. J. Arnold, P. G. Sammes, and T. W. Wallace, *J. Chem. Soc. Perkin Trans. I*, 409 (1974).

 14. J. Wolinsky and R. B. Login, *J. Org. Chem.* **35**, 3205 (1970).

 15. H. Gotthardt, R. Huisgen, and H. O. Bayer, *J. Am. Chem. Soc.* **92**, 4340 (1970).

 16. R. J. McClure, G. A. Sim, P. Coggon, and A. T. McPhail, *Chem. Commun.*, 128 (1970).

 17a. R. A. Carboni and R. V. Lindsey, Jr., *J. Am. Chem. Soc.* **81**, 4342 (1959).

 b. L. A. Carpino, *J. Am. Chem. Soc.* **84**, 2196 (1962); **85**, 2144 (1963).

 18. R. J. Crawford and G. L. Erickson, *J. Am. Chem. Soc.* **89**, 3907 (1967).

 19a. T. J. Barton and E. Kline, *J. Organometal. Chem.* **42**, C21 (1972).

 b. C. H. Clapp and F. H. Westheimer, *J. Am. Chem. Soc.* **96**, 6710 (1974).

 20. D. Bichan and M. Winnik, *Tetrahedron Lett.*, 3857 (1974).

 21a. P. G. Gassman and K. T. Mansfield, *Org. Synth.* **V**, 96 (1973).

 b. E. L. Allred and J. C. Hinshaw, *J. Am. Chem. Soc.* **90**, 6885 (1968).

 c. C. D. Smith, *Org. Synth.* **51**, 133 (1971).

 d. J. S. McKennis, L. Brener, J. S. Ward, and R. Pettit, *J. Am. Chem. Soc.* **93**, 4957 (1971).

 e. W. G. Dauben and F. G. Willey, *Tetrahedron Lett.*, 893 (1962).

 f. K. B. Wiberg, G. J. Burgmaier, and P. Warner, *J. Am. Chem. Soc.* **93**, 246 (1971).

 22a. D. J. Faulkner and M. R. Peterson, *J. Am. Chem. Soc.* **95**, 553 (1973).

 b. N. A. Le Bel, N. D. Ojha, J. R. Menke, and R. J. Newland, *J. Org. Chem.* **37**, 2896 (1972).

 c. G. Büchi and H. Wüest, *J. Am. Chem. Soc.* **96**, 7573 (1974).

 d. C. A. Henrick, F. Schaub, and J. B. Siddall, *J. Am. Chem. Soc.* **94**, 5374 (1972).

 e. R. E. Ireland and R. H. Mueller, *J. Am. Chem. Soc.* **94**, 5897 (1972).

 f. E. J. Corey, R. B. Mitra, and H. Uda, *J. Am. Chem. Soc.* **86**, 485 (1964).

 g. J. E. McMurry and L. C. Blaszczak, *J. Org. Chem.* **39**, 2217 (1974).

 h. W. Sucrow, *Angew. Chem. Int. Ed. Engl.* **7**, 629 (1968).

 i. O. P. Vig, K. L. Matta, and I. Raj, *J. Ind. Chem. Soc.* **41**, 752 (1964).

 j. W. Nagata, S. Hirai, T. Okumura, and K. Kawata, *J. Am. Chem. Soc.* **90**, 1650 (1968).

 k. H. O. House, J. Lubinkowski, and J. J. Good, *J. Org. Chem.* **40**, 86 (1975).

 23. N. Shimizu, M. Ishikawa, K. Ishikura, and S. Nishida, *J. Am. Chem. Soc.* **96**, 6456 (1974).

 24. R. Schug and R. Huisgen, *Chem. Commun.*, 60 (1975).

 25. W. L. Howard and N. B. Lorette, *Org. Synth.* **V**, 25 (1973).

 26a. L. F. Fieser, *Org. Synth.* **V**, 604 (1973).

 b. D. H. Miles, P. Loew, W. S. Johnson, A. F. Kluge, and J. Meinwald, *Tetrahedron Lett.*, 3019 (1972).

 c. J. P. Marino and T. Kaneko, *J. Org. Chem.* **39**, 3175 (1974).

 27. L. N. Mander and J. V. Turner, *J. Org. Chem.* **37**, 2915 (1972).

Chapter 7

 1b. E. C. Taylor, F. Kienzle, R. L. Robey, and A. McKillop, *J. Am. Chem. Soc.* **92**, 2175 (1970).

c. G. F. Hennion and S. F. de C. McLeese, *J. Am. Chem. Soc.* **64**, 2421 (1942).

d. J. Koo, *J. Am. Chem. Soc.* **75**, 1889 (1953).

e. E. C. Taylor, F. Kienzle, R. L. Robey, A. McKillop, and J. D. Hunt, *J. Am. Chem. Soc.* **93**, 4845 (1971).

f. G. A. Ropp and E. C. Coyner, *Org. Synth.* **IV**, 727 (1963).

2. G. R. Elling, R. C. Hahn, and G. Schwab, *J. Am. Chem. Soc.* **95**, 5659 (1973).

3a. R. J. Sundberg and D. E. Blackburn, *J. Org. Chem.* **34**, 2799 (1969).

b. M. Rosenblum, *J. Am. Chem. Soc.* **82**, 3796 (1960).

4. H. A. Bruson and H. L. Plant, *J. Org. Chem.* **32**, 3356 (1967).

5. C. F. Bernasconi, *J. Am. Chem. Soc.* **93**, 6975 (1971).

6. R. Foster and C. A. Fyfe, *J. Chem. Soc.* **B**, 53 (1966).

7. J. C. Sheehan and G. D. Daves, Jr., *J. Org. Chem.* **30**, 3247 (1965).

8. E. J. Corey, S. Barcza, and G. Klotmann, *J. Am. Chem. Soc.* **91**, 4782 (1969).

9. E. C. Taylor, F. Kienzle, R. L. Robey, and A. McKillop, *J. Am. Chem. Soc.* **92**, 2175 (1970).

10. P. B. D. de la Mare, O. M. H. el Dusouqui, J. G. Tillett, and M. Zeltner, *J. Chem. Soc.*, 5306 (1964).

11. A. A. Khalaf and R. M. Roberts, *J. Org. Chem.* **36**, 1040 (1971).

12a. J. H. Boyer and R. S. Buriks, *Org. Synth.* **V**, 1067 (1973).

b. H. P. Schultz, *Org. Synth.* **IV**, 364 (1963); F. D. Gunstone and S. H. Tucker, *Org. Synth.* **IV**, 160 (1963).

c. D. H. Hey and M. J. Perkins, *Org. Synth.* **V**, 51 (1973).

d. K. Rorig, J. D. Johnston, R. W. Hamilton, and T. J. Telinski, *Org. Synth.* **IV**, 576 (1963).

e. K. G. Rutherford and W. Redmond, *Org. Synth.* **V**, 133 (1973).

f. M. M. Robison and B. L. Robison, *Org. Synth.* **IV**, 947 (1963).

g. R. Adams, W. Reifschneider, and A. Ferretti, *Org. Synth.* **V**, 107 (1973).

13a. C. L. Perrin and G. A. Skinner, *J. Am. Chem. Soc.* **93**, 3389 (1971).

b. R. A. Rossi and J. F. Bunnett, *J. Am. Chem. Soc.* **94**, 683 (1972).

c. M. Jones, Jr., and R. H. Levin, *J. Am. Chem. Soc.* **91**, 6411 (1969).

15a. G. H. Cleland, *Org. Synth.* **51**, 1 (1971).

b. G. Smolinsky, *J. Am. Chem. Soc.* **82**, 4717 (1960).

c. J. T. Koo, *Org. Synth.* **V**, 550 (1973).

d. A. B. Galun and A. Kalir, *Org. Synth.* **48**, 27 (1968).

16a. R. E. Ireland, C. A. Lipinski, C. J. Kowalski, J. W. Tilley, and D. M. Walba, *J. Am. Chem. Soc.* **96**, 3333 (1974).

b. J. J. Korst, J. D. Johnston, K. Butler, E. J. Bianco, L. H. Conover, and R. B. Woodward, *J. Am. Chem. Soc.* **90**, 439 (1968).

Chapter 8

1b. E. Schmitz, D. Habisch, and A. Stark, *Angew. Chem. Int. Ed. Engl.* **2**, 548 (1963).

g. R. Breslow and H. W. Chang, *J. Am. Chem. Soc.* **83**, 2367 (1961).

h. P. A. Bartlett and W. S. Johnson, *J. Am. Chem. Soc.* **95**, 7501 (1973).

3. W. J. Baron, M. E. Hendrick, and M. Jones, Jr., *J. Am. Chem. Soc.* **95**, 6286 (1973).

4. C. D. Poulter, E. C. Friedrich, and S. Winstein, *J. Am. Chem. Soc.* **91**, 6892 (1969).

5. D. M. Lemal and E. H. Banitt, *Tetrahedron Lett.*, 245 (1964).

6a. R. K. Hill and D. A. Cullison, *J. Am. Chem. Soc.* **95**, 2923 (1973).

b. R. Oda, Y. Ito, and M. Okano, *Tetrahedron Lett.*, 7 (1964).

c. R. A. Carboni, J. C. Kauer, J. E. Castle, and H. E. Simmons, *J. Am. Chem. Soc.* **89**, 2618 (1967).

d. M. E. Peek, C. W. Rees, and R. C. Storr, *J. Chem. Soc. Perkin Trans. I*, 1260 (1974).

e. K. Biemann, G. Büchi, and B. H. Walker, *J. Am. Chem. Soc.* **79**, 5558 (1957).

f. J. A. Marshall and J. L. Belletire, *Tetrahedron Lett.*, 871 (1971).

g. L. A. Paquette and R. W. Houser, *J. Am. Chem. Soc.* **91**, 3870 (1969).

h. D. C. Richardson, M. E. Hendrick, and M. Jones, Jr., *J. Am. Chem. Soc.* **93**, 3790 (1971).

i. J. C. Sheehan and U. Zoller, *J. Org. Chem.* **39**, 3415 (1974).

j. J. E. McMurry and A. P. Coppolino, *J. Org. Chem.* **38**, 2821 (1973).

k. L. A. Paquette, R. H. Meisinger, and R. E. Wingard, Jr., *J. Am. Chem. Soc.* **95**, 2230 (1973).

7. P. S. Skell and R. R. Engel, *J. Am. Chem. Soc.* **88**, 3749 (1966).

8. P. L. Barieli, G. Berti, B. Macchia, and L. Monti, *J. Chem. Soc. C*, 1168 (1970).

9. H. O. House and G. A. Frank, *J. Org. Chem.* **30**, 2948 (1965).

10. W. M. Jones, R. C. Joines, J. A. Myers, T. Mitsuhashi, K. E. Krajca, E. E. Waali, T. L. Davis, and A. B. Turner, *J. Am. Chem. Soc.* **95**, 826 (1973); G. G. Vander Stouw, A. R. Kraska, and H. Schechter, *J. Am. Chem. Soc.* **94**, 1655 (1972); P. Schissel, M. E. Kent, D. J. McAdoo, and E. Hedaya, *J. Am. Chem. Soc.* **92**, 2147 (1970); W. J. Baron, M. Jones, Jr., and P. P. Gaspar, *J. Am. Chem. Soc.* **92**, 4739 (1970); E. Hedaya and M. E. Kent, *J. Am. Chem. Soc.* **93**, 3283 (1971).

11. D. M. Lemal and K. S. Shim, *Tetrahedron Lett.*, 3231 (1964).

12. G. Büchi and J. D. White, *J. Am. Chem. Soc.* **86**, 2884 (1964).

13a. F. Fisher and D. E. Applequist, *J. Org. Chem.* **30**, 2089 (1965).

   b. B. Boyer, P. Dubreuil, G. Lamaty, J. P. Rogue, and P. Geneste, *Tetrahedron Lett*, 2919 (1974).

   c. S. Winstein and J. Sonnenberg, *J. Am. Chem. Soc.* **83**, 3235 (1973).

   d. B. M. Trost, R. M. Cory, P. H. Scudder, and H. B. Neubold, *J. Am. Chem. Soc.* **95**, 7813 (1973).

14a. E. J. Corey, J. F. Arnett, and G. N. Widiger, *J. Am. Chem. Soc.* **97**, 430 (1975).

   b. W. H. Staas and L. A. Spurlock, *J. Org. Chem.* **39**, 3822 (1974).

   c. P. G. Gassman and T. J. Atkins, *J. Am. Chem. Soc.* **94**, 7748 (1972).

   d. J. A. Marshall and J. A. Ruth, *J. Org. Chem.* **39**, 1971 (1974).

   e. K. P. Dastur, *J. Am. Chem. Soc.* **96**, 2605 (1974).

15. J. Hoffman, *Org. Synth.* **V**, 818 (1973).

16a. E. Ciganek, *J. Am. Chem. Soc.* **93**, 2207 (1971).

   b. M. Miyashita and A. Yoshikoshi, *J. Am. Chem. Soc.* **96**, 1917 (1974).

   c. N. J. Leonard and J. C. Coll, *J. Am. Chem. Soc.* **92**, 6685 (1970).

   d. C. D. Gutsche and H. R. Zandstra, *J. Org. Chem.* **39**, 324 (1974).

   e. U. Biethan, U. v. Gizycki, and H. Musso, *Tetrahedron Lett.*, 1477 (1965).

   f. A. C. Cope and A. S. Mehta, *J. Am. Chem. Soc.* **86**, 1268 (1964).

   g. E. J. Corey, Z. Arnold, and J. Hutton, *Tetrahedron Lett.*, 307 (1970).

   h. E. J. Corey, M. Ohno, R. B. Mitra, and P. A. Vatakencherry, *J. Am. Chem, Soc.* **86**, 478 (1964).

   i. M. Ohno, N. Naruse, and J. Terasawa, *Org. Synth.* **V**, 266 (1973).

   j. B. M. Jacobson, *J. Am. Chem. Soc.* **95**, 2579 (1971).

17. E. W. Warnhoff, C. M. Wong, and W. T. Tai, *J. Am. Chem. Soc.* **90**, 514 (1968).

Chapter 9

1a. W. T. Smith and G. L. McLeod, *Org. Synth.* **IV**, 345 (1963).

   b. L. F. Fieser and J. Szmuszkovicz, *J. Am. Chem. Soc.* **70**, 3352 (1948).

   c. F. Freeman and K. W. Arledge, *J. Org. Chem.* **37**, 2656 (1972).

   d. E. J. Corey and R. S. Glass, *J. Am. Chem. Soc.* **89**, 2600 (1967).

   e. J. K. Crandall and L. C. Crawley, *Org. Synth.* **53**, 17 (1973).

   f. T. Uematsu and R. J. Suhadolnik, *J. Org. Chem.* **33**, 726 (1968).

2. C. S. Foote and R. W. Denny, *J. Am. Chem. Soc.* **93**, 5162 (1971).

3. B. Rickborn and S.-Y. Lwo, *J. Org. Chem.* **30**, 2212 (1965).

4a. J. J. Plattner, R. D. Gless, and H. Rapoport, *J. Am. Chem. Soc.* **94**, 8613 (1972).

   b. R. N. Mirrington and K. J. Schmalzl, *J. Org. Chem.* **37**, 2877 (1972).

   c. T. G. Clarke, N. A. Hampson, J. B. Lee, J. R. Morley, and B. Scanlon, *Tetrahedron Lett.*, 5685 (1968).

   d. K. B. Sharpless and R. F. Lauer, *J. Org. Chem.* **39**, 429 (1974); *J. Am. Chem. Soc.* **95**, 2697 (1973).

   e. J. W. Burnham, W. P. Duncan, E. J. Eisenbraun, G. W. Keen, and M. C. Hamming, *J. Org. Chem.* **39**, 1416 (1974).

5a. A. McKillop, J. D. Hunt, F. Kienzle, E. Bigham, and E. C. Taylor, *J. Am. Chem. Soc.* **95**, 3635 (1973).

   b. A. McKillop, B. P. Swann, and E. C. Taylor, *J. Am. Chem. Soc.* **95**, 3340 (1973).

   c. J. E. Byrd and J. Halpern, *J. Am. Chem. Soc.* **95**, 2586 (1973).

6a. H. C. Brown, J. H. Kawakami, and S. Ikegami, *J. Am. Chem. Soc.* **92**, 6914 (1970).

   b. T. Sakan and K. Abe, *Tetrahedron Lett.*, 2471 (1968).

   c. K. J. Clark, G. I. Fray, R. H. Jaeger, and R. Robinson, *Tetrahedron* **6**, 217 (1959).

   d. C. S. Foote, *Acc. Chem. Res.* **1**, 104 (1968).

   e. S. Moon and H. Bohm, *J. Org. Chem.* **37**, 4338 (1972).

   f. W. G. Dauben and D. A. Cox, *J. Am. Chem. Soc.* **85**, 2130 (1963).

7a. R. B. Miller and R. D. Nash, *J. Org. Chem.* **38**, 4424 (1973).

b. R. Grewe and I. Hinrichs, *Chem. Ber.* **97**, 443 (1964).

c. W. Nagata, S. Hirai, K. Kawata, and T. Okumura, *J. Am. Chem. Soc.* **89**, 5046 (1967).

d. W. G. Dauben, M. Lorber, and D. S. Fullerton, *J. Org. Chem.* **34**, 3587 (1969).

e. E. E. van Tamelen, M. Shamma, A. W. Burgstahler, J. Wolinsky, R. Tamm, and P. E. Aldrich, *J. Am. Chem. Soc.* **80**, 5006 (1958).

8. E. L. Eliel, S. H. Schroeter, T. J. Brett, F. J. Biros, and J.-C. Richer, *J. Am. Chem. Soc.* **88**, 3327 (1966).

9a. F. G. Bordwell and A. C. Knipe, *J. Am. Chem. Soc.* **93**, 3416 (1971).

b. C. S. Foote, S. Mazur, P. A. Burns, and D. Lerdal, *J. Am. Chem. Soc.* **95**, 586 (1973).

c. J. P. Marino, K. E. Pfitzner, and R. A. Olofson, *Tetrahedron* **27**, 4181 (1971).

d. H. Hart and L. R. Lerner, *J. Org. Chem.* **32**, 2669 (1967).

e. W. H. Dennis, Jr., L. A. Hull, and D. H. Rosenblatt, *J. Org. Chem.* **32**, 3783 (1967).

10. E. J. Corey, N. W. Gilman, and B. E. Ganem, *J. Am. Chem. Soc.* **90**, 5616 (1968).

11. W. S. Trahanovsky, J. R. Gilmore, and P. C. Heaton, *J. Org. Chem.* **38**, 760 (1973).

12a. J. A. Marshall and R. A. Ruden, *J. Org. Chem.* **36**, 594 (1971).

b. J. A. Marshall and G. M. Cohen, *J. Org. Chem.* **36**, 877 (1971).

c. A. C. Cope, S. Moon, and C. H. Park, *J. Am. Chem. Soc.* **84**, 4843 (1962).

d. L. A. Paquette and J. H. Barrett, *Org. Synth.* **V**, 467 (1973).

e. J. V. Paukstelis and B. W. Macharia, *J. Org. Chem.* **38**, 646 (1973).

13a. W. S. Johnson, T. Li, C. A. Harbert, W. R. Bartlett, T. R. Herrin, B. Staskun, and D. H. Rich, *J. Am. Chem. Soc.* **92**, 4461 (1970); W. S. Johnson, M. F. Semmelhack, M. U. S. Sultanbawa, and L. A. Dolak, *J. Am. Chem. Soc.* **90**, 2994 (1968).

b. E. J. Corey, J. A. Katzenellenbogen, N. W. Gilman, S. A. Roman, and B. W. Erickson, *J. Am. Chem. Soc.* **90**, 5618 (1968).

c. D. Taub, R. D. Hoffsommer, C. H. Kuo, H. L. Slates, Z. S. Zelawski, and N. L. Wendler, *Chem. Commun.*, 1258 (1970).

d. J. F. Bagli, T. Bogri, and R. Deghenghi, *Tetrahedron Lett.*, 465 (1966).

e. D. Taub, R. D. Joffsommer, C. H. Kuo, H. L. Slates, Z. S. Zewlawski, and N. L. Wendler, *Chem. Commun.*, 1258 (1970).

f. D. P. Strike and H. Smith, *Tetrahedron Lett.*, 4393 (1970).

g. E. J. Corey, N. H. Andersen, R. M. Carlson, J. Paust, E. Vedejs, I. Vlattas, and R. E. K. Winter, *J. Am. Chem. Soc.* **90**, 3245 (1968).

14. L. M. Stephenson, D. E. McClure, and P. Sysak, *J. Am. Chem. Soc.* **95**, 7888 (1973).

15. N. M. Hasty and D. R. Kearns, *J. Am. Chem. Soc.* **95**, 3380 (1973).

16. K. B. Sharpless, R. F. Lauer, and A. Y. Teranishi, *J. Am. Chem. Soc.* **95**, 6137 (1973).

17. D. P. Higley and R. W. Murray, *J. Am. Chem. Soc.* **96**, 3330 (1974).

18. R. Criegee and P. Günther, *Chem. Ber.* **96**, 1564 (1963).

19. A. P. Schaap and G. R. Faler, *J. Am. Chem. Soc.* **95**, 3381 (1973).

20. J. D. Albright and L. Goldman, *J. Am. Chem. Soc.* **89**, 2416 (1967).

21. A. J. Whitworth, R. Ayoub, Y. Rousseau, and S. Fliszár, *J. Am. Chem. Soc.* **91**, 7128 (1969).

22. W. P. Keaveney, M. G. Berger, and J. J. Pappas, *J. Org. Chem.* **32**, 1537 (1967).

23. K. Gollnick and G. Schade, *Tetrahedron Lett.*, 2335 (1966).

24. W. V. McConnell and W. H. Moore, *J. Org. Chem.* **28**, 822 (1963).

25. E. E. Royals and J. C. Leffingwell, *J. Org. Chem.* **31**, 1937 (1966).

26. M. Doyle, R. J. Swedo, and J. Roček, *J. Am. Chem. Soc.* **95**, 8352 (1973).

Chapter 10

1a. D. M. Simonović, A. S. Rao, and S. C. Bhattacharyya, *Tetrahedron* **19**, 1061 (1963).

b. R. E. Ireland and L. N. Mander, *J. Org. Chem.* **32**, 689 (1967).

c. G. Büchi, W. D. MacLeod, Jr., and J. Padilla O., *J. Am. Chem. Soc.* **86**, 4438 (1964).

d. P. Doyle, I. R. Maclean, W. Parker, and R. A. Raphael, *Proc. Chem. Soc.*, 239 (1963).

e. J. C. Sheehan and K. R. Henery-Logan, *J. Am. Chem. Soc.* **84**, 2983 (1962).

f. E. J. Corey, M. Ohno, R. B. Mitra, and P. A. Vatakencherry, *J. Am. Chem. Soc.* **86**, 478 (1964).

2a. A. B. Smith, III, and W. C. Agosta, *J. Am. Chem. Soc.* **96**, 3289 (1974).

b. R. S. Cooke and U. H. Andrews, *J. Am. Chem. Soc.* **96**, 2974 (1974).

c. L. A. Hulshof and H. Wynberg, *J. Am. Chem. Soc.* **96**, 2191 (1974).

d. J. P. Marino and T. Kaneko, *J. Org. Chem.* **39**, 3175 (1974).

3a. T. Hylton and V. Boekelheide, *J. Am. Chem. Soc.* **90**, 6987 (1968).

b. B. W. Erickson, *Org. Synth.* **53**, 189 (1973).

c. H. Paulsen, V. Sinnwell, and P. Stadler, *Angew. Chem. Int. Ed. Engl.* **11**, 149 (1972).

d. S. Torii, K. Uneyama, and M. Isihara, *J. Org. Chem.* **39**, 3645 (1974).

e. R. Noyori, S. Makino, T. Okita, and Y. Hayakawa, *J. Org. Chem.* **40**, 806 (1975).

4a. R. B. Miller and E. S. Behare, *J. Am. Chem. Soc.* **96**, 8102 (1974).

b. S. Iwaki, S. Marumo, T. Saito, M. Yamada, and K. Katagiri, *J. Am. Chem. Soc.* **96**, 7842 (1974).

c. G. Büchi, W. Hofheinz, and J. V. Paukstelis, *J. Am. Chem. Soc.* **91**, 6473 (1969).

d. M. Brown, *J. Org. Chem.* **33**, 162 (1968).

5. A. Eschenmoser, D. Felix, and G. Ohloff, *Helv. Chim. Acta* **50**, 708 (1967).

6a. J. M. Harless and S. A. Monti, *J. Am. Chem. Soc.* **96**, 4714 (1974).

b. P. E. Eaton and G. H. Temme, III, *J. Am. Chem. Soc.* **95**, 7508 (1973).

c. D. M. Lemal and J. P. Lokensgard, *J. Am. Chem. Soc.* **88**, 5934 (1966).

d. W. von E. Doering, B. M. Ferrier, E. T. Fossel, J. H. Hartenstein, M. Jones, Jr., G. Klumpp, R. M. Rubin, and M. Saunders, *Tetrahedron* **23**, 3943 (1967).

e. E. Vogel, W. Klug, and A. Breuer, *Org. Synth.* **54**, 11 (1974).

f. L. A. Paquette and J. S. Ward, *J. Org. Chem.* **37**, 3569 (1972).

7a. J. A. Marshall and A. E. Greene, *J. Org. Chem.* **36**, 2035 (1971).

b. P. S. Wharton, C. E. Sundin, D. W. Johnson, and H. C. Kluender, *J. Org. Chem.* **37**, 34 (1972).

c. E. J. Corey, B. W. Erickson, and R. Noyori, *J. Am. Chem. Soc.* **93**, 1724 (1971).

d. R. E. Ireland and J. A. Marshall, *J. Org. Chem.* **27**, 1615 (1962).

e. W. S. Johnson, T. J. Brocksom, P. Loew, D. H. Rich, L. Werthemann, R. A. Arnold, T. Li, and D. J. Faulkner, *J. Am. Chem. Soc.* **92**, 4463 (1970).

f. L. Birladeanu, T. Hanafusa, and S. Winstein, *J. Am. Chem. Soc.* **88**, 2315 (1966); T. Hanafusa, L. Birladeanu, and S. Winstein, *J. Am. Chem. Soc.* **87**, 3510 (1965).

8a. E. H. White, F. McCapra, and G. F. Field, *J. Am. Chem. Soc.* **85**, 337 (1963).

b. E. J. Corey and S. Nozoe, *J. Am. Chem. Soc.* **85**, 3527 (1963).

c. R. Howe and F. J. McQuillin, *J. Chem. Soc.*, 2423 (1955).

d. J. R. Dyer, W. E. McGonigal, and K. C. Rice, *J. Am. Chem. Soc.* **87**, 654 (1965).

9a. Z. G. Hajos and D. R. Parrish, *J. Org. Chem.* **39**, 1615 (1974).

b. L. Hub and H. S. Mosher, *J. Org. Chem.* **35**, 3691 (1970).

c. S. Musierowicz, A. Wróblewski, and H. Krawczyk, *Tetrahedron Lett.*, 437 (1975).

10a. S. Ohki and T. Nagasaka, *Chem. Pharm. Bull.* **19**, 545 (1967).

b. A. Guyer, A. Bieler, and E. Pedrazzetti, *Helv. Chim. Acta* **39**, 423 (1956).

c. S. F. Thames and H. C. Odom, Jr., *J. Heterocycl. Chem.* **3**, 490 (1966).

d. H. C. Beyerman and G. J. Heiszwolf, *Recl. Trav. Chim. Pays-Bas* **84**, 203 (1965).

e. S. W. Pelletier, R. L. Chappell, and S. Prabhakar, *J. Am. Chem. Soc.* **90**, 2889 (1968); S. W. Pelletier and S. Prabhakar, *J. Am. Chem. Soc.* **90**, 5318 (1968).

f. P. Doyle, I. R. MacLean, W. Parker, and R. A. Raphael, *Proc. Chem. Soc.*, 239 (1963).

g. M. Kato, H. Kosugi, and A. Yoshikoshi, *Chem. Commun.*, 185 (1970).

11. A. I. Meyers and E. W. Collington, *J. Am. Chem. Soc.* **92**, 6676 (1970).

12a. I. Fleming, *Selected Organic Syntheses*, Wiley, London, 1973, pp. 3–6; J. E. McMurry and J. Melton, *J. Am. Chem. Soc.* **93**, 5309 (1971).

b. R. M. Coates and J. E. Shaw, *J. Am. Chem. Soc.* **92**, 5657 (1970).

c. W. S. Johnson, T. J. Brocksom, P. Loew, D. H. Rich, L. Werthemann, R. A. Arnold, T. Li, and D. J. Faulkner, *J. Am. Chem. Soc.* **92**, 4463 (1970); E. E. van Tamelen and J. P. McCormick, *J. Am. Chem. Soc.* **92**, 737 (1970); S. Tanaka, H. Yamamoto, H. Nozaki, K. B. Sharpless, R. C. Michaelson, and J. D. Cutting, *J. Am. Chem. Soc.* **96**, 5254 (1974); K. Imai, S. Marumo, and K. Mori, *J. Am. Chem. Soc.* **96**, 5925 (1974); D. J. Faulkner and M. R. Petersen, *J. Am. Chem. Soc.* **95**, 553 (1973); J. A. Findlay and W. D. MacKay, *Chem. Commun.*, 733 (1969); H. Schulz and I. Sprung, *Angew. Chem. Int. Ed. Engl.* **8**, 271 (1969); K. Mori, B. Stalla-Bourdillon, M. Ohki, M. Matsui, and W. S. Bowers, *Tetrahedron* **25**, 1667 (1969); W. S. Johnson, T. Li, D. J. Faulkner, and S. F. Campbell, *J. Am. Chem. Soc.* **90**, 6225 (1968); R. Zurflüh, E. N. Wall, J. B. Siddall, and A. Edwards, *J. Am. Chem. Soc.* **90**, 6224 (1968); E. J. Corey, J. A. Katzenellenbogen, N. W. Gilman, S. A. Roman, and B. W. Erickson, *J. Am. Chem. Soc.* **90**, 5618 (1968); E. J. Corey, H. Yamamoto, D. K. Herron, and K. Achiwa, *J. Am. Chem. Soc.* **92**, 6635 (1970); K. H. Dahm, B. M. Trost, and H. Röller, *J. Am. Chem. Soc.* **89**, 5292 (1967); P. Loew and W. S. Johnson, *J. Am. Chem. Soc.* **93**, 3765 (1971).

d. J. H. Babler, D. O. Olsen, and W. H. Arnold, *J. Org. Chem.* **39**, 1656 (1974); R. J. Crawford, W. F. Erman, and C. D. Broaddus, *J. Am. Chem. Soc.* **94**, 4298 (1972).

e. E. J. Corey and R. D. Balanson, *J. Am. Chem. Soc.* **96**, 6516 (1974).

f. J. L. Herrmann, M. H. Berger, and R. H. Schlessinger, *J. Am. Chem. Soc.* **95**, 7923 (1973).

g. R. F. Romanet and R. H. Schlessinger, *J. Am. Chem. Soc.* **96**, 3701 (1974); R. A. LeMahieu, M. Carson, and R. W. Kierstead, *J. Org. Chem.* **33**, 3660 (1968); G. Büchi, D. Minster, and J. C. F. Young, *J. Am. Chem. Soc.* **93**, 4319 (1971).

h. E. J. Corey, M. Ohno, R. B. Mitra, and P. A. Vatakencherry, *J. Am. Chem. Soc.* **86**, 478 (1964).

13a. H. M. Walborsky, T. Sugita, M. Ohno, and T. Inouye, *J. Am. Chem. Soc.* **82**, 5255 (1960).

b. H. M. Walborsky, L. Barash, and T. C. Davis, *J. Org. Chem.* **26**, 4778 (1961); *Tetrahedron* **19**, 2333 (1963).

c. W. S. Johnson, C. A. Harbert, and R. D. Stipanovic, *J. Am. Chem. Soc.* **90**, 5279 (1968).

d. M. Mousseron, M. Mousseron, J. Neyrolles, and Y. Beziat, *Bull. Soc. Chim. Fr.*, 1483 (1963); Y. Beziat and M. Mousseron-Canet, *Bull. Soc. Chim. Fr.*, 1187 (1968).

Chapter 11

4. B. W. Erickson and R. B. Merrifield, *J. Am. Chem. Soc.* **95**, 3757 (1973).

8a. P. Sieber and B. Iselin, *Helv. Chim. Acta* **51**, 614 (1968).

b. M. Smith, D. H. Rammler, I. H. Goldberg, and H. G. Khorana, *J. Am. Chem. Soc.* **84**, 430 (1962).

c. P. G. Pietta, P. Cavallo, and G. R. Marshall, *J. Org. Chem.* **36**, 3966 (1971).

10. S.-S. Wang and R. B. Merrifield, *J. Am. Chem. Soc.* **91**, 6488 (1969); S.-S. Wang, *J. Am. Chem. Soc.* **95**, 1328 (1973).

11. B. F. Gisin and R. B. Merrifield, *J. Am. Chem. Soc.* **94**, 3102 (1972).

13. S. A. Narang, O. S. Bhanot, J. Goodchild, R. H. Wightman, and S. K. Dheer, *J. Am. Chem. Soc.* **94**, 6183 (1972).

14. E. Ohtsuka, M. Ubasawa, and M. Ikehara, *J. Am. Chem. Soc.* **92**, 5507 (1970).

15. J. D. Glass, I. L. Schwartz, and R. Walter, *J. Am. Chem. Soc.* **94**, 6209 (1972).

16. H. K. Hall, Jr., and M. J. Steuk, *J. Polym. Sci. Polym. Chem. Ed.* **11**, 1035 (1973); H. K. Hall, Jr., L. J. Carr, R. Kellman, and F. De Blauwe, *J. Am. Chem. Soc.* **96**, 7265 (1974).

17a. T. Saegusa, Y. Kimura, K. Sano, and S. Kobayashi, *Macromolecules* **7**, 546 (1974).

b. P. M. Hergenrother, *Macromolecules* **7**, 575 (1974).

c. K. Teranishi, M. Iida, T. Araki, S. Yamashita, and H. Tani, *Macromolecules* **7**, 421 (1974).

d. J. K. Stille and G. K. Noren, *Macromolecules* **5**, 49 (1972).

18. K. E. Pfitzner and J. G. Moffatt, *J. Am. Chem. Soc.* **87**, 5661 (1965).

Subject Index

509